ANNUAL
REPORTS IN
MEDICINAL
CHEMISTRY
Volume 31

ANNUAL REPORTS IN MEDICINAL CHEMISTRY
Volume 31

Sponsored by the Division of Medicinal Chemistry of the American Chemical Society

EDITOR-IN-CHIEF:

JAMES A. BRISTOL
PARKE-DAVIS PHARMACEUTICAL RESEARCH
DIVISION OF WARNER-LAMBERT COMPANY
ANN ARBOR, MICHIGAN

SECTION EDITORS

DAVID W. ROBERTSON • ANNETTE M. DOHERTY • JACOB J. PLATTNER
WILLIAM K. HAGMANN • WINNIE W. WONG • GEORGE L. TRAINOR

EDITORIAL ASSISTANT
LISA BAUSCH

ACADEMIC PRESS
San Diego London Boston New York Sydney Tokyo Toronto

This book is printed on acid-free paper. ∞

Copyright © 1996 by ACADEMIC PRESS

All Rights Reserved.
No part of this publication may be reproduced or transmitted in any form or by any
means, electronic or mechanical, including photocopy, recording, or any information
storage and retrieval system, without permission in writing from the publisher.

Academic Press, Inc.
525 B Street, Suite 1900, San Diego, California 92101-4495, USA
http://www.apnet.com

Academic Press Limited
24-28 Oval Road, London NW1 7DX, UK
http://www.hbuk.co.uk/ap/

International Standard Serial Number: 0065-7743

International Standard Book Number: 0-12-040531-8

PRINTED IN THE UNITED STATES OF AMERICA
96 97 98 99 00 01 MM 9 8 7 6 5 4 3 2 1

CONTENTS

III.   CANCER AND INFECTIOUS DISEASES

IV.   IMMUNOLOGY, ENDOCRINOLOGY, AND METABOLIC DISEASES

V. TOPICS IN BIOLOGY

Section Editor: Winnie W. Wong, BASF Bioresearch Corporation,
                   Worcester, Massachusetts

VI. TOPICS IN DRUG DESIGN AND DISCOVERY

Section Editor: George L. Trainor, DuPont Merck Pharmaceutical Company,
                   Wilmington, Delaware

VII.   TRENDS AND PERSPECTIVES

Section Editor: James A. Bristol, Parke-Davis Pharmaceutical Research,
                   Ann Arbor, Michigan

CONTRIBUTORS

PREFACE

Annual Reports in Medicinal Chemistry continues to strive to provide timely and critical reviews of important topics in medicinal chemistry together with an emphasis on emerging topics in the biological sciences which are expected to provide the basis for entirely new future therapies.

Volume 31 retains the familiar format of previous volumes, this year with 35 chapters. Sections I - IV are disease oriented and generally report on specific medicinal agents, with updates from volume 30 on anti-infectives, antivirals, endothelin, and thrombin. We have continued the trend of reducing the number of annual updates in favor of specifically focussed and mechanistically oriented chapters, where the objective is to provide the reader with the most important new results in a particular field. To this end, chapters on topics not reported in at least two years include: neuropeptide Y, gonadal steroid receptors, P2 purinoreceptors, anesthetics, PDE inhibitors, GPIIb/IIIa inhibitors, NK antagonists, malaria, estrogen receptor modulators, cell adhesion integrins, obesity, osteoporosis, and MMP inhibitors.

Sections V and VI continue to emphasize important topics in medicinal chemistry, biology, and drug design as well as the critical interfaces among these disciplines. Included in Section V, Topics in Biology, are chapters on cell cycle, apoptosis, JAKs and STATs, proteosome, and MAP kinase. Some of the topics of research reviewed in this section will appear in a chapter related to medicinal agents in a future volume, once sufficient time has passed to allow new compounds to be developed from a biological strategy. Chapters in Section VI, Topics in Drug Design and Discovery, reflect the current focus on mechanism-directed drug discovery and newer technologies. These include chapters on new NMR methods, combinatorial chemistry from the perspectives of both synthesis and analysis, and plasma protein binding.

Volume 31 concludes with Trends and Perspectives, with chapters on NCE introductions worldwide in 1995, and an interesting reflection on the Protein Project. In addition to the chapter reviews, a comprehensive set of indices has been included to enable the reader to easily locate topics in volumes 1-31 of this series.

Over the past year, it has been my pleasure to work with 6 highly professional section editors and 78 authors, whose critical contributions comprise this volume.

James A. Bristol
Ann Arbor, Michigan
June 1996

SECTION I.   CENTRAL NERVOUS SYSTEM DISEASES

Editor: David W. Robertson
Ligand Pharmaceuticals, San Diego, California

Chapter 1. Neuropeptide Y: At The Dawn Of Subtype Selective Antagonists

Philip A. Hipskind and Donald R. Gehlert
Central Nervous System Research; Lilly Research Laboratories
Lilly Corporate Center, Indianapolis, IN 46285.

Introduction - Neuropeptide Y (NPY) is a 36 amino acid peptide and a member of the family of "PP-fold" peptides that includes peptide YY (PYY) and pancreatic polypeptide (PP). Originally isolated from extracts of porcine brain using a procedure that identified C-terminally amidated peptides, NPY is named for the presence of an N-terminal tyrosine and a C-terminal tyrosine amide (1,2). NPY is the most abundant peptide neurotransmitter in the brain, and its receptors are prominent in both central and peripheral nervous systems.

$$
\begin{array}{l}
\quad\quad\quad\;\; 1 \quad\quad\quad\quad\quad\quad\quad\quad\quad 10 \\
\textbf{NPY:} \quad \text{Tyr-Pro-Ser-Lys-Pro-Asp-Asn-Pro-Gly-Glu-Asp-Ala-} \\
\quad\quad\quad\quad\quad\quad\quad\quad\quad 20 \\
\quad\quad\quad \text{Pro-Ala-Glu-Asp-Leu-Ala-Arg-Tyr-Tyr-Ser-Ala-Leu-} \\
\quad\quad\quad\quad\quad\quad\quad 30 \\
\quad\quad\quad \text{Arg-His-Tyr-Ile-Asn-Leu-Ile-Thr-Arg-Gln-Arg-Tyr-NH}_2
\end{array}
$$

One of the fundamental difficulties of defining a functional role of NPY has been the absence of specific receptor ligands. The synthesis of peptide analogs or chimeric peptides has provided some selective agonists, but few useful peptide antagonists are available. Recently, some progress has been made in the discovery of specific non-peptide NPY antagonists. In this chapter, we will review essential elements of NPY pharmacology, receptor distribution and molecular biology as well as recent developments in NPY medicinal chemistry.

DISTRIBUTION OF NPY

Significant NPY levels are found in most brain regions including cerebral cortex, hippocampus, thalamus, hypothalamus and brainstem (3). Dense clusters of NPY-containing cell bodies are found in the arcuate nucleus of the hypothalamus, locus coeruleus and nucleus of solitary tract (4,5). Dense NPY-staining nerve terminals are found in hypothalamic, brainstem and certain limbic regions. The localization of these nerve terminals and cell bodies suggests potential roles for NPY in somatic, sensory and cognitive brain functions. Animal studies suggest a direct role for NPY in regulation of food intake, memory retention and anxiolysis (6-9).

Outside the CNS, NPY is found in both peripheral nerves and in the circulation. NPY containing nerve plexuses surround blood vessels in a variety of organs (10,11). NPY is an important cotransmitter with norepinephrine (NE) and is costored and coreleased in sympathetic fibers (12). Enteric, cardiac nonsympathetic, nonadrenergic perivascular and certain parasympathetic nerves also contain NPY (13). These localizations suggest potentially important roles for NPY in sensory and autonomic function. The effects of NPY in the periphery as well as the central nervous system have been reviewed in detail (14,15).

The related peptides, PYY and PP, are considered to be endocrine peptides. In contrast to NPY, there are few neurons that contain PYY in the central or peripheral nervous system. However, some PYY positive neurons are found in the brainstem suggesting this peptide could play a limited role in autonomic regulation (16,17). In humans, PP is a hormone secreted from the islets of Langerhans of the pancreas (18). Less is known about the function of PP though it is well established that PP can inhibit pancreatic exocrine secretion (19). A human receptor with high affinity for PYY and PP (hPP1 or Y4) has recently been cloned (20,21).

NPY RECEPTOR PHARMACOLOGY, DISTRIBUTION AND MOLECULAR BIOLOGY

Since the peptide sequences of the PP-fold family of peptides are similar, it is not surprising that there is extensive overlap in receptors that respond to each peptide. For instance, PYY can be used to replicate many effects of NPY. The first evidence for receptor heterogeneity was made using a preparation of the sympathetic nervous system (22). In this study, C-terminal fragments of NPY and PYY (NPY13-36 and PYY13-36) effectively suppressed electrically evoked twitches of the vas deferens. Unlike NPY and PYY, these fragments were unable to produce a contraction of the guinea pig iliac vein. Therefore, the prejunctional receptor was termed Y2 and the postjunctional was termed Y1 (22). Recent work has examined in more detail the agonistic effects of NPY and PYY peptides and their C-terminal fragments on various Y1 and Y2 cell lines and tissue preparations (23,24). At this time, evidence for additional receptors is less well established, NPY receptor subtypes are summarized in Table 1 and discussed in more detail below.

TABLE 1. NPY Receptor Subtypes Based on Peptide Potency

Receptor Subtype	Peptide Potency	Non-peptide Antagonist
Y1	NPY = PYY = $Pro^{34}NPY$ > NPY2-36 >> NPY-13-36	Benextramine BIBP3226 SR120107A SR120819A PD160170
Y2	NPY = PYY > NPY13-36 = NPY2-36 >> $Pro^{34}NPY$	Benextramine
Y3	NPY = $Pro^{34}NPY$ = NPY13-36 >> PYY	-
PP1 / Y4	PP >> PYY > $Leu^{31}Pro^{34}NPY$ > NPY > NPY13-36	-
"Feeding"	NPY2-36 > NPY = PYY ≈ $Pro^{34}NPY$ > NPY13-36[a]	-

a) refers to potency in food consumption assays.

Y1 Receptor - The Y1 receptor is considered to be postsynaptic and mediates many peripheral actions of NPY. Originally, this receptor was described as having high affinity for both NPY and PYY but poor affinity for C-terminal fragments of NPY such as NPY 13-36 (22,25). Pro^{34} substitution leads to Y1 selectivity in NPY and PYY (26-28). These receptor probes have helped establish a role for Y1 receptors in a variety of functions. The receptor is considered to be coupled to adenylate cyclase in an inhibitory manner in cerebral cortex (29), vascular smooth muscle cells (13) and SK-N-MC cells (29). The Y1 receptor also mediates mobilization of intracellular calcium in porcine vascular smooth muscle cells (30) and human erythroleukemia cells (31) .

The human Y1 receptor was cloned by two independent groups of investigators (32-34). This receptor is a member of the seven transmembrane domain G-protein coupled receptor family, and consists of an open reading frame of 384 amino acids. Interestingly, this receptor can couple to either phosphatidyl inositol hydrolysis or the inhibition of adenylate cyclase depending on the type of cell in which the receptor is expressed (33). The importance of the existance of multiple signal transduction pathways in normal cell function is unknown, however, the Y1 receptor has been reported to couple to either second messenger system when studied using tissue preparations or cell lines naturally expressing the receptor (31,35). Therefore, the Y1 receptor cannot be distinguished solely on the basis of coupling to a single second messenger. Antisense oligonucleotides directed at the mRNA for this receptor reduced the vasoconstriction in response to NPY, and produced an "anxiogenic" response in rats (36,37)

Y2 Receptor - As with the Y1 receptor, this receptor subtype was first delineated using tissue bath preparations (22). Pharmacologically, the Y2 receptor is distinguished from Y1 by exhibiting affinity for C-terminal fragments of NPY. The receptor can be differentiated by use of NPY 13-36, though the 3-36 fragments of NPY and PYY provide improved affinity and selectivity (38). Several discontinuous NPY analogs have been synthesized that are missing the "PP-fold" and these also have selectivity for the Y-2 receptor (39,40). The Y2 receptor binds NPY and PYY with similar affinity, but has low affinity for [Leu^{31}-Pro^{34}]-NPY (26). Like the Y1 receptor, this receptor is coupled to inhibition of adenylate cyclase, though in some preparations it may not be sensitive to pertussis toxin, suggesting potential subtypes of the Y2 receptor (41,42). In several types of neuronal preparations, a pertussis toxin-sensitive Y2 receptor mediates a decrease in the peak amplitude of calcium currents (43-45). In the dorsal root ganglion, stimulation of the Y2 receptor reduces intracellular calcium levels by selective inhibition of N-type calcium channels (46). Presently it is unclear whether this action is related to the inhibition of adenylate cyclase. Thus, the Y2 receptor may exhibit differential coupling to second messengers.

Central Y2 receptors are found in the hippocampus, substantia nigra-lateralis, thalamus, hypothalamus and brainstem (47,48). Outside the brain, Y2 is found in the peripheral nervous system such as sympathetic (49), parasympathetic (50) and sensory (51) neurons. In all these tissues, Y2 mediates a decrease in the release of neurotransmitters. The Y2 receptor has been cloned using expression cloning techniques (52-54). The cDNA encodes a 381 amino acid, seven transmembrane domain receptor that can couple to the inhibition of adenylate cyclase.

Y3 Receptor - The Y3 receptor has high affinity for NPY with lower affinity for PYY. In the CA3 region of the hippocampus, a Y3 receptor has been identified using electrophysiological techniques. This receptor potentiates the excitatory response of these neurons to NMDA (55) though another group was unable to repeat these results (56). Direct binding to this subtype has been performed using [^{125}I]-NPY as the radioligand with membranes from rat heart (57) and bovine adrenal chromaffin cells (58). In these preparations, NPY, [Leu^{31}-Pro^{34}]-NPY and NPY 13-36 potently displaced [^{125}I]-NPY while PYY and PP were significantly less potent. In addition, a Y3 functional bioassay has been described using the rat distal colon (59). The Y3 receptor in bovine adrenal chromaffin cells is coupled to the influx of calcium and does not appear to affect adenylate cyclase (58).

The presence of the Y3 receptor is best established in the nucleus tractus solitarius of the rat brainstem. Application of NPY to this region produces a dose-dependent reduction in blood pressure and heart rate (60,61). This area of the brain also may have significant populations of the Y1 and Y2 receptor, and therefore cardiovascular regulation by NPY via the nucleus tractus solitarius is quite complex.

Feeding Receptor - Intracerebroventricular (i.c.v.) or hypothalamic administration of NPY to rats results in a profound increase in food consumption (62,63). Chronic administration of peptide results in a significant increase in body weight and produces many of the metabolic and hormonal abnormalities associated with obesity (64,65). The feeding response is greatest when the peptide is infused into the periformical region of the hypothalamus (66). While the pharmacology of this response resembles that of the Y1 receptor, the 2-36 fragment of NPY was significantly more potent than NPY (67,68). In addition, intracerebroventricular NPY 2-36 fully stimulates feeding but does not reduce body temperature as does NPY (69). In a guinea pig tracheal preparation, NPY 2-36 behaves like a competitive antagonist of NPY (70). Thus the unique activity of NPY 2-36 may be a combination of agonist and antagonist activities at several NPY receptor subtypes.

The localization of NPY receptors in the hypothalamus also fails to support a role for the Y1 receptor in feeding. Only low levels of Y2 binding are found in the paraventricular and perifornical areas of the hypothalamus (47,48). Messenger RNA encoding for the Y1 receptor is found in high density in the arcuate nucleus (71,72) while being relatively absent from the paraventricular and perifornical regions of the hypothalamus. In addition, direct labeling of rat brain sections with $[^{125}I]$-NPY 2-36 has failed to reveal appreciable quantities of binding in these regions (73). Futher illustrating the complexity of these systems, it was found that antisense inhibition of Y1 receptor expression actually increased feeding (74). Much of the difficulty in establishing the feeding receptor as a separate Y1-like receptor is the limited pharmacological tools available to study Y1 receptors. The recent discovery of selective non-peptide Y1 antagonists will be critical to further study of this putative receptor. In addition, the cloning and identification of this receptor would be useful.

NPY ANTAGONISTS

One of the challenges in the study of NPY has been the lack of suitable antagonists to evaluate the precise role of NPY in normal and pathological function. Peptide analogs, chimeric peptides and random mixtures of peptides have been tried with limited success. Recently, some progress has been made in the discovery of synthetic molecules that antagonize effects of NPY. The structures of these molecules and their pharmacologic effects are discussed in the paragraphs below.

Peptide Antagonists - Efforts have been to made identify potent and selective NPY peptide analogs that are NPY antagonists. The first endeavor was initiated by screening mixtures of peptides, and, when an active peptide mixture was identified, the active component was purified and sequenced. This procedure yielded PYX-2, a peptide with the sequence Ac-[3-(2,6-dichlorobenzyl)Tyr^{27}, D-Thr^{32}]NPY27-36 (75). PYX-2 blocked NPY-induced mobilization of intracellular calcium in human erythroleukemia (HEL) cells (76). PYX-2 supresses both natural and neuropeptide Y-induced carbohydrate feeding in rats (77). A D-Trp^{32} was incorporated by other workers into the full-length sequence of NPY, yielding a peptide (D-Trp^{32}NPY) which was shown to be a competitive antagonist for NPY in an isoproteranol-stimulated hypothalamic membrane preparation (78). Similar effects with [D-Trp^{32}]-NPY have been noted in response to feeding induced by NPY (78). Paradoxically, these peptides have poor affinity for Y1 receptor binding and hypothalamic $[^{125}I]$-PYY binding (79).

Based on prior work with centrally truncated analogs of NPY as well as the importance of the D-Trp^{32} substitution, a new peptide analog, bis(31 31')[Cys^{31}, Trp^{32}, Nva^{34}]NPY31-36 was discovered which exhibited antagonistic activity at Y1 receptors (40,80). Simultaneously, a series of cyclic and/or dimeric peptides, including 1229U91 (**1**), appeared (81,82). Compound **1** had binding affinities in Y1 and Y2 receptor preparations of 10.9 nM and 7.9 nM, and was selective verses a variety of other neurotransmitter or peptide receptor assays (81,83). In addition, **1** antagonized the *in vivo* elevation of arterial blood pressure in rats produced by NPY while exhibiting no effect towards the pressor responses to norepinephrine or angiotensin II (83). Interestingly, in a functional Y2 assay (rat vas deferens), **1** showed no activity up to 3 μM. Another recently disclosed analog, [D-Tyr^{27,36}, D-Thr^{32}]-NPY 27-36, has been described as a selective antagonist for the feeding response to NPY (84) though *in vitro* data supporting this have not yet been published. Collectively, the activity of these compounds suggest amino acid substitution in position 32 is critical for agonist/antagonist functionality at Y1 and Y1-like receptors.

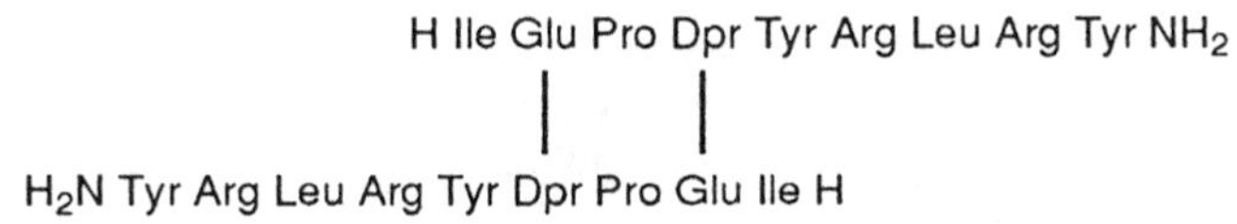

H Ile Glu Pro Dpr Tyr Arg Leu Arg Tyr NH$_2$

H$_2$N Tyr Arg Leu Arg Tyr Dpr Pro Glu Ile H

1

Benextramine and Its Analogs - Benextramine (**2**, BXT), a tetramine disulfide originally described as a non-specific, irreversible antagonist of α_1-adrenergic receptors (85), was discovered to have a similar effect at a subpopulation of NPY receptors in the rat brain (86). Compound **2** maximally inhibited 61% of the N-[propionyl-^{3}H]-NPY ([^{3}H]-NPY) specific binding to rat brain membranes with an IC$_{50}$ of 55 μM., however the inhibition was irreversible since [^{3}H]-NPY specific binding decreased 44% in membranes pretreated with **2**. The same authors later concluded that the BXT-sensitive [^{3}H]-NPY binding sites in the rat brain displayed pharmacology characteristic of a Y1-like receptor, whereas the BXT-insensitive sites appeared to be Y2-like.(87). Interestingly **2** appeared to block the constrictive effects of NPY, [Leu^{31}, Pro^{34}]NPY and NPY(13-36) in the rat femoral artery (88). Recently, studies comparing [^{3}H]-NPY displacement in rat whole brain, frontoparietal cortex and brainstem preparations revealed the existence of two sites, one with micromolar affinity and another with picomolar affinity for **2**. In rat hippocampus, however, only the micromolar BXT-sensitive site was observed. BXT-insensitive sites were also characterized at all loci with the exception of frontoparietal cortex binding where only BXT-sensitive sites were observed (89). This would be consistent with the predominant localization of Y1 in the rat cerebral cortex, while the hippocampus contains both Y1 and Y2 receptors (47,48). It has also been shown that **2** blocks NPY mediated effects *in vivo*. The observed increase in NPY driven rat locomotor and exploratory behaviors was totally abolished by pretreatment with i.c.v. **2**. In the same model, prazosin, a specific α1-adrenergic receptor antagonist also blocked the NPY response. This implies that indirect activation of α1-adrenergic receptor might play a role in the observed NPY behavioral effects (90). Unfortunately **2** also has α1-adrenergic effects, and this complicates mechanistic interpretation of observations using this agent (85).

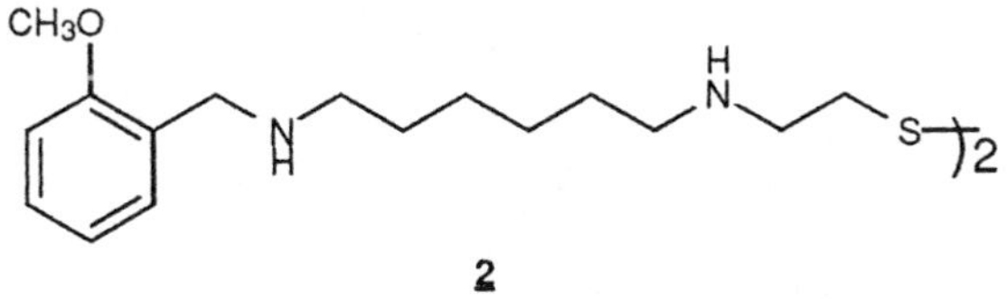

2

Due to its non-selective and irreversible properties, use of **2** as a pharmacological tool has been limited. However, it has served as a medicinal chemistry lead for the design of more potent and selective, non-peptide NPY receptor antagonists. Considering the distance between aryl and quanidinyl groups found in both the N- and C-terminal sections of NPY and the extent to which similar groups in BXT mimic these distances, several new NPY antagonists have been prepared. Compounds SC3117 (**3**) and SC3199, (**4**) both of this genus, displace [^{3}H]-NPY from rat brain binding sites (IC$_{50}$ values of 19 and 45 μM, respectively), and exhibit antagonistic effects in rat femoral artery preparations (IC$_{50}$ values of 43 and 414 nM, respectively) (91). This discrepancy between binding potency vs. *in vitro* efficacy may be indicative of a difference between peripheral post-synaptic and central NPY receptors. Additional pharmacological differences were observed between **3** and **4**: 1) **3** appeared to select only for BXT-sensitive sites, whereas **4** displaced [^{3}H]-NPY from **4** BXT-sensitive and -insensitive sites; and 2) **3** was irreversible (like **2**) whereas **4** was reversible. This is consistent with the notion that the irreversibility of **3** is due to the disulfide linkage present in this molecule (92). This disulfide structural feature is not present in **4**. Both **3** and **4** were selective for NPY verses α-adrenergic receptors. Additional medicinal chemistry uncovered CC217 (**5**) and CC2137 (**6**), both bis(*N,N*-dialkylguanidine) analogs (93). These analogs are similar to **2** in that they are selective for the BXT-sensitive [^{3}H]-NPY binding populations in rat brain membranes (IC$_{50}$ values ranging from 15 to 18 μM). Compound **6**, however, competitively antagonized the effects of the Y2 selective peptide, NPY(13-36) with a pA$_2$ of 9.2 and K$_d$ = 0.63 nM and had no effect on the vasoconstriction produced by the Y1 selective agonist [Leu^{31}, Pro^{34}]NPY.

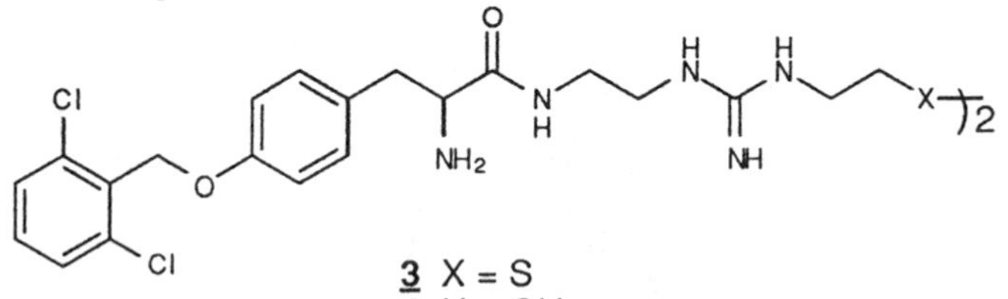

3 X = S
4 X = CH$_2$

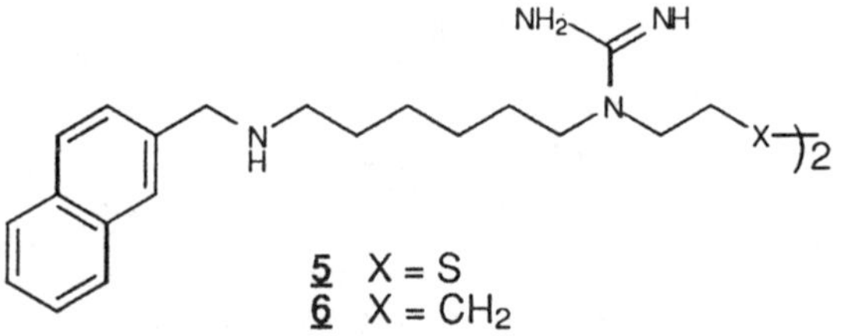

5 X = S
6 X = CH$_2$

BIBP3226 - A low molecular weight, highly selective and potent, non-peptide NPY antagonist, BIBP3226 (**7**), was designed by mimicking the C-terminal part of NPY (94,95). Compound **7** enantiospecifically and competitively displaced [^{125}I]BH-NPY in human SK-N-MC cells (K$_i$ = 5.1 nM), a rat cortex (K$_i$ = 6.8 nM) Y1 preparation and in pig hypothalamus and renal artery preparations (IC$_{50}$ ≈ 10 nM). In a human clonal cell line (CHO cells), **7** was reportedly more potent (K$_i$ = 0.5 nM) (96). Affinities of **7** at Y2 receptors (SMS-KAN cells), as well as 60 other receptor preparations and 15 enzyme systems were >200 fold less potent. It antagonized the increase in intracellular Ca^{2+} induced by NPY with a pK$_b$ of approximately 7.3, and in the isolated perfused rat kidney assay, inhibited the NPY-induced increase in perfusion pressure with an IC$_{50}$ value of 26 nM (79). In the rabbit vas deferens Y1 assay, a dose - response effect was observed (pK$_b$ = 6.98), while antagonistic effects were not observed in a Y2 rat vas deferens assay system. Also a dose dependent inhibition of the pressor response of NPY in pithed rats with **7** was observed. Further work using the pig showed that **7** inhibited the vasoconstrictive effects of a Y1 receptor agonist without influencing the response to a Y2 agonist, noradrenaline, α–adrenergic agonists, β-methylene ATP or angiotensin II. In studies involving human cerebral vessels, **7** caused a parallel and rightward shift in the NPY dose-response curves without any significant change in the maximal contractile response. The calculated pA$_2$ was 8.52, a value compatible with the reported affinity at the rodent and human Y-1 receptor (97,98). Additionally, a 50% reduction in the vasoconstrictive response to sympathetic nerve stimulation was observed with **7** upon i.v. administration (99).

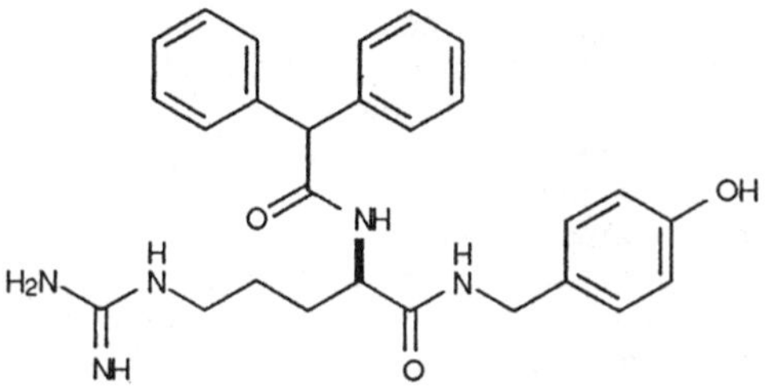

7

SR 120107A and SR 120819A - Two synthetic compounds, SR 120107A (**8**) and SR 120819A (**9**) were recently disclosed as potent and selective Y1 antagonists (100). Both were competitive inhibitors of $[^{125}I]$-NPY in SK-N-MC cells with IC_{50} values of approximately 92 nM and 42 nM, respectively, and to be selective for rat, guinea pig and human Y1 verses Y2 receptors. Compound **9** antagonized the NPY inhibition of cAMP accumulation induced by forskolin in SK-N-MC cells with an IC_{50} value of 92 nM. In a rabbit Y1 vas deferens preparation, competitive antagonism of the selective Y1 agonist, [Leu^{31}, Pro^{34}]NPY, was observed with both compounds (pA$_2$ values of 6.9 and 7.2 respectively). Both agents blocked the hypertensive effects of [Leu^{31}, Pro^{34}]NPY in anesthetized guinea-pigs for a period of 3 to 4 hours (101).

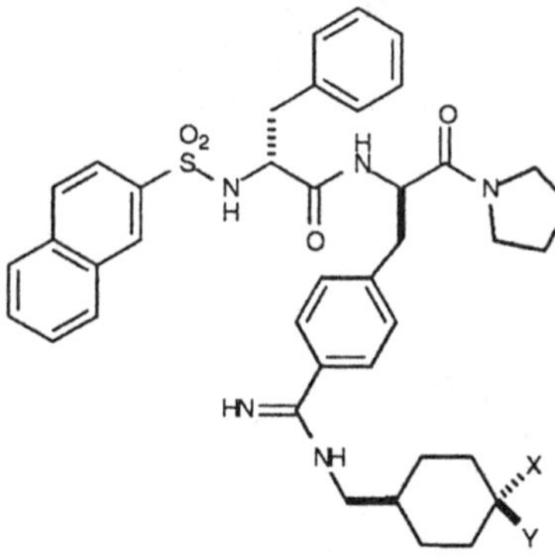

8   X = CH$_2$N(CH$_3$)$_2$; Y = H
9   X = H; Y = CH$_2$N(CH$_3$)$_2$

Others - A series of quinoline Y1 antagonists appeared recently. The most potent of this series was PD160170 (**10**) (K_i = 48 nM) (102). This series was 200-fold selective for Y1 verses Y2, and was a functional antagonist in the NPY inhibition of forskolin stimulated cAMP production. A series of substituted benzylamines, represented by **11**, were recently disclosed in the patent literature (103). Potency in an assay using the SK-N-MC cell line was reported to be in the range of 67 to 525 nM, IC50. No other additional information has been published to date on either series.

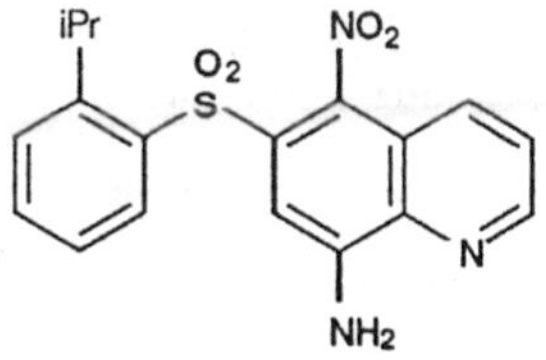

10 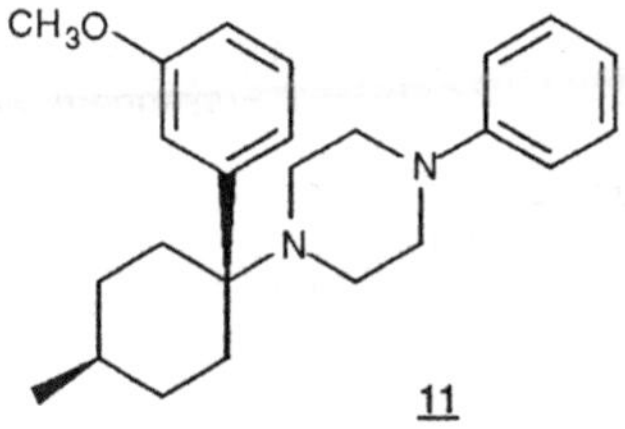   **11**

Conclusion - Much of the pharmaceutical interest in NPY is based on the hypothesis that disturbances in NPY-mediated processes may play a role in a variety of human diseases. Application of exogenous NPY produces effects consistent with a role for this peptide in obesity and diabetes as well as psychiatric and cardiovascular disorders. Alterations in the endogenous levels of NPY have been observed in both animal models of disease and, in several cases, human diseases. While the importance of NPY and related peptides in biological function is well established, the discovery and development of specific receptor antagonists will be a critical element to determine the relevance of this peptide system in human disease. Through advances in molecular biology and medicinal chemistry this goal is beginning to be met at the first member of this family of receptors.

References

1.   K. Tatemoto, M. Carlquist and V. Mutt, Nature, 296, 659, (1982).
2.   K. Tatemoto, Proc. Natl. Acad. Sci. USA, 79, 5485, (1982).
3.   D.A. DiMaggio, B.M. Chronwall, K. Buchman and T.L. O'Donohue, Neuroscience, 15, 1149, (1985).
4.   B.M. Chronwall, D.A. Dimaggio, V.J. Massari, V.M. Pickel, D.A. Ruggerio and T.L. O'Donohue, Neuroscience, 15, 1159, (1985).
5.   M.E. de Quidt and P.C. Emson, Neuroscience, 18, 545, (1986).
6.   M. Heilig and E. Widerlov, Critical Reviews In Neurobiology, 9, 115, (1995).
7.   W.F. Colmers and D. Bleakman, Trends Neurosci., 17, 373, (1994).
8.   S. Dryden, H. Frankish, Q. Wang and G. Williams, Eur. J. Clin. Invest., 24, 293, (1994).
9.   C. Wahlestedt and D.J. Reis, Annu. Rev. Pharmacol. Toxicol, 33, 309, (1993).
10. Z. Zukowska-Grojec and C. Wahlestedt in "Biol. Neuropept. Y Relat. Pept.," W. F. Colmers and C. Wahlestedt, Ed., Humana, Totowa, N. J., 1993, p. 315.
11. L. Grundemar and R. Hakanson, General Pharmacology, 24, 785, (1993).
12. J. Clarke, N. Benjamin, S. Larkin, D. Webb, A. Maseri and G. Davies, Circulation, 83, 774, (1991).
13. B.J. McDermott, B.C. Millat and H.M. Piper, Cardiovascular Res., 27, 893, (1993).
14. D.R. Gehlert, Life Sciences, 55, 551, (1994).
15. Y. Dumont, J.C. Martel, A. Fournier, S. St-Pierre and R. Quirion, Prog. Neurobiol., 38, 125, (1992).
16. R. Ekman, C. Wahlestedt, G. Bottcher, F. Sundler, R. Håkanson and P. Panula, Regul. Pept., 16, 157, (1986).
17. M. Broomé, T. Hökfelt and L. Terenius, Acta Physiol. Scand., 125, 349, (1985).
18. L.-I. Larsson, F. Sundler and R. Håkanson, Cell Tissue Res., 156, 167, (1975).
19. T.-M. Lin in "Gastrointestinal Hormones," G. B. Jerzy Glass, Ed., Raven Press, 1980, p. 275.
20. J.A. Bard, M.W. Walker, T.A. Branchek and R.L. Weinshank, J. Biol. Chem., 270, 26762, (1995).
21. I. Lundell, A.G. Blomqvist, M. Berglund, D.A. Schober, D. Johnson, M.A. Statnick, R. Gadski, D.R. Gehlert and D. Larhammar, J. Biol. Chem., 270, 29123, (1995).
22. C. Wahlestedt, N. Yanaihara and R. Håkanson, Regul. Peptides, 13, 307, (1986).
23. H.A. Wieland, K. Willim and H.N. Doods, Peptides, 0016, 01389, (1995).
24. A.G. Beck-Sickinger and G. Jung, Biopolymers, 37, 123, (1995).
25. C. Wahlestedt, L. Edvinsson, E. Ekblad and R. Håkanson in "Neuronal Messengers in Vascular Function," A. Nobin, C. Owman and B. Arneklo-Nobin, Ed., Elsevier, 1987, p. 231.
26. J.U. Fuhlendorff, U. Gerther, L. Aakerlund, N.L. Johansen, H. Thøgersen, S.G. Melberg, U.B. Olsen, O. Thastrup and T.W. Schwartz, Proc. Natl. Acad. Sci. USA, 87, 182, (1990).
27. Y. Dumont, A. Cadieux, L.-H. Pheng, A. Fournier, S. St-Pierre and R. Quirion, Mol. Brain Res., 26, 320, (1994).
28. E.K. Potter, J. Fuhlendorff and T.W. Schwartz, Eur. J. Pharmacol., 193, 15, (1991).
29. A. Westlind-Danielsson, A. Unden, J. Abens, S. Andell and T. Bartfai, Neurosci. Lett., 74, 237, (1987).
30. S. Mihara, Y. Shigeri and M. Fujimoto, FEBS Lett., 259, 79, (1989).
31. L. Aakerlund, U. Gether, J. Fuhlendorff, T.W. Schwartz and O. Thastrup, FEBS Lett., 260, 73, (1990).
32. D. Larhammar, A.G. Blomquist, F. Yee, E. Jazin, H. Yoo and C. Wahlestedt, J. Biol. Chem., 267, 10935, (1992).
33. H. Herzog, Y.J. Hort, H.J. Ball, G. Hayes, J. Shine and L.A. Selbie, Proc. Natl. Acad. Sci. USA, 89, 5794, (1992).

34. H. Herzog, Y.J. Hort, J. Shine and L.A. Selbie, DNA Cell Biol., 12, 465, (1993).
35. J. Hinson, C. Rauh and J. Coupet, Brain Res., 446, 379, (1988).
36. C. Wahlestedt, E.M. Pich, G.F. Koob, F. Yee and M. Heilig, Science, 259, 528, (1993).
37. D. Erlinge, L. Edvinsson, J. Brunkwall, F. Yee and C. Wahlestedt, Eur. J. Pharmacol., 240, 77, (1993).
38. Y. Dumont, A. Fournier, S. St-Pierre and R. Quirion, Society for Neuroscience Abstracts, 19, 726, (1993).
39. A. Beck, G. Jung, W. Gaida, H. Koppen, R. Lang and G. Schnorrenberg, FEBS Lett., 244, 119, (1989).
40. J.L. Krstenansky, T.J. Owen, S.H. Buck, K.A. Hagaman and L.R. McLean, Proc. Natl. Acad. Sci. U S A, 86, 4377, (1989).
41. S. Foucart and H. Majewski, Naunyn-Schmeideberg's Arch. Pharmacol., 340, 658, (1989).
42. W.F. Colmers and Q.J. Pittman, Brain Res., 489, 99, (1989).
43. D.A. Ewald, P.C. Sternweis and R.J. Miller, Proc. Natl. Acad. Sci. U S A, 85, 3633, (1988).
44. D. Bleakman, N.L. Harrison, W.F. Colmers and R.J. Miller, Br. J. Pharmacol., 107, 334, (1992).
45. J.W. Wiley, R.A. Gross and R.L. MacDonald, J. Neurophysiol., 70, 324, (1993).
46. P.T. Toth, V.P. Bindokas, D. Bleakman, W.F. Colmers and R.J. Miller, Nature, 364, 635, (1993).
47. D.R. Gehlert, S.L. Gackenheimer and D.A. Schober, Neurochem Int., 21, 45, (1992).
48. Y. Dumont, A. Fournier, S. St-Pierre and R. Quirion, J. Neurosci., 13, 73, (1993).
49. C. Wahlestedt and R. Håkanson, Med. Biol., 64, 85, (1986).
50. M. Stjernquist and C. Owman, Acta Physiol. Scand., 138, 95, (1990).
51. L. Grundemar, N. Grundström, I.G.M. Johansson, R.G.G. Andersson and R. Håkanson, Br. J. Pharmacol., 99, 473, (1990).
52. P.M. Rose, P. Fernandez, J.S. Lynch, S.T. Frazier, S.M. Fisher, K. Kodukula, B. Kienzle and R. Seetha, J. Biol. Chem., 270, 22661, (1995).
53. C. Gerald, M.W. Walker, P.J.J. Vaysse, C.G. He, T.A. Branchek and R.L. Weinshank, J. Biol. Chem., 0270, 26758, (1995).
54. D.R. Gehlert, L.S. Beavers, D. Johnson, S.L. Gackenheimer, D.A. Schober and R.A. Gadski, Molecular Pharmacology, 49, 224, (1996).
55. F.P. Monnet, G. Dubonnel and C. de Montigny, Eur. J. Pharmacol., 182, 207, (1990).
56. A.R. McQuiston and W.F. Colmers, Neurosci. Lett., 138, 261, (1992).
57. A. Balasubramaniam, S. Sheriff, D.F. Rigel and J.E. Fischer, Peptides, 11, 545, (1990).
58. C. Wahlestedt, S. Regunathan and R. Håkanson, Life Sci., 50, PL7, (1992).
59. Y. Dumont, H. Satoh, A. Cadieux, M. Taoudi-Benchekroun, L.H. Pheng, S. St-Pierre, A. Fournier and R. Quirion, Eur. J. Pharmacol., 238, 37, (1993).
60. L. Grundemar, C. Wahlstedt and D.J. Reis, J. Pharmacol. Exp. Therap., 258, 633, (1991).
61. L. Grundemar, C. Wahlstedt and D.J. Reis, Neurosci. Lett., 122, 135, (1991).
62. J.T. Clark, P.S. Kalra, W.R. Crowley and S.P. Kalra, Endocrinology, 115, 427, (1984).
63. B.G. Stanley and S.F. Leibowitz, Life Sci., 35, 2635, (1984).
64. N. Zarjevski, I. Cusin, R. Vettor, F. Rohner-Jeanrenaud and B. Jeanrenaud, Endocrinology, 133, 1753, (1993).
65. B.G. Stanley, S.E. Kyrkouli, S. Lampert and S.F. Leibowitz, Peptides, 7, 1189, (1986).
66. B.G. Stanley, W. Magdalin, A. Seirafi, W.J. Thomas and S.F. Leibowitz, Brain Res., 604, 304, (1993).
67. S.P. Kalra and W.R. Crowley, Front. Neuroendocrinol., 13, 1, (1992).
68. B.G. Stanley, W. Magdalin, A. Seirafi, M.M. Nguyen and S.F. Leibowitz, Peptides, 13, 581, (1992).
69. F.B. Jolicoeur, J.N. Michaud, D. Menard and A. Fournier, Brain Res. Bull., 26, 309, (1991).
70. M.T. Benchekroun, S. St-Pierre, A. Fournier and A. Cadieux, Br. J. Pharmacol., 109, 902, (1993).
71. C. Eva, K. Keinänen, H. Monyer, P. Seeburg and R. Sprengel, FEBS Lett., 271, 80, (1990).
72. J.D. Mikkelsen and P.J. Larsen, Neuroscience Lett., 148, 195, (1992).
73. D.A. Schober and D.R. Gehlert, Society for Neuroscience Abstracts, 19, 727, (1993).
74. M. Heilig, Regulatory Peptides, 59, 201, (1995).
75. K. Tatemoto, Ann. NY Acad. Sci., 611, 1, (1990).
76. K. Tatemoto, M. Mann and M. Shimizu, Proc. Natl. Acad. Sci., 89, 1174, (1992).
77. S.F. Leibowitz, M. Xuereb and T. Kim, Neuroreport, 3, (1992).
78. A. Balasubramaniam, S. Sheriff, M.E. Johnson, M. Prabhakaran, Y. Huang, J.E. Fischer and W.T. Chance, J. Med. Chem., 37, 811, (1994).
79. H.A. Wieland, K.D. Willim, M. Entzeroth, W. Wienen, K. Rudolf, W. Eberlein, W. Engel and H.N. Doods, J. Pharmacol. Exp. Therap., 275, 143, (1995).
80. A. Balasubramaniam, W. Zhai, S. Sheriff, Z. Tao, W.T. Chance, J.E. Fischer, P. Eden and J. Taylor, J. Med. Chem., 39, 811, (1996).
81. A.J. Daniels, J.E. Matthews, R.J. Slepetis, M. Jansen, O.H. Viveros, A. Tadepalli, W. Harrington, D. Heyer and A. Landavazo, Proc. Natl. Acad. Sci. U. S. A., 92, 9067, (1995).

82.   A.J. Daniels, D. Heyer, A. Landavazo, J.J. Leban and A. Spaltenstein, Wo 9400486 A1 940106 (9403), (1994).
83.   S.S. Hegde, D.W. Bonhaus, W. Stanley, R.M. Eglen, T.M. Moy, M. Loeb, S.G. Shetty, A. Desouza and J. Krstenansky, J. Pharmacol. Exp. Therap., 275, 01261, (1995).
84.   R.D. Myers, M.H. Wooten, C.D. Ames and J.W. Nyce, Brain Res. Bull., 37, 237, (1995).
85.   B.G. Benfey, Trends Pharmacol. Sci., 3, 470, (1982).
86.   M.B. Doughty, S.S. Chu, D.W. Miller, K. Li and R.E. Tessel, Eur. J. Pharmacol., 185, 113, (1990).
87.   M.B. Doughty, K. Li, L. Hu, S.S. Chu and R. Tessel, Neuropeptides, 23, 169, (1992).
88.   R.E. Tessel, D.W. Miller, G.A. Misse, X. Dong and M.B. Doughty, J. Pharmacol. Exp. Ther., 265, 172, (1993).
89.   C. Melchiorre, P. Romualdi, b.M. L., A. Donatini and S. Ferri, Eur. J. Pharmacol., 265, 1, (1994).
90.   M. Smialowska, L. Gastollewinska and K. Tokarski, Neuropeptides, 26, 225, (1994).
91.   M.B. Doughty, S.S. Chu, G.A. Misse and R. Tessel, Bioorg. Med. Chem. Lett, 2, 1497, (1992).
92.   C. Melchiorre, Trends Pharmacol. Sci., 2, 209, (1981).
93.   C. Chaurasia, G. Misse, R. Tessel and M.B. Doughty, J. Med. Chem., 37, 2242, (1994).
94.   K. Rudolf, W. Eberlein, W. Engel, H.A. Wieland, K.D. Willim, M. Entzeroth, W. Wienen, A.G. Beck-Sickinger and H.N. Doods, Eur. J. Pharmacol., 271, R11, (1994).
95.   M. Sautel, K. Rudolf, H. Wittneben, H. Herzog, R. Martinez, M. Munoz, W. Eberlein, W. Engel, P. Walker and A.G. Beck-Sickinger, Mol Pharmacol., in press, (1996).
96.   H.N. Doods, M. Entzeroth, W. Wienen, K. Rudolf, W. Engel, W. Eberlein and H.A. Wieland, Pharmacol. Toxicol., 76, 76, (1995).
97.   T. Nilsson, L. Cantera and L. Edvinsson, Neuroscience Letters, 0204, 00145, (1996).
98.   R. Abounader, J.G. Villemure and E. Hamel, Br. J. Pharmacol., 0116, 02245, (1995).
99.   J. Lundberg and A. Modin, Pharmacol. Toxicol., 76, S13, (1995).
100. C. Serradeil-Le Gal, G. Valeet, P.E. Rouby, A. Pellet, G. Villanova, L. Foulon, L. Lespy, G. Neliat, J.P. Chambon, J.P. Maffrand and G. LeFur, Neuroscience Abstracts, 20, 907, (1994).
101. C. Serradeil-Le Gal, G. Valeet, P.E. Rouby, A. Pellet, F. Oury-Donat, G. Brossard, L. Lespy, E. Marty, G. Neliat, P. de Cointet, J.P. Maffrand and G. LeFur, FEBS Lett., 362, 192, (1995).
102. J.L. Wright, J.A. Bikker, D.M. Downing, T.G. Heffner, R.G. MacKenzie, J.R. Rubin and L.D. Wise, 211th ACS National Meeting, New Orleans, LA, (1996)
103   J.M. Peterson, C.A. Blum, G. Cai, A. Hutchison, WO9614307 A (1996).

Chapter 2. Gonadal Steroid Receptors: Possible Roles In The Etiology And Therapy Of Cognitive And Neurological Disorders

Rajesh C. Miranda and Farida Sohrabji
Texas A&M University, Dept. Human Anatomy & Medical Neurobiology
Reynold's Medical Bldg., College Station TX 77843

Introduction - Gonadal hormones (the estrogens [e.g., estradiol-17β, **1**], androgens [e.g., testosterone, **2**] and progestins [e.g., progesterone, **3**]) have long been known to regulate sexual behavior and to integrate the hypothalamic-pituitary-gonadal axis. However, gonadal hormone receptors are also localized to regions of the forebrain that mediate cognition and affect. The important role of these steroids in the management of neurological and psychiatric disorders has recently come to be appreciated. For example, clinical trials suggest that estrogen replacement therapy may be effective in the treatment of some types of Alzheimer's disease (1,2). Sex differences have been observed in the incidence of psychiatric and neurological disorders including schizophrenia (3) and autism (4). Furthermore, progesterone antagonists have been effective in treating nervous system neoplasms including meningiomas (5). Thus gonadal steroids may play an effective and plieotropic role in disease management. However, a basic conundrum is how to accentuate the central nervous system actions of the gonadal steroids, while attenuating their harmful effects in non-neural target tissues such as the uterine endometrium, breast epithelial tissue and the prostate (6,7). There is a paucity of drugs that can specifically target brain gonadal steroid receptors. Furthermore, estrogen receptor (ER) antagonists, used routinely in the management of breast cancer, have poorly understood or unknown actions in the brain. The answer to this issue of tissue specificity may well lie in the basic biology of the brain and of the specific gonadal steroid receptors. This chapter will focus on neural-selective steroid receptor ligands and on alternate mechanisms of gonadal steroid receptor activation as strategies in the management of neuro-psychiatric disorders. Emphasis in this review, will be placed on the estrogen receptor, as the best understood gonadal steroid regulator of neural function.

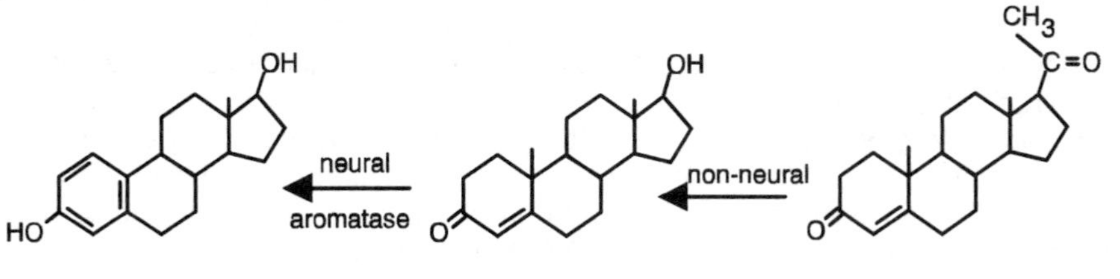

1   **2**   **3**

THE GONADAL HORMONE RECEPTORS AND THEIR LOCALIZATION IN THE BRAIN:

Gonadal hormone receptors are nuclear transcription factors and members of a superfamily [containing approximately 150 members] of thyroid/steroid/vitamin-D3/retinoic acid receptors. This superfamily is divided into steroid [gonadal and adrenal] and non-steroidal [thyroid/ vitamin-D3/retinoic/peroxisomal proliferator] sub-families. The classical model of signal transduction in this superfamily suggests that ligand binding results in receptor phosphorylation, a conformational change in the steroid receptor and binding of the ligand-receptor complex to target DNA sequences (hormone response elements) within nuclear chromosomes. DNA binding in turn activates the transcription initiation complex. Excellent recent reviews of the biology of steroid hormone receptors have been published (8,9).

Neural localization of gonadal hormone receptors - Estrogen (for references, see (10-12)), androgen (13,14) and progesterone (15,16) receptors are widely expressed throughout the central nervous system both during development and in the adult (a composite of data obtained from rodents and primates). In addition to their well-known localization to the hypothalamus, gonadal steroid receptors are also localized to the nuclei of the septum and diagonal band of Broca, hippocampus, allocortex, isocortex and amygdaloid complex. These limbic and isocortical regions are the neural foci for cognition and affect and are the targets of neuronal degeneration characteristic of disorders such as Alzheimer's disease. Though gonadal hormone receptors are expressed throughout the forebrain, there are temporal variations in the peak receptor levels within individual regions both during development and aging. The cellular actions of the estrogens and the androgens often overlap since estrogens can be synthesized intra-neurally from testosterone.

HORMONE REPLACEMENT THERAPY (HRT) IN NEUROLOGICAL AND PSYCHIATRIC DISORDERS

Although the benefits and risks of estrogen replacement therapy (ERT) continue to be debated (6,7), this debate proceeds almost without reference to the actions of estrogen on the brain. Prescribed overwhelmingly for peri-menopausal vasomotor problems (reviewed in (17)) and depression, HRT may provide unexpected benefits by lowering the risk for cognitive dementias.

Hormone replacement therapy and Alzheimer's disease - Peri-menopausal loss of estrogen titers has been associated with anxiety, depression and loss of memory. Studies in the late seventies and eighties established the beneficial effects of estrogen treatment on cognitive functioning and mood in populations of post-menopausal women. The initial studies on ERT for women with Alzheimer's disease, using oral estradiol or (2 mg/day, (1)) or conjugated equine estrogen (estrone sulfate **4**, 1.25 mg/day) (2), attributed a moderate protective role for estrogen. Recent evidence from two large scale retrospective studies of estrogen usage among women with senile dementia differed widely in their conclusions. The first study, based on a population of women listed in the Alzheimer's Disease Patient Registry, reported no association between estrogen prescriptions (in either oral or parenteral form, data obtained from pharmacy records) and the incidence of Alzheimer's disease (18). In contrast, another major study using a sample of women recruited from the community (19), found that while control and AD patients did not differ in total number of prescriptions, current use of estrogen was significantly lower in patients with Alzheimer's disease. Moreover, among those women with AD, performance on a cognitive screening test (the Mini Mental Status exam) was significantly better among current users of estrogen. The relationship between AD and ERT was further underscored in a case control study nested in a prospective study of women in a retirement community (20). This study concluded that the risk of Alzheimer's disease was significantly less in estrogen users than non-users. Moreover the risk of AD declined with increased dosage of the oral estrogen and increased duration of all (oral or parenteral) estrogens.

Despite the correlation between ERT and Alzheimer's disease, the mechanism by which estrogen may protect against AD is unclear at this time. A recent prospective study on a cohort of older non-demented women indicates that estrogen use *per se* does not enhance performance on cognitive tests (7) though estrogen may protect cognitive neural circuits against disease. In this context, it is intriguing that the Ullrich-Turner Syndrome, where the primary pathogenic target is the female reproductive system (involution of the ovaries), is also accompanied by specific cognitive losses (21) including impaired visual-spatial-perceptual abilities and performance IQ. Some of the contradiction between the epidemiological studies may be eventually related to differences in the experimental approach, the type of HRT and the lack of good measures for patient compliance. Recent clinical reports indicate that transdermal or sublingual hormone patches may greatly increase patient compliance (see (22)) and future studies on AD and other dementias using these technologies may resolve some of the current debate on gonadal hormonal regulation of cognition. At the present time, there are no reports available on the effects of gonadal hormones for male patients of Alzheimer's disease. While males have a readily available source of estrogen from testosterone (**2**), the amount of free testosterone and the diurnal rhythm of testosterone production decreases with age (23). In view of testosterone's ability to enhance spatial cognition in older men (23), HRT may be beneficial to male patients of Alzheimer's disease. Furthermore combined estrogens and androgens, when appropriately titered (24), may be of therapeutic value to women patients as well.

Gonadal hormones and other psychiatric disorders - HRT may also be relevant to the management of affective and perceptual disorders, such as depression and schizophrenia. HRT has long been used to treat depression and mood swings related to menopause (25) and more recently, to treat postnatal depression. For example, a recent study (26) suggests that estradiol-17β (200μg/day, transdermal) leads to a significant improvement in depressive symptoms related to childbirth, as measured by the Edinburgh postnatal depression scale and by clinical psychiatric interview.

A related issue concerns lithium therapy, one of the common prescriptions for bipolar disorders (manic depressive psychoses). At therapeutic doses (1 to 5 mM), lithium appears to stimulate proliferation of the breast carcinoma-derived cell line MCF-7, using mechanisms similar to that of estradiol-17β (27). Intriguingly, this effect was specific to lithium and not evoked by other cations, and was mainly seen in estrogen-dependent cell lines. While this study indicates a potential collateral risk associated with lithium therapy, it also raises the possibility that, in the brain, lithium may act on hormone sensitive neurons and *via* signaling pathways used by gonadal steroids.

The onset and severity of psychoses such as schizophrenia may also be related to hormonal milieu. Several studies have also indicated that female schizophrenics have a much later mean age of onset of schizophrenic symptoms than in males (reviewed in (3)). Furthermore, increased susceptibility to schizoaffective psychosis occurs during periods of lowered levels of circulating gonadal hormones, observed in women, at menopause (for review see (3)), and postpartum (28). This suggests that gonadal hormones protect against the symptoms of schizophrenia.

In rodent model systems, estrogen appears to decrease behaviors mediated by dopamine, a key neurotransmitter system implicated in schizophrenia. In neonatal rats for example, estradiol-17β led to a decrease in the affinity of ^{3}H-Sulpride (**5**, a D2 receptor-selective antagonist) binding in the striatum. This decrease in affinity was correlated with an estradiol-related decrease in haloperidol (**6**)-induced cataleptic behavior and apomorphine (**7**)-induced stereotypic (oral, grooming) behaviors. Conversely, postpartum female patients with psychotic symptoms demonstrated an increased response on a dopamine agonist test (increased growth hormone secretion in response to apomorphine administration) as compared to non-affected women (28). An imbalance in estrogen-dopamine interactions may therefore be an integral aspect of the etiology of schizophrenia.

STRATEGIES FOR THE THERAPEUTIC TARGETING OF NEURAL GONADAL HORMONE RECEPTORS

The effective management of neurological and psychiatric disorders with HRT requires the development of drug delivery systems and protocols that specifically target the nervous system. There are at least three potentially effective strategies to accomplish this aim: The first strategy is to design hormones that are biologically active selectively in neural tissue; A second strategy is to induce neural specific activation of the gonadal hormone receptors in a ligand independent manner; A third strategy is to target tissue-selective transcription intermediary factors that are down stream of the gonadal hormone signaling pathway, The latter strategies bypass or attenuate the need for hormone delivery. The following section will address these approaches.

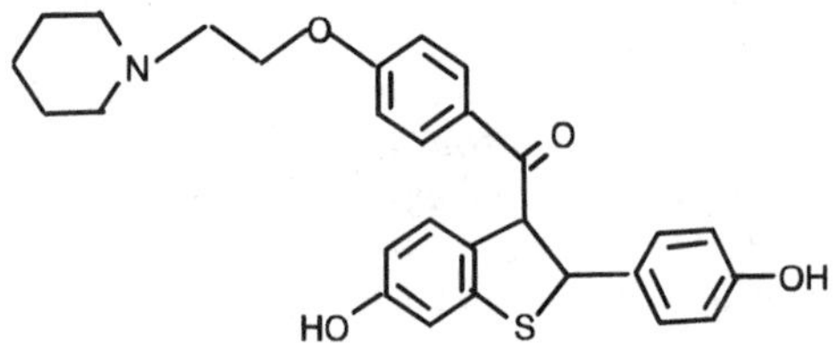

8

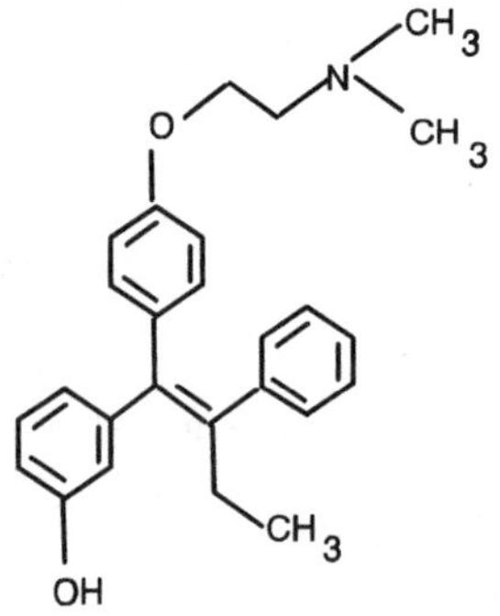

9

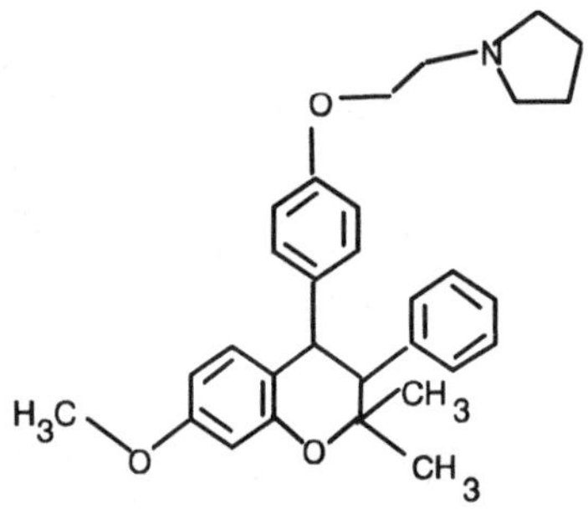

NEURAL-SELECTIVE HORMONE REPLACEMENT STRATEGIES

Tissue-specific estrogenic ligands - In general estrogen has a strong beneficial effect on the circulatory system, on bones and, when taken for long periods of time, on mortality. The risks of unopposed estrogen on uterine tissue (endometrial disease) may be reduced by HRTs that include progesterone (29). The more problematic risks such as breast cancer, vaginal bleeding and other idiosyncratic reactions (thrombosis, hypertension, allergies, somatic complaints of bloating and breast tenderness) need to be resolved before HRT can be commonly used. Such considerations have been instrumental in the development of synthetic estrogens (**8,9,10**) adapted for a specific target or ligands that are

10

agonists in some tissues and antagonists in others (29). Many of these selective estrogen-receptor modulators (SERMs) have been directed towards bone, since peri-menopausal estrogen depletion has been recognized as an important risk factor for osteoporosis. The antiestrogen, raloxifene (**8**) and the tamoxifene analog, droloxifene (3-hydroxytamoxifene, **9**) reportedly protect against ovariectomy-induced cancellous bone loss and bone turnover, while acting as antagonist in the uterus (30,31) and in prostate adenocarcinomas (32). Another bone-specific estrogen antagonist, Centchroman (**10**), prevents osteoclast-mediated bone resorption (33) in *in vitro* assays, with an IC_{50} (0.1μM) that is 10% of the dose of estradiol-17β required to produce the same inhibition. **These data suggest** that the dosages of at least some SERMs may be titrated downward as compared to estradiol, while yet maintaining selective, therapeutic advantages. While these new SERMs are aggressively tested to develop alternates to HRT, little is known of their actions on the brain. Given the cluster of syndromes that occur menopausally, effects of new SERMs on cognition and affect becomes increasingly urgent.

<u>Neural-selective chemical delivery systems</u> - A second promising strategy (termed a

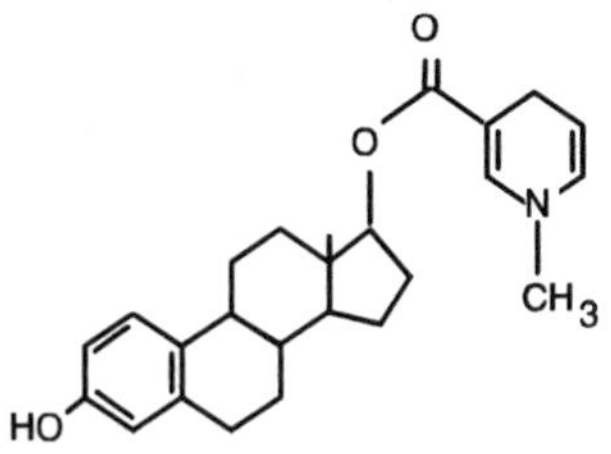

11 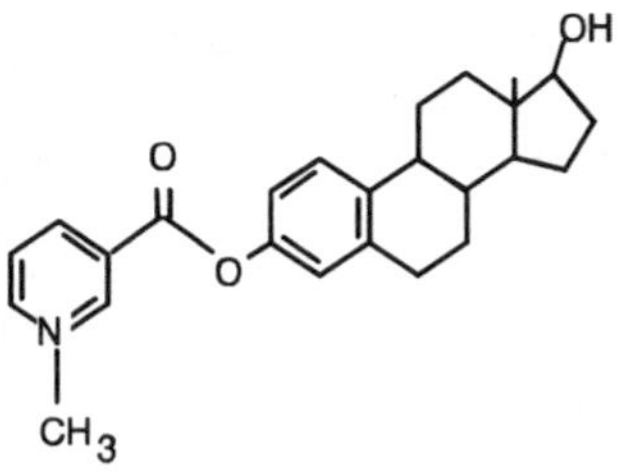   **12**

chemical delivery system or CDS) for selectively targeting hormones to the brain, makes use of redox reactions and the blood-brain barrier to selectively sequester hormones to the brain while eliminating them from other organs (34,35). In this strategy, 1-methyl-1,4-dihydronicotinate (the CDS) can be conjugated either to the 17-alcohol position (i.e., **11**, 17-E2-CDS, (35)) or the 3-phenol (i.e., **12**, 3-E2-CDS, (34)) of estradiol-17β. Modification of the 17-alcohol position (**11**) renders the estradiol biologically inactive, but lipophilic. Intra-organ reduction to the nicotinate salt, renders this compound hydrophilic. While it is rapidly cleared from all other organ systems, 17-E2-CDS salt is trapped within the brain by the blood brain barrier. Hydrolysis of 17-E2-CDS leads to higher levels and sustained release of biologically active estradiol (with a half-life of 8 days) within the brain as compared to levels in plasma, fat and liver (36). The neural-selective action of the E2-CDS system is further supported by experiments in rats, demonstrating prolonged suppression of plasma leutinizing hormone, follicle-stimulating hormone and weight gain, as well as a prolonged upregulation of plasma prolactin levels following a single dose of E2-CDS as compared to estradiol-17β or controls (35,37). Furthermore, 17-E2-CDS has been used in female rats, to reverse castration-induced decreases in high affinity choline uptake in the brain (35). This data is relevant to the treatment of Alzheimer's disease which is characterized by a loss of function of the basal forebrain cholinergic system.

LIGAND-INDEPENDENT ACTIVATION OF THE GONADAL HORMONE RECEPTORS

Steroid-independent mechanisms represent an alternate route for therapeutic intervention. Classically, ligand binding has been suggested as the key event in receptor activation. However, recent evidence suggests the ligand is not the only activator of gonadal steroid receptors. Two important alternative activators are growth factors and neurotransmitters. The activation of neural-specific growth factors and neurotransmitters along with HRT using tissue-selective hormones may be the key to neural-specific therapeutic interventions. Such ligand independent activation of gonadal steroid receptors is dependent on receptor phosphorylation.

Phosphorylation of gonadal steroid hormone receptors - The estrogen (38,39), androgen (40) and progesterone (41) receptors exist uniformly as phospho-proteins and demonstrate basal levels of phosphorylation. Binding of the ligand leads to hyper-phosphorylation of these transcription factors, a feature that is critical for activation. The human estrogen receptor, for example, is phosphorylated on both serine$^{118, 167}$ and tyrosine^{537} residues as a critical component of receptor activation (42,43). Phosphorylation of serine^{167} is ligand (estradiol)-dependent (42,44). However, phosphorylation of serine^{118} and tyrosine^{167} is ligand-independent and the result of the activation of mitogen-activated protein kinase (MAPK) and src- or tissue-specific tyrosine kinases (TK) (42,45,46) respectively. Such ligand-independent phosphorylation by MAPK and TKs suggests that nuclear steroid receptor activation may alternatively be regulated by signaling agents (i.e., either growth factors or neurotransmitters) acting at cell surface receptors.

Growth factor-mediated, ligand-independent activation of steroid hormone receptors - Growth factors including insulin-like growth factor (IGF)-1 (43,47,48), transforming growth factor α (TGFα, (49))and epidermal growth factor (EGF, (43)) appear to activate the estrogen receptor in a ligand-independent manner (Figure 1). Recent evidence (43) indicates that this activation is mediated by a Ras-Raf-MAPK pathway (for review, see (50)) of kinases. The potential for neural interactions of growth factors like IGF-1 with the estrogen receptor, is supported by early reports suggesting that insulin synergizes with estradiol-17β, to induce neurite outgrowth in cerebral cortical explant cultures (51).

The neurotrophins (nerve growth factor [NGF], brain derived growth factor [BDNF], neurotrophin-3 [NT-3] and neurotrophin-4 [NT-4]) are another important and relatively neural-specific growth factor family that may regulate the estrogen receptor. Neurotrophins are neuroprotective in the central nervous system, both during development and in the adult (reviewed in (52)). Indeed neurotrophic factor therapy has already been suggested for the management of a wide range of neurological disorders including Alzheimer's and Parkinson's disease, amyotrophic latreral sclerosis and peripheral neuropathies among others (52). These neuroprotective actions of the neurotrophins are transduced by specific, cell surface, receptor tyrosine kinases (trkA, trkB and trkC) (53). The neurotrophins and their receptors are preferentially expressed in the nervous system (54,55) and cardiovascular systems (56), two targets that are also protected by estrogens. NGF and estrogen receptors co-localize to the basal forebrain (12,57), a region that degenerates in Alzheimer's disease. NGF up-regulates the density of nuclear estrogen binding sites in vitro, in cerebral cortical explant cultures (58) and in a prototypic neurotrophin target, rat pheochromocytoma (PC12) cells (59). This increase in nuclear estrogen binding sites may be a function of hormone receptor phosphorylation, since the neurotrophins are known to phosphorylate and activate other nuclear transcription factors (e.g., CREB, (60) Fos (61), Pax (62)) including those (e.g., Nur77 (63)) related to the steroid receptor family. The binding of NGF to its cognate receptor (TrkA) activates a Ras-Raf kinase pathway (64). This in turn activates serine/threonine kinases such as MAPK (64), that are known to lead to the ligand-independent phosphorylation and activation of the estrogen receptor (43).

While the neurotrophins regulate nuclear estrogen receptor levels, estradiol in turn up-regulates mRNA for the NGF-specific receptor, trkA *in vivo*, in adult dorsal root ganglia (65) and *in vitro*, in PC12 cells (59). Estradiol also upregulates the levels of BDNF mRNA in adult cerebral cortex and olfactory bulb (66). Recent evidence suggests that the neurotrophins may be regulated by the binding of the estrogen receptor to a hormone response element within the BDNF gene (66). Collectively, these data suggest that neural-selective growth factors such as the neurotrophins may activate brain estrogen receptors specifically, as part of their neuroprotective role (Figure 1). Furthermore, the presence of reciprocal estrogen-neurotrophin regulatory interactions in neural tissue favors combination therapies involving both trophic factor activation and gonadal steroid replacement. Such therapeutic regimens may permit a reduction in the hormone dose, and a consequent reduction in the incidence of un-desirable, extraneural side effects.

Neurotransmitter activation of gonadal steroid hormone receptors - In cell culture transcription assays, the neurotransmitter dopamine appears to specifically activate both the progesterone receptor and the estrogen receptor (but not the glucocorticoid receptor) in the absence of hormone (Figure 1) (49,67). This ligand-independent activation of the estrogen and progesterone receptors appears to be mediated specifically by ligands (SKF-38393, **13**) that activate of the D1 receptor subtype (49). In the human brain, D1 receptors are extensively expressed in limbic and neocortical regions of the forebrain, regions that also express high levels of gonadal hormone receptors (68,69). Experiments in rats (70) indicate that the specific DI receptor agonist SKF-38393 but not the D2 receptor agonist (Quinpirole, **14**) induces a dose-related potentiation of estrogen (estradiol benzoate, 10μg, **15**)-induced lordotic (sexual

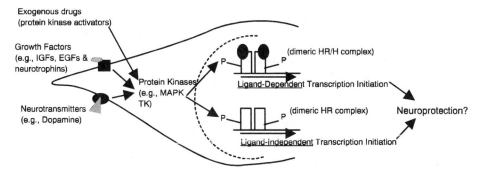

Figure 1: Growth factors and neurotransmitters are alternate activators of the steroid hormone receptors. **HR**, hormone receptor; **HR/H**, hormone receptor/hormone complex.

receptive) behavior. D1 receptor antagonists (SCH 23390, **16**) as well as progesterone antagonists (RU 38486 [**17**] and ZK 98299 [**18**]) specifically block the D1 agonist and estrogen-induced sexual behavior. The results of these studies suggest that combination therapies with HRT and specific D1 receptor agonists may be a mechanism to target the nervous system selectively. As with trophic factors, such neural-selective activation of the gonadal hormone receptors would decrease the negative impact of more traditional HRT by permitting a decrease in the effective dose of gonadal steroids required to produce a neuroprotective response.

REGULATION OF THE TRANSCRIPTIONAL ACTIVATION FUNCTIONS OF
GONADAL HORMONE RECEPTORS

A third approach to HRT may be to target intermediary factors that couple gonadal hormone receptors to the transcription machinery (Figure 2). Gonadal hormone receptors posses two transactivational (transcriptional activating) domains, a constitutively active N-terminal domain (AF-1) and a ligand-dependent domain (AF-2) (reviewed in (9)). The AF1 and AF2 transactivating domains interact indirectly with the TATA-binding protein (TBP) of the TFIID complex *via* transcriptional intermediary factors (TIFs, Figure 2). Recent research has identified several estrogen receptor-interacting TIFs. Some, such as TAF(TBP associated factor)$_{II}$30 interact with AF-2 domain of the estrogen receptor (71) and promote ligand-independent transcription in *in vitro* assays. In contrast, other TIFs such as TIF1 (72), RIP (receptor interacting protein)140 (73) and RIP160 (ERAP160 (74)) interact with the AF-2 domain of the estrogen receptor to modulate ligand-dependent transcription. The linking of the transactivating domains of gonadal hormone receptors to the TFIID complex by TIFs is a critical requirement for hormone mediated transcription. While neural-specific, gonadal hormone receptor-interacting TIFs have yet to be identified, the modulation of these specific intermediaries may be an additional tool that can supplement HRT in the management of disease.

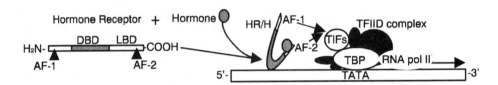

Figure 2: In the presence of ligand, transactivational domains (**AF-1** & **AF-2**) of hormone receptors (HR/H) interact with the TFIID complex *via* specific transcription intermediary factors (**TIFs**), to promote transcription. **DBD**, DNA binding domain; **LBD**, ligand binding domain; **TBP**, TATA binding protein. (The assembly of the transcription initiation complex is reviewed in (75)).

Conclusions - Hormone replacement therapy represents a promising avenue for the management of neurological and psychiatric diseases including Alzheimer's disease, depression and schizophrenia. The challenge however, is to design drug delivery systems and treatment protocols that target neural tissue specifically, while minimizing harmful effects on breast and uterine tissue. The use of ligands with tissue specific agonist or antagonist effects, specialized chemical delivery systems, neural specific growth factor- and neurotransmitter- mediated gonadal hormone receptor activation systems, as well as the direct targeting of transcriptional machinery, all represent reasonable strategies to achieve tissue specificity and therefore, effective therapy.

References

1. H. Fillit, H. Weinreb, I. Cholst, B. Luine, B. McEwan, R. Amador, and J. Zabriskie, Psychoneuro 11, 337 (1986).
2. H. Honjo, Y. Ogino, K. Naitoh, M. Urabe, J. Kitawaki, J. Yasuda, T. Yamamoto, S. Ishihara, H. Okada, T. Yonezawa, K. Hayashi, and T. Nambara, J Steroid B 34, 521 (1989).
3. H. Hafner, A. Riecher-Rossler, W. An Der Heiden, K. Maurer, B. Fatkenheuer, and W. Loffler, Psychological Med 23, 925 (1993).

4. F. Volkmar, P. Szatmari, and S. Sparrow, J. Autism Dev. Disord. 23, 571 (1993).
5. S. Lamberts, H. Tanghe, C. Avezaat, R. Braakman, R. Wijngaarde, J. Koper, and F. de Jong, J Ne Ne Psy 55, 486 (1992).
6. R. Lobo, Am J Obst Gyn 173, 982 (1995).
7. E. Barrett-Connor, and D. Kritz-Silverstain, JAMA 269, 2637 (1993).
8. D. Mangelsdorf, C. Thummel, M. Beato, P. Herrlich, G. Schutz, K. Umesono, B. Blumberg, P. Kastner, M. Mark, P. Chambon, and R. Evans, Cell 83, 835 (1995).
9. M. Beato, P. Herrlich, and G. Schutz, Cell 83, 851 (1995).
10. R. Miranda, and C.D. Toran-Allerand, Cerebral Cortex 2, 1 (1992).
11. C.D. Toran-Allerand, R. Miranda, R.B. Hochberg, and N.J. MacLusky, Brain Res 576, 25 (1992).
12. R. Miranda, F. Sohrabji, and C.D. Toran-Allerand, Mol Cell Neurosci 4, 510 (1993).
13. J. Kerr, R. Allore, S. Beck, and R. Handa, Endocrinol 136, 3213 (1995).
14. R. Handa, P. Connolly, and J. Resko, Endocrinol 122, 1890 (1988).
15. J. Kato, S. Hirata, A. Nozawa, and N. Yamada-Mouri, Hormone Beh 28, 454 (1994).
16. K. Hagihara, S. Hirata, T. Osada, M. Hirai, and J. Kato, J Steroid B 41, 637 (1992).
17. J. Stanford, J NIH Res 8, 40 (1996).
18. D. Brenner, W. Kukull, A. Stergachis, G. van Belle, J. Bowen, W. McCormick, L. Teri, and E. Larson, Am J Epidemiol 140, 262 (1994).
19. V. Henderson, A. Paganini-Hill, C. Emanuel, M. Dunn, and J. Buckwalter, Arch Neurol 51, 896 (1994).
20. A. Paganini, and V. Henderson, Am J Epidem 140, 256 (1994).
21. J. Ross, G. Stefanatos, D. Roeltgen, H. Kushner, and G. Cutler, Am J Med Genetics 58, 74 (1995).
22. S. Gordon, Am J Obst Gyn 173, 998 (1995).
23. J. Janowsky, S. Oviatt, and E. Orwoll, Behav Neurosci 108, 325 (1994).
24. R. Sands, and J. Studd, Am J Med 98, 1A-76s (1995).
25. L. Palinkas, and E. Barrett-Connor, Obstet Gyn 80, 30 (1992).
26. A. Gregoire, R. Kumar, B. Everitt, A. Henderson, and J. Studd, Lancet 347, 930 (1996).
27. W. Welshons, K. Engler, J. Taylor, L. Grady, and E. Curran, J Cell Phys 165, 134 (1995).
28. A. Wieck, R. Kumar, A. Hirst, M. Marks, I. Campbell, and S. Checkley, Br Med J 303, 613 (1991).
29. G. Evans, and R. Turner, Bone 17, 181s (1995).
30. H. Chen, H. Ke, C. Lin, Y. Ma, H. Qi, D. Crawford, C. Pirie, H. Simmons, W. Jee, and D. Thompson, Bone 17, 175s (1995).
31. M. Anzano, C. Peer, J. Smith, L. Mullen, M. Shrader, D. Logsdon, C. Driver, C. Brown, A. Roberts, and M. Sporn, J Nat Canc 88, 123 (1996).
32. B. Neubauer, K. Best, D. Counts, R. Goode, D. Hoover, C. Jones, M. Sarosdy, C. Shaar, L. Tanzer, and R. Merriman, Prostate 27, 220 (1995).
33. T. Hall, H. Nyugen, M. Schaueblin, and B. Fournier, Biochem Biophys Res Commu. 216, 662 (1995).
34. M. Brewster, P. Druzgala, W. Anderson, M. Huang, N. Bodor, and E. Pop, J Pharm Sci. 84, 38 (1995).
35. M. Brewster, S. Bartruff, W. Anderson, P. Druzgala, N. Bodor, and E. Pop, J Med Chem 37, 4237 (1994).
36. M. Rahimy, J. Simpkins, and N. Bodor, Drug Design and Delivery 6, 29 (1990).
37. M. Rahimy, J. Simpkins, and N. Bodor, Pharm Res 7, 1107 (1990).
38. S. Arnold, J. Obourn, M. Yudt, T. Carter, and A. Notides, J Steroid B 52, 159 (1995).
39. H. Lahooti, R. White, P. Danielian, and M. Parker, Mol Endocri 8, 182 (1994).
40. A. Brinkmann, G. Jenster, G. Kuiper, C. Ris, J. Van Larr, J. Van Der Korput, H. Degenhart, M. Trifiro, L. Pinsky, G. Romalo, H. Schweikert, J. Veldscholte, E. Mulder, and J. Trapman, J Steroid B 41, 361 (1992).
41. M. Bagchi, S. Tsai, M. Tsai, and B. O'Malley, Proc. Natl. Acad. Sci. 89, 2664 (1992).
42. S. Arnold, J. Obourn, H. Jaffe, and A. Notides, J Steroid B 55, 163 (1995).
43. S. Kato, H. Endoh, Y. Masuhiro, T. Kitamoto, S. Uchiyama, H. Sasaki, S. Masushige, Y. Gotoh, E. Nishida, H. Kawashima, D. Metzgar, and P. Chambon, Science 270, 1491 (1995).
44. S. Arnold, J. Obourn, H. Jaffe, and A. Notides, Mol Endocr 8, 1208 (1994).
45. S. Arnold, J. Obourn, H. Jaffe, and A. Notides, Mol Endocr 9, 24 (1995).

46. G. Castoria, A. Miglaiccio, S. Green, M. Di Domenico, P. Chambon, and F. Auricchio, Biochem 32, 1740 (1993).

47. Z. Ma, S. Santagati, C. Patrone, E. Vegeto, and A. Maggi, Mol Endocr 8, 910 (1994).

48. S. Aronica, and B. Katzenellenbogen, Mol Endocr 7, 743 (1993).

49. B. O'Malley, W. Schrader, S. Mani, C. Smith, N. Weigel, O. Conneely, and J. Clark, Recent Progress in Hormone Research 50, 333 (1995).

50. J. Downward, Cell 83, 831 (1995).

51. C.D. Toran-Allerand, L. Ellis, and K.H. Pfenninger, Dev Brain Res 41, 87 (1988).

52. F. Hefti, J Neurobiol 25, 1418 (1994).

53. M. Barbacid, J Neurobiol. 25, 1386 (1994).

54. R. Miranda, F. Sohrabji, and C.D. Toran-Allerand, Proc Natl Acad Sci 90, 6439 (1993).

55. Z. Kokaia, J. Bengzon, M. Metsis, M. Kokaia, H. Persson, and O. Lindvall, Proc Natl Acad Sci 90, 6711 (1993).

56. M. Donovan, R. Miranda, R. Kraemer, T. McCaffrey, L. Tessarollo, D. Mahadeo, D. Kaplan, P. Tsoulfas, L. Parada, D. Toran-Allerand, D. Hajjar, and B. Hempstead, Am J Path 147, 309 (1995).

57. C.D. Toran-Allerand, R. Miranda, W. Bentham, F. Sohrabji, T.J. Brown, R.B. Hochberg, and N.J. MacLusky, Proc Natl Acad Sci 89, 4668 (1992).

58. R. Miranda, F. Sohrabji, M. Singh, and C.D. Toran-Allerand, J Neurobiol 31, in press (1996).

59. F. Sohrabji, L. Greene, R. Miranda, and C.D. Toran-Allerand, J Neurobiol 25, 974 (1994).

60. D. Ginty, D. Bonni, and M. Greenberg, Cell 77, 713 (1994).

61. L.K. Taylor, K.D. Swanson, J. Kerigan, W. Mobley, and G.E. Landreth, J Biol Chem 269, 308 (1994).

62. C. Kioussi, and P. Gruss, J Cell Biol 125, 417 (1994).

63. J. Yoon, and L. Lau, Mol Cell Biol 14, 7731 (1994).

64. D. Kaplan, and R. Stephens, J Neurobiol 25, 1404 (1994).

65. F. Sohrabji, R. Miranda, and C.D. Toran-Allerand, J Neurosci 14, 459 (1994).

66. F. Sohrabji, R. Miranda, and D. Toran-Allerand, Proc Natl Acad Sci 92, 11110 (1995).

67. C. Smith, O. Conneely, and B. O'Malley, Proc Natl Acad Sci 90, 6120 (1993).

68. H. Hall, G. Sedvall, O. Magnusson, J. Kopp, C. Halldin, and L. Farde, Neuropsycho 11, 245 (1994).

69. J. Meador-Woodruff, D. Grandy, H. Van Tol, S. Damask, K. Little, O. Civelli, and S. Watson, Neuropsycho 10, 239 (1994).

70. S. Mani, J. Allen, J. Clark, J. Blaustein, and B. O'Malley, Science 265, 1246 (1994).

71. X. Jacq, C. Brou, Y. Lutz, I. Davidson, P. Chambon, and L. Tora, Cell 79, 107 (1994).

72. B. Le Douarin, C. Zechel, J. Garnier, Y. Lutz, L. Tora, B. Pierrat, D. Heery, H. Gronemeyer, P. Chambon, and R. Losson, EMBO J 14, 2020 (1995).

73. V. Cavilles, S. Dauvois, F. L'Horset, G. Lopez, S. Hoare, P. Kushner, and M. Parker, EMBO J 14, 3741 (1995).

74. S. Halachmi, E. Marden, G. Martin, H. MacKay, C. Abbondanza, and M. Brown, Science 264, 1455 (1994).

75. R. Tjian, and T. Maniatis, Cell 77, 5 (1994).

Chapter 3. P₂ Purinoceptors: A Family of Novel Therapeutic Targets

Michael Williams and Shripad S. Bhagwat
Neuroscience Discovery, Pharmaceutical Products Division,
Abbott Laboratories, Abbott Park, IL 60064 - 3500

Introduction: In the past two decades, the neurotransmitter / neuromodulator function of the purine nucleotide, adenosine triphosphate (ATP, **1**) has been clearly delineated from its role as an intracellular building block and energy source. ATP modulates CNS, immune and systemic function (1 - 3), acting via a family of cell surface receptors, the P₂ receptor superfamily (Table 1). ATP and related nucleotides also act as trophic or mitotic factors, alone or in combination with other pleiotropic / mitotic agents to affect DNA synthesis, cell differentiation and apoptosis (4 - 6). A physiological role for ATP was described in 1929 (7) and a neurotransmitter role in the 1950s (8), leading, in 1972, to the seminal purinergic nerve hypothesis (9, 10). Adenosine diphosphate (ADP, **2**) stimulates platelet aggregation (11) while uridine triphosphate (UTP, **3**), guanosine triphosphate (GTP, **4**) and diadenosine tetraphosphate (Ap4A, **5**) also modulate cellular function via receptor - mediated mechanisms (4 - 6, 12).

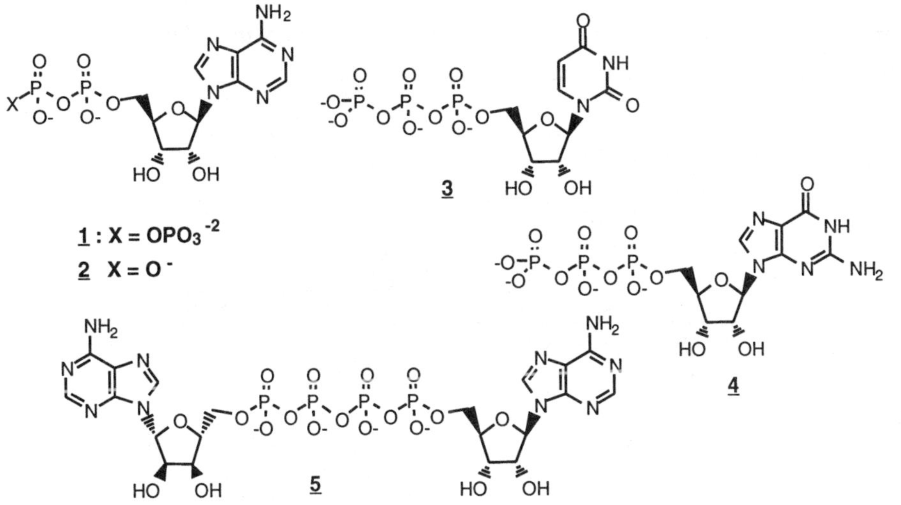

1 : X = OPO₃⁻²

2 X = O⁻

ATP AS A NEUROTRANSMITTER

ATP is an unusual candidate for a neurotransmitter, this role being conceptually viewed as an unorthodox, if not inappropriate, use of cellular energy stores (1,13). For each ATP molecule used extracellularly, three high energy phosphate bonds are hydrolyzed resulting in an energy "cost" of approximately -30 Kcal/mol. In comparison peptide neuromodulators use one ATP molecule to form each peptide bond at a "cost" of -10 Kcal/mol. The degradation of a 36 amino acid peptide like neuropeptide Y to terminate its neuromodulatory activity results in a minimal net energy debit of -350

Kcal/mol. Thus peptide neuromodulators potentially expend more energy than ATP. Energy usage is obviously not the sole criteria used to assess the role of a molecule as an intercellular mediator and must be viewed in the overall context of cell-to-cell communication in maintaining cell function. As well as interacting with members of the P_2 receptor family, ATP can be broken down via synaptic ectonucleotidase action to form adenosine (ADO), a neuromodulator in its own right (14). ATP can also modulate ADO interactions at the A_1 receptor (15). The nucleotide is potentially a multifunctional effector agent acting in a cascade - like manner to produce effects that may be mutually antagonistic. While ATP is a fast excitatory neurotransmitter at P_{2x} receptors (16, 17), ADO is a potent and selective inhibitor of excitatory transmitter release via the activation of presynaptic ADO A_1 receptors (14). The degree to which such mechanisms are physiologically important depends on tissue ectonucleotidase activity (18) as well as the array of purinergic receptors and cell types present (19). An additional facet of the neurotransmitter role of ATP is that its release, as well as degradation and resynthesis (the latter potentially in cells different from that from which ATP is released) may contribute to changes in the dynamics of the cellular "energy charge" (20) providing yet another nuance to the messenger role of ATP. The cellular energy charge and ATP have been linked to cellular apoptosis (21, 22).

P_2 PURINOCEPTORS

Purinoceptors were originally divided into P_1 (ADO) and P_2 (ATP) subclasses (10). Pharmacological evaluation led to the classification of P_{2x} and P_{2Y} classes (23) with the addition of P_{2T} and P_{2Z} receptors (24). Receptor characterization is limited by the

Table 1: Pharmacological classification of P_2 receptors (26)

Receptor	Agonist potency	Representative tissues
P_{2X1}	2-(4-nitrophenylethylthio) ATP > 3'-amino-3'deoxyATP = 2-hexylthioATP.	Guinea pig vas deferens
P_{2X2}	5-Fluoro-UTP > 2-hexylthioATP > 3'-acetylamino-3'-deoxy-ATP	Guinea pig and rat urinary bladder, rabbit bladder detrussor, rat colon longitudinal muscle
P_{2X3}	No known selective agonists	Vascular smooth muscle
P_{2X4}	No known selective agonists	Peripheral and central neurons
P_{2Y1}	2 - MeSATP $\geq$ ATP >> ADP > $\alpha\beta$ – meATP	Chick brain, rat brain cortex
P_{2Y2}	ATP $\geq$ UTP = ATPγS	Rat and guinea pig brain, rat renal mesangial cells, hepatocytes, osteoblasts, myocardial endothelium, skeletal muscle, aorta, liver, rabbit neutrophils, PC12 cells
P_{2Y3}	2- MeSADP > ADP	Platelets, ? brain endothelium
P_{2Y4}	2 - MeSATP >> ATP = ADP = $\alpha\beta$ meATP > $\beta\gamma$ meATP	Guinea-pig taenia coli, turkey erythrocytes
P_{2Y5}	2 - MeSATP >> ATP = ADP >> $\alpha\beta$ - meATP. $\beta\gamma$meATP inactive	Rabbit aorta, cortical and dorsal spinal cord astrocytes
P_{2Y6}	2 - MeSATP > ATP > ADP?	rabbit coronary, hepatic and mesenteric arteries
P_{2Y7}	Ap4A	Rat neurons, chromaffin cells

dependence on the rank order potencies of a series of ATP agonists and a lack of potent and selective antagonists.

Agonist potency can be tissue dependent, reflecting the receptor present and intrinsic ectonucleotidase activity (18). The paucity of potent and selective ligands for the various P$_2$ receptor subtypes has led to their potential function being deduced on the basis of localization and *in situ* hybridization studies, rather than the classical approach of using potent, selective compounds to determine *in vivo* function.

P$_2$ receptor nomenclature is currently in a state of considerable flux, evolving from a pharmacological basis (25, 26) to one relating receptor cloning to function (27). IUPHAR Receptor Nomenclature guidelines (28) require that pharmacologically defined receptors be described in italics e.g. *P$_{2Y1}$*. Cloned receptors for which <u>only</u> the amino acid sequence is known are shown in lower case e.g. p2y(x). Receptors for which <u>both</u> sequence <u>and</u> pharmacology are known are shown in upper case, e.g. P2Y$_1$. Clones for the two major P$_2$ receptor families are numbered in the sequence identified, based on structural, rather than cladistic, relationships (29). P$_2$ will be used generically to describe the receptor family.

The P2X is a ligand gated ion channel (LGIC) receptor family (30, 31), seven of which, P2X$_{1-7}$ have been cloned. The P2Y is a G-protein coupled receptor (GPCR) family with seven reported members designated P2Y$_{1-7}$ (32, 33). Two UTP- sensitive, ATP - insensitive pyrimidinoceptors (34 - 37) have been cloned as P2Y$_4$ and P2Y$_6$ receptors (36, 37) and the *P$_{2Z}$* receptor, an ATP - gated ion pore sensitive to 2'- and 3'-O- (4-benzoylbenzoyl) ATP (4), as the P2X$_7$ receptor (38). *Ap4A* and *P$_{2T}$* receptors (39) have yet to be cloned.

The evolution of P$_2$ receptor nomenclature has been confusing and is compounded by the random order in which receptors have been assigned a letter of the alphabet (40). The older nomenclature (x, y, t, z; 23, 24), the LGIC/GPCR nomenclature (Table 1; 26) and the evolving cloned receptor nomenclature (27) are used concurrently. For instance, the *P2$_U$* receptor is also known as the P2Y$_2$ (28, 41) and the *Ap4A* as the *P$_{2D}$* receptor (39). The first cloned P$_2$ receptors were members of the GPCR/P2Y family. The P2Y$_1$ receptor was cloned from chick brain (32, 42) and the P2Y$_2$ receptor from NG 108-15 neuroblastoma/glioma cells (41). The P2Y$_2$ receptor is responsive to both ATP and UTP. The P2Y$_4$ receptor is unique in being the only P$_2$ receptor lacking an RGD integrin binding domain (32). The P2Y$_3$ receptor responsive to ADP is developmentally expressed (32). The P2Y$_5$ receptor has been cloned from activated chick T cells (36, 43), the P2Y$_6$ from human placenta, the P2X$_1$ from rat vas deferens (44), the P2X$_2$ from PC12 neuroblastoma cells (45), the P2X$_3$ from rat dorsal root ganglion (DRG; 46,47), the P2X$_4$ from rat hippocampus (48) and the P2X$_5$ and P2X$_6$ from rat celiac and cervical ganglion, respectively (49). Other P2X receptors have been tentatively identified in rat brain (50).

The structure of P2X receptors is unlike other LGICs, consisting of two transmembrane spanning regions related to the epithelial amiloride - sensitive sodium channel. The P2X$_1$ is sensitive to α,β-meATP (<u>6</u>) and rapidly desensitizes. P2X$_2$ and receptors are insensitive to <u>6</u> and do not readily desensitize (30). The P2X$_1$ receptor is identical to RP- 2, a cDNA clone found in apoptotic thymocytes that potentially may play a role in programmed cell death (44, 45). P2X receptors exist as multimers in transfected oocytes (47, 51) with their pharmacological diversity and function being defined on the basis of whether and which homo- and hetero-oligomeric forms are formed. Sensory neurons, for instance, express LGICs composed of $P2X_2$ and $P2X_3$ subunits (47). At the present time, the relationship between hetero-oligomeric structure and receptor function has not been defined.

Table 2: P_2 receptor classification based on functional cloning (27)

Receptor	Tissue of Origin *GenBank Accession #*	Pharmacology	Transduction system
$P2X_1$	rat vas deferens - *X80477* Human urinary bladder *X 83688*	2 - MeSATP$\geq$ ATP > α,β meATP ATP > α,β meATP	$I_{Na/K/Ca}$ $I_{Na/K/Ca}$
$P2X_2$	Rat PC12 cell - *U14414*	2 - MeSATP > ATP, α,β meATP inactive	$I_{Na/K/Ca}$
$P2X_3$	rat dorsal root ganglion - *X90651*	2 - MeSATP > ATP > α,β meATP	$I_{Na/K}$
$P2X_4$	rat hippocampus *X91200* rat brain - *U32497/ X93565*	ATP > 2 - MeSATP > α,β meATP ATP > 2 - MeSATP > α,β meATP	$I_{Na/K}$
$P2X_5$	rat celiac ganglia - *X92069*	ATP > 2 - MeSATP > ADP	
$P2X_6$	rat brain, rat cervical ganglion - *X92070*	ATP > 2 - MeSATP > ADP	
$P2X_7$	mouse macrophage - *X95882*	BzATP>ATP>UTP	$I_{Na/K}$ or I_{ca}
$P2Y_1$	chick brain - *X73268* turkey brain - *U09842* insulinoma cell mouse - *U22829* - rat *U22830* bovine endothelium - *X87628*	2 - MeSATP > ATP> ADP, UTP inactive 2 - MeSATP > ATP> ADP, UTP inactive 2 - MeSATP $\geq$ 2Cl-ATP $\geq$ ATP, α,β meATP inactive 2 - MeSATP > ATP >> UTP	PLCβ/IP$_3$/Ca^{2+} PLCβ/IP$_3$/Ca^{2+} PLCβ/IP$_3$/Ca^{2+} PLCβ/IP$_3$/Ca^{2+}
$P2Y_2$	Mouse NG-108 -1 5 - *L14751* Human CT/43 cells - *U07225* rat lung - *U09402*	ATP = UTP >> 2 - MeSATP ATP = UTP >> 2 - MeSATP ATP = UTP	PLCβ/IP$_3$/Ca^{2+} PLCβ/IP$_3$/Ca^{2+} PLCβ/IP$_3$/Ca^{2+}
$P2Y_3$	chick brain	ADP > UTP > ATP = UDP	PLCβ/IP$_3$/Ca^{2+}
$P2Y_4$	Human placenta rat brain - *X91852*	UTP = UDP > ATP = ADP	PLCβ/IP$_3$/Ca^{2+}
$P2Y_5$	Activated chick T cells *P32250*	ATP > ADP > 2 - MeSATP >> UTP, α,β- meATP	
$P2Y_6$	Human placenta	UDP >5- Br- UTP > UTP > ADP >2 - MeSATP > ATP	PLCβ/IP$_3$/Ca^{2+}
p2y7	HEL cells - *U41070*	ATP > ADP = UTP	

I = LGIC cation current; PLC = phospholipase C; IP$_3$ = inositol-1, 4, 5 - trisphosphate. Bz ATP = 2' and 3' O - (4 - benzoylbenzoyl) ATP.

From a drug discovery perspective, it is a major step from the cloning and expression of the members of a new receptor family to defining their function (52). As noted, much of what is known regarding cloned P_2 receptor function is either structurally inferred or based on localization. The presence of the P2X$_3$ receptor in the rat DRG and its localization to small nociceptive neurons using anti-peripherin staining (46) suggests a role for ATP in pain perception. The assignment of function based on structural homology and/or localization also requires substantiation by physiological data. The latter can be derived by developing transgenic knockout or overexpression models or by genomic analysis of cDNA libraries from diseased human tissues. However, in the few instances where these molecular biological approaches have been successful, selective ligands are still required to focus drug discovery efforts.

Diadenosine Polyphosphate Receptors. The P_{2D}/P_{2Y7} receptor is a GPCR activated by diadenosine polyphosphates like Ap4A. It is present in the nervous (39) and cardiovascular (53) systems. Ap4A is an "alarmone" (54), a molecule mediating cellular responses to stress via gene transcription and translation mechanisms. Ap4A is equipotent to ATP at the P_{2U} receptor (55) and blocks platelet aggregation (P_{2T} receptor; 56) suggesting that it is not selective for a single class of P_2 receptor.

Pyrimidine Nucleotide Receptors? UTP mimics ATP effects in a number of tissues suggesting the existence of a P_{2U} or P_{2n} receptor, 'n' for nucleotide (1). A P_{2U} receptor in the C6-2B rat glioma cell activated by UDP and UTP but not by ATP (34) suggested that unique "pyrimidinoceptors" distinct from the various P_{2U} receptors might exist. Two human placental pyrimidinoceptors, P2Y$_4$ and P2Y$_6$ at which UTP and UDP are the preferred agonists respectively, are members of the P2Y receptor family (36, 37). ATP can antagonize the actions of UTP at the human uridine nucleotide receptor (35).

LIGANDS FOR P_2 PURINOCEPTORS

Agonists. All known agonist ligands for the P_2 receptor family are variations on the purine nucleotide pharmacophore (57- 60). SAR efforts have focused on increasing the stability of the polyphosphate side chain using various bioisotere strategies (58) and on modifying the purine nucleus to enhance selectivity and potency (59). ATP analogs like ARL 67156 (7) are potent ectonucleotidase inhibitors and can thus alter the rank order agonist potency of ATP analogs as well as endogenous ATP (18).

Bioisoteric substitution of the phosphoester bridging oxygens with methylene or dihalomethylene groups or the introduction of a terminal thio group increases ligand stability (13, 58). α,β-meATP (6) is an agonist at all P$_{2x}$ receptors except the P2X$_2$. 3'-Deoxy-3'-benzylaminoATP (8) is a more potent P2X agonist in the guinea pig vas deferens and urinary bladder. Attachment of substituted alkylthio chains at the 2-thio position of adenine nucleotides can enhance activity as in the 2 - (6 - cyanohexyl) analog (9), of the potent P$_{2Y}$ receptor agonist, 2 - methylthio ATP (2-MeSATP, 10). While AMP is inactive at P$_{2Y}$ receptors, 2-thioether AMP analogs like 11, have greater potency than ATP at erythrocyte P$_{2Y}$ receptors (13). This retention of affinity appears to occur as a result of adding a long chain at a site distal to the triphosphate group that reduces nucleotidase activity. The 2-thioether may also act as a secondary anchor for binding (13). These analogs are inactive at P$_{2x}$ receptors.

N^6-modification increases P$_{2Y}$ selectivity. N^6-methyl ATP is equipotent to ATP at taenia coli P$_{2Y}$ receptors and inactive at both vascular P$_{2Y}$-and smooth muscle P$_{2x}$-purinoceptors. N^6-methyl-2-(5-hexenylthio)-ATP (12) is active at erythrocyte and taenia coli P_{2Y} receptors and inactive at P2X receptors. AdoCAsp4, (**13**), where the labile phosphate groups are replaced with peptide residues like aspartate, glutamate and γ-carboxyglutamate is functionally active at C6 glioma cell P_2 receptors (60).

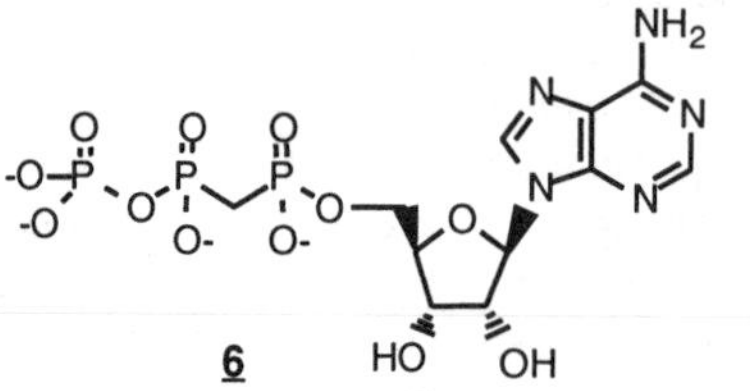

6 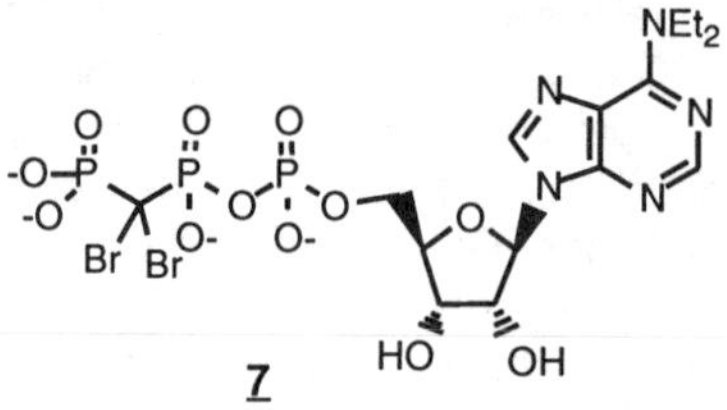   **7**

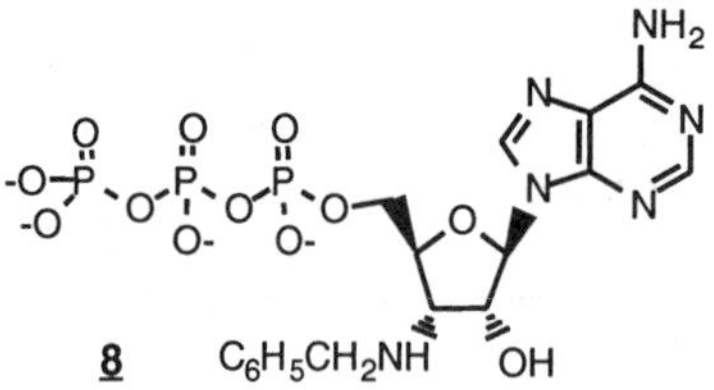

8 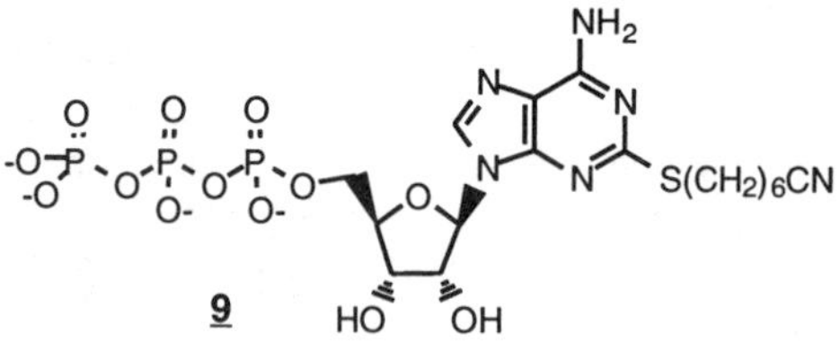   **9**

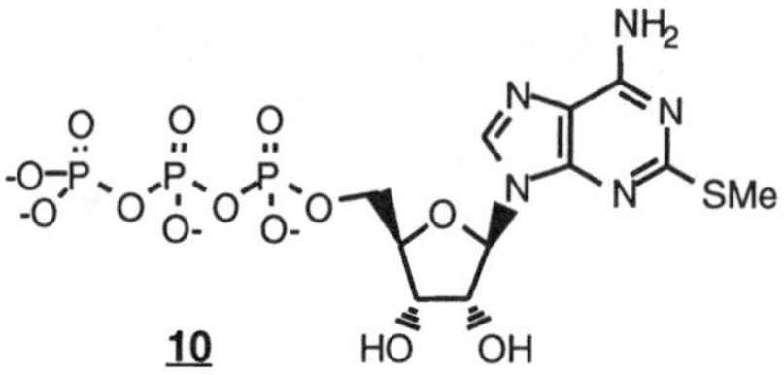

10 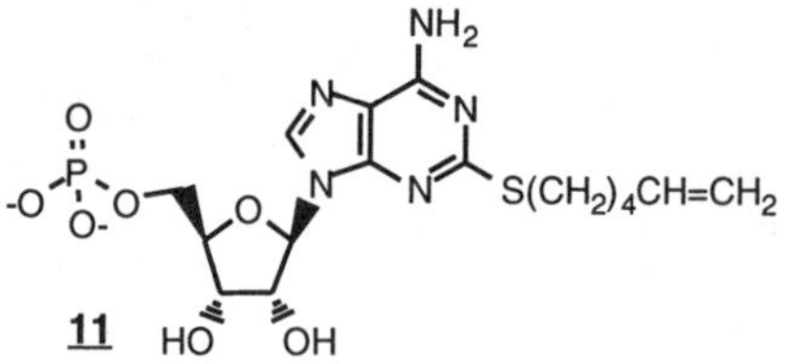   **11**

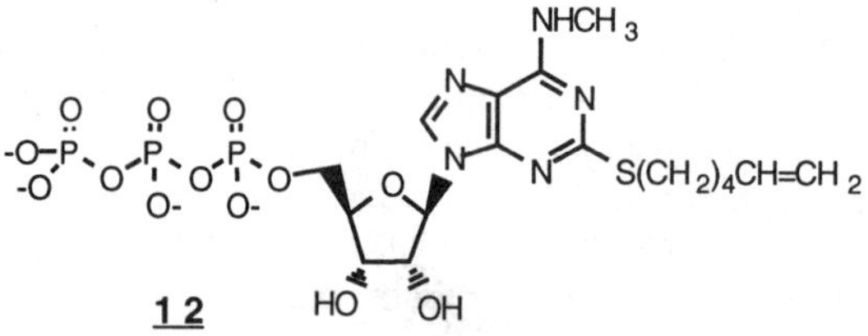

12 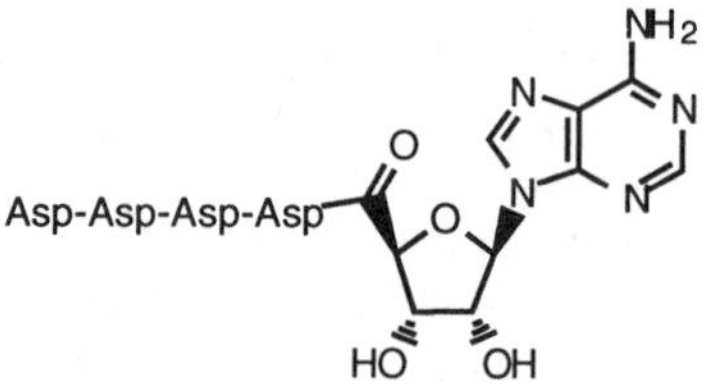   **13**

Antagonists. Antagonist ligands like suramin (**14**), PPADS (**15**) , and reactive blue 2 (**17**) evidence some degree of selectivity and efficacy at P_2 receptors. However, they have other activities that limit their use, especially *in vivo*. For instance, suramin, a polysulfonated aromatic antitrypanocidal drug, is a competitive P_2 receptor antagonist but also interacts with bFGF and NMDA receptors, ectonucleotidases and protein kinases. XAMR 0721 (**16**) is a suramin analog equipotent to the parent that lacks ectonucleotidase inhibitory activity (61). The identification of novel pharmacophores selective for P_2 receptors that can be used to unambiguously characterize the various members of the receptor superfamily is critical to advancing the field.

Allosteric Modulators. There is good precedent from NMDA, GABA$_A$/ benzodiazepine and nicotinic LGICs for physiologically relevant allosteric modulation of receptor function. PIT (2,2'-pyridylisatogen, **18**), originally described as a P_2 receptor antagonist, can enhance P2Y$_1$ receptor responses in oocytes (62).

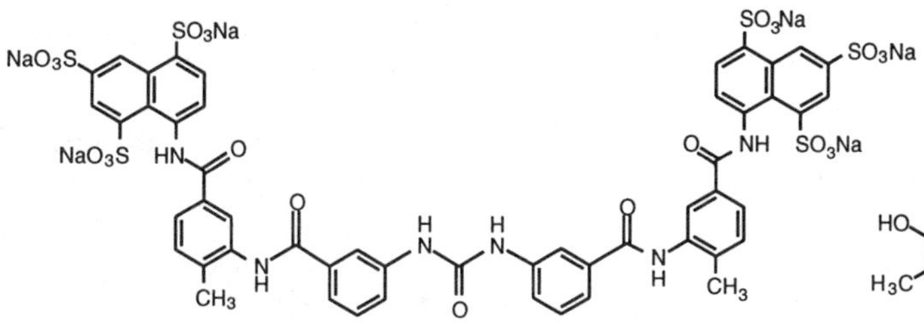

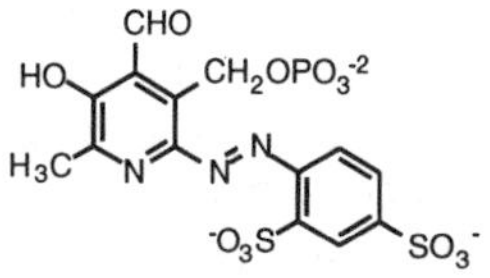

P_2 RECEPTOR CHARACTERIZATION

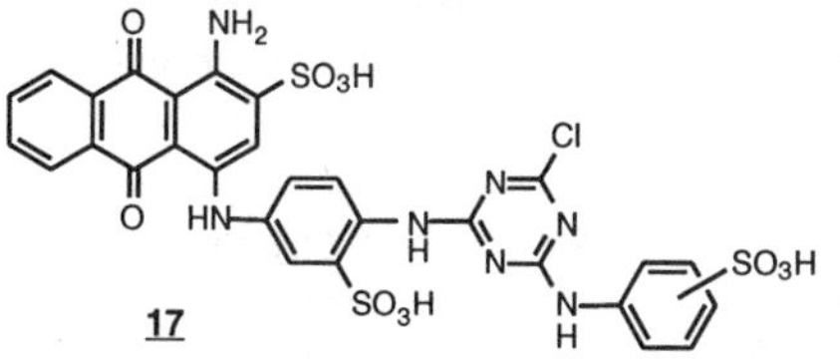

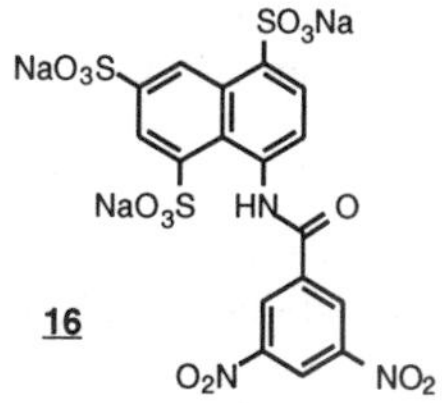

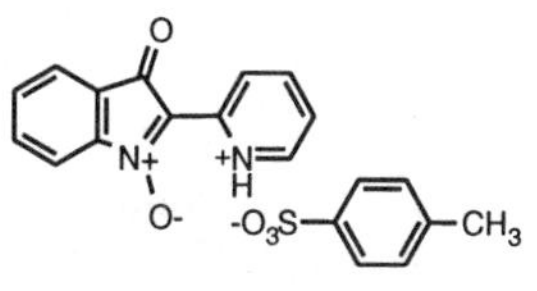

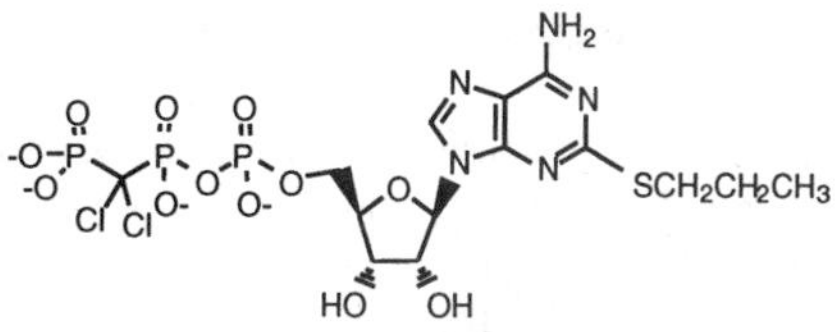

Binding Assays: Current P_2 receptor binding assays use agonist ligands (3), [^{3}H]α, β meATP and [^{35}S]ATPγS for P_{2X} and [^{32}P]BzATP and [^{35}S]dATPγS for P_{2Y}. (3, 31, 63 - 67).The density of brain P_{2Y} receptors (Bmax = 38 pmol/mg protein; 32) is 25 - 400 times greater than ADO receptors and greater than the density of CNS excitatory amino acid receptors. This suggests that brain P_{2Y} receptors subserve a critical role in CNS function. Improved radioligand binding assays are critical to the use of high throughput screening to identify novel, non - purine, non - nucleotide pharmacophores using chemical and combinatorial library compound sources.

Signal Transduction: P2Y receptors, like other GPCRs, are coupled to phospholipase C activation, adenylate cyclase modulation and Ca $^{2+}$ translocation (Table 2). P2Y receptor coupling can be used to differentiate receptor subtypes (3, 67). P_2 receptor activation can inhibit guanylate cyclase (68) and nitric oxide synthase (NOS) activity (1, 6). P2X receptors modulate cell function by gating Na$^+$, K$^+$ and Ca^{2+} permeability (3).

POTENTIAL THERAPEUTIC APPLICATIONS OF P_2 RECEPTOR LIGANDS

Cancer. ATP is an effective and long - lasting inhibitor of human tumor cell growth, an effect involving tumor cell membrane permeabilization via the opening of Ca^{2+} and Na$^+$ channels (69). These cytotoxic actions of ATP involve the gap junction protein, connexin- 43, and may thus be mechanistically related to the ATP-stimulated, P_{2Z} / P2Y$_7$ receptor-mediated apoptosis seen in macrophages (4). Exercise - related reductions in tumor volume have been associated with increases in tumor energy charge (70). A Phase II trial of ATP for the treatment of small cell lung cancer showed increased weight gain, improved performance status and quality of life (71).

Trophic and Apoptotic Effects. P_2 receptor agonists promote astrocyte maturation (6) and the proliferation of rat mesengial cells (5). ATP acts synergistically with bFGF to enhance mitotic processes, activating MAP kinase to modulate early response gene expression (e.g. *c-jun, c-fos*, AP-1 etc.) and cell growth, differentiation and stress responses (6). The apoptotic role of ATP has been most clearly defined for the macrophage P_{2Z} / $P2Y_7$ receptor (4). The structural homology between the $P2X_2$ receptor and the thymocyte apoptotic RP-2 receptor (44, 45) further supports a critical role for P_2 receptors in regulating events related to programmed cell death. Cellular ATP levels and mitochondrial ATP potential may also influence apoptotic events (21, 22).

Anesthesia. ATP can be used as an adjunct to inhalation anesthetics (72) reducing typical anesthetic doses, providing an improved safety margin, decreasing the risk associated with overdosage and toxicity and improving post surgical recovery time. ATP elicits controlled hypotension, decreases the possibility of surgical bleeding, induces analgesia without respiratory depression and blunting autonomic responses.

Cystic Fibrosis. ATP and UTP promote airway defense by stimulating mucin secretion, activating chloride channel function and increasing ciliary beat frequency (12). ATP and UTP have been evaluated as treatments for cystic fibrosis, UTP being preferred to ATP to avoid the bronchoconstrictor effects of the ADO formed from ATP. UTP alone, or in combination with amiloride, can restore normal function to the peripheral airways of the lung in cystic fibrosis patients.

Pain. The selective destruction of the rat DRG $P2X_3$ receptor by capcasin (45) suggests a role for ATP in pain processing. Antagonists selective for this receptor may represent novel, non-opioid analgesic agents (73).

Septic Shock. 2 - MeSATP blocks endotoxin-stimulated TNF-α and interleukin-1 (IL-1) release in a mouse model of septic shock (74) and prevents endotoxin lethality. 2 - MeSATP also attenuates LPS-induced expression of inducible macrophage NOS (75).

Thrombosis. ARL 67085 (**19**) is an ATP bioisostere with ADP antagonist activity (76) that is 30,000 - fold selectivity for the P_{2T} receptor. It had antithrombotic activity in dog without affecting cardiovascular parameters, was safer than fibrinogen receptor antagonists and superior in efficacy to aspirin (76).

Miscellaneous: P_2 receptor mechanisms may also be involved in cell adhesion (67), which in lymphocytes occurs via L-selectin shedding (77), chondrocyte and osteoblast function (78), the latter via potentiation of PDGF effects (79), diabetes (80), auditory (81) and renal cell (82) function. The fast transmitter role of ATP (16, 17) has yet to be linked to a distinct CNS function although mast cell P_2 receptors in the medial habenula may be involved in stress and neural-endocrine interactions (83).

FUTURE DIRECTIONS

    With an increased focus on rational drug design using cloned human molecular targets (52), the inability to relate a defined molecular target to a given therapeutic utility can significantly limit the prioritization and pace of a drug discovery program.

The identification of specific ligands, agonists, antagonists and allosteric modulators, to define the physiological / pharmacological function of a particular P$_2$ receptor thus remains a key element in the discovery of new therapeutic agents. An additional area of interest is whether the ATP-modulated potassium inward rectifier channel (K$_{IR}$s) family might not also be classified as a P$_2$ receptor subclass.

As knowledge in the area of P$_2$ purinoceptor function continues to accumulate and as medicinal chemistry and screening efforts are focused on the various discrete receptor targets, it is anticipated that novel P$_2$ receptor ligands, particularly antagonists, will represent innovative therapeutic modalities for the treatment of disease states in a variety of tissue systems including the CNS.

References

1. G. Burnstock in "P$_2$ Purinoceptors: Localization, Function and Transduction Mechanisms," Ciba Foundation Symposium, 198, John Wiley, Chichester, U.K., 1996, p.1.
2. H. Zimmerman, Trends Neurosci., 17, 420 (1994).
3. T.K. Harden, J.L. Boyer and R.A. Nicholas, Ann.Rev.Pharmacol.Toxicol., 35, 541 (1995).
4. F. DiVigilio, Immunol.Today, 16, 524 (1995)
5. E. Schulze-Lohoff, S. Zanner, A.Olgivie and R.B. Sterzel, Hypertension, 26, 899 (1995).
6. J.T. Neary, M.P. Rathbone, F. Cattabeni, M.P. Abbracchio and G. Burnstock, Trends Neurosci., 19, 13 (1996).
7. A.N. Drury and A. Szent-Gyorgi, J.Physiol (Lond.), 68, 214 (1929).
8. R.A. Holton and P. Holton, J.Physiol. (Lond.), 126, 124 (1954).
9. G. Burnstock, Pharmacol.Rev., 24, 509 (1972).
10. G. Burnstock in "Cell Membrane Receptors for Drugs and Hormones," L. Bolis and R.W. Straub, Eds., Raven, New York, 1978, p.107.
11. S.M.O. Hourani and D.A. Hall, Trends Pharmacol.Sci., 15, 103 (1994).
12. R.C. Boucher, M.R. Knowles, K.N. Olivier, W. Bennett, S.J. Mason and M.J. Stutts in "Adenosine and Adenine Nucleotides Nucleotides:From Molecular Biology to Integrative Physiology," L. Bellardinelli and A. Pelleg, Eds., Kluwer, Boston, 1995, p. 525.
13. M. Williams and K.A. Jacobson, Exp.Opin.Ther.Patents, 4, 925 (1995).
14. M Williams in "Psychopharmacology, The Fourth Generation of Progress," F.E. Bloom and D.J. Kupfer, Eds., Raven, New York, 1995, p. 643.
15. M.Williams and A. Braunwalder, J.Neurochem., 47, 88 (1986).
16. R.J. Evans, V. Derkach and A. Surprenant, Nature, 357, 505 (1992).
17. F.A .Edwards, A.J. Gibb and D. Colquhoun, Nature, 359, 144(1992).
18. C. Kennedy and P. Leff, Trends Pharmacol.Sci., 16, 168 (1995).
19. Y. Zhang, J. Palmblad and B.B. Fredholm, Biochem.Pharmacol., 51, 957 (1996).
20. D.E. Atkinson, "Cellular Energy Metabolism and Its Regulation" Academic, San Diego, CA, 1977.
21. L.M. Zheng, A. Zychlinsky, C.-C. Liu, D.M. Ojcius and J.D. Young, J.Cell.Biol., 112, 279 (1991)
22. C. Richter, M. Schweizer, A. Cossarizza and C. Francheschi, FEBS Lett., 378, 107 (1996).
23. G. Burnstock and C. Kennedy, Gen.Pharmacol., 16, 433 (1985).
24. J. Gordon, Biochem.J., 233, 309 (1986).
25. G. R. Dubyak and C. El- Moatassim, Am.J.Physiol., 265, 577 (1993).
26. M.P. Abbracchio and G.Burnstock, Pharmacol.Ther., 64, 445 (1994).
27. B. King and G. Burnstock (1996) "Purinoceptor Update Newssheet" available on the World Wide Web from the NIH Purine Home Page http//mgddk1.niddk.nih.gov.8000.
28. P.M. Vanhoutte, P.P.A. Humphrey and M. Spedding, Pharmacol.Rev., 48, 1 (1996).
29. J. Linden, M.A. Jacobson, C. Hutchins and M. Williams in "Handbook of Receptors and Channels. G -Protein Coupled Receptors," S.J. Peroutka, Ed., CRC Press, Boca Raton, FL., 1994, p. 29.
30. A. Surprenant, G. Buell and R.A. North, Trends Neurosci., 18, 224 (1995).
31. P.P.A. Humphrey, G.Buell, I. Kennedy, B. S. Khakh, A.D. Michel, A. Surprenant and D.J. Trezise, Naunyn-Schmideberg's Arch.Pharmacol., 352, 585 (1995).
32. J. Simon, T.E. Webb and E.A. Barnard, Pharmacol.Toxicol., 76, 302 (1995).
33. M.R. Boarder, G.A. Weisman, J.T. Turner and G.F. Wilkinson, Trends Pharmacol.Sci., 16, 133 (1995).
34. E.R. Lazarowski and T.K. Hardin, J.Biol.Chem., 269, 11830 (1994).
35. T. Nguyen, L. Erb, G.A. Weisman, A. Marchese, H.H.Q. Heng, R.C. Garrad, S.R. George, J.T. Turner and B.F. O'Dowd, J.Biol.Chem., 270, 30845 (1995).
36. D. Communi, S. Pirotton, M. Parmentier and J.M. Boeynaems, J.Biol.Chem., 270, 30849 (1995).
37. D. Communi, M. Parmentier and J.M. Boeynaems, Biochem.Biophys.Res.Comm., in press, (1996).

38. A. Surprenant, F.A. Rassendren, E. Kawashima, R.A. North and G. Buell, Science, 272, 735 (1996).
39. J. Pintor and M.T. Miras-Portugal, Gen.Pharmacol., 26, 229 (1995).
40. M.P. Abbracchio, F. Cattabeni, B.B. Fredholm and M. Williams, Drug Dev.Res., 28, 207 (1993).
41. K.D. Lustig, A.K. Shiau, A.J. Brake and D. Julius, Proc.Natl.Acad.Sci.U.S.A., 90, 5113 (1993).
42. T.E. Webb, J. Simon, B.J. Krishek, A.N. Bateson, T.G. Smart, B.J. King, G. Burnstock and E.A. Barnard, FEBS Lett., 324, 219 (1993).
43. T.E. Webb, M.G. Kaplan and E.A. Barnard, Biochem.Biophys.Res.Comm., 219, 105 (1996).
44. S. Valera, N. Hussy, R.J. Evans, N. Adami, R.A. North, A. Surprenant and G. Buell, Nature, 371, 516 (1994).
45. A.J. Brake, M.J. Wagenbach and D. Julius, Nature, 371, 519 (1994).
46. C.C. Chen, A.N. Akopian, L. Sivilotti, D. Colquhoun, G. Burnstock and J.N. Wood, Nature, 377, 428 (1995).
47. C. Lewis, S. Neldgart, C. Holy, R.A. North, G. Buell and A. Surprenant. Nature, 377, 432 (1995).
48. X. Bo, Y. Zhang, M. Nassar, G. Burnstock and R. Shoepfer, FEBS Lett., 375, 129 (1995).
49. G. Collo, R.A. North, E. Kawashima,E. Merlo-Pich,S. Neidhart.,A. Surprenant and G. Buell, J.Neurosci., 16, 2495 (1996)
50. E.J. Kidd, B.A. Grahames, J. Simon, A.D. Michel, E.A. Barnard and P.P.A. Humphrey, Mol. Pharmacol., 48, 178 (1995).
51   R.J. Evans, C. Lewis, G. Buell, S. Valera, R.A. North and A. Surprenant, Mol.Pharmacol., 48, 178 (1995).
52. M.J. Ashton, M. Jaye and J.S. Mason, Drug Disc.Today, 1, 11 (1996).
53. J. Walker, T.E. Lewis, E.P. Pivorun and R.H. Hilderman, Biochem., 32, 1264 (1993).
54. A. Varsgavasky, Cell, 34, 711 (1983).
55. E.R. Lazarowksi, W.C. Watt, M.J. Stuts, R.C. Boucher and T.K. Harden, Brit.J.Pharmacol., 116, 1619 (1995).
56. P.C. Zamecnik, K. Byung, M.J. Gao, G. Taylor and M. Blackburn, Proc.Natl.Acad.Sci.U.S.A., 89, 2370 (1992).
57. N.J. Cusack, Drug Dev.Res., 28, 244 (1993).
58. K.A. Jacobson, B. Fischer, M. Malliard, J.L. Boyer, C.H.V. Hoyle, T.K. Hardin and G Burnstock in "Adenosine and Adenine Nucleotides:From Molecular Biology to Integrative Physiology," L. Bellardinelli and A. Pelleg, Eds., Kluwer, Boston, 1995, p. 149.
59. B. Fischer, J.L. Boyer, C.H.V. Hoyle, A.U. Ziganshin, A.L. Brizzolara, G.E. Knight, J. Zimmet, G. Burnstock, T.K. Harden and K.A. Jacobson, J.Med.Chem., 36, 3937 (1993).
60. A. Uri, L. Jarlebark, I. von Kugelgen, T. Schonberg, A. Unden and E. Heilbronn, Bioorg.Med. Chem., 2, 1099 (1994).
61. A.M. van Rhee, M.P.A. van der Heijden, M.W. Beukers, A.P. Ijzerman, W. Soudijn and P. Nickel, Eur.J.Pharmacol., 268, 1 (1994).
62. B.F. King, C, Dacquet, A.U. Ziganshin, D.F. Weetman, G. Burnstock, P.M. Vanhoutte, M. Spedding and G. Burnstock, Br.J.Pharmacol., 117, 1111 (1996).
63. X. Bo, J. Simon, G. Burnstock and E.A. Barnard, J.Biol.Chem., 267, 17581 (1992).
64. A.D. Michel and P.P.A. Humphrey. Naunyn-Schmiederberg's Arch.Pharmacol., 348, 608 (1993).
65. A.D. Michel and P.P.A. Humphrey. Br.J.Pharmacol., 117, 63 (1996).
66. J.L. Boyer, C.L. Cooper, T.K. Harden, J.Biol.Chem., 265, 13515 (1990).
67. M.A. Ventura and P. Thomopoulos, Mol.Pharmacol., 47, 104 (1995).
68. S.J. Parkinson, S.L. Carrithers and S.A. Waldman, J.Biol.Chem., 269, 22683 (1994).
69. E. Rapaport, Drug Dev.Res., 28, 428 (1993).
70. P. Daneryd, T. Westlin, S. Edstrom and B. Soyssi, Eur.J.Cancer, 31A, 2309 (1995).
71. Bioworld, 6 (247) 12/28/95.
72. A.F. Fukunaga, T.A. Miyamoto, Y.Kikuta, Y. Kaneko and T. Ichinohe in "Adenosine and Adenine Nucleotides: From Molecular Biology to Integrative Physiology," L. Bellardinelli and A. Pelleg, Eds., Kluwer, Boston, 1995, p. 511.
73. C. Kennedy and P. Leff, Nature, 377, 385 (1995).
74. R.A. Proctor, L. Denlinger, P.S. Leventhal, S.K. Daugherty, J.-W. Van Der Loo, T. Tanake, G.S. Firestein and P.J. Bertics, Proc.Natl.Acad.Sci. U.S.A., 91, 6017 (1994).
75. L.C. Denlinger, P.L. Fisette, K.A. Garis, G. Kwon, A. Vazquez-Torres, A.D. Simon, B. Nguyen, R.A. Proctor, P.J. Bertics and J.A. Corbett, J.Biol.Chem., 271, 337 (1996).
76. R.G. Humphries, M.J. Robertson and P. Leff, Trends Pharmacol.Sci., 16, 179 (1995).
77. G.P. Jamieson, M.B. Snook, P.J. Thurlow and J.S. Wiley, J.Cell.Physiol., 166, 637 (1996).
78. H. Yu and J. Ferrier, Biochem.Biophys.Res.Comm., 191, 357 (1993).
79. S. Shimegi, Calcif. Tissue Inter., 58, 109 (1996).
80. D. Hillaire-Buys, G. Bertrand, J. Chapal, R. Peuch, G. Ribes and M.M. Loubatieres- Mariani, Br. J.Pharmacol., 109, 183 (1993).
81. D.J.B. Munoz, P.R. Thome, G.D. Housley, T.E. Billett and J.M. Battersby, Hear.Res., 90, 106 (1995).
82. P.C. Churchill and V.R. Ellis, J.Pharmacol.Exp.Ther., 266, 190 (1993)
83. R. Silver, A.-J. Silverman, L. Vitkovic and I.I. Lederhendler, Trends Neurosci., 19, 25 (1996).

Chapter 4. The Metabotropic Glutamate Receptors

David J. Madge and Andrew M. Batchelor
The Cruciform Project, 140 Tottenham Court Road, London, W1P 9LN

Introduction - L-Glutamate (**1**) is the neurotransmitter that mediates communication at the majority of excitatory synapses in the mammalian brain. Glutamate receptors are responsible for mediating and modulating synaptic transmission, and also for regulating the way in which electrical signals reaching a nerve cell are integrated. It is now well established that over-activation of glutamate receptors, or a failure of the cellular systems that regulate the activities of these receptors, plays a part in many CNS disease states such as epilepsy, ischaemia, neuropathic pain and neurodegenerative disorders. Glutamate elicits its effects at a variety of receptors and these can be broadly classified into the ionotropic glutamate receptors (iGluRs) and the metabotropic glutamate receptors (mGluRs). Ionotropic glutamate receptors are ligand-gated, cation-permeable channels and they may be further classified, according to their selective agonists, into the NMDA and the AMPA/kainate receptors (1).

The metabotropic glutamate receptors are characterised by the observation that they exert their effects *via* a G-protein. These receptors conform to the traditional 7-transmembrane (7-TM) spanning model of G-protein coupled receptors (GPCRs), but have little or no homology to other GPCRs. The existence of mGluRs was first postulated in 1985 when it was demonstrated that glutamate (GLU) could stimulate phospholipase C (PLC), in cultured striatal neurones, *via* a receptor that was not blocked by antagonists of the NMDA or AMPA receptors, and therefore did not belong to the iGluR family (2). Similar effects were also demonstrated, using various non-selective ligands such as quisqualate (QUIS, **2**), in other brain tissues and cultures, leading to the conclusion that glutamate, like GABA, 5-HT and acetylcholine, acts at G-protein coupled receptors as well as ligand-gated ion channels (3,4). Around the same time another glutamate analogue, L-2-amino-4-phosphonobutyrate (L-AP4, **3**) was recognised as a ligand for a new glutamate receptor sub-type (5), mediating a depression of glutamatergic transmission at many synapses (6). Similar receptors, coupled to cyclic GMP, have been discovered in ON bipolar cells in the retina (7). Developments in the mGluR area were accelerated by the discovery of the rigid glutamate analogue (+/-) *trans*-1-aminocyclopentane-1,3-dicarboxylic acid (*trans*-ACPD, **4**) (8), which was shown to be a selective agonist for mGluRs (9).

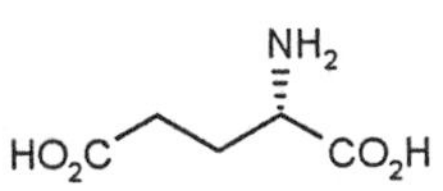

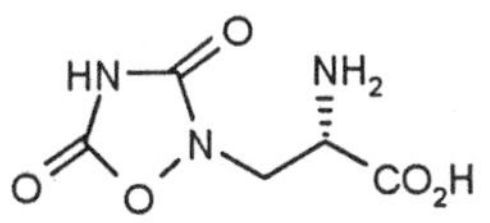

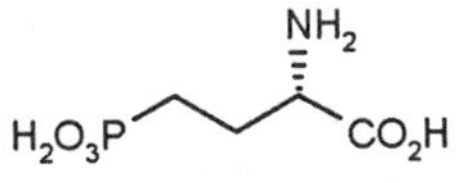

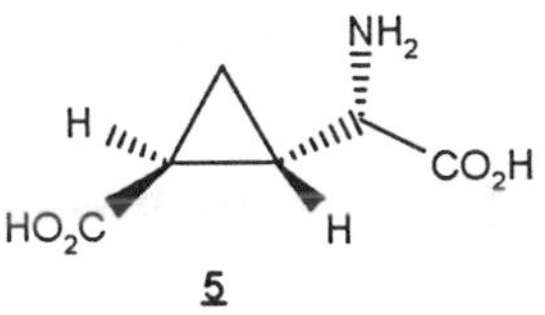

intracellular, carboxy-terminus domain, which is variable in length and not well conserved between members of this class of receptors, is involved with the selection of interactions with G-proteins (15,16). Computational analysis has revealed a high degree of homology between the extracellular domain of mGluRs (particularly mGluR1) and the leucine/isoleucine/valine bacterial periplasmic binding protein (LIVBP). The crystal structure of LIVBP, in conjunction with a sequence alignment with the extracellular amino terminal domain (ATD) of mGluR1 provided an indication of the possible glutamate binding site in the mGluR1 ATD, and this was supported by mutational studies in which the key serine-165 and threonine-188 residues in the mGluR1 ATD were converted to alanine residues (17). This model predicts a binding site for glutamate on one of a pair of globular lobes, with a connecting hinge region which closes the two lobes together upon binding of glutamate. The structure and transduction mechanisms of the mGluRs have been reviewed in depth recently (11,12,18).

DEVELOPMENT OF mGLUR AGONISTS

At present little is known about the physicochemical characteristics of the binding of glutamate and its analogues to the mGluR receptors. Consequently the search for new ligands has depended on either a screening approach or the use of established ligands as structural templates. The mGluR ligands reported to date all consist of an S-amino acid function linked through various chain or ring systems to a distal acidic function; clearly these ligands closely mimic the binding of glutamate to the mGluR receptors. The selective mGluR agonist, ACPD, in its most active 1S,3R conformer, adopts an extended relationship of the two acid functions, and this has been shown to be a common feature of mGluR agonists. In contrast, ligands for the iGluRs classically mimic a more folded conformer of glutamate.

Group-I Receptor Agonists - Quisqualic acid (4) remains the most potent agonist at these receptors to be reported so far, but it shows no selectivity between mGluR1 and mGluR5 (19), or over the ionotropic glutamate receptors, and its use is occaisionally complicated by an ability to sensitise certain Group-III receptors to increased activation by L-AP4 (the QUIS effect) (20). Phenylglycines have been widely reported as antagonists at metabotropic glutamate receptors (see later) but one such compound, 3,5-dihydroxyphenylglycine (6), has been found to act as a moderately potent agonist of Group-I receptors. This compound provided elevated phosphoinositide (PI) hydrolysis in adult and neonatal rat hippocampus (21) with a potency greater than that of ACPD. No effects on cAMP formation could be detected with this compound and it has recently been resolved and the mGluR activity found to reside, as would be expected, in the S enantiomer (22). The rigid glutamate analogue trans-ADA (7) has been reported to be able to activate Group-I receptors, as measured by inositol phosphate (IP) formation, in cerebellar granule neurons (23). However, a recent profiling of this compound in cells expressing various mGluRs showed no activity at the Group-I or Group-III receptors and only weak activity at the Group-II receptor (24). It remains to be determined whether the IP formation noted is mediated indirectly by a Group-II receptor or via an unidentified Group-I receptor.

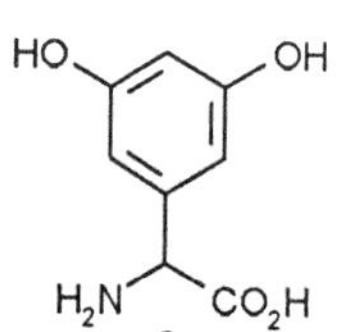

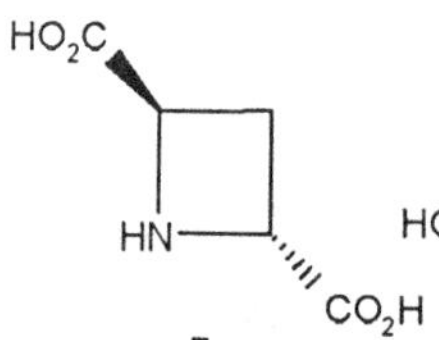

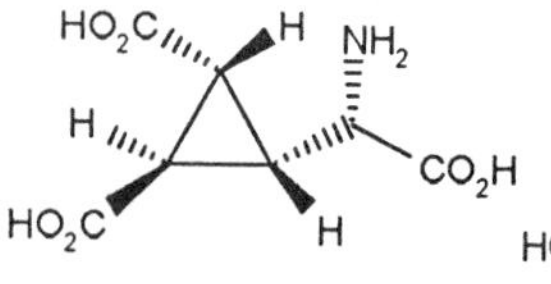

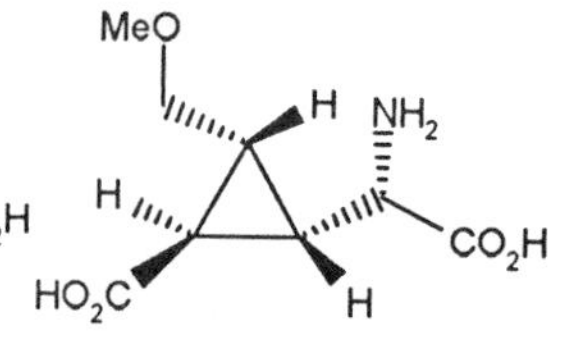

It was soon clear that the effects of ACPD were being mediated by multiple mGluR subtypes (10). Cross-hybridisation and PCR amplification have led to the discovery of 8 mGluR subtypes, and several splice variants. To date ACPD remains the only broad-spectrum mGluR agonist with no activity at iGluRs, but this compound shows little selectivity among the sub-types of mGluR (11).

CLASSIFICATION OF mGLURs

Traditional classification techniques for new receptor groups are based on the use of selective pharmacological tools. However, the advent of rapid molecular biology techniques for the discovery of receptor subtypes, and the range of mGluRs found using these techniques, means that selective ligands have yet to be developed for each mGluR subtype. Consequently the classification of mGluRs has been based mainly on the homology of the receptor proteins and their observed second messenger coupling (Table 1). The mGluRs have been divided into two main groups, according to the principal mode of second-messenger coupling identified in mammalian cells expressing the respective mGluR sub-types. The first group couple to up-regulation of the PLC cascade, and quisqualate is the most potent agonist for this class of receptor (Group-I). The remaining two groups both couple to down regulation of stimulated adenylate cyclase (AC) activity, but differ in their sensitivity to the glutamate analogue L-AP4. Group-II receptors are not responsive to this ligand (but are very sensitive to the cyclopropyl analogue of glutamate, L-CCG-I, (**5**) whereas at Group-III receptors L-AP4 is more potent than glutamate (12). Screening of cDNA libraries has revealed a significant homology of the mGluRs to a bovine parathyroid calcium sensing receptor (BoPCaR1), which is also coupled to PLC activation (13).

Table 1. Classification of Metabotropic Glutamate Receptors

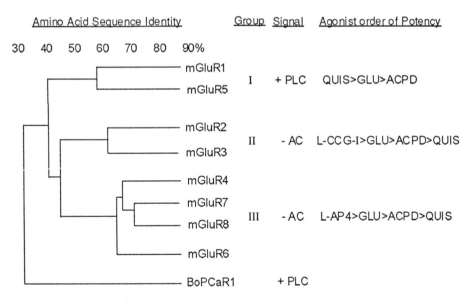

Although the classical 7-TM structural motif is retained in members of the mGluR family of receptors, the third intracellular loop is shorter than usual while the extracellular, N-terminal domain and the second intracellular loop are longer than usual, prompting suggestions that the latter may be critical in second messenger selectivity determination (14). There is also evidence accumulating that the

Group-II Receptor Agonists - L-CCG-I (**5**) is significantly more potent than either GLU or ACPD for the inhibition of forskolin-stimulated cAMP formation mediated by Group-II receptors, and is only weakly active at Group-III receptors. The dicarboxycyclopropyl analogue of **5**, (DCG-IV **8**), has been found to have enhanced potency at Group-II receptors (25), while demonstrating no effect on basal PI hydrolysis (26). An enhancement of QUIS-stimulated PI hydrolysis, however, has been observed with DCG-IV (27). Although it is not yet clear how this effect is mediated, involvement of a novel mGluR coupled to phospholipase D is a possibility (28). Replacing the 3-carboxyl moiety in DCG-IV with a methoxymethyl group gives rise to another L-CCG-I analogue, MCG-I (**9**), with quite potent activity at Group-II receptors (29). MCG-I, in cultured rat cortical cells, inhibited forskolin-stimulated cAMP formation with a potency slightly lower than DCG-IV and L-CCG-I, but greater than that of ACPD. MCG-I does not affect basal PI hydrolysis (30). A detailed conformational analysis of this compound, along with L-CCG-I, was published recently (31).

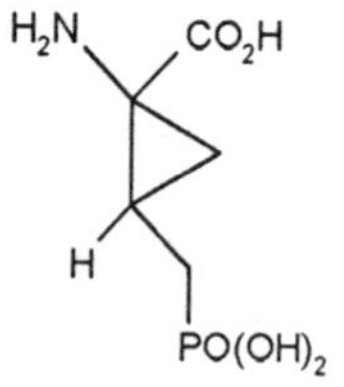

10 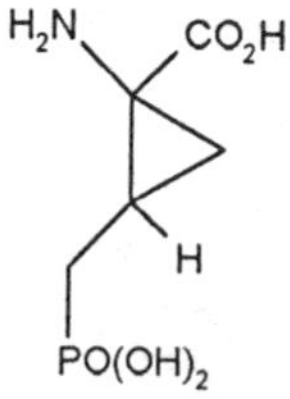   **11** 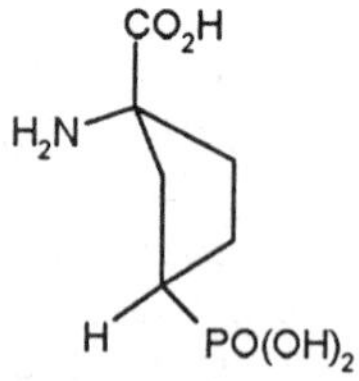   **12** 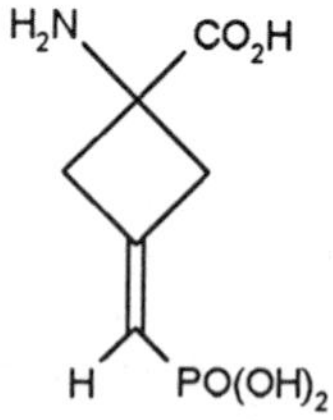   **13**

Group-III Receptor Agonists - The Group-III receptors are characterised by their sensitivity to L-AP4. The agonist selectivity and regional distribution of these receptors has led to the suggestion that one or more of this Group may be a candidate for the classical L-AP4 receptor. L-AP4 is a potent agonist for mGluR4 and mGluR6 but is much weaker at mGluR7 (32). The discovery of L-AP4 sensitive mGluRs led to a re-evaluation of a range of compounds that were originally synthesised in an attempt to characterise the L-AP4 receptor. A recent study compared the activity of a series of L-AP4 analogues at mGluR4a-expressing BHK570 cells to their potency as inhibitors of neurotransmission in the lateral perforant pathway (LPP) (33), one of the pathways in which L-AP4 has classically been shown to be active (34). The results obtained suggest some correlation between the two activities, with several L-AP4 analogues showing good activity in both assays (e.g. **10-13**). However, the mGluR antagonist MCPG (**20**) blocked the effect of L-AP4 on transmission in the LPP but did not block the inhibition of forskolin-stimulated cAMP accumulation produced by L-AP4. This suggests that another member of the Group-III receptors, rather than mGluR4a, mediates the effects of L-AP4 in this model. Some of these compounds have also been evaluated as agonists for the postsynaptic, cGMP-coupled L-AP4 receptor in retinal ON bipolar cells (mGluR6), although none were as active as L-AP4 itself (35). The agonist activity of a range of phenylglycines at Group-III receptors in the retina have been measured and several were found to be active, but with a different rank order of potency to that observed in spinal cord (36).

DEVELOPMENT OF mGLUR ANTAGONISTS

In the early days of mGluR research, several phosphonates (e.g. L-AP3, L-AP4) were identified as antagonists of glutamate-evoked PI hydrolysis in rat brain slices or in neuronal cultures. However, in mammalian cell lines transfected with individual mGluRs, L-AP3 and L-AP4 were found to be incapable of blocking glutamate stimulated PI hydrolysis even at millimolar concentrations of the putative antagonists

(37,38,39). The antagonist actions of these compounds may be the result of an indirect effect on PI hydrolysis in the case of L-AP3 (40), or *via* the mediation of a presynaptic Group-III receptor in the case of L-AP4 (41). However, recent data suggest that in earlier experiments, at least in the case of L-AP3, a weak antagonist activity was being masked by the use of high (competing) concentrations of glutamate (42). It is becoming increasingly clear that antagonist activity at mGluRs can often be achieved by the addition of an α-alkyl group to the amino acid moiety of a known agonist. Several examples of this are shown below.

Group-I Receptor Antagonists - The antagonist pharmacology of the Group-I receptors is based almost entirely around the wide range of phenylglycines synthesised by Watkins *et.al.* (43) (e.g. **14**-**20**). In all cases reported so far the mGluR activity has resided in the *S* or (+) isomer of the amino acid (44). Many of the earlier, α-hydrogen, phenylglycines, eg. CHPG, **14**, also demonstrated weak partial agonist activity at Group-II receptors (45). The α-alkyl analogues of these compounds show only antagonist activity, but they demonstrate only limited selectivity (46). The first of this class of ligand to be fully exemplified was **20,** an α-methyl phenylglycine (47). A conformationally constrained analogue of **20** that shows potent mGluR1a antagonist activity, and complete selectivity over Group-II and Group-III receptors, is (+/-)-1-aminoindan-1,5-dicarboxylic acid (**21**). This compound provides a useful insight into the preferred rotational conformation of the amino acid side chain in this class of antagonist (48).

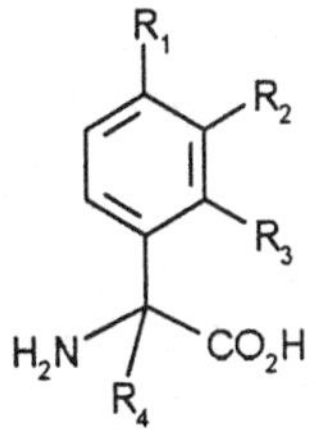

		R_1	R_2	R_3	R_4	
	14	COOH	OH	H	H	(CHPG)
	15	OH	COOH	H	Me	
	16	H	CH_2COOH	H	Me	
	17	CH_2COOH	H	H	Me	
	18	COOH	H	H	Et	
	19	COOH	H	I	H	
	20	COOH	H	H	Me	(MCPG)

Group-II Receptor Antagonists - Several of the phenylglycines described above show antagonist activity at these receptors, but unfortunately they are not selective over the Group-III receptors (46). LY302427 (**22**) reversed the ACPD induced inhibition of forskolin stimulated cAMP accumulation in rat hippocampal slices, and in cells transfected with the mGluR2 cDNA, whilst showing no activity at Group-I receptors (49). The α-methyl analogue of L-CCG-I, MCCG (**23**), was found to antagonise the synaptic depression caused by ACPD and L-CCG-I in a competitive manner, while having minimal effect on L-AP4 induced depression (50, 51).

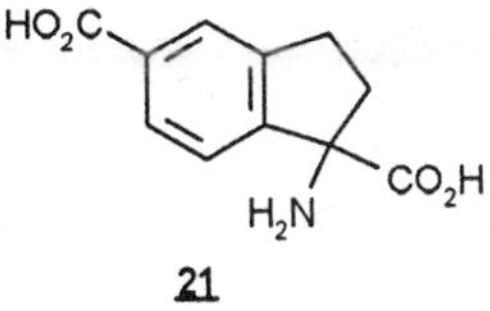

21 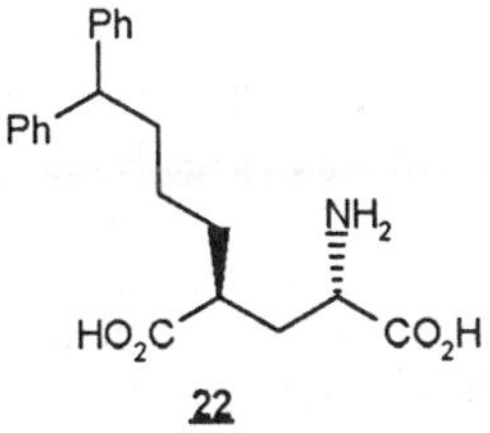　　　　　　　　**22** 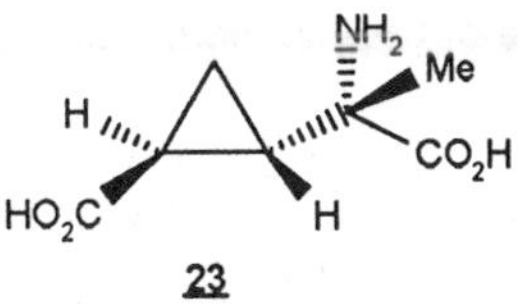　　　　　　　　**23**

Group-III Receptor Antagonists - The antagonist pharmacology of these receptors has only recently begun to be developed. In common with the Group-II receptors, the α-methyl analogue of a known agonist has frequently been shown to be an antagonist at these receptors. Thus MAP4 (**24**), the methyl analogue of L-AP4, is an antagonist at mGluR4a (52), and also blocks the synaptic depressant effects of L-AP4 in the LPP (53). Most of the phenylglycines that show antagonist activity at Group-III receptors also show similar potency at Group-II receptors (46) but **25**, the 4-phosphonic acid analogue of **20**, has been reported to be around 10-fold selective for inhibiton of L-AP4 mediated depression of monosynaptic responses in neonatal motoneurones, with respect to a similar depression induced by ACPD (54). In addition, α-methyl-3-carboxyphenylalanine, **26**, has been shown to be a potent and selective inhibitor of L-AP4 mediated inhibition of forskolin-stimulated cAMP accumulation (55).

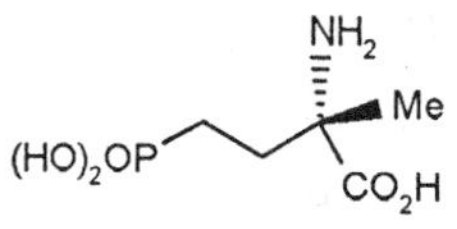

| 24 | 25 | 26 |

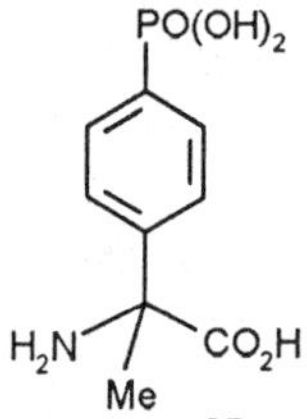

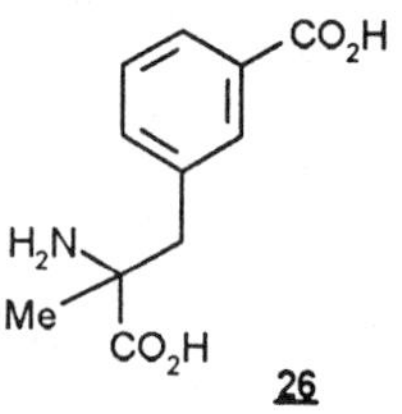

PHYSIOLOGICAL FUNCTIONS OF mGLURs

The availability of cloned cDNA sequences for distinct receptor subtypes, *via* their expression in cell lines, has enabled the full characterisation of mGluR ligands, and hence the identification of those that show some selectivity among the various mGluR subtypes. However, there is no guarantee that a receptor expressed in a cell type other than its usual host is normal, both in terms of its pharmacology and especially in terms of the linkage to second-messenger cascades. It is therefore imperative to study the mGluRs in their natural environment. Metabotropic glutamate receptors can be situated on pre- or post-synaptic membranes, and exert their influence through a range of G-proteins. In general, activation of postsynaptic mGluRs is excitatory (increasing the likelihood of the cell firing) and numerous studies show that activation of Group-I receptors can mediate this excitation. Activation of presynaptic mGluRs has been shown to inhibit transmitter release. Depending on whether the transmitter released is excitatory or inhibitory the effect of reducing its release can be either inhibitory or excitatory, respectively (11).

In order to determine the function of mGluRs it is important to identify the conditions under which they are activated. It is apparent that mGluRs generally require a high frequency burst of synaptic firing in order to become activated (56,57,58). In accordance with their ability to detect high frequency firing, mGluRs have been shown to participate in synaptic plasticity in several brain regions. The best studied model of synaptic plasticity is long term potentiation (LTP) of afferents into the CA1 pyramidal cells of the hippocampus. In common with considerable research in this area the data is controversial, with some groups reporting an essential role of mGluRs (59) and some indicating the opposite view (60). In the cerebellum, the parallel fibre input to Purkinje cells undergoes a long term depression (LTD) which, in the vast majority of reports, requires activation of mGluRs (61). It is early days for a full understanding of the capabilities fo mGluRs in brain function, but already it is clear that they are an interesting target for the development of therapeutic compounds.

POSSIBLE PATHOLOGICAL ROLES OF mGLURs

Malfunctioning glutamate transmission has been implicated in many disorders of the nervous system, including epilepsy, ischaemia, chronic pain and slow degenerative disorders.

Epilepsy - Epilepsy is a common neuronal disorder affecting about 1% of the population. It is characterised by abnormal patterns of neuronal activity where large populations of neurones fire synchronously. This can occur in a small localised region of the brain (focal epilepsy) or the whole brain (generalised epilepsy), and, depending on which parts of the brain are affected, it can result in stereotypic movements, sensory illusions, or even loss of consciousness. In some cases the underlying cause is a tumour, traumatic head injury, infection or an area damaged by oxygen starvation, but in many other cases the cause is unknown. What is common to all forms of epilepsy is that the normal balance between inhibitory and excitatory processes in the brain becomes biased towards excitation. Current anti-epileptic treatments act by widespread dampening of neuronal activity via an inhibitory interaction with the sodium channels responsible for action potentials (e.g. phenytoin or carbamazepine) or enhance inhibitory (GABAergic) transmission (e.g. phenobarbitone or the barbiturates). The other logical possibility, specific reduction of excitatory transmission, has been avoided because inhibition of the postsynaptic ionotropic glutamate receptors is prone to unwanted side effects. Currently available treatments are only effective in 50-80% of cases (62) and side effects are problematic for the majority of these; there is clearly room for improved therapies.

Metabotropic glutamate receptors could provide a new avenue for anti-epileptic treatment. Activation of mGluRs on the terminals of glutamate-releasing neurones, or blockade of those on inhibitory terminals or postsynaptic cells, would reduce excitation and might therefore be expected to be beneficial in epilepsy. This approach could be successful if different mGluR subtypes are found to be expressed on inhibitory terminals, excitatory terminals, and on the postsynaptic neurone, although the development of selective ligands is still required. Even with the limited knowledge and pharmacological tools available at the moment, there is already evidence for this differential expression of mGluR subtypes in the hippocampus (63), and also in the cerebellum (64).

There have been several reports of the actions of mGluR ligands using in vitro models of epilepsy. One such model is the hippocampal slice which can be made to show epileptiform activity by various means. In this model, when ionotropic glutamate receptors and $GABA_A$ receptors are blocked, application of the K^+-channel blocker 4-aminopyridine causes long bursts of synchronised firing that originate in the CA3 region. MCPG-sensitive mGluRs seem to be responsible for the formation of these bursts, and also for their propagation to the CA1 pyramidal cells (65). In the same preparation a different pattern of bursting behaviour is evoked by adding the $GABA_A$ blocker picrotoxin; under these conditions MCPG decreases the frequency of the bursts but has no effect on the shape of an individual burst. This suggests that endogenous glutamate activates mGluRs (probably the mGluR2/3 subclass) and that this is partially responsible for controlling burst firing in the disinhibited hippocampus (66). Inhibitory interneurones in the hippocampus are also activated via mGluRs, both synaptically and by exogenous agonists (possibly via mGluR1/5, 63) and in addition there are mGluRs on the terminals of inhibitory interneurones (64) and on excitatory terminals (67). In addition, the non-selective mGluR agonist, ACPD (4), has been shown to reduce the frequency of spontaneous epileptiform activity in the rat cortex (68). More recently the phenylglycine CHPG (14), which shows both agonist activity at

Group-II (usually presynaptic) and antagonist activity at Group-I (usually postsynaptic) receptors, has been shown to be capable of blocking audiogenic seizures (69).

Ischaemia - One of the most common forms of damage to the brain results from ischaemia. It is known that during an ischaemic attack, and for quite some time afterwards, levels of glutamate are elevated and this is thought to cause the excitotoxic damage that results in long term impairment of brain function. One or more mGluR may be activated by the elevated glutamate levels observed during ischaemia, and therefore contribute to the excitotoxicity, as has been implied from models in cultured neurones (70,71). In addition, there is growing evidence that vulnerable neurons can react to increased extracellular glutamate levels, following an ischaemic incident, by instigating changes in the expression of the mRNA for certain mGluRs (72). This suggests a role for these receptors in the modulation of excitability. It would be anticipated that a reduction in neuronal excitability could be achieved through the blockade of the postsynaptic, calcium mobilising, Group-I mGluRs, and some data for **20** in hypoxia/hypoglycemia models supports this (73). Depression of synaptic transmission can also be achieved by activation of the presynaptic (Group-II and/or Group-III) receptors, which appear to act as autoreceptors, responding to elevated glutamate levels by reducing glutamate release from the presynaptic terminal (74,75). Initial studies using **4** provided mixed results, almost certainly because of the lack of selectivity of this compound and the resulting mixed (and therefore not optimal for inhibition of neurodegeneration) pattern of mGluR activation (76,77).

Pain - The sensation of pain is a protective mechanism involving a specific neuronal pathway. There are two broad categories of pain: sharp/acute pain and dull/chronic pain. Stimuli such as heat, mechanical stress or chemical stress are detected in the skin by the free nerve endings of neurones situated in the dorsal root ganglion, which send axons to the dorsal horn region of the spinal cord. Here the primary sensory neurones synapse onto the spinothalamic neurones which in turn send axons up to the brain. The two sorts of pain are kept distinct by separate pathways. Glutamate is highly involved in synaptic communication in the dorsal horn (78) and the thalamus (79). There is already evidence of mGluRs being involved in both areas (80,81). Particularly interesting is their involvement in the plasticity of pain response, such as hyperalgesia and the phenomenon of 'wind-up'. Group-I mGluRs seem to be involved in both cases and in situ hybridisation shows mGluR5 in the correct place in the dorsal horn (82).

Slow Neurodegeneration - There has been a link postulated between the slow neurodegenerative diseases and glutamate, for example in Parkinson's disease, amyotrophic lateral sclerosis (ALS or motor neurone disease), Alzheimer's disease, cerebellar atrophies and Huntington's disease (83,84). Evidence suggests that glutamate uptake could be disrupted, leading to toxic levels in the cerebrospinal fluid of ALS patients (85,86). It would seem feasible that manipulation of mGluR activity could be used to inhibit glutamate release, without interfering with normal glutamatergic transmisssion, perhaps in a specific locations in the brain.

Conclusions - It is only in the last three years that the heterogeneity of mGluRs, and the complexity of their involvement in neurotransmission, has begun to be more fully explored. Clearly there is a pressing need for more potent and, vitally, selective ligands for these receptors in order to more concisely study their individual functions. The breadth of involvement of mGluRs in pathological functions that has already been elucidated will certainly fuel the search for new ligands.

References

1.   B. Sommer and P.H.Seebeurg, Trends. Pharmacol. Sci., 13, 291 (1992).
2.   F. Sladeczek, J.-P. Pin, M. Recasens, J. Bockaert and S. Weiss, Nature, 317, 717 (1985).
3.   F. Nicoletti, M.J. Iadarola, J.T. Wroblewski and E. Costa, Proc. Natn. Acad. Sci. U.S.A., 83, 1931 (1986).
4.   B. Pearce, J. Albrecht, C. Morrow and S. Murphy, Neurosci. Lett., 72, 335 (1986).
5.   A.C. Foster and G.E. Fagg, Brain Res. Rev., 7, 103 (1984).
6.   J.F. Koerner and R.L. Johnson in 'Excitatory Amino Acid Receptors; Design of Agonists and Antagonists,' Krogsgaard-Larsen, P. and Hansen, J.J. Eds, Ellis Horwood Limited, West Sussex, U.K., 1992, p.308.
7.   S. Nawy and C.E. Jahr, Nature, 346, 269 (1990).
8.   E. Palmer, D.T. Monaghan, and C.W. Cotman, Eur. J. Pharmacol., 166, 585 (1990).
9.   A.J. Irving, J.G. Schofield, J.C. Watkins, D.C. Sunter, and G.L. Collingridge, Eur. J. Pharmacol., 186, 363 (1990).
10.   D.D. Schoepp and P.J. Con, Trends Pharmacol. Sci., 14, 13 (1993).
11.   J.-P. Pin and R. Duvoisin, Neuropharmacol., 34, 1 (1995).
12.   T. Knopfel, R. Kuhn and H. Allgeier, J. Med. Chem., 38, 1417 (1995).
13.   E.M. Brown, G. Gamba, D. Riccardi, M. Lombardi, R. Butters, O. Kifor, A. Sun, M.A. Hediger, J. Lytton and S.C. Hebert, Nature, 366, 575 (1993).
14.   J.-P. Pin, J. Gomeza, C. Joly and J. Bockaert, Biochem. Soc. Trans., 23, 91 (1995).
15.   D.R. Hampson, E. Theriault, X.-P. Huang, P. Kristensen, D.S. Pickering, J.E. Franck and E.R. Mulvihill, Neuroscience, 5, 325 (1994).
16.   N. Gabellini, R.M. Manev, P. Candeo, M. Favaron and H. Manev, NeuroReport, 4, 531 (1993).
17.   P.J. O'Hara, P.O. Sheppard, H. Thogersen, D. Venezia, B.A. Haldeman, V. McGrane, K.M. Houamed, C. Thomsen, T.L. Gilbert and E.R. Mulvihill, Neuron, 11, 41 (1993).
18.   J.-P. Pin and J. Bockaert, Curr. Opin. Neurobiol., 5, 342 (1995).
19.   D.D. Schoepp, J. Bockaert and J. Sladeczek, Trends Pharmacol. Sci., 11, 508 (1990).
20.   E.R. Whittemore and J.F. Koerner, Brain Res., 489, 146 (1989).
21.   D.D. Schoepp, J. Goldsworthy, B.G. Johnson, C.R. Sakoff and S.R. Baker, J. Neurochem., 63, 769 (1994).
22.   S.R. Baker, J. Goldsworthy, R.C. Harden, C.R. Sakoff and D.D. Schoepp, Bioorg. Med. Chem. Lett., 5, 223 (1995).
23.   M. Favaron, R.M. Manev, P. Candeo, R. Arban, N. Gabellini, A.P. Kozikowski and H. Manev, NeuroReport, 4, 967 (1993).
24.   T. Knopfel, J. Sakaki, P.J. Flor, P. Baumann, A.I. Sacaan, G. Velicelebi, R. Kuhn and H. Allgeier, Eur. J. Pharmacol. (Mol. Pharmacol. Sect.), 288, 389 (1995).
25.   M. Ishida, T. Saitoh, K. Shimamoto, Y. Ohfune and H.Shinozaki, Br. J. Pharmacol., 109, 1169 (1993).
26.   F. Nicoletti, G. Cosabana, A.A. Genazzani, M.R. L'Episcopo and H. Shinozaki, Eur. J. Pharmacol. (Mol. Pharmacol. Sect.), 245, 297 (1993).
27.   A.A. Genazzani, M.R. L'Episcopo, G. Cosabana, H. Shinozaki and F. Nicoletti, Brain Res., 659, 10 (1994).
28.   T. Holler, E. Cappel, J. Klein and K. Loffelholz, J. Neurochem., 61, 1569 (1993).
29.   M. Ishida, T. Saitoh, Y. Nakamura, K. Kataoka and H.Shinozaki, Eur. J. Pharmacol. (Mol. Pharmacol. Sect.), 268, 267 (1994).
30.   M. Ishida, T. Saitoh and H. Shinozaki, Poster 251.8, presented at 25th Annual Meeting of the Society for Neuroscience, San Diego, California, 13th. Nov. 1995.
31.   K. Shimamoto and Y. Ohfune, J. Med. Chem., 39, 407 (1996).
32.   N. Okamoto, H. Seiji, C. Akazawa, Y. Hayashi, R. Shigemoto, N. Mizuno and S. Nakanishi, J. Biol. Chem., 269, 1231 (1994).
33.   P.A. Johansen, L.A. Chase, A.D. Sinar, J.F. Koerner, R.L. Johnson and M.B. Robinson, Mol. Pharmacol., 48, 140 (1995).
34.   J.F. Koerner, R.L. Johnson, N.L. Peterson and H.B. Kroona, Brain Res., 571, 162 (1992).
35.   W.B. Thoresen and J.S. Ulphani, Brain Res., 676, 93 (1995).
36.   W.B. Thoresen, T.J. Velte and R.F. Miller, Neuropharm., 34, 27 (1995).
37.   I. Aramori and S. Nakanishi, Neuron, 8, 757 (1992).
38.   D.S. Pickering, C. Thomsen, P.D. Suzdak, E.J. Fletcher, R. Robitaille, M.W. Salter, J.F. MacDonald, X.-P. Huang and D.R. Hampson, J. Neurochem., 61, 85 (1993).
39.   T. Abe, H. Sugihara, H. Nawa, R. Shigemoto, N. Mizuno and S. Nakanishi, J. Biol. Chem., 267, 361 (1992).
40.   M. Ikeda, Neurosci. Lett., 157, 87 (1993).
41.   P. Kristensen, P.D. Suzdak and C. Thomsen, Neurosci. Lett., 155, 159 (1993).
42.   J.A. Saugstad, T.P. Segerson and G.L. Weatbrook, Eur. J. Pharmacol. (Mol. Pharmacol. Sect.) 289, 395 (1995).
43.   J.C. Watkins and G. Collingridge, Trends. Pharmacol. Sci., 15, 333 (1994).

44. C. Thomsen, E. Boel and P.D. Suzdak, Eur. J. Pharmacol. (Mol. Pharmacol. Sect.), 267, 77 (1994).
45. Y. Hayashi, N. Sekiyama, S. Nakanishi, D.E. Jane, D.C. Sunter, E.F. Birse, P.M. Udvarhelyi and J.C. Watkins, J. Neuroscience 14, 3370 (1994).
46. J.S. Bedingfield, M.C. Kemp, D.E. Jane, H.-W. Tse, P.J. Roberts and J.C. Watkins, Br. J. Pharmacol., 116, 3323 (1995).
47. S.A. Eaton, D.E. Jane, P.L. St.J.Jones, R.H. P.Porter, P.C.-K. Pook, D.C. Sunter, P.M. Udvarhelyi, P.J. Roberts, T.E. Salt and J.C. Watkins, Eur. J. Pharmacol. (Mol. Pharmacol. Sect.), 244, 195 (1993).
48. R. Pellicciari, R. Luneia, G. Constantino, M. Marinozzi, B. Natalini, P. Jakobsen, A. Kanstrup, G. Lombardi, F. Moroni and C. Thomsen, J. Med. Chem., 38, 3717 (1995).
49. D.D. Schoepp, R.A. Wright, B.G. Johnson, N.G. Mayne, J.P. Burnett and A. Mann, Poster 251.7, 25th Meeting of the American Society for Neuroscience, San Diego, California, 13th Nov. 1995.
50. D.E. Jane, P.L.St.J. Jones, P.C.-K. Pook, H.-W. Tse and J.C. Watkins, Br. J. Pharmacol., 112, 809 (1994).
51. T.E. Salt and S.A. Eaton, Neuroscience, 65, 5 (1995).
52. P.A. Johansen and M.B. Robinson, Eur. J. Pharmacol. (Mol. Pharmacol. Sect.), 290, R1 (1995).
53. T.J. Bushell, D.E. Jane, H.-W. Tse, J.C. Watkins, C.H. Davies, J. Garthwaite and G.L. Collingridge, Neuropharm. 34, 239 (1995).
54. N.K. Thomas, D.E. Jane, K. Pittaway, D.C. Sunter and J.C. Watkins, Br. J. Pharmacol., 114 (Proc. Suppl., 9P, 1995).
55. M.C. Kemp, D.E. Jane, H.-W. Tse, P.J. Roberts and J.C. Watkins, Br. J. Pharmacol., 113 (Proc. Suppl., 147P, 1994).
56. A.M. Batchelor, D.J. Madge and J.G. Garthwaite, Neuroscience, 63, 911 (1994).
57. S. Charpak and B.H. Gahwiler, Proc. R. Soc. Lond., B243, 221 (1991).
58. D.A. McCormick and M.V. Krosigk, Proc. Natn. Acad. Sci. U.S.A., 89, 2774 (1992).
59. Z.A. Bortolotto, Z.I. Bashir, C.H. Davies and G.L. Collingridge, Nature, 368, 740 (1994).
60. M.J. Thomas and T.J. O'Dell, Brain Res., 695, 45 (1995).
61. D.J. Linden, Neuron, 12, 457 (1994).
62. H.P. Rang, M.M. Dale and J.M. Ritter, in 'Pharmacology'. 3rd Edition, Churchill-Livingstone, New York, 1995.
63. J.-C. Poncer, H. Shinozaki and R. Miles, J. Physiol., 485, 121 (1995).
64. I. Llano and A. Marty, J. Physiol., 486, 163 (1995).
65. R. Bianchi and R.K.S. Wong, J. Physiol., 487, 663 (1995).
66. L.R. Merlin, G.W. Taylor and R.K.S. Wong, J. Neurophysiol., 74, 896 (1995).
67. A. Baskys and R.C. Malenka, J. Physiol, 444, 687 (1991).
68. M.J. Sheardown, NeuroReport, 3, 916 (1992).
69. C. Thomsen, H. Klitgaard, M. Sheardown, H.C. Jackson, K. Eskesen, J. Jacobsen, S. Treppendahl and P.D. Suzdak, J. Neurochem., 62, 2492 (1994).
70. S. Miller, J.P. Kesslak, C. Romano and C.W. Cotman, Ann. N.Y. Acad. Sci., 757, 460 (1995).
71. T. Opitz, Neuropharm., 33, 715 (1994).
72. L. Iversen, E. Mulvihill, B. Haldeman, N.H. Diemer, F. Kaiser, M. Sheardown and P. Kristensen, J. Neurochem., 63, 625 (1994).
73. T. Opitz, P. Richter and K.G. Reymann, Neuropharm., 33, 715 (1994).
74. M. Vignes, V.R. Clarke, C.H. Davies, A. Chambers, D.E. Jane, J.C. Watkins and G.L. Collingridge, Neuropharm., 34, 973 (1995).
75. S.J. East, M.P. Hill and J.M. Brotchie, Eur. J. Pharmacol., 277, 117 (1995).
76. R. Siliprandi, M. Lipartiti, E. Fadda, J. Sautter and H. Manev, Eur. J. Pharmacol., 219, 173 (1992).
77. D.D. Schoepp, J.P. Tizzano, R.A. Wright and A.S. Fix, Neurodegeneration, 4, 71 (1995).
78. L. Urban, S.W.N. Thompson and A. Dray, Trends Neurosci., 17, 432 (1994).
79. T.E. Salt and S.A. Eaton, Neurochem. Int., 24, 451 (1994).
80. V. Neugebauer, T. Lucke and H.-G. Schaible, Eur. J. Neurosci., 6, 1179 (1994).
81. T.E. Salt and S.A. Eaton, Eur. J. Neurosci., 3, 1104 (1991).
82. Z. Vidnyanszky, J. Hamori, L. Negyessy, D. Ruegg, T. Knopfel, R. Kuhn and T.J. Gorcs, NeuroReport, 6, 209 (1994).
83. C. Ikonamidou and L. Turski, Curr. Opin. Neurol., 8, 487 (1995).
84. B. Meldrum and J. Garthwaite, Trends. Pharmacol. Sci., 11, 379 (1990).
85. J.D. Rothstein, G.Tsai, R. Kunel, L. Clawson, D.R. Cornblath, D.B. Drachman, A. Pestronk, B.L. Stauch and J.T. Coyle, Ann. Neurol., 28, 18 (1990).
86. A. Plaitakis, E. Constantakakis and J. Smith, Ann. Neurol., 24, 446 (1988).

Chapter 5. Drugs in Anesthetic Practice

David C. Rees and David R. Hill
Organon Laboratories Ltd, Newhouse, ML1 5SH, Scotland, UK.

Introduction - Anesthesia and the drugs that induce it has helped transform surgical interventions from being painful and, in many cases unsuccessful, to procedures that are predictable, safe and which bring about major benefit to patients. Advances in surgical procedures together with cost containment provide the impetus for discovering new anesthetic drugs. The current need is for fast acting drugs with minimal side effects and rapid patient recovery characteristics.

Surgical anesthesia is not simply a lack of consciousness. It is a state of reversible insensibility consisting of three separate components; hypnosis with amnesia, analgesia and skeletal muscle relaxation. The preservation of normal physiology (cardiovascular, acid-base and endocrine stability) is also desirable. No single molecule provides all three components of anesthesia at doses which do not impair critical functions. Consequently, the trend in recent years has been towards the combined use of individual agents, each providing one component of anesthesia. Thus, the concept of 'balanced anesthesia' has emerged.

Although there are many excellent reviews covering individual classes of anesthetic drugs, there is to our knowledge no recent publication which surveys the three pillars of anesthesia (hypnosis, analgesia and skeletal muscle relaxation). Anesthetics have not been reviewed in *Annual Reports in Medicinal* Chemistry for over 20 years.

In many respects the profile of agents used during anesthetic practice is markedly different from that which characterizes most CNS (and other) drugs. Some of these differences are listed below in order to introduce generic properties of anesthetic drugs and to highlight some of the atypical challenges facing medicinal chemists involved in discovering new anesthetic substances:

- administration is commonly by injection or inhalation (not oral).
- rapid onset of clinical effect is required (within a few seconds or minutes).
- short duration of effect with rapid and predicable patient recovery is desired.
- usually administered in hospitals with close medical supervision.
- used acutely (not chronically).
- drug combinations used to produce desired effect
- in vivo animal models predictive of the desired clinical effects are well described.

A number of chemical properties are claimed to be involved in obtaining the desirable pharmacokinetic properties of rapid onset and short duration of action. These include: the incorporation of ester groups susceptible to rapid metabolism by esterases (soft-drug deactivation approach) e.g. remifentanil, atracurium; decreasing the potency of muscle relaxants to increase the speed of onset; changing the relative solubility of volatile anesthetics in blood and tissues to change the recovery rate; utilizing the Hofmann degradation reaction to inactivate atracurium and redistribution of steroidal muscle relaxants within the body to terminate their effects. Further details are given in the appropriate sections below.

GENERAL ANESTHETICS (HYPNOTICS)

From the earliest days it was apparent that induction of anesthesia by inhalation of a volatile liquid (diethyl ether, chloroform) was unsatisfactory and there has been a continual search for better anesthetic agents. Rapid induction by the intravenous route is the most popular method today but maintenance of anesthesia has by contrast, remained largely dependent upon inhalation of gases. Recently gaseous agents with very low blood solubilities have been developed that provide excellent control over depth of anesthesia. These are described in more detail later in this section. The development of potent short acting intravenous agents suitable for both induction *and* maintenance (Total Intravenous Anesthesia, TIVA) now challenges the dominance of the inhalational anesthetics.

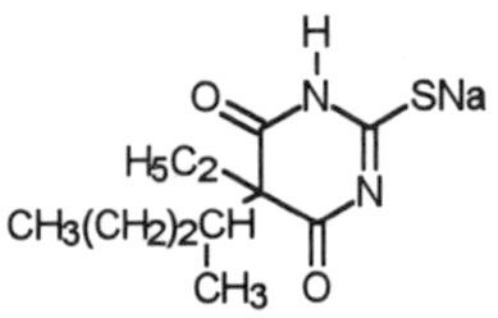

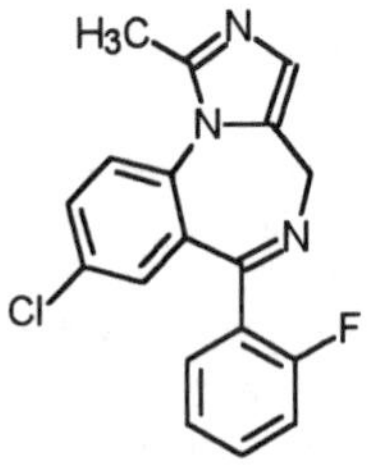

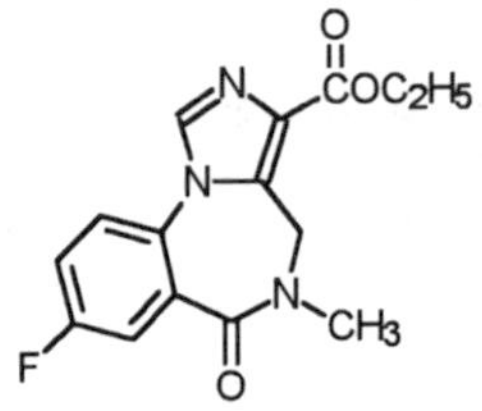

Intravenous Agents - Older agents such as thiopentone **1** and midazolam **2**, are still used extensively today. Thiopentone offers an inexpensive water-soluble agent which, when given as a bolus dose, induces a rapid and smooth hypnosis with minimal CNS excitatory phenomena. Sub-anesthetic brain concentrations occur quickly and result from a rapid redistribution throughout the body. The well-defined site of action of midazolam at the GABA$_A$ receptor has permitted its use to be combined with an antagonist, flumazenil **3** to provide rapid reversal from the sedative effects of the benzodiazepine (1,2). Propofol **4** has established itself as a popular induction *and* maintenance agent, becoming the drug of choice for ambulatory surgery in outpatients (3,4). The drug is rapidly redistributed from brain to peripheral tissues where it may persist for some time. Propofol is not water soluble and so is formulated in intralipid, an emulsion of soya bean oil, egg phosphatide and glycerol. Etomidate **5** is a short acting anesthetic agent which may be used either as part of a balanced anesthesia technique or to provide TIVA (5). Although the salt is soluble in water, it is unstable and so the drug is available for use as the base in propylene glycol/water or ethanol.

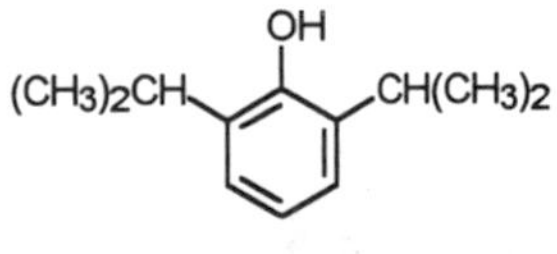

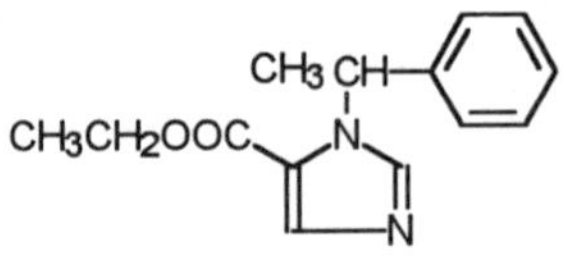

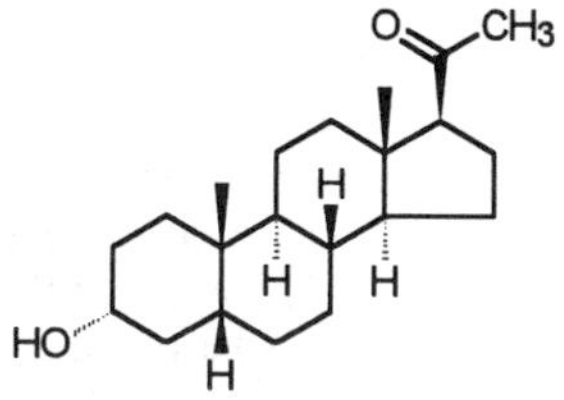

6 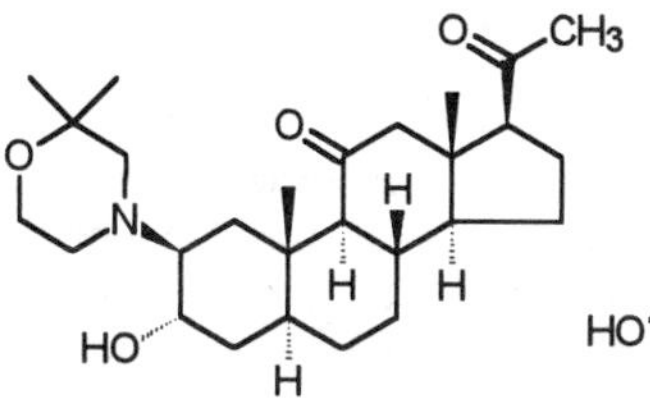   **7** 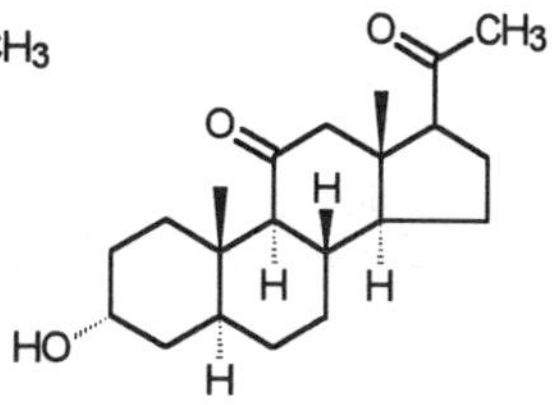   **8**

The endogenous steroid pregnanolone **6** (eltanolone) and Org 21465, a synthetic steroid **7** have recently been tested clinically (6-8). These are structurally related to the older anesthetic alphaxalone **8**. Org 21465 has the advantage of an amino substituent with water soluble salts. The clinical pressure to provide water soluble agents contrasts with the highly lipophilic nature of many anesthetic agents and the need to maintain lipophilicity in order to permit rapid access to the CNS.

Inhalational Agents - The general pharmacology of the inhalational agents halothane **9**, desflurane **10**, sevoflurane **11** and isoflurane **12** is similar but the compounds differ kinetically because of their varying solubilities in blood; a feature due primarily to the degree and type of halogen substitution. It has been proposed that the speed of equilibration of an anesthetic in the lung alveolus, which in turn determines the potential for speed of onset, is inversely related to the blood/gas partition coefficient. For the less soluble compounds with lower blood/gas partition coefficients and lower tissue/blood partition coefficients, the alveolar concentration more rapidly approaches that of the inspired gas and the more rapid is onset (9). It also follows that elimination of the anesthetic will be more rapid and so then will offset of anesthesia. The properties of newer inhalational agents are summarised below .

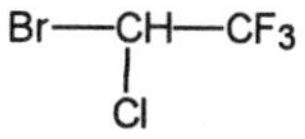

9 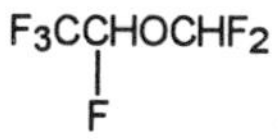   **10** 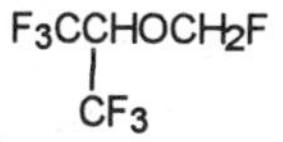   **11** 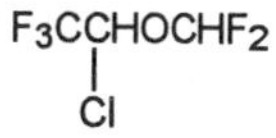   **12**

Table 1.   Comparison of two new inhalational anesthetics (9)

	Desflurane 10	**Sevoflurane 11**
Blood/gas partition coefficient	Best (very low)	Good (low)
Suitability for induction	Poor (pungent)	Good
Recovery characteristics	Very rapid	Rapid
In vitro stability	Normally stable	Difluorovinyls (toxic)
In vivo stability	Very stable	Fluoride ion
Cost compared to existing agents	Competitive	Increased

Although only recently introduced into the clinic, neither sevoflurane nor desflurane are new molecules. Both were first synthesised almost two decades ago but were not developed for a variety of reasons including low potency, difficulties in synthesis and fears over toxicity. Their introduction into the clinic now underlines how changing trends in surgery have driven the introduction of new molecules which permit greater precision of control and rapid recovery characteristics.

Because six of the seven F atoms in **11** are magnetically identical, this drug is a good candidate for in vivo magnetic resonance imaging. ^{19}F NMR spectroscopy at 4.7 Tesla has been used to observe the uptake, distribution and metabolism of this drug in rats. At anesthetic doses, it is detected in two compartments, brain tissue and neighbouring adipose/muscle tissue (10).

The biological and chemical transformations of volatile anesthetics leading to potentially toxic species is an issue of some concern. It has been proposed that metabolism may produce traces of reactive acyl fluorides (11) or F^- (12). Furthermore, there may be a chemical reaction between the volatile agent and the strong base (e.g. soda-lime) used to absorb CO_2 gas from the drug delivery apparatus generating small amounts of CO (13) or potentially toxic difluorovinyl species (14). It may be possible to alleviate this by using molecular sieves to remove the CO_2 (15).

Mechanisms - Theories of general anesthesia tend to fall into two main groups: the lipid hypothesis and the membrane-bound protein hypothesis, with present opinion tending towards the protein hypothesis (16). The lipid hypothesis is based upon the observation that there is a strong positive correlation between a compound's octanol:water partition coefficient and its ability to induce anesthesia. The protein hypothesis gains support from the finding that relevant concentrations of anesthetic agents preferentially interact in vitro with receptors for inhibitory neurotransmitters present throughout the nervous system.

Many anesthetics currently used are powerful positive allosteric modulators of $GABA_A$ (16-20). The gaseous agents halothane **9**, isoflurane **12** and the intravenous agents, propofol **4**, alphaxalone **8** and midazolam **2** together with barbiturates all potentiate the effects of low concentrations of GABA using in vitro assays. Anesthetics may influence strychnine-sensitive glycine receptors (18,21) in a similar way and voltage-gated calcium channels have also been implicated (22). Further support for a role of proteins and particularly $GABA_A$ receptors in anesthetic action comes from the finding of stereoselectivity in isoflurane's **12** anesthetic effect (23,24).

MUSCLE RELAXANTS

Skeletal muscle relaxants (neuromuscular blockers, NMB's) are used widely in the clinic particularly to paralyse laryngeal muscles prior to airway intubation and also to paralyse other skeletal muscles before general surgery. One of the most significant advances in the safety of anesthetic practice occurred in the early 1940's with the first use of a drug (curare) to produce muscle relaxation during surgery. An ideal modern agent should induce rapid block with minimal side effects, be inherently short acting or reversible e.g. by an anticholinesterase.

All currently used muscle relaxing drugs in surgery today achieve their effect by blocking the physiological effects of acetylcholine **13** at nicotinic acetylcholine receptors (nAChR) on the muscle. Like the $GABA_A$ receptors described above, nAChR's are membrane-bound hetero-pentameric ligand-gated ion channels (25). Poorly selective drugs may produce undesirable side effects by also blocking the effects of acetylcholine at other receptors, particularly on autonomic ganglia and the heart (26).

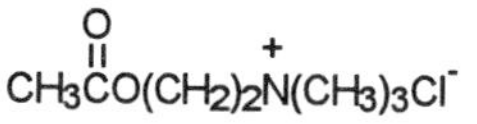

$$CH_3\overset{O}{\overset{\|}{C}}O(CH_2)_2\overset{+}{N}(CH_3)_3Cl^-$$

13

$$\begin{array}{c} COOCH_2CH_2\overset{+}{N}(CH_3)_3 \\ (CH_2)_2 \\ COOCH_2CH_2\overset{+}{N}(CH_3)_3 \end{array}$$

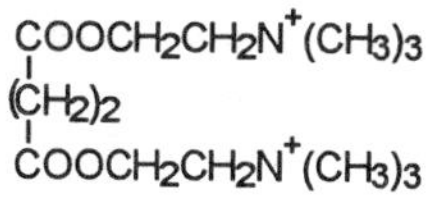

14

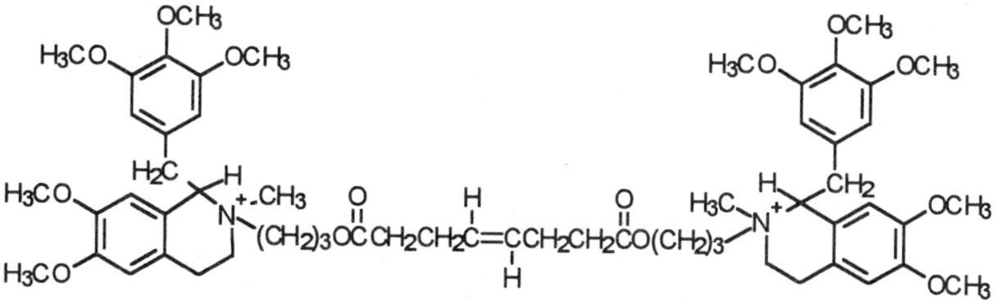

15

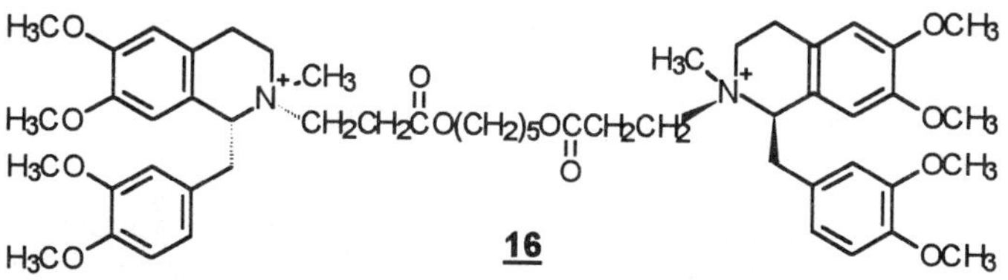

16

NMB drugs may be divided into two classes: depolarising (e.g. suxamethonium **14**) and the non-depolarising (e.g. mivacurium **15** cis-atracurium **16**, vecuronium **17** and rocuronium **18**). Depolarisers and non-depolarisers may be considered as agonists and antagonists respectively. The non-depolarising agents act in a purely competitive manner and unlike the depolarising blocker, their action may be reversed by an anticholinesterase e.g. neostigmine (27). The agonist suxamethonium **14** produces rapid muscle relaxation by receptor desensitization (tachyphylaxis). However it causes mechanism-related unwanted side effects such as muscle pain and autonomic activation (through ganglionic receptors). Such serious adverse effects have prompted the search for non-depolarising (competitive) neuromuscular blocking agents possessing the rapid onset and short duration of **14**. This has resulted in the introduction of compounds that have either rapid onset (**18**) or short duration (**15**). One other agent, still in the development phase **19** (Org 9487) may combine both rapid onset and fairly short duration (28). The aminosteroid **17** is a remarkably selective muscle type nAChR antagonist with negligible affinity for ganglionic or muscarinic receptors (26); properties which have made it a widely used drug.

Most clinically used NMB's contain two alkylated nitrogen atoms with at least one of these quaternised resulting in a permanently charged ('onium) species. Many of the SAR data are from whole animal experiments, frequently cats, and the two most widely accepted conclusions are: 1) NMB activity is loosely related to the inter-nitrogen distance, very approximately 1.2 nm is optimal and 2) non-depolarisers tend to be more sterically bulky and less conformationally flexible than depolarisers (29,30). The two most extensively studied series are represented by **17**, **18**, **19**, the aminosteroids (30) and **15**, **16**, the bisbenzylisoquinolinium diesters (31). Although completely different chemically, the conotoxin peptide muscle relaxants e.g. alpha-conotoxin have a very similar in vitro profile (32).

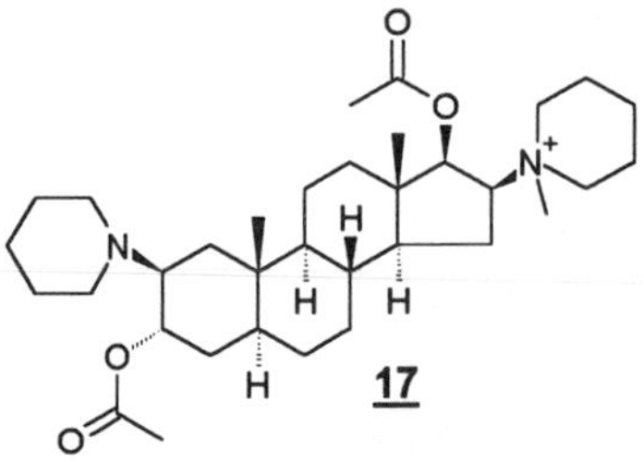

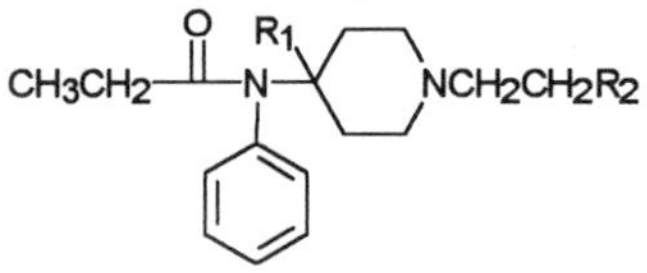

$R_1 = H$ $R_1 = CH_2OCH_3$ $R_1 = CO_2Me$
$R_2 = Ph$ $R_2 = $ ethyltetrazolone $R_2 = CO_2Me$

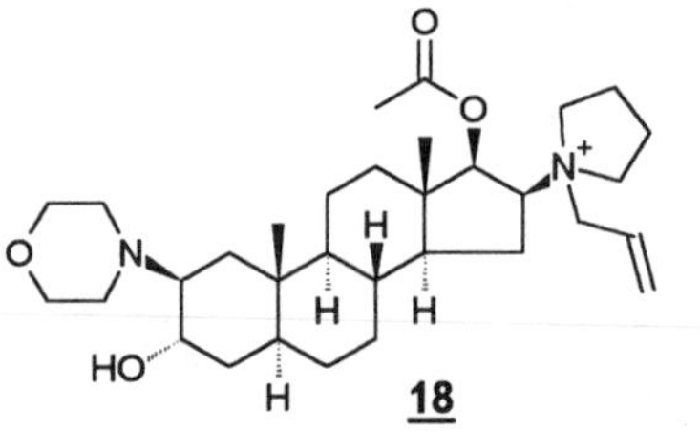

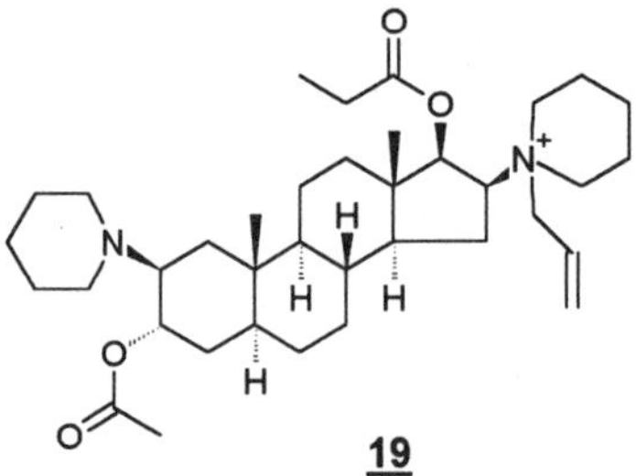

17   **18**   **19**   **20**   **21**   **22**

    In the cat, fast onset and brief duration may only be produced by compounds of rather low potency (33). The same inverse relationship between potency and onset time has also been observed in man for a range of different neuromuscular blockers (34). Accordingly, the newer agent **18**, which has an ED_{95} in man of 0.3 mg/kg (compared to 0.05 mg/kg for **17**) has a more rapid onset of block than other commonly used blockers (35). In the case of **16** duration is thought to be terminated by a combination of non-specific plasma esterase hydrolysis and non enzymatic base-catalysed Hofmann elimination of the quaternary amine. It was developed as the *1 R-cis, 1' R-cis* configuration of the mixture of stereoisomers known as atracurium. On a molar, basis **16** is 3 times more potent than atracurium so it is an intermediate-acting relaxant with a clean cardiovascular side effect profile (36). Compound **15** is a short acting non-depolarising blocker and has a duration of action in man between suxamethonium and atracurium or vecuronium (35).

ANALGESICS

    The analgesic properties of opioid agonists comprise one of the three key components of surgical anesthesia. In addition they have intrinsic hypnotic properties and may reduce the requirement for other general anesthetics. The clinical pharmacology of opioids is based upon over 2000 years of experience with the poppy plant extract, opium, which contains amongst others the alkaloid morphine. During the last two decades research into the pharmacology of opioid receptors has led to the identification of at least three G-protein coupled opioid receptor subtypes, designated mu, kappa and delta (37).

    The 4-anilidopiperidine, fentanyl **20** is the prototype of a series which includes the faster onset, shorter duration compound alfentanyl **21** which has over a decade of clinical use as an injectable analgesic. In surgical practice, opioids based on **20** can be used in conjunction with **4** or **2** to achieve total intravenous anesthesia (38). These, and all other currently used analgesic anesthetic opioids act as agonists at the mu-opioid receptor subtype and hence suffer from a number of predictable side-effects. In the context of anesthesia, respiratory depression and muscle rigidity cause the most common concern. The fentanyl chemical series continues to attract the interest of medicinal chemists (39). For example, trefentanil **23** which is a structural hybrid of **20**, **21** and the 4-phenylpiperidine pharmacophore of morphine, was investigated in clinical trials as an analgesic with a reduced cardiovascular side-effect profile (40). This compound was withdrawn but remifentanil **22** appears to be more promising. The methyl ester attached to the piperidine nitrogen is susceptible to rapid metabolism by blood and tissue esterases and since the resulting carboxylic acid metabolite is almost inactive it can be regarded as a soft-drug. It has a rapid onset and appears to be ultra-short acting in man and suitable for administration by continuous infusion (41-43).

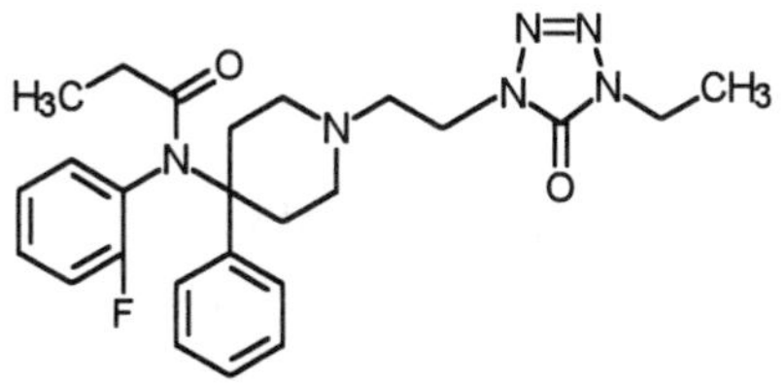

23 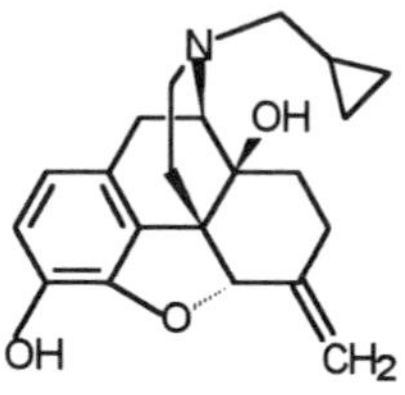　　　　　　　　　　　　　　　　**24**

Opioid (side) effects can be reversed by administering an opioid antagonist such as nalmefene **24** which has a longer elimination half-life than the structurally-related prototype antagonist naloxone, hence reducing the risk of the agonist effects reappearing (44).

The utility of kappa or delta receptor selective opioids in anesthetic practice has not been reported but this may reflect the lack of appropriate ligands. Recently, both kappa-selective (45) and delta-selective (46) agonists have been shown to have analgesic effects in animal studies and spinally administered combinations of delta and mu (but not kappa) agonists and an α_2 adrenergic agonist may be synergistic in the control of visceral pain (47).

In addition to opioids, N_2O is still widely used in anesthetic practice, particularly for its analgesic properties (48). Alpha$_2$ adrenoceptor agonists such as medetomidine and its enantiomer dexmedetomidine **25** have potential for clinical use as non-opioid analgesics (49).

LOCAL ANESTHETICS

The use of local anesthetics in medical practice can be traced back to 1884 when the alkaloid cocaine was first used in this context. Their anesthetic mechanism is believed to be due to (intracellular) block of voltage-dependent Na+ channels although they also interact with other channels such as voltage-dependent K+ channels, Ca2+ activated K+ channels and glibenclamide-sensitive K+ channels (50). Local anesthetics are particularly popular for regional surgery (epidural injections used during childbirth and spinal injections for surgery of the lower limbs and abdomen) and outpatient surgery where they offer a less complicated and less expensive alternative to general anesthesia. Local anesthetics are also widely available for non-surgical use (e.g. throat lozenges, topical preparations) but this aspect is not described here.

Many local anesthetics (e.g. bupivacaine (+) 26,) are used clinically as racemic mixtures (51). In 1996 S-ropivacaine 27 was launched as the first single enantiomer local anesthetic and this compound also has some post operative analgesic properties (52). Levobupivacaine 26, the S-enantiomer of bupivacaine, is in advanced clinical trials and, on the basis of preclinical and clinical studies, it is claimed to have less cardiotoxicity and similar anesthetic potency than the racemate (53) hence making it safer for use in patients with compromised cardiac function. Sameridine 28 provides a potentially useful combination of spinal anesthesia plus subsequent analgesia (54).

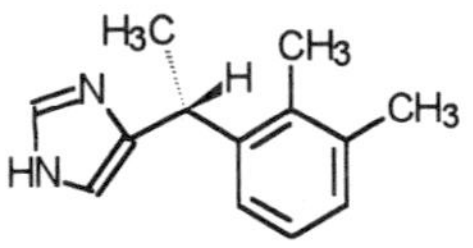

25

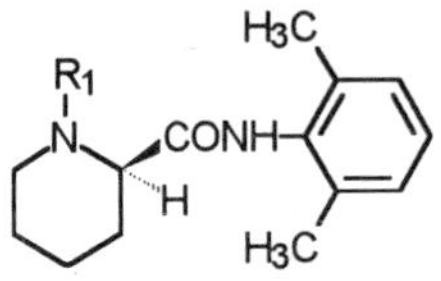

26 R₁ = nBu
27 R₁ = nPr

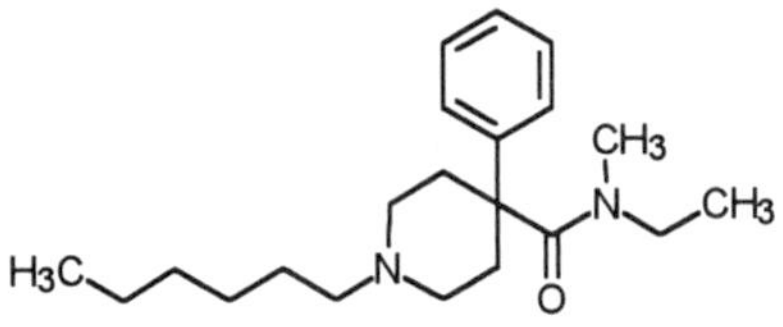

where 26 R_1 = nBu and 27 R_1 = nPr

28

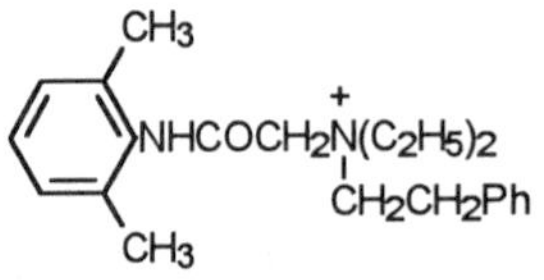

29

The structures 26 and 27 illustrate the groups classically associated with local anesthetic activity; a tertiary amine linked by an amide to an aromatic ring. In a slight modification of this pharmacophore, the linker group has been lengthened by two carbon atoms and the aromatic ring replaced by benzotriazole to give compounds with similar potency but with improved therapeutic index to lidocaine (55). The permanently charged quaternary derivative of lidocaine, N-phenethyl lidocaine 29, is a potent Na^+ channel blocker in vitro and is a local anesthetic in rats with a longer duration and greater potency than the parent tertiary amine (56). This enhanced potency may be due to the phenethyl group interacting with a lipophilic site on the sodium channel or it may be due to the absorption/diffusion of the permanently charged molecule.

Conclusions - Drugs used in anesthetic practice comprise three main classes: hypnotics, muscle relaxants and analgesics. Research and development of new drugs has been characterised by a trend towards identifying fast onset, short duration compounds with rapid patient recovery characteristics and minimal side-effect profiles which offer opportunities to reduce health care costs. This pharmacokinetic profile is atypical for CNS drugs, and may be achieved by incorporating metabolically labile bonds, by modifying the partition coefficient or by redistribution. The original hypothesis that general anesthetics exert their action by non-specific interactions with lipids has been superseded by current understanding that their mechanism actually involves direct interaction with membrane bound proteins and particularly ligand-gated ion channels.

References

1. B.R.P. Birch and R.A. Miller, J.Psychopharmacol., 9,103 (1995).

2. L. Claffey, G. Plourde, J. Morris, M. Trahan and D.M. Dean, Can.J.Anaesth., <u>41</u>, 1084 (1994).
3. B. Fulton and E.M. Sorkin, Drugs, <u>50</u>, 636 (1995).
4. H.M. Bryson, B.R. Fulton and D. Faulds, Drugs, <u>50</u>, 513 (1995).
5. A. Rajah and M. Morgan in "Intravenous Anaesthesia - What is new?", J.W. Dundee and J.W. Sear, Eds., Balliere Tindall, London, 1991, p.425.
6. P. Carl, S. Hogskilde, T. Langjensen, V. Bach, J. Jacobsen, M.B. Sorensen, M. Gralls and L. Widlund, Acta Anaesth.Scand., <u>38</u>, 734 (1994).
7. D.K. Gemmell, A. Byford, A. Anderson, R.J. Marshall, D.R. Hill, A.C. Campbell, N. Hamilton, C. Hill-Venning and J.J. Lambert, Br.J.Pharmacol., <u>116</u>, 443P (1995).
8. C.J. Daniell, M. Cross, J. R. Sneyd, P. Thompson, M.M. Weideman and C.J. Andrews, Br.J.Anaesthesia, 586P (1996).
9. E.I. Eger, Anesthesiology, <u>80</u>, 906 (1994).
10. Y. Xu, P. Tang, W. Zhang, L. Firestone and P.M. Winter, Anesthesiology, <u>83</u>, 766 (1995).
11. J.L. Martin, D.J. Plevak, K.D. Flannery, M. Charlton, J.J. Poterucha, C.E. Humphreys, G. Derfus and L.R. Pohl, Anesthesiology, <u>83</u>, 1125 (1995).
12. E.D. Kharasch, M.D. Karol, C. Lanni and R. Sawchuk, Anesthesiology, <u>82</u>,1369 (1995).
13. Z.X. Fang, E.I. Eger, M.J. Laster, B.S. Chortkoff, L. Kandel and P. Ionescu, Anesth.Analg., <u>80</u>, 1187 (1995).
14. M. Morio, K. Fujii, N. Satoh, M. Imai, U. Kawakami, T. Mizuno, Y. Kawai, Y. Ogasawara, T. Tamura, A. Negishi, Y. Kumagai and T. Kawai, Anesthesiology, <u>77</u>, 1155 (1992).
15. J.P.H. Fee, J.M. Murray and S.R. Luney, Anaesthesia, <u>50</u>, 841 (1995).
16. N.P. Franks and W.R. Lieb, Nature, <u>367</u>, 607 (1994).
17. D.L. Tanelian, P. Kosek, I. Mody and M.B. MacIver, Anesthesiology, <u>78</u>, 757 (1993).
18. R.A. Harris, S.J. Mihic, J.E. Dildy-Mayfield and T.K. Machu, FASEB J., <u>9</u>, 1454 (1995).
19. S.J. Mihic, S.J. Mcquilkin, E.I. Eger, P. Ionescu and R.A. Harris, Mol.Pharmacol., <u>46</u>, 851 (1994).
20. M. Hara, Y. Kai and Y. Ikemoto, Anesthesiology, <u>81</u>, 988 (1994).
21. N.L. Harrison, J.L. Kugler, M.V. Jones, E.P. Grenblatt and D.B. Pritchett, Mol.Pharmacol., <u>44</u>, 628 (1993).
22. K. Hirota and D.G. Lambert, Br.J.Anaesthesia, <u>76</u>, 344 (1996).
23. E.J. Moody, B.D. Harris and P. Skolnick, Trends in Pharmacol.Sci., <u>15</u>, 387 (1994).
24. A.C. Hall, W.R. Lieb and N.P. Franks, Br.J.Pharmacol., <u>112</u>, 906 (1994).
25. N. Unwin, Nature, <u>373</u>, 37 (1995).
26. I.G. Marshall, A.J. Gibb and N.N. Durant, Br.J.Anaesthesia, <u>55</u>, 703 (1983).
27. F.G. Standaert, in "Anesthesia", Vol. 2, R.D. Miller Ed., Churchill Livingstone, New York, 1986 p835.
28. L. Vandenbroek, J.M.K.H. Wierda, N.J. Smeulers and J.H. Proost, Br.J.Anaesthesia, <u>73</u>, 331 (1994).
29. M. Kimura, I. Kimura, M. Muroi, K. Tanaka, H. Nojima, T. Uwano and T. Koizumi, Biological and Pharmaceutical Bulletin, <u>17</u>, 1224 (1994).
30. S.A. Hill, R.P.F. Scott and J.J. Savarese, Bailliere's Clin.Anaesth., <u>8</u>, 317 (1994).
31. N.C. Dhar, R.B. Maehr, L.A. Masterson, J.M. Midgley, J.B. Stenlake and W.B. Wastila, J.Med.Chem., <u>39</u>, 556 (1996).
32. I.G. Marshall and A.L. Harvey, Toxicon., <u>28</u>, 231 (1990).
33. W.C. Bowman, Anesthesiology, <u>69</u>, 57 (1988).
34. A.F. Kopman, Anesthesiology, <u>70</u>, 915 (1989).
35. J.M.K.H. Wierda, F.D.M. Hommes, H.J.A. Nap and L. Vandenbroek, Anaesthesia, <u>50</u>, 393 (1995).
36. I.H. Littlejohn, K. Abhay, A. Elsayed, C.J. Broomhead, P. Duvaldestin and P.J. Flynn, Anaesthesia, <u>50</u>, 499 (1995).
37. A.B. Reitz, M.C. Jetter, K.D. Wild, R.B. Raffa, Annu.Repts.Med.Chem., <u>30</u>, 11 (1995).
38. J.G. Bovill, in "Intravenous Anaesthesia - What is new?", J.W. Dundee and J.W. Sear, Eds., Balliere Tindall, London 1991, p283.
39. L.V. Kudzma, S.M. Evans, S.P. Turnbull, S.A. Severnak and E.F. Ezell, BioMed. Chem.Lett., <u>5</u>, 1177 (1995).
40. L.V. Kudzma, S.A. Severnak, M.J. Benvenga, E.F. Ezell, M.H. Ossipov, V.V. Knight, F.G. Rudo, H.K. Spencer and T.C. Spaulding, J.Med.Chem., <u>32</u>, 2534 (1989).
41. T.D. Egan, H.J.M. Lemmens, P. Fiset, D.J. Hermann, K.T. Muir, D.R. Stanski and S.L. Shafer, Anesthesiology , <u>79</u>, 881 (1993).
42. C.L. Westmoreland, J.F. Hoke, P.S. Sebel, C.C. Hug and K.T. Muir, Anesthesiology, <u>79</u>, 893 (1993).
43. J.P. Thompson and D.J. Rowbotham, Br.J.Anaesthesia, <u>76</u>, 341 (1996).

44. J.A. Wilhelm, P. Vengpedersen, T.B. Zakszewski, E. Osifchin and S.J. Waters, Int. J.Clin.Pharmacol.&Ther., 33, 540 (1995).
45. D.C. Rees in "Progress in Medicinal Chemistry", Vol. 29, G.P. Ellis Ed., Elsevier, Amsterdam, 1992, p109.
46. E.J. Bilsky, S.N. Calderon, T. Wang, R.N. Bernstein, P. Davis, V.J. Hruby, R.W. McNutt, R.B. Rothman, K.C. Rice and F. Porreca, J.Pharmacol.Exp.Ther., 273, 359 (1995).
47. Y. Harada, K. Nishioka, L.M. Kitahata, K. Kishikawa and J.G. Collins, Anesthesiology, 83, 344 (1995).
48. P.J. Armstrong and A.A. Spence in "Intravenous Anaesthesia - what is new?", J.W. Dundee and J.W. Sear, Eds., Balliere Tindall, London, 1991, p453.
49. R. Aantaa and M. Scheinin, Acta Anaesth.Scand., 37, 433 (1993).
50. I. Yoneda, S. Hidenari, K. Okamoto and Y. Watanabe, Eur.J.Pharmacol., 247, 267 (1993).
51. T.N. Calvey, Acta Anaesth.Scand., 39, 83 (1995).
52. G. S. Leisure and C.A. DiFazio, Seminars in Anesthesia, 15, 1,(1996)
53. R. Gristwood, H. Bardsley, H. Baker and J. Dickens, Exp.Opin.Invest.Drugs, 3, 1209 (1994).
54. SCRIP, 2081, 22 (1995).
55. G. Caliendo, G. Greco, G. Mattace Raso, R. Meli, E. Novellino, E. Perissutti and V. Santagada, Eur.J.Med.Chem., 31, 99 (1996).
56. G.K. Wang, C. Quan, M. Vladimirov, W.M. Mok and J.G. Thalhammer, Anesthesiology, 83, 1293 (1995).

SECTION II. CARDIOVASCULAR AND PULMONARY DISEASES

Editor: Annette M. Doherty
Parke-Davis Pharmaceutical Research, Ann Arbor, MI 48105

Chapter 6. Thrombin and Factor Xa Inhibition

Jeremy J. Edmunds and Stephen T. Rapundalo
Parke-Davis Pharmaceutical Research, Ann Arbor, MI, USA,
M. Arshad Siddiqui, BioChem Therapeutic, Laval, Quebec, Canada

Introduction - The homeostasis of internal body fluids following tissue injury is maintained by interdependent enzymatic processes comprising the blood coagulation cascade in which thrombin (factor IIa) plays a key role (1). This serine protease catalyses the cleavage of soluble plasma fibrinogen at four Arg-Gly bonds to form fibrin, which readily polymerizes to generate an interconnected gel network. In situations where abnormal clot formation is associated with cardiovascular disease such as deep vein thrombosis, coronary artery thrombosis, pulmonary embolism, and angina, the indirect thrombin inhibitors heparin and warfarin have demonstrated clinical utility (2-4). Following the promising antithrombotic efficacy afforded by the direct thrombin inhibitors hirudin and hirulog, recent attention has focused on the design of low molecular weight thrombin active site inhibitors (5, 6). These inhibitors address the limitations of previous antithrombotics including bleeding propensity, serine protease selectivity, lack of significant oral bioavailability, and inability to inhibit clot-bound thrombin (7, 8). An alternative antithrombotic strategy has also emerged in the form of inhibitors of factor Xa, the penultimate serine protease in the blood coagulation cascade. This enzyme, as part of a prothrombinase complex composed of non enzymatic co-factor Va, calcium ions and a phospholipid membrane surface, regulates the generation of thrombin from its zymogen prothrombin. Polypeptide factor Xa inhibitors have demonstrated that specific inhibition of factor Xa is a valid antithrombotic strategy and hence low molecular weight agents that selectively inhibit factor Xa, or clot-bound factor Xa, are currently being developed (9).

THROMBIN INHIBITORS

Naturally Occurring Thrombin Inhibitors - Recent additions to this class of inhibitors include the cyclic peptides isolated from the marine sponge *Theonella swinhoei*, and presumed microbial symbionts. These cyclotheonamides C-E (CT-C,D,E) (IIa iC_{50}, 8.4, 5.2, and 28 nM respectively) **1-3** differ principally in the P1' and P3 regions of the inhibitor, but are not specific serine protease inhibitors (10). Another naturally occurring non-specific serine protease cyclic peptide A90720A (IIa IC_{50}, 260 nM) **4** was isolated from blue-green algae *Microchaete loktakensis* (11).

Non Covalent Inhibitors - Even though there are substantial structural differences between the prototypical agents of this class, argatroban **5** and NAPAP **6** (12, 13), the mode of binding of these inhibitors to thrombin is similar, which has allowed molecular modeling experiments to aid in the design of inhibitors such as L-NAPAMP **7** (IIa K_i 2.5 nM, trypsin K_i 42 nM), **8**, CRC 220 **9** (IIa K_i 2.5 nM, trypsin K_i 312 nM), **10** (IIa K_i 1.6 nM, trypsin K_i 35.2 nM) and **11** (14-18). An alternative structure based drug design process which relied upon the selection of thrombin selective S1 compounds culminated in the synthesis of Ro 46-6240 **12**, (napsagatran) (IIa K_i 0.27 nM, trypsin K_i 1.9 µM) (19, 20). While this compound is potent and selective, the short half life of 15 minutes has led to the continued exploration of SAR and the patenting (21) of a series of compounds exemplified by **13**. Other approaches to the design of thrombin inhibitors have originated from peptide screening leads, which led to the discovery of a retro-binding tripeptide **14** (IIa K$_i$ 17.2 nM, trypsin K$_i$ 189.2 nM) (22, 23). Further refinement of inhibitory activity resulted in the particularly selective inhibitor BMS-189090 **15** (IIa K$_i$ 3.64 nM, trypsin K$_i$ 47 µM) (24).

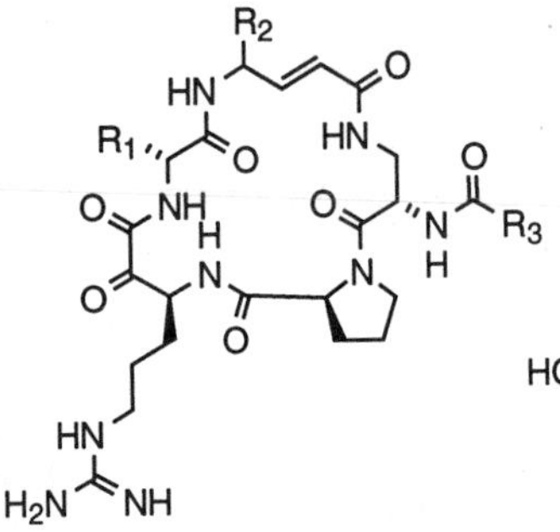

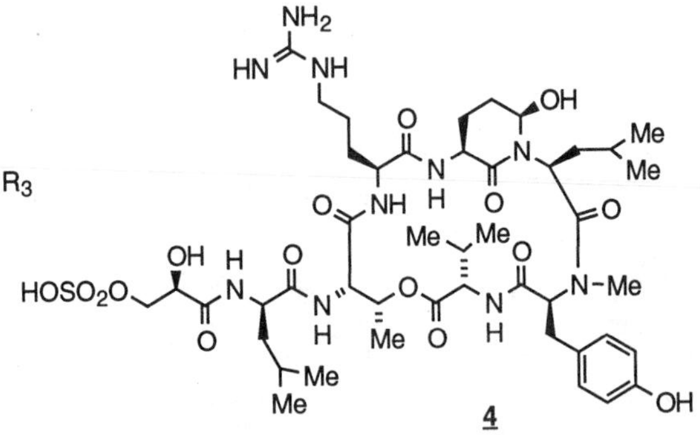

1 CT-C R$_1$ = CH$_2$Ph, R$_2$ = $\overset{Z}{\equiv}$CH(4-OH-Ph), R$_3$ = H

2 CT-D R$_1$ = CH$_2$CH(Me)$_2$, R$_2$ = CH$_2$(4-OH-Ph), R$_3$ = H

3 CT-E R$_1$ = CH(Me)(Et), R$_2$ = CH$_2$(4-OH-Ph), R$_3$ = CH(Me)NH(C=O)CH$_2$Ph

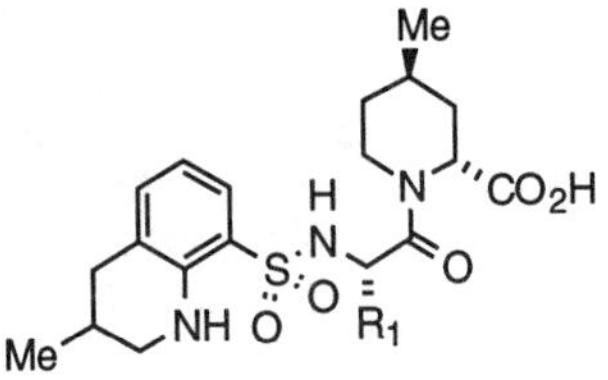

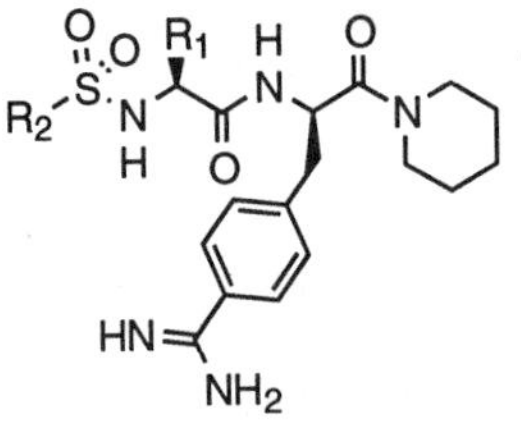

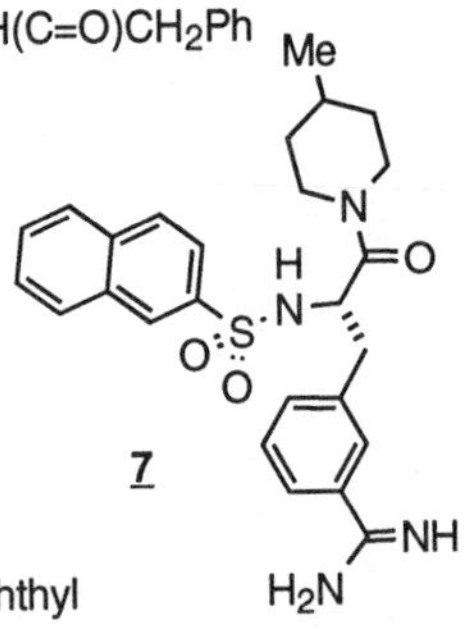

5   R$_1$ = (CH$_2$)$_3$NHC(=NH)NH$_2$

8   R$_1$ = (CH$_2$)$_3$-imidazole

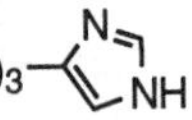

6   R$_1$ = H, R$_2$ = β-Naphthyl

9   R$_1$ = CH$_2$CO$_2$H, R$_2$ = 4-methoxy-2,3,6-trimethylphenyl

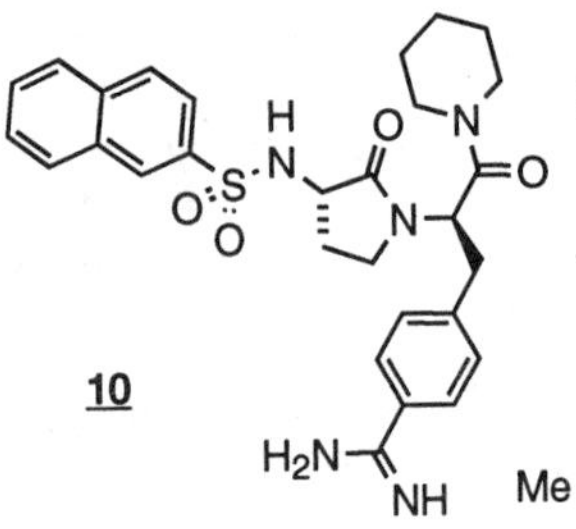

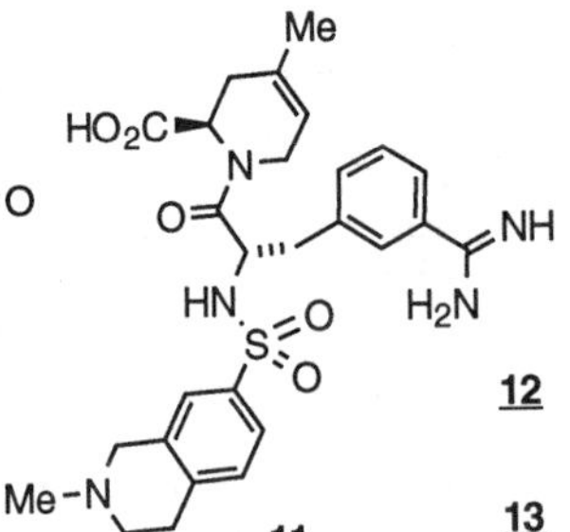

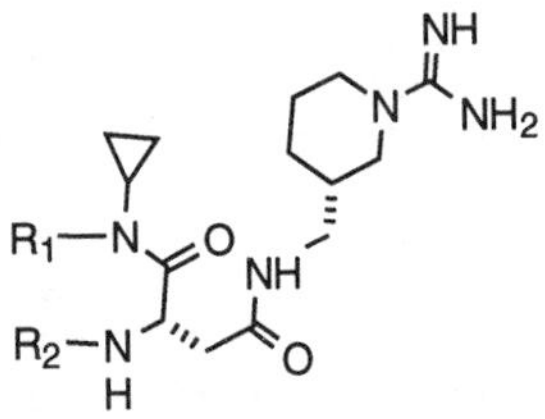

12 R$_1$ = CH$_2$CO$_2$H
     R$_2$ = β-Naphthylsulphonyl

13 R$_1$ = (CH$_2$)$_2$NH(C=O)-2-Pyrazine
     R$_2$ = (C=O)CH$_2$S-2-Pyrimindine

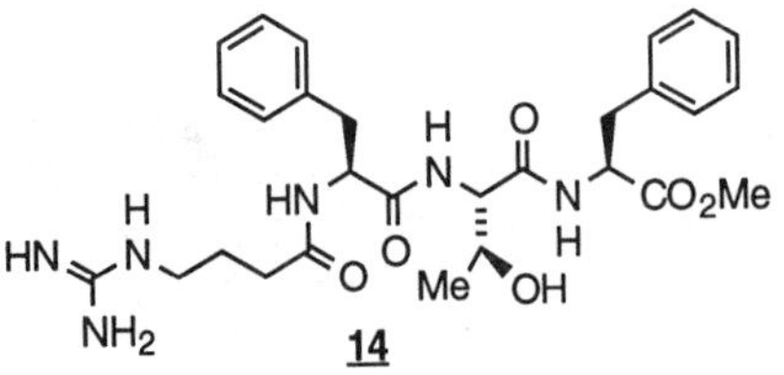

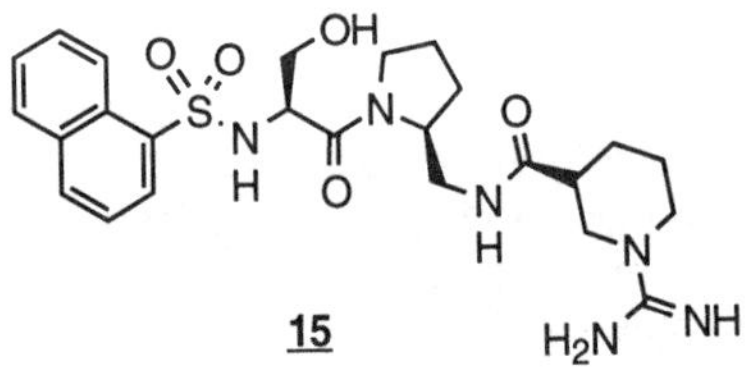

A combination of X-ray crystallography, molecular modeling and readily accessible 1,3-dipolar cycloaddition chemistry identified a rigid tricyclic phenylamidine based thrombin inhibitor **16** (IIa K$_i$ 90 nM) with 7.8 fold selectivity versus trypsin (25). Finally the non covalent inhibitor inogatran **17** (IIa K$_i$ 15 nM, trypsin K$_i$ 45 nM) was reported to be a potent and selective thrombin inhibitor (26-29). It has a substantial oral bioavailability of 32% and 51% in male and female rats respectively, compared with 44% and 34% in male and female dogs. Further SAR exploration has led to the patenting of **18**, with the P1 modification of a (4-amidinophenyl)methyl moiety (30).

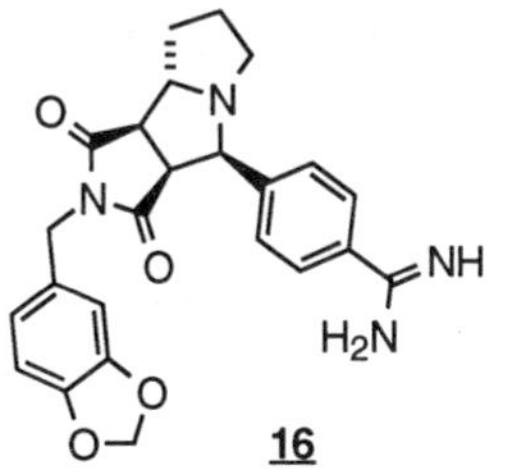

16

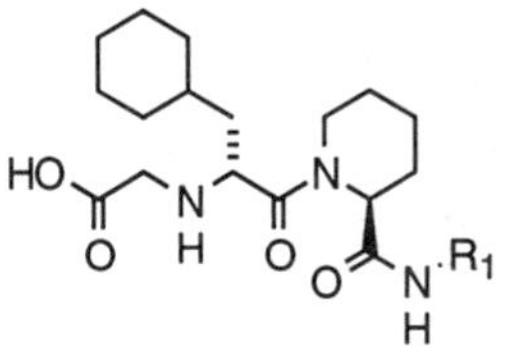

17 R$_1$ = (CH$_2$)$_3$NHC(=NH)NH$_2$

18 R$_1$ = (4-amidinophenyl)methyl

<u>Transition State Inhibitors</u> - This class of agents is characterized by those that incorporate an electrophilic moiety that interacts covalently with the active site Ser-195 in a reversible manner. These inhibitors have typically been considered to exhibit slow tight binding kinetics and poor serine protease selectivity. However, a number of fluoroalkyl ketones such as **19**, and **20** have demonstrated that the kinetics of association is affected by the nature of the alkyl ketone substituent (31, 32). Furthermore the poor serine protease selectivity of **21** was modified by incorporation of substituents at for example P3 **22**, P1 **23** (IIa IC$_{50}$ 3.11 nM, trypsin IC$_{50}$ 150 nM) and a combination of P2 and P1 substituents **24** (IIa IC$_{50}$ 0.8 nM, trypsin IC$_{50}$ > 25000 nM) (33-36). In addition a series of lysyl-alpha-keto-amide thrombin inhibitors further demonstrate that serine protease selectivity may be affected by the nature of the P1' substituent, with compound **25** being over 100 fold selective versus trypsin (31, 37). The related compound **26** demonstrates between 10 and 19% oral bioavailability, similar to H-D-Phe-Pro-Arg-H in mice. Other compounds containing primary amines at P1 include **27** with over 12,000 fold selectivity versus trypsin and the boroarginyl DuP 714 based inhibitors **28** (IIa K$_i$ 0.46 nM), and **29** (IIa K$_i$ 0.07 nM) (38-40). Finally a series of hexapeptides, containing a keto methylene replacement for the scissile amide bond, were prepared to specifically explore the S prime sites of thrombin (41). The most potent compound, **30** (IIa K$_i$ 29 nM), through molecular modeling, suggests the existence of hydrophobic S2' pocket and potential of the inhibitor to selectively hydrogen bond to Arg-35 and Lys-60F.

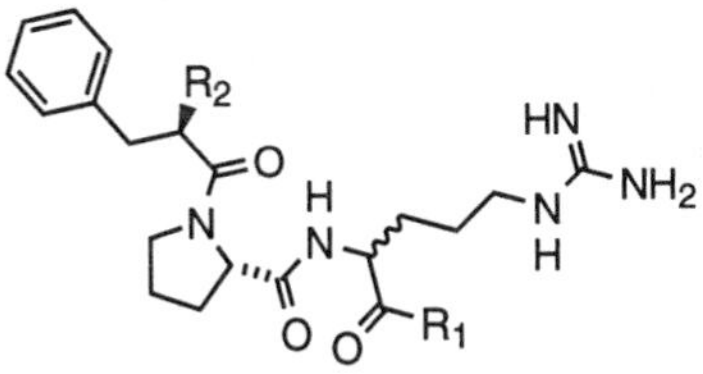

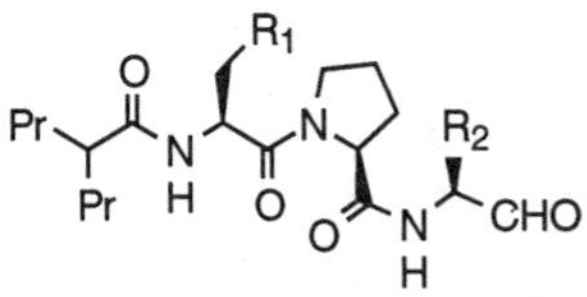

19 R$_1$ = CF$_2$CF$_2$CF$_3$ R$_2$ = NH$_2$

20 R$_1$= CH$_2$OCH$_2$CF$_3$ R$_2$ = NH$_2$

40 L-isomer, R$_1$= H R$_2$ = NH(Me)

41 L-isomer, R$_1$= H R$_2$ = H

21 R$_1$ = CO$_2$Me R$_2$ = (CH$_2$)$_3$NHC(=NH)NH$_2$

22 R$_1$ = tetrazole R$_2$ = (CH$_2$)$_3$NHC(=NH)NH$_2$

23 R$_1$ = tetrazole R$_2$ = -3(1-amidino)piperidinylmethyl

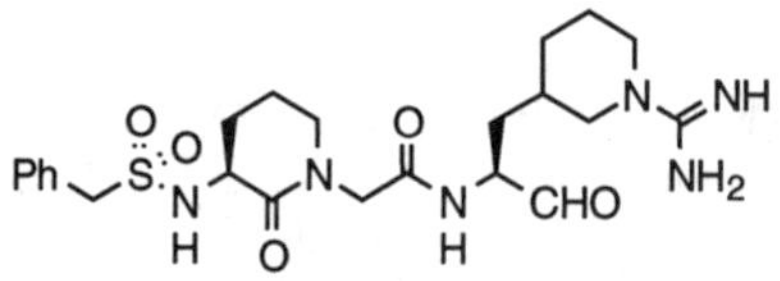

24

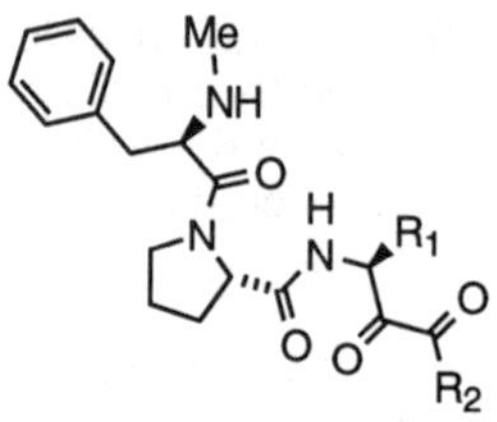

25 R_1 = (CH$_2$)$_4$NH$_2$ R_2 =1- pyrrolidine
26 R_1 = (CH$_2$)$_4$NH$_2$ R_2 = NH(Me)
27 R_1 = (4-amino)cyclohexane R_2 = NH(Me)

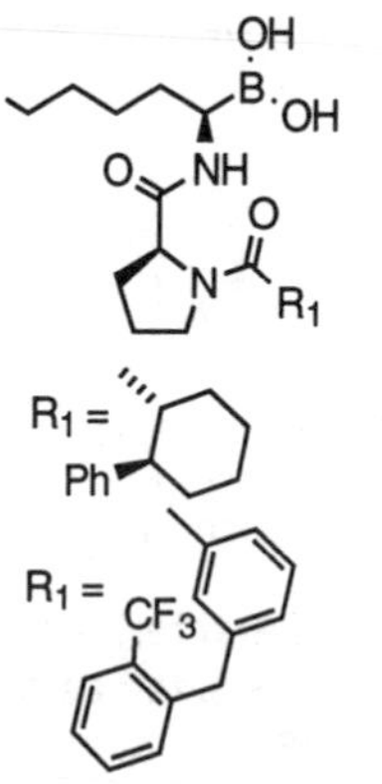

28 R_1 =

29 R_1 =

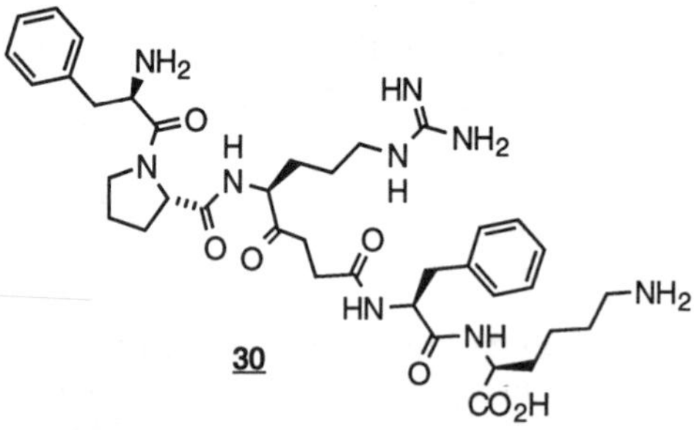

30

FACTOR Xa INHIBITORS

The factor Xa inhibitors (42, 43) known to date can be classified into three types namely, i) endogenous inhibitors that are present in the blood including antithrombin III (ATIII) and tissue factor pathway inhibitor (TFPI) (44, 45), ii) inhibitors isolated from the salivary glands of blood sucking insects (46, 47) and animals (48) and iii) rational drug designed inhibitors comprising of peptides and non peptides.

Polypeptide Inhibitors - Detailed spectroscopic studies have only recently confirmed that inhibition of factor Xa by the recombinant tick anticoagulant peptide (rTAP) is achieved by a low affinity exosite association followed by a high affinity interaction with the active site (49-51). While rTAP demonstrates slow kinetics of association, a poly peptide isolated from the human hookworm, *Ancylostoma caninum* (AcAP) significantly and dose-dependently prolonged prothrombin time with rapid onset of factor Xa inhibition (52). Recently it has been determined that the protease nexin-2/amyloid b-protein precursor (PN-2/AbPP), which inhibits factors XIa (K_i 6.9 nM) and IXa (K_i 0.15 nM), also inhibits factor Xa (K_i 45 nM) as an isolated protein and as part of the prothrombinase complex (53). The complex formed between factor Xa and PN-2/AbPP does not appear to be active-site directed and the slow tight kinetics are characteristic of Kunitz-type inhibitors. Ecotin, an *E. coli* periplasmic protein was recently demonstrated to be a potent inhibitor of factor Xa (K_i 50 pM) with a fast K_{on} rate, and slow K_{off} rate, but has previously been shown to inhibit trypsin. Replacement of the P1 Met-84 residue in ecotin with either Arg or Lys led to factor Xa inhibitors with K_i's of 11 and 21 pM respectively (54-57).

Synthetic Inhibitors of Factor Xa - DX9065a **31** remains the prototypical low molecular weight active-site factor Xa competitive inhibitor (Xa K_i 41 nM) with no activity against thrombin (58). Its discovery resulted from the systematic structure activity relationship studies of the weak factor Xa inhibitor, 1,2-di(5-amidino-2-benzofuranyl)ethane (DABE) **32** (Xa K_i 570 nM) and amidinophenyl pyruvic acid, and presumably bis(amidinobenzyl)cycloheptanone (BABCH) **33** (Xa K_i 13 nM, trypsin K_i 720 nM, IIa K_i 8300 nM) (59-61). Further structure based drug design may be aided by the recently disclosed crystal structure of DX9065a with trypsin (Ki 620 nM), a serine protease highly homologous to factor Xa (62-64).

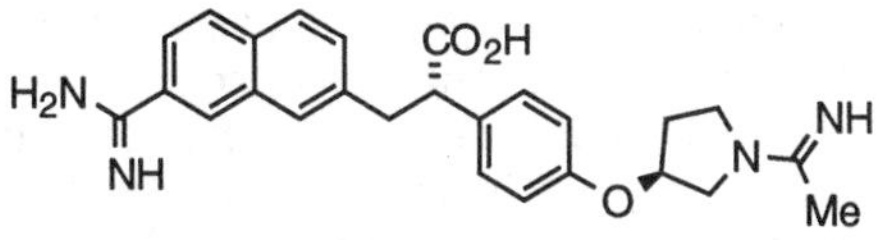

31

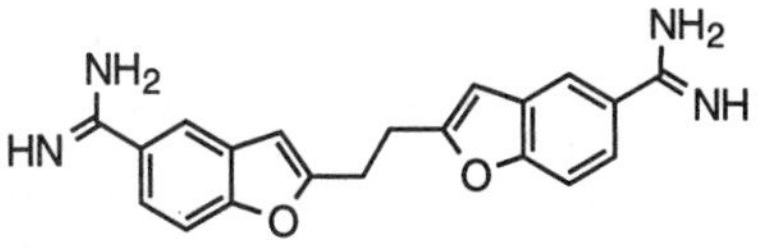

32

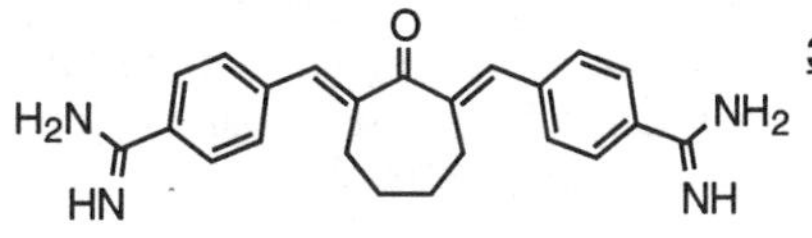

33

DX9065a, like other factor Xa inhibitors, shows species dependent inhibition of factor Xa activity (65, 66). The hydrolysis of a synthetic chromogenic substrate S2222 by factor Xa purified from human (Xa K_i 78.8 nM), rabbit (Xa K_i 102 nM), and rat (Xa K_i 198 nM) plasma demonstrated different sensitivity to DX9065a. The relative potency of factor Xa inhibition also paralleled the concentration of DX9065a that was required to double prothrombin time (dPT) between species (human dPT= 0.52 μM, rabbit dPT = 1.5 μM, rat dPT = 22.2 μM).

There are a number of peptides that are potent inhibitors of factor Xa including dansyl-Glu-Gly-Arg-chloromethyl ketone **34**, **35** (Xa IC_{50} 30 nM), **36** (Xa IC_{50} 9.7 nM), CV-12296 **37** (Xa IC_{50} 23 nM) and **38** (Xa IC_{50} 376 nM) (67, 68). A series of Tyr-Ile-Arg (YIR) peptides have also been reported which include Ac-Y-I-R-L-P-NH₂ (Xa K_i 500 nM), with no activity against thrombin, plasmin, trypsin and elastase. The most potent Xa inhibitor reported is Ac-(iBu)Y-I-R-L-P-NH₂ **39** (Xa K_i 40 nM) (69). At this time it is not known whether these peptides are active site directed inhibitors of factor Xa or indirect inhibitors preventing assembly of the prothrombinase complex.

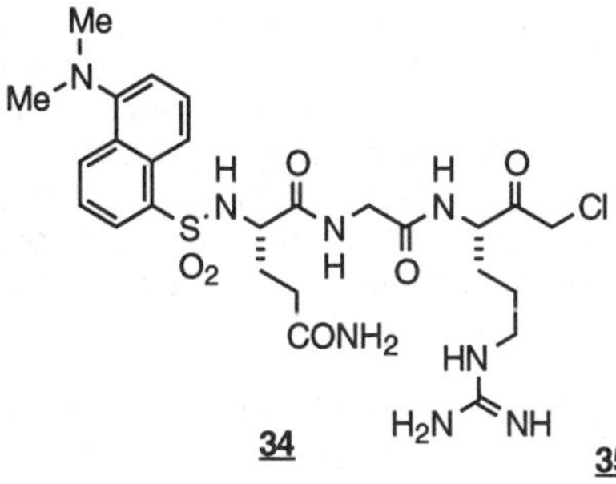

34

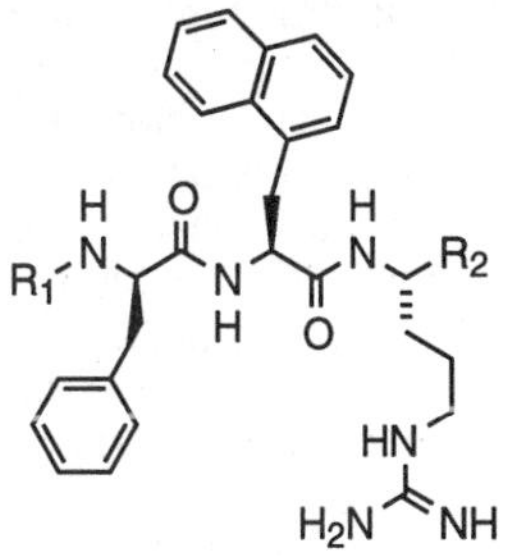

35   R₁ = BOC   R₂ = CHO
36   R₁ = Ac   R₂ = B(OH)₂
37   R₁ = C(O)CH₂CH₂CO₂H   R₂ = CHO

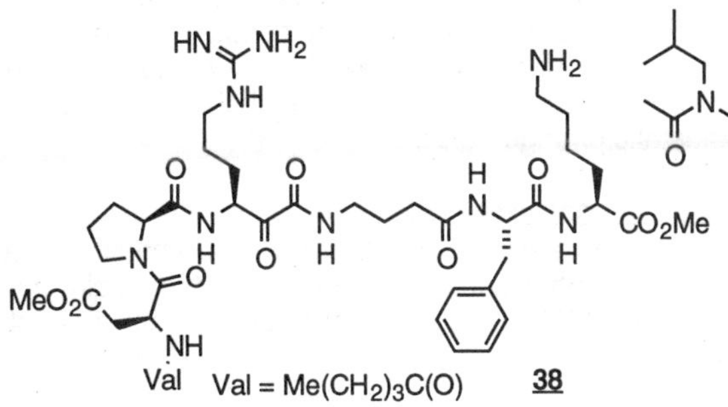

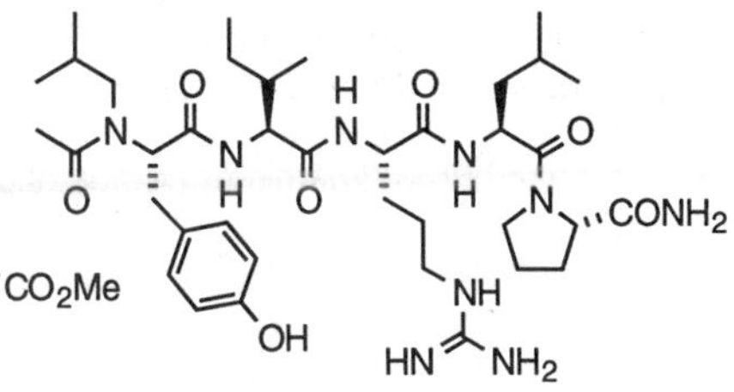

39

38   Val = Me(CH₂)₃C(O)

RECENT STUDIES ON EXPERIMENTAL AND CLINICAL PHARMACOLOGY OF THROMBIN AND FACTOR Xa INHIBITORS

In Vitro Plasma Anti-Coagulant Activity - Several recent studies have utilized defined biochemical systems, particularly plasma, to characterize mechanism of action profiles for a variety of antithrombin agents (70-73). Evidence has emerged demonstrating the ability of some thrombin inhibitors like argatroban **5** to inactivate not only plasma clot bound thrombin but also free and fibrin bound thrombin (71). Other studies have revealed that efegatran (LY 294468) **40** and Ro 46-6240 **12**, unlike argatroban and other antithrombin agents, are effective in blocking extrinsic and intrinsic generation of thrombin and factor Xa in human plasma (72, 73).

Anti-Coagulant Activity in Healthy Individuals - The specific mode of action and relative effect on coagulation markers of a number of anti-coagulant agents such as low molecular weight heparin (LMWH), unfractionated heparin (UH), r-hirudin and dermatan sulphate (Desmin 370), have recently been evaluated in normal volunteers (74-77).

Venous Thrombosis - Several clinical and experimental studies have recently described the anticoagulant effects of UH and LMWHs in deep vein thrombosis (78), although no clinical studies have been reported on the potential effectiveness of synthetic active-site inhibitors (79-81). Argatroban was found to be less active on a weight basis than heparin when given i.v. bolus in a rat model of thromboplastin-induced venous thrombosis (82). However, when adminstered as a continuous i.v. infusion, argatroban had the same antithrombotic potency as heparin, indicating different durations of action of the agents. In a rabbit model of venous thrombosis, administration of intrathrombic argatobran during pulse-spray pharmacomechanical thrombolysis with tPA significantly improved clot lysis, especially compared with tPA and intrathrombic heparin treatment (83). In a tissue factor-induced model of venous thrombosis, SDZ 217-766 treatment was found to be superior to an equivalent dose of heparin in preventing thrombus formation, and had lower aPTT prolongation times (84). The synthetic polypeptide thrombin inhibitor, CVS995, induced a sustained antithrombotic effect in a rabbit venous thrombosis model, along with r-hirudin (CGP 39393) and hirulog-1, as compared with either fraxiparine, a LMWH, or rTAP (85). Additionally, CVS995 demonstrated an enhanced degree of endogenous fibrinolysis. A study by Yokohama demonstrated in baboons that parenteral and oral administration of the specific factor Xa inhibitor, DX9065a, dose-dependently reduced platelet deposition and fibrin accumulation in the formation of fibrin-rich venous thrombi (86). However, Biemond has reported that rTAP treatment is ineffective at preventing thrombus growth in an rabbit venous thrombosis model (85).

Arterial Thrombosis - The antithrombotic effects of r-hirudin and synthetic thrombin active-site inhibitors such as BMY 44621 **41**, CRC 220, and Ro 46-6240 have been shown to be promising therapy in recent studies utilizing various animal models of arterial thrombosis (16, 87-90). The efficacy of the factor Xa inhibitor rTAP was examined in a canine coronary artery electrolytic lesion model and on acute arterial thrombus formation triggered by tissue factor/factor VIIa or collagen in an *ex vivo* model of shear-dependent human thrombogenesis (91, 92). Intravenous infusion of rTAP markedly reduced residual thrombus mass and time to occlusion in the absence of dramatic hemodynamic or hemostatic changes (92).

Myocardial Infarction and Coronary Thrombolysis - Both hirulog and hirudin treatment have each been compared with heparin treatment as adjuncts to streptokinase and aspirin therapy (93-95). Hirulog yielded higher early coronary patency rates along with less adverse events including bleeding (96). The utility of hirudin as an adjunct to thrombolytic therapy in acute myocardial infarction has also been evaluated in several pilot trials (97). A comparison between fraxiparine and UH was made as an adjunct to coronary thrombolysis with alteplase and aspirin in a dog model of myocardial infarction which indicated that both UH and the higher two doses of fraxiparine were associated with higher coronary patency rates (98). In a similar study, Chen found that the direct thrombin inhibitor inogatran modulated coronary artery reocclusion and reduced time to reperfusion (99).

Coronary Angioplasty - rTAP and heparin, or local delivery of r-hirudin, have recently been shown to be capable of reducing or preventing reocclusion following coronary artery angioplasty in various experimental models (100, 101). Clinically, several thrombin inhibitors have now been evaluated in the setting of adjunctive treatment in angioplasty (102), or secondary to restenosis (103) or unstable angina (97, 104-106).

Restenosis - Hirudin and rTAP have been evaluated in a rabbit atherosclerotic femoral artery injury model of restenosis and were found to have no preventive effects on neointimal hyperplasia (107).

Cerebral Ischemia - Evidence is now emerging describing the efficacy of argatroban in experimental models of cerebral ischemia evolving from embolism or thrombosis (108, 109).

Disseminated Intravascular Coagulation (DIC). - Oral administration of DX9065a was observed to reduce the severity of both endotoxin- and thromboplastin-induced DIC in rats, without any effect on bleeding time (110, 111).

Conclusions - A number of new direct thrombin inhibitors have emerged as potential therapeutic antithrombotic agents. For thrombin inhibitors to show pharmacological efficacy, rapid and complete inhibition of thrombin is required to suppress the powerful positive feedback loop for thrombin generation. The tight binding electrophilic thrombin inhibitors typically demonstrate slow binding behavior and hence may not be well suited to prevent the local rapid increase in thrombin generation upon tissue injury. It should also be noted that upon discontinuation of treatment with thrombin inhibitors, particularly the reversible non covalent inhibitors, a rebound increase of thrombin is observed which leads to a procoagulant state. Thus the failure to sufficiently and chronically suppress thrombin and its generation may ultimately limit the clinical utility of thrombin inhibitors (73). Factor Xa inhibitors offer the attractive potential to suppress thrombin generation while, as competitive inhibitors, allowing some platelet aggregation (112). This translates to a reduced bleeding tendency and presents an exciting area for drug development.

References

1.   L. Luchtman-Jones and G.J. Broze Jr., Ann. Med., 27, 47 (1995).
2.   J. Fareed, D.D. Callas, D. Hoppensteadt, W. Jeske and J.M. Walenga, Exp. Opin. Invest. Drugs., 4(5), 389 (1995).
3.   D. Fitzgerald, Ann. N. Y. Acad. Sci., 714, 41 (1994).
4.   J. Lefkovits and E.J. Topol, Circulation, 90, 1522 (1994).
5.   S.D. Kimball, Curr. Pharm. Des., 1, 441 (1995).
6.   J. Das and S.D. Kimball, Bioorg. Med. Chem., 3(8), 999 (1995).
7.   S.D. Kimball, Blood Coagul. Fibrin., 6(6), 511 (1995).
8.   M. Verstraete and P. Zoldhelyi, Drugs, 49(6), 856 (1995).
9.   M. Iino, H. Takeya, T. Takemitsu, T. Nakagaki, E.C. Gabazza and K. Suzuki, Eur. J. Biochem., 232(1), 90 (1995).
10. Y. Nakao, S. Matsunaga and N. Fusetani, Bioorg. Med. Chem., 3(8), 1115 (1995).
11. R. Bonjouklian, T.A. Smitka, A.H. Hunt, J.L. Occolowitz, T.J. Perun Jr., L. Doolin, S. Stevenson, L. Knauss, R. Wijayaratne, S. Szewczyk and G.M.L. Patterson, Tetrahedron, 52, 395 (1996).

12. L.R. Bush, Cardiovasc. Drug Rev., 9(3), 247 (1991).
13. H. Brandstetter, D. Turk, W.H. Hoeffken, D. Grosse, J. Stuerzebecher, P.D. Martin, B.F.P. Edwards and W. Bode, J. Mol. Biol., 226(4), 1085 (1992).
14. A. Bergner, M. Bauer, H. Brandstetter, J. Stuerzebecher and W. Bode, Enzyme Inhib., 9(1), 101 (1995).
15. G. Lassalle, T. Purcell, D. Galtier, P.H. Williams and F. Galli, US 5453430 (1995).
16. G. Dickneite, D. Seiffge, K.H. Diehl, M. Reers, J. Czech, E. Weinmann, D. Hoffmann and W. Stuber, Thromb. Res., 77(4), 357 (1995).
17. H. Mack, T. Pfeiffer, W. Hornberger, H.-J. Boehm and H.W. Hoeffken, Enzyme Inhib, 9(1), 73 (1995).
18. J.C. Danilewicz, D. Ellis and R.J. Kobylecki, WO 9513274 (1995).
19. K. Hilpert, J. Ackermann, D.W. Banner, A. Gast, K. Gubernator, G. Schmid, T.B. Tschopp and H. van de Waterbeemb, Eur. J. Med. Chem., 30, 131 (1995).
20. K. Hilpert, J. Ackermann, D.W. Banner, A. Gast, K. Gubernator, P. Hadvary, L. Labler, K. Muller, G. Schmid, T.B. Tschopp and H. van de Waterbeemd, J. Med. Chem., 37, 3889 (1994).
21. J. Ackermann, D. Banner, K. Gubernator, K. Hilpert and G. Schmid, EP 641779 (1995).
22. L. Tabernero, C.Y. Chang, S.L. Ohringer, W.F. Lau, E.J. Iwanowicz, W.-C. Han, T.C. Wang, S.M. Seiler, D.G.M. Roberts and J.S. Sack, J. Mol. Biol., 246(1), 14 (1995).
23. E.J. Iwanowicz, W.F. Lau, J. Lin, D.G.M. Roberts and S.M. Seiler, J. Med. Chem., 37, 2122 (1994).
24. M.F. Malley, L. Tabernero, C.Y. Chang, S.L. Ohringer, D.G.M. Roberts, J. Das and J.S. Sack, Protein Sci., 5, 221 (1996).
25. U. Obst, V. Gramlich, F. Diederich, L. Weber and D.W. Banner, Angew. Chem., Int. Ed. Engl., 34(16), 1739 (1995).
26. A-C. Teger-Nilsson and R.E Bylund, WO 9311152 (1993).
27. A-C. Teger-Nilsson, E. Gyzander, S. Andersson, M. Englund, C. Mattsson, J-C. Ulvinge and D. Gustafsson, Thromb. Haemostasis, 73, 1325 (1995).
28. D. Gustafsson, M. Elg, S. Lenfors, I. Borjesson and A-C. Teger-Nilsson, Thromb. Haemostasis, 73, 1319 (1995).
29. U.G. Eriksson, L. Renberg, C. Vedin and M. Strimfors, Thromb. Haemostasis, 73, 1318 (1995).
30. K.T. Antonsson, R.E. Bylund, N.D. Gustafsson and N.O.I. Nilsson, WO 9429336 (1994).
31. D.M. Jones, B. Atrash, H. Ryder, A.-C. Teger-Nilsson, E. Gyzander and M. Szelke, Enzyme Inhib., 9(1), 43 (1995).
32. D.M. Jones, B. Atrash, A.-C. Teger-Nilsson, E. Gyzander, J. Deinum and M. Szelke, Lett. Pept. Sci., 2(3/4), 147 (1995).
33. G.P. Vlasuk, T.R. Webb and D.A. Pearson, WO 9315756 (1993).
34. G.P. Vlasuk, T.R. Webb, D.A. Pearson and M.M. Abelman, WO 9417817 (1994).
35. O.E. Levy, S.Y. Tamura, R.F. Nutt and W.C. Ripka, WO 9535312 (1995).
36. J.E. Semple, O.E. Levy, R.F. Nutt and W.C. Ripka, WO 9535313 (1995).
37. S.D. Lewis, A.S. Ng, E.A. Lyle, M.J. Mellott, S.D. Appleby, S.F. Brady, K.J. Stauffer, J.T. Sisko, S.-S. Mao, D. F. Veber, R. F. Nutt, J. J. Lynch, J. J. Cook, S. J. Gardell and J. A. Shafer, Thromb. Haemostasis, 74(4), 1107 (1995).
38. D.F. Veber, S.D. Lewis, J.A. Shafer, D-M. Feng, R.F. Nutt, S.F. Brady, WO 9425051 (1994).
39. J.M. Fevig, M.M. Abelman, D.R. Brittelli, C.A. Kettner, R.M. Knabb and P.C. Weber, Biorg. Med. Chem. Lett., 6, 295 (1996).
40. J. Cacciola, J.M. Fevig, R.S. Alexander, D.R. Brittelli, C.A. Kettner, R.M. Knabb and P.C. Weber, Biorg. Med. Chem. Letts., 6, 301 (1996).
41. M. Jetten, C.A.M. Peters, A. Visser, P.D.J. Grootenhuis, J.W. van Nispen and H.C.J. Ottenheijm, Bioorg. Med. Chem., 3(8), 1099 (1995).
42. B. Kaiser and J. Hauptmann, Cardiovasc. Drug Rev., 12, 225 (1994).
43. M. Yamazaki, Drugs of the Future, 20(9), 911 (1995).
44. G.J. Broze, Jr., Blood Coagul. Fibrin., 6(1), S7 (1995).
45. G.J. Broze, Jr., Annu. Rev. Med., 46, 103 (1995).
46. A.M. Joubert, J.C. Crause, A.R.M.D. Gaspar, F.C. Clarke, A.M. Spickett and A.W.H. Neitz, Exp. Appl. Acarol., 19, 79 (1995).
47. A.R.M.D. Gaspar, J.C. Crause and A.W.H. Neitz, Exp. Appl. Acarol., 19, 117 (1995).
48. R. Apitz-Castro, S. Beguin, A. Tablante, F. Bartoli, J.C. Holt and H.C. Hemker, Thromb. Haemostasis, 73(1), 94 (1995).
49. M.S.L. Lim-Wilby, K. Hallenga, M. De Maeyer, I. Lasters, G.P. Vlasuk and T.K. Brunck, Protein Sci., 4(2), 178 (1995).

50. G. Sanyal, D. Marquis-Omer, L. Waxman, H. Mach, J.A. Ryan, J. O'Brien Gress and C.R. Middaugh, Biochim. Biophys. Acta., 1249(10), 100 (1995).
51. S.-S. Mao, J. Huang, C. Welebob, M.P. Neeper, V.M. Garsky and J.A. Shafer, Biochem., 34(15), 5098 (1995).
52. M. Cappello, G.P. Vlasuk, P.W. Bergum, S. Huang and P.J. Hotez, Proc. Natl. Acad. Sci. USA, 92(13), 6152 (1995).
53. F. Mahdi, W.E. van Nostrand and A.H. Schmaier, Biol., 270, 23468 (1995).
54. M.E. McGrath, S.A. Gillmor and R.J. Fletterick, Protein Sci., 4(2), 141 (1995).
55. J.L. Seymour, R.N. Lindquist, M.S. Dennis, B. Moffat, D. Yansura, D. Reilly, M.E. Wessinger and R.A. Lazarus, Biochem., 33(13), 3949 (1994).
56. S.A. Gillmor, M.E. McGrath and R.J. Fletterick, Perspect. Drug Discovery Des., 2(3), 475 (1995).
57. R.A. Lazarus, M.S. Dennis and J.S. Ulmer, WO 9507986 (1995).
58. T. Nagahara, Y. Yokoyama, K. Inamura, S-i. Katakura, S. Komoriya, H. Yamaguchi, T. Hara and M. Iwamoto, J. Med. Chem., 37, 1200 (1994).
59. T. Nagahara, S.-i. Katakura, Y. Yokoyama, N. Kanaya, K. Inamura, S. Komoriya, K. Yamada, H. Yamaguchi, K. Tanabe, Y. Morishima, H. Ishihara, T. Hara, K. -I. Yamazaki, S. Kunitada and M. Iwamoto, Eur. J. Med. Chem., 30, 139s (1995).
60. P. Walsmann, H. Horn, F. Markwardt, P. Richter, J. Sturzebecher, H. Vieweg and G. Wagner, Acta. Biol. Med. Ger., 35, K1 (1976).
61. J. Sturzebecher, U. Sturzebecher, H. Vieweg, G. Wagner, J. Hauptmann and F. Markwardt, Thromb Res, 54, 245 (1989).
62. M.T. Stubbs and W. Bode, Curr. Opin. Struct. Biol., 4, 823 (1994).
63. M.T. Stubbs, R. Huber and W. Bode, FEBS Lett, 375, 103 (1995).
64. M.T. Stubbs and W. Bode, Trends Cardiovasc. Med., 5(4), 157 (1995).
65. R.R. Tidwell, W.P. Webster, S.R. Shaver and J.D. Geratz, Thromb. Res., 19, 339 (1980).
66. T. Hara, A. Yokoyama, Y. Morishima and S. Kunitada, Thromb. Res., 80(1), 99 (1995).
67. T.K. Brunck, T.R. Webb and W.C. Ripka, WO 9413693 (1994).
68. W. Ripka, T. Brunck, P. Stanssens, Y. LaRoche, M. Lauwereys, A.-M. Lambeir, I. Lasters, M. DeMaeyer, G. Vlasuk, O. Levy, T. Miller, T. Webb, S. Tamura and D. Pearson, Eur. J. Med. Chem., 30, 87s (1995).
69. F. Al-Obeidi, M. Lebl, J. Ostrem, P. Safar, A. Stierandova, P. Strop and A. Walser, WO 9529189 (1995).
70. F.A. Ofosu, J.C. Lormeau, S. Craven, L. Dewar and N. Anvari, Thromb. Haemostasis, 72(6), 862 (1994).
71. C.N. Berry, C. Girardot, C. Lecoffre and C. Lunven, Throm. Haemostasis, 72(3), 381 (1994).
72. D.D. Callas, D. Hoppensteadt and J. Fareed, Semin. Thromb. Hemostasis, 21(2), 177 (1995).
73. A. Gast and T.B. Tschopp, Blood Coag. Fibrinolysis, 6(6), 553 (1995)
74. C. Legnani, G. Palareti, R. Biagi, S. Ludovici, L. Maggiore, M.R. Milani and S. Coccheri, Eur. J. Clin. Pharmacol., 47(3), 247 (1994).
75. S. Eichinger, M. Wolzt, M. Nieszpaur-Los, B. Schneider, K. Lechner, H.-G. Eichler and P.A. Kyrle, Thromb. Haemostasis, 72(6), 831 (1994).
76. S. Eichinger, M. Wolzt, B. Schneider, M. Nieszpaur-Los, H. Heinrichs, K. Lechner, H.-G. Eichler and P.A. Kyrle, Arterioscler. Thromb. Vasc. Biol., 15(7), 886 (1995).
77. B.I. Eriksson, K. Soederberg, L. Wildlund, B. Wandeli, L. Tengborn and B. Risberg, Thromb. Haemostasis, 73(3), 398, (1995).
78. E.E. Weinmann and E.W. Salzman, N. Engl. J. Med., 331, 1630 (1994).
79. P. Lindmarker, M. Holmstrom, S. Granqvist, H. Johnsson and D. Lockner, Thromb. Haemostasis, 72(2), 186 (1994).
80. M. Alhenc-Gelas, C. Jestin-Le Guernic, J.F. Vitoux, A. Kher, M. Aiach and J.N. Fiessinger, Thromb. Haemostasis, 71(6), 698 (1994).
81. A. Diquelou, D. Dupouy, R. Cariou, K.S. Sakariassen, B. Boneu and Y. Cadroy, Thromb. Haemostasis, 74, 1286 (1995).
82. C.N. Berry, D. Girard, S. Lochot and C. Lecoffre, Br. J. Pharmacol., 113(4), 1209 (1994).
83. K. Valji, K. Arun and J.J. Bookstein, J. Vasc. Interv. Radiol., 6(1), 91 (1995).
84. C. Tapparelli, R. Metternich, P. Gfeller, B. Gafner and M. Powling, Thromb. Hemostasis, 73, 641 (1995).
85. B.J. Biemond, P.W. Friederich, M. Levi, G.P. Vlasuk, H.R. Buller and J.W. ten Cate, Circulation, 93, 153 (1996).

86. T. Yokoyama, A.B. Kelly, U.M. Marzec, S.R. Hanson, S. Kunitada and L.A. Harker, Circulation, 92(3), 485 (1995).

87. H.F. Kotze, S. Lamprecht and P.N. Badenhorst, Blood, 85(11), 3158 (1995).

88. J.P. Carteaux, A. Gast, T.B. Tschopp and S. Roux, Circulation, 91(5), 1568 (1995).

89. Y. Takiguchi, F. Asai, K. Wada, H. Hayashi and M. Nakashima, Br. J. Pharmacol., 116, 3056 (1995).

90. W.A. Schumacher, N. Balasubramanian, D.R. St Laurent and S.M. Seiler, Eur. J. Pharmacol., 259(2), 165 (1994).

91. J.J. Lynch, G.R. Sitko, E.D. Lehman and G.P. Vlasuk, Thromb. Haemostasis, 74, 640 (1995).

92. U. Orvim, R.M. Barstad, G.P. Vlasuk and K.S. Sakariassen, Arterioscler. Thromb. Vasc. Biol., 15, 2188 (1995).

93. P.R. Eisenberg, J. Thromb. Thrombolysis, 1(3), 237 (1995).

94. R.M. Lidon, P. Theroux and R. Bonan, Circulation, 89, 1567 (1994).

95. L.V. Lee, Am. J. Cardiol., 75, 7 (1995).

96. P. Theroux, F. Perez Villa, D. Waters, J. Lesperance, F. Shabani and R. Bonan, Circulation, 91(8), 2132 (1995).

97. C.P. Cannon and E. Braunwald, J. Am. Coll. Cardiol., 25(7 Suppl), 30s (1995).

98. L. Jun, J. Arnout, P. Vanhove, F. Dol, J.C. Lormeau, J.M. Herbert, D. Collen and F. Van de Werf, Coron. Artery. Dis., 6(3), 257 (1995).

99. L.Y. Chen, W.W. Nichols, C. Mattsson, A.C. Tegernilson, R. Wallin, T.G.P. Saldeen and J.L. Mehta, Cardiovasc. Res., 30, 866 (1995).

100. E.M. Lyle, T. Fujita, M.W. Conner, T.M. Connolly, G.P. Vlasuk and J.L. Lynch, Jr., J. Pharmacol. Toxicol. Methods, 33(1), 53 (1995).

101. B.J. Meyer, A. Fernandez-Ortiz, A. Mailhac, E. Falk, L. Badimon, A.D. Michael, J.H. Chesebro, V. Fuster and J.J. Badimon, Circulation, 90(5), 2474 (1994).

102. P.W. Serruys, J.P. Herrman, R. Simon, W. Rutsch, C. Bode, G.J. Laarman, R. van Dijk, A.A. van den Bos, V.A. Umans, K.A. Fox, P. Close and J.W. Deckers, N. Engl. J. Med., 333(12), 757 (1995).

103. S. Suzuki, S. Sakamoto, K. Adachi, K. Mizutani, M. Koide, N. Ohga, T. Miki and T. Matsuo, Thromb. Res., 77(4), 369 (1995).

104. J. Fuchs and C.P. Cannon, Circulation, 92(4), 727 (1995).

105. J.A. Bittl, Am. Heart. J., 130, 658 (1995).

106. J.A. Bittl, J. Strony, J.A. Brinker, W.H. Ahmed, C.R. Meckel, B.R. Chaitman, J. Maraganore, E. Deutsch and B. Adelman, N. Engl. J. Med., 333(12), 764 (1995).

107. Y. Jang, L.A. Guzman, M. Lincoff, M. Gottsauner-Wolf, F. Forudi, C.E. Hart, D.W. Courtman, M. Ezban, S.G. Ellis and E.J. Topol, Circulation, 92, 3041-3050 (1995).

108. T. Hara, M. Iwamoto, M. Ishihara and M. Tomikawa, Haemostasis, 24, 351 (1994).

109. K. Umemura, H. Kawai, H. Ishihara and M. Nakashima, Jpn. J. Pharmacol., 67(3), 253 (1995).

110. M. Yamazaki, H. Asakura, K. Aoshima, M. Saito, H. Jokaji, C. Uotani, I. Kumabashiri, E. Morishita, T. Ikeda and T. Matsuda, Thromb. Haemostasis, 72(3), 393 (1994).

111. T. Hara, A. Yokoyama, K. Tanabe, H. Ishihara and M. Iwamoto, Thromb. Haemostasis, 74(2), 635 (1995).

112. J.M. Herbert, A. Bernat, F. Dol, J.P. Herault, B. Crepon and J.C. Lormeau, J. Pharmacol. Exp. Ther., 276, 1030, 1996.

Chapter 7. Inhibitors of Types I and V Phosphodiesterase: Elevation of cGMP as a Therapeutic Strategy

Michael Czarniecki, Ho-Sam Ahn and Edmund J. Sybertz
Schering-Plough Research Institute
2015 Galloping Hill Road, Kenilworth, NJ 07033

Introduction - cGMP plays a key role in the modulation of vascular function. Therapeutic manipulation of cGMP has been exploited in angina pectoris, congestive heart failure, myocardial infarction, and hypertension. The evolving biology of cGMP and understanding of its mechanisms of regulation are opening up new opportunities for discovery of drugs which can produce therapeutically desirable effects through alteration of cGMP levels. It is becoming apparent that at least two isoforms of cyclic nucleotide phosphodiesterases participate in the hydrolysis and inactivation of cGMP. Inhibitors of these isoforms have the potential to offer novel approaches to treating cardiovascular diseases. The field of cyclic nucleotide phosphodiesterase (PDE) inhibition as a therapeutic target has received considerable attention over the years and has been the subject of numerous reviews (1 - 8). This chapter will focus on recently discovered cyclic GMP phosphodiesterase inhibitors and will examine the biochemistry and pharmacology of selected inhibitors of Type I and/or V PDE.

Biochemistry of cGMP Phosphodiesterases - Steady state levels of cGMP are maintained by the rate of formation and degradation of the nucleotide. Formation is governed by the state of activation of the various receptor-linked and soluble guanylate cyclases. Degradation can be regulated by the various cyclic nucleotide phosphodiesterases. This field has been the subject of several recent review articles (2, 8 - 12).

At least seven classes of phosphodiesterases have been identified to date (Table I). The classes share several common structural features and the amino acid sequences in the putative hydrolytic sites are highly conserved. The classes have been distinguished by their specificity for cAMP (cyclic adenosine monophosphate) and cGMP, by the mechanisms of regulation and their sensitivity to various pharmacological agents. The classes are encoded by distinct genes and subclasses have been formed through alternative splicing mechanisms.

PDE I is a calmodulin (CaM) dependent phosphodiesterase which hydrolyzes both cAMP and cGMP (9). The enzyme is stimulated several fold by calcium and CaM. The cDNAs for several CaM PDE's have been cloned and the primary sequences identified (9, 11 - 13). These enzymes share a high degree of sequence homology. PDE I is distributed in the cytosolic fractions of various tissues (9). It has been identified in heart, lung, brain, testes, kidney, pancreas and circulating blood cells. In addition it is found in vascular smooth muscle cells where it plays a significant role in cyclic nucleotide hydrolysis in this tissue (14,15). Inhibitors should have a major influence on cardiovascular function in vitro and in vivo.

Table I.   Major Classes of Cyclic Nucleotide Phosphodiesterases.

PDE	Substrate	Gene Products	Regulatory Mechanisms	Tissue Distribution
I	cGMP cAMP	3	CaM dependent	Vascular tissue, brain, kidney, pancreas, circulating blood cells
II	cGMP cAMP	1	Stimulated by cGMP	Adrenal cortex, platelet, vascular tissue, heart, lung
III	cAMP	2	Inhibited by cGMP	Heart, vascular tissue, liver, platelet, adipocyte
IV	cAMP	4	cAMP specific	Heart, kidney, brain, GI tract, liver, lung, circulating blood cells
V	cGMP	2	cGMP specific	Platelet, vascular tissue, lung
VI	cGMP	3	cGMP specific	Retinal rods, cones, kidney
VII	cAMP	1	Unknown	Skeletal muscle, T cells

PDE V binds and selectively hydrolyzes cGMP (9,16). Its cDNA has been cloned and shown to be divergent from that of the retinal cGMP specific PDE (11,17,18). The cDNA encodes an 875 amino acid polypeptide with a homologous catalytic segment that is conserved across PDE types. It is distributed in lung, kidney, spleen, platelets, endothelial cells and smooth muscle cells. The enzyme plays a key role in the hydrolysis of cGMP in different tissues (5,15,19).

Structure Activity Relationships of cGMP PDE Inhibitors - Recent investigations of inhibitors of cGMP phosphodiesterases can be divided between those which inhibit the Type I PDE and Type V PDE isozymes. This chapter will cover only those new types cGMP PDE inhibitors which have recently been published. For a discussion of the historical literature the reader is referred to the most recent comprehensive review of the medicinal chemistry of PDE inhibition (20). In general much of the information is derived from the patent literature with a smaller amount published in refereed scientific articles. This may reflect the early stage of PDE selective isozyme research. Most of the work has used known, generally non selective, PDE inhibitors, e.g. zaprinast (**1**), papavarine (**2**), IBMX (**3**), sulmazole (**4**), and griseolic acid (**5**), as the conceptual starting point for the design of new compounds.

Analogs of **1** have received the greatest attention in the design of inhibitors of cGMP hydrolysis (21 - 29). Significant structural latitude is possible while retaining potent inhibition of Type V PDE. Benzopyrimidine **6** (IC_{50} = 6.5 nM), isomeric pyrazolopyrimidines **7** (IC_{50} = 15 nM), and **8** (IC_{50} = 3.7 nM), imidazopyrimidine **9** (IC_{50}= 6.4 nM), and pyridylpyrimidine **10** (IC_{50} = 6.5 nM), are all potent inhibitors of Type V. In addition there appears to be a wide tolerance for substitution on the pendant benzoxy group or even its replacement with nitrogen substituted heterocycles such as pyridine. This class of molecules shows remarkable selectivity for PDE V relative to PDE III. In the examples shown this ranges from 10^2 to 10^4 fold selective for PDE V.

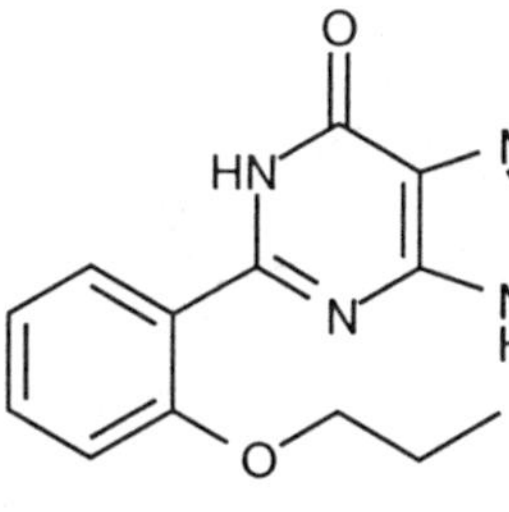

The non-selective PDE inhibitor **2** has been optimized to give compounds with high selectivity for PDE V (30 - 37). Conservative changes gave **11** with only modest potency and selectivity, IC_{50} = 5.5, 8.7, >100, >100 and 0.36 µM for PDE's Type I, II, III, IV and V respectively. Changes in both the aryl substitution pattern and at the quinazoline 2-position yielded inhibitors with substantially enhanced PDE V inhibition. IC_{50}'s of 19, 0.53, and 2.6 nM have been reported for **12**, **13** and **14** respectively. In general these improvements in potency for PDE V result in corresponding enhancements in selectivity relative to other isozymes. Interestingly, compound **15** is also a potent inhibitor of Type V PDE with an IC_{50} reported of 0.4 nM (38). The selectivity of **15** against other PDE isozymes has not been reported.

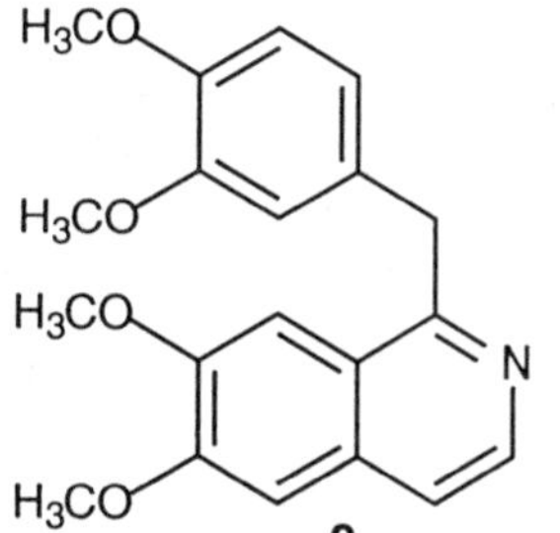

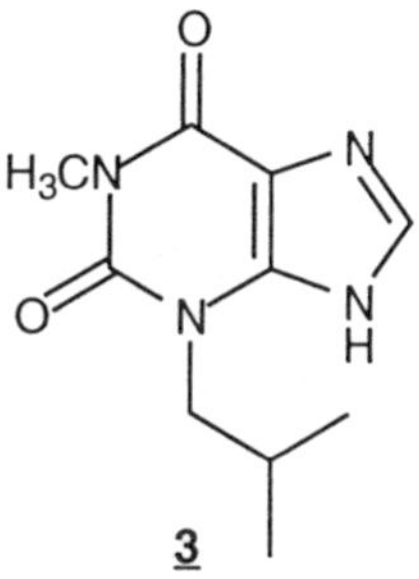

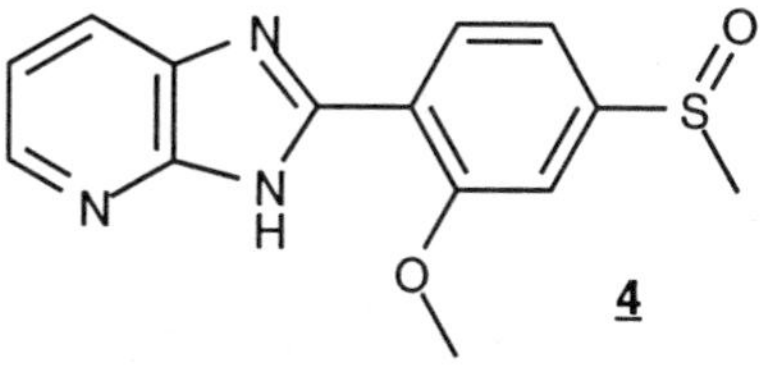

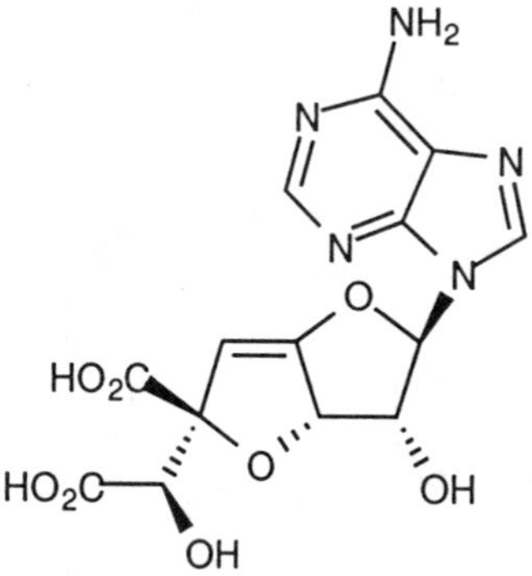

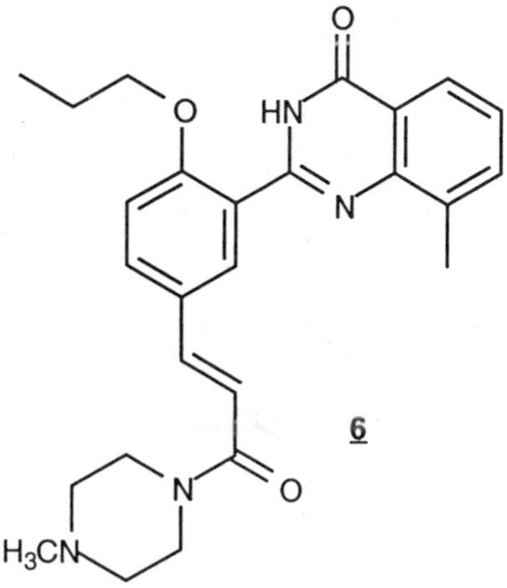

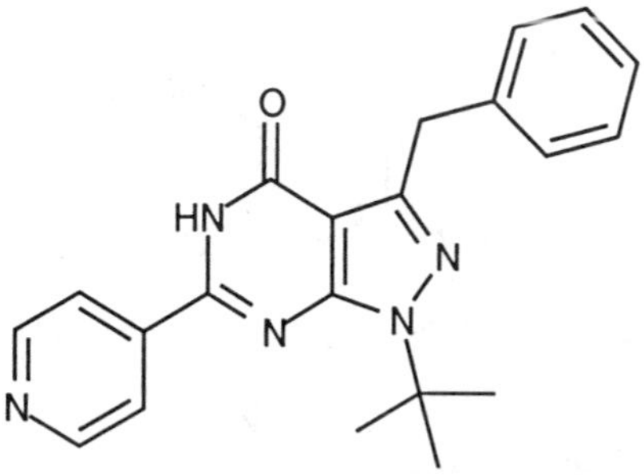

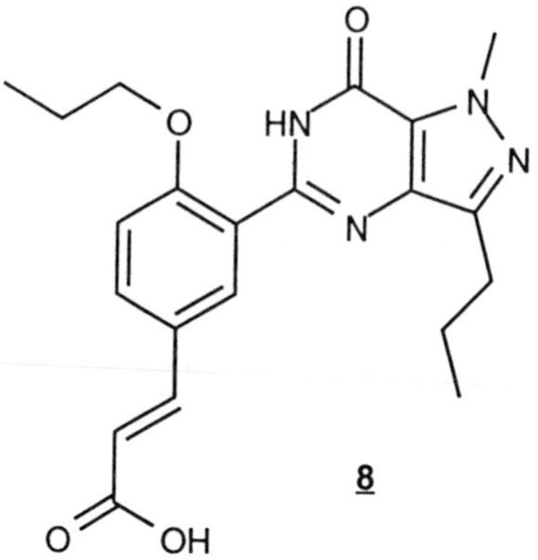

8

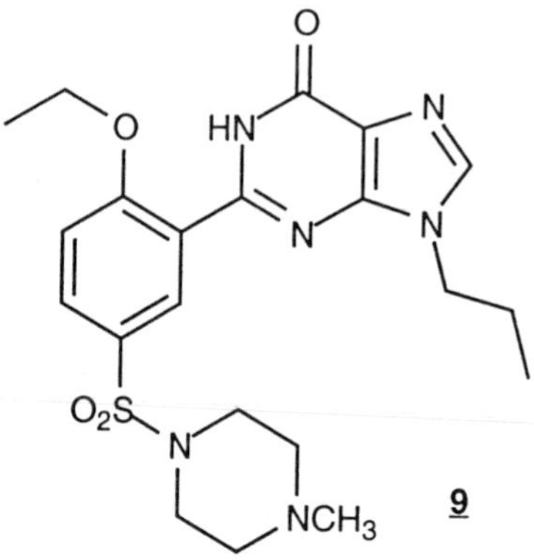

9

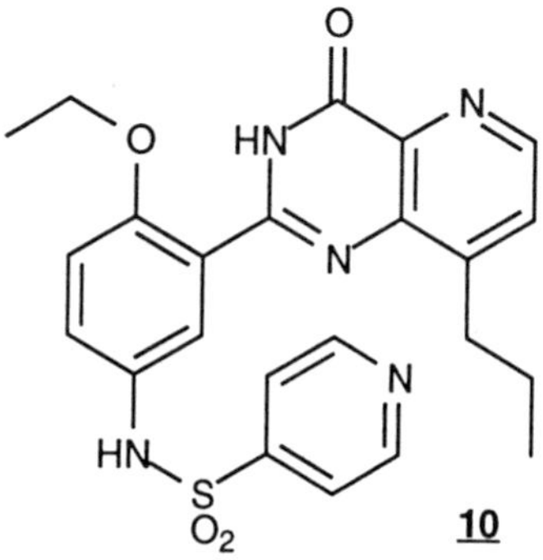

10

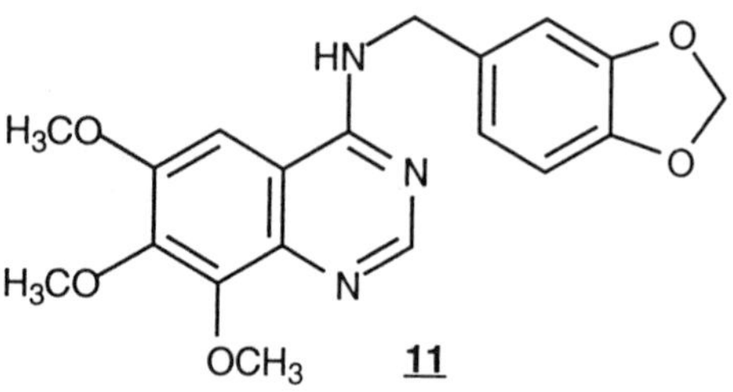

11

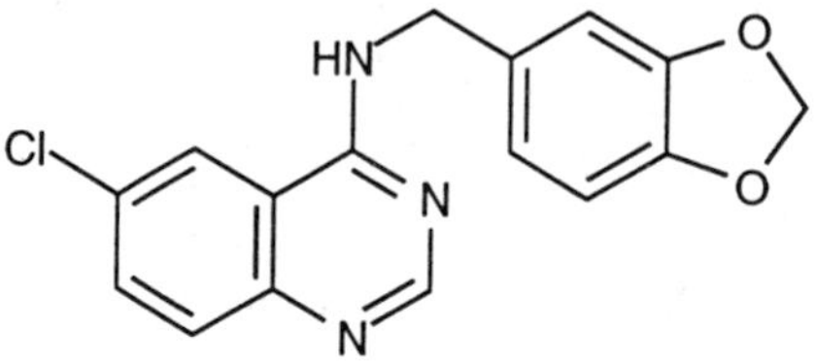

12

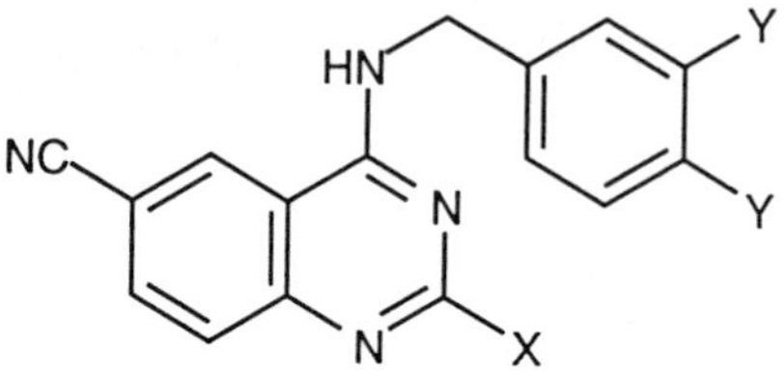

13 X = 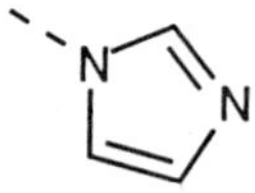   Y = H

14 X = 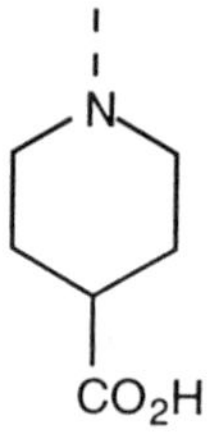   Y =

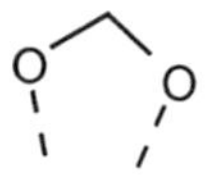

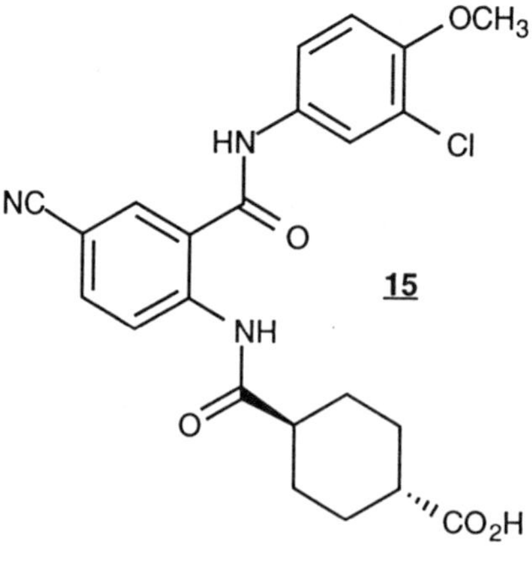

15

The imidazopyridine sulmazole (**4**) was originally developed as a positive inotropic agent. A recent report, however, has shown that **4** is actually a selective inhibitor of Type V PDE with an IC_{50} of 21 µM compared to its inhibition of Types III and IV with inhibitions of 35% and 31% at 100 µM respectively (39). This series also contains potent dual inhibitors of Type IV and V PDE's. Compound **16** inhibits Type IV PDE with an IC_{50} of 6 µM and Type V with an IC_{50} of 1 µM. Type III PDE is not inhibited by **16** at 10 µM. A recent patent application has disclosed the unprecedented diaminobutenedione class of Type V PDE inhibitors (**17**). Although only modest activity is claimed (IC_{50}'s in the low µmolar range) they do represent a very significant structural departure compared to known classes of PDE inhibitors (40).

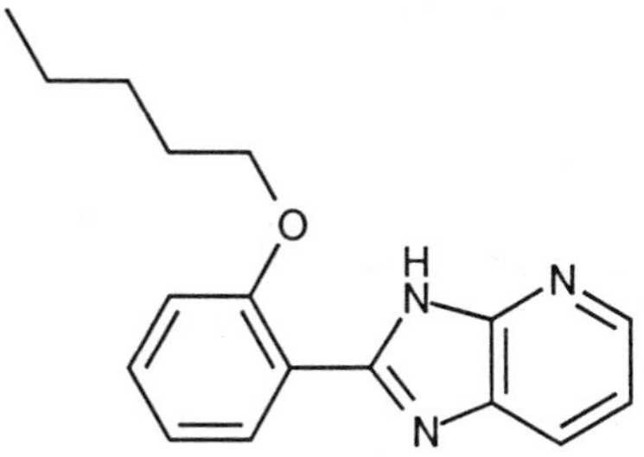

16 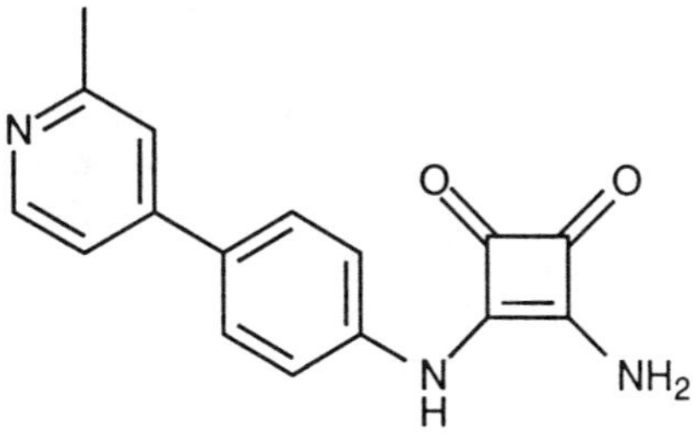   **17**

Inhibition of Type I PDE as a target has received less attention by medicinal chemists. One of the early examples in this area were analogs of the natural product griseolic acid A (**5**). Since **5** is an adenosine analog it was reasoned that selectivity for cGMP hydrolyzing enzymes could be attained by guanine analogs of **5**. This was confirmed by example **18** which had an IC_{50} = 0.3 µM for Type I PDE but was only a very weak inhibitor of Type III PDE with an IC_{50} of 94 µM (41). Using a similar concept

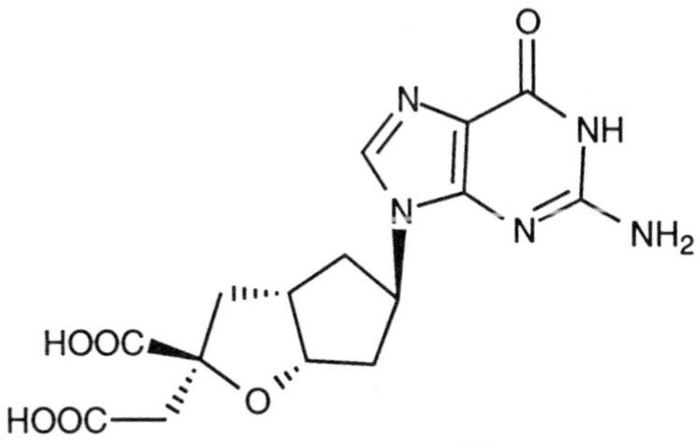

18 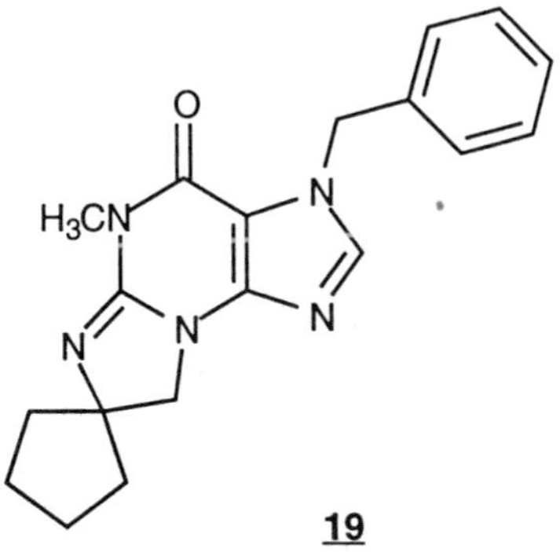   **19**

the nonselective xanthine **3** was "hybridized" to make it more guanine like resulting in a series of tetracyclic guanines, an example of which is represented by **19**. Compound **19** shows dual inhibition of both Type I and Type V PDE's with IC_{50}'s for both enzymes of 100 nM (42).

A recent patent application has claimed Type I PDE inhibitors related to the quinazolines represented by **11**. Compound **20** and related structures are potent inhibitors of Type I PDE with an IC_{50} = 170 nM for **20**. Since no data have been reported for Type V PDE inhibition of **20** it is not known if these structural modifications result in selective inhibitors of Type I PDE (43).

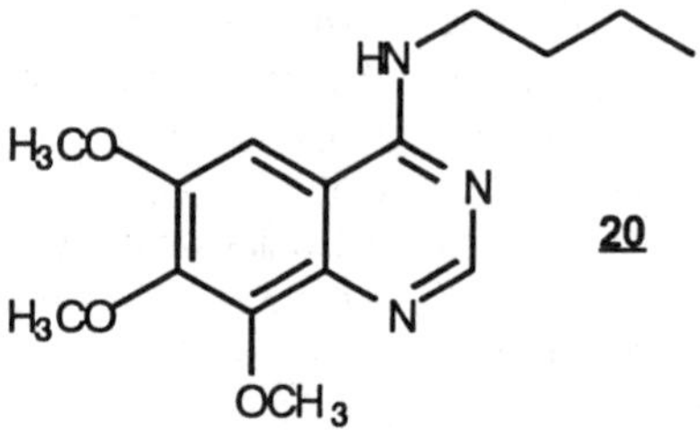

20

Pharmacology of Type I/V PDE Inhibitors - A member of the quinazoline PDE inhibitors which has been extensively studied is E 4021 (**21**). This molecule displays a very high level of potency and selectivity for inhibiting Type V PDE (44 - 46). E 4021 increases cGMP levels in porcine coronary arteries without affecting cAMP. It produces a potent concentration dependent relaxation of isolated coronary arteries. This response is partially dependent upon an intact endothelium. *In vivo*, the molecule induces modest coronary artery vasodilatation in the pig without affecting systemic hemodynamics. Its most prominent vascular action is on the pulmonary circulation when administered *in vivo*. The compound displays good bioavailability in animals. Recent studies suggest that **21** exerts salutary effects in a porcine model of cardiac pacing-induced congestive heart failure. Intravenous administration of **21** to dogs resulted in significant reductions in pulmonary capillary wedge pressure, left atrial pressure and systemic resistance while improving cardiac output and stroke volume. Given the potency, selectivity and oral bioavailability of this molecule, it represents an useful investigational tool to probe the full potential of Type V PDE inhibition in vascular diseases.

PDE	IC_{50} (nM)
I	35700
II	8700
III	45100
IV	10000
V	3.9

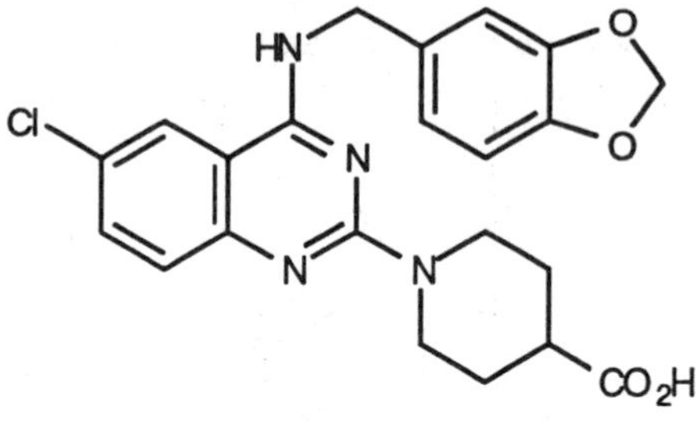

21

Quinazolines such as **13** and **22** appear to manifest a completely unexpected activity as inhibitors of thromboxane A_2 synthesis inhibitors (47). Although compound **22** is only a modestly potent inhibitor of Type V PDE with an IC_{50} = 0.31 μM, the presence of the pyridyl group (or an imidazolyl group) imparts the property of TXA$_2$ synthesis inhibition (47). In washed human platelets **22** inhibited the synthesis of TXA$_2$ with an IC_{50} = 0.83 μM and was selective relative to cyclooxygenase synthesis (IC_{50} = 14 μM).

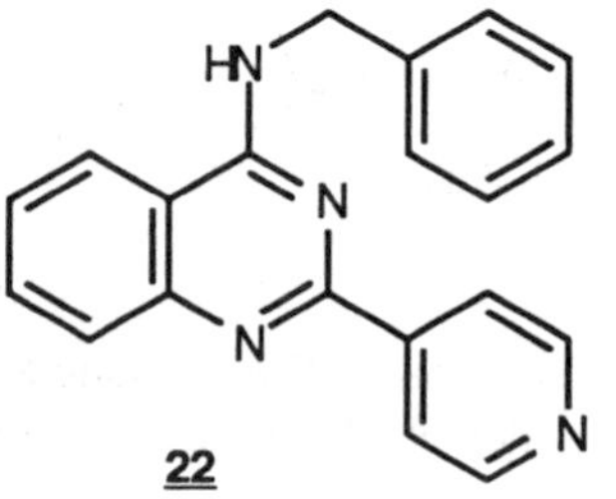

22

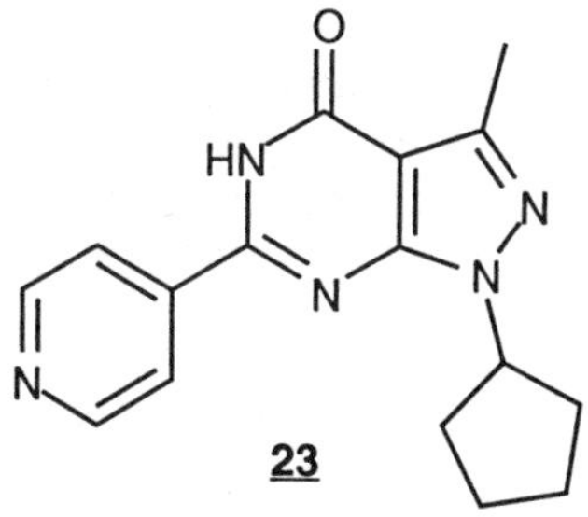

23

The zaprinast analog, Win 58237 (**23**), is a submicromolar inhibitor of PDE V with a K_i of 170 nM (48). The compound shows only modest selectivity for PDE V in that it also inhibits PDE I and PDE IV at 20 X and 2 X the concentrations which inhibit PDE V, respectively. In spite of this lack of selectivity, **23** shows a PDE V like biological profile. The compound potentiates vascular responses to nitroprusside and atrial natriuretic peptide in an endothelium-dependent fashion. *In vivo*, **23** lowers blood pressure when it is administered intra-venously, and this response is inhibited by NO synthase inhibitors. The ability of **23** to potentiate. responses to sodium nitroprusside is likely due to its PDE V activity since a related compound, which inhibited only PDE I, did not enhance nitroprusside responsiveness.

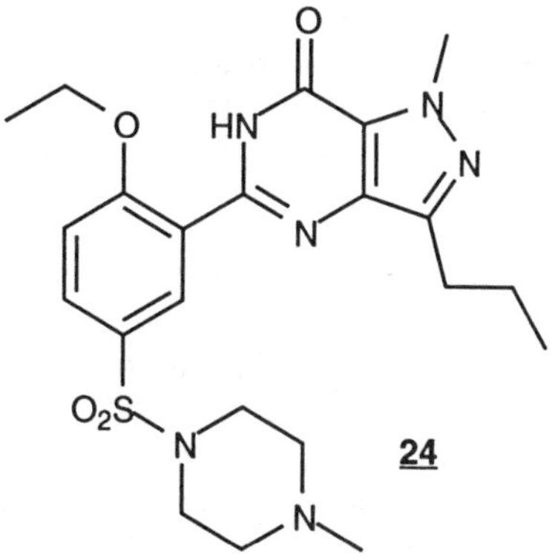

24

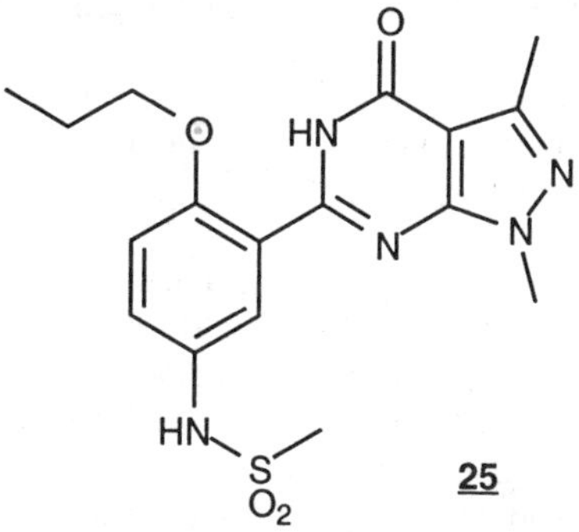

25

The pyrazolopyrimidine UK-92480 (**24**) inhibited rabbit platelet phosphodiesterases Types I, III and V with IC_{50} values of 260, 65000, and 3.6 nM, respectively (49). When evaluated against human *corpus cavernosum* Type V PDE it inhibited with an IC_{50} = 3.0 nM. UK-92480 is being developed to treat male erectile dysfunction (49). The pyrazolo[3,4-d]pyrimidone **25** inhibited PDE Types I, II, III, IV and V with IC_{50} values of 1000, 3000, 10000, 22000, and 3 nM respectively (50). In rat smooth muscle cells **25** potentiated the ANF induced elevation of intracellular cGMP with an EC_{50} of 0.35 µM. Upon oral administration in spontaneously hypertensive rats at 5 mg/kg, **25** reduced mean blood pressure approximately 20 mm Hg. This effect was sustained over a period of 6 hours.

SCH 51866 (**26**) is a prototypical Type I/V inhibitor related to **19** which has undergone extensive pharmacological characterization. SCH 51866 inhibited PDE I and V with IC_{50} values of 70 and 63 nM, respectively (51). It inhibited Types II, III and IV at IC_{50}'s of 1800, 5000, and 30000 nM, respectively. SCH 51866 increased cGMP, but not cAMP levels in vascular tissue at concentrations of 0.3 - 10 µM. The increase in cGMP was greater in tissues where the endothelium was present.

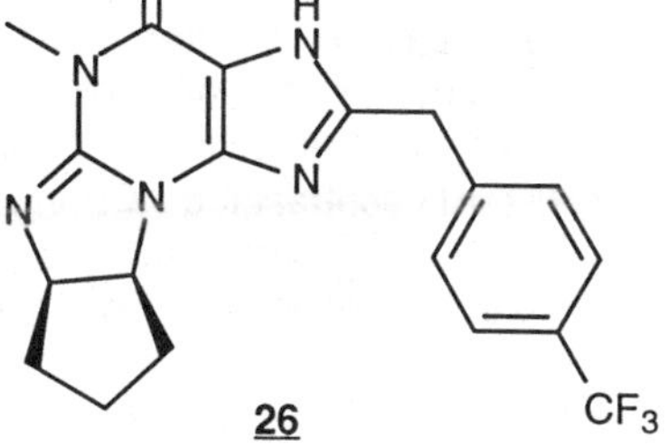

26

Consistent with its PDE inhibitory properties, **26** markedly enhanced the increase in cGMP that is produced by sodium nitroprusside. In isolated rat aorta, **26** produced a concentration-dependent relaxation with an IC_{50} of approximately 1 µM. Its effect was dependent upon NO and an intact endothelium since the response was reduced in the presence of l-NAME (l-nitroarginine methyl ester), an inhibitor of NO synthesis (51).

In vivo studies demonstrate that **26** is an orally active antihypertensive agent in the spontaneously hypertensive rat at doses of 1 - 10 mg/kg, reducing blood pressure by 39 mm Hg at a dose of 10 mg/kg. The antihypertensive actions were abolished by inhibitors of NO synthesis, indicating that enhancement of NO participates in the response (51).

Studies in dogs have evaluated the coronary vascular actions of **26**. Intravenous administration of 0.03 - 1 mg/kg produced significant increases in coronary blood flow by up to 100 percent. Analysis of flow distribution within the myocardium indicated that the drug elicits equivalent vasodilatation in the endocardium and epicardium. SCH 51866 did not decrease blood flow in ischemic myocardium in the dog, suggesting that the compound is unlikely to create a condition of coronary steal (51).

cGMP inhibits the proliferation of vascular smooth muscle cells that is stimulated by growth factors and vasoactive substances (52). Since the proliferation of vascular smooth muscle cells is a hallmark of the late-stage of restenosis which accompanies balloon catheter angioplasty, the effects of **26** were evaluated in both rodent and porcine models of angioplasty. Balloon catheter injury of the rat carotid artery results in a marked intimal hyperplasia which is mediated by the migration of smooth muscle cells from the media to the intima and the proliferation of cells within the intima. Over a two week time period, this results in a marked enhancement of the intimal to medial area ratio of the injured vessel wall. Treatment of rats with 3 and 10 mg/kg of **26**, once-a-day over a two week period suppressed intimal/medial area ratios by 35% - 50%, respectively (51). The response was associated with a significant inhibition of smooth muscle cells in the intima and overall DNA content of the blood vessel, indicative of an inhibition of cell migration and cell proliferation, respectively. The response was slightly less than the 70% inhibition which was achieved with the ACE inhibitor spirapril in these studies.

The rodent model of balloon catheter injury has been shown to be a poor model of the vascular restenosis post angioplasty which occurs in humans (53). Agents which are active in this model have failed in human trials (53). A conspicuous example is the failure of ACE inhibitors in clinical trials (54, 55). For this reason, a model of vascular restenosis in pigs has been developed in which the iliac and femoral arteries are injured by balloon catheterization and intimal hyperplasia which develops over a one month period. Treatment of pigs with **26** in the diet at 17 mg/kg/day over one month resulted in a 55% suppression of the intimal/medial ratio in balloon catheter injured pigs (51). In contrast the ACE inhibitor spirapril which is very effective in the rat model was inactive. These data suggest that PDE I/V inhibition may be effective in situations of vascular injury which are not driven by ACE or sensitive to ACE inhibitors.

<u>Perspective</u> - Given the important roles described for cGMP in cardiovascular regulation, it is evident that agents which modulate cGMP through inhibition of cGMP PDE's can have desirable effects in cardiovascular diseases. Moreover, many

cardiovascular disorders, including hypertension, dyslipidemia, coronary artery disease and congestive heart failure are associated with impaired endothelial cell NO responses and drugs which can restore this impaired function by enhancing the activity of cGMP may reverse this basic vascular defect.   The studies described above with newer, more potent and selective PDE inhibitors suggest that these agents may have utility in treating a variety of cardiovascular disorders. Although this review has considered predominantly the potential beneficial pharmacolgy of cGMP PDE inhibition, one must also be wary of the risks.   There are at least seven distinct classes of PDE with many isoforms.   Tissue distribution is broad and there is considerable regulation of cyclic nucleotide metabolism in different tissues.   In addition, NO is now recognized as a mediator of cellular signaling in many cell types. The safety consequences of PDE I and V inhibition have not been completely elucidated.   However, NO and cGMP are important regulators of gastrointestinal motility and electrolyte transport.   NO, cGMP and PDE I form a novel signaling pathway in many neurons. Furthermore, PDE V and VI are very closely related enzymes and selective inhibitors which inhibit Type V but not VI have not been discovered. The potential impact of newer inhibitors on these processes will need to be assessed and considered together with the potential clinical benefit of PDE inhibition.

References

1.   D. Raeburn, J. Souness, A. Tomkinson, J. Karlsson, Prog. Drug. Res., 40, 9 (1993).
2.   J. A. Beavo, Physiological Rev., 75, 725 (1995).
3   C.D. Nicholson, R.A.J. Challiss, M. Shahid, Trends Pharmacol. Sci., 12, 19 (1991).
4.   W. J. Thompson, Pharamc. Ther., 51, 13 (1991).
6.   I.P. Hall, Br. J. Clin. Pharmacol., 35, 1 (1993).
7.   K.J. Murray and P.J. England, Biochem. Soc. Trans., 20, 460 (1992).
8.   E.J. Sybertz, M. Czarniecki, H.-S. Ahn, Curr. Pharmaceutical Design, 1, 373 (1996).
9.   W.K. Sonnenburg and J.A. Beavo, Adv. Pharmacol., 26, 87 (1994).
10.   G.B. Bolger, Cell Signal., 6, 851 (1994).
11.   J.A. Beavo, M. Conti, R.J. Heaslip, Mol. Pharmacol., 46, 399 (1994).
12.   J. Beltman, W.K. Sonnenburg, J.A. Beavo, Mol. Cell. Biochem., 127/128, 239 (1993).
13.   W.K. Sonnenberg, P.J. Mullaney, J.A. Beavo, J. Biol. Chem., 268, 645 (1993).
14.   H.S. Ahn, W. Crim, B. Pitts, E.J. Sybertz, Adv. Sec. Mess. Phospho. Res., 25, 271 (1992).
15.   T. Saeki and I. Saito, Biochem. Pharmacol., 46, 833 (1993).
16.   M.K. Thomas, S.H. Francis, J.D. Corbin, J. Biol. Chem., 265, 14964 (1990).
17.   S.H. Francis, M.K. Thomas, J.D. Corbin, in "Cyclic Nucleotide Phosphodiesterases: Structure, Regulation and Drug Action." J. A. Beavo, M.D. Houslay, eds. Wiley; Chichester, 1990, Vol. 2, p. 117.
18.   L.M. McAllister-Lucas, W.K. Sonnenberg, A. Kadlecek, D. Seger, H.L.Trong , J.L. Colbran, M.K.Thomas, K.A. Walsh, S.H. Francis, J.D. Corbin, J. A. Beavo, J. Biol. Chem., 268, 22863 (1993).
19.   R.T. MacFarland, B.D. Zelus, J.A. Beavo, J. Biol. Chem., 266, 136 (1991).
20.   R. E. Weishaar, J. A. Bristol, in "Comprehensive Medicinal Chemistry" C. Hansch, P. G.; Sammes, J. B. Taylor, eds.; Pergamon Press: Oxford, 1990, Vol. 2, Chapter 8.5.
21.   A. Bell, D. Brown, N. Terrett, European Patent 463 756 A1 (1992).
22.   A. Bell and N.Terrett, European Patent 526 004 A1 (1993).
23.   D. Brown and N. Terrett, World Patent 9306104 A1 (1993).
24.   A. Bell and N.Terrett, World Patent 9307149 A1 (1993).
25.   N.Terrett, World Patent 9312095 A1 (1993).
26.   N.Terrett, World Patent 9400453 A1 (1994).
27.   A. Bell and N.Terrett, World Patent 9405661 A1 (1994).
28.   B. Dumaitre and N. Docic, European Patent 636 626 A1 (1995).
29.   E. Bacon, S. Baldev, G. Lesher,. US Patent 5 294 612 (1994).
30.   Y. Takase, N. Watanabe, H. Adachi, K. Kodoma,T. Saeki, S. Souda, F. Furuya, World Patent 9507267 A1 (1995).
31.   Y. Takase, T. Saeki, M. Fujimoto, I. Saito, J. Med. Chem., 36, 3765 (1993).
32.   Y. Takase, T. Saeki, N. Watanabe, H. Adachi, S. Souda, I. Saito, J. Med. Chem., 37, 2106 (1994).

33.   Y. Takase, N. Watanabe, M. Matsui, H. Ikuta, T. Kimura, T. Saeki, H. Adachi, T. Tokumura, H. Mochida, Y. Akita, S. Souda, South African Patent 927465 (1992).
34.   Y. Takase, N. Watanabe, H. Adachi, K. Kodoma, H. Ishihara, T. Saeki, S. Souda, World Patent 9422855 A1 (1994).
35.   S. Lee, Y. Konishi, O. Macina, K. Kondo, D. Yu, European Patent 579 496 A1 (1993).
36.   S. J. Lee, Y. Konishi, O.T. Macina, K. Kondo, D.T. Yu, U.S. Patent 5436233-A (1995).
37.   S. J. Lee, Y. Konishi, O.T. Macina, K. Kondo, D.T. Yu, U.S. Patent 5439895-A (1995).
38.   F. Ozaki, K. Ishibashi, H. Ikuta, H. Ishihara, World Patent 9518097-A1 (1995).
39.   W. Coates, B. Connolly, D. Dhanak, S. Flynn, A. Worby, J. Med. Chem., $\underline{36}$, 1387 (1993).
40.   W. Coates and D. Rawlings, World Patent 9429277 A1 (1994)
41.   D. Tulshian, M. Czarniecki, R. Doll, H.-S. Ahn, J. Med. Chem., $\underline{36}$, 1210 (1993).
42.   B. Neustadt, N. Lindo, B. McKittrick, U.S. Patent 5393755 (1995).
43.   Y. Takase, H. Adachi, K. Kodama, H. Ishihara, T. Saeki, S. Souda, K. Furuya, World Patent 9507267 (1995).
44.   T. Saeki, H. Adachi, Y. Takase, S. Yoshitake, S. Souda, I. Saito, J. Pharmacol. Exp. Ther., $\underline{272}$, 825 (1995).
45.   H. Adachi, S.Yoshitake, I. Saito, 68th Ann. Meeting Jpn. Pharmacol. Soc. March 25 - 28, 1995; Nagoya, Japan .
46.   K. Kodama H. Adachi, S. Yoshitake, I. Saito, 68th Ann. Meeting Jpn. Pharmacol Soc. March 25 - 28, 1995; Nagoya, Japan.
47.   S. J. Lee, Y. Konishi, D.T. Yu, T.A. Miskowski, C.M. Riviello, O.T. Macina, M.R.Frierson, K. Kondo, M. Sugitani, J.C. Sircar, K. M. Blazejewski, J. Med. Chem., $\underline{38}$, 3547 (1995).
48.   P. J. Silver, R. L. Dundore, D.C. Bode, L. de Garavilla, R.A. Bucholz, G. Van Aller, L. Hamel, E. Bacon, B. Singh, G.Y. Lesher, J. Pharmacol. Exp. Ther., $\underline{271}$, 1143 (1994).
49.   N. K. Terret, A. S. Bell, D. Brown, P. Ellis, 210th American Chemical Society Meeting, August 20 - 24, 1995; Chicago , Illinois MEDI 229.
50.   B. Dumaitre and N. Dodic, 8th RSC-SCI Medicinal Chemistry Symposium, September 10 - 13, 1995; Cambridge, U.K.
51.   R.W. Watkins, H.R. Davis, A. Fawzi, H.-S. Ahn, J. Cook, R. Cleven, L. Hoos, D. McGregor, R. McLeod, K. Pula, R. Tedesco, D. Tulshian, FASEB J., $\underline{9}$, 1984 (1995).
52.   U.C. Garg and A. Hassid, J. Clin. Invest., $\underline{83}$, 1774 (1989).
53.   L. A. Shaw, M. Rudin, N. S. Cook, Trends Pharm. Sci. $\underline{16}$, 401 (1996).
54.   Mercator Study Group, Circulation, $\underline{86}$, 100 (1992).
55.   W. Desmet, M. Vrolix, I. Descheerder, J. Vanlierde, J.L. Willems, J. Piessens, Circulation, $\underline{89}$, 385 (1994).

Chapter 8. Chronic Pulmonary Inflammation and Other Therapeutic Applications of PDE IV Inhibitors

Jeffrey A. Stafford and Paul L. Feldman
Glaxo Wellcome Research
Research Triangle Park, NC 27709

Introduction - Current treatment for asthma involves relaxation of the airway smooth muscle (bronchodilation) and inhibition of the underlying pulmonary inflammatory events (1,2). Present clinical agents, however, inadequately produce both pharmacological effects. β_2-Adrenoceptor agonists elevate cAMP levels in smooth muscle cells promoting muscle relaxation and bronchdilation, however, these agents provide only moderate antiinflammatory effects. Conversely, the glucocorticosteroids possess antiflammatory properties; however, chronic administration of these drugs is often not well tolerated. Finally, the benefits of the dual antiinflammatory/bronchodilator, theophylline, which is both a nonselective phosphodiesterase (PDE) inhibitor and a nonselective adenosine receptor antagonist, are limited due to its dose-limiting side effects.

The cyclic nucleotide PDEs comprise a family of enzymes whose role is to regulate cellular levels of the second messengers, cAMP and cGMP, by their hydrolysis to inactive metabolites. Elevated levels of cAMP are associated with the suppression of cell activation in a wide range of inflammatory and immune cells, including lymphocytes, macrophages, basophils, neutrophils, and eosinophils (3-7). cAMP-specific, rolipram-sensitive phosphodiesterase (PDE IV) is the predominant PDE isozyme in these cells and thus regulates a major pathway of cAMP degradation (8). The attraction of PDE IV inhibition as a therapy for asthma derives from the potential of elevating cAMP levels selectively in the airway smooth muscle and the inflammatory cells involved in the asthmatic response, thereby achieving both of the desired pharmacological effects with a single agent and without the side effects that result from nonselective drug interactions (9).

This chapter highlights some of the important, recent developments in the medicinal chemistry and pharmacology of PDE IV inhibition, especially from the period beginning 1994 to early 1996. Other reviews on the subject, including discussions of PDE nomenclature, have appeared during this period (10-13).

PDE IV AS A THERAPEUTIC TARGET FOR ASTHMA

PDE IV Inhibitors as Antiinflammatory Agents - Pulmonary eosiniphilia is a characteristic feature of bronchial asthma (2). The well-documented findings that PDE IV is the predominant cAMP PDE isozyme in human eosinophils have triggered numerous studies that examine the effects of selective PDE IV inhibitors on eosiniphilia. Other pharmacological models of antiinflammatory activity relevant to the pathology of asthma have also been examined with PDE IV inhibitors.

In guinea pigs rolipram (**1**) dosed at 75 µg/kg, i.p., reduced allergen-induced bronchial hyperreactivity at a dose that only slightly inhibited histamine-induced bronchoconstriction (14). Theophylline (**2**), and the dual PDE III/IV inhibitor, Org 20241 (**3**), showed a similar effect; however, bronchoalveolar lavage studies on the treated animals showed different selectivities of these inhibitors with respect to their antiinflammatory activities. The rolipram-treated guinea pigs did not show a significant reduction in the number of eosinophils 24 hours after a second allergen provocation, although the numbers of neutrophils and lymphocytes were reduced. Previous studies on PDE IV inhibition of eosinophil influx have used higher doses (≥1 mg/kg), where histamine-induced bronchoconstriction is also inhibited (15). In a study using Brown-Norway rats, **1** or **3**, each dosed at 30 µmol/kg (i.p.), prevented accumulation of both eosinophils and neutrophils into the lungs (16).

Acute and chronic exposure of *Ascaris suum* aerosol antigen to atopic cynomolgus monkeys produces effects similar to those observed in bronchial asthma, including bronchoconstriction, neutrophil and eosinophil influx, and airway hyperresponsiveness (AHR). Subcutaneous treatment with **1** (10 mg/kg) caused a significant reduction in the antigen-induced increase in both neutrophils and eosinophils (17). *Ex vivo* measurements four hours after antigen exposure showed that rolipram-treated monkeys had significantly reduced plasma levels of TNFα and IL-8, however IL-1 and IL-6 levels were unaffected. Further rolipram treatment protected the monkeys against a chronic response to antigen and prevented the development of AHR.

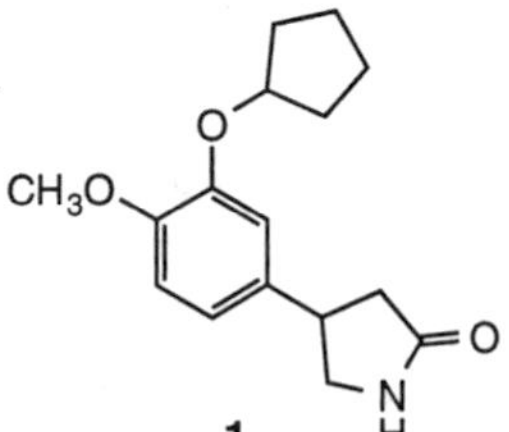

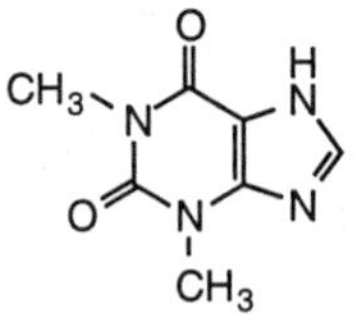

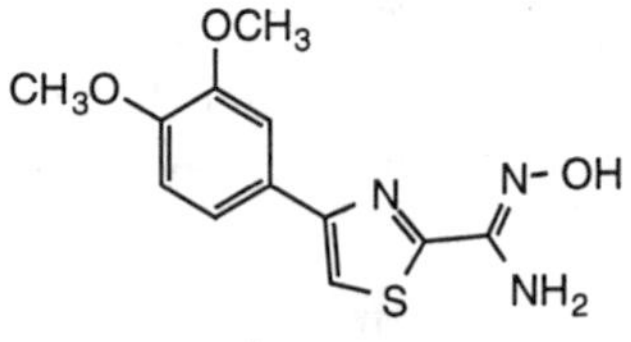

The effect of intracellular cAMP levels on human eosinophil function was measured in a chemotaxis model (18). Forskolin and PGE$_1$, both activators of adenylate cyclase, and the selective PDE IV inhibitor, RS-25344 (**4**), were each measured at 10 µM for their ability to inhibit PAF- and C5a-stimulated eosinophil chemotaxis. In keeping with its greater ability to elevate cAMP in eosinophils, **4** was also found to be more potent at inhibiting chemotaxis than forskolin and PGE$_1$, showing a mean percent inhibition of eosinophil chemotaxis in the PAF- and C5a-stimulated groups of 35% and 40%, respectively.

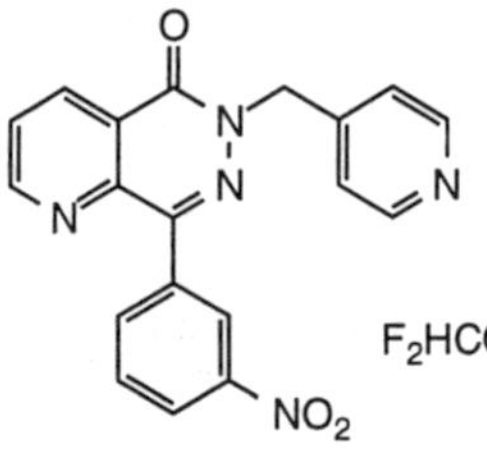

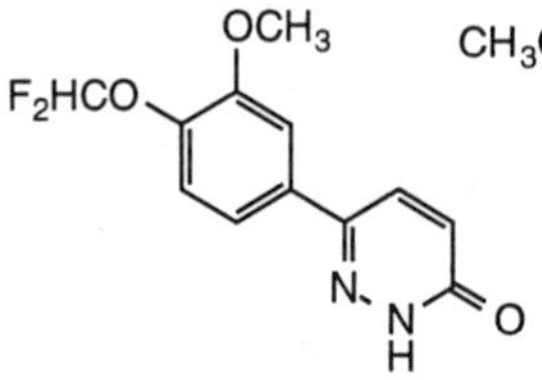

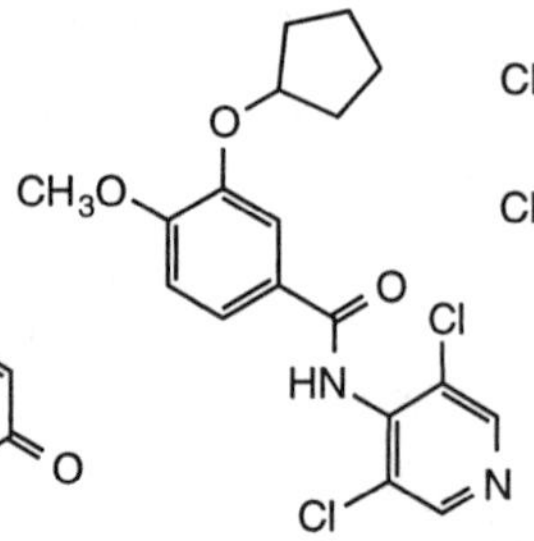

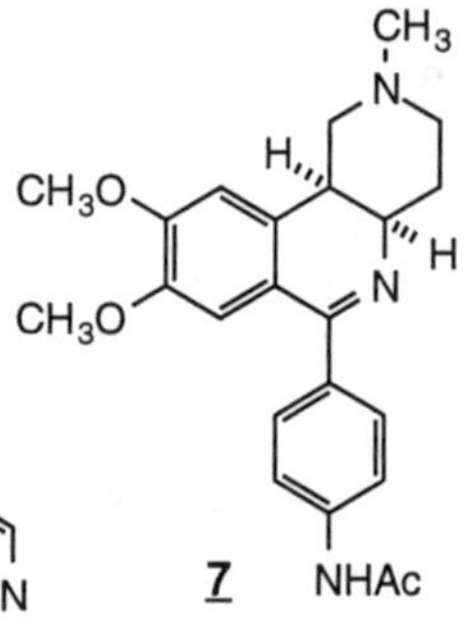

Human eosinophil respiratory burst in opsonized zymosan-stimulated cells was dose-dependently suppressed by rolipram and the dual PDE III/IV inhibitor, zardaverine (**5**), with IC$_{50}$'s of 40 µM and 30 µM, respectively. No effect was observed by the selective PDE III inhibitor, SK&F 94120 (19). In the presence of albuterol, a β$_2$-adrenoceptor agonist and itself an inhibitor of eosinophil burst (i.e., superoxide anion generation), the inhibitory effects of rolipram and **5** on respiratory burst were additive. These observations are in contrast to the findings of other researchers who reported that each of the nonselective, methylxanthine PDE inhibitors, theophylline and IBMX, suppressed the functions of C5a-stimulated human eosinophils (including respiratory burst), which were not significantly inhibited by the selective PDE IV inhibitiors rolipram and RP 73401 (**6**), each at concentrations up to 10 µM (20). The underlying reasons for the different findings in the two studies are unclear.

The dual PDE III/IV inhibitor, AH 21-132 (**7**), at a dose of 3 mg/kg (i.p.), significantly inhibited antigen-induced eosinophilia in mice (21). Monocyte accumulation was also suppressed although to a lesser extent. In guinea pigs, antigen-induced immediate asthmatic response, measured as a decrease in both respiratory rate and volume as compared to pre-challenge levels, was inhibited by **7** at a dose of 30 mg/kg (p.o.). Rolipram showed "clear and almost complete inhibition" of these responses at a dose of 10 mg/kg. Compound **7** was more potent than rolipram at inhibiting antigen-induced contraction of isolated guinea pig trachea.

PDE IV Inhibitors Undergoing Clinical Evaluation - CDP 840 (**8**) reportedly showed "pronounced effects in preclinical models of bronchoconstriction, eosinophilia, and airway hyperresponsiveness (22)." However, **8** was recently dropped from its phase II trials because it demonstrated no "significant therapeutic advance" over current asthma therapies (23).

The N-pyridyl benzamide **6** was identified from a series of rolipram-related benzamides as a potent and selective inhibitor of PDE IV (IC_{50} = 1 nM, pig aorta PDE IV) with 19,000-fold selectivity over the other PDE isozymes (24). Compound **6** is reported to have low oral bioavailibility and is being developed as an inhaled therapy for asthma (12). The compound is 3- to 14-fold more potent than rolipram at inhibiting guinea pig eosinophil function (25). At doses that are significantly bronchodilating, **6** is reported to exert little effect on the cardiovascular system (26). Further side effect profiles on **6** have not been disclosed.

Cyanocyclohexane carboxylic acid **9** (SB 207499) has been compared with (R)-rolipram in a panel of biochemical assays including the suppression of inflammatory cell functions (27). Among the data provided, **9** inhibits PDE IV with a K_i = 92 nM and inhibits the production of TNFα from activated human monocytes with an IC_{50} = 110 nM. The ability of **9** to induce gastric acid secretion from isolated rabbit parietal glands was 200-fold less than that of (R)-rolipram.

The emetic profile of the development candidate, WAY-PDA-641 (**10**), was compared with the other selective PDE IV inhibitors, rolipram and denbufylline (**11**) (28). When administered i.v. to conscious dogs, rolipram, **10**, and **11** displayed mean emetic doses of 0.3, 1.0, and 1.2 mg/kg, respectively. The "stepping behaviors and apparent anxiety" associated with rolipram and denbufylline were much less pronounced with **10**, which has an ED_{30} of 1.6 µg/kg for bronchodilation in an anesthetized dog. Cardiovascular effects were not observed at this dose.

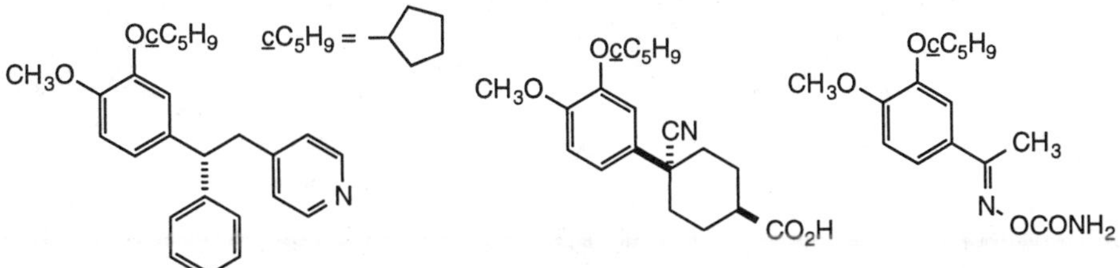

8   **9**   **10**

The *in vitro* and *in vivo* pharmacology of the selective PDE IV inhibitor, CP-80,633 (**12**) was recently disclosed. Compound **12** exhibited no isoform selectivity between PDE IV$_{A-D}$ (10), and suppressed eosinophil superoxide production (IC_{50} = 2.7 µM) and chemotaxis (IC_{50} = 0.2-0.4 µM) (29). In the monkey and guinea pig **12** showed significant antiinflammatory effects (30); however, in fasted ferrets **12** (3 mg/kg, p.o.) produced vomiting and retching in 20% of the animals, an effect prevented by prior feeding (31). Clinical development of **12** for asthma has been discontinued (32).

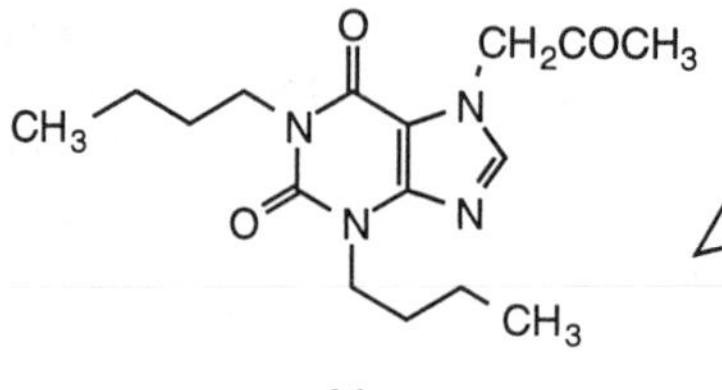

11 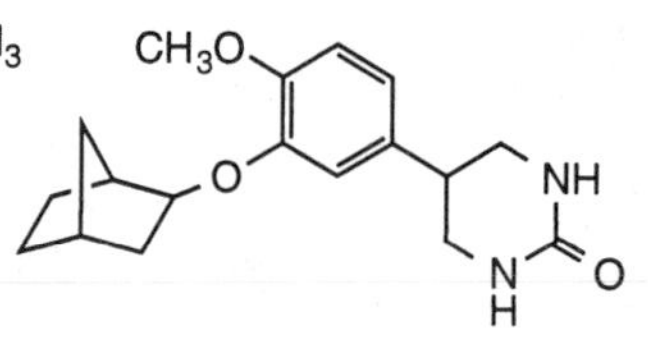   **12**

NEW PDE IV INHIBITORS

There continues to be significant research in the discovery of novel and selective PDE IV inhibitors. The atomic structure of the PDE IV catalytic site remains elusive, and discovery efforts have been guided by traditional structure-activity methods.

Rolipram and Catechol Ether Analogs - A series of aryl-substituted pyrrolidines was described in which the carbonyl group of rolipram was translocated from an endocyclic position to an exocyclic position (33). A number of N-substituted derivatives were prepared and their PDE IV inhibition data were reported. Carbamate **13** is a representative member of this class and is approximately 10-fold more potent at inhibiting human PDE IV than rolipram. Subsequently, a patent application describing this class of molecules appeared (34). Structure-activity studies within the pyrrolidine class led to the discovery of the more potent inhibitors, methyl ketones **14** and **15** (35-37). Ketone **15** is a low-nanomolar inhibitor of PDE IV (K_i = 1 nM), which is also a potent inhibitor of TNFα secretion both *in vitro* and *in vivo*. On the basis of proton NMR and computer modelling studies a pharmacophore model of **15** was proposed (36). In this model the ketone carbonyl oxygen makes an important interaction within the PDE IV active site. Interestingly, the pharmacophore model proposed for **15** encompasses the structural features of ketone **16**, a member of a class of arylcyclopropane PDE IV inhibitors (38). A homologous series of achiral N-substituted piperidines (e.g., **17**) was also the subject of a patent application (39). The 3-aryl-2-isoxazoline carboxylic acid **18** was also described albeit no biological data on this compound were provided (40).

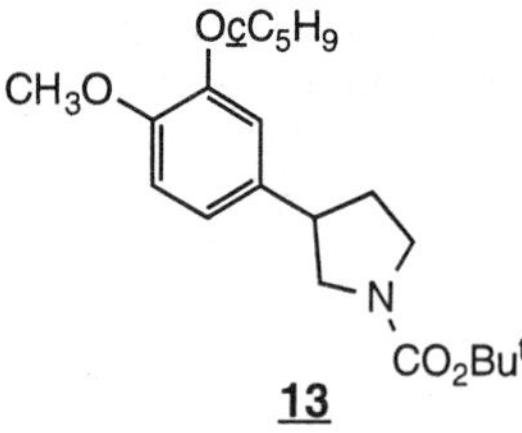

13 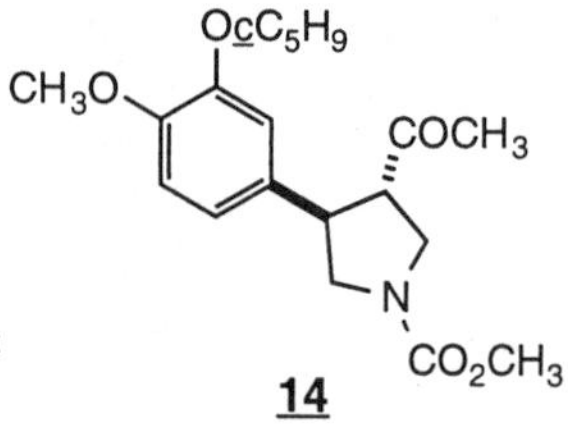   **14** 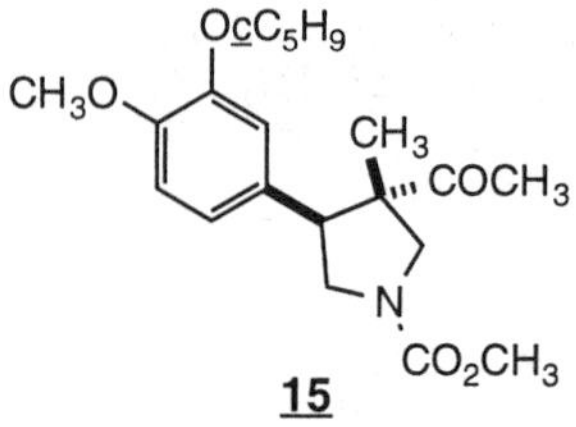   **15**

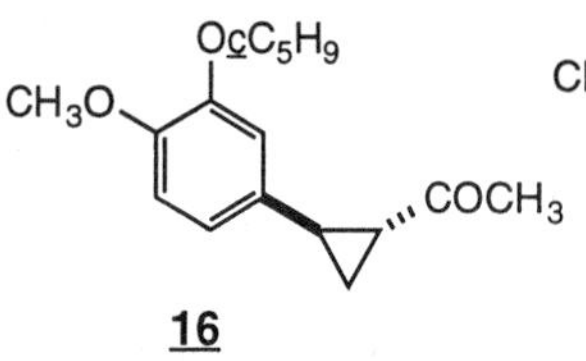

16 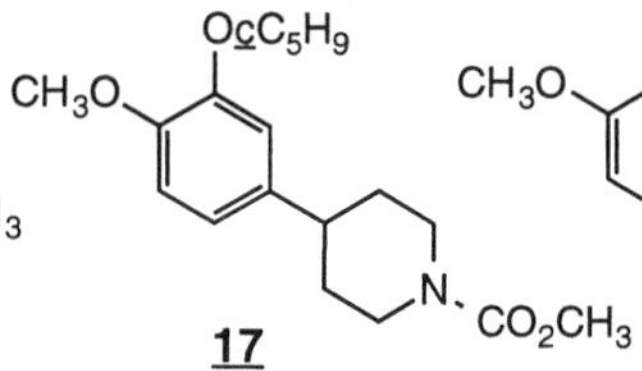   **17** 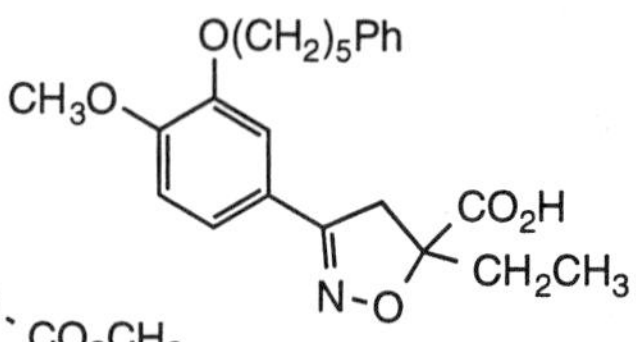   **18**

Heterobicyclic Derivatives - Naphthyridinone **19** belongs to a series of PDE IV inhibitors that is structurally related to the weak and nonselective PDE IV inhibitor, imidazonaphthyridone KF-17625 (**20**) (41). Compound **19**, which has an IC_{50} of 3.6 µM against porcine cardiac ventricular PDE IV, showed 84% inhibition of antigen-induced bronchoconstriction in anesthetized guinea pigs when dosed at 10 mg/kg (i.v). All of the naphthyridinones in this study lack the theophylline-like structural motif found in **20** and also show greater PDE IV selectivity than **20**. The most potent inhibitor reported in this series was **21** (IC_{50} = 0.14 µM), however **21** was less efficacious than **19** in a variety of functional assays.

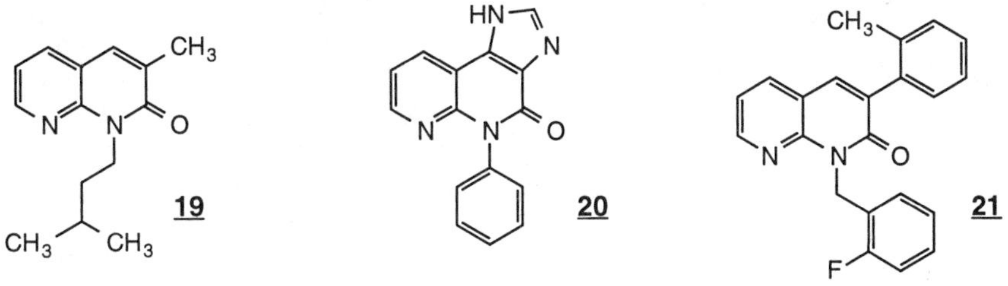

PDE IV High-Affinity Rolipram Binding - In addition to the cAMP-hydrolyzing activity of PDE IV that can be inhibited by (±)-rolipram (K_i = 220 nM), human PDE IV also expresses a high-affinity binding site for (±)-rolipram (K_d = 10 nM). The functional relevance of this so-called rolipram binding site remains unclear, although it has been proposed that the well-publicized side effects of emesis, nausea, and gastrointestinal disturbance that plagued the clinical development of rolipram stem from its high-affinity binding (42,43).

Selective PDE IV inhibitors were compared for their abilities to suppress superoxide anion production from guinea pig eosinophils, to inhibit the catalytic activity of human PDE IV_A, and to bind to the high-affinity site (44). Suppression of eosinophil function was better correlated to inhibition of catalytic activity than to high-affinity binding. Carboxylic acid **9** displayed an IC_{30} for inhibition of O_2^- production, IC_{50} for catalytic activity, and K_d of rolipram binding of 0.34, 0.14, and 0.12 µM, respectively. The novel cyanocyclohexane derivative **22** was the most potent compound in these three assays measuring 0.045, 0.02, and 0.025 µM, respectively.

In a similar study, the secretion of acid from isolated rabbit gastric glands was measured and correlated to the high-affinity binding of PDE IV inhibitors (45). (R)-rolipram, pyrrolidinone **23**, and imidazolidinone **24** increased acid secretion with EC_{50}'s of 0.004 µM, 0.017 µM, and 0.036 µM, respectively. This is in agreement with their rank affinities for the rolipram binding site. In contrast, the IC_{50}'s of PDE IV inhibition for (R)-rolipram, **23**, and **24** are 1.4 µM, 0.044 µM, and 8.5 µM, respectively.

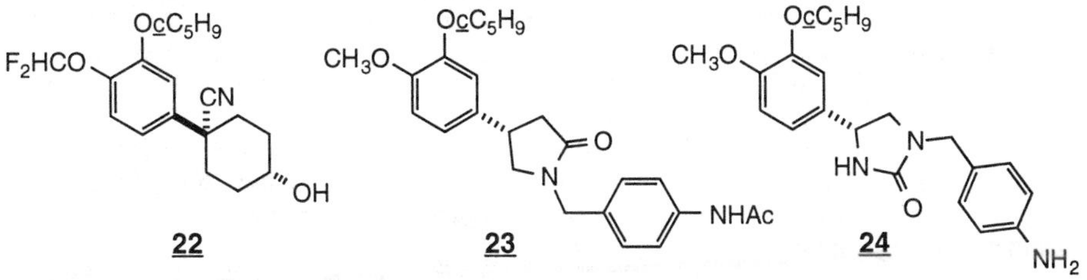

The speculation that the adverse side effect profile associated with PDE IV inhibitors derives from an interaction with the high affinity binding site has prompted researchers to target compounds in which the relative activity for enzyme inhibition and high-affinity binding is reversed with respect to rolipram, i.e., compounds which are potent PDE IV inhibitors yet display concomitant weak affinity for the rolipram binding site (46). The oxindole analog **25** has an IC_{50} for inhibition of 0.48 µM and an IC_{50} for rolipram binding of 4.6 µM (47). In a series of benzimidazole carboxylates, modification of the catechol ether moiety influenced the PDE IV inhibition/rolipram binding ratio (48). Thus, the cyclopentyl ether **26** is equipotent for inhibition and for binding with IC_{50}'s of 0.071 µM and 0.078 µM, respectively. However, the 2-indanyl ether **27** shows a fifteen-fold difference in these two activities, with IC_{50}'s of 0.002 µM and 0.032 µM, respectively. The observation that the nature of the alkyl groups on the catechol ether moiety influences the high-affinity binding profile within a series of closely related analogues has been previously documented (49). In this regard the benzofuran derivative of rolipram, **28**, inhibits PDE IV catalytic activity equal to rolipram, while the binding affinity of **28** decreases approximately thirty-fold (50).

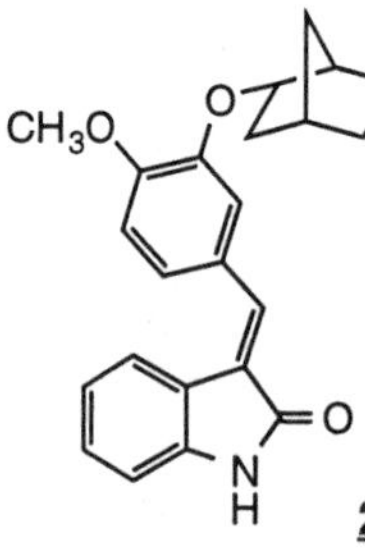

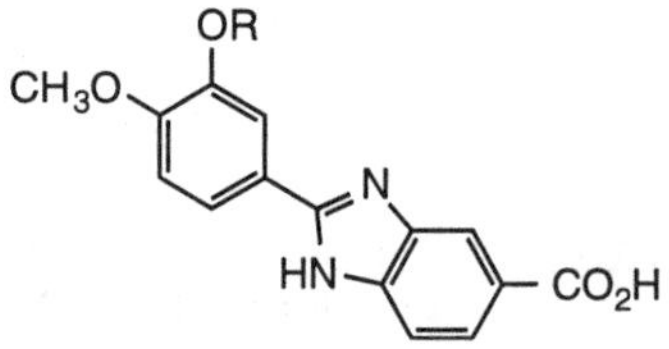

26: R = cyclopentyl
27: R = 2-indanyl

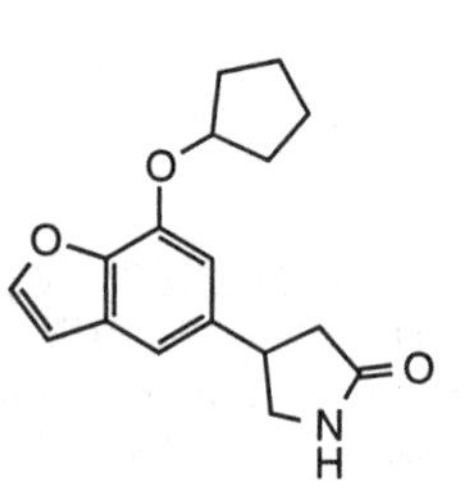

25

28

The biaryl carboxylate **29** has an IC_{50} for PDE IV inhibition of 0.41 µM and an IC_{50} for rolipram binding of 1.68 µM (51). Also in this series, a dependence on the nature of the catechol ether with respect to the PDE IV inhibition/rolipram binding ratio was observed. The derivative **30**, which possesses the same catechol ethers as rolipram, also shares with rolipram an enhanced affinity for rolipram binding over PDE IV inhibition. Compound **29** caused prodromal or emetic behavior in 3 of 15 ferrets when given orally at 30 mg/kg, a dose that provided a drug concentration of 24 µM one hour postdose. By comparison, rolipram caused prodromal or emetic behavior in 2 of 10 ferrets when administered orally at only 0.03 mg/kg.

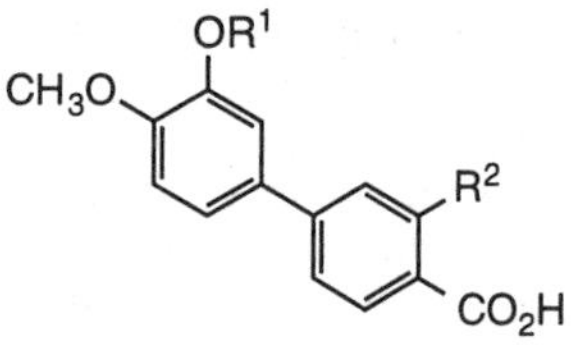

29: R^1 = 5-phenylpentyl;
R^2 = CH_3
30: R^1 = cyclopentyl; R^2 = H

OTHER THERAPEUTIC APPLICATIONS FOR PDE IV INHIBITORS

The ability of PDE IV inhibitors to increase intracellular cAMP levels, which subsequently inhibits the production and/or release of cytokines such as TNFα, has led to an increased interest in their use as therapeutic agents for pathologies in which cytokine overexpression and/or deregulation has been implicated (52,53).

Arthritis - Elevated levels of TNFα are found in both the serum and synovial fluid of arthritic patients. The recently reported positive clinical data using TNFα-neutralizing antibodies for the treatment of arthritis has stimulated investigation on the use of PDE IV inhibitors for rheumatoid arthritis (54-56).

Rolipram (**1**) and the structurally distinct, heterobicyclic PDE IV inhibitor, CP-77059 (**31**), were measured for their ability to inhibit LPS-induced elevation of serum TNFα levels in mice (57). Thirty minutes prior to a sublethal i.p. administration of LPS, mice were dosed orally with either rolipram or **31**. The minimum effective dose (ED_{50}) of **31** was approximately 0.1 mg/kg, compared with approximately 3 mg/kg for rolipram. Further, in a three-week rat adjuvant arthritis model, rolipram inhibited ankle swelling by 32% at 3 mg/kg p.o., while **31** inhibited ankle swelling by 40% and 42% at 1 and 3 mg/kg p.o., respectively. Subsequent radiological analysis of the ankles assessed the following parameters: bone demineralization, bone erosion, periostitis, cartilage space reduction , and soft tissue swelling. At 3 mg/kg, rolipram showed a 24% reduction in radiological changes, while **31** showed 34-62% reduction at 0.3, 1, and 3 mg/kg.

In a model of T cell-dependent arthritis induced by mercuric chloride in Brown-Norway rats, tissue injury is associated with cytokine production from the Th2 subset of CD4$^+$ T cells and an increase in TNFα production. When administered intraperitoneally, the nonselective PDE inhibitor, pentoxifylline (**32**), which is a selective inhibitor of the Th1 response and an inhibitor of TNFα production, demonstrated a protective effect against arthritis (58). Soluble TNF receptor was similarly protective against arthritis in this model. These data suggest that inhibition of TNFα production, and not Th2-dependent cytokine production, is the important mechanism of action of **32** in this model.

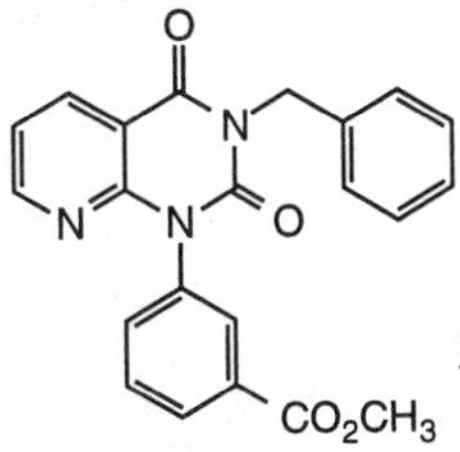

31

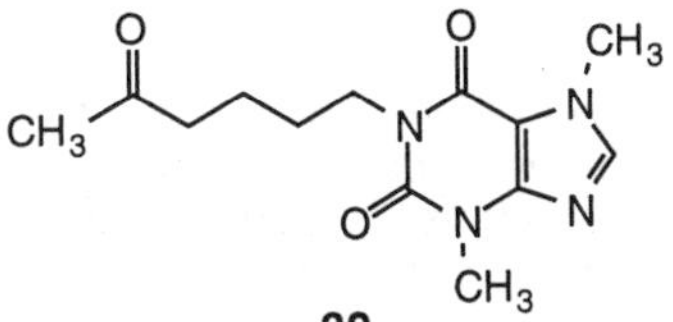

32

Experimental Autoimmune Encephalomyelitis (EAE) - Multiple sclerosis (MS), a human demyelinating disorder of the central nervous system (CNS), is widely considered to be a T cell-mediated autoimmune disease. Experimental autoimmune encephalomyelitis (EAE) is an animal model believed to be predictive for MS, wherein susceptible animals are immunized with human CNS tissue, eliciting an inflammatory response that resembles MS (59).

TNFα has been implicated in the pathogenesis of MS and EAE. Due to the observation that PDE IV inhibitors, such as rolipram, inhibit TNFα secretion from activated monocytes and macrophages, and that PDE IV is expressed in physiologically relevant concentrations in the CNS, two independent studies were reported in which rolipram was examined in EAE (60,61). In one of these studies three marmosets were treated with rolipram (10 mg/kg s.c. every 48 hr) seven days following immunization. In all three animals onset of clinical EAE was completely prevented. By contrast, in the control group, two of the three animals developed EAE 15 and 17 days following immunization, while the third animal unexpectedly died. Cessation of rolipram treatment led to eventual onset of EAE, while animals that continued treatment remained asymptomatic. Measurement of the brain levels of TNFα mRNA revealed that the two animals that did not develop EAE had reduced levels compared to the animal that did not receive rolipram for the duration. A similar study using Lewis rats showed equally promising results. A patent application has been filed for the use of rolipram and its analogs for the prevention and treatment of multiple sclerosis (62).

Human Immunodeficiency Virus and AIDS - TNFα levels are elevated in the serum of AIDS patients and it has been suggested that the cytokine may actually play a role in contributing to and exacerbating the clinical signs of the disease (63-65). In fact, it has been shown that TNFα increases the replication of HIV-1 *in vitro* by stimulation of HIV expression (66). The non-specific PDE inhibitor **32** has been in clinical trials for the treatment of HIV-infected individuals due to its ability to down-regulate TNFα activity through inhibition of TNFα synthesis (67). A phase I/II study of **32** and zidovudine (AZT) on HIV-1 growth in AIDS patients suggests that a **32**/AZT combination therapy is more beneficial than either drug alone (68).

Rolipram (**1**) is a more potent inhibitor of TNFα production in human peripheral blood monocytes than **32**. The effects of rolipram on acutely and chronically HIV-infected cells were also recently compared to **32**, and it was shown in both models that rolipram is 10-600 times more potent at inhibiting HIV replication and TNFα production (69). The mechanism by which rolipram inhibits HIV replication is thought to be the same as that with **32**, i.e., interference of the activation of the nuclear transcription factor, NF-κB, via a decrease in TNFα (70). In support of this hypothesis, the dose-dependent inhibition of viral protein p24 antigen paralleled the dose-dependent inhibition of TNFα production.

<u>Type-II Diabetes (NIDDM)</u> - Elevation of intracellular cAMP in pancreatic β-cells as a response to glucose stimulation potentiates insulin secretion (71,72). The deficiency in insulin secretion that is characteristic of Type II diabetics has prompted the recent investigation of PDE inhibitors for use as glucose-dependent insulin secretagogues.

Although the identity of the relevant PDE isozyme expressed in β-cells remains uncertain, the emergence of selective inhibitors of the various PDE isozymes has allowed researchers to examine the effect that a selective PDE inhibitor has on glucose-dependent insulin secretion. The arylpiperazine derivative, L-686,398 (**33**), was shown to be an insulin secretagogue (EC$_{50}$ = 50 μM) when incubated, in the presence of glucose, with pancreatic islet cells isolated from *ob/ob* mice (73). Compound **33** was further shown to inhibit PDE activity from an islet homogenate with an IC$_{50}$ of approximately 50 μM. Using recombinant PDE's **33** inhibited PDE III and PDE IV with K$_i$'s of 27 and 5 μM, respectively.

In another study rolipram showed no effect on glucose-stimulated insulin secretion when incubated with isolated rat islets at 100 μM. At a concentration of 1 mM, rolipram *inhibited* insulin secretion 55%. In addition, rolipram at 10-100 μM failed to consistently inhibit homogenate PDE activity (74). By contrast, four selective PDE III inhibitors each increased insulin secretion. The most potent effect was shown by the PDE III inhibitor, Org 9935, which also showed significant inhibition of total PDE activity in the islet homogenate. These studies, as well as one that used human islets, point to a greater functional role of the PDE III isozyme in pancreatic β-cells.

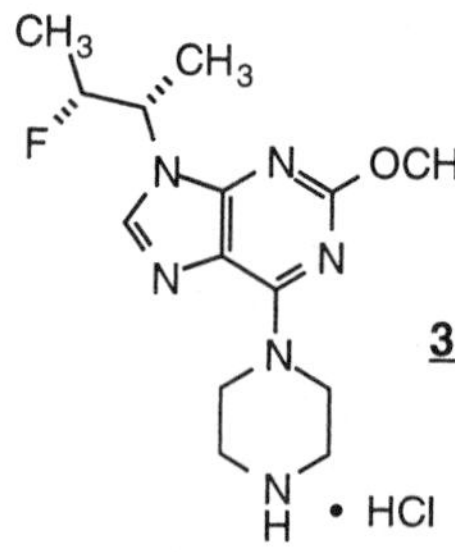

<u>Conclusions</u> - Within the last few years, extensive medicinal chemistry research has resulted in the discovery of potent and selective inhibitiors of specific cyclic nucleotide PDEs. Among the PDEs that has received recent attention, PDE IV continues to be an attractive target for the treatment of inflammatory diseases. The hypothesis that selective PDE IV inhibition can be an effective therapy for asthma is presently undergoing clinical evaluation, and it remains to be seen whether a favorable therapeutic index can be achieved. Notwithstanding these formidable hurdles, the importance of cytokine regulation and/or dysfunction in various disease states continues to be uncovered. It is therefore likely that selective PDE IV inhibitors, with their ability to suppress TNFα secretion from immune cells, will continue to be evaluated for their therapeutic potential.

References

1. Asthma, 3rd ed.; T.J.H. Clark, S. Godfrey and T.H. Lee, Eds., Chapman and Hall, London, 1992.
2. D.R. Buckle and H. Smith. Development of Anti-Asthma Drugs; Butterworths, London, 1984.
3. A. Lerner, B. Jacobson and A. Miller, J. Immunol., 140, 936 (1988).
4. M. Bachelet, M.J.P. Adolfs, J. Masliah, G. Bereziat, B.B. Vargaftig and I.L. Bonta, Eur. J. Pharmacol., 149, 73 (1988).
5. J.A. Warner, D.W. MacGlashan, S.P. Peters, A. Kagey-Sobotka and L.M. Lichtenstein, J. Allergy Clin. Immunol., 82, 432 (1988).
6. F.A. Kuehl, Jr., M.E. Zanetti, D.D. Soderman, D.K. Miller and E.A. Ham, Am. Rev. Respir. Dis., 136, 210 (1987).
7. G. Dent, M.A. Giembycz, K.F. Rabe and P.J. Barnes, Br. J. Pharmacol., 103, 1339 (1991).
8. T.J. Torphy and B.J. Undem, Thorax, 46, 512 (1991).
9. T.J. Torphy, G.P. Livi, J.M. Balcarek, J.R. White, F.H. Chilton and B.J. Undem in "Advances in Second Messenger and Phosphoprotein Research," Vol. 25, S.J. Strada and H. Hidaka, Eds., Raven Press, New York, N.Y., 1992, p. 289.
10. J.A. Beavo, M. Conti and R.J. Heaslip, Mol. Pharmacol., 46, 399 (1994).
11. S.B. Christensen and T.J. Torphy, Ann. Rep. Med. Chem., 29, 185 (1994).
12. M.N. Palfreyman, Drugs of the Future, 20, 793 (1995).
13. D. Cavalla and R. Frith, Curr. Med. Chem., 2, 561 (1995).
14. R.E. Santing, C.G. Olymunder, K. Van der Molen, H. Meurs and J. Zaagsma, Eur. J. Pharmacol., 275, 75 (1995).
15. D.C. Underwood, R.R. Osborn, L.B. Novak, J.K. Matthews, S.J. Newsholme, B.J. Undem, J.M. Hand and T.J. Torphy, J. Pharmacol. Exp. Ther., 266, 306 (1993).
16. W. Elwood, J. Sun, P.J. Barnes, M.A. Giembycz and K.F. Chung, Inflamm. Res., 44, 83 (1995).
17. C.R. Turner, C.J. Andresen, W.B. Smith and J.W. Watson, Am. J. Respir. Crit. Care Med., 149, 1153 (1994).
18. T. Kaneko, R. Alvarez, I.F. Ueki and J.A. Nadel, Cell. Signalling, 7, 527 (1995).
19. G. Dent, M.A. Giembycz, P.M. Evans, K.F. Rabe and P.J. Barnes, J. Pharmacol. Exp. Ther., 271, 1167 (1994).
20. A. Hatzelmann, H. Tenor and C. Schudt, Br. J. Pharmacol., 114, 821 (1995).
21. H. Nagai, H. Takeda, T. Iwama, S. Yamaguchi and H. Mori, Jpn. J. Pharmacol., 67, 149 (1995).
22. G.J. Warrellow, R.P. Alexander, E.C. Boyd, M.A.W. Eaton, J.C. Head, G.A. Higgs, B. Hughes, M.J. Perry, R. Allen and R. Owens, 8th RSC-SCI Medicinal Chemistry Symposium, Churchill College, Cambridge, UK, September 10-13, 1995.
23. Financial Times (London), February 2, 1996.
24. M.J. Ashton, D.C. Cook, G. Fenton, J.-A. Karlsson, M.N. Palfreyman, D. Raeburn, A.J. Ratcliffe, J.E. Souness, S. Thurairatnam and N. Vicker, J. Med. Chem., 37, 1696 (1994).
25. J.E. Souness, C. Maslen, S. Webber, M. Foster, D. Raeburn, M.N. Palfreyman, M.J. Ashton and J.-A. Karlsson, Br. J. Pharmacol., 115, 39 (1995).
26. D. Raeburn, S.L. Underwood, S.A. Lewis, V.R. Woodman, C.H. Battram, A. Tomkinson, S. Sharma, R. Jordan, J.E. Souness, S.E. Webber and J.-A. Karlsson, Br. J. Pharmacol., 113, 1423 (1994).
27. M.S. Barnette, S.B. Christensen, D.M. Essayan, K.M. Esser, M. Grous, S.-K. Huang, C.D. Manning, U. Prabhakar, J. Rush and T.J. Torphy, Am. J. Resp. Crit. Care Med., 149 (Suppl.), A209 (1994).
28. R.J. Heaslip, D.Y. Evans and B.D. Sickels, FASEB, 8, A371 (1994).
29. V.L. Cohan, H.J. Showell, E.R. Pettipher, D.A. Fisher, C.J. Pazoles, J.W. Watson, C.R. Turner and J.B. Cheng, J. Allergy Clin. immun., 95, Abstract 840 (1995).
30. C.R. Turner, J.B. Cheng, V.L. Cohan, H.J. Showell, C.J. Pazoles and J.W. Watson, J. Allergy Clin. Immun., 95, Abstract 842 (1995).
31. J.W. Watson, R. Beck, V.L. Cohan, H.J. Showell, C.J. Pazoles, J.B. Cheng,T. Liston and C.R. Turner, J. Allergy Clin. Immun., 95, Abstract 850 (1995).
32. H. Masamune, J.B. Cheng, K. Cooper, J.F. Eggler, A. Marfat, S.C. Marshall, J.T. Shirley, J.E. Tickner, J.P. Umland and E. Vazquez, Biomed. Chem. Lett., 5, 1965, ref. 6 (1995).
33. P.L. Feldman, M.F. Brackeen, D.J. Cowan, B.E. Marron, F.J. Schoenen, J.A. Stafford, E.M. Suh, P.L. Domanico, D. Rose, M.A. Leesnitzer, E.S. Brawley, A.B. Strickland, M.W. Verghese, K.M. Connolly, R. Bateman-Fite, L.S. Noel, L. Sekut and S.A. Stimpson, J. Med. Chem., 38, 1505 (1995).
34. T. Tanaka, A. Yamamoto and A. Amenomori, European Patent Application, 0671389A1.
35. J.A. Stafford, N.L. Valvano, P.L. Feldman, E.S. Brawley, D.J. Cowan, P.L. Domanico, M.A. Leesnitzer, D.A. Rose, S.A. Stimpson, A.B. Strickland, R.J. Unwalla and M.W. Verghese, Biomed. Chem. Lett., 5, 1977 (1995).
36. J.A. Stafford, J.M. Veal, P.L. Feldman, N.L. Valvano, P.G. Baer, M.F. Brackeen, E.S. Brawley, K.M. Connolly, P.L. Domanico, B. Han, D.A. Rose, R.D. Rutkowske, L. Sekut, A.B. Strickland and M.W. Verghese, J. Med. Chem., 38, 4972 (1995).
37. P.L. Feldman and J.A. Stafford, WO9508534 (1995).
38. G. Fenton, J.S. Mason, M.N. Palfreyman and A.J. Ratcliffe, WO9427947 (1994).

39. L.J. Lombardo, WO9425437 (1994).
40. E.F. Kleinman, WO9514680 (1995).
41. A. Matsuura, N. Ashizawa, R. Asakura, T. Kumonaka, T. Aotsuka, T. Hase, C. Shimizu, T. Kurihara and F. Kobayashi, Biol. Pharm. Bull., 17, 498 (1994).
42. R. Schiechen, H.H. Schneider and H. Wachtel, Psychopharmacology, 102, 17 (1990).
43. M.S. Barnette, Challenges for Drug Discovery. New Drugs for Asthma-III, Montebello, Quebec, July, 1994.
44. M.S. Barnette, C.D. Manning, L.B. Cieslinski, M. Burman, S.B. Christensen and T.J. Torphy, J. Pharmacol. Exp. Ther., 273, 674 (1995).
45. M.S. Barnette, M. Grous, L.B. Cieslinski, M. Burman, S.B. Christensen and T.J. Torphy, J. Pharmacol. Exp. Ther., 273, 1396 (1995).
46. M.S. Barnette, T.J. Torphy and S.B. Christensen, WO9500139 (1995).
47. H. Masamune, J.B. Cheng, K. Cooper, J.F. Eggler, A. Marfat, S.C. Marshall, J.T. Shirley, J.E. Tickner, J.P. Umland and E. Vazquez, Biomed. Chem. Lett., 5, 1965 (1995).
48. J.B. Cheng, K. Cooper, A.J. Duplantier, J.F. Eggler, K.G. Kraus, S.C. Marshall, A. Marfat, H. Masamune, J.T. Shirley, J.E. Tickner and J.P. Umland, Biomed. Chem. Lett., 5, 1969 (1995).
49. B.K. Koe, L.A. Lebel, J.A. Nielsen, L.L. Russo, N.A. Saccomano, F.J. Vinick and I.H. Williams, Drug Dev. Res., 21, 135 (1990).
50. J.A. Stafford, P.L. Feldman, B.E. Marron, F.J. Schoenen, N.L. Valvano, R.J. Unwalla, P.L. Domanico, E.S. Brawley, M.A. Leesnitzer, D.A. Rose and R.W. Dougherty, Biomed. Chem. Lett., 4, 1855 (1994).
51. A.J. Duplantier, M.S. Biggers, R.J. Chambers, J.B. Cheng, K. Cooper, D.B. Damon, J.F. Eggler, K.G. Kraus, A. Marfat, H. Masamune, J.S. Pillar, J.T. Shirley, J.P. Umland and J.W. Watson, J. Med. Chem., 39, 120 (1996).
52. J. Semmler, H. Wachtel and S. Endres, Int. J. Immunopharmacol., 15, 409 (1993).
53. T. Kambayashi, C.O. Jacob, D. Zhou, N. Mazurek, M. Fong and G. Strassmann, J. Immunol., 155, 4909 (1995).
54. S.J. Hopkins and A. Meager, Clin. Exp. Immunol., 73, 88 (1988).
55. T. Saxne, M.A. Palladino, D. Heinegard, N. Talal and F.A. Wollheim, Arthritis Rheum., 31, 1041 (1988).
56. M.J. Elliott, R.N. Maini, M. Feldmann, J.R. Kalden, C. Antoni, J.S. Smolen, B. Leeb, F.C. Breedveld, J.D. Macfarlane, H. Bijl and J.N. Woody, Lancet, 344, 1105 (1994).
57. L. Sekut, D. Yarnall, S.A. Stimpson, L.S. Noel, R. Bateman-Fite, R.L. Clark, M.F. Brackeen, J.A. Menius Jr. and K.M. Connolly, Clin. Exp. Immunol., 100, 126 (1995).
58. P.D.W. Kiely, K.M. Gillespie and D.B.G. Oliveira, Eur. J. Immunol., 25, 2899 (1995).
59. R.H. Swanborg, Methods Enzymol., 162, 413 (1988).
60. C.P. Genain, T. Roberts, R.L. Davis, M.-H. Nguyen, A. Uccelli, D. Faulds, Y. Li, J. Hedgpeth and S.L. Hauser, Proc. Natl. Acad. Sci., 92, 3601 (1995).
61. N. Sommer, P.-A. Löschmann, G.H. Northoff, M. Weller, A. Steinbrecher, J.P. Steinbach, R. Lichtenfels, R. Meyermann, A. Riethmüller, A. Fontana, J. Dichgans and R. Martin, Nature (Med), 1, 244 (1995).
62. H. Wachtel, P.-A. Löschmann, H. Graf and J. Hedgpeth, WO9528926 (1995).
63. J. Lähdevirta, C.P.J. Maury, A.-M. Teppo and H. Repo, Am. J. Med., 85, 289 (1988).
64. S. Kobayashi, Y. Hamamoto, N. Kobayuashi and N. Yamamoto, AIDS, 4, 169 (1990).
65. G. Darling, D.L. Fraker, J.C. Jensen, C.M. Gorschboth and J.A. Norton, Cancer Res., 50, 4008 (1990).
66. K. Kitano, C.I. Rivas, G.C. Baldwin, J.C. Vera and D.W. Golde, Blood, 82, 2742 (1993).
67. B.J. Dezube, A.B. Pardee, B. Chapman, L.A. Beckett, J.A. Korvick, W.J. Novick, J. Chiurco, P. Kasdan, C.M. Ahlers, L.T. Ecto, C.S. Crumpacker and the NIAID AIDS Clinical Trials Group, J. Acquir. Immune Defic. Syndr., 6, 787 (1993).
68. D.R. Luke, B.J. McCreedy, T.P. Sarnoski, J.B. Bookout, A.M. Johnston, J.E. Lell, E.B. Wiggan, N. Bell, R.A. Limjuco, K.I. Guthrie, C.A. Rogers-Phillips and S. Dennis, Int. J. Clin. Pharm. Ther. Tox., 31, 343 (1993).
69. J.B. Angel, B.M. Saget, S.P. Walsh, T.F. Greten, C.A. Dinarello, P.R. Skolnik and S. Endres, AIDS, 9, 1137 (1995).
70. D.K. Biswas, B.J. Dezube, C.M. Ahlers and A.B. Pardee, J. Acquir. Immune Defic. Syndr., 6, 778 (1993).
71. W.J. Malaisse, F. Malaisse-Lagae and D. Mayhew, J. Clin. Invest., 46, 1724 (1967).
72. W.S. Zawalich and H. Rasmussen, Mol. Cell. Endocrinol., 70, 119 (1990).
73. M.D. Leibowitz, C. Biswas, E.J. Brady, M. Conti, C.A. Cullinan, N.S. Hayes, V.C. Manganiello, R. Saperstein, L. Wang, P.T. Zafian and J. Berger, Diabetes, 44, 67 (1995).
74. R. Shafiee-Nick, N.J. Pyne and B.L. Furman, Br. J. Pharmacol., 115, 1486 (1995).

Chapter 9. Endothelin Antagonists

M. A. Lago, J. I. Luengo, C. E. Peishoff and J. D. Elliott

SmithKline Beecham Pharmaceuticals, P.O. Box 1539, King of Prussia, PA. 19406

Introduction - Endothelin-1 (ET-1) is a 21 amino acid peptide produced by endothelial cells which exhibits vasoconstricting activity a full order of magnitude greater (EC_{50} of 0.1-1 nM) than angiotensin II, vasopressin or norepinephrine (1). There are three closely related ET isoforms, ET-1, ET-2 and ET-3, each encoded by a distinct gene. Circulating levels of ETs are low, ~1 pM, well below the EC_{50} and it is now apparent that the ETs function locally as autocrine or paracrine agents. ET-1, the only one of the isopeptides produced by endothelial cells, plays an important role in cardiovascular regulation. In addition, ET-1 elicits a number of other biological actions, including mitogenesis, chemotaxis, regulation of renal tubule reabsorption, modulation of neurotransmitter and hormone release as well as hyperplasia and hypertrophy. Consequently ET-1 has been linked to a number of pathophysiological conditions such as hypertension, myocardial dysfunctions, vascular restenosis, atherosclerosis, renal failure and cerebrovascular disease.

Progress in the discovery of ET receptor antagonists has been reviewed in previous volumes in this series (2,3). Several reviews covering different aspects of ET research have also appeared in the literature in the last year (4-10). Likewise, a book on ET receptors and physiology as well as a full supplement in the *Journal of Cardiovascular Pharmacology* containing the proceedings of the Fourth International Conference on Endothelin (London, April 1995) have been published (11,12).

Endothelin Biosynthesis - ET-1 is ubiquitous, being produced in numerous organs by a variety of cell types including endothelial, epithelial, endometrial, mesangial, smooth muscle, neural, and mast cells as well as macrophages. Distribution of both ET-2 and ET-3 is much more restricted and sites of production are still uncertain. ET-1 production is regulated at the transcription level by a variety of agents (PDGF, TGF-β, TNF-α, angiotensin II, vasopressin, thrombin, oxidized LDL, cyclosporin, radiocontrast media) as well as by biophysical stimuli (shear stress, hypoxia). Human ET-1 is derived from the 212-amino acid preproendothelin-1, through cleavage by a dibasic specific endopeptidase, reported to be the mammalian convertase furin, to produce a 38-amino acid inactive precursor peptide, termed "big ET-1" (13). Further processing to mature ET-1 is catalyzed by endothelin-converting enzyme (ECE) through specific cleavage of a Trp^{21}-Val^{22} bond. Two types of ECE enzymes have been recently identified, ECE-1 and ECE-2 (14-20). In addition, two ECE-1 isoforms, ECE-1a and ECE-1b, encoded by a single gene, but differing in their N-terminal cytosolic domains have been reported (21). ECE-1 and ECE-2 are membrane-associated, zinc-binding metalloproteases containing a single transmembrane domain, N-glycosylation sites and a short N-terminal cytoplasmic tail. These enzymes have homology to both neutral endopeptidase (NEP) and Kell blood group antigen. ECE-1 and ECE-2 have an overall identity of 59% but differ in their localization and their sensitivities to pH and phosphoramidon (1). ECE-1 has a sharp activity optimum at neutral pH, is located on the plasma membrane and is 250 times less sensitive to phosphoramidon than ECE-

2. In contrast, ECE-2 has an optimal pH of 5.5 (inactive at neutral pH) and, being located in the trans-Golgi network, is presumably the enzyme that processes the endogenously synthesized big ET-1 within the endothelial cell.

Endothelin Receptors - The ETs elicit their effects through activation of two receptor subtypes, ET_A and ET_B, that belong to the seven transmembrane spanning, G-protein coupled superfamily. The ET_A receptor is characterized by high affinity and selectivity for ET-1 and ET-2 over ET-3 whereas the ET_B receptor has equivalent, high affinity for all three peptides. The ET_A receptor mediates vasoconstriction and vascular smooth muscle proliferation, whereas the ET_B subtype mediates vasodilation promoted by release of nitric oxide from endothelial cells as well as vasoconstriction in certain vascular beds. A recent report has implicated the ET_B receptor as the mediator of the autoinduction of ET-1 in rat mesangial cells (22).

ET Gene Disruption - Quite unexpected results are obtained from ET gene disruption studies using "knockout mice". Homozygous $ET-1^{-/-}$ mice die at birth of respiratory failure exhibiting severe craniofacial abnormalities and disordered development of the pharyngeal arches, heart and great vessels (23). $ET-1^{+/-}$ heterozygous mice develop normally, but have reduced plasma and tissue levels of ET-1 and mild hypertension. Interestingly, the ET_A receptor knockout mice exhibit morphological abnormalities nearly identical to the ET-1 deficient mice (24). The phenotype of these mice resembles the human congenital conditions, Pierre-Robin and Treacher-Collins diseases, for which no genes have been implicated. Targeted disruption of either the ET-3 peptide or the ET_B receptor gene results in homozygous mice that share almost identical phenotypes (aganglionic colon and coat color abnormalities) which are related to Hirschsprung's disease, a hereditary human condition characterized by a mis-sense mutation in the ET_B gene (25-27). These studies demonstrate that, in addition to vasoactive properties, the ETs play an important role in the regulation of development during embryogenesis. In particular, ET-1 and ET-3, acting through the ET_A and ET_B receptors, respectively, seem to be essential to the normal development of tissues derived from the neural crest in mice and humans.

ECE INHIBITORS

Inhibition of ECE activity has the potential to provide therapeutic benefit in those clinical conditions where ET-1 is overproduced. The two enzymes characterized so far, ECE-1 and ECE-2, are zinc proteases that are inhibited by phosphoramidon, **1**, (IC_{50} of 1 μM and 4 nM, respectively) and the quinone FR 901533, **2**, (IC_{50} of 2 μM and 3 μM, respectively) but not by other metalloprotease inhibitors such as thiorphan (**3**) or captopril (20,28). In contrast, NEP is very efficiently inhibited by both **1** (IC_{50} = 30 nM) and **3** (IC_{50} = 5 nM), but not by **2**. Both **1** and **2** have been shown to inhibit big ET-1 induced pressor response, in a dose dependent manner, when injected (1-10 mg/kg i.v.) in male Sprague-Dawley rats prior to big ET-1 challenge (29,30). In addition, infusion of **1** (30 nmol/min) into the human brachial artery inhibits both the rise in plasma ET-1 and the associated vasoconstriction caused by exogenously administered big-ET-1 (50 pmol/min) (31). The use of **1** in *in vivo* studies is somewhat limited due to its lack of selectivity, low potency and poor pharmacokinetics and a number of groups have developed SAR studies on **1** to improve its activity profile.

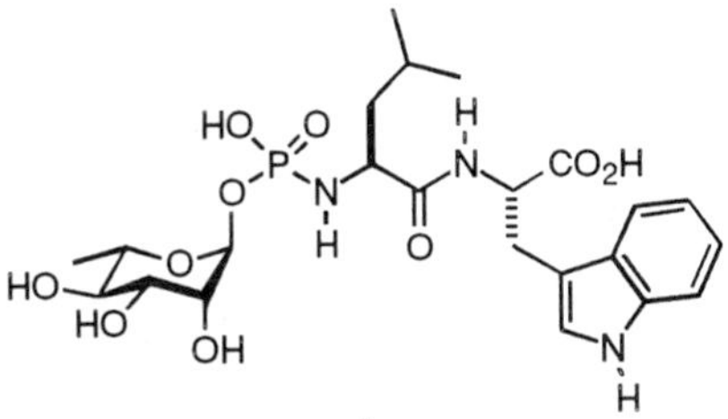

1

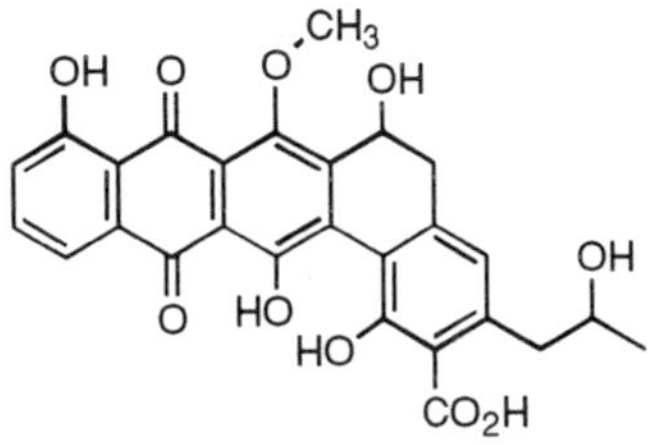

2

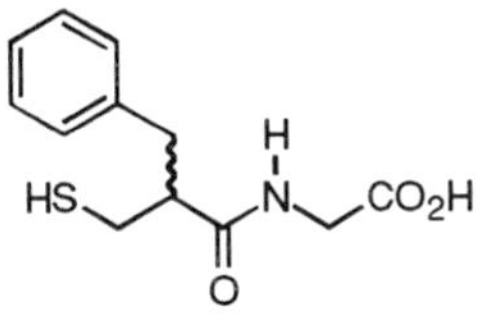

3

Phosphoramidon Analogs - Conflicting results have been reported about the role of the carbohydrate ring in **1**. Thus the des-rhamnosyl derivative **4** has been reported to be 3-fold more potent (IC$_{50}$ = 0.96 µM) than **1** as an inhibitor of the ECE activity of a preparation from the microsomal fraction of bovine cultured endothelial cells, but 20-fold less potent (IC$_{50}$ = 70 µM) than **1** in a partially purified ECE preparation from porcine primary aortic endothelial cells (32,33). Compound **4** shows greater inhibitory activity towards NEP, angiotensin-converting enzyme and thermolysin than the parent phosphoramidon (**1**). Derivative **4** is active *in vivo*, showing an effect similar to that of **1** in the attenuation of the pressor response induced by exogenously administered big ET in the rat (34). Replacement of the rhamnose ring of **1** by simple alkyl groups, such as ethyl (**5**), results in a loss of potency (IC$_{50}$ in a rabbit lung ECE preparation of 109 µM for **5**, compared with 2 µM for **1**) (35). The original activity of **1** can be recovered by replacement of the phosphoramide group in **5** by a phosphonamide as in **6** (IC$_{50}$ = 2 µM); the corresponding phosphate (**7**) and phosphonate (**8**), however, are devoid of ECE inhibitory activity. Likewise, replacement of the phosphoramide of **4** by metal-chelating functionalities, such as in the thiol **9** or hydroxamate **10**, results in compounds with reduced inhibitory potency of an ECE preparation from rabbit lung homogenate (IC$_{50}$ of 12 and 24 µM, respectively) (36).

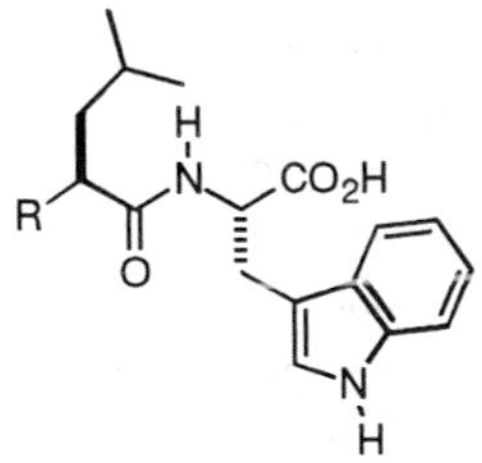

		R			R
4	-NH-PO$_3$H$_2$		**8**	-O-P(O)(OH)-Pr	
5	-NH-P(O)(OH)-OEt		**9**	(RS)-CH$_2$SH	
6	-NH-P(O)(OH)-Pr		**10**	(RS)-CONHOH	
7	-O-P(O)(OH)-OEt		**11**	-NHP(O)(OH)(CH$_2$)$_2$-2-naphthyl	

Introduction of a P$_1$ lipophilic unit adjacent to the phosphoryl group of **1** results in an analog with increased potency and selectivity, i.e. the naphthyl derivative **11** (IC$_{50}$ = 0.55 µM). Alternatively, the lipophilic P$_1$ group can be introduced between the phosphoryl and leucinyl residues through homologation of the phosphoramide to an α-amino phosphonic acid (32). Analogs such as **12** and **13**, show increased ECE potencies (3 and 10-fold, respectively) and ECE/NEP selectivities (10 and 400-fold, respectively) relative to **1** (33,37). As anticipated, greater activity is only seen in one of the two epimers at the amino-phosphoryl center.

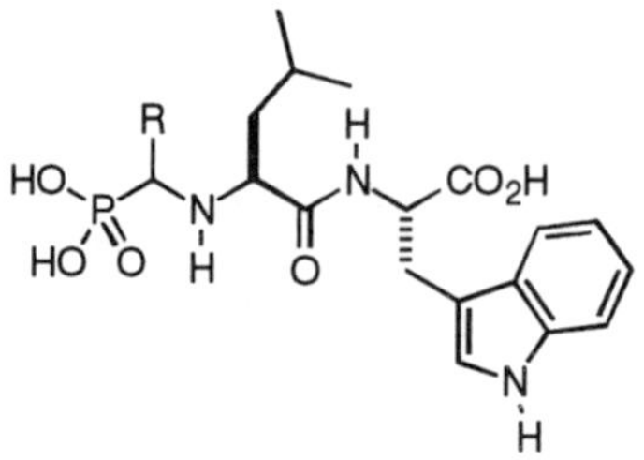

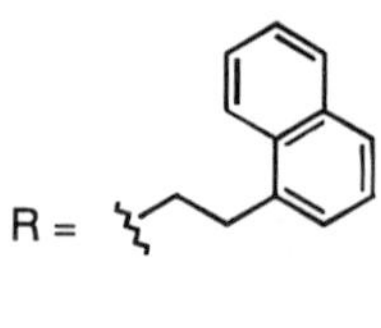

Other Inhibitors - The phosphonic acid tetrazole CGS 26303 (**14**), a potent inhibitor of NEP (IC_{50} = 0.9 nM), has been found to possess modest ECE inhibitory activity (IC_{50} = 1.1 µM) (38). Compound **14** (30 mg/kg i.v.) blocked the pressor response produced by challenge with big ET in anesthetized rats. In addition, compound **14** inhibits the enzymatic degradation of exogenously administered atrial natriuretic peptide (ANP) and reduces the mean arterial pressure of spontaneously hypertensive rats during chronic administration (5 mg/kg/d continuous infusion). Compound **14** does not show biological activity after oral administration, but the corresponding diphenylphosphonate CGS 26393 (**15**), an orally bioavailable prodrug of **14**, shows a long-lasting antihypertensive effect in the deoxycorticosterone acetate (DOCA) salt rat (30 mg/kg p.o.)(39).

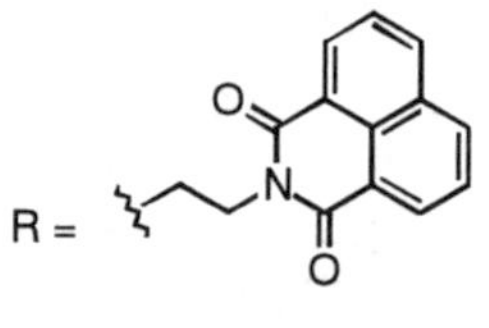

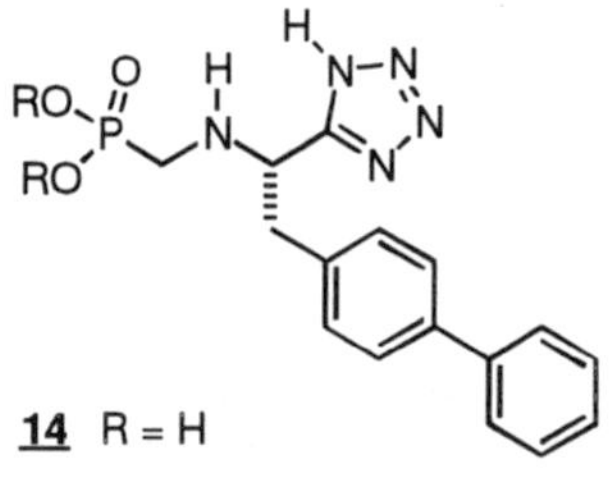

Hydroxamic acids derived from malonyl amino acids have recently been reported as extremely potent inhibitors of an ECE-like activity in human bronchiolar smooth muscle (HBSM) cells (40). Interestingly, the authors found that compound **10**, a weak inhibitor of partially purified ECE from rabbit lung (see above) is one of the more potent inhibitors of HBSM ECE (IC_{50} = 0.25 nM). Hydroxamate **16** shows remarkable potency, with an IC_{50} of 0.01 nM. The HBSM ECE preparation, however, has properties markedly distinct from those reported for ECE-1, ECE-2 or endothelial cell homogenates as it is potently inhibited by both phosphoramidon (IC_{50} = 0.8 nM) and thiorphan (IC_{50} = 8 nM) at an optimum activity pH of 7.2. Potency of the hydroxamate inhibitors against NEP (with big ET-1 as substrate) correlates strongly with their potencies against HBSM ECE. The possibility of HBSM ECE actually being NEP was eliminated by the authors based on efficiency of big ET-1 conversion to ET-1 (100 fold greater than that of NEP, without subsequent cleavage of mature ET-1) and lack of immunoprecipitation by anti-NEP monoclonal antibody. No studies on the ability of these extremely potent malonyl hydroxamates to inhibit big ET processing *in vivo* have yet been reported.

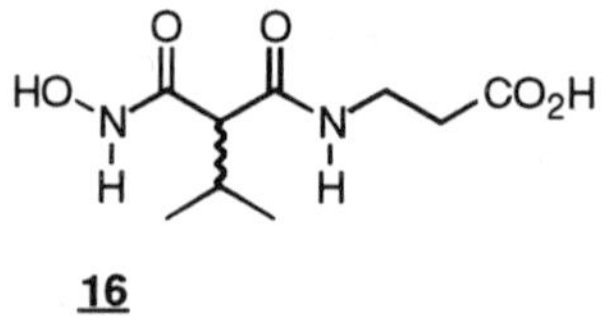

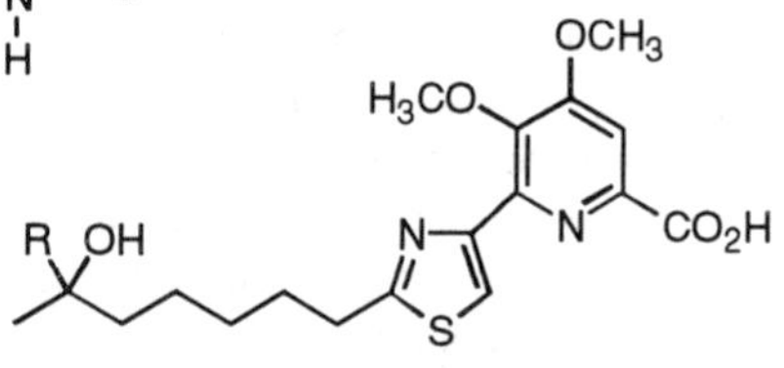

Two pyridyl thiazoles, **17** and **18**, isolated from fermentation of *Saccarothrix* sp. have recently been reported as selective ECE inhibitors (IC_{50} values of 0.03 µg/mL for

ECE and 1.25 µg/mL for NEP). Compound **17** also shows *in vivo* activity, blocking the hypertensive effect of a big ET-1 bolus (1 nmol/kg i.v.) when compound was infused at 30 mg/kg (41). The development of potent and selective ECE inhibitors remains a considerable challenge, despite the progress in recent years. A number of questions remain unanswered, such as the possibility of additional endothelin converting enzymes, physiological relevance of ECE isoforms and ECE/NEP inhibitor selectivity.

ENDOTHELIN RECEPTOR ANTAGONISTS

Peptide ET Antagonists - The earliest reports of receptor antagonists in the endothelin area describe compounds peptidic in nature and much of this work has been previously reviewed (42,43). The majority of peptide ET receptor antagonists thus far described fall into the categories of ET_A selective or dual ET_A/ET_B affinity compounds. A very significant breakthrough in the search for compounds that could elucidate a role for the endothelins in pathophysiology came with the discovery of the ET_A selective antagonist BQ-123 (**19**), which has been widely utilized in many biological studies (44,45). In man, local intraarterial administration of **19** (or phosphoramidon) causes forearm blood vessel dilation, suggesting that ET-1 contributes to the regulation of basal forearm vascular tone in humans.

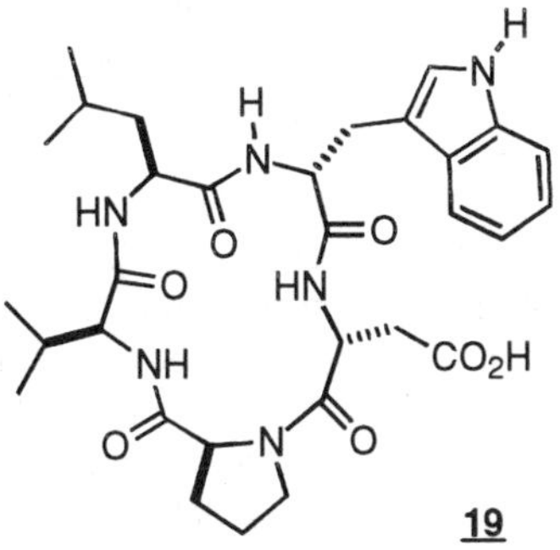

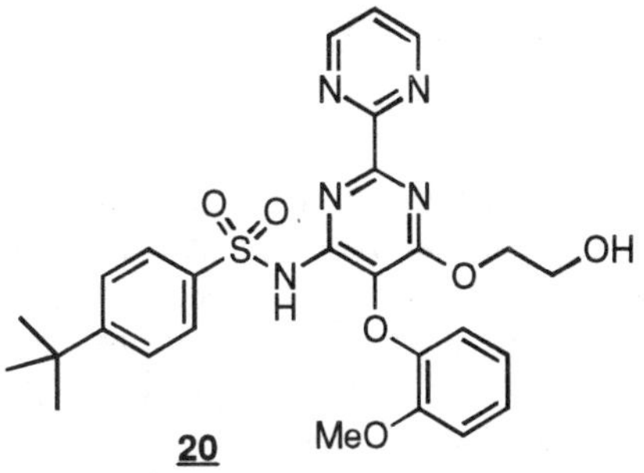

Recent data for RES 701-1 questioned the potency of this previously described, ET_B selective antagonist, (IC_{50} = 10 nM and >5 mM, respectively, against human ET_B and ET_A receptors stably transfected in CHO cells) (46). The apparent discrepancy has been attributed to a marked conformational difference between the molecules of synthetic and microbial origin (47-49).

Nonpeptide ET Antagonists - A very significant step toward the discovery of non-peptide dual ET_A/ET_B endothelin receptor antagonists of therapeutic potential came with the disclosure of bosentan (**20**), the first orally bioavailable ET antagonist (50). Compound **20** has played an important role in the implication of endothelin in pathophysiology. Clinical studies have been recently disclosed in three patient groups: congestive heart failure, hypertensive and gastric ulcer sufferers (51,52).

The indane-2-carboxylic acid series of mixed ET_A/ET_B antagonists are exemplified by SB 209670 (**21**), now in early development as an i.v. agent, and by the orally active analog SB 217242 (**22**). When administered intravenously, **21** has beneficial effects in a number of animal models of disease including hypertension, acute renal failure, restenosis following percutaneous transluminal coronary angioplasty and blockade of radiocontrast-induced nephrotoxicity (53-56). Compound **22**, a more orally bioavailable analog of **21**, was identified with the aid of intestinal permeability screening (57). Compound **22** binds to human endothelin receptors with high affinity (K_i values: ET_A = 1.1 nM, ET_B = 111 nM), and it is a potent, competitive, functional antagonist of ET_A receptors (rat aorta, K_b = 4.4 nM) with potential as a therapeutic agent for chronic indications and with demonstrated efficacy in stroke and pulmonary hypertension models (58). As a result of molecular modeling comparisons with the indane compounds, a new series of antagonists have been designed, which includes indole **23** and dihydroisoindole **24** (59). Site directed mutagenesis results suggest that these compounds interact with a lysine residue on transmembrane spanning domain III, which is conserved between receptor subtypes (ET_A K182, ET_B K166) (58).

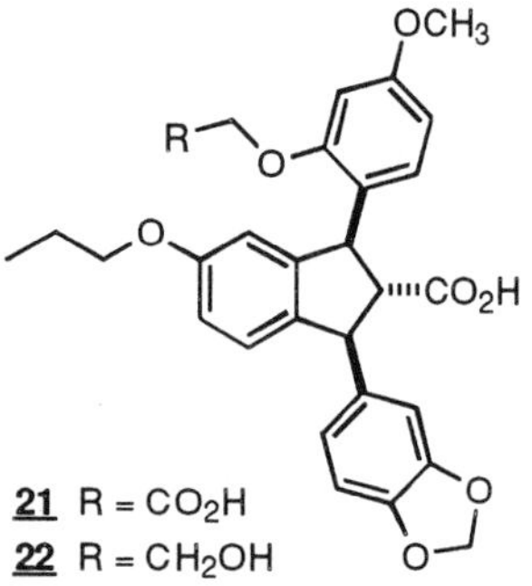

21 R = CO$_2$H
22 R = CH$_2$OH

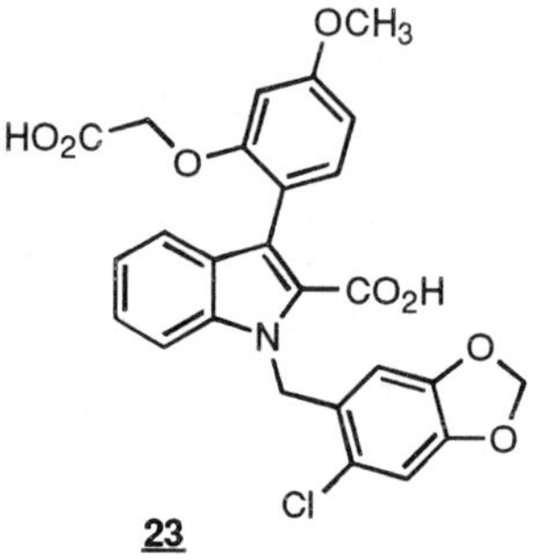

23

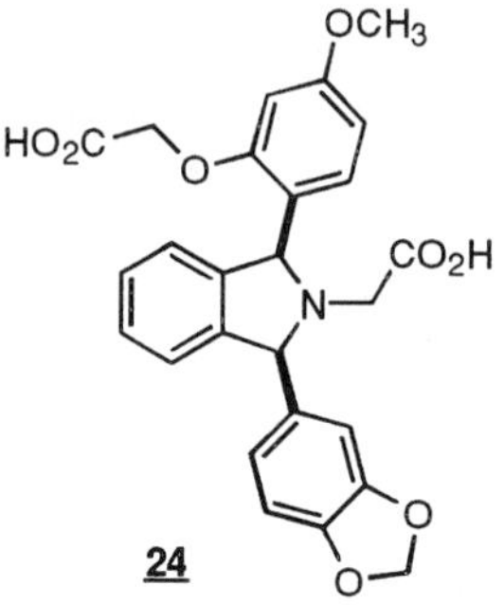

24

Recently, an account of the optimization of the butenolide series of endothelin antagonist has been published (60). This series of compounds arose from a lead obtained through high-throughput screening. This lead was then used in SAR studies guided by a Topliss decision tree protocol. One of the highlighted compounds, PD 156707 (**25**), is a potent ET_A selective antagonist active against human receptors (IC_{50}'s: ET_A = 0.3 nM, ET_B = 420 nM) and is reported to be 50% bioavailable in rats. Compound **25** has proven efficacious in a cat model of stroke and in a rat model of hypoxic pulmonary hypertension (61). Further structural modifications to the butenolide series provided compounds of mixed receptor profile, as exemplified by butenolide **26** (IC_{50}'s: ET_A = 13 nM, ET_B = 240 nM)(62).

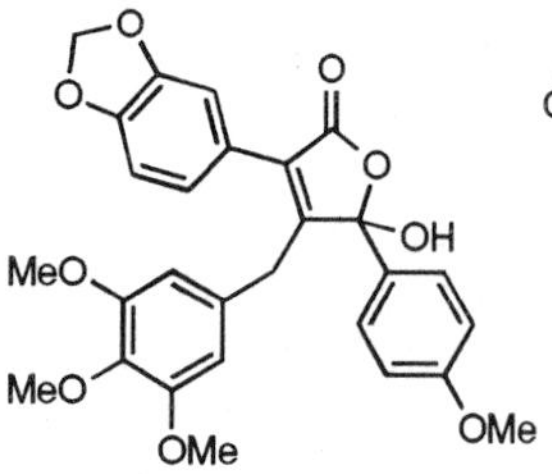

25

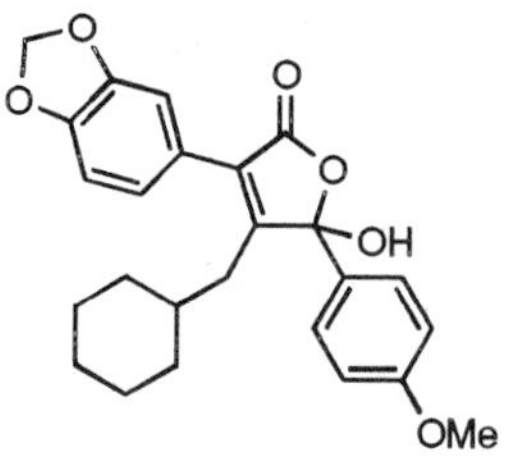

26

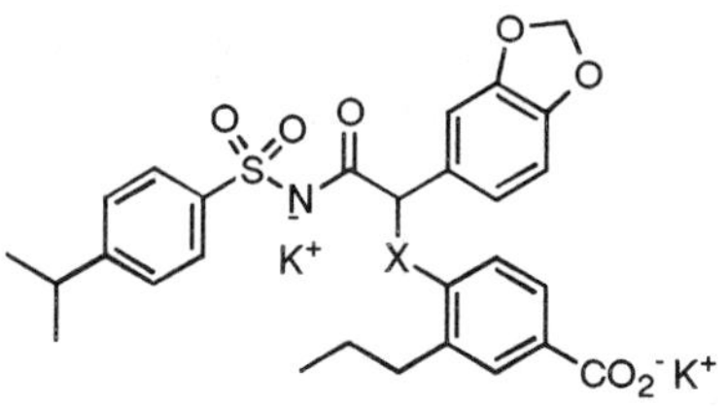

27 X = O
28 X = CH$_2$

SAR studies relating to the acylsulfonamide antagonist, L-754,142 (**27**), have also been recently disclosed (63). The lead structure for this series was obtained by directed database screening and was originally prepared as an angiotensin II receptor antagonist. Compound **27** is a mixed antagonist of ET_A/ET_B receptors (K_i's: ET_A = 0.06 nM, ET_B = 2.25 nM) and is efficacious when dosed orally in several animal models. Concerns centering around the possible configurational and metabolic stability of the ether oxygen linkage of **27** led to studies directed at modification of this moiety. While these concerns now appear to have been unfounded, this work resulted in the identification of compound **28** as a potent mixed ET_A/ET_B antagonist (64).

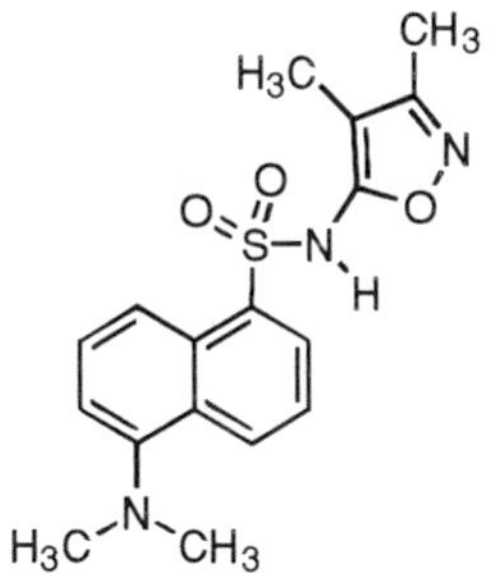

29 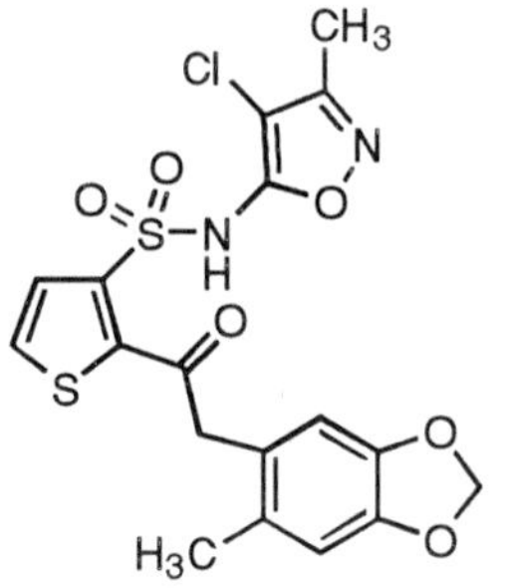   **30** 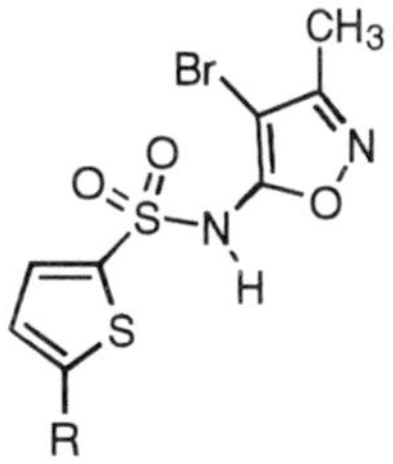   **31** R =H
                           32 R = p-MeC$_6$H$_4$

A detailed SAR account on the ET_A selective isoxazolyl sulfonamide antagonists, represented by BMS 182874 (**29**), has recently been published (65). Further investigation into this compound series by a different group has provided the thiophene **30**, a potent ET_A selective antagonist (IC_{50}'s: ET_A =1.7 nM, ET_B = 11,000 nM) (66,67). Compound **30** has proven to be efficacious in rat models of acute pulmonary hypertension (68). ET_A/ET_B selectivity can be structurally modulated in this series of compounds. Thus the ET_A selective compound **31** (IC_{50}'s: ET_A =3.4 nM, ET_B = 40,000 nM) can be turned into an ET_B selective analog by simple addition of an aryl substituent at the 4-position of the thiophene ring as in **32** (IC_{50}'s: ET_A =72,000 nM, ET_B = 36 nM).

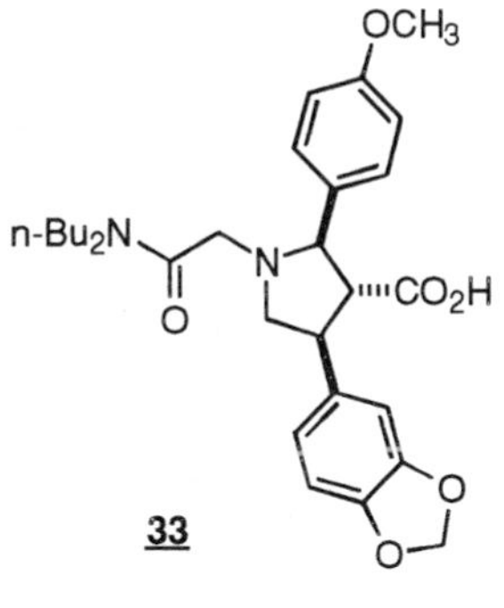

33 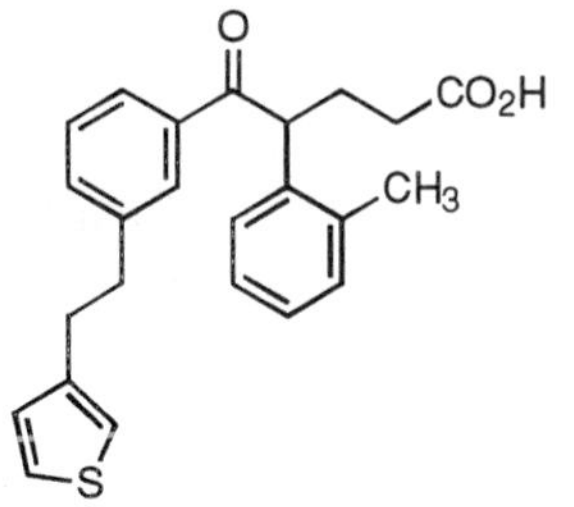   **34** 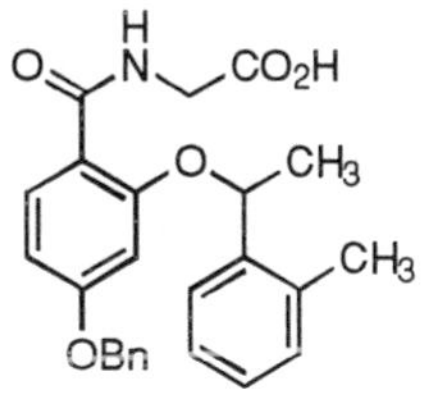   **35**

The 2,4-diaryl-pyrrolidine-3-carboxylic acid **33** has recently been reported as a new class of ET_A selective antagonists (69). Design of these compounds was based on the pharmacophore analysis of **22** in which the indane nucleus was replaced by an N-substituted pyrrolidine. The activity of **33** resides in the *RRS* isomer as shown, which has the same absolute configuration as that of **22**. Compound **33** has IC_{50}'s of 0.36 nM and 515 nM for inhibition of ET-1 radioligand binding at the ET_A and ET_B receptors, respectively. Functionally, **33** is a potent inhibitor of phosphoinositol hydrolysis mediated by ET-1 (IC_{50} = 0.16 nM) and antagonizes the ET-1 induced contraction of the rabbit aorta with a pA2 = 9.20. Compound **33** is reported to have 70% oral bioavailability in rats.

A two point pharmacophore analysis (acid group to aromatic ring) of **19** and SG 50-235, the caffeoyl ester of a pentacyclic triterpenoid isolated from *Myrica cerifera,* was used to search a 3D database search leading, after optimization, to the ET_A selective ketopentanoic acid **34** (IC_{50}'s: ET_A = 90 nM, ET_B > 30 µM) and benzoic acid derivative **35** (IC_{50}'s: ET_A =76 nM, ET_B > 30 µM) (70).

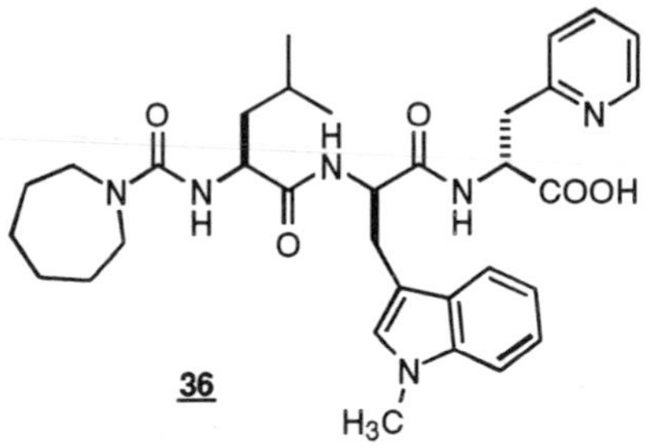

Backbone modification of the potent and ET_A selective antagonist FR-13937 (**36**) led to the development of a modified pseudotetrapeptide which showed improved *in vivo* stability and oral pharmacokinetics (71-73). A strong correlation between Δlog P and drug absorption was found in this series and, after extensive modification, carbamate **37** was found to be optimal in terms of activity (IC_{50}'s: ET_A = 4.6 nM, ET_B = 90,000 nM) and oral bioavailability (approximately 13% in rat).

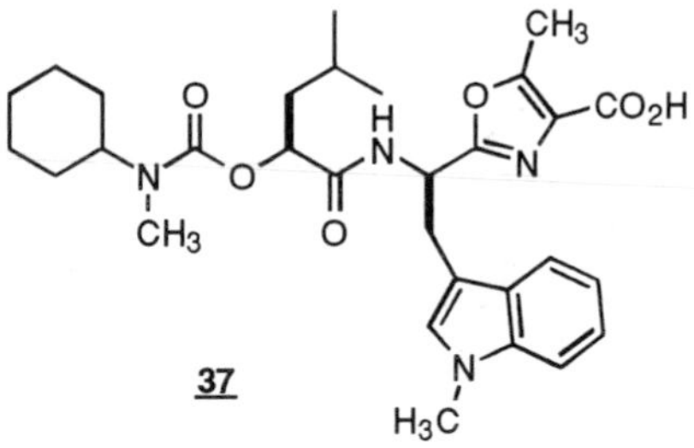

<u>Summary</u> - The development of endothelin receptor antagonists has progressed rapidly since the identification of ET-1. Most significantly, highly potent, non-peptide receptor antagonists have been advanced by several groups to the point at which potential therapeutic candidates have been identified. Multiple strategies have proven effective both in lead generation and compound development. These include random screening, QSAR analyses, directed database screening of compounds developed for other GPCRs, and/or peptidomimetic strategies based on features of ET-1 important to either binding or function. In the near future, data from clinical trials should be available to establish whether or not a role exists for endothelin in the aetiology of human disease as well as to provide valuable information concerning the optimal receptor subtype selectivity for therapeutic agents.

<u>References</u>

1. M. Yanagisawa, H. Kurihara, S. Kimura, Y. Tomobe, M. Kobayashi, Y. Mitsui, Y. Yazaki, K. Goto and T. A. Masaki, Nature, <u>332</u>, 411 (1988).
2. D.C. Spellmeyer, Annu. Rep. Med. Chem., <u>29</u>, 65 (1994).
3. T. F. Walsh, Annu. Rep. Med. Chem., <u>30</u>, 91 (1995).
4. M. Sokolovsky, Pharmac. Ther., <u>68</u>, 435 (1995).
5. T. Masaki, Annu. Rev. Pharmacol. Toxicol., <u>35</u>, 235 (1995).
6. T. Inagami and M. Naruse, R. Hoover, Annu. Rev. Physiol., <u>57</u>, 171 (1995).
7. T. J. Opgenorth, Adv. Pharmacol., <u>33</u>, 1 (1995).
8. W. L. Cody and A. M. Doherty, Biopolymers, <u>37</u>, 89 (1995).
9. A. J. Turner and L. J. Murphy, Biochem. Pharmacol., <u>51</u>, 91 (1996).
10. C. J. Ferro and D. J. Webb, Drugs, <u>51</u>, 12 (1996).
11. J. Cardiovasc. Pharmacol., <u>26</u> (Suppl. 3), S1-S521 (1995).
12. "Endothelin Receptors: From the Gene to the Human" R.R. Ruffolo, Jr., Ed., CRC Press, Boca Raton, Fla. 1995.
13. J.-B. Denault, P. D'Orléans-Juste, T. Masaki and R. Leduc, J. Cardiovasc. Pharmacol., <u>26</u> (Suppl. 3), S47 (1995).
14. K. Shimada, M. Takahashi and K. Tanzawa, J. Biol. Chem., <u>269</u>, 18275 (1994).

15. D. Xu, N. Emoto, A. Giaid, C. Slaughter, S. Kaw, D. deWit and M. Yanagisawa, Cell., 78, 473 (1994).
16. T. Ikura, T. Sawamura, T. Shiraki, H. Hosokawa, T. Kido, H. Hoshikawa, K. Shimada, K. Tanzawa, S. Kobayashi, S. Miwa and T. Masaki, Biochem. Biophys. Res. Commun., 207, 807 (1995).
17. M. Schmidt, B. Kroger, E. Jacob, H. Seulberger, T. Subkowski, R. Otter, T. Meyer, G. Schmalzing and H. Hillen, FEBS Letters, 356, 238 (1994).
18. K. Yorimitsu, K. Moroi, N. Inagaki, T. Saito, Y. Masuda, T. Masaki, S. Seino and S. Kimura, Biochem. Biophys. Res. Commun., 208, 721 (1995).
19. K. Shimada, Y. Matsushita, K. Wakabayashi, M. Takahashi, A. Matsubara, Y. Iijima and K. Tanzawa, Biochem. Biophys. Res. Commun., 207, 807 (1995).
20. N. Emoto and M. Yanagisawa, J. Biol. Chem., 270, 15262 (1995).
21. O. Valdenaire, E. Rohrbacher and M. G. Mattei, J. Biol. Chem., 270, 29794 (1995).
22. S. Iwasaki, T. Homma, Y. Matsuda and V. Kon, J. Biol. Chem., 270, 6997 (1995).
23. Y. Kurihara, H. Kurihara, H. Suzuki, T. Kodama, K. Maemura, R. Nagai, H. Oda, T. Kuwai, W. H. Cao, N. Kamada, K. Jishage, Y. Ouchi, S. Azuma, Y. Toyoda, T. Ishikawa, M. Kumada and Y. Yazaki, Nature, 368, 703 (1994).
24. K. Hosoda, A. Giaid, R. E. Hammer and M. Yanagisawa, Circulation, 90, 634 (1994).
25. A. G. Baynash, K. Hosoda, A. Giaid, J. A. Richardson, N. Emoto, R. E. Hammer and M. Yanagisawa, Cell, 79, 1277 (1994).
26. K. Hosoda, R. E. Hammer, J. A. Richardson, A. G. Baynash, J. C. Cheung, A. Giaid and M. Yanagisawa, Cell, 79, 1267 (1994).
27. E. G. Puffenberger, K. Hosoda, S. S. Washington, K. Nakao, D. deWit, M. Yanagisawa and A. Chakravarti, Cell, 79, 1257 (1994).
28. Y. Tsurumi, N. Ohhata, T. Iwamoto, N. Shigematsu, K. Sakamoto, M. Nishikawa, S. Kiyoto and M. Okuhara, J. Antibiot., 47, 619 (1994).
29. T. Bennett and S. M. Gardiner and J. Hum, Hypertens., 8, 587 (1994).
30. Y. Tsurumi, K. Fujie, M. Nishikawa, S. Kiyoto and M. Okuhara, J. Antibiot., 48, 169 (1995).
31. C. Plumpton, W. G. Haynes, D. J. Webb and A. P. Davenport, Br. J. Pharmacol., 116, 1821 (1995).
32. P. J. Kukkola, P. Savage, Y. Sakane, J. C. Berry, N. A. Bilci, R. D. Ghai and A. Y. Jeng, J. Cardiovasc. Pharmacol., 26 (Suppl. 3), S65 (1995).
33. T. Fukami, T. Hayama, Y. Amano, Y. Nakamura, Y. Arai, K. Matsuyama, M. Yano and K. Ishikawa, Bioorg. Med. Chem. Lett., 4, 1257 (1994).
34. D. M. Pollock, K. Shiosaki, G. M. Sullivan and T. J. Opgenorth, Biochem. Biophys. Res. Commun., 186, 1146 (1992).
35. S. R. Bertenshaw, R. S. Rogers, M. K. Stern, B. H. Norman, W. M. Moore, G. M. Jerome, L. M. Branson, J. F. McDonald, E. G. MacMahon and M. A. Palomo, J. Med. Chem., 36, 173 (1993).
36. S. R. Bertenshaw, J. J. Talley, R. S. Rogers, J. S. Carter, W. M. Moore, L. M. Branson, and C. M. Koboldt, Bioorg. Med. Chem. Lett., 3, 1953 (1993).
37. C. P. Lee, P. M. Keller, W. E. DeWolf. E. H. Ohlstein and J. D. Elliott, 210th American Chemical Society National Meeting, Chicago, IL, Aug 1994, Abstract MEDI 40.
38. S. DeLombaert, R. D. Ghai, A. Y. Jeng, A. J. Trapani and R. L. Webb, Biochem. Biophys. Res. Commun., 204, 407 (1994).
39. S. DeLombaert, L. Blanchard, C. Berry, R. D. Ghai and A. J. Trapani, Bioorg. Med. Chem. Lett., 5, 145 (1995).
40. R. Bihovsky, B. L. Levinson, R. C. Loewi, P. W. Erhardt, and M. A. Polokoff, J. Med. Chem., 38, 2119 (1995).
41. Y. Tsurumi, H. Ueda, K. Hayashi, S. Takase, M. Nishikawa, S. Kiyoto and M. Okuhara, J. Antibiot., 48, 1066 (1995).
42. T. D Warner, B. Battistini, A. M Doherty and R. Corder, Biochem. Pharmacol., 48, 625 (1994).
43. A. M. Doherty, Chemical and Structural Approaches to Rational Drug Design, CRC Press, Boca Raton, 85 (1995).
44. M. Ihara, R. Yamanaka, K. Ohwaki, S. Ozaki, T. Fukami, K. Ishikawa, P. Towers and M. Yano Eur. J. Pharm., 274, 1 (1995).
45. B. Battistini and R. Botting, Drug News & Perspectives, 8, 365 (1995).
46. Y. Morishita, S. Chiba, T. Tsukuda, T. Tanaka, T. Ogawa, M. Yamasaki, M. Yoshida, I. Kawamoto and Y. Matsuda, J. Antibiot., 47, 269 (1994).
47. R. Katahira, K. Shibata, M. Yamasaki, Y. Matsuda and M. Yoshida, Bioorg. Med. Chem. Lett., 5, 1595 (1995)
48. R. Katahira, K. Shibata, M. Yamasaki, Y. Matsuda and M. Yoshida, Bioorg. Med. Chem. 3, 1273 (1995)
49. J. X. He, W. L. Cody M. A. Flynn, K. M. Welch, E. E. Reynolds and A. M. Doherty, Bioorg. Med. Chem. Lett., 5, 621 (1995).

50. M. Clozel, V. Breau, G. A. Gray, B. Kalina, B.-M. Loffler, K. Burri, J.-M. Cassal, G. Hirth, M. Muller, W. Neidhart and H. Ramuz, J. Pharmacol. Exp. Ther., 270, 228, (1994).

51. W. Kiowski, G. Sutsch, P. Hunziker, P. Muller, J. Kim, E. Oechslin, R. Schmitt, R. Jones and O. Bertel, Lancet, 346, 732 (1995).

52. C. R. Jones "Human Experience with Bosentan" Endothelin Inhibitors: Advances in Therapeutic Application & Development , Coronado, CA, Feb 5-7, 1996.

53. J. M. Gellai, M. Jugus, T. Fletcher, P. Nambi, E. H. Ohlstein, J. D. Elliott and D. P. Brooks, J. Pharmacol. Exp. Ther., 275, 200 (1995).

54. S. A Douglas, C. Louden, L. M. Vickery-Clark, B. L. Storer, T. Hart, G. Z. Feuerstein, J. D. Elliott and E. H. Ohlstein, Circ. Res., 75, 190 (1994).

55. J. D Elliott, M. Ezekiel, M. Gellai, S. A. Douglas, E. H. Ohlstein, FASEB J. 8(4), A7 (1994).

56. D. P. Brooks and P.D. DePalma, Nephron, 72, 629, (1996).

57. M. A. Lago, R. D. Cousins, A. Gao, J. D. Leber, C. E. Peishoff, E. H. Ohlstein and J. D. Elliott, Fourth International Conference on Endothelin, C44 (abstract) (1995).

58. J. D. Elliott, "The Design and Synthesis of Non-Peptide Endothelin Receptor Anatagonist" Endothelin Inhibitors: Advances in Therapeutic Application & Development , Coronado, CA, Feb 5-7, (1996).

59. A. Gao, M. A. Lago, P. Nambi, E.H. Ohlstein and J. D. Elliott, 211th American Chemical Society National Meeting, 141 (abstract), New Orleans March 24-28 (1996).

60. A. M. Doherty, W. C. Patt, J. J. Edmunds, K. A. Berryman, B. R. Reisdorph, M. S Plummer, A. Shahripour, C. Lee, X.-M. Cheng, D. M. Walker, S. J. Haleen, J. A. Keiser, M. A . Flynn, K. M. Welch, H. Hullak, D. G. Taylor and E. E. Reynolds, J. Med. Chem., 38, 1259 (1995).

61. R. Bialecki, "Endothelin's Pathologic Role in Pulmonary Hypertension". Endothelin Inhibitors: Advances in Therapeutic Application & Development, Coronado, CA, Feb 5-7, (1996).

62. A. M. Doherty, "Non-peptide Endothelin ET_A selective and ET_A/ET_B Receptor Antagonists" Endothelin Inhibitors: Advances in Therapeutic Application & Development, Coronado, CA, Feb 5-7 (1996).

63. T. F. Walsh, K. J. Fitch, D. L. Williams, K. L. Murphy, N. A. Nolan, D. J. Pettibone, R. S. L Chang, S. S. O'Malley, B. V. Clineschmidt, D. F. Veber and W. J. Greenlee, Bioorg. & Med. Chem. Lett., 5, 1155 (1995).

64. T. F. Walsh, "The Discovery of Potent, Orally Active, Acylsulfonamide-Based Non-Peptidic ET_A/ET_B Receptor Antagonists", Endothelin Inhibitors: Advances in Therapeutic Application & Development, Coronado, CA, Feb 5-7 (1996).

65. P. D. Stein, D.M. Floyd, S. Bisaha, J. Dickey, R.N. Gitotra, J.Z. Gougoutas, M. Kozlowski, V.G. Lee, E. C. Liu, M.F. Malley, C. McMullen, C. Mitchell, S. Moreland, N. Murugesan, R. Serafino, M.L. Webb, R. Ahang and J.T. Hunt, J. Med. Chem., 38, 1344 (1995).

66. M. F. Chan, B. Raju, A. Kois, R. S. Castillo, E. J. Verner, C. Wu, Y. V. Venkatachalapathi, I. Okun, E. Hwang, F. D. Stavros and V. N. Balaji, 209th American Chemical Society National Meeting, 167,168 (abstracts), Anaheim, CA, April 2-6 (1995).

67. M. F. Chan, "The design and Synthesis of Non-Peptide antagonists Selective for Both the ET_A and ET_B Receptor subtypes" Endothelin Inhibitors: Advances in Therapeutic Application & Development, Coronado, CA, Feb 5-7 (1996).

68. T. A. Brock, "Therapeutic Uses of Selective ET_A Receptor Antagonists in Vascular Diseases" Endothelin Inhibitors: Advances in Therapeutic Application & Development, Coronado, CA, Feb 5-7 (1996).

69. M. Winn, T. W. von Gelden, T. J. Opgenorth, H. S. Jae, A. S. Tasker, S. A. Boyd, J. A. Kester, R. A. Mantei, R. Bal, B. K. Sorensen, J. R. Wu-Wong, W. J. Chiou, D. B. Dixon, E. I. Novosad, L. Hernandez and K. C. Marsh, J. Med. Chem., 39, 1039 (1996).

70. P. C. Astles, N. V. Harris, M. F. Harper, R. Lewis, C. McCarthy, I. McLay, T. N. Majid,B. Porter, C. Smith and R. J. A. Walsh, 8th SCI-RSC Medicinal Chemistry Symposium, Cambridge, UK 10-13 September (1995).

71. T. W. von Geldern, C. Hutchins, J. A. Kester, J. R. Wu-Wong, W. Chiou and D. B. Dixon and T. J. Opgenorth, J. Med. Chem., 39, 957 (1996).

72. T. W. von Geldern, J. A. Kester, R. Bal, J. R. Wu-Wong, W. Chiou and D. B. Dixon and T. J. Opgenorth, J. Med. Chem., 39, 968 (1996).

73. T. W. von Geldern, D. Hoffman, J. A. Kester, H. N. Nellans, B. D. Dayton, S. V. Calzadilla, K. C. Marsh, L. Hernandez, W. Chiou, D. B. Dixon, J. R. Wu-Wong and T. J. Opgenorth, J. Med. Chem., 39, 982 (1996).

Chapter 10. GPIIb/IIIa Antagonists

James Samanen

SmithKline Beecham Pharmaceuticals, King of Prussia, PA 19406

Introduction - When this area was last reviewed in 1993, the evolution of GPIIb/IIIa antagonists from proteins to peptides to nonpeptides was still under way (1). The hypothesis that GPIIb/IIIa antagonists could provide more reliable control of thrombosis than cyclooxygenase inhibitors had been demonstrated in animal models, but the therapeutic potential of these novel antithrombotic agents was just beginning to be recognized. By now, many potent orally active GPIIb/IIIa antagonists have been reported. An intravenous GPIIb/IIIa antagonist, abciximab has proven that GPIIb/IIIa antagonists are efficacious in humans. Abciximab (ReoPro) is now commercially available, and several other GPIIb/IIIa antagonists are on the horizon. This chapter will rely on recent reviews that sufficiently survey the field up to mid-1994 (1-10). It will focus upon recent literature regarding current and potential products, intravenous and oral agents in both clinical and preclinical research, and current important issues.

MARKETED PRODUCTS

ReoPro (abciximab, c7E3) - The first demonstrations of *in vitro* and *in vivo* GPIIb/IIIa antagonist activity were performed with monoclonal antibodies 10E5 and 7E3, respectively, from the Coller group in the early 1980s (11,12). 7E3 has evolved into abciximab, a chimeric Fab fragment of 7E3 approved by the FDA on January 22, 1994. Abciximab is indicated for the reduction of acute cardiac complications in patients undergoing angioplasty procedures who are at high risk for abrupt artery closure. Two current trials seeking to expand the indications to refractory unstable angina and low risk angioplasty patients were stopped due to the positive findings at each first interim analysis (13). In the large-scale trial (EPIC) that evaluated abciximab for the prevention of cardiovascular complications after percutaneous transluminal coronary angioplasty (PTCA), abciximab significantly reduced the 30-day incidence of major ischemic events relative to placebo, and significantly decreased the need for repeat revascularization during a six month follow-up period (14). In 14% of the patients receiving abciximab in the EPIC trial, twice the number of bleeding episodes occurred over placebo, usually at the site of vascular puncture (4). Current trials intend to explore strategies for practical use of abciximab that minimize bleeding risk in the larger patient population (14,15). The high initial cost of abciximab therapy ($1350 per treatment) may provide an opportunity for agents that could be marketed at lower cost, such as peptide and nonpeptide GPIIb/IIIa antagonists (16).

AGENTS IN PRECLINICAL AND CLINICAL DEVELOPMENT

Most of the agents that are undergoing preclinical and clinical investigation to compete against abciximab, are not other antibodies, but peptides and nonpeptides. The peptide and nonpeptide agents have demonstrated *in vivo* efficacy via platelet aggregation *ex vivo* in a number of species (17-23) and efficacy in a variety of animal models of thrombosis (17,18,20,24-31). Consistent with their proposed mechanism of action, these compounds have shown superior efficacy to other antiplatelet drugs: aspirin (24,25,27,30); dazoxiben (a thromboxane synthetase inhibitor) (18); and ticlopidine (27). Yet to be reported are head-to-head comparisons with clopidogrel, which demonstrated superior efficacy over ticlopidine in animals and is currently completing Phase III trials (32,33). The impressive array of *in vivo* studies firmly establish the antithrombotic efficacy of GP IIb/IIIa antagonists in animals.

<u>Agents for Intravenous Antithrombotic Therapy</u> - Besides abciximab, three peptides and two nonpeptides have demonstrated efficacy in humans. Integrelin is an undisclosed cyclic peptide developed from the snake venom protein barbourin, which contains a unique KGD sequence that in disclosed peptide analogs confers a high degree of selectivity towards GPIIb/IIIa (34). In Phase I trials with integrelin almost no effect on bleeding time was observed. Platelet aggregation, which was inhibited during the infusion of 1-1.5 mg/kg/min, returned to normal 2-4 hr after the infusion was stopped (35). This potent and reversible inhibition allowed for integrelin to be evaluated in patients undergoing angioplasty, unstable angina, and myocardial infarction under the umbrella trial IMPACT. Preliminary results appear encouraging (4). An NDA was expected to be filed in the US in the first quarter of 1996 (36). While the short term effects of integrelin in angioplasty and other thrombotic crises appear promising, benefits at the 6-month follow-up in the angioplasty trial were not significant (4,37). Aggrastat, tirofiban, (MK-383, L 700,462, **1**) displayed potent antiaggregatory activity in humans in Phase I (38-42). The compound is currently in Phase III to evaluate the reduction of cardiac events with balloon angioplasty and for treatment of unstable angina (4,32). Lamifiban, (Ro 44-9883, **2**) displays a unique property over other analogs in its ability to bind to GPIIb/IIIa without inducing a change in

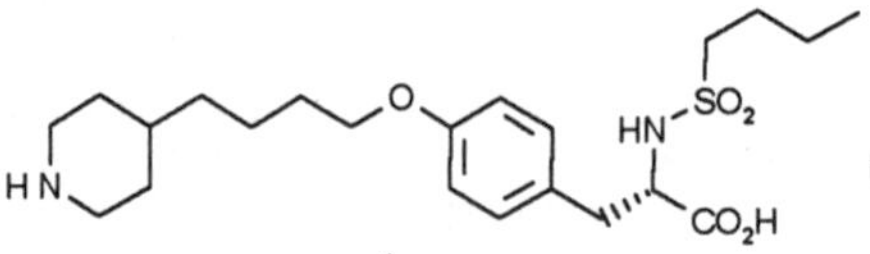

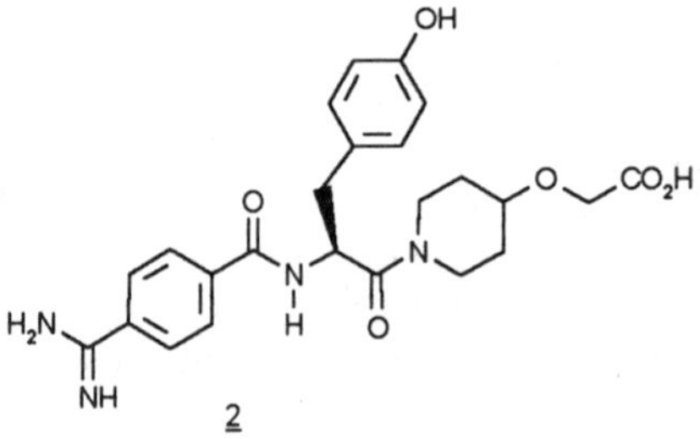

Ac-Cys-Asn-Dtc-Amp-Gly-Asp-Cys-OH

3

Ac-Cys-Asn-Pro-Arg-Gly-Asp-Tyr(OMe)-Arg-Cys-NH$_2$

4

Ac-Cys-(NMe)Arg-Gly-Asp-Pen-NH$_2$

5

Mba-(NMe)Arg-Gly-Asp-Man

6

CO-D-Tyr-Arg-Gly-Asp-Cys-OH

7

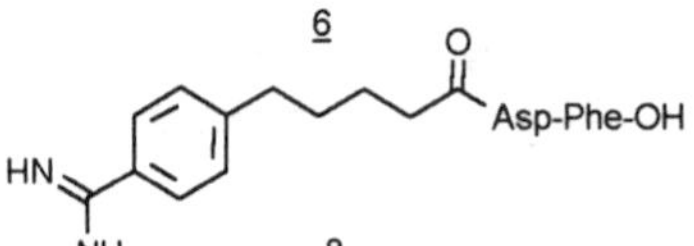

8

conformation that exposes the LIBS (ligand-induced binding site) epitope (43,44). Preliminary Phase I and II results have been published, and the compound was reported to be in Phase III (4,45-47). MK-852, **3**, has been studied in clinical trials through Phase II (48-52). TP-9201, **4** was selected for Phase I trials for its ability to minimally augment template bleeding (31,54,55). The penultimate arginine residue was determined to be critical for this effect (55). The fact that this compound works well in citrated blood *in vitro*, but not heparinized blood, led to the hypothesis that calcium ions, and not selectivity against αvβ5, may play a role in this phenomenon (17). Compounds that have been evaluated preclinically in animal models of thrombosis include an antibody YM 337 (56), three highly constrained cyclic peptides SK&F-106760 **5**, (25,26,30) SK&F-107260 **6**, (25,26) G-4120 **7**, (17) and a nonpeptide SC-49992, **8** (24).

<u>Agents for Oral Antithrombotic Therapy</u> - Compounds that display antithrombotic activity by oral administration provide an attractive alternative to solely intravenous agents for acute indications and could potentially provide a treatment for chronic indications, such as the prevention of myocardial infarction, stroke, or surgical intervention (47). Most of the GPIIb/IIIa antagonists that display potent oral activity are nonpeptides. All of these compounds inhibit platelet aggregation *ex vivo* at oral doses of $\leq$ 10 mg/kg, but with varying duration of action, designated below by <u>W</u> percent inhibition of *ex vivo* platelet aggregation at <u>X</u> hr with <u>Y</u> mg/kg po/id, or by the pharmacokinetic half-life ($T^{1/2}$, <u>Z</u> min). The obligatory requirements of the polar hydrophilic cationic and anionic groups in high affinity GPIIb/IIIa antagonists challenge the discovery of agents with high oral bioavailability. The first two oral GPIIb/IIIa antagonists to reach clinical trials contain groups that mask one or both of these polar groups. Fradafiban (BIBU-104, **9**) is a bis-prodrug of the compound BIBU-52, **10** (57,58). Phase I and II results have been published, and Phase III trials are underway (59-63). The oral efficacy of this bis-prodrug is consistent with rapid metabolism to **10**, since either mono-prodrug would be expected to lack affinity for GPIIb/IIIa. It displays moderate duration in rhesus monkeys (49% inhibition at 8 h with 1 mg/kg po) (57).

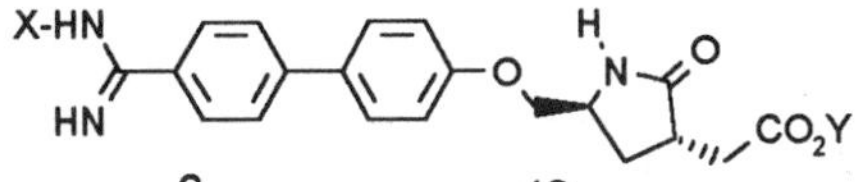

$$\textbf{9} \qquad \qquad \textbf{10}$$
$$X = CH_3O_2\overset{O}{C}\text{-}; \; Y = \text{-}CH_3 \quad X,Y = H$$

Xemlofiban, (SC-54684, **11**) is an ethyl ester prodrug which metabolizes rapidly to the acid SC-54701, **12** (64). Compound **11** displays 22.7% oral bioavailability in dogs (64). It has demonstrated potent antiaggregatory activity (30% inhibition at 24 h with 2.4 mg/kg po, bid) in dogs, and in both Phase I and Phase II pilot studies (65-67).

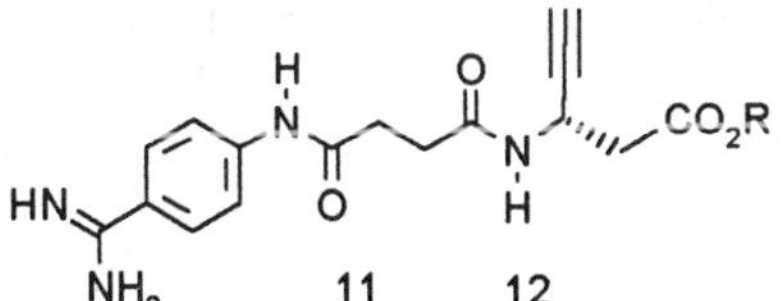

$$\textbf{11} \qquad \textbf{12} \qquad \qquad \qquad \textbf{13}$$
$$R = Et \quad R = H \qquad \qquad R = Et$$

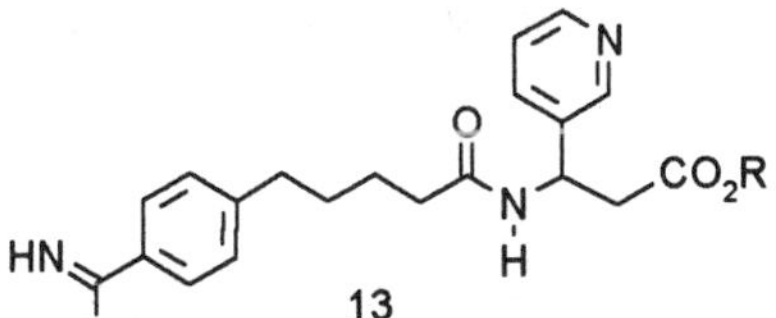

Compounds such as **13** lacking a linking amide show diminished efficacy and bioavailability in dogs (68). G-6788, **14**, an ethyl ester of G-6249, **15**, was recently reported to display 2 and 21% oral bioavailability in dogs and rhesus monkeys, respectively (69,70). S-1762 was reported to display long duration ($T^{1/2}$ 6.8 h) and 47% oral bioavailability in dogs (23). Ro-44-3657, a bis-prodrug of Ro-44-3888 related to **2**, was also reported to display long duration in dogs ($T^{1/2}$ 10-12 h) and rhesus monkeys ($T^{1/2}$ 5.2 h) and 33% oral bioavailability in rhesus monkeys (71,72).

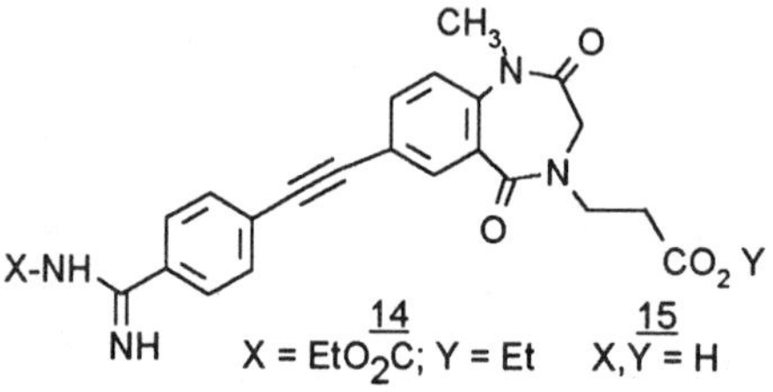

14 X = EtO$_2$C; Y = Et   **15** X,Y = H

A number of compounds have now shown oral activity without prodrug modification. Ro-43-8857, **16** was one of the first high affinity GPIIb/IIIa antagonist nonpeptides to appear in the patent literature (73). Compound **16** was the first nonpeptide for which oral activity was reported, albeit with low potency (80% inhibition at 4 h with 10 mg/kg, po) and low bioavailability (3%) (74,75). The peptide DMP-728, **17** displays long duration of action (60% at 24 h with 1 mg/kg po) and moderate oral bioavailability (10-15% in dogs) (76,77). It has been evaluated in Phase I trials (78).

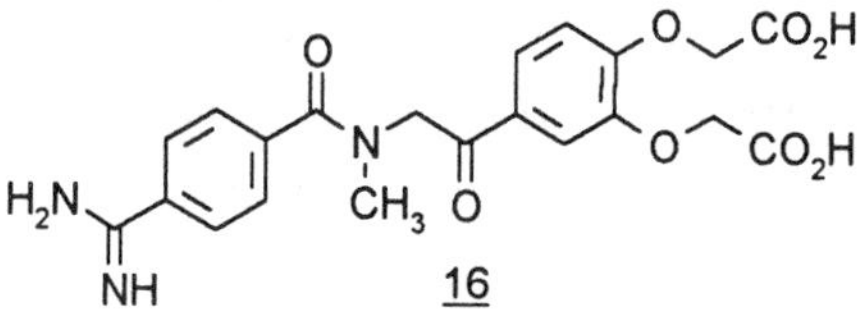

(NMe)Arg-Gly-Asp
   | |
D-Abu———————Amb

16   **17**

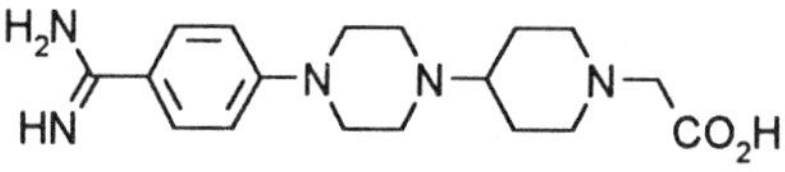

18

GR-144053, **18** is a potent compound (100% inhibition at 4 h with 3 mg/kg po), displaying long duration (T$^{1/2}$ 9 h) and ~30% oral bioavailability in cynomolgous monkeys (22,79). SB-214857, **19** arose from a search for compounds with enhanced oral duration over SB-208651, **20**, an orally active analog of **21**. Compound **19** displays considerably enhanced oral duration (80% inhibition at >7 h with 1 mg/kg id) in dogs over **20** (80% inhibition at 1 h with 3 mg/kg id) (80-82). Potent *in vivo* activity

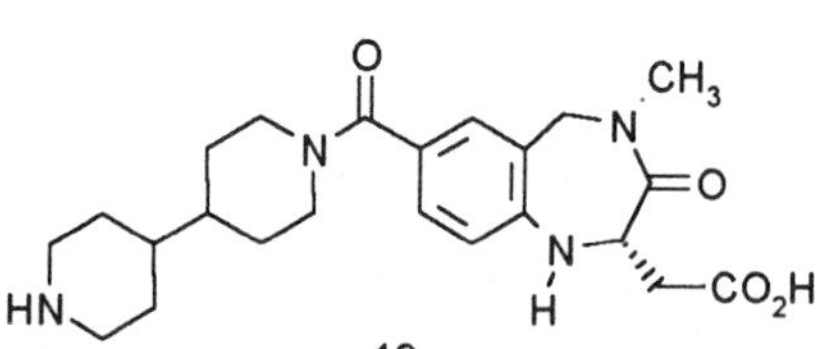

19 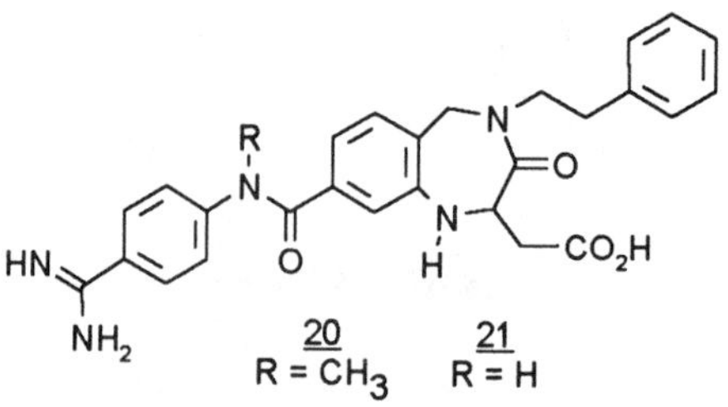   **20** **21**
              R = CH$_3$ R = H

in dogs has been described for L-703,014, **22** (80% inhibition at 9 h with 2 mg/kg po) (19,83,84), L-709780, **23** (50% inhibition at 3 h with 4 mg/kg po) (85-87), and L-734,217, **24** (63% inhibition at 8 h with 1 mg/kg po) (88,89). The achiral (oxodioxolenyl)methyl carbamate prodrug **25** was shown to undergo quantitative reconversion to **24** in dog plasma and displayed comparable bioavailability, ~23%, as **24** (89). Thus, good oral bioavailability can be achieved in certain GPIIb/IIIa antagonists without the masking of ionic groups. Compound **26** is from a series of compounds reported to be orally active (90).   RWJ-50042, **27** (91) is reported to have led to oral compounds (92).   Several other orally active compounds have been reported, including the following, for which structures are unavailable: TAK-029, which is reported to be in Phase I in Japan (93); DMP-757, (94,95); the oral prodrug of EF-5077 (28); FR-144653 (96); RPR-110173 (97); and SDZ-GPI-562 (98).

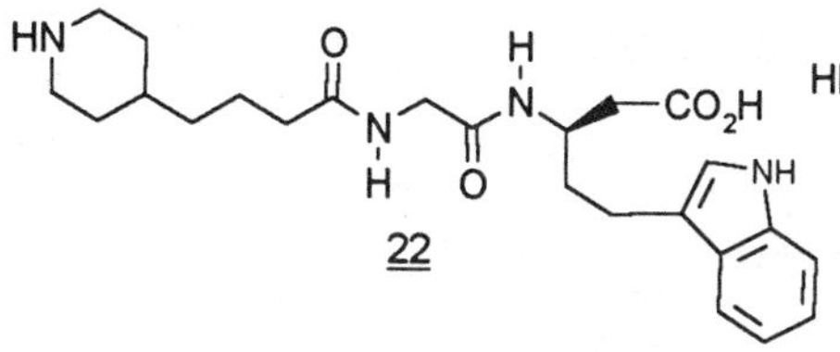

22

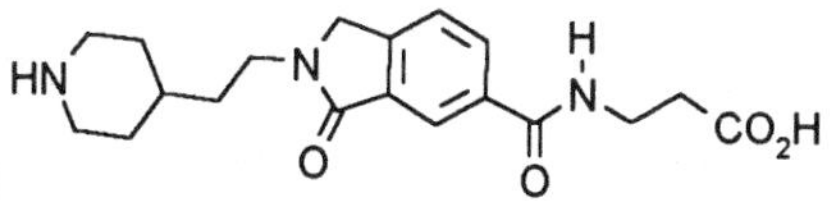

23

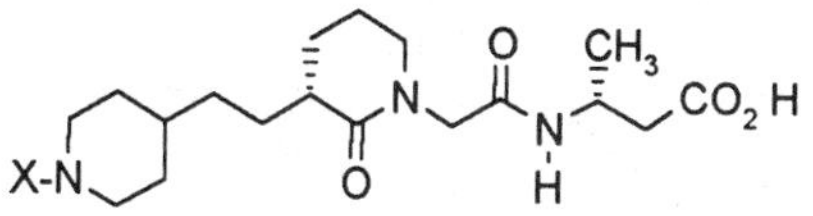

24   25
X = H   X =

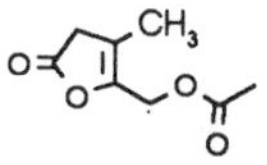

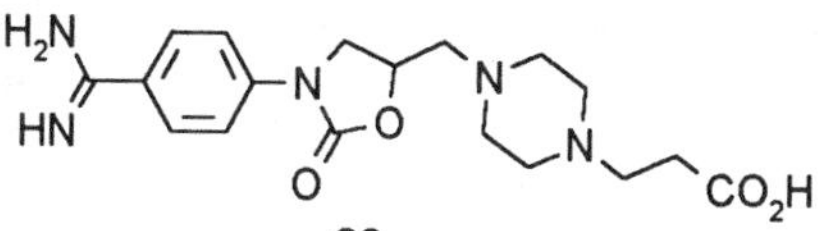

26

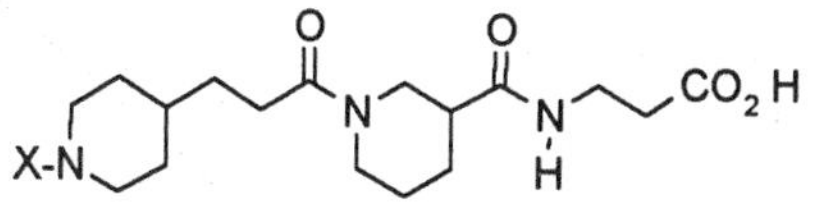

27

These compounds represent a flavor of the intense competition in the arena of oral GPIIb/IIIa antagonists.

ISSUES, UNSOLVED PROBLEMS, AND NEW AREAS OF EXPLORATION

GPIIb/IIIa Structure - The tertiary structure of the heterodimeric GPIIb/IIIa complex has yet to be reported. Calvete has attempted to review and correlate the diverse structural and functional data into a coherent structural model (99). Recent reviews provide further insight into the state of understanding of GPIIb/IIIa structure (100,101).

GPIIb/IIIa and Signal Transduction - That integrins are signal transducing receptors is now widely accepted (102,103). Ligand binding to GPIIb/IIIa has been shown to induce conformational changes in the receptor,(101) but additional post-receptor occupancy events appear to be necessary for "signal transduction" (104).

Pharmacophore Hypotheses - The structural requirements for high affinity ligands have been reported and summarized in a two-dimensional figure (7). It has been proposed that additional regions of space where the cation may be positioned arise from either an expanded or second cation binding site (105). Most of the three dimensional pharmacophore information comes from the structural and SAR studies of the high affinity highly constrained cyclic peptides, such as **5** and **6** (106-108), GR 83895 (109), **17** (77,110,111), and **7** (112-114). Interestingly, the conformations of all but **7** are highly homologous, and show a "turn-extended-turn" conformation about the critical tripeptide, as determined by X-ray crystal structure and nmr studies. A cyclic hexapeptide GPIIb/IIIa antagonist bearing this conformation has also been reported (6). Peptides that lack the turn-extended-turn conformation typically display diminished affinity for GPIIb/IIIa (104,115). An exception is **7**, which adopts a more puckered or "cup-shaped" conformation about the critical tripeptide. Compounds, such as **20** and **21**, directly designed from **6** displayed high affinity (81,115). The compounds directly designed from **7** were inactive (69,112), however, subsequent SAR studies gave rise to the high affinity nonpeptide **15**. Overlays of the more rigid nonpeptides with the cyclic peptides have helped to locate the spatial positions of some of the pharmacophoric elements (79,105).

The Value of Extended GPIIb/IIIa Antagonist Therapy - Patients at risk of a life-threatening thrombotic event should benefit from extended GPIIb/IIIa antagonist therapy (47). Following a surgical procedure, such as balloon angioplasty, to remove a stenotic lesion produced by atherosclerosis, the inner surface of the repaired vessel is highly thrombogenic and GPIIb/IIIa antagonist therapy is important immediately after(15). From the EPIC trial it appears that passivation (loss of thrombogenicity) takes 2-8 days after angioplasty, and that a bolus dose of abciximab only protected patients from urgent PTCA for about 4 h (15). The extended protection afforded by a bolus plus 12 h infusion of abciximab, however, provided a >80% decrease in the need for repeat revascularization over a 30 day follow-up period. Thus, more extended GPIIb/IIIa antagonist therapy could offer even greater protection in the hours or days following angioplasty (15).

Is Bleeding an Issue? - In animal thrombosis models, GPIIb/IIIa antagonists have typically displayed modest enhancements (2-3 fold) of experimental cutaneous bleeding in doses that achieve 80-100% GPIIb/IIIa receptor occupancy (3,7). The inclusion of anticoagulant heparin in most protocols is an important factor. Analysis of EPIC trial results revealed that the bleeding risk in abciximab-treated patients was directly proportional to the weight-adjusted heparin dose, and patients receiving the lowest weight-adjusted heparin dose had the "fewest endpoint events" (15). Therapeutic efficacy, furthermore, did not correlate with weight-adjusted heparin dose (47). In a recently completed trial with abciximab, weight adjustment and heparin dose lowering did not alter efficacy but significantly reduced bleeding complications (47). Major bleeding events in patients were not significantly greater than placebo in recent trials with Integrelin and **2** (47). Since activated platelets can play a role in coagulation, GPIIb/IIIa antagonist intervention with platelet aggregation might be expected to impact upon the platelet component of coagulation and influence the anticoagulant action of heparin (15). Since coagulation and platelet aggregation are involved in thrombus formation, it might be expected that anticoagulant and antiaggregatory agents could synergistically block thrombus formation. The fact that **4** has shown only a small effect on template bleeding, suggests that GPIIb/IIIa antagonists may not inherently exacerbate cutaneous bleeding. The fact that template bleeding enhancements beyond 2-3 fold may be seen with certain GPIIb/IIIa antagonists when the dose exceeds GPIIb/IIIa receptor occupancy, argues that certain compounds influence bleeding by a mechanism not involving GPIIb/IIIa (74,116).

$\alpha v\beta 3$, Restenosis and Selectivity - In the days following angioplasty the wounded vessel may begin to develop a lesion that could progress to occlusion, in a process called restenosis (117). The integrin $\alpha v\beta 3$, containing the same beta subunit as GPIIb/IIIa ($\alpha IIb\beta 3$), is found on migrating smooth muscle cells during neointimal hyperplasia (118). The recent demonstration of neointimal hyperplasia inhibition by an $\alpha v\beta 3$ -specific peptide in animals (119), led to the proposal that $\alpha v\beta 3$ inhibition may have contributed to the reduction in restenosis among patients receiving abciximab in the EPIC trial (120). Abciximab binds $\alpha v\beta 3$ and Mac-1, which is found on leukocytes (118,121), besides $\alpha IIb\beta 3$. Compounds **1-4,9-27** display high selectivity towards $\alpha IIb\beta 3$, but many RGD-containing GPIIb/IIIa antagonists, e.g. the earlier non-selective compounds, also display high affinity for $\alpha v\beta 3$ (7). Pure $\alpha v\beta 3$

antagonists have also been identified (122,123). The investment in the development of high affinity nonpeptide ligands for αIIbβ3 has contributed to the rapid identification of a subset of compounds with affinity for αvβ3 and subsequent development of high affinity αvb3 antagonists which could be developed for this indication (124).

Conclusions - Clinical research in the next few years will continue to reveal the extent to which acute GPIIb/IIIa antagonist therapy helps to reduce morbidity in antithrombotic crises. The risk-benefit ratio of GPIIb/IIIa antagonists in various thrombotic disorders remains to be determined for both acute and chronic therapies.

References

1   B.K. Blackburn, and T.R. Gadek, In "Ann. Rep. in Med. Chem." J.A. Bristol, Ed., Academic Press, New York 1993, p. 79.
2   V. Austel, F. Himmelsbach, and T. Muller, Drugs of the Future, 19, 757 (1994).
3   N. S. Cook, G. Kottirsch, and H-.G. Zerwes, Drugs of the Future, 19, 135 (1994).
4   W.H. Frishman, B. Burns, N. Alturk, and B. Altajar, Am. Heart J., 130, 877 (1995).
5   R.J. Gould Pers. Drug Discovery Design, 1, 537 (1994).
6   G. Muller, M. Gurrath, and H. Kessler, J. Computer-Aided Mol. Des., 8, 709 (1994).
7   I. Ojima, S. Charkravarty, and Q. Dong, Biorg. Med. Chem., 1995, 3, 337 (1995).
8   J.M. Stadel, A.J. Nichols, D.R. Bertolini, and J.M. Samanen, In "Integrins: Molecular and Biological Response to the Extracellular Matrix", D. Cheresh, R.P. Mecham, Eds., Academic Press, New York, 1994, p. 237.
9   T. Weller, L.H. Alig, M.H. Muller,and W.C. Kouns, Drugs of the Future, 19, 461 (1994).
10   J.A. Zablocki, N.S. Nicholson, and L.P. Feigen, Exp. Opinion Invest. Drugs, 3, 437 (1994).
11   B.S. Coller, E.I. Peerschke, L.E. Scudder, and C.A. Sullivan, J. Clin. Invest., 72, 325 (1983).
12   B.S. Coller , J. Clin. Invest., 76, 101 (1985).
13   J.J. Ferguson, Circulation, 93, 637 (1996).
14   E.J. Topol, Am. Heart J., 130, 666 (1995).
15   B.S. Coller, K. Anderson, and H. Weisman, Thromb. Haem., 74, 302 (1995).
16   Mark Simon, Robertson Stephens & Co. Feb. 8, 1995.
17   D. Collen, H.R. Lu, J.-M. Stassen, I. Vreys, T. Yasuda, S. Bunting, and H.K. Gold, Thromb. Haem., 71, 95 (1994).
18   B.D. Guth, U. Gerster, V. Koch, T.H. Muller, Thromb. Haem., 69, (Abst. 1887) 1072 (1993).
19   R.J. Gould, J.J., Lynch Jr., and L. Schaffer, Thromb. Haem., 69, (Abst. 3) 539 (1993).
20   J. J. Lynch, J. J., Cook G.R. Sitko, M.A. Holahan, D.R. Ramjit, M.J. Mellott, M.T. Stranieri, I.I. Stabilito, G. Zhang, and R.J. Lynch, J. Pharmacol. Exp. Ther., 272, 20 (1995).
21   T.H. Muller, U. Gerster, V. Koch, F. Himmelsbach, and B.D. Guth, Thromb. Haem., 69, (Abst. 1889) 1072 (1993).
22   N.B. Pike, M.R. Foster, and E.J. Hornby, Thromb. Haem. 69, (Abst. 1886) 1071 (1993).
23   M. Just, M. Hropot, B. Jablonka, and Thromb. Haem., 73, (Abst. 2080) 1444 (1995).
24   L.P. Feigen, N.S. Nicholson, L.W. King, J.G. Campion, F.S. Tjoeng, and S.G. Panzer-Knodle, J. Pharmcol. Exp. Ther., 267, 1191 (1993).
25   A.J. Nichols, J.A. Vasko, P.F. Koster, R.E. Valocik, and J.M. Samanen, In "Cellular Adhesion: Molecular Definition to Therapeutic Potential". B.W. Metcalf, B.J. Dalton, and G. Poste, Ed., Plenum Press, New York, 1994, p. 213.
26   A.J. Nichols, J.A. Vasko, P.F. Koster, R.E. Valocik, G.R. Rhodes, C. Miller-Stein, V. Boppana, and J.M. Samanen, J. Pharmacol. Exp. Ther., 270, 614 (1994).
27   S.A. Mousa, J.M. Bozarth, M.S. Forsythe, S.M. Jackson, A. Leamy, M.M. Diemer, R.P. Kapil, R.M. Knabb, M.C. Mayo, and S.K. Pierce, Circulation, 89, 3 (1994).
28   H. Iida, K. Asai, E. Hatsushiba, M. Ishikawa, T. Ishizuka, T. Kurosawa, S. Ohuchi, K. Katano, T. Tsuruoka, and S. Omoto, Thromb. Haem., 73, (Abst. 1595) 1315 (1995).
29   D.R. Ramjit, J.J. Lynch, Jr, and G.R. Sitko, J. Pharmacol. Exp. Ther., 266, 1501 (1993).
30   R.N. Willette, C.F. Sauermelch, R. Rycyna, S. Sarkar, G.Z. Feuerstein, A.J. Nichols, and E.H. Ohlstein, Stroke, 23, 703 (1992).
31   J.F. Tschopp, E.M. Driscoll, E.-X. Mu, S.C. Black, M.D. Pierschbacher, B.R. and Lucchesi, Coronary Artery Dis., 4, 809 (1993).
32   The Pink Sheet, 57, No. 51, December 18, 1995.
33   S.J. Gardell, Perspectives in Drug Discovery and Design, 1, 521 (1994).

34   R.M. Scarborough, M.A. Naughton, W. Teng, J.W. Rose, D.R. Phillips, L. Nannizzi, A. Arfsten, A.M. Campbell, and I.F. Charo, J. Biol. Chem., 268, 1066 (1993).

35   I.F. Charo, R.M. Scarborough, C.P. duMee, D. Wolf, D.R. Phillips, and R.L. Swift, Circulation, 86, (Supp. I) I-260 (1992).

36   Business Wire, Jan. 5, 1996.

37   Business Wire, June, 14, 1995.

38   M.S. Egbertson, T. Chang, M.E. Duggan, R.J. Gould, W. Halczenko, and G.D. Hartman, J. Med. Chem., 37, 2537 (1994).

39   G.D. Hartman, M.S. Egbertson, W. Halszenko, W.L. Laswell, M.E. Duggan, and R.L. Smith, J. Med. Chem., 35, 4640 (1992).

40   K. Peerlinck, I. De Lepeiere, M. Goldberg, D. Panebianco, J. Vermylen, and J. Arnout, Circulation, 86, (Supp. I) I-260 (1992).

41   K. Peerlinck, I. DeLepeleire, and M. Goldberg, Circulation, 88, 1512 (1993).

42   K. Peerlinck, J. Arnout, I. DeLepeleire, M. Goldberg, V. Fitzpatrick, and J. Vermylen, Thromb. Haem., 69, (Abst.2) 560 (1993).

43   L. Alig, A. Edenhofer, P. Hadvary, M. Hurzeler, D. Knopp, M. Muller, and B. Steiner, J. Med. Chem. 1992, 35, 4393 (1992).

44   B. Steiner, P. Haring, L. Jennings, and W.C. Kouns, Thromb. Haem., 69, (Abst. 860) 782 (1993).

45   C.R. Jones, R.J. Ambros, H.J. Rapold, B. Steiner, T. Weller, P. van Heiningen, and H.J.M.J. Crijins, Thromb. Haem., 69, a65 (1993).

46   P. Theroux, S. Kouz, and M. Knudtson, Circulation, 90, I-232 (1994).

47   J. Lefkovits, and E. J. Kopol, Eur. Heart J. 17, 9-18 (1996).

48   R.F. Nutt, S.F. Brady, C.D. Colton, J.T. Sisko, T. Ciccarone, M.R. Levy, M.E. Duggan, I.S. Imagire, and R.J. Gould, in "Peptides, Chemistry and Biology, Proc. 12th Amer. Peptide Symp. " J.A. Smith, and J.E. Rivier, Ed., Leiden: ESCOM, 1992, p. 914.

49   P. Theroux, N. Kleiman, and P.K. Shah, Circulation, 88, I-201 (1993).

50   H.D. White, B. Charbonnier, F. van de Werf, L. Erhardt, D. David, W.S. Hillis, G. von der Lippe, F.W.A. Verheugt, G. Jennings, D. Braun, and V. Fitzpatrick, Circulation, 88, I-201 (1993).

51   J.D. Rossen, and T. Palabrica, J. Amer. Coll. Cardiol. 21, (Supp. 2): Abst. 91 (1993).

52   P. Wissel, T.D. Bjornsson, and A. Barchowsky, Clin. Pharmacol. Ther., 1992, 51, 149 (1992).

53   A.S. Yuan, E.L. Hand, and M. Hichens, J. Pharmaceut. Biomed. Anal., 11, 427 (1993).

54   N.B. Modi, S.A. Baughman, B.D. Paasch, A. Celniker, and S. Smith, J. Cardiovasc. Pharmacol., 25, 888 (1995).

55   S. Cheng, W.S. Craig, D. Mullen, J.F. Tschopp, D. Dixon, and M.D. Pierschbacher, J. Med. Chem. 37, 1 (1994).

56   S. Kaku, T. Kawasaki, N. Hisamichi, Y. Sakai, Y. Taniuchi, O. Inagaki, S. Yano, K-I. Suzuki, C. Terazaki, and Y. Masuho, Thromb. Haem., 73, (Abst. 1589) 1314 (1995).

57   T.H. Muller, H. Schurer, L. Waldmann, E. Bauer, F. Himmelsbach, and K. Binder, Thromb. Haem., 69, 975 (1993).

58   F. Himmelsbach, A. Volkhard, B. Guth, G. Linz, T.H. Muller, H. Pieper, E. Seewaldt-Becker, and H. Weisenberger, Eur. J. Med. Chem., Suppl. Vol. 30, 244s (1995).

59   N.H. Weisenberger, T. H. Muller, G. Deichsel, and J. Krause, Thromb. Haem., 73, (Abst. 1592) 1315 (1995).

60   M. Maher, N. Moran, D. O'Callaghan, D. Coyle, and D. Fitzgerald, Thromb. Haem., 73, (Abst. 2088) 1446 (1995).

61   T.H. Muller, H. Weisenberger, R. Brickl, K. Ruhr, H. Narjes, and F. Himmelsbach, Thromb. Haem., 73, (Abst. 2091) 1447 (1995).

62   T.H. Muller, H. Weisenberger, R. Brickl, M. Kirchner, H. Narjes, and F. Himmelsbach, Thromb. Haem., 73, (Abst. 2081) 1445 (1995).

63   F. van de Werf, Eur. Heart J., 17, 325 (1996).

64   J.A. Zablocki, J.G. Rico, R.B. Garland, T.E. Rogers, K. Williams, and L.A. Schretzman, J. Med. Chem., 38, 2378 (1995).

65   R.J. Anders, J.C. Alexander, G.L. Hantsbarger, D.M. Burns, S.D. Oliver, G. Cole, and D.J. Fitzgerald, J. Am. Coll. Cardiol., 25, (Abst. 931-113), 117A (1995).

66   K. Kottke-Marchant, C. Simpfendorfer, M. Lowrie, D. Burns, and R.J. Anders, Circulation, 92, Suppl. 1, (Abst. 2331) I-988 (1995).

67   C. Simpfendorfer, K. Kottke-Marchant, and E.J. Topol, J. Am. Coll. Cardiol., 27, Suppl. A, (Abst. 969-110) 242A (1996).

68   J. A. Zablocki, F.S. Tjoeng, P.R. Bovy, M. Miyano, R.B. Garland, and K. Williams, Biorg. Med. Chem., 3, 539 (1995).

69   K.D. Robarge, A.G. Olivero, B.K. Blackburn, M. Baier, and A. Lee, 210th National Meeting of the American Chemical Society, August 20-24, Chicago, Abstract MEDI53 (1995).

70   R.S. McDowell, B.K. Blackburn, T.R. Gadek, L.R. McGee, T. Rawson, and M.E. Reynolds, J. Amer. Chem. Soc., 116, 5077 (1994).

71   B. Steiner, L. Alig, B. Blackburn, S. Bunting, M. Hurzeler, M. Schmitt, S. Weiss, and T. Weller, Blood, 86, Supp. I, (Abst. 351) 91a (1995).

72   C. Refino, N. Modi, S. Bullens, C. Pater, T. Lipari, D. Wu, B. Blackburn, T. Weller, B. Steiner, and S. Bunting, Blood, 86, Supp. I, (Abst. 348) 90a (199?).

73   L. Alig, A. Edenhofer, M. Muller, A. Trzeciak, and T. Weller, European Patent Application , EP 0,372, 468. (1989); b) L. Alig, A. Edenhofer, M. Muller, A. Trzeciak, and T. Weller, European Patent Application, EP 0,381,033. (1990).

74   N.S. Cook, O. Bruttger, C. Pally, and A. Hagenbach, Thromb. Haem., 70, 838-847 (1993).

75   N.S. Cook, G. Kottirisch, and H.-G. Zerwes, Drugs of the Future, 19, 135 (1994).

76   S. Jackson, W. DeGrado, A. Dwivedi, A. Parthasarathy, A. Higley, J. Krywko, A. Rockwell, J. Markwalder, G. Wells, and R. Wexler, J. Amer. Chem. Soc., 116, 3220 (1994).

77   S.A. Mousa, W.F. DeGrado, D.-X. Mu, R.P. Kapil, B.R. Lucchesi, and T.M. Reilly, Circulation, 93, 537-543 (1996).

78   DuPont Merck communication, March, 1995.

79   C.D. Eldred, B. Evans, S. HIndley, B.D. Judkins, H.A. Kelly, J. Kitchin, P. Lumley, B. Porter, B.C. Ross, K.J. Smith, and N. Taylor, J. Med. Chem., 37, 3882 (1994).

80   W.E. Bondinell, R.M. Keenan, W.H. Miller, F.E. Ali, A.C. Allen, and C.W. DeBrosse, Bioorg. Med. Chem., 2, 897 (1994).

81   J. Samanen, F.E. Ali, L. Barton, W. Bondinell, J. Burgess, J. Callahan, and R. Calvo, In "Peptides: Chemistry, Structure, and Biology (Proc. 14th Amer. Peptide Symp.)" P.T.P. Kaumaya, and R.S. Hodges, Ed., ESCOM, Leiden, 1996, in press.

82   W.H. Miller, T. Ku, F.E. Ali, W.E. Bondinell, R.C. Calvo, L.D. Davis, K.F. Erhard, and L.B. Hall, Tetrahedron Lett., 36, 9433 (1996).

83   R.J. Gould, J.S. Barrett, J.D. Ellis, M.A. Holahan, M.T. Stranieri, and A.D.. Theoharides, Thromb. Haem., 69, (Abst 2) 539.

84   J.S. Barrett, R.J. Gould, J.D. Ellis, M.M. Holahan, M. T. Stranieri, J.J. Lynch Jr., G.D. Hartman, N. Ihle, M. Duggan, and O.A. Moreno, Pharm. Res., 11, 426 (1994).

85   M.S. Egbertson, A.M. Naylor, G.D. Hartman, J.J. Cook, and R.J. Gould, American Chemical Society National Meeting, San Diego, March 13-17, Abst. MEDI-16 (1994).

86   M.S. Egbertson, A.M. Naylor, G.D. Hartman, J.J. Cook, R.J. Gould, M.A. Holahan, J.J. Lynch, Jr., R.J. Lynch, and M.T. Stranieri, Biorg. Med. Chem. Lett., 4, 1835 (1994).

87   A.M. Naylor, M.S. Egbertson, L.M. Vassalo, L. A. Birchenough, G.X. Zhang, R.J. Gould, and G.D. Hartman, Biorg. Med. Chem. Lett., 4, 1841 (1994).

88   M.E. Duggan, A.M. Naylor-Olsen, J.J. Perkins, P.S. Anderson, and C.T.-C. Chang, J. Med. Chem., 38, 3332 (1995).

89   J. Alexander, D.S. Bindra, J.D. Glass, M.A. Holahan, M.L. Renyer, G.S. Rork, G. R. Sitko, M.T. Stranieri, R.F. Stupienski, and H. Veerapanane, J. Med. Chem. 39, 480 (1996).

90   J. Gante, H. Juraszyk, H. Wurziger S. Bernotat-Danielowski, G. Melzer, and F. Rippman, Lett. Peptide Sci., 2, 135 (1995).

91   W.J. Hoekstra, M.P. Beavers, P. Andrade-Gordon, M.F. Evangelisto, P.M. Keane, and J.B. Press, J. Med. Chem., 38, 1582 (1995).

92   Drug News and Perspectives, 7, 353 (1994).

93   K. Nakamura, Scrip no. 2002, Feb. 24, 1995, p. 25.

94   S.A. Mousa W.F. DeGrado, S. Jackson, M. Flint, S. Forsythe, and T.M. Reilly, Circulation, 92, (Abst. 2332) I-488 (1995).

95   C.-B. Xue, M. Rafalski, J. Roderick, C.J. Eyermann, S. Mousa, R.E. Olson, and W.F. DeGrado, Biorg. Med. Chem. 6, 339-344 (1996).

96   D. Cox, T. Aoki, J. Seki, and Y. Motoyama, Thromb. Haem., 69, (Abst. 598), p. 706 (1993).

97   R.J. Leadley, J.S. Bostwick, C.J. Kasiewski, V. Chu, S. Klein, M. Czekaj, C.J. Gardner, and M.H. Perrone, Circulation, 92, Supp. I, (Abst. 2329) I-488 (1996).

98   H. Demrow, C. Batley, and J.D. Folts, J. Am. Coll. Cardiol., 25, (Abst. 914-68) 70A (1995).

99   J.J. Calvete, Thromb. Haem., 72, 1 (1994).

100 J.W. Smith, In "Integrins, Molecular and Biological Response to the Extracellular Matrix" D.A. Cheresh, and R.P. Mecham, Ed., Academic Press, New York, 1994, p. 1.

101 M.H. Ginsberg, D. Xiaping, T.E. O'Toole, and J.C. Loftus, Thromb. Haem., 74, 352 (1995).

102 M.A. Schwartz, In "Integrins, Molecular and Biological Response to the Extracellular Matrix," D. Cheresh, and R. Mecham, Ed., Academic Press, New York, 1994, p. 33.

103 J. Ashkenas, C.H. Damsky, M.J. Bissell, and Z. Werb, In "Integrins, Molecular and Biological

Response to the Extracellular Matrix," D.A. Cheresh, and R.P. Mecham, Ed., Academic Press, New York, 1994, p. 79.

104 T.J. Kunicki, and Z.M. Ruggeri, In "Integrins, Molecular and Biological Response to the Extracellular Matrix," D.A. Cheresh, and R.P. Mecham, Ed., Academic Press, New York, 1994, p. 195.

105 T.W. Ku, W.H. Miller, W.E. Bondinell, K.F. Erhard, R.M. Keenan, and A.J. Nichols, J. Med. Chem., 38, 9 (1995).

106 K.D. Kopple, P.W. Baures, J.W. Bean, C.A. D'Ambrosio, J.L. Hughes, C.E. Peishoff, and D.S. Eggleston, J. Amer. Chem. Soc. 1992, 114, 9615 (1992).

107 C.E. Peishoff, F.E. Ali, J.W. Bean, R. Calvo, C.A. D'Ambrosio, D.S. Eggleston, and S.M. Hwang, J. Med. Chem., 35, 3962 (1992).

108 F.E. Ali, D.B. Bennett, R.R. Calvo, J.D. Elliott, S.-M. Hwang, T.W. Ku, M.A. Lago, and A.J. Nichols, J. Med. Chem., 37, 769-780 (1994).

109 M.M. Hann, B. Carter, J. Kitchin, P. Ward, A. Pipe, J. Broomhead, E. Hornby, M. Forster, and C. Perry, In "Molecular Recognition: Chemical and Biochemical Problems II" S.M . Roberts, Ed., The Royal Society of Chemistry, Cambridge, 1992, p. 145.

110 A.C. Bach, C.J. Eyermann, J.D. Gross, M.J. Bower, R.L. Harlow, P.C. Weber, and W.F. DeGrado, J. Amer. Chem. Soc., 116, 3207 (1994).

111 R.L. Harlow, J. Amer. Chem. Soc., 115, 9838 (1993).

112 R.S. McDowell, T.R. Gadek, P.L. Barker, D.J. Burdick, K.S. Chan, C.L. Quan, and N. Skelton, J. Amer. Chem. Soc., 116, 5069 (1994).

113 P.L. Barker, S. Bullens, S., Bunting, D.J. Burdick, K.S. Chan, T. Deisher, C. Eigenbrot, and T.R. Gadek, J. Med. Chem., 35, 2040 (1992).

114 R.S. McDowell, and T.R. Gadek, J. Amer. Chem. Soc., 114, 9245 (1992).

115 T.W. Ku, F.E. Ali, L.S. Barton, J.W. Bean, W.E. Bondinell, J.L. Burgess, and J.F. Callahan, J. Amer. Chem. Soc., 115, 8861 (1993).

116 S. Roux, S. Bunting, W. C. Kouns and B. Steiner, Thromb. Haem., 69, (Abst. 66) 560 (1993).

117 V. Fuster, E. Falk, J.T. Fallon, and L. Badimon, Thromb. Haem., 74, 552 (1995).

118 I.F. Charo, L.S. Bekeart, and D.R. Phillips, J. Biol. Chem., 262, 9935 (1987).

119 E.T. Choi, L. Engel, and A.D. Callow, J. Vasc. Surg., 19, 125 (1994).

120 J. Lefkovits, E.F. Plow, and E.J. Topol, New Engl. J. Med., 332, 1553 (1995).

121 D.I. Simon, H. Xu, N.K. Rao, Circulation 92, Supp. I, (Abst. 0519), I-110 (1995).

122 M. Pfaff, K. Tangemann, B. Muller, M. Gurrath, G. Muller, H. Kessler, R. Timpl, and J. Engel, J. Biol. Chem., 269, 20233 (1994).

123 T. Siahaan, L.R. Lark, M. Pierschbacher, E. Ruoslahti, and L.M. Gierasch, In "Peptides: Chemistry, Structure and Biology, Proceedings of the 11th American Peptide Symposium" J.E. Rivier, and G.R. Marshall, Ed., ESCOM, Leiden, 1990, p. 699.

124 R. Keenan, W. Miller, F. Ali, L. Barton, and W. Bondinell, 211th National Meeting of the American Chemical Society, March 24-28, New Orleans, Abst. MEDI 236 (1996).

Chapter 11. Atherothrombogenesis

Peter Charlton[1] and Michael Sumner[2]

[1]Xenova Limited, Slough, UK and [2]GlaxoWellcome, Stevenage, UK

Introduction - Arterial thrombosis rarely occurs in healthy vessels but is, in general, associated with atherosclerosis. The development of atherosclerotic plaques is not usually associated with clinical signs of thrombosis. However, the disruption of an atherosclerotic plaque accompanied by resultant intraluminal thrombosis plays a fundamental role in the pathogenesis of acute coronary (unstable angina, acute myocardial infarction, sudden death) and cerebral (transient ischaemic attack, stroke) syndromes. These diseases are the leading cause of death in the Western world and consequently are the subject of intense investigation (1).

Although atherosclerotic plaque disruption and thrombosis can lead to such catastrophic clinical events, there is growing evidence for an ongoing, dynamic interplay between the processes of thrombosis and atherosclerosis in arteries. These interactions are the subject of this review (see Figure 1) which will also consider how an understanding of these processes can be translated into new approaches to the treatment of arterial diseases.

THROMBOSIS AND ATHEROSCLEROTIC PLAQUE DEVELOPMENT

Plaque Initiation - Atherosclerotic lesions can be induced to form in two ways: by enriching the diet with cholesterol or by repeated injury of the wall of an artery of normocholesterolemic animals, the latter being exacerbated by cholesterol-rich diets. In either case, the earliest lesions of atherosclerosis are focal and occur at vessel orifices and branches, points where normal arterial shear forces are disrupted by turbulent flow conditions. Such areas are associated with endothelial cell dysfunction, evidenced by an increased permeability, loss of "protective factors"(e.g., nitric oxide, prostacyclin), low density lipoprotein (LDL) accumulation and monocyte adhesion. The latter migrate into the intima where, together with vascular smooth muscle cells (VSMCs), they accumulate lipid to become foam cells, giving rise to fatty streaks, the characteristic feature of early atherosclerotic lesions (2).

In hyperlipidemic models, frank endothelial cell denudation only occurs after several weeks at which point platelets adhere to exposed connective tissue matrix proteins and ensuing platelet activation and aggregation occurs. Thus the consensus view is that platelet adhesion and aggregation do not play a part in plaque initiation. However, microthrombi and fibrin are frequently found on the surface of macroscopically normal vessels and have been associated with areas of focal increases in endothelial cell permeability (3). Cytokine and oxidized-LDL-stimulated endothelial cells and monocytes express procoagulant tissue factor (TF) and anti-fibrinolytic plasminogen activator inhibitor-1 (PAI-1), agents which favor thrombosis and fibrin deposition onto the luminal surface of the vessel. This may explain why microthrombi occur even in the absence of developed lesions and may reflect an important contribution of thrombotic events to early plaque development.

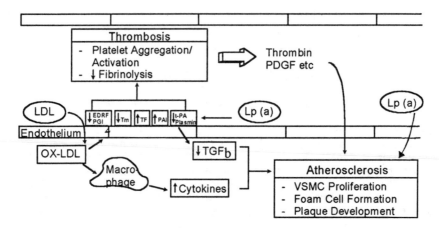

Figure 1 Simplified scheme illustrating the interactions between thrombosis and atherosclerosis Endothelial cell 'dysfunction' leads to monocyte adhesion and accumulation in the vessel wall. In a lipid-rich environment macrophages and smooth muscle cells (SMCs) engulf oxidized LDL (OX-LDL) to become foam cells, the characteristic early phase of atherosclerosis. OX-LDL, lipoprotein(a) (Lp(a)) and the cytokines released from endothelial cells and macrophages induce the expression and release of prothrombotic factors (TF, PAI-1) and inhibition of anticoagulant and profibrinolytic elements (thrombomodulin (Tm) and tissue plasminogen activator (tPA)). This prothrombotic environment favors activation of platelets, generation of thrombin, increased fibrin deposition and further lipid accumulation in the vessel wall. Thrombin, together with platelet derived growth factor (PDGF), promotes VSMC proliferation and a proinflammatory environment. Lp(a) accumulation in the vessel wall also reduces transforming growth factor-β (TGF-β) levels thus promoting SMC proliferation and plaque growth.

Thrombosis and Plaque Progression - Endothelial loss or damage results in the adhesion of platelets to the vessel wall. This leads to their activation and secretion of granule contents and provides a focus for the assembly of the procoagulant prothrombinase complex. The result is the generation of Factor Xa (FXa), thrombin and platelet-derived mitogens, such as PDGF, which are mitogenic for SMCs (2). The importance of platelet adhesion for the SMC proliferative response, which is characteristic of atherosclerotic plaque progression, is illustrated by experiments in von Willebrand's disease pigs (4). These animals are deficient in von Willebrand's Factor (vWF) and show defective platelet adhesion. Moreover, they are relatively resistant to atherosclerosis, an observation which has been ascribed to a defect in platelet adhesion and which has been interpreted as confirming the key role of platelets in the growth of atherosclerotic plaques (3,4).

Plaque growth is mediated largely by the proliferation of SMCs and the elaboration of extracellular matrix proteins (2). Clearly there is an important contribution to this response from thrombotic elements. In addition, however, there is clear evidence for the ongoing incorporation of fibrotic thrombi into the developing plaque, thereby contributing to its bulk and providing an ongoing source of mitogens (3).

Thrombosis and Plaque Rupture - The clinical sequalae of plaque rupture and ensuing thrombosis are evident as acute coronary or cerebral syndromes. There is overwhelming evidence that such episodes are precipitated by formation of a large thrombus (3).

Approximately 25% of thrombi occur on the surface of the plaque - these are usually associated with a high grade stenosis and surface infiltration by foam cells leading to intimal injury. In contrast, 75% of coronary thrombi are associated with small, ruptured atherosclerotic plaques, which tend to be lipid-rich, have a high macrophage and a low smooth muscle cell content and a thin fibrous cap. The rupture of such plaques exposes a large mass of prothrombotic material, including TF and matrix proteins. As a result, a massive thrombus occurs which initially forms deep within the soft interior of the plaque itself. If the thrombus is sufficiently large, this will occlude the affected vessel leading to acute ischaemia. Whether the thrombus grows sufficiently to be occlusive will depend upon the haemostatic status of the individual. The risk of developing acute myocardial infarction is related to plasma levels of fibrinogen, activated Factor VII (FVIIa) and PAI-1; elevation of one or more of these factors will favor intraluminal thrombosis (5).

KEY MEDIATORS OF ATHEROTHROMBOGENESIS

Lipoprotein (a) - Lipoprotein (a), a low density lipoprotein-like particle that contains apolipoprotein-a (Apo(a)), is a primary risk factor for premature atherosclerosis (6,7). Apo(a) has been found within atherosclerotic plaques but not in other regions of the vessel wall and increased Lp(a) concentrations are associated with atherosclerotic vascular disease including coronary artery disease, myocardial infarction (MI), peripheral vascular disease, retinal vascular occlusion and occlusive arterial thrombosis (7). In addition, expression of human Apo(a) in transgenic mice results in an increase in atheroma. Survivors of MI who fail to recanalise have significantly higher Lp(a) than those patients with a patent artery and Lp(a) is also the only significant predictor of restenosis after percutaneous transluminal coronary angioplasty (PTCA) (8).

Lp(a) is a variant of LDL particle in which Apo(a) is connected by a disulphide bond to apolipoprotein B-100. The finding that Lp(a) is an independent cardiovascular risk factor distinct from LDL suggests that risk is also associated with the Apo(a) moiety (9). Apo(a) is a glycoprotein of variable molecular mass in the range 280 - 800 kDa which is structurally related to proteins involved in fibrinolysis, coagulation, lipid transport and mitogenesis (7,8). In particular, Apo(a) has a significant primary sequence homology with plasminogen and, in common with plasminogen, contains a number of kringle domains (9). The lysine binding sites of plasminogen are associated with the kringle structure and mediate binding to fibrin(ogen) (10). Lp(a) may contain a functional lysine binding site in one of the kringle 4 motifs and antibodies to kringle 4 of plasminogen inhibit the interaction of Lp(a) with lysine (9,11). This similarity between the Apo(a) and plasminogen may help explain some of the pathophysiological effects of elevated plasma Lp(a) concentrations by interference with several of the functions of plasminogen (6,12). These include activation of plasminogen activators, binding and proteolysis of fibrin, plasmin-mediated activation of TGF-β and binding of plasminogen to cells. Thus through inhibition of plasminogen binding to its receptor and interference with the functions of plasminogen, Lp(a) impedes fibrinolysis and reduces plasmin generation. In addition, once bound to plasminogen receptors and binding sites the cell and matrix associated Lp(a) is oxidized and contributes to the development of atherosclerotic lesions. Furthermore, Lp(a) also promotes a thrombogenic state by increasing PAI-1 activity (6,13).

Both recombinant Apo(a) and native Lp(a) inhibit the binding of plasminogen to fibrin (14). Small isoforms of Lp(a) (≤ 500 kDa) are more efficient inhibitors of plasminogen activation in vitro than larger isoforms (> 500 kDa) and low molecular mass isoforms of Apo(a) are consistently associated with coronary disease (15). Mutations of kringle 4-37 of Apo(a) may also affect the lysine binding properties and be associated with coronary disease as a met[66] → thr mutation was found in 40% of patients with normal or high plasma Lp(a) and a history of cardiovascular disease (16). Thus mutations may also contribute to the atherothrombogenic potential of Lp(a).

Such relationships provide links between lipoprotein metabolism and thrombosis (see Figure 2). In addition, there is growing evidence that TGF-β may be a key regulator of atherosclerosis by inhibiting both the migration and proliferation of VSMCs. Importantly, TGF-β is produced in a latent form which is activated by the action of plasmin. Thus by inhibiting plasmin activity, Lp(a) and PAI-1 block the activation of TGF-β and allow VSMC proliferation to proceed unchecked, an observation confirmed in a transgenic mouse model (17). In addition, TGF-β itself can stimulate PAI-1 production, which in turn decreases plasmin generation thereby providing a negative feedback control mechanism. TGF-β may thus play a pivotal role in preventing atherogenesis, by maintaining the architecture of the normal vessel (18). Reduction in TGF-β levels allows damage of the endothelium by increased leucocyte binding. Uptake of lipids and SMC proliferation and migration results in growth of the lesion. The potential importance of TGF-β is supported by a recent study which demonstrated that a population of patients with advanced atherosclerotic lesions have five-fold less active TGF-β in their sera than patients with normal arteries (19). Interestingly, this study also showed that women had a higher proportion of active TGF-β than men, perhaps accounting for their relative greater resistance to coronary artery disease.

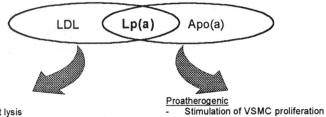

Prothrombotic	Proatherogenic
- Impaired clot lysis	- Stimulation of VSMC proliferation
- Competition with Plasminogen for binding sites	- Foam cell generation
- Increased PAI-1 synthesis	- Increased lipid delivery to cells

Figure 2 Prothrombotic and Proatherogenic Role of Lp(a)

Thrombin - A healthy endothelium regulates thrombosis through the expression of a wide variety of mediators (eg., vWF, prostacyclin, nitric oxide, Tm, tPA, PAI-1, tissue factor pathway inhibitor) (20). Thrombin shifts the balance from an antithrombotic to a prothrombotic state through a number of actions including modulation of these mediators, activation of platelets, generation of fibrin and activation of various coagulation factors (FXIII, FVIII and FV). Thus thrombin has a major role in thrombogenesis and is an important trigger of thrombosis in acute coronary syndromes (21).

Protein C Pathway - The protein C anticoagulant pathway controls excessive thrombus extension through the control of thrombin formation. The protein C pathway is initiated on binding of thrombin to Tm on endothelial cells (22). This complex then cleaves the protein C zymogen to activated protein C (APC). Once complexed with Tm, the prothrombotic activities of thrombin are down-regulated as thrombin is prevented from cleaving fibrinogen or activating platelets (23). APC then down-regulates prothrombin activation and thrombin generation through inactivation of Factor VIIIa and Factor Va (24). APC is also profibrinolytic through the attenuation of TAFI (thrombin-activatable fibrinolysis inhibitor) activation and may further enhance a profibrinolytic state by inhibiting PAI-1 activity (24-26). *In vivo* studies support an antithrombotic role for APC as APC is beneficial in both animal models of thrombosis and in patients with disseminated intravascular coagulation (DIC) (24,26). In a canine model of coronary artery thrombosis APC prevented reocclusion after recanalization with tPA, thus APC may be useful as an adjunctive treatment to enhance thrombolytic therapy in AMI patients (24). As PAI-1 may be elevated in acute MI patients the concurrent inhibition of PAI-1 by APC may also be beneficial.

PAI-1 - Clearance of a thrombus and restoration of normal blood flow requires the activation of the fibrinolytic system. A key event in this activation is the conversion by tPA of the zymogen plasminogen to its active form plasmin which then degrades fibrin into soluble degradation products (27). The primary physiological regulator of tPA activity is the fast-acting inhibitor, PAI-1 which is a member of the serine proteinase inhibitor (serpin) family (28). PAI-1 is a glycoprotein with a predicted mass of 48kDa which acts as a pseudosubstrate inhibiting tPA through a "bait" residue (Arg 346 - Met 347) (29,30). PAI-1 is produced by endothelial cells and platelets and in circulation active PAI-1 rapidly complexes with tPA with a rate constant of the order of $10^7 M^{-1}s^{-1}$ (31).

The correct level of PAI-1 is critically important in regulating the correct fibrinolytic capacity. Increased PAI-1 concentration is associated with familial venous thrombosis and an inherited deficiency of PAI-1 can result in a hyperfibrinolytic state but reports of congenital deficiencies or increased levels of PAI-1 are both extremely uncommon (32,33). Transgenic mice which express high levels of human PAI-1 suffer severe venous thrombosis and PAI-1 deficient mice exhibit a mild hyperfibrinolytic state and a greater resistance to venous thrombosis (34,35). Release of PAI-1 from activated platelets plays a major role in resistance to thrombolysis and promotes reocclusion after thrombolytic therapy and several groups report that the spontaneous lysis of thrombi is dependent on PAI-1 content (36-38). There is also a rebound increase in PAI-1 activity following local intra-arterial tPA infusion which may lead to reocclusion (39). Balloon injury in rabbit carotid arteries induces intra-mural expression of PAI-1 and decreased surface fibrinolytic activity which may result in the development of mural thrombosis (40).

Elevated PAI-1 has been reported in a variety of disorders including deep vein thrombosis, unstable angina, coronary artery disease and acute MI (13,27,31,41). In unstable angina there is a clear correlation between elevated PAI-1 activity at presentation and patients who suffered subsequent cardiovascular events (ie., acute MI) (42). Elevated PAI-1 is also a predictor of recurrent infarction in young survivors of a first MI and is a risk factor for coronary atherosclerosis in survivors of MI with glucose intolerance (35,43). In diabetic patients increased PAI-1 levels may predispose individuals to MI and raised PAI-1 levels may inhibit spontaneous reperfusion and so contribute to a poor outcome. Increased PAI-1 activity induced by proinsulin and insulin contributes to a prothrombotic state resulting in the accelerated atherosclerosis and impaired thrombolysis in non-insulin dependent diabetes mellitus patients (44).

Several studies have linked an increase in PAI-1 expression with the progression and severity of atherosclerotic disease (45-47). Increased PAI-1 in atheromatous arterial wall favors fibrin deposition which promotes plaque development and release of PAI-1 may contribute to thrombotic complications observed following plaque rupture (13,48). The large excess of PAI-1 over tPA in the endothelium intima and media of diseased coronary arteries may predispose to thrombosis (49). In situ analysis of atherosclerotic plaques indicates that the increased PAI-1 expression in intimal SMCs and macrophages is coincident with the expression of TF, thrombin, fibrin and vitronectin. This balance of growth factors, cytokines and lipids within plaques induces PAI-1 activity and contributes to impaired fibrinolysis and increased matrix deposition (50).

Interest in fibrinolysis results from the proposal that impaired fibrinolysis facilitates thrombus growth and plaque formation (51). The proposed link between intracellular lipid metabolism and regulation of the fibrinolytic system is supported by results with the lipid lowering agents gemfibrozil and niacin (52). Gemfibrozil decreases plasma PAI-1 in patients with elevated triglycerides and niacin attenuates PAI-1 synthesis in Hep G2 cells. Inhibition of HMG-CoA reductase may also lower PAI-1 levels (53).

In addition to its role in fibrinolysis PAI-1 may play an important role in a variety of other biological processes including trauma, tumour metastasis, angiogenesis, cellular migration and ovulation (54-56).

STRATEGIES FOR THERAPEUTIC INTERVENTION

Monocyte Adhesion and Accumulation - Since one of the earliest events in atherogenesis is monocyte adhesion to, and infiltration of, the dysfunctional endothelium, this would represent a potential target for therapeutic intervention. Targets might include adhesion molecules expressed on the surface of endothelial cells, some of which may be unique to early atherosclerotic lesions, or monocyte chemoattractants, such as chemokines, which are expressed by "dysfunctioning" endothelial cells (57).

Oxidized-LDL Accumulation - Strategies designed to lower plasma LDL-cholesterol levels have so far proven to be the most effective approach to the management of atherosclerotic diseases (58). Alternative strategies include the further inhibition of LDL synthesis, perhaps by inhibiting apolipoprotein B synthesis, upregulation of the LDL receptor or increased formation of protective HDL fractions(59-61). Inhibition of LDL oxidation may also be an attractive target since OX-LDL has been shown to promote endothelial cells to express molecules which favor monocyte accumulation and to induce expression of tissue factor and PAI-1 by endothelial cells or monocytes (62,63).

Lp(a) - Due to the central role of Lp(a) in both thrombosis and atheroclerosis there is a clear potential for agents that lower plasma Lp(a). Some studies with nicotinic acid and N-acetylcysteine have been reported but further studies are warranted in this area (7). A two-step model for Lp(a) formation has been proposed and, as Apo(a) is rapidly cleared, inhibitors of Lp(a) assembly should reduce plasma Lp(a) concentration (64).

Thrombin and other Coagulation Enzymes - OX-LDL and cytokine-stimulated endothelial cells and macrophages express high levels of TF (65). Binding of FVIIa (a recognised risk factor for coronary artery disease) to form a TF/FVIIa complex initiates the coagulation cascade leading to the generation of thrombin and the formation of fibrin. Thrombin has multiple actions which promote thrombosis and atherogenesis (66). Thus thrombin converts fibrinogen to fibrin to complete the coagulation cascade, promotes PAI-1 release (hence inhibits fibrinolysis), is a powerful activator of platelets, promoting their aggregation and granule content secretion, and is chemotactic for monocytes, neutrophils and lymphocytes. In addition, thrombin promotes vWF secretion, P-selectin expression, cell retraction (leading to increased permeability) and proliferation in endothelial cells and is mitogenic for lymphocytes, fibroblasts and VSMC. Recent reports establishing thrombin receptor expression in macrophages and VSMC in atheromatous tissues, but not in normal vasculature (where thrombin receptors are only found on the endothelium), demonstrate that thrombin has both the plethora of actions and the necessary receptors to be a major factor in atherothrombogenesis (67). Thus antagonism of the actions of thrombin either by direct inhibition or reduction in thrombin formation (eg., inhibition of FXa activity) remains an important therapeutic target and these approaches are reviewed in Chapter 6 of this volume.

Thrombomodulin and Protein C - The endogenous anti-thrombotic mechanisms include both anti-platelet agents such as nitric oxide and prostanoids and anti-coagulants. In view of the potential major contribution of thrombin to atherothrombogenesis, endogenous anti-coagulant mechanisms which limit thrombin formation and action may have therapeutic value. Thrombin specificity is switched from fibrinogen to protein C upon binding to Tm, an endothelial cell surface proteoglycan. The thrombin/Tm complex is a potent activator of protein C which subsequently cleaves and inactivates FVa and FVIIIa to terminate thrombin formation *via* the coagulation pathways. In addition to having anti-coagulant actions, APC is also anti-inflammatory and hence may be an important endogenous modulator of thrombosis and atherogenesis (22,68).

The change in specificity of thrombin consequent upon binding to Tm has recently been mimicked by a single substitution resulting in a thrombin mutant which lacked procoagulant activity but was still able to activate protein C (69). An intriguing possibility arising out of this work is that it may be possible to design a "chemical thrombomodulin" which could bind to, and alter, the specificity of thrombin to enhance its anti-coagulant function.

Thrombin Receptor Antagonist - As outlined above and reviewed elsewhere, thrombin has a multitude of actions which are pro-coagulant, pro-inflammatory and pro-mitogenic (70). Many (but not all) of these actions are mediated through a novel type of protease-activated receptor, of which more than one type may exist (71). The now classical thrombin receptor characterised by Coughlan and co-workers has a widespread distribution and may represent a key target for the design of antagonists (72). Such agents would be expected to block many of the actions of thrombin which contribute to atherothrombogenesis with perhaps less compromise of the haemostatic system than with direct thrombin inhibitors. As such, thrombin receptor antagonists may be well suited to the prophylactic prevention of coronary and cerebral artery diseases.

Platelet Aggregation and Adhesion Inhibitors - There is a clear platelet involvement in the thrombotic response following plaque rupture and strategies to inhibit platelet aggregation (eg., GPIIb/IIIa antagonists) or reduce platelet adhesion to the vessel wall are reviewed in Chapter 10 of this volume.

PAI Antagonists - PAI-1 contributes to the prothrombotic state and is implicated in the progression of atherosclerotic disease. Thus inhibition of PAI-1 production or action should decrease extracellular matrix accumulation and promote fibrinolysis by enhancing plasmin generation and intramural proteolysis. Antisense oligonucleotides targeting specific sequences of the PAI-1 gene were prepared and shown to inhibit PAI-1 production in human vascular endothelial and smooth muscle cells (73). This decreased PAI-1 synthesis correlated with an increase in cell-associated plasmin activity.

An alternative approach to regulating PAI-1 activity has been the generation of inhibitory antibodies to the protein. PRAP-1, a fragment of a rabbit polyclonal antibody against human PAI-1, inhibited PAI-1 activity and enhanced clot lysis *in vitro* (74). In a rat *in vivo* model where animals were pretreated with endotoxin to elevate PAI-1 then given the coagulant enzyme batroxobin an infusion of PRAP-1 increased fibrinolysis and decreased the occurrence of fibrin clots in the lungs (75). The effects of MAI-12, a murine monoclonal antihuman PAI-1 antibody, have been studied in *in vivo* thrombolysis and thrombus extension in a rabbit thrombosis model (76). Incorporation of the inhibitory antibody into human whole blood thrombi formed in the jugular vein of the rabbit significantly enhanced endogenous thrombolysis and partially prevented thrombus extension in this model. These studies have now been extended to evaluate the systemic administration of a second PAI-1 inhibitory murine monoclonal antibody, CLB-2C8 (77). In the rabbit jugular vein model intravenous administration of CLB-2C8 (1mg/kg) significantly enhanced endogenous thrombolysis and significantly reduced thrombus extension. CLB-2C8 was also evaluated in a canine coronary artery thrombosis model (77). In combination with tPA the anti-PAI-1 antibody significantly reduced the time to reperfusion and delayed the occurrence of reocclusion.

Although the problem of immunogenicity would hinder the clinical development of murine antibodies for use in man the data generated with MAI-12 and CLB-2C8 have stimulated the search for small non-peptide inhibitors of PAI-1. Recently the first low molecular weight inhibitors have been described (78,79). These diketopiperazine-based compounds, including XR5082 (1) modulated PAI-1 activity *in vitro* and, following intravenous bolus administration dose-dependently increased the rate and extent of *ex vivo* blood clot lysis in the rat. The compounds also significantly prolonged the time to thrombus formation in the rat electrically stimulated carotid artery thrombosis model but had no effect on human platelet aggregation *in vitro* or other global tests of haemostasis.

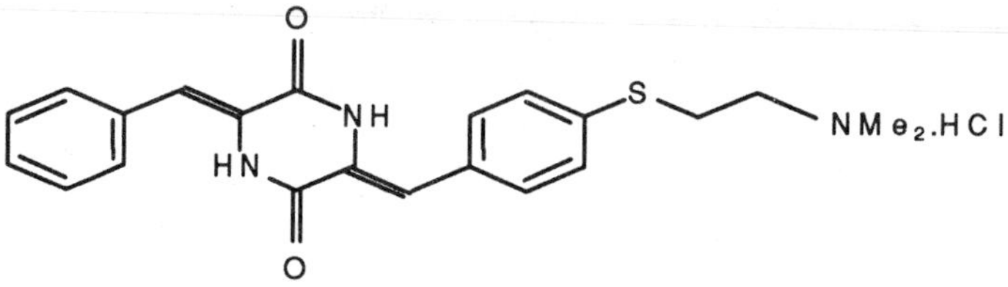

1

One further approach to inhibiting PAI-1 activity is by generating peptides based on the PAI-1 amino acid sequence that interfere with the binding of PAI-1 to tPA. In a recent study a tetradecapeptide corresponding to P_1 . P_{14} of the mobile loop of PAI-1 was shown to rapidly inactivate PAI-1 and to increase clot lysis *in vitro* (80,81).

To date, antibodies or small molecule inhibitors of PAI-1 activity have only been evaluated in acute thrombosis models, but this approach, or down regulation of PAI-1 expression, may have utility in controlling the development of atherothrombotic lesions as well as protecting against thrombosis.

Summary and Conclusions - The catastrophic event of atherosclerotic plaque rupture and occlusive thrombus formation in a coronary or cerebral artery remains the single major cause of death in the Western world. There is, however, an ongoing dynamic interplay between the processes of haemostasis and atherosclerosis which contributes to all stages of plaque formation. Key players in these interactions are monocytes, which act as a source of TF, cytokines and mitogens and platelets which provide mitogens and thrombotic elements. A number of potential points for intervention can be identified which seek either to affect the atherogenic process directly or indirectly via inhibition of thrombosis. Clearly, as our understanding of atherothrombogenesis grows we may be better able to identify agents which can more dramatically impact upon this major unmet medical need.

References

1.   Heart and Stroke Facts, 1994, Statistical Supplement, American Heart Association.
2.   R. Ross, Nature, 362, 801 (1993).
3.   N. Woolf and M.J. Davies in "Thrombosis in Cardiovascular Disorders," V. Fuster, M. Verstraete, Ed.,W.B. Saunders Company, p. 41 (1992).
4.   V. Fuster, E.J.W. Bowie, J.C. Lewis, D.N. Fass, C.A. Owen and A.L. Brown, J. Clin. Invest. 61, 722 (1978).
5.   P.A. Sytkowski, W.B. Kannel and R.B. D'Agostino, New Eng. J. Med., 322, 1635 (1990).
6.   G.Y.H. Lip and A.F. Jones, Q. J. Med., 88, 529 (1995).
7.   R.L. Desmarais, I.J. Sarembock, C.R. Ayers, S.M. Vernon, E.R. Powers and L.W. Gimple, Circulation, 91, 1403 (1995).
8.   P.C. Harpel, A. Hermann, X. Zhang, I. Ostfeld and W. Borth, Thromb. Haemost., 74, 382 (1995).

9.   E.F. Plow, T. Herren, A. Redlitz, L.A. Miles and J.L. Hoover-Plow, FASEB J., 9, 939 (1995).
10.   A. Ernst, M. Helmholds, C. Brunner, A. Pethö-Schramm, V.W. Armstrong and HJ. Müller, J. Biol. Chem., 270, 6227 (1995).
11.   J.L. Hoover-Plow, L.A. Miles, G.M. Fless, A.M. Scanu and E.F. Plow, Biochemistry, 32, 13681 (1993).
12.   A.M. Scanu, Biochem. Pharmacol., 46, 1675 (1993).
13.   J. Aznar and A. Estellés, Haemostasis, 24, 243 (1994).
14.   L. Hervio, V. Durlach, A. Girard-Globa and E. Anglés-Cano, Biochemistry, 34, 13353 (1995).
15.   M.J. Chapman, T. Huby, F. Nigon and J. Thillet, Atherosclerosis, 110 (Suppl.), S69 (1994).
16.   A.M. Scanu, Am. J. Cardiol. 75, 58B (1995).
17.   D.J. Grainger, P.R. Kemp, A.C. Liu, R.M. Lawn and J.C. Metcalfe, Nature, 370, 460 (1994).
18.   D.J. Grainger and J.C. Metcalfe, Biol. Rev., 70, 571 (1995).
19.   D.J. Grainger, B.R. Kemp, J.C. Metcalfe, A.C. Liu, R.M. Lawn, N.R. Williams, A.A. Grace, P.M. Schofield and A. Chauhan, Nature Medicine, 1, 74 (1995).
20.   J.D. Pearson, British Medical Bulletin, 50, 776 (1994).
21.   J.I. Weitz, Am. J. Cardiol., 75, 18B (1995).
22.   C.T. Esmon and K. Fukudome, Seminars in Cell Biology, 6, 259 (1995).
23.   C.T. Esmon, FASEB J., 9, 946 (1995).
24.   T. Sakamoto, H. Ogawa, H. Yasue, Y. Oda, S. Kitajima, K. Tsumoto and H. Mizokami, Circulation, 90, 427 (1994).
25.   L. Bajzar, R. Manuel and M.E. Nesheim, J. Biol. Chem., 270, 14477 (1995).
26.   Y. Katsuura, K. Aoki, H. Tanabe, M. Kiyoki and A. Funatsu, Thromb. Res., 76, 353 (1994).
27.   P.J. Declerck, I. Juhan-Vague, J. Felez and B. Wiman, Journal of Internal Medicine, 236, 425 (1994).
28.   J. Potempa, E. Korzus and J. Travis, J Biol Chem, 269, 15957 (1994).
29.   E. Sancho, P.J. Declerck, N.C. Price, S.M. Kelly and N.A. Booth, Biochemistry, 34, 1064 (1995).
30.   T.L. Lindahl, P.I. Ohlsson and B. Wiman, Biochem. J., 265, 109 (1990).
31.   S. Dawson and A. Henney, Atherosclerosis, 95, 105 (1992).
32.   G.M. Patrassi, M.T. Sartori, G. Saggiorata, G. Boeri and A. Girolami, Fibrinolysis, 6, 99 (1992).
33.   M.H. Lee, E. Vosburgh, K. Anderson and J. McDonagh, Blood, 81, 2357 (1993).
34.   L.A. Erickson, G.J. Fici, J.E. Lund, T.P. Boyle, H.G. Polites and K.R. Marotti, Nature, 346, 74 (1990).
35.   P. Carmeliet, J.M. Stassen, L. Schoonjans, B. Ream, J.J. van den Oord, M. De Mol, R.C. Mulligan and D. Collen, J. Clin. Invest., 92, 2756 (1993).
36.   H.A.R. Stringer, P. van Swieten, H.F.G. Heijnen, J.J. Sixma and H. Pannekoek, Arterioscler. Thromb., 14, 1452 (1994).
37.   I.M. Lang, J.J. Marsh, M.A. Olman, K.M. Moser, D.J. Loskutoff and R.R. Schleef, Circulation, 89, 2715 (1994).
38.   B.J. Potter van Loon, D.C. Rijken, E.J.P. Brommer and A.P.C. van der Maas, Thromb. Haemost., 67, 101 (1992).
39.   B. Åstedt, B. Casslén, I. Lecander, L. Nilsson and L. Norgren, Blood Coagulation and Fibrinolysis, 4, 563 (1993).
40.   H. Sawa, C. Lundgren, B.E. Sobel and S. Fujii, J. Am. Coll. Cardiol., 24, 1742 (1994).
41.   R.P. Gray, V. Mohamed-Ali, D.L.H. Patterson and J.S. Yudkin, Thromb. Haemost., 73, 261 (1995).
42.   I. Wieczorek, C.A. Ludlam and K.A.A. Fox, Am. J. Cardiol., 74, 424 (1994).
43.   A. Hamsten, U. de Faire, G. Walldius, G. Dahlen, A.Szamosi, C. Landou, M. Blombäck and B. Wiman, Lancet, 2, 3 (1987).
44.   T.K. Nordt, H. Sawa, S. Fujii and B.E. Sobel, Circulation, 91, 764 (1995).
45.   P.N. Raghunath, J.E. Tomaszewski, S.T. Brady, R.J. Caron, S.S. Okada and E.S. Barnathan, Arterioscler. Thromb. Vasc. Biol., 15 1432 (1995).
46.   J. Schneiderman, M.S. Sawdey, M.R. Keeton, G.M. Bordin, E.F. Bernstein, R.B. Dilley and D.J. Loskutoff, Proc. Natl. Acad. Sci. USA, 89, 6998 (1992).
47.   I. Juhan-Vague and M.C. Alessi, Thromb. Haemost., 70, 138 (1993).
48.   F. Lupu, G.E. Bergonzelli, D.A. Heim, E. Cousin, C.Y. Genton, F. Backmann and E.K.O. Kruithof, Arteriosclerosis and Thrombosis, 13, 1090 (1993).
49.   T. Padró, J.J. Emeis, M. Steins, K.W. Schmid and J. Kienast, Arterioscler. Thromb. Vasc. Biol., 15, 893 (1995).
50.   P. Carmeliet and D. Collen, FASEB J., 9, 934 (1995).
51.   A. Hamsten and P. Eriksson, Fibrinolysis, 8 (Suppl. 1), 253 (1994).
52.   S.L. Brown, B.E. Sobel and S. Fujii, Circulation, 92, 767 (1995).

53. H. Wada, Y. Mori, T. Kaneko, Y. Wakita, T. Nakase, K Minamikawa, M. Ohiwa, S. Tamaki, M. Tanigawa, S. Kageyama, K. Deguchi, T. Nakano, S. Shirakawa and K Suzuki Am. J. Hematol., 44, 112 (1993).

54. H. Pappot, H. Gårdsvoll, J. Rømer, A. Navrsted Pederson, J. Grøndahl-Hansen, C. Pyke and N. Brünner, Biol. Chem. Hoppe-Seyler, 376, 259 (1995).

55. X.-R. Peng, A.J.W. Hsueh and T. Ny, Eur. J. Biochem., 214, 147 (1993).

56. J.J. Emeis, R. Hoekzema and A.F. de Vos, Blood, 85, 115 (1995).

57. J.S. Friedland, Clin. Sci. 88, 393 (1995).

58. M.F. Oliver, Lancet, 346, 1378 (1995).

59. R.L. Hamilton, Trends Cardiovasc. Med., 4, 131 (1994).

60. S.L. Hoffmann, D.L. Eaton, M.S. Brown, W.J. McConathy, J.L. Goldstein and R.E. Hammer, J. Clin. Invest. 85, 1542 (1990).

61. R. Fears, H. Ferres and A.W.R. Tyrell in "Pharmacolgical Control of Hypermipidemia", Prous Science Publishers 1986, p.353.

62. J.A. Berliner, M. Navab, A.M. Fogelman, J.S. Frank, L.L. Demer, P.A. Edwards, A.D. Watson and A.J. Lusis, Circulation, 91, 2488 (1995).

63. P. Holvoet and D. Collen, FASEB J., 8, 1279 (1994).

64. V.N. Trieu and W.J. McConathy, J. Biol. Chem., 270, 15471 (1995).

65. N. Mackman, FASEB J., 9, 883 (1995).

66. P.M. Dennington and M.C. Berndt, Clin. Exp. Pharmacol. Physiol., 21, 349 (1994).

67. N.A. Nelkin, S.J. Soifer, J. O'Keefe, T.K.H. Vu, I.F. Charo and S.R. Coughlin, J. Clin. Invest., 90, 1614 (1992).

68. C.T. Esmon and H.P. Schwarz, Trends Cardiovasc. Med., 5, 141 (1995).

69. C.S. Gibbs, S.E. Coutre, M. Tsiang, W.X. Li, A.K. Jain, K.E. Dunn, V.S. Law, C.T. Mao, S.Y. Matsumura, S.J. Mejza, L.R. Paborsky and L.L.K. Leung, Nature, 378, 413 (1995).

70. S.R. Coughlin, Trends Cardiovasc. Med., 4, 77 (1994).

71. M.D. Hollenberg, Trends Pharmacol. Sci., 17, 3 (1996).

72. T.K.H. Vu, D.T. Hung, V.I. Wheaton and S.R. Coughlin, Cell, 64, 1057 (1991).

73. H. Sawa, B.E. Sobel and S. Fujii, J. Biol. Chem., 269, 14149 (1994).

74. T. Abrahamsson, P. Bjorquist, J. Deinum, J. Ehnebom, M. Hulander, A. Legnehed, C. Matsson, V. Nerme, A. Westin-Eriksson and B. Åkerblom, Fibrinolysis, 8 (Suppl. 2), 55 (1994).

75. T. Abrahamsson, V. Nerme, M. Strömqvist, B. Åkerblom, A. Legnehed, K. Pettersson and A. Westin Eriksson, Thromb. Haemost., 75, 118 (1996).

76. M. Levi, B.J. Biemond, A. van Zonneveld, J.W. ten Cate and H. Pannekoek, Circulation, 85, 305 (1992).

77. B.J. Biemond, M. Levi, R. Coronel, M.J. Janse, J.W. ten Cate and H. Pannekoek, Circulation, 91, 1175 (1995).

78. P. Charlton, R. Faint, F. Bent, J. Bryans, I. Chicarelli-Robinson, I. Mackie, S. Machin and P. Bevan, Thromb. Haemost., 75, in press (1996).

79. P. Charlton, C. Barnes, F. Bent, J. Bryans, R. Faint, A. Folkes, S. Machin, I. Mackie and P. Bevan, Circulation, 92, 486 (Abstract), (1995).

80. D.T. Eitzman, W.P. Fay, D.A. Lawrence, A.M. Francis-Chmura, J.D. Shore, S.T. Olson and D. Ginsburg, J. Clin. Invest., 95, 2416 (1995).

81. M. van Meijer and H. Pannekeok, Fibrinolysis, 9, 263 (1995).

Chapter 12. Neurokinin Receptor Antagonists

Christopher J Swain and Richard J Hargreaves

Merck, Sharp and Dohme Research Laboratories, Neuroscience Research Centre, Terlings Park, Harlow, Essex. CM20 2QR

Introduction Three extensive reviews have been published which cover much of the current literature on receptor distribution, *in vitro* and *in vivo* pharmacology and potential clinical utilities of neurokinin (NK) receptor antagonists that include analgesia, anxiety, arthritis, asthma, emesis, migraine and schizophrenia (1-3). Receptors for tachykinins are currently divided into three main populations. The preferred mammalian tachykinin agonist for each of these sites is substance P (SP) for NK_1, neurokinin A for NK_2 and neurokinin B for NK_3. All three human receptors have been cloned and expressed in cell lines facilitating rapid screening of novel agents for receptor selectivity. Substance P and the related tachykinins neurokinin A and neurokinin B are mainly found in neurons particularly in small diameter sensory somatic and visceral fibres, in enteric sensory neurons and in a number of pathways within the brain. The release of tachykinins from the peripheral ends of these neurons may play an important role in the neurogenic inflammatory responses to local injury and inflammation by promoting the release of histamine from mast cells and the release of cytokines from invading white cells as well as acting directly upon blood vessels to produce vasodilation and plasma extravasation. Neurogenic inflammation within the dura mater, a pain-producing intracranial tissue, has been hypothesised to be the source of migraine headache pain since it is inhibited by current clinically effective anti-migraine agents. Within the CNS at the central terminals of primary sensory neurons in the brainstem cranial nerve nuclei and spinal cord dorsal horn substance P may function as a sensory neurotransmitter or neuromodulator particularly with regard to the central input to pain pathways from unmyelinated C-fibres involved in nociception. This central antinociceptive site of action may be critical to the speed of onset and the effectiveness of NK_1 antagonists for the acute treatment of migraine since they would be able to block headache pain rapidly and directly. Substance P containing afferent nerve fibers also innervate the brainstem in the region of the nucleus tractus solitarius, a CNS area involved in the control of emesis suggesting a role for central NK_1 receptor antagonists as antiemetics. It has also been hypothesised that tachykinins within the CNS may have a role in modulating the activity of norepinephrine and dopamine pathways that are involved in affective disorders and responses to stress.

NK_2 receptor antagonists have been claimed to have anxiolytic properties in preclinical assays. However, until very recently it had not been possible to demonstrate the existence of NK_2 receptors in brain tissue from adult rats (although present in neonatal tissue) using radioligand binding techniques (4). In December 1995, new autoradiographic evidence for the existence of NK_2 receptors in adult rat brain using localisation of a radiolabelled NK_2 receptor selective antagonist, [^{3}H]-SR48968 was presented (5). A low abundance of receptors (< 30% specific binding) was detected in basal ganglia structures. Early studies using [^{125}I]Bolton-Hunter labeled neurokinin A, the preferred endogenous ligand at NK_2 receptors, did not detect the presence of NK_2 receptors in *human* brain (6) but the properties of this ligand for

111

autoradiography are suboptimal compared to [^{3}H]-SR48968 and so the studies need to be repeated.

NK$_3$ receptors have been shown to modulate central monoamine function suggesting that agonists or antagonists may be clinically useful for the treatment of schizophrenia, Parkinson's disease and depression. NK$_3$ receptors have been shown to modulate the activity of dopaminergic cell bodies in the ventral tegmental area and substantia nigra in rodents suggesting that an NK$_3$ receptor *antagonist* might exhibit antipsychotic activity (7). On the other hand, central infusion of the peptide NK$_3$ receptor agonist senktide in rats elicits the 5-HT behavioural syndrome (8) and increases release of norepinephrine suggesting that NK$_3$ *agonists* may have utility as antidepressants.

NK$_1$ Receptor Selective Non-peptide Antagonists- Pfizer discovered the first non-peptide NK$_1$ receptor antagonist (CP-96,345; **1**) in 1991 (9). An investigation of the structure-activity relationships within this series of quinuclidines highlighted the crucial importance of the *ortho*-methoxy substitution of the benzylamine side-chain (10). Whilst **1** was a potent antagonist of substance P *in vitro* and *in vivo* it also had significant affinity for a number of ion channels which compromised its use *in vivo* (11).

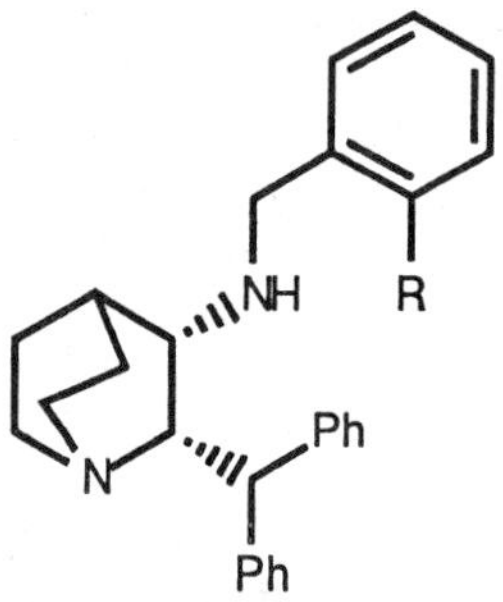

	R	IC$_{50}$ (nM)
1	OMe	0.77
2	H	16
3	Cl	33
4	Et	17

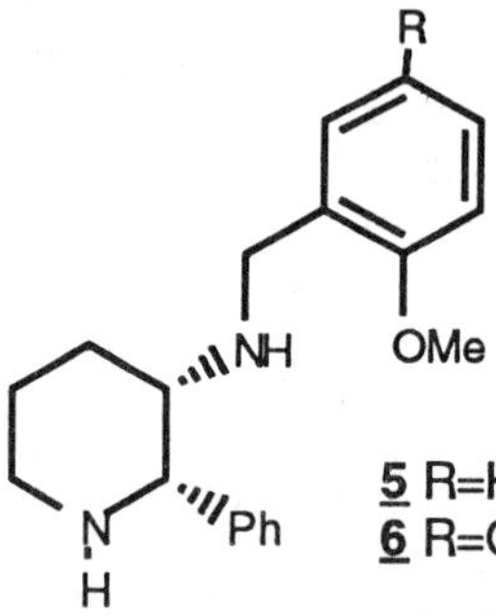

5 R=H
6 R=OCF3

More recently the piperidine analogue (CP-99,994; **5**) was reported which *in vitro* has excellent affinity and selectivity [human NK$_1$ receptor (hNK$_1$), human NK$_2$ receptor (hNK$_2$), human NK$_3$ receptor (hNK$_3$)] (hNK$_1$ IC$_{50}$ 0.6 nM; hNK$_2$ IC$_{50}$ >1000 nM; hNK$_3$ IC$_{50}$ >1000 nM) and considerably reduced ion channel activity compared to **1** (12). *In vivo* CP-99,994 showed potent activity after i.v. administration but was poorly active after oral administration (13). The clinical trials with **5** have not yet been published in full. To date two abstracts have described firstly the lack of activity in preventing pain associated with peripheral neuropathy, a condition with a highly heterogeneous patient population and that is currently very poorly controlled (14), and secondly in the treatment of asthma (15). Clinical studies with CP-99,994 did however appear to be complicated by its lack of oral bioavailability and consequent need to administer the test compound by the intravenous route where the dose regime appears to have been somewhat limited. The poor oral bioavailability of **5** would now appear to have been solved with the discovery of (CP-122,721; **6**), which is a potent and selective NK$_1$ antagonist (hNK$_1$ IC$_{50}$ 0.14 nM) but has significant affinity for the L-type calcium channel (390 nM) (15). Compound **6** was active in assays of dermal extravasation (0.01-0.05 mg/kg p.o.), substance P induced locomotor hyperactivity (0.2 mg/kg) and against cisplatin induced emesis (0.3 mg/kg i.v.). Glaxo postulated that the 5-position of the aromatic ring, *para* to the methoxy, was a likely site for hydroxylation. Introduction of a variety of substituents into the 5-position resulted in the discovery of the tetrazole (GR 203,040 ; **7**) which had an improved *in vivo* profile over CP-99,994 (13). Compound **7** has excellent *in vitro* affinity (hNK$_1$ IC$_{50}$ 0.06nM) and good oral bioavailability in dogs (76%). In ferrets **7** (0.1 mg/kg s.c.) significantly reduced emesis induced by a wide range of emetogens including X-irradiation, cisplatin and morphine. Interestingly, while the isomeric tetrazole **8** and the pyridine **9** afford excellent affinity they have reduced *in vivo* potency (13). Compound **7** was initially selected for clinical evaluation in emesis and migraine but has now been replaced by GR 205,171 **10** (17). The tetrazole has also been utilised in the corresponding benzofuran analogue **11** which is reported to be active against radiation induced emesis in ferrets (0.3 mg/kg s.c.) (18).

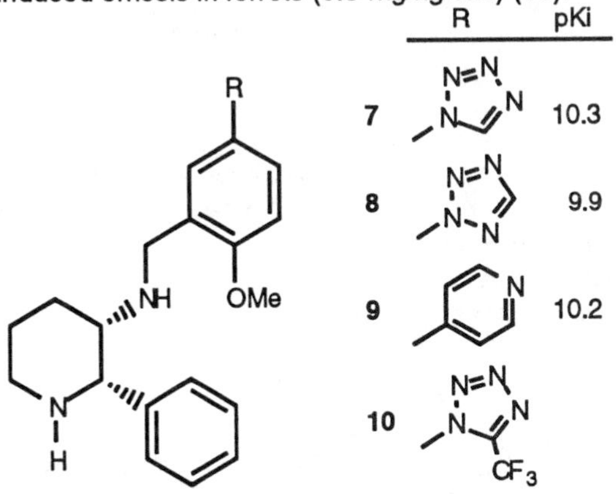

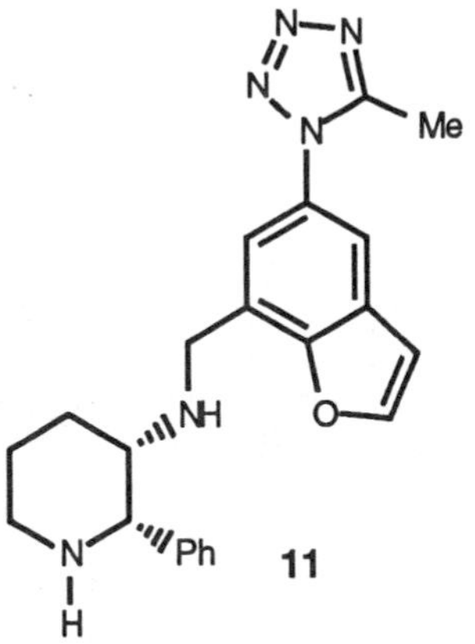

	R	pKi
7	$N{\equiv}N$ ring	10.3
8	$N{\equiv}N$ ring	9.9
9	pyridine	10.2
10	$N{\equiv}N$ ring, CF_3	

Since the first publication disclosing a novel series of amino ether based NK_1 antagonists **12**, the structure activity in this series together with a summary of the mutagenesis studies that have helped to define the pharmacophore have been reviewed (19).

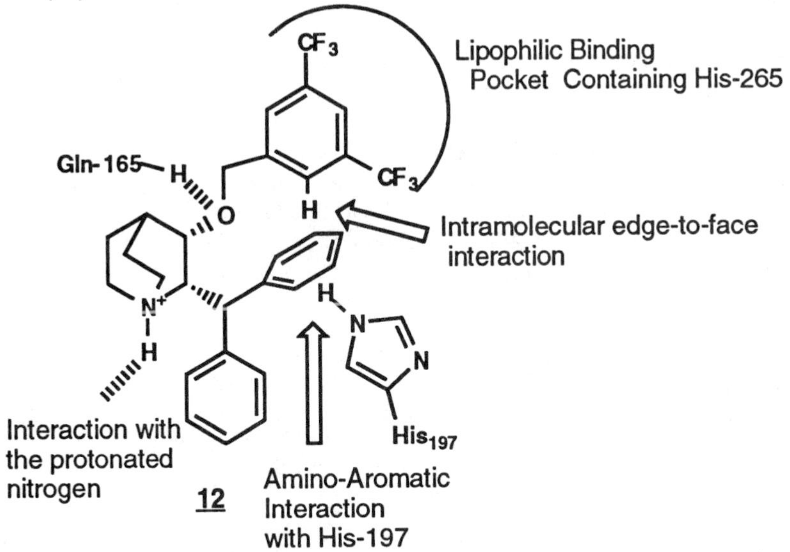

CF₃

Lipophilic Binding Pocket Containing His-265

Gln-165

Intramolecular edge-to-face interaction

Interaction with the protonated nitrogen

12

Amino-Aromatic Interaction with His-197

His₁₉₇

The 3,5-bis(trifluoromethyl)benzyl ether is also present in the N-substituted 2-phenyl-3-benzyloxypiperidine series, of note are the triazole **13** and the triazolinone **14** (20). Both compounds have excellent affinity (**13**, hNK_1 IC_{50} 0.18 nM; **14**, hNK_1 IC_{50} 0.1 nM), and block the extravasation induced by substance P (0.5 pmol) injected into the dorsal skin in the guinea-pig (**14**, ID_{50} 0.007 mg/kg p.o.). They are also active in the trigeminal dural extravasation model of migraine (10-100 µg/kg) and block cisplatin induced emesis in ferrets (ID_{50} 0.3 mg/kg). The corresponding morpholine **15** has also been described and have similar activity (21).

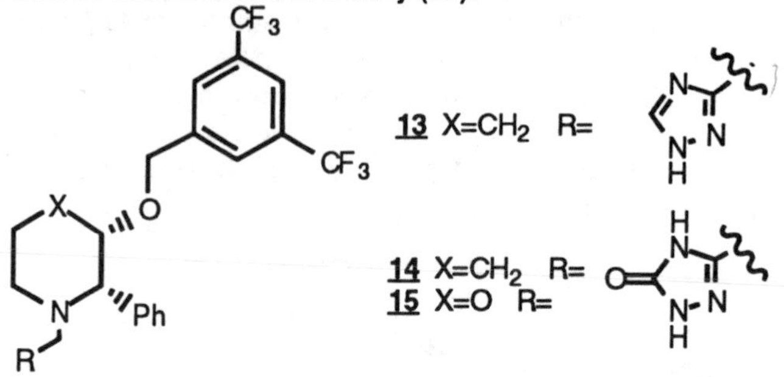

13 X=CH_2 R=

14 X=CH_2 R=
15 X=O R=

Fujisawa have replaced the cyclic peptide FK-224 previously in development for asthma with a highly modified tri-peptide FK-888 **16** because of its superior pharmacological profile (22). FK 888 inhibited substance P binding in guinea-pig lung with a Ki of 0.69 nM and antagonised the airway constriction induced by 10 nmol/kg substance P with an ED_{50} of 0.0032 mg/kg i.t. but was inactive p.o. (23).

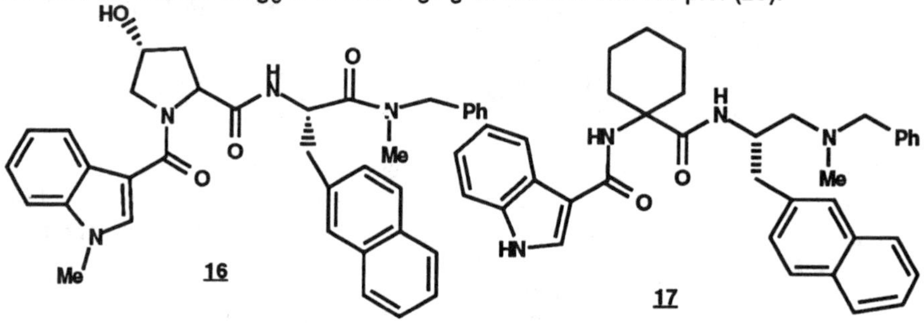

The structure activity relationships of a related series of compounds based on 1-aminocycloalkylcarboxylic acid residues leading to MEN 10930 **17** has been reported (24,25). This compound has good affinity for the NK_1 receptor (hNK_1 IC_{50} 1 nM) with good selectivity over the other receptors (hNK_2 IC_{50} 1 mM, hNK_3 IC_{50} 10 mM) as well as verapamil sensitive Ca^{2+} sites (K_i 1.5 mM). Since the discovery of the perhydroisoindole-based NK_1 antagonist **18** (RP 67580) that showed excellent affinity for the rat NK_1 receptor but 25-fold reduced affinity for the human NK1 receptor, continued exploration of this series has identified compounds with excellent affinity at the human NK_1 receptor (26).

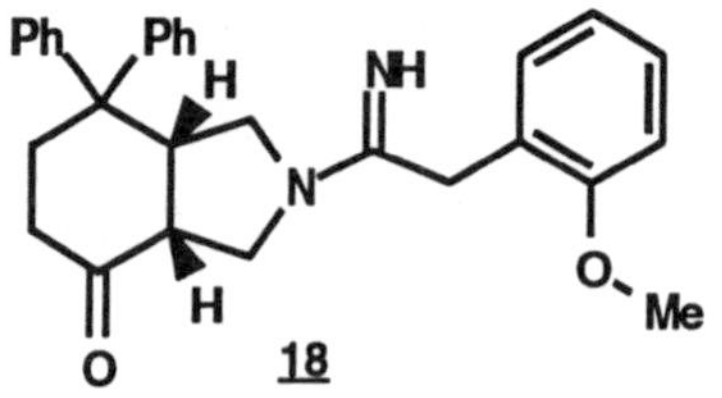

18

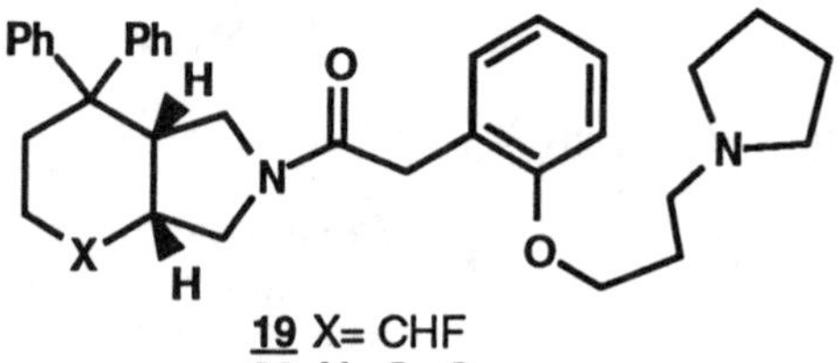

19 X= CHF
20 X= S=O

The key structural changes that confer increased affinity for the human NK_1 receptor appear to be modification of the ketone functionality and replacement of the amidine linking group by amide. The *ortho* methoxy has also been changed to an alkyl amino but this may be intended to improve water solubility . These compounds are claimed to be active in models of analgesia (**19** ED_{50} 3 mg/kg p.o. **20** ED_{50} 1.7 mg/kg p.o.) and inhibit septide induced extravasation (**19** ED_{50} 0.7 mg/kg s.c. **20** ED_{50} <0.1 mg/kg s.c.) (27). Further modifications of the ketone functionality have now been disclosed in which a fourth aryl ring has been introduced to afford the tertiary alcohol (RPR 100893; **21**) (INN erispant). Interestingly, it is now the opposite enantiomer that has high affinity. The *in vitro* and *in vivo* pharmacological profiles have been extensively evaluated (28,29). In IM9 cells **21** has an affinity for the hNK_1 receptor of 30 nM, whilst in rat brain the affinity is 1417 nM. Compound **21** was active in models of analgesia (formalin paw ED_{50} 3.1 mg/kg s.c. ED_{50} 11 mg/kg p.o.), inflammation (septide induced extravasation ED_{50} 30.1 mg/kg i.v. ED_{50} 0.33 mg/kg p.o.), and migraine (dural extravasation ED_{50} 2.5 ng/kg i.v. ED_{50} 0.5 mg/kg p.o.). Oral doses of 1, 5 and 20 mg (in an undisclosed specialised formulation) **21** was reported to be no better than placebo in Phase II clinical trials for the treatment of migraine. No plasma drug levels nor evidence of functional activity at the doses used were however disclosed. On the assumption that the doses used produced reasonable plasma drug levels, this lack of activity against migraine may be related to the poor brain penetrability of **21**. This result suggests that a peripheral NK_1 antagonist action alone may not be insufficient to give headache relief especially within the 2h time frame used for efficacy evaluation in clinical trials. Thus, NK_1 receptor antagonists may rely for their anti-migraine effects on inhibition of central nociceptive pathways in the trigeminal nuclei where headache pain could be blocked rapidly and directly (30). It is however noteworthy that in preclinical models these pathways can be inhibited only 50% by NK_1 receptor antagonists and should a greater level of blockade be required then even highly brain penetrant compounds may be ineffective against migraine (31).

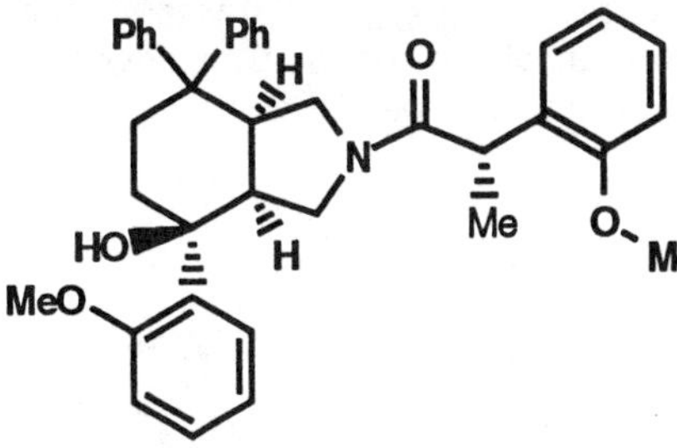

21

A series of 4-aminopiperidine amides have been disclosed, and a compound from this series (CGP-49823; **22**) which has relatively modest affinity (11nM) was reported to be active in the rat social interaction and forced swim tests for anxiolytic and antidepressant activity respectively (32,33). Compound **22** is reported to be in Phase I clinical trials as a potential treatment for anxiety disorders (34).

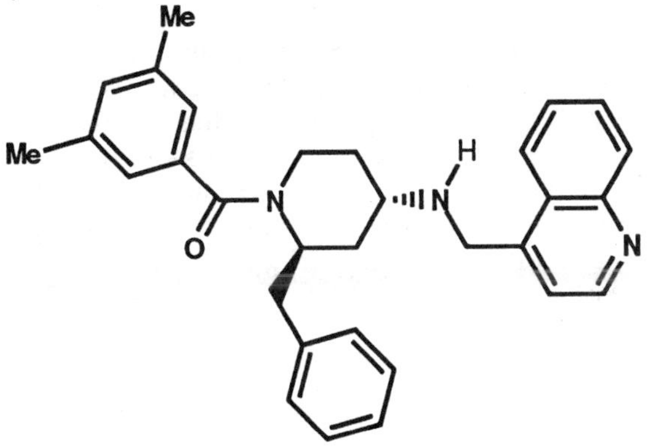

22

A new class of amide based NK$_1$ antagonists have been described, again bearing the 3,5-bis(trifluoromethyl) substitution (**23,24**). The detailed SAR in this series has recently been published (35). While a variety of benzyl substitutions gave excellent *in vitro* affinity (**24** hNK$_1$ IC$_{50}$ 0.21 nM) only the 3,5-bis(trifluoromethyl) yielded good oral activity in assays of capsaicin induced extravasation (**24** 0.017 mg/kg i.v., 0.068 mg/kg p.o compared to CP 99,994 **2** 0.017 mg/kg i.v., 8.7 mg/kg p.o.)

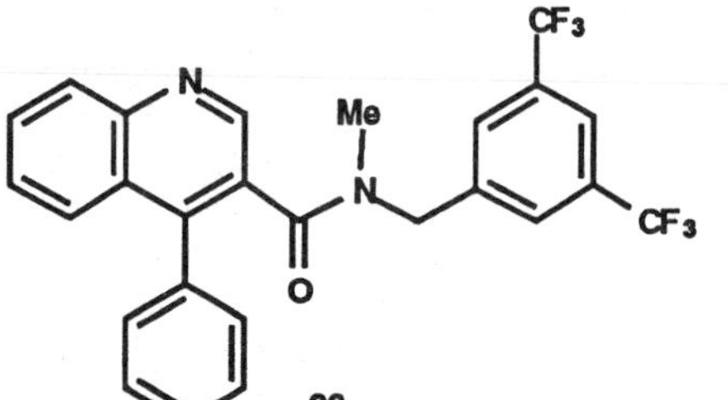

Directed screening of compound libraries has identified several new structural classes of NK$_1$ antagonists. The structure activity of a novel series of tryptophan esters **25** has been described in detail in which the 3,5-bis(trifluoromethyl) substituted ring again gives the highest NK$_1$ affinity and *in vivo* activity (36). Subsequent work showed that it was not the benzyl ether oxygen that was required for binding to the receptor but the carbonyl oxygen of the ester. Consequently the corresponding ketones, with a variety of groups linked to the nitrogen **26-29**, have now been claimed (37).

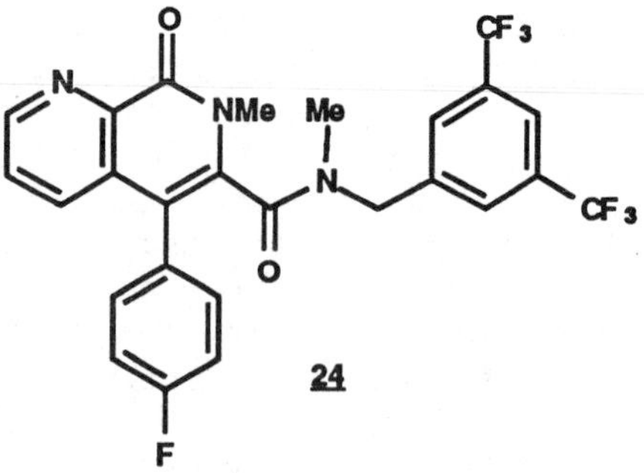

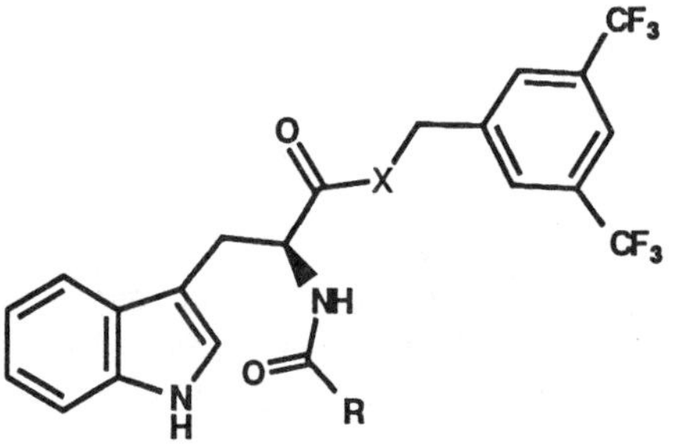

		hNK$_1$ IC$_{50}$ (nM)
25 R=Me	X=O	1.6
26 R= -(CH$_2$)$_3$NMe$_2$	X=CH$_2$	0.7
27 R= 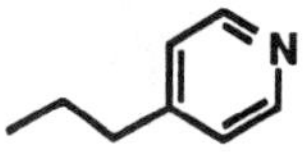	X=CH$_2$	0.47
28 R= 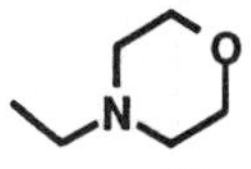	X=CH$_2$	0.17
29 R= 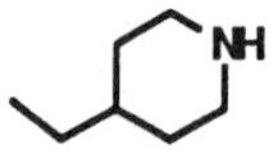	X=CH$_2$	0.37

The discovery and *in vivo* evaluation of (LY303870; **30**), a potent NK$_1$ receptor antagonist (hNK$_1$ 0.2 nM) with good selectivity (inactive vs. 70 other receptors and channels, including Ca^{2+} channels) has been reported (38). Using a new technique to assess dural extravasation in guinea-pigs, **30** was said to be equipotent with sumatriptan; however, the active dose range for sumatriptan (0.01-0.1 mg/kg i.v.) was considerably lower than that previously reported by other groups. LY303870 was more potent than sumatriptan when given orally (ID$_{50}$ 0.1 mg/kg vs 3 mg/kg), and given as an aerosol, **30** completely inhibited dural extravasation at 10 mg/kg.

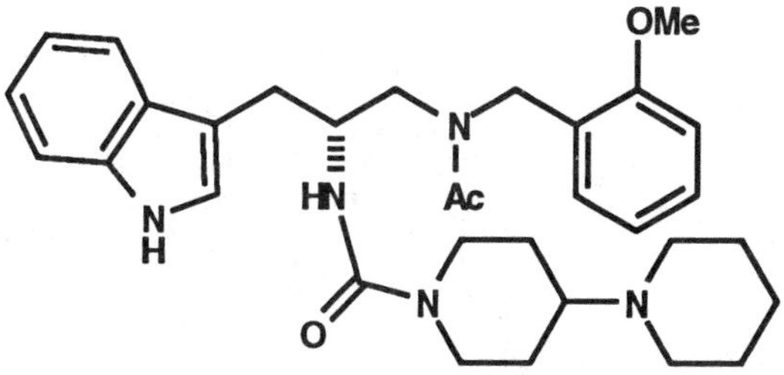

30

In contrast to the high activity of **30** in peripheral assays of NK$_1$ receptor function, very high doses were required to obtain activity in CNS assays. Doses of up to 30 mg/kg were required to inhibit NK$_1$ agonist induced hyperalgesia and formalin induced nociception in rats, and even higher doses (up to 60 mg/kg) were needed to inhibit amphetamine induced locomotor activity in mice. **30** has completed phase I studies for migraine and was well tolerated up to a 1g total oral dose in man

A novel N-acylated 3-(3,4-dichlorophenyl)piperidine (SR 140,333; **31**) that displays high affinity both for rat and human NK$_1$ receptors (rat NK$_1$ IC$_{50}$ 0.02 nM, human NK$_1$ IC$_{50}$ 0.01 nM) has been identified (39). This compound is structurally related to NK$_2$ and NK$_3$ antagonists that have also been disclosed by the same group but shows excellent selectivity for the NK$_1$ receptor. In the rat **31** given intravenously potently inhibited the plasma extravasation induced by sciatic nerve stimulation, mustard oil application and substance P (40). Compound **31** is reported to be in Phase I trials for inflammation and migraine but will have poor CNS availability due to its quaternary nitrogen moeity. The issue of receptor selectivity within this class of compounds is highlighted by (MDL 105212; **32**) reported by Marion Merrell Dow, a novel non-selective neurokinin receptor antagonist, (hNK$_1$ IC$_{50}$ 3.1 nM, hNK$_2$ IC$_{50}$ 8.4 nM, hNK$_3$ IC$_{50}$ 21 nM). MDL 105212 is being developed for the treatment of asthma (41).

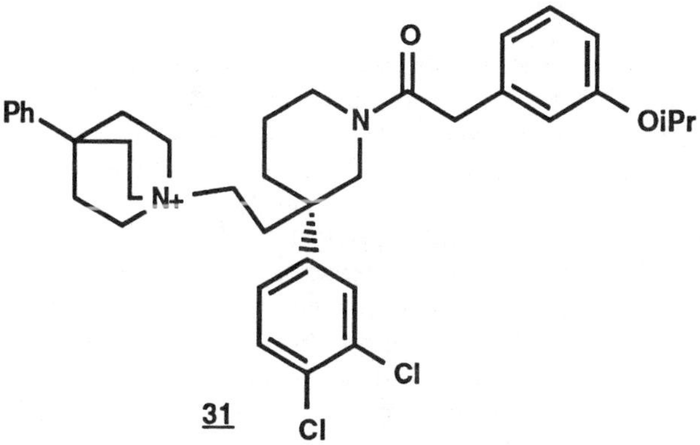

31

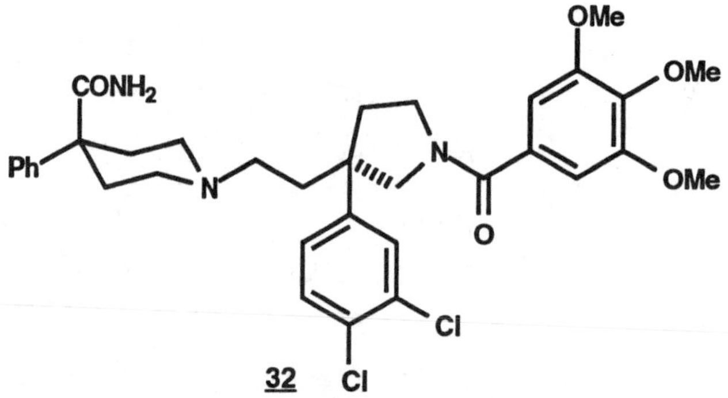

NK₂ Receptor Selective Non-peptide Antagonists- A novel class of NK₂ antagonists have been reported by Glaxo and an example of this class has been evaluated in some detail (42) (GR 159,897 **33**). This compound shows high affinity and good NK₂ receptor selectivity with the affinity of the R enantiomer of the sulphoxide being some 10 fold higher than the S enantiomer. From the assignment of NK₂ subtype selectivity based on the Menarini tissue assays this compound would be a NK₂b selective agent. *In vitro* the compound showed excellent activity rat colon pA₂=10, human bronchus=8.6, human ileum=9.5.

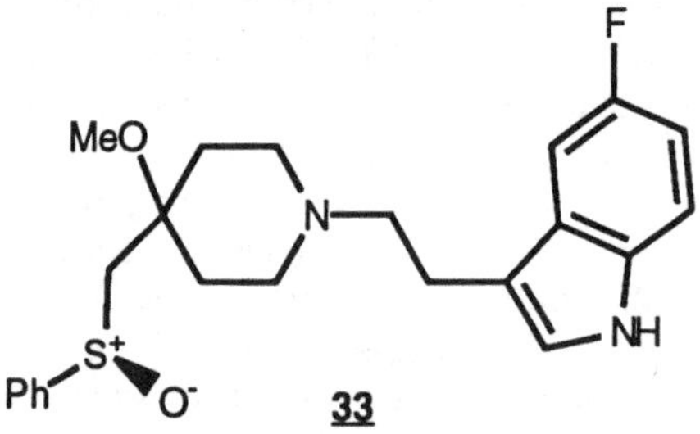

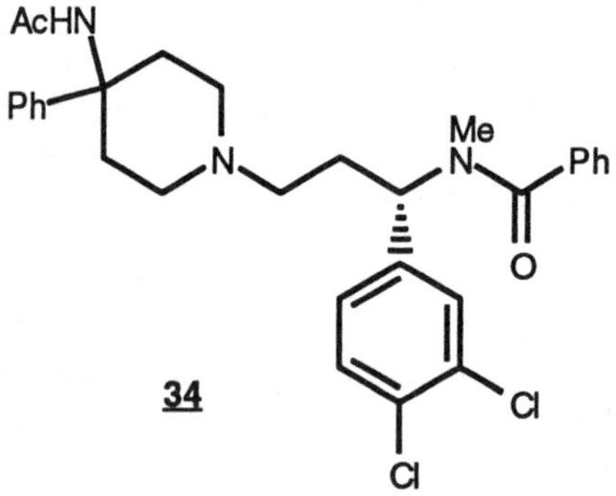

In vivo **33** (3 mmol/kg i.v.) blocks bronchoconstriction induced by the selective NK₂ agonist GR 64349 (43). Claims for anxiolytic activity with NK₂ receptor antagonists in preclinical assays were originally made by Glaxo (5). Glaxo have continued to assess the role of NK₂ receptors in anxiety paradigms and find that in the social interaction test injection of the NK₂ agonist GR-64349 (100pmol) into the dorsal raphe nucleus of the brain gave a reduction in activity under low light conditions that was taken to be predictive of an anxiogenic action. They have also shown in the elevated plus maze that **33**, given directly into the dorsal raphe nucleus of the brain at doses of 3-300 pmol, produced dose related anxiolysis with an effect as large as that produced by systemic benzodiazepines. The black and white box was also used to test for anxiolysis and in this test both **33** and the Sanofi NK₂ antagonist (SR 48968; **32**) (44) were active at systemic doses of <1ug/kg s.c. In the marmoset human threat test both **33** and **34** were anxiolytic at 50μg/kg sc, with a ceiling of effect 60% of that produced by a benzodiazepine. Independent support for the involvement of NK₂ receptors in anxiogenesis came recently from Calixto and colleagues (45) who showed that the NK₂ receptor antagonists SR 48968 and GR100679 had anxiolytic-like effects when administered either by direct central injection into the raphe nucleus, or sytemically at very low doses and across a wide dose range (0.02 to 200 μg/kg s.c.). The behavioural data with these NK₂ antagonists is however confusing since the putative anxiolytic dose range $(\mu g/kg)$ is considerably lower than that required to block NK_2 agonist-mediated effects in peripheral tissues (0.6 to 1.2 mg/kg) (46-48).

NK_3 Receptor Selective Non-peptide Antagonists The first non-peptide NK_3 antagonist reported is (SR 142,801; **35**) [49]. *In vitro* **35** has excellent affinity and receptor selectivity (hNK_3 0.21 nM) and *in vivo* shows CNS activity blocking the turning behaviour induced by intrastriatal injection of the the NK_3 selective agonist senktide in gerbils.

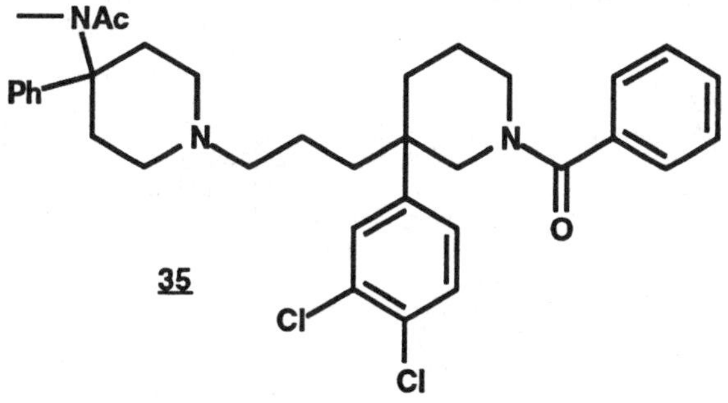

35

In conclusion it is fascinating to note the diversity of structural classes that have now been disclosed as high affinity NK_1 antagonists since the report of the first non-peptide antagonist. The outcome of clinical trials with these novel agents is now eagerly awaited and should be forthcoming in the near future.

References

1. C.A. Maggi, R. Patacchini and P. Rovero, and A. Giachetti, J. Auton. Pharmacol., 13, 23 (1993).
2. D. Regoli, A. Boudon, and J-L. Fauchere, Pharmacol Rev, 46(4), 551 (1994).
3. J. Longmore, C.J. Swain, and R. Hill, Drug News and Perspectives, 8(1), 5 (1995).
4. R.M. Hagan, I.J.M. Beresford, J. Stables, J. Dupere, C. M. Stubbs, P.J. Elliott, R.L.G. Sheldrick, A. Chollet, E. Kawashima, A.B. McElroy,and P. Ward, Regul. Pept ., 46, 9 (1993).
5. S.C. Stratton, I.J.M. Beresford, and R.M. Hagan, Br. J. Pharmacol., 112, 49P (1994).
6. M.M. Dietl, and J.M. Palacios, Brain Res ., 539, 211 (1991).
7. P.J. Elliott, G.S. Mason, M. Stephens-Smith, and R.M. Hagan, Neuropeptides, 19, 119 (1991).
8. A.J. Stoessl, C.T. Dourish,and S.D. Iversen, Br. J. Pharmacol., 94, 285 (1988).
9. R.M. Snider, J.W. Constantine, J.A. Lowe III, K.P. Longo, W.S. Lebel, H.A. Woody, E. Drozda, M.C. Desai, F.J. Vinick, R.W. Spencer, and H-J Hess, Science, 251, 435 (1991).
10. J.A. Lowe III, S.E. Drozda, R.M. Snider, K.P. Longo, S.H. Zorn, J. Morrone, E.R. Jackson, S. McLean, D.K. Bryce, J. Bordner, A. Nagahisa, Y. Kanai, O. Suga, and M. Tsuchiya, J. Med.Chem., 35, 2591 (1992).
11. M. Caeser, G.R. Seabrook, and J.A. Kemp, Brit. J. Pharmacol., 109, 918 (1993).
12. T. Rosen, T.F. Seegar, S. McLean, M.C. Desai, K.J. Guarino, D. Bryce, K. Pratt, and J. Heym, J.Med.Chem, 36, 3197 (1993).
13. P. Ward, D.R. Armour, D.E. Bays, B. Evans, G.M.P. Giblin, N. Heron, T. Hubbard, K. Liang, D. Middlemiss, J. Mordaunt, A. Naylor, N.A. Pegg, V. Vinader, S.P. Watson, C. Bountra, and D.C. Evans, J.Med.Chem., 38, 4985 (1995).
14. G.A. Suarez, T.L. Opfer-Gehring, D. MacLean, and P.A. Low, Neurology, 44 (supp2), A220 (1994).
15. J.V. Fahy, H.F. Wong, P. Geppetti, J.M. Reis, S.C. Harris, D.B. Maclean, and J.A. Nadel, Boushey, Amer. J. Resp. and Critical Care Med., 152, 879 (1995).
16. S. McLean, A. Fossa, S. Zorn,and T. Rosen, Presented at Tachykinins '95, Florence Oct 16-18 (1995).
17. C.J. Gardner, D.J. Twissell, P. Ward, Presented at British Pharmacology Society Meeting, Univ. of Leicester, 17-19 April P29 (1996).
18. A. Naylor,and B. Evans, WO 95/00536 (1995).

19. C.J. Swain, E.M. Seward, M.A. Cascieri, T. Fong, R. Herbert, D.E. MacIntyre, K. Merchant, S.N. Owen, A. P. Owens, V. Sabin, M. Teall, M.B. VanNiel, B.J. Williams, S. Sadowski, C. Strader, R. Ball, and R. Baker, J. Med. Chem., 38, 4793 (1995).

20. C.J. Swain, Presented at Tachykinins '95, Florence Oct 16-18 (1995).

21. J. Hale, S. Mills, M. MacCoss, S. Shah, H. Qi, D. Mathre, M. Cascieri, S. Sadowski, C. Strader, D. MacIntyre, and J. Metzger, Presented at Tachykinins '95, Florence Oct 16-18 (1995).

22. M. Ichinose, N Nakajima, T Takahashi, H. Yamauchi, H. Inoue, and T. Takishima, Lancet, 340, 1248 (1992).

23. M. Murai, Y. Maeda, M. Yamaoka, D. Hagiwara, H. Miyake, M. Matsuo, and T. Fujii, Regulatory Peptides, 46, 335 (1993).

24. A. Sisto, F. Centini, C.I.Fincham, P. Lombardi, E. Monteagudo, E. Potier, R. Terracciano, A. Giolitti, M. Venanzi, B. Pispisa, Presented at Tachykinins '95, Florence Oct 16-18 (1995).

25. M. Astolfi, C. Goso, M. Maggi, and S. Manzini, Presented at Tachykinins '95, Florence Oct 16-18 (1995).

26. C. Garret, A. Carruette, V. Fardin, S. Moussaoui, J-F. Peyronel, J-C. Blanchard, and P.M. Laduron, Proc. Natl. Acad. Sci. USA, 88, 10208 (1991).

27. D. Achard, A. Truchon, and J-F. Peyronel, Bioorg. and Med. Chem. Lett., 4, 669 (1994).

28. M. Tabart, and J-F. Peyronel, Bioorg. and Med. Chem. Lett., 4 (5) 673 (1994).

29. S. M. Moussaoui, F. Monttier, M.A. Carruette, V. Fardin, C. Floch, and C. Garret, Neuropeptides, 26, 35 (1994).

30. S.L. Shepheard, D.J. Williiamson, J. Williams, R.G. Hill, and R. Hargreaves, Neuropharmacology, 34, 255 (1995).

31. J.S.Polley, P.J. Gaskin, D.T.Beattie, H.E. Conner, Presented at British Pharmacology Society Meeting, Univ. of Leicester, 17-19 April P104 (1996).

32. N. Subramanian, C. Ruesch, G.P. Anderson, and W. Schilling, J. Physiol. Pharmacol., 72 Supp 1P (1994).

33. A. Vassout, M. Schaub, C. Getsch, S. Ofner, W. Schilling, and S. Veenstra, Neuropeptides, 26, S38 (1994).

34. Scrip No 2111, 7, March 15th (1996)

35.. H. Natsugari, Y. Ikeura, Y. Kiyota, Y. Ishichi, T. Ishimaru, O. Saga, H. Shirafuji, T. Tanaka, I. Kamo, T. Doi, and M. Otsuka, J. Med. Chem., 38, 3106-3120 (1995).

36. A.M. Macleod, K.J. Merchant, F. Brookfield, R. Lewis, F. Kelleher, G. Stevenson, A. Owens, C.J. Swain, R. Baker, M.A. Cascieri, S. Sadowski, E. Ber, D.E. MacIntyre, J. Metzger, and R Ball, J. Med. Chem., 36, 2044 (1993).

37. A.M. MacLeod, M.A. Cascieri, K.J. Merchant, S. Sadowski, S. Hardwicke, R.T. Lewis, D.E. MacIntyre, J.M. Metzger, T.M. Fong, S. Shepheard, F.D. Tattersall, R. Hargreaves, and R. Baker, J. Med. Chem., 38, 934 (1995).

38. B.D. Gitter, R.F. Bruns, J.J. Howbert, P.W. Stengel, D.R. Gehlert, S. Iyengar, D.O. Calligaro, D. Regoli, and P.A. Hipskind, Presented at Tachykinins '95, Florence Oct 16-18 (1995).

39. X. Emonds-Alt, J.D. Doutremepuich, M. Healulme, G. Neliat, V. Santucci, R. Steinberg, P. Vilain, D. Bichon, J.P. Ducoux, V. Proietto, D. van Brock, P. Soubrie, G. Le Fur, and J.C. Breliere, Eur J Pharmacol., 250, 403 (1993).

40. I. Juranek, and F. Lembeck, Proc. Br. Pharmacol., 13-16 Dec P189 (1994).

41. T.P. Burkholder, T-B. Le, G.D. Maynard, S.A. Shatzer, R.W. Knippenberg, and E. Kudlacz, Presented at Tachykinins '95, Florence Oct 16-18 (1995).

42. W.J. Cooper, H.S. Adams, R. Bell, P.M. Gore, A.B. McElroy, J.M. Pritchard, P.W. Smith, and P. Ward, Bioorg. and Med. Chem. Lett., 4, 1951 (1994).

43. D.I. Ball, G.P.A. Wren, Y.D. Pendry, J.R. Smith, L. Piggott, and R.L.G. Sheldrick, Neuropeptides, 24, 190 (1993).

44. J.D. Emonds-Alt Xdoutremepuich, M. Jung, E. Proietto, V. Santucci, D. Van Broeck, P. Vilian, P. Soubrie, G. LeFur, and J.C. Breliere, Neuropeptides, 24, 231 (1993).

45. R.M. Teixeira, A.R.S. Santos, G.A. Rae, J.B. Calixto, and T.C.M. de Lima, Presented at Tachykinins '95 Florence Oct 16-18 (1995).

46. C-C. Chan, C. Tousignant, E. Ho, C. Brideau, C. Savoie, and I.W. Rodger, Can. J. Physiol. Pharmacol., 72, 11 (1993).

47. C. Tousignant, C-C. Chan, D. Guevremont, C. Brideau, J.J. Hale, M. MacCoss, and I.W. Rodger, Br. J. Pharmacol., 108, 383 (1993).

48. D.I. Ball, I.J.M. Beresford, G.P.A. Wren, Y.D. Pendry, R.L.G. Sheldrick, D.M. Wask, M.P. Turpin, R.M. Hagan, and R.A. Coleman, Br. J. Pharmacol., 112, 48P (1994).

49. X. Emonds-Alt, D. Bichon, J.P. Ducoux, M. Heaulme, B. Miloux, M. Poncelet, V. Proietto, D. Van Broeck, P. Vilain, G. Neliat, P. Soubrie, G. Le Fur, and J.C. Breliere, SR 142,801, Life Sciences, 56, (1), 27 (1995).

SECTION III. CANCER AND INFECTIOUS DISEASES

Editor: Jacob J. Plattner
Abbott Laboratories, Abbott Park, IL 60064

Chapter 13. Antibacterial Agents

Burton H. Jaynes*, John P. Dirlam* and Scott J. Hecker†
*Pfizer Inc, Central Research Division, Groton, CT 06340
†Microcide Pharmaceuticals Inc., 850 Maude Ave., Mountain View, CA 94043

<u>Introduction</u> - There is a continuing need for new antibacterial agents as infectious bacteria have proven to be incredibly adaptable in their fight for survival (1). Bacteria are mutating and evolving in ways that make them resistant in commonly occurring bacterial infections (2), and better drugs are needed for emerging diseases (3), such as those caused by invasive (so-called flesh eating) streptococci, as well as drug resistant enterococci and *Mycobacteria* (see Chapters 14 and 17, respectively). To meet this challenge, research has focused on finding 1) new analogs of existing classes of antibacterial agents, 2) novel classes lacking cross-resistance, and 3) novel antibacterial screening targets. This review summarizes important developments of the past year related to ß-lactams, macrolides, quinolones, oxazolidinones and tetracyclines, in addition to several less-developed or new areas that offer exciting potential including novel screening targets.

<u>ß-Lactams</u> - Recently, ß-lactam permeation and transport mechanisms have been reviewed (4). In the cephalosporin area, a review has appeared describing pharmaco-kinetic studies of newer compounds (5) and new information has appeared on agents previously described in *Annu. Rep. Med. Chem.*, including S-1090 (6), E1100 (7), "dual-action" cephalosporin-quinolone conjugates (8), RU-59863 (9) and 2-oxaisocephems (10).

Among anti-MRSA cephalosporins, the structure and biological profiling of previously reported (11) TOC-39 (**1**) has appeared (12). This compound displays the best activity (MIC$_{90}$ = 3.13 µg/ml) against methicillin-resistant *S. aureus* (MRSA) of a series of compounds resulting from an exploration of SAR surrounding the ß-lactam skeleton (cephem, 2-thioisocephem, 2-oxaisocephem), the alkoxyimino group on the 7-substituent, and the linking group between the pyridiniumthio moiety and the cephem skeleton (13,14). The discovery and biological profiling of MC-02,306 (**2**), a new cephem containing a novel solubilizing 7-substituent, was described (15). This agent displays good activity against both MRSA (MIC$_{90}$ = 8 µg/ml) and enterococci, including strains of *E. faecium* showing high-level resistance to ampicillin (MIC$_{90}$ = 16 µg/ml). Efficacy *in vivo* against MRSA in mouse models was demonstrated for **1** and **2**.

New broad spectrum cephalosporins have also been described. The product of an SAR study on 3-vinyl cephems, Ro 25-6833 (**3**) displays good potency *in vitro* against methicillin-susceptible *S. aureus* (MSSA), *E. faecalis* and *E. coli* (MIC$_{90}$ = 1, 2 and 0.5 µg/ml, respectively), as well as a long half-life in several species. The elimination half-life in man was 3.8 hours (16). The SAR and activity surrounding CP6679 (**4**) was reported (17). This agent displays modest activity against MSSA, MRSA, penicillin-resistant *S. pneumoniae*, *E. coli* and *P. aeruginosa* (MIC$_{90}$ = 3.13, 12.5, 1.56, 0.1 and 12.5 µg/ml, respectively), but lesser activity against *E. faecalis* and imipenem-resistant *P. aeruginosa* (MIC$_{90}$ = 25 µg/ml for each).

Reports of other broad spectrum cephalosporins where less data are provided (no MIC_{90}s) include: 3-acryloxymethyl cephems, such as **5**, designed to be more stable to metabolic esterase hydrolysis than the corresponding 3-acetoxy compounds (18); a series of benzotriazolium cephalosporins, exemplified by **6** (19); a series of 3'-quaternary ammonium cephems, such as **7**, which are structurally related to previously reported FK-037 (20); and a group of benzothiopyranylthiomethyl cephems (21), exemplified by **8**.

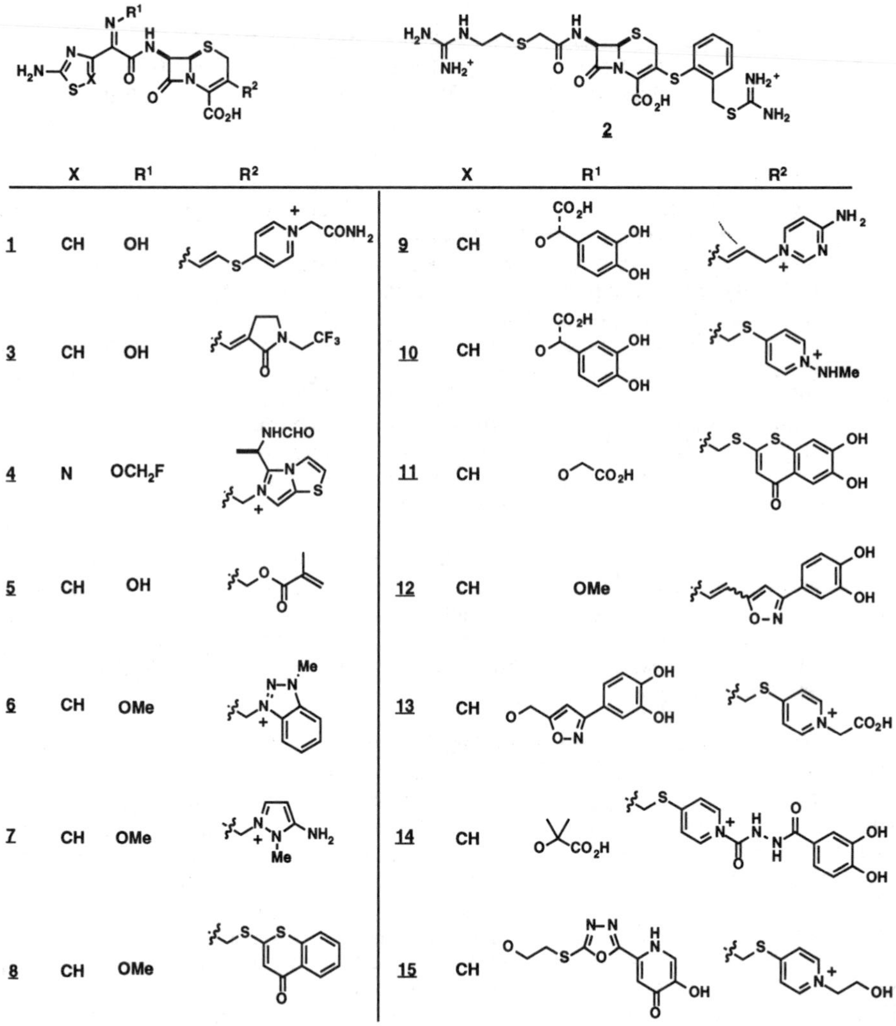

	X	R¹	R²
1	CH	OH	[structure: vinylthio-pyridinium-CONH₂]
3	CH	OH	[structure: CF₃-pyrrolidinone]
4	N	OCH₂F	[structure: NHCHO thiazolium]
5	CH	OH	[structure: O-methacryloyl]
6	CH	OMe	[structure: Me benzotriazolium]
7	CH	OMe	[structure: Me, NH₂ pyrazolium]
8	CH	OMe	[structure: benzothiopyranylthio]

	X	R¹	R²
9	CH	[structure: CO₂H catechol-O]	[structure: allyl pyrimidinium NH₂]
10	CH	[structure: CO₂H catechol-O]	[structure: thio-pyridinium-NHMe]
11	CH	[structure: O–CO₂H]	[structure: dithio benzothiopyranyl OH, OH]
12	CH	OMe	[structure: isoxazole catechol OH, OH]
13	CH	[structure: isoxazole catechol OH, OH, O]	[structure: thio-pyridinium-CO₂H]
14	CH	[structure: O-C(CH₃)₂-CO₂H]	[structure: thio-pyridinium-N-NH-CO-catechol OH, OH]
15	CH	[structure: O-ethylthio-oxadiazole-pyridone OH]	[structure: thio-pyridinium-OH]

Several new examples of cephalosporins containing a catechol moiety to enhance uptake into Gram-negative bacteria have appeared. Two of these reports contain results of extensive evaluation *in vitro*. LB10522 (**9**) displays excellent potency (22) against *P. aeruginosa*, *E. coli* and MSSA (MIC_{90}s of 2, 0.25 and 0.5 µg/ml, respectively). A study that varied the oxyimino substituent in cephems containing a 3-methylaminopyridiniumthiomethyl group resulted in the discovery of BRL 57342 (**10**), which also displays excellent potency (23) against *P. aeruginosa*, *E. coli* and MSSA (MIC_{90}s of 4, 0.06 and 0.25 µg/ml, respectively). Both **9** and **10** show excellent β-lactamase stability, efficacy *in vivo*, and long half-lives.

Reports of other catechol-substituted cephalosporins where less data are provided (no MIC_{90}s) include: catechol versions (24) of compound **8** above, such as AM-1647 (**11**); isoxazole-substituted 3-vinyl cephems (25), such as **12**; oxyimino-substituted cephems (26) exemplified by **13**; 3-pyridiniumthio cephems (27) such as **14**; and hydroxypyridone derivatives (28), represented by **15**.

A review of activity and resistance in the carbapenem arena has appeared (29). SM-17466 (**16**) contains the 2-mercaptothiazole substituent (30) that previously afforded impressive anti-MRSA potency when appended to the cephem (31) and carbacephem (32) nuclei. This agent displays compelling activity *in vitro* against MRSA, penicillin-resistant *S. pneumoniae*, *E. faecalis* and *E. faecium* (MIC_{90}s of 3.13, 0.1, 3.13 and 12.5 μg/ml, respectively), as well as activity against MRSA *in vivo*. An independent report of an SAR study surrounding 2-mercaptobenzothiazolyl carbapenems, exemplified by **17**, has also appeared (33). New reports detailing the *in vitro* activity of previously reported L-695,256 (34) have appeared (35-37). In addition, two new series of carbapenems have been reported which are structurally related to this agent, represented by biphenyl analog **18** (38) and carbolinyl derivative **19** (39).

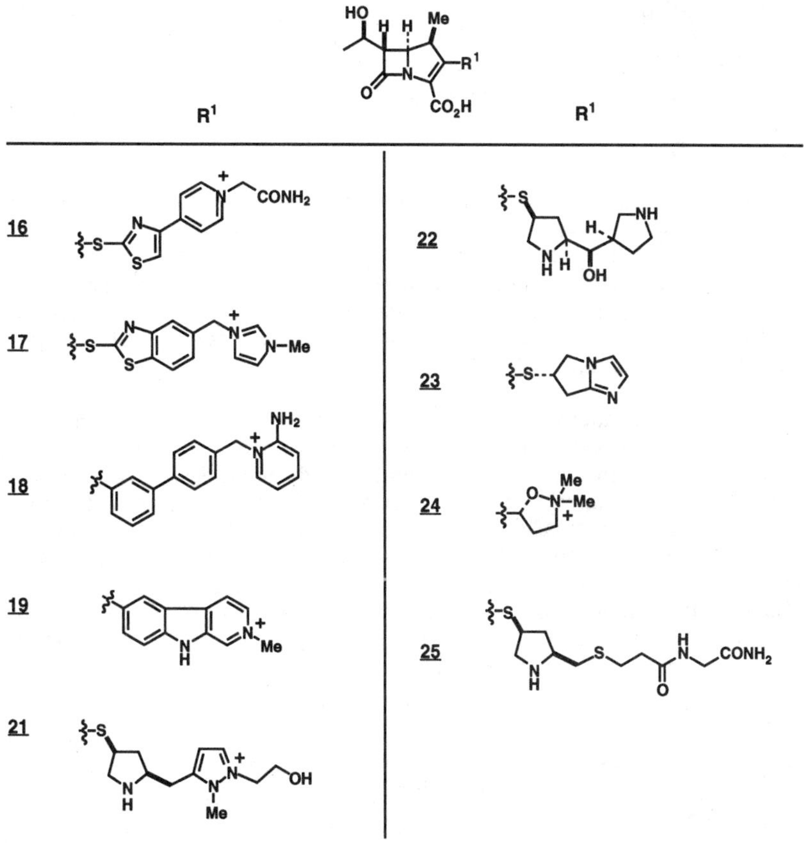

New carbapenems have also been reported with potent antipseudomonal activity. BMS-181139 (**20**) compares favorably (40,41) with imipenem and mero-penem against *P. aeruginosa* and *E. coli*, and has useful activity against MSSA and *E. faecalis* (MIC_{90}s of 2, 0.03, 1 and 8 μg/ml, respectively). In addition, this agent is 30-fold more stable than imipenem and 7-fold more stable than meropenem to hydrolysis by hog kidney dehydropeptidase. FR-21818 (**21**) displays MIC_{90}s of 3.13, 0.1, 0.1 and 12.5 μg/ml against *P. aeruginosa*, *E. coli*, MSSA and *E. faecalis*, respectively. Efficacy against *P. aeruginosa* has been demonstrated *in vivo,* as well as low acute toxicity and acceptable pharmacokinetics (41). The discovery and bioprofiling of ER-35786 (**22**) has been reported (42). This new carbapenem displays good activity against *P. aeruginosa* (MIC$_{90}$ = 6.25 µg/ml), even against imipenem- and meropenem-resistant strains. MIC$_{90}$s against *E. coli*, MSSA and *E. faecalis* are 0.05, 0.1 and 12.5, respectively. Efficacy *in vivo,* low acute toxicity, and acceptable pharmacokinetics were also demonstrated.

The discovery and bioprofiling of DU-6681 (**23**) and its orally-bioavailable pivaloyloxymethyl ester prodrug DZ-2640 were described (43). Compound **23** has potent broad spectrum activity (MIC$_{90}$s against MSSA, penicillin-resistant *S. pneumoniae*, and *E. coli* of 0.1, 0.2 and 0.025 µg/ml, respectively), but weak activity against *P. aeruginosa* (MIC$_{90}$ = 50 µg/ml). Oral bioavailability of 33% was observed in rats. Reports on new carbapenems in which less data are provided (no MIC$_{90}$s) include isoxazolidinium derivatives (44) modeled after thienamycin, such as **24**, and new carbamoylpyrrolidinylthio carbapenems (45), exemplified by **25**.

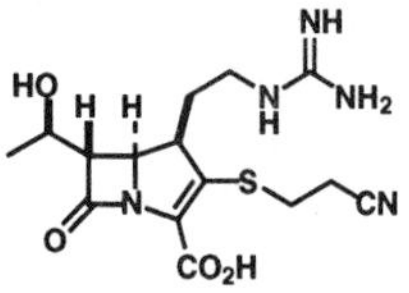

20	**26**	**27**

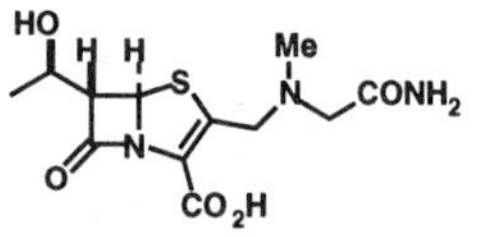

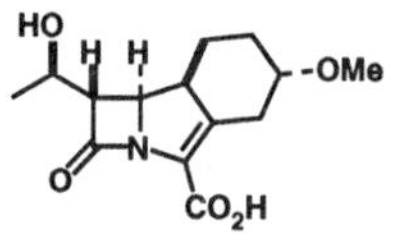

In other classes, reviews covering the utility of penems (46) and tribactams (47) have appeared. Impressive potency has been observed (48) with new penem Men 10700 (**26**), showing *in vitro* activity equal to or better than imipenem against many isolates. In particular, this compound displays exceptional activity against MRSA, with an MIC$_{90}$ of 4 µg/ml; no *in vivo* data were provided. The synthesis and *in vitro* evaluation of 5a-methoxy-tribactam **27**, which shows moderate broad spectrum activity, have also been reported (49).

New reviews concerning ß-lactamases and their inhibitors have appeared (50, 51). An approach to substituted derivatives of tazobactam yielded the compound **28**, which exhibits excellent activity in combination with piperacillin against organisms producing plasmid-mediated enzymes. However, it has insufficient activity against those producing chromosomally mediated class I enzymes (52). The preparation and evaluation of a series of bridged ß-lactams, including Ro 48-1256 (**29**), resulted in compounds capable of rendering susceptible the majority of imipenem-resistant *P. aeruginosa* isolates tested (53). The synthesis of a set of 7-alkylidene cephems produced analogs such as **30**, which displays activity superior to tazobactam against enzymes derived from *E. cloacae* (54,55).

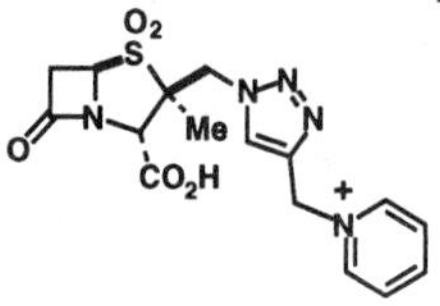

28	**29**	**30**	**31**	**32**

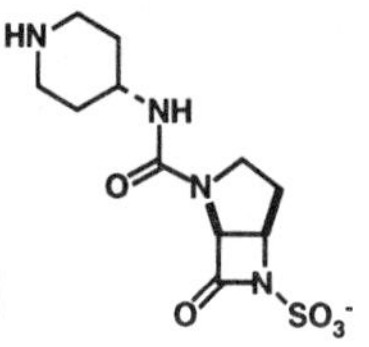

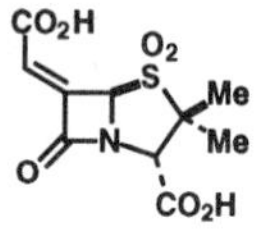

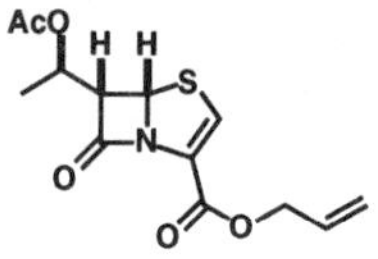

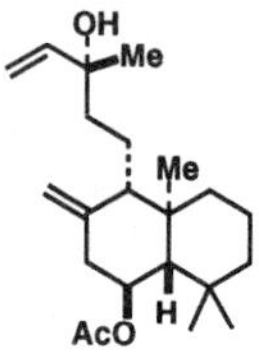

With regard to novel mechanisms of action in the ß-lactam area, penem **31** was found to be a potent inhibitor of *E. coli* signal peptidase (56,57). Also, two new agents were reported to abolish PBP 2a-mediated resistance in MRSA. MC-207,252 and MC-

200,616 (**32**) potentiate the activity of a range of ß-lactams against MRSA, reducing the MIC_{90} of imipenem from 64 to 0.125 µg/ml by addition of 10 µg/ml of the potentiator. Specificity of action against the resistance pathway is evidenced by the fact that no potentiation is observed against MSSA or other organisms, nor is it observed with non-ß-lactam antibiotics (58). Activity *in vivo* against MRSA by both subcutaneous and oral administration, in combination with imipenem, was demonstrated in a murine soft tissue model (59).

Macrolides - Having now witnessed the commercial introduction of several second generation macrolides (i.e., clarithromycin, azithromycin, roxithromycin, and dirithromycin), pharmaceutical scientists are poised to develop the next wave of improved macrolides, best exemplified by the ketolides. An international symposium devoted to macrolides, azalides, and streptogramins reflected the current level of activity (60).

Several reviews have been published describing the successes and failures of chemistry leading from older macrolides to the recently introduced agents mentioned above (61-63). While some early objectives including improved stability and pharmacokinetics have been achieved, other challenges such as broader spectrum of activity and overcoming macrolide resistance remain to be conquered. Topics related to the former continued to appear over the last year: *Helicobacter pylori*, *Mycobacteria*, and *Chlamydia* susceptibilities to macrolides have been the subject of recent studies (64-66). New analogs in the 16-membered tylosin series have emerged (67,68) and a series of 9a-azalides bearing N-9a acyl groups (e.g., **33**) was synthesized (69), although neither group offered significant biological advantages over the parent series. Three *in vivo* active 8a-azalides (70) were evaluated for acid stability and exhibited improvements over erythromycin (71). NMR-based conformation studies of erythromycin, clarithromycin and azithromycin in a weakly bound state with bacterial ribosomes revealed superimposition of the three macrolides except in the cladinose sugar region, which showed varied orientations (72).

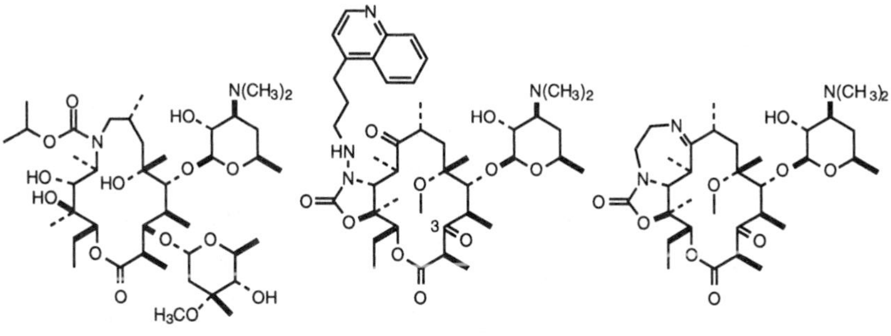

33   34   35

The ketolide class represents an important step in understanding the SAR of macrolides by virtue of the replacement of cladinose, a sugar long thought to be essential for activity, with a C-3 ketone. Bicyclic hydrazono-carbamate RU 004 (**34**) and tricyclic imine TE-802 (**35**) represent the two main subclasses that have been reported (73-74). The latter was prepared in seven steps from clarithromycin and demonstrates good MICs against Gram-positive bacteria, including some strains of erythromycin-resistant *S. aureus*. In addition, **35** was effective against systemic infections in mice when administered orally. Compound **34** displays good *in vitro* activity against multiresistant pneumococci (e.g., erythromycin-resistant MIC_{90} = 0.25 µg/ml) and other respiratory pathogens, which successfully translates into potent mouse protection in various infection models. Both series of ketolides have remarkable acid stability.

Quinolones - Fluoroquinolones, first introduced in the early 1980s, have the advantage of being broad spectrum, orally active and possess a unique mechanism of action as inhibitors of DNA gyrase (75). These drugs, e.g., ciprofloxacin (**36**) and ofloxacin, are indicated in the treatment of both respiratory and urinary tract infections, bacterial diarrhea, sexually transmitted diseases, and Gram-negative bacilli infections (76). Although they are used against some staphylococcal infections caused by methicillin-resistant isolates of *S. aureus* and *S. epidermidis*, newer quinolones are being sought with improved potency. The limited spectrum of activity against *Streptococcus* and anaerobes remains a problem in the clinical utility of this class of compounds; in addition, poor activity against many Gram-positive bacterial pathogens limits their use. Finally, an increasing number of *P. aeruginosa* and *Enterobacteriaceae* isolates have become resistant to **36** since its introduction.

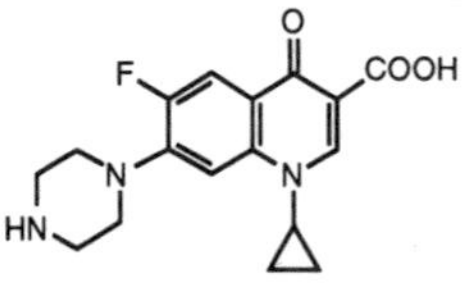

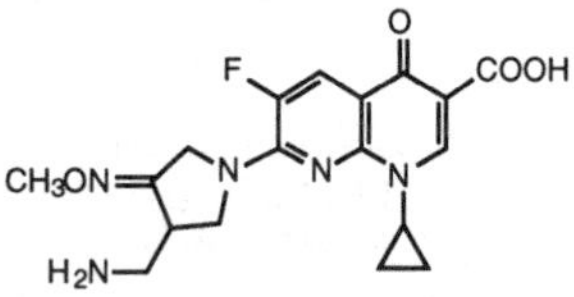

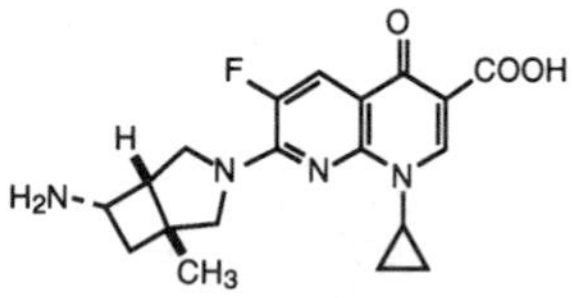

New quinolones with improved properties in the above mentioned areas would expand the clinical use of this class. The antimicrobial and pharmacokinetic properties of most of the quinolones under development, which include a number of very promising candidates, have been reviewed (76). New information pertaining to the mechanisms of quinolone activity has appeared (77), and the toxicity of this class has been reviewed (78). Although it has proven difficult to achieve all of the desirable properties in a single compound, such drugs are being actively sought. A new fluoronaphthyridone, LB20304 (**37**), showed favorable pharmacokinetics in rodents and dogs, along with potent *in vitro* activity against Gram-positive bacteria (79). It was 8- to 64-fold more active than **36** against MSSA, MRSA, and methicillin-sensitive and methicillin-resistant *S. epidermidis* (MRSE). A related compound with a C-7 bicyclic amine moiety, CFC-222 (**38**), was particularly effective against MRSA and penicillin-resistant *S. pneumoniae* (80).

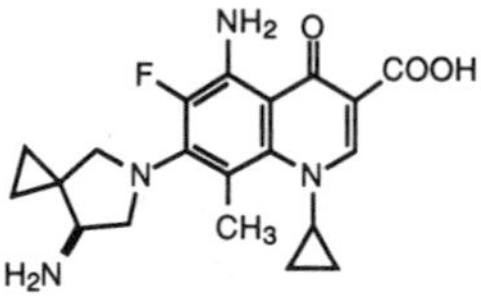

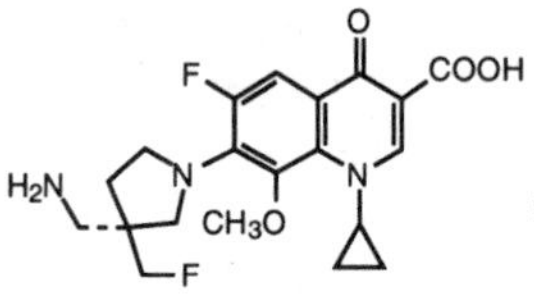

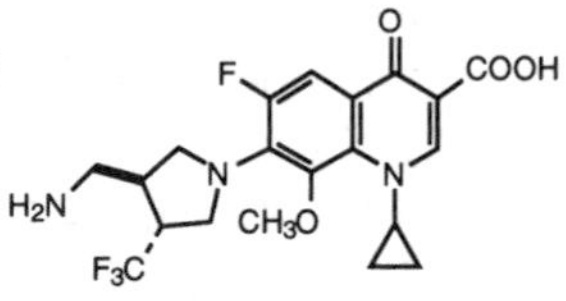

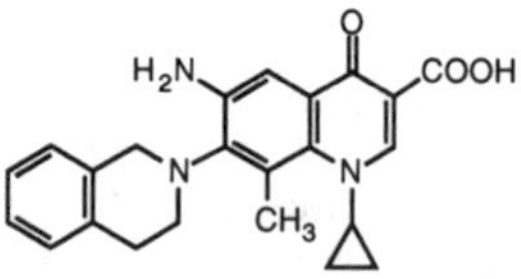

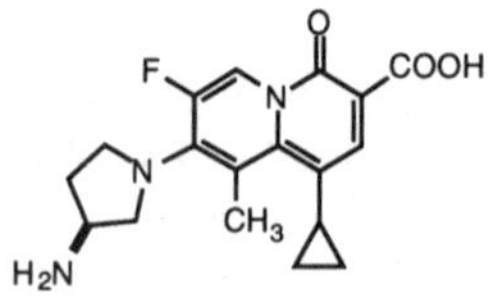

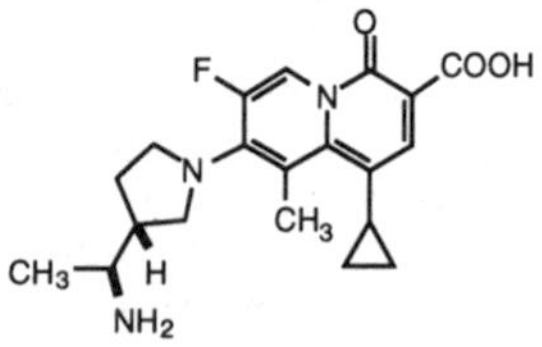

HSR-903 (**39**) exhibited MICs for clinical isolates that are 8- to 64-fold more potent against Gram-positive cocci and 2- to 16-fold more potent against Gram-negative bacilli relative to **36** (81). No phototoxicity of **39** was observed in a study in guinea pigs. Introduction of a methoxy group at the 8-position of the quinolone ring improved the photostability of Y-688 (**40**), and excellent activity was observed against quinolone-resistant *S. aureus*, *S. epidermidis*, and enterococci (82). S-32730 (**41**), another 8-methoxy analog with low phototoxicity, possesses superior Gram-positive activity including penicillin- and quinolone-resistant strains (83). Medicinal chemistry studies have been reported with a number of interesting quinolones (84-89), including the MRSA-active 6-aminoquinolone **42** (90).

The 2-pyridones, a novel class of DNA gyrase inhibitors, exemplified by ABT-719 (previously A-86719.1; **43**), were first reported in 1994 (91). The synthesis, antibacterial activity, and pharmacokinetics of **43** have been reviewed (92). As part of a continuing effort in this area, two new series of compounds were reported at the 35th ICAAC (93). In general, bicyclic pyridones were more active than the tricyclic pyridones. In comparative studies, the potency of A-104954 (**44**), a bicyclic pyridone, against *S. aureus*, *P. aeruginosa*, and *K. pneumoniae* was superior to all other quinolones tested.

Oxazolidinones - The discovery of potent, orally active oxazolidinone antibacterial agents, exemplified by DuP 721 (**45**), was reported in 1987 (94). Compound **45** possessed potent *in vitro* and *in vivo* activity against Gram-positive bacteria including MRSA and MRSE, as well as activity against Gram-negative anaerobes and *Mycobacterium tuberculosis*. The mechanism of action of **45** shows it is a bacterial protein synthesis inhibitor, with inhibition occurring at an early event in the initiation phase of protein synthesis (95). Recently, two novel analogs of **45**, U-100592 (**46**) and U-100766 (**47**), in which a substituted piperazine moiety or a morpholine ring has been incorporated, were reported that are currently in clinical development for treatment of serious human infections involving multiply-resistant strains of staphylococci, streptococci, or enterococci (96,97). The *in vitro* and *in vivo* (po and iv) activities of **46** and **47** against representative strains are similar to those of vancomycin.

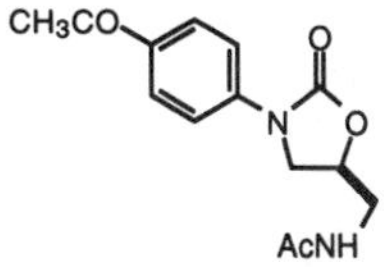

45 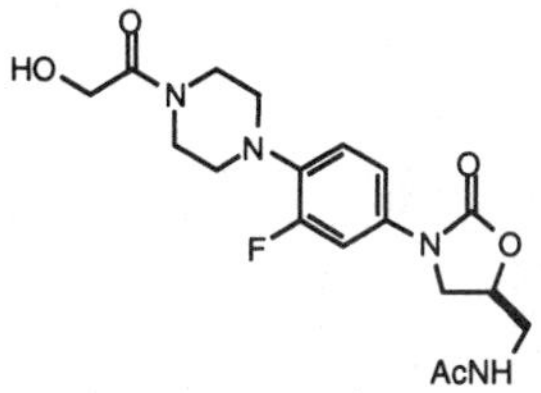   **46** 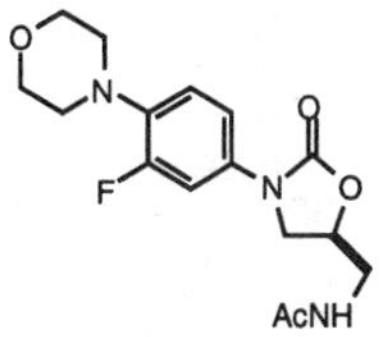   **47**

Tetracyclines - Several reviews of tetracycline antibiotics have appeared over the last year (98,99), including an account of the newest generation of tetracyclines, the glycylcyclines (100). Additional biological data have been reported on the two most widely profiled glycylcyclines, DMG-MINO (CL 329,998; **48**) and DMG-DMDOT (CL 331,002; **49**) (101,102). Both agents demonstrated good activity against several hundred clinical strains of enterococci (103), 203 isolates of *Neisseria gonorrhoeae* (104), 102 strains of MRSA, and 55 *S. pneumoniae* isolates representing penicillin-susceptible, -intermediate, and -resistant strains (105).

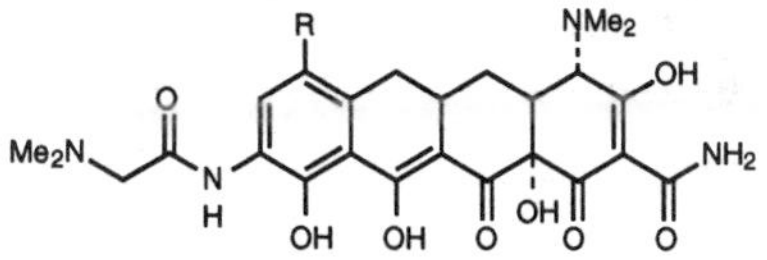

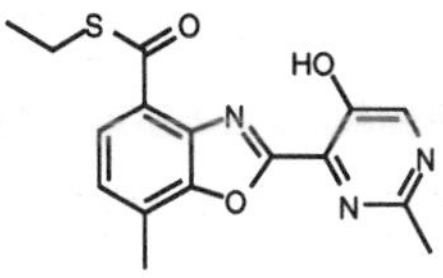

48: R = NMe₂
49: R = H   **50**

<u>Novel Areas and Screening Targets</u> - A program of mechanism-based screening yielded several benzoxazole/benzimidazole leads (106). These compounds were structurally related to a new class of benzoxazole-containing natural products, the boxazomycins. From a series of novel boxazomycins prepared by total synthesis, PD-155392 (**50**) was found to be the most active against Gram-positive organisms including MRSA. Novel screening targets were reported at a symposium at the 35th ICAAC (107). These ranged from a genetic approach using a set of conditional lethal mutants of *Salmonella* and *S. aureus* that will be screened against compounds to identify novel antibiotics specific for new targets (108), to a program in which 20 aminoacyl-tRNA synthetases were cloned from a diverse range of pathogens for the development of a high-throughput screening effort to identify compounds that target tRNA (109).

References

1. P.B. Fernandes, ASM News, <u>62</u>, 21 (1996).
2. Report of the ASM Task Force on Antibiotic Resistance, Suppl. to Antimicrob. Agents Chemother., 1 (1995).
3. JAMA, <u>275</u>, 181 (1996).
4. K. Bush, Infect. Dis. Ther., <u>17</u>, 133 (1995).
5. M.E. Klepser, M.N. Marangos, K.B. Patel, D.P. Nicolau, R. Quintiliani and C. H. Nightingale, Clin. Pharmacokinet., <u>28</u>, 361 (1995).
6. M. Tsuji, Y. Ishii, A. Ohno, S. Miyazaki and K. Yamaguchi, Antimicrob. Agents Chemother., <u>39</u>, 2544 (1995); 35th ICAAC (1995), Abs. Nos. F32-F37.
7. 35th ICAAC (1995), Abs. Nos. F38-F40.
8. 35th ICAAC (1995), Abs. Nos. F24-F27.
9. 35th ICAAC (1995), Abs. Nos. F59 and F60.
10. H. Tsubouchi, K. Tsuji, K. Yasumura, M. Matsumoto, T. Shitsuta and H. Ishikawa, J. Med. Chem., <u>38</u>, 2152 (1995).
11. H. Hanaki, H. Akagi, C. Shimizu, A. Hyodo, N. Unemi and M. Yasui, 33rd ICAAC, 889 (1993).
12. H. Hanaki, H. Akagi, Y. Masaru, T. Otani, A. Hyodo and K. Hiramatsu, Antimicrob. Agents Chemother., <u>39</u>, 1120 (1995).
13. H. Hanaki, H. Akagi, T. Otani and A. Hyodo, 35th ICAAC, F58 (1995).
14. H. Hanaki, H. Akagi, M. Yasui and T. Otani, J. Antibiot., <u>48</u>, 901 (1995).
15. 35th ICAAC (1995), Abs. Nos. F48-F50.
16. 35th ICAAC (1995), Abs. Nos. F51-F57.
17. 35th ICAAC (1995), Abs. Nos. F43-F45.
18. J.B. Harbridge, G. Burton and J.H. Bateson, BioMed. Chem. Lett., <u>5</u>, 657 (1995).
19. J.C. Jung, M.G. Kim, M.J. Sung, Y.K. Choi, S.G. An, S.H. Oh, S.S. Yim and C.J. Moon, J. Antibiot., <u>48</u>, 530 (1995).
20. H. Ohki, K. Kawabata, S. Okuda, T. Kamimura and K. Sakane, J. Antibiot., <u>48</u>, 1049 (1995).
21. K. Obi, T. Saito, H. Fukuda, K. Hirai and S. Suzue, J. Antibiot., <u>48</u>, 274 (1995).
22. 35th ICAAC (1995), Abs. Nos. F41 and F42.
23. R.G. Adams, E.G. Brain, C.L. Branch, A.W. Guest, F.P. Harrington, L. Mizen, J.E. Neale, M.J. Pearson, I.N. Simpson, H. Smulders and I.I. Zomaya, J. Antibiot., <u>48</u>, 417 (1995).
24. K. Obi, T. Saito, A. Kojima, H. Fukuda, K. Hirai and S. Suzue, J. Antibiot., <u>48</u>, 278 (1995).
25. K.I. Choi, J.H. Cha, A.N. Pae, Y.S. Cho, H.-Y. Kang, H.Y. Koh and M.H. Chang, J. Antibiot., <u>48</u>, 1371 (1995).
26. K.I. Choi, J.H. Cha, A.N. Pae, Y.S. Cho, H.-Y. Kang, H.Y. Koh and M.H. Chang, J. Antibiot., <u>48</u>, 1375 (1995).
27. K. Tsuji, K. Yasumura and H. Ishikawa, BioMed. Chem. Lett., <u>5</u>, 963 (1995).
28. K. Obi, A. Kojima, H. Fukuda and K. Hirai, BioMed. Chem. Lett., <u>5</u>, 2777 (1995).
29. J. Blahova, M. Hupkova, K. Kralikova and V. Krcmery, Antiinfect. Drugs Chemother., <u>31</u>, 49 (1995).
30. Y. Sumita, H. Nouda, K. Kanazawa and M. Fukasawa, Antimicrob. Agents Chemother., <u>39</u>, 910 (1995).
31. M. Tsushima, A. Tamura, T. Hara, K. Iwamatsu and S. Shibahara, 32nd ICAAC 394 (1992).
32. R.J. Ternansky, S.E. Draheim, A.J. Pike, F.W. Bell, S.J. West, C.L. Jordan, C.Y.E. Wu, D.A. Preston, W. Alborn, Jr., J.S. Kasher and B.L. Hawkins, J. Med. Chem., <u>36</u>, 1971 (1993).

33. S.T. Waddell, R.W. Ratcliffe, S.P. Szumiloski, K.J. Wildonger, R.R. Wilkening, R.A. Blizzard, J. Huber, J. Kohler, K. Dorso, E. St. Rose, J.G. Sundelof and G.G. Hammond, BioMed. Chem. Lett., 5, 1427 (1995).
34. 34th ICAAC (1994), Abs. Nos. B85, F50, F52, F54, F56, F58, F60, F62, F64 and F66.
35. H.F. Chambers, Antimicrob. Agents Chemother., 39, 462 (1995).
36. G. Malanoski, L. Collins, C.T. Eliopoulos, R.C. Moellering, Jr., and G.M. Eliopoulos, Antimicrob. Agents Chemother., 39, 990 (1995).
37. M. Rylander, J.Rollof, K. Jacobsson and S.R. Norrby, Antimicrob. Agents Chemother., 39, 1178 (1995).
38. F. DiNinno, D.A. Muthard and T.N. Salzmann, BioMed. Chem. Lett., 5, 945 (1995).
39. L.D. Meurer, R.N. Guthikonda, J.L. Huber and F. DiNinno, BioMed. Chem. Lett., 5, 767 (1995).
40. J. Banville, C. Bachand, J. Corbeil, J. Desiderio, J.F. Tomc, P. Lapointe, A. Martel, V.S. Rao, R. Remillard, M. Menard, R.E. Kessler and R.A. Partyka, 31st ICAAC, 828 (1991).
41. 35th ICAAC (1995), Abs. Nos. F145 and F146.
42. 35th ICAAC (1995), Abs. Nos. F151-F155.
43. 35th ICAAC (1995), Abs. Nos. F133-F135.
44. K. Nishi, M. Imuta, Y. Kimura and H. Miwa, J. Antibiot., 48, 1481 (1995).
45. H.-W. Lee, E.-N. Kim, H.-J. Son, K.-K. Kim, J.-K. Kim, C.-R. Lee and J.-W. Kim, J. Antibiot., 48, 1046 (1995).
46. A. Bryskier, Exp. Opin. Invest. Drugs, 4, 705 (1995).
47. G. Gaviraghi, Eur. J. Med. Chem., 30, Suppl., Proc. of the 13th Int. Symp. on Med. Chem., 1994 (1995).
48. M. Altamura, E. Perrotta, P. Sbraci, V. Pestellini, F. Arcamone, G. Cascio, L. Lorenzi, G. Satta, G. Morandotti and R. Sperning, J. Med. Chem. 38, 4244 (1995).
49. S. Hanessian, M.J. Rozema, G.B. Reddy and J.F. Braganza, BioMed. Chem. Lett., 5, 2535 (1995).
50. K. Coleman, Exp. Opin. Invest. Drugs, 4, 693 (1995).
51. J.-M. Frere, Mol. Microbiol., 16, 385 (1995).
52. E.L. Setti, C. Fiakpui, O.A. Phillips, D.P. Czajkowski, K. Atchison, R.G. Micetich and S.N. Maiti, J. Antibiot., 48, 1320 (1995).
53. 35th ICAAC (1995), Abs. Nos. F147, F148, F149 and F150.
54. J.D. Buynak, B. Geng, B. Bachmann and L. Hua, BioMed. Chem. Lett, 5, 1513 (1995).
55. J.D. Buynak, K. Wu, B. Bachmann, D. Khasnis, L. Hua, H.K. Nguyen and C.L. Carver, J. Med. Chem., 38, 1022 (1995).
56. A.E. Allsop, G. Brooks, G. Bruton, S. Coulton, P.D. Edwards, I.K. Hatton, A.C. Kaura, S.D. McLean, N.D. Pearson, G.C. Smale and R. Southgate, BioMed. Chem. Lett, 5, 443 (1995).
57. A.E. Allsop, S. McLean, C. Perry, T. Smale and R. Southgate, 35th ICAAC, F129 (1995).
58. S. Chamberland, J. Blais, A.F. Boggs, Y. Bao, F. Malouin, S.J. Hecker and V.J. Lee, 35th ICAAC, F144 (1995).
59. D. Griffith, T. Annamalai, R. Williams and T.R. Parr, Jr., 35th ICAAC, F143 (1995).
60. The Third International Conference on the Macrolides, Azalides and Streptogramins, Lisbon, Portugal; January 24-27, 1996.
61. M.S. Marriott, Exp. Opin. Invest. Drugs, 4, 61 (1995).
62. D.T.W. Chu, Exp. Opin. Invest. Drugs, 4, 65 (1995).
63. H.A. Kirst, Drugs of Today, 31, Suppl. C, 1 (1995).
64. A. Markham and D. McTavish, Drugs, 51, 161 (1996).
65. L.E. Bermudez and L.S. Young, Curr. Opin. Infect. Dis., 8, 428 (1995).
66. L. Welsh, C. Gaydos and T.C. Quinn, Antimicrob. Agents Chemother., 40, 212 (1996).
67. A. Narandja, Z. Kelneric, L. Kolacny-Babic and S. Djokic, J. Antibiot., 48, 248 (1995).
68. S. Bobillot, T. Bakos, P. Sarda, T.T. Thang, L. Ming, A. Olesker and G. Lukacs, J. Antibiot., 48, 667 (1995).
69. N. Kujundzic, G. Kobrehel, Z. Banic, Z. Kelneric and B. Koic-Prodic, Eur. J. Med. Chem., 30, 455 (1995).
70. R.R. Wilkening, R.W. Ratcliffe, G.A. Doss, K.F. Bartizal, A.C. Graham and C.H. Herbert, BioMed. Chem. Lett., 3, 1287 (1993).
71. C.J. Gill, G.K. Abruzzo, A.M. Flattery, J.G. Smith, J. Jackson, L. Kong, R. Wilkening, K. Shankaran, H. Kropp and K. Bartizal, J. Antibiot., 48, 1141 (1995).
72. A. Awan, R.J. Brennan, A.C. Regan and J. Barber, J. Chem. Soc., Chem. Commun., 1653 (1995).
73. 35th ICAAC (1995), Abs. Nos. F157-F175.
74. 35th ICAAC (1995), Abs. Nos. F176 and F177.
75. L.L. Shen in "Quinolone Antibacterial Agents," 2nd Edition, D.C. Hooper and J.S. Wolfson Eds., ASM, Washington, D.C., 1993, p. 77.
76. T. D. Gootz and P. R. McGuirk, Expert Opin. Invest. Drugs, 3, 93 (1994).
77. J.R. Spitzner, I.K. Chung, T.D. Gootz, P.R. McGuirk and M.T. Muller, Mol. Pharm., 48, 238 (1995).
78. J.M. Domagala, J. Antimicrob. Chemother., 33, 685 (1994).
79. 35th ICAAC (1995), Abs. Nos. F204 and F205.

80.   35th ICAAC (1995), Abs. Nos. F198-F201.
81.   35th ICAAC (1995), Abs. Nos. F202 and F203.
82.   35th ICAAC (1995), Abs. Nos. F190 and F191.
83.   T. Maejima, H. Senda, W. Iwatani, Y. Tatsumi, T. Arika, H. Fukui, T. Shibata, J. Nakano and T. Naito, 35th ICAAC, F189 (1995).
84.   V. Cecchetti, S. Clementi, G. Cruciani, A. Fravolini, P.G. Pagella, A. Savino and O. Tabarrini, J. Med. Chem., 38, 973 (1995).
85.   J. Frigola, D. Vano, A. Torrens, A. Gomez-Gomar, E. Ortega and S. Garcia-Granda, J. Med. Chem., 38, 1203 (1995).
86.   M. Reuman, S.J. Daum, B. Singh, M.P. Wentland, R.B. Perni, P. Pennock, P.M. Carabateas, M.D. Gruett, M.T. Saindane, P.H. Dorff, S.A. Coughlin, D.M. Sedlock, J.B. Rake and G.Y. Lesher, J. Med. Chem., 38, 2531 (1995).
87.   T.E. Renau, J.P. Sanchez, M.A. Shapiro, J.A. Dever, S.J. Gracheck and J.M. Domagala, J. Med. Chem., 38, 2974 (1995).
88.   J.P. Sanchez, R.D. Gogliotti, J.M. Domagala, S.J. Gracheck, M.D. Huband, J.A. Sesnie, M.A. Cohen and M.A. Shapiro, J. Med. Chem., 38, 4478 (1995).
89.   T.E. Renau, J.P. Sanchez, J.W. Gage, J.A. Dever, M.A. Shapiro, S.J. Gracheck and J.M. Domagala, J. Med. Chem., 39, 729 (1996).
90.   V. Cecchetti, A. Fravolini, M.C. Lorenzini, O. Tabarrini, P. Terni and T. Xin, J. Med. Chem., 39, 436 (1996).
91.   34th ICAAC (1994), Abs. Nos. F41, F43, F45, F47, F49, F51, F53 and F55.
92.   R.A. Fromtling and J. Castaner, Drugs of the Future, 20, 1103 (1995).
93.   35th ICAAC (1995), Abs. Nos. F8 and F9.
94.   A.M. Slee, M.A. Wuonola, R.J. McRipley, I. Zajac, M.J. Zawada, P.T. Bartholomew, W.A. Gregory and M. Forbes, Antimicrob. Agents Chemother., 31, 1791 (1987).
95.   D.C. Eustice, P.A. Feldman, I. Zajac and A.M. Slee, Antimicrob. Agents Chemother., 32, 1218 (1988).
96.   35th ICAAC (1995), Abs. Nos. F206-F230.
97.   S.J. Brickner, D.K. Hutchinson, M.R. Barbachyn, P.R. Manninen, D.A. Ulanowicz, S.A. Garmon, K.C. Grega, S.K. Hendges, D.S. Toops, C.W. Ford and G.E. Zurenko, J. Med. Chem., 39, 673 (1996).
98.   V.J. Lee, Exp. Opin. Ther. Patents, 5, 787 (1995).
99.   D.E. Taylor and A. Chau, Antimicrob. Agents Chemother., 40, 1 (1996).
100. F.T. Tally, G.A. Ellestad and R.T. Testa, J. Antimicrob. Chemother., 35, 449 (1995).
101. R.A. Fromtling and J. Castaner, Drugs of the Future, 20, 1116 (1995).
102. R.A. Fromtling and J. Castaner, Drugs of the Future, 20, 1001 (1995).
103. A.P. Fraise, N. Brenwald, J.M. Andrews and R. Wise, J. Antimicrob. Chemother., 35, 877 (1995).
104. W.L. Whittington, M.C. Roberts, J. Hale and K.K. Holmes, Antimicrob. Agents Chemother., 39, 1864 (1995).
105. W.J. Weiss, N.V. Jacobus, P.J. Petersen and R.T. Testa, J. Antimicrob. Chemother., 36, 225 (1995).
106. 35th ICAAC (1995), Abs. Nos. F21-F23.
107. 35th ICAAC (1995), Abs. Nos. S42-S47.
108. M.B. Schmid, 35th ICAAC, S47 (1995).
109. C.L. Quinn, 35th ICAAC, S45 (1995).

Chapter 14. Semisynthetic Glycopeptide Antibiotics

Robin D.G. Cooper and Richard C. Thompson
Infectious Diseases Research
Lilly Research Laboratories
Eli Lilly & Co., Indianapolis, IN 46285

Introduction - Vancomycin, 1, isolated in the early 1950's, was the first example of the presently large family of naturally occurring glycopeptide antibiotics (1). Vancomycin has been in clinical use now for over thirty five years and during the past twenty years its use has increased steadily, such that it is now the first line of therapy against serious infections due to staphylococci and enterococci. As the importance of vancomycin as an agent to treat resistant Gram-positive bacteria has increased, resistance to vancomycin has also made its first appearance, affording potentially serious clinical consequences (2,3).

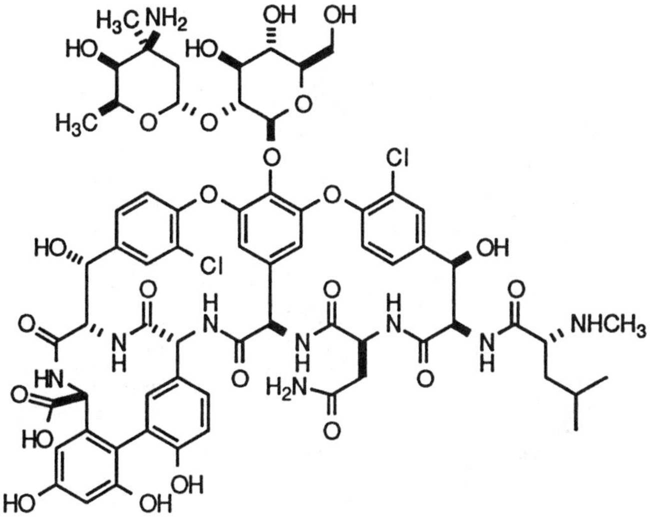

1

Indeed, the increasing prevalence and importance of hospital-acquired infections due to Gram-positive infections has been a major trend in hospitals over the last decade (4). This, coupled with ever-increasing antibiotic resistance, has made vancomycin a critically important drug. A remarkable property of vancomycin has been the long delay in the emergence of resistance. The unusual mechanism of action of vancomycin, the inhibition of bacterial cell wall synthesis, is often cited as the reason that mutational resistance is essentially unknown.

As vancomycin use has grown, interest in the glycopeptide class has increased. Systematic efforts to find new naturally occurring glycopeptides have been successful although none of these has yet to be developed clinically, primarily because they have not been found to show any substantive improvements over vancomycin or teicoplanin (5), the only two glycopeptide antibiotics approved for clinical use. A new development over the past few years has been interest in semisynthetic glycopeptides (6,7) as it has been demonstrated that activity of the class can be expanded to include vancomycin-resistant enterococci and even some Gram-negative bacteria.

MECHANISM OF ANTIBIOTIC ACTIVITY

The antibacterial effect exhibited by the glycopeptide class of antibiotics is attributed to their ability to inhibit the biosynthesis of peptidoglycan, a key structural component of the bacterial cell wall (8,9). Vancomycin and other glycopeptide antibiotics have been shown to form complexes with the C-terminal D-Ala-D-Ala sequence of the bacterial cell wall precursor, thus preventing the subsequent transglycosylation and transpeptidation reactions necessary for construction of the peptidoglycan. In addition to an affinity for D-Ala-D-Ala, other factors may also play an important role in the activity exerted by the glycopeptide antibiotics (10). Specifically, evidence has recently been reported that suggests dimerization of the antibiotic may work cooperatively with ligand binding (11-13). Figure 1 illustrates the two recognition sites of the glycopeptide eremomycin. A hydrogen-bond network is observed between two molecules of eremomycin to form an asymmetric dimer, as well as to the cell wall precursor D-Ala-D-Ala. The ability to form dimeric complexes may partially explain why certain vancomycin-type glycopeptides, such as eremomycin and A82846B, have excellent antimicrobial activity while displaying relatively weak binding to synthetic ligands (10).

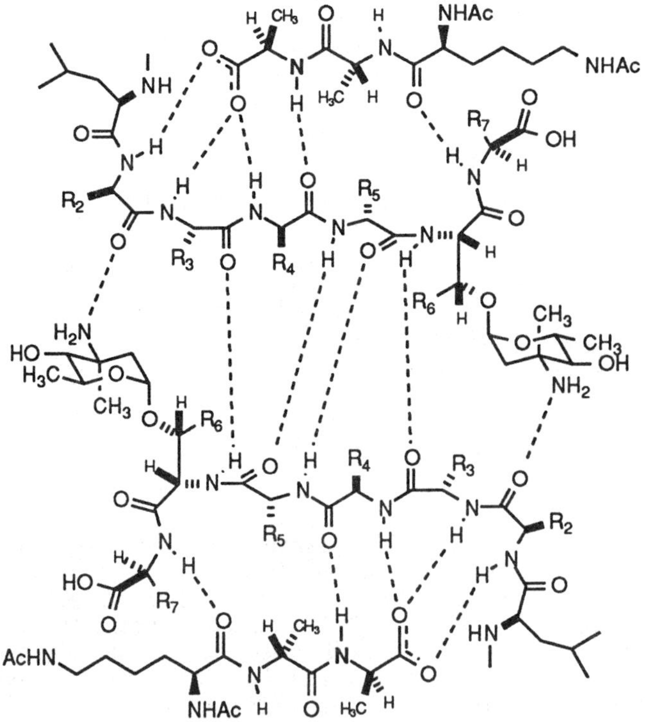

Figure 1. Hydrogen-bond interactions between eremomycin dimer
and cell wall precursor

The molecular basis of resistance to glycopeptide antibiotics is well understood (14-17). The bacteria have acquired the necessary enzymes to synthesize an altered cell wall precursor, terminating in D-lactate rather than D-alanine. The replacement of an amide linkage with an ester linkage results in the loss of one hydrogen-bond interaction between vancomycin and the ligand. The binding of vancomycin to D-Ala-D-Lact is approximately 1000-fold weaker, resulting in loss of activity (16,18).

SEMI-SYNTHETIC GLYCOPEPTIDE DERIVATIVES

The variety of naturally occurring glycopeptide antibiotics has facilitated a greater understanding of structure-activity relationships within this class which was reviewed recently (6).   The complexity of the structure has limited the variety of possible chemical modifications, and until recently most efforts failed to improve activity. However, improving the activity of this group of antibiotics by chemical modification is now yielding interesting results (6,7). Perhaps the most clinically important developments are the extension of the spectrum of glycopeptide antibiotics to include vancomycin-resistant enterococci, Gram-negative activity, and enhanced potency afforded by certain amide modifications of the terminal carboxyl (7).

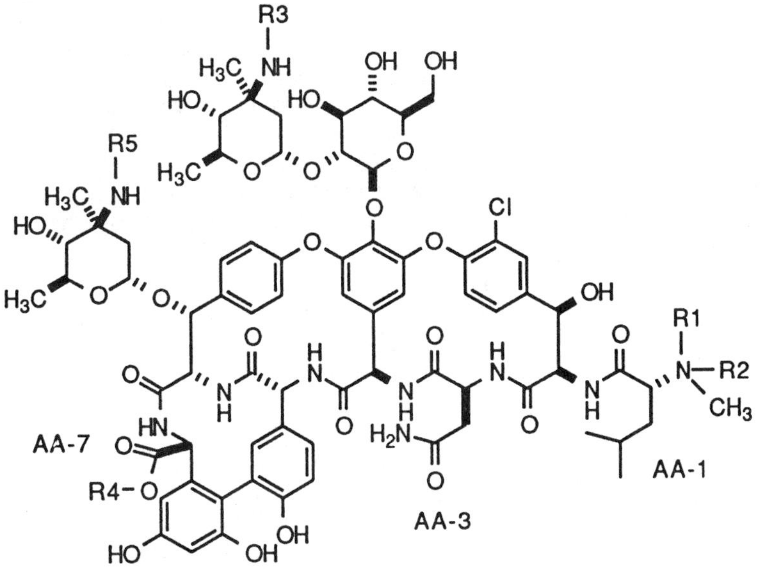

2	R1=R2=R3=R4=R5=H (**Eremomycin**)
3	R1=Boc, R2=R3=R4=R5=H
4	R1=Cbz, R2=R3=R4=R5=H
5	R1=CONH$_2$, R2=R3=R4=R5=H
6	R1=R3=CONH$_2$, R2=R4=R5=H
7	R1=NO, R2=R3=R4=R5=H
8	R1=NO, R3=R5=CH$_2$Ph, R2=R4=H
9	R1=CH$_3$, R2=R3=R4=R5=H
10	R1=CH$_2$CH=CH$_2$, R2=R3=R4=R5=H
11	R1=R2=CH$_3$, R3=R4=R5=H
12	R1=R2=CH$_2$CH=CH$_2$, R3=CH$_3$, R4=R5=H
13	R4=Pr, R1=R2=R3=R5=H
14	R1=R2=R4=CH$_3$, R3=R5=H
15	R1=R3=R4=CH$_2$Ph, R2=R5=H

N-Alkyl Derivatives - The substitution of eremomycin has been explored (19,20). While these efforts did not result in more active compounds, they illustrate the problem of selective reactions. Eremomycin (A82846A; **2**) has three basic nitrogen atoms, nevertheless a selective acylation reaction of the amino group of AA-1 (AA-1 refers to amino acid 1, or the N-methylleucine) was accomplished using standard procedures to yield either the N-Boc (**3**) or the N-Cbz (**4**) derivatives. Carbamoylation also gave the

N-carbamate at AA-1 (**5**), however on further reaction the di-substituted derivative (**6**) was obtained in which the second group was located on the disaccharide of AA-4. Nitrosation also gave the AA-1 derivative (**7**), in the case of both eremomycin and vancomycin. Reductive alkylation of the N-NO compound yielded the AA-4, AA-6 dialkylated product (**8**). The N-NO compound (**7**) did retain most of its antibacterial activity against methicillin-resistant *S. aureus* (MRSA), while the carbamoyl compounds (**3**, **4**, **5**, **6**) were devoid of activity.

Further work studied the alkylation of eremomycin with alkyl and arylalkyl halides (20). The most reactive halides [methyl iodide and allyl iodide] reacted initially at the N-Me group of AA-1, resulting in both the tertiary (**9**, **10**) and quaternary derivatives (**11**, **12**), followed then by alkylation of the carboxyl group (**14**). With the bulkier benzyl chloride, no quaternary compound was observed, but alkylation now afforded the tertiary derivative, followed by esterification and then by reaction at the amino group of the disaccharide of AA-4 (**15**). Alkylations on AA-1 with a small alkyl group had little effect on activity, both the tertiary and quaternary derivatives (**9**, **11**) possessed somewhat superior activity to vancomycin and comparable activity to eremomycin. However, the overall size of the functionality on this nitrogen did appear to play a role in that the quaternary compound having one methyl group and two allyl groups (**12**) was less active than the compound having three methyl groups (**11**), which was also less active than the parent eremomycin (**2**). Esterification of the carboxyl group of eremomycin, i.e., (**13**) caused little change to the activity. These compounds were inactive against both vancomycin-resistant enterococci and Gram-negative organisms.

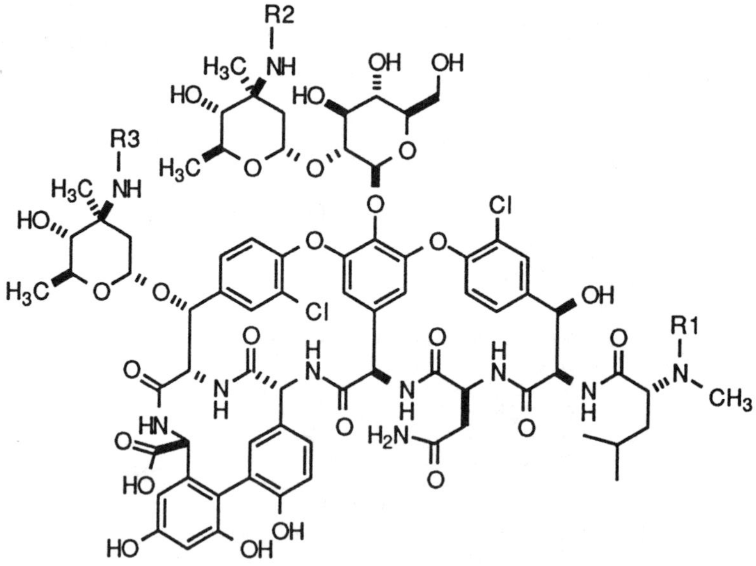

16	R1=R2=R3=H (**A82846B**)
17	R1=R3=H, R2=n-C_8H_{17}
18	R1=R3=H, R2=*p*-chlorobenzyl
19	R1=R3=H, R2=*p*-phenylbenzyl
20	R1=R3=H, R2=chloro-biphenylmethyl

It had been noted that N-alkylation of vancomycin produced derivatives which were more active and had a longer serum elimination half-life than vancomycin (21,22). The regiochemistry of the alkylation was also critical in that the alkylation on the disaccharide of AA-4 displayed higher potency than that on the N-Me leucine of

AA-1. When this chemistry was applied to A82846B (LY264826; **16**), where there are now three potential sites of reaction, it was again observed that derivatization of AA-4 was critical and that monoalkylation was preferred over di- and tri- alkylation for enhancement of activity (6). Such compounds showed both interesting activity and alterations in pharmacokinetics.

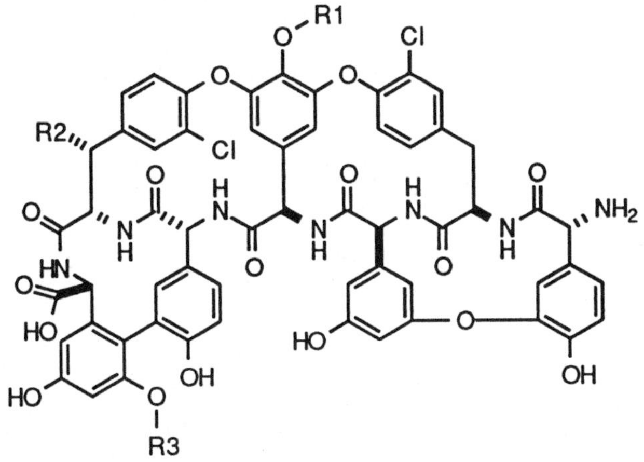

21 R1 =

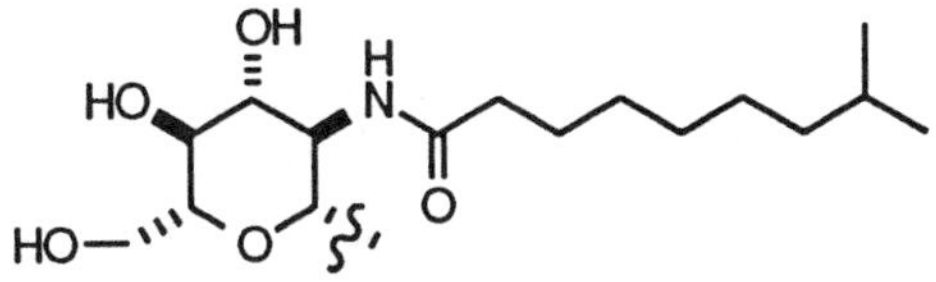

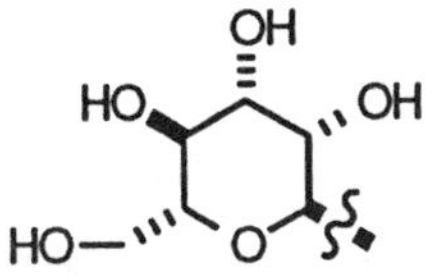

22 R1 = H, R2 = OH, R3 =

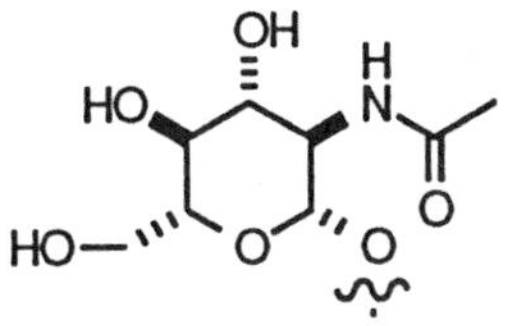

R2 =   R3 =

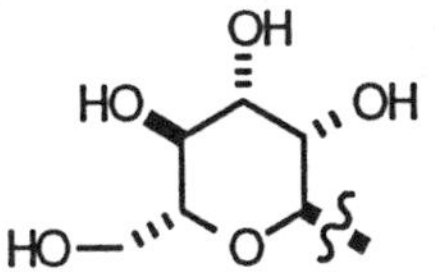

23 R1 = R3 = H, R2 =

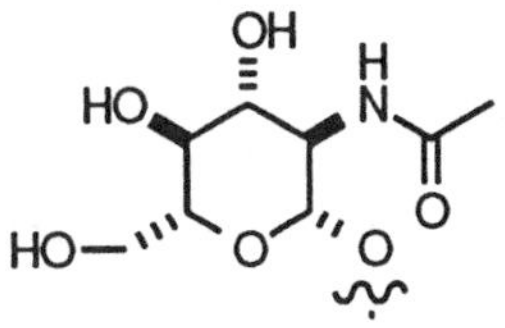

24 R1 = R2 = R3 = H

Some of the N-alkyl derivatives showed interesting activity against strains of vancomycin-resistant enterococci (VRE). Certain A82846B derivatives were the most active and demonstrated a dramatic improvement over vancomycin. The best of these compounds included the N-octyl (**17**) and the N-p-chlorobenzyl (**18**) derivatives of A82846B (6,23,24). Further modification of the alkyl group gave compounds with high potency against the vancomycin resistant strains of enterococci of both VanA and VanB phenotypes. These new compounds retain excellent activity against methicillin-resistant *S. aureus* (MRSA), methicillin-resistant *S. epidermidis* (MRSE), and the pneumococci (25). Highly active side chains include the phenylbenzyl (LY307599; **19**) and chloro-biphenylmethyl (LY333328; **20**) whose MIC's for VanA enterococci are typically 0.5 to 1 μg/ml (26,27). It has been speculated that these derivatives (**19** and **20**) incorporate all the essential structural elements in order to take full advantage of the mechanism of action (28). The sugars and chlorine atoms promote dimerization and ligand binding (29), while the sidechain may act as a membrane anchor, concentrating the antibiotic at its site of action (30). These compounds have the valuable property of bactericidal activity against some enterococci, good activity against MRSA and coagulase-negative staphylococci, and remarkable potency against the streptococci (25,26).

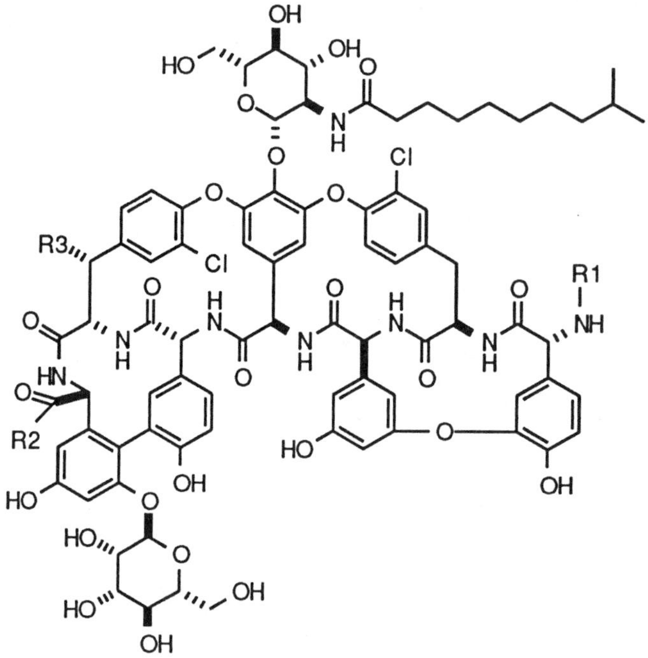

25 MDL62,873 (mideplanin)

R1 = H, R2 =

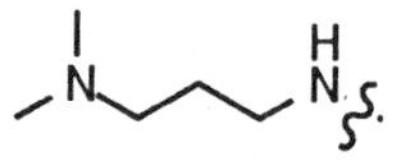

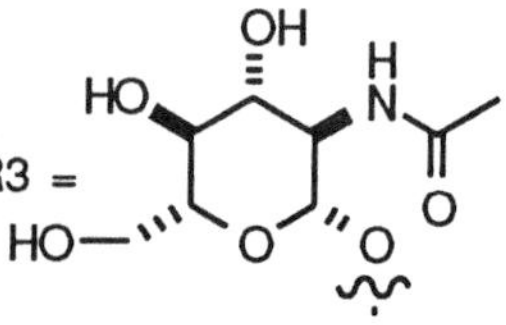

26 MDL62,600

R1 = R3 = H, R2 =

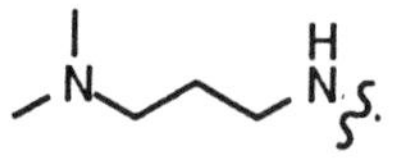

27 MDL63,246

R1 = CH$_3$, R3 = OH, R2 =

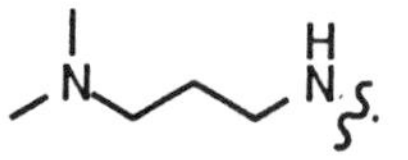

28 MDL63,042

R1 = CH$_3$, R3 = OH, R2 = H$_2$N

R3 =

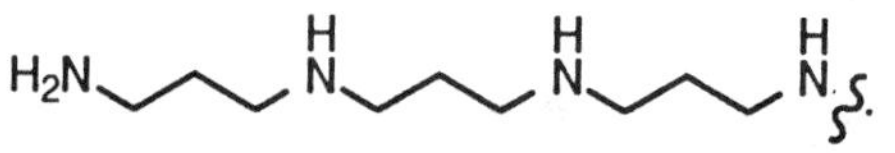

Amide Modifications of Carboxyl Group - The condensation of amines with teicoplanin (**21**), the pseudoaglycones (**22, 23**), and its aglycone (**24**) furnished a series of new amide derivatives which exhibited improved activity over the parent molecule. The best improvement in activity was obtained when basic amino groups were included in the sidechain (31,32). One of these derivatives, mideplanin (MDL62,873; **25)** has considerable improvements over teicoplanin especially against the coagulase-negative staphylococci, against which teicoplanin is somewhat weak and inconsistent (33). However, this new compound has no improvement against the vancomycin and teicoplanin VanA resistant enterococci.

Reductive removal of the sugar unit on AA-6 from mideplanin resulted in a compound, MDL62,600 (**26**), which now began to show some activity against VanA enterococci, albeit of insufficient magnitude to be considered clinically useful (MIC's against five isolates in the range of 16-32 µg/ml) (34). Using an alternative natural glycopeptide, A-40,926 as starting material, a series of amide derivatives were synthesized on a nucleus that had the sugar on AA-6 removed but which still retained the hydroxyl substitution (35). Two examples of these compounds are MDL63,246 (**27**) and MDL63,042 (**28**). Although the overall activity of these compounds are very similar to that of mideplanin, they show some interesting activity against the VanA enterococci, with MDL63,042 having a MIC_{90} against 20 strains of 16 µg/ml (35-37).

Similar derivatives were also prepared on the aglycone of teicoplanin (**24**) (7,38). The two most interesting ones (**29** and **30**) again had amide side chains having amino-functionality. What was clearly striking about these derivatives was their unusual activity against Gram-negative organisms in sharp distinction to mideplanin and all the other glycopeptides. This appears to be related to the ability of the cationic side chain to promote permeablility of the Gram-negative outer membrane (39). MDL62,708 (**29** demonstrated a rapid bactericidal activity against *E. coli*. This activity could also be demonstrated in an *in vivo* septicemia model in mice against an *E. coli* challenge (38).

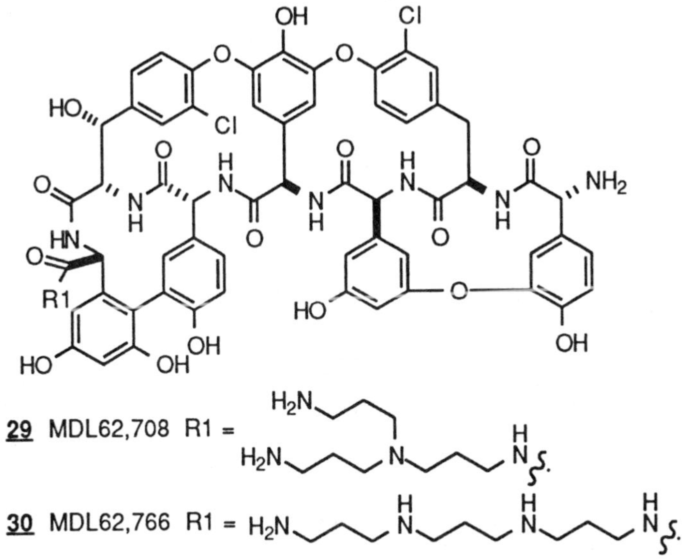

29 MDL62,708 R1 =

30 MDL62,766 R1 =

An alternate approach to enhancing activity of vancomcyin-like glycopeptides is under study (40). The overall objective is eventually to build a molecule that will have catalytic as well as binding activity. Vancomycin derivatives bearing propyl, histaminyl and 3-aminopropyl groups attached to the C terminus [AA-7] have been reported. To date, the group has successfully taken a first step in this direction by making derivates capable of catalytic carbamate hydrolysis. Also reported are amide derivatives of vancomycin (41) that included some peptidyl-type (e.g. the L-Ala-L-Ala amide) but no biological activity was given.

Nucleus Variations - Although the glycopeptides are a large and reasonably diverse group of naturally occurring antibiotics, to date, all of the differences in the heptapeptide backbone have been limited to amino acids 1 and 3. Recently a chemical methodology has been developed which could allow a far more systematic investigation of the effect upon binding and activity that can be obtained by changes in this part of the nucleus (42,43).

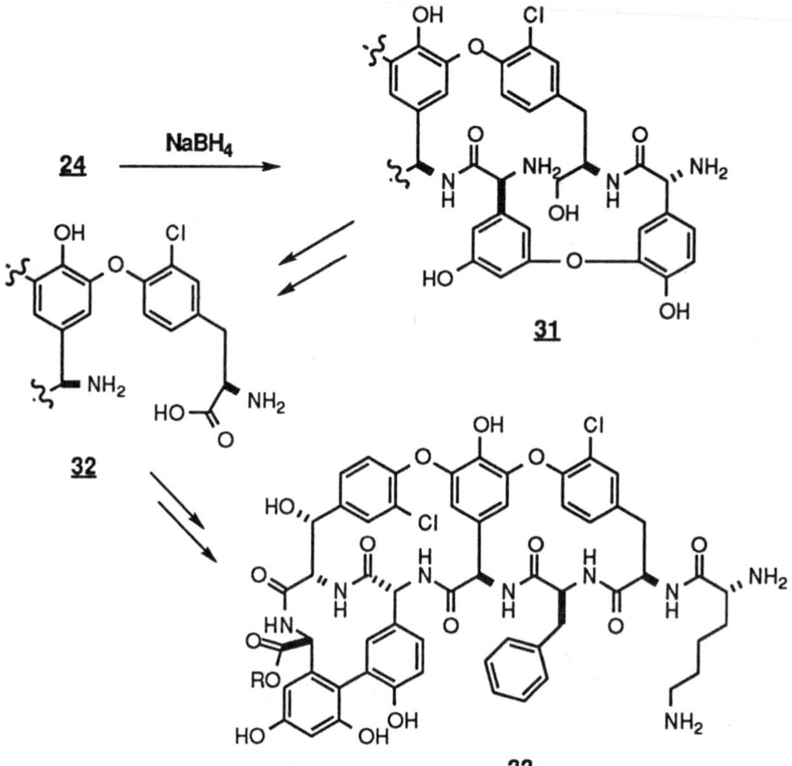

Reduction of teicoplanin aglycone (24) with sodium borohydride in aqueous alcohol solution resulted in a specific reductive cleavage of the amide bond between AA-2 and AA-3 to afford the amino-alcohol (31). After oxidation of the resultant primary alcohol, a double Edman degradation protocol removed both amino acids 1 and 3. This new tetrapeptide (32) represents a synthon from which glycopeptides varying at amino acids 1 and 3 can be generated (42,44,45).

The reconstruction of a new heptapeptide nucleus (33) has recently been demonstrated (46). Condensation of a differentially protected tetrapeptide derivative with N-Boc-L-phenylalanine followed by deprotection and cyclization provided the hexapeptide intermediate. Subsequent addition of D-lysine and removal of protecting groups furnished the novel glycopeptide aglycone MDL63,166 (33) where AA-1 and AA-3 are now D-phenylalanine and L-lysine. This new derivative had comparable activity to the original teicoplanin aglycone, with some interesting additional activity against a teicoplanin-resistant strain of *E. faecalis*. Overall, this methodology will allow the synthesis of highly novel compounds, and is a very promising approach to modulation of activity of glycopeptides.

Future Directions - The highly specific and elegant interaction between the glycopeptide antibiotics and their molecular target, the D-Alanyl-D-Alanine moiety of peptidoglycan precursor once afforded speculation that this was an entity optimized by evolution. As our knowledge expands and chemical modifications of both older and newly discovered natural products is explored, it now seems that the potential of the class for highly potent and broad spectrum activity is greater than might have been previously imagined. New compounds are not likely to be exempt from the toxicities found in the glycopeptide class, and this, as well as their pharmacodyanamic complexity, will remain a restriction to their future development. Some of the most interesting semisynthetic molecules now being reported are intrinsically complex to make, and cost may limit the attractiveness of development. Nonetheless, the class provides an instructive example of the effectiveness of the marriage of natural product research and medicinal chemistry applied to new medical needs.

References

1. M.H. McCormick, W.M. Stark, G.E. Pittenger and J.R. McGuire, Antibiot. Annual, 601 (1955).
2. A.H.C. Uttley, C.H. Collins, J. Naidoo and R.C. George, Lancet, 1, 57 (1988).
3. R. Leclercq, E. Derlot, J. Duval and P. Courvalin, N. Eng. J. Med., 319, 157 (1988).
4. S.N. Banerjee, T.G. Emori, D.H. Culver, R.P. Gaynes, W.R. Jarvis, T. Horan, J.R. Edwards, J. Tolson, T. Henderson and W.J. Martone, Am. J. Med., 91, 86S (1991).
5. R.N. Brogden and D.H. Peters, Drugs, 47, 823 (1994).
6. R. Nagarajan, J. Antibiot., 46, 1181 (1993).
7. R. Ciabatti and A. Malabarba, La Chimica e L'industria, 76, 300 (1994).
8. J.C.J. Barna and D.H. Williams, Annual Review of Microbiology, 38, 339 (1984).
9. P.E. Reynolds, European Journal of Clinical Microbiology and Infectious Disease, 8, 789 (1989).
10. D.A. Beauregard, D.H. Williams, M.N. Gwynn and D.J.C. Knowles, Antimicrob. Agents Chemother., 39, 781 (1995).
11. W.G. Prowse, A.D. Kline, M.A. Skelton and R.J. Loncharich, Biochemistry, 34, 9632 (1995).
12. J.P. Mackay, U. Gerhard, D.A. Beauregard, R.A. Maplestone and D.H. Williams, J. Am. Chem. Soc., 116, 4573 (1994).
13. J. P. Mackay, U. Gerhard, D.A. Beauregard, M.S. Westwell, M.S. Searle and D.H. Williams, J. Am. Chem. Soc. 116, 4581 (1994).
14. M. Arthur and P. Courvalin, Antimicrob. Agents Chemother., 37, 1563 (1993).
15. C.T. Walsh, Science, 261, 308 (1993).
16. C.T. Walsh, S.L. Fisher, I.-S. Park, M. Prahalad and Z. Wu, Chemistry and Biology, 3, 21 (1996).
17. M. Arthur, P.E. Reynolds, F. Depardieu, S. Evers, S. Dutka-Malen, R. Quintiliani Jr. and P. Courvalin, J. Infect., 32, 11 (1996).
18. T.D.H. Bugg, G.D. Wright, S. Dutka-Malen, M. Arthur, P. Courvalin and C.T. Walsh, Biochemistry, 30, 10408 (1991).
19. A.Y. Pavlov, T.F. Berdnikova, E.N. Olsufyeva, E.I. Lazhko, I.V. Malkova, M.N. reobrazhenskaya, R.T. Testa and P.J. Petersen, J. Antibiot., 46, 1731 (1993).
20. A.Y. Pavlov, E.N. Olsufyeva, T.F. Berdnikova, I.V. Malkova, M.N. Preobrazhenskaya and G.D. Risbridger, J. Antibiot., 47, 225 (1994).
21. R. Nagarajan, A.A.Schabel, J.L. Occolowitz, F.T. Counter, J.L. Ott and A.M. Felty-Duckworth, J. Antibiot., 42, 63 (1989).
22. R. Nagarajan, A.A. Schabel, J.L. Occolowitz, F.T. Counter and J.L. Ott, J. Antibiot., 41, 1430 (1988).
23. T.I. Nicas, D.L. Mullen, J.E. Flokowitsch, D.A. Preston, N.J. Snyder, R.E. Stratford and R.D.G. Cooper, Antimicrob. Agents Chemother., 39, 2585 (1995).
24. T.I. Nicas, Can. J. Infect. Dis., 6 Supp. C, 207C (1995).
25. T.I. Nicas, J.E. Flokowitsch, D.A. Preston, D.L. Mullen, J. Grissom-Arnold, N.J. Snyder, M.J. Zweifel, S.C. Wilkie, M.J. Rodriguez, R.C. Thompson and R.D.G. Cooper, Interscience Conference on Antimicrobial Agents and Chemotherapy, San Fransisco, CA, session 152, F248 (1995).
26. M.J. Zweifel, N.J. Snyder, S.C. Wilkie, D.L. Mullen, T.F. Butler, Y. Lin, T.I. Nicas, M.J. Rodriguez, R.C. Thompson and R.D.G. Cooper, Interscience Conference on Antimicrobial Agents and Chemotherapy, San Fransisco, CA, session 152, F245 (1995).

27.   M.S. Marriot, Expert Opinions Investigational Drugs, 5, 1017 (1995).
28.   D.H. Williams and M.S. Westwell, Chemtech, 17 (1996).
29.   U. Gerhard, J.P. Mackay, R.A. Maplestone and D.H. Williams, J. Am. Chem. Soc., 115, 232 (1993).
30.   M.S. Westwell, U. Gerhard and D.H. Williams, J. Antibiot., 48, 1292 (1995).
31.   A. Malabarba, A. Trani, P. Strazzolini, G. Cietto, P. Ferrari, G. Tarzia, R. Pallanza and M. Berti, J. Med. Chem., 32, 2450 (1989).
32.   C. Altomare, A. Carotti, S. Cellamare, A. Carrieri, R. Ciabatti and A. Malabarba, J. Pharm. Pharmacol., 46, 994 (1994).
33.   M. Berti, G Candiani, M Borgonovi, P Landini, F Ripamonti, R Scotti, L Cavenaghi, M Denaro and B.P. Goldstein., Antimicrob. Agents Chemother., 36, 446 (1992).
34.   A. Malabarba, R. Ciabatti, J. Kettenring, P. Ferrari, R. Scotti, B.P. Goldstein and M. Denaro, J. Antibiot., 47, 1493 (1994).
35.   A. Malabarba, R. Ciabatti, R. Scotti, B.P. Goldstein, P. Ferrari, M. Kurz, B.P. Andreini and M, Denaro, J. Antibiot., 48, 869 (1995).
36.   B.P. Goldstein, G. Candiani, T.M. Arain, G. Romanò, I. Ciciliato, M. Berti, M. Abbondi, R. Scotti, M. Mainini, F. Ripamonti, A. Resconi and M. Denaro, Antimicrob. Agents Chemother., 39, 1580 (1995).
37.   M.T. Kenny, M.A. Brackman and J.K. Dulworth, Antimicrob. Agents Chemother., 39, 1589 (1995).
38.   A. Malabarba, R. Ciabatti, J. Kettenring, R. Scotti, G. Candiani, R. Pallanza, M. Berti and B.P. Goldstein, J. Med. Chem., 35, 4054 (1992).
39.   R.E.W. Hancock and S. Farmer, Antimicrob. Agents Chemother., 37, 453 (1993).
40.   Z. Shi and J.H. Griffin, J. Am. Chem. Soc., 115, 6482 (1993).
41.   U.N. Sundram and J.H. Griffin, J. Org. Chem., 60, 1102 (1995).
42.   A. Malabarba and R. Ciabatti, J. Med. Chem., 37, 2988 (1994).
43.   A. Malabarba, Int. Patent WO9426780 (1994).
44.   A. Malabarba, R. Ciabatti, J. Kettenring, P. Ferrari, K. Vekey, E. Bellasio and M. Denaro, J. Org. Chem., 61, 2137 (1996).
45.   A. Malabarba, R. Ciabatti, M. Maggini, P. Ferrari, L. Colombo and M. Denaro, J. Org. Chem., 61, 2151 (1996).
46.   A. Malabarba and R. Ciabatti, Interscience Conference on Antimicrobial Agents and Chemotherapy, San Francisco, CA, session 152, F257 (1995).

Chapter 15.   Chemotherapy of Malaria

Marianne C. Murray and Margaret E. Perkins
Pfizer Inc, Central Research Division, Groton, CT 06340,
Columbia University School of Public Health, New York, New York 10032

INTRODUCTION

Malaria remains a serious endemic disease in more than 100 countries in Africa, Asia, Oceania, Latin and South America (1). Approximately 300 million people are affected with the parastic disease and reports of 1-2 million deaths per year, mostly African children, are attributed to malaria. Travelers from nonendemic areas are at risk of exposure and thousands of cases are reported in the US and in Europe annually (2).

Malaria is caused by the protozoan parasite, *Plasmodium*, a member of the phylum Apicomplexan. There are four species that infects humans: *P. falciparum, P. vivax, P. malariae* and *P. ovale* of which *P. falciparum* and *P. vivax* are the most important (3). Immature young erythrocytes are invaded by *P. vivax* and *P. ovale* malaria, whereas *P. malariae* infects mature cells and *P. falciparum* infects cells of all ages (4).   *P. falciparum* and *P. vivax* are widespread; and in the case of *P. falciparum*, the parastic disease results in severe infections and is responsible for the malarial-related deaths. *P. malariae* occurs widely and causes the least severe but persistent infections, whereas *P. ovale* is mainly confined to Africa (5).

Malaria is mainly transmitted by the female Anopheline mosquito and in rare cases by transfusion of infected blood. The life cycle of plasmodia involves two stages: 1) a sexual reproductive stage with multiplication (sporogony) which occur in the gut of the mosquito and 2) an asexual reproduction phase with multiplication (schizogeny) which takes place in the host (6). Sporozoites, the infective stage, are injected during a blood meal from the mosquito and circulate to the liver where there is a period of pre-erythrocytic development into tissue schizonts (7).   Once the tissue schizonts rupture, each releases thousands of merozoites which enter the circulation and invade erythrocytes. In *P. falciparum* and *P.malariae*, all tissue schizonts rupture at the same time and none persist in the liver. In contrast, *P. vivax* and *P. ovale* can persist in a dormant exoerythrocytic form called hypnozoites that remain in the liver for months before rupturing, which then results in relapses of erythrocyte infections (2, 5). The exoerythrocytic phase is clinically asymptomatic and its duration depends on the species.

BIOLOGY OF THE MALARIAL PARASITE

Hepatocyte Invasion- Infective sporozoites injected by the mosquito invade hepatic parenchymal cells. The circumsporozoite protein on the sporozoite surface attaches to the hepatocyte receptor which has been identified to be a proteoglycan (8).

Erythrocyte Invasion- Invasion of the malarial merozoites involves a complex series of events that depend on receptor interactions between the surface of both the erythrocyte and merozoite. The invasion process has been studied by interference microscopy and electron microscopy (EM) (9-11). Initial attachment of the merozoite to the erythrocyte occurs randomly in which any point on the merozoite surface can attach to the erythrocyte. Following the initial attachment the merozoite reorients so that the apical protuberance (where the rhoptery and microneme organelles are located) is in apposition to the erythrocyte. The major erythrocyte receptor for *P. falciparum* has been identified as glycophorin A and the receptor for *P. vivax* is the Duffy blood group antigen (12,13). Specific proteins of the apical organelles, rhoptry and microneme, are secreted during invasion and are believed to play a major role in the invasion process. As apical attachment occurs, the erythrocyte undergoes a wave of deformation followed by merozoite entry into the erythrocyte and then resealing of the erythrocyte membrane.

Growth and Development- Once the merozoites invade the erythrocytes, they begin to undergo further development and asexually reproduce yielding schizonts. The process of intracellular maturation to schizonts is called schizogony. All developmental erythrocytic stages can be distinguished by their morphology on Giemsa-stained blood smears. The development of the schizont progresses from the initial ring-form stage to the trophozoite stage to the schizont stage (containing 6-24 merozoites). The merozoites rupture out of the mature schizont-infected erythrocyte and within seconds attach to and invade uninfected erythrocytes (2,6).

Schizogony requires 72 hours for *P. malariae* (quartan malaria) and 48 hours for the other three species of malaria (tertian malaria) (2). *P. falciparum* infected erythrocytes undergoing schizogony sequester in capillary and venular beds. Generally, the mature stages (trophozoites and schizonts) are not detected in the peripheral circulation (6).

Merozoites never return to the liver, therefore the blood stage parasite is not responsible for relapses which are seen in *P. vivax* and *P. ovale*. Some merozoites in the erythrocytic stage differentiate into the sexual stage known as gametocytes (male and female) which can be transmitted to an uninfected mosquito during a blood meal completing the parasite life cycle.

Plasmodium depends on non-hemoglobin sources of amino acids that are inadequately represented (Met, Cys, Gln, Glu, Ile) in hemoglobin and therefore the parasite must obtain these amino acids from the host environment (14,15). This is seen when medium supplemented with the above amino acids is able to support normal parasite growth (16). Host hemoglobin is a major energy source for intra-erythrocytic *Plasmodium*. Interferring with hemoglobin catabolism is toxic to the parasite. Anti-malarial drugs such as chloroquine are known to accumulate in the food vacuoles of trophozoites and thereby prevent hemoglobin degradation in the organelle resulting in selective toxicity (17,18). Various classes of anti-malarial drugs target the intraerythrocytic growth and development of the parasite and is the subject of this review.

Transport in the Parasite Infected Cell- The intracellular location of the malarial parasite affords it protection from immune attack but also presents a weak link in its armory since it must obtain at least some of its nutrients from the external environment. The nutrient pathways in the parasite-infected red blood cell are very complex. Metabolites such as glucose, nucleotides and amino acids must be transported across the plasma membrane of the red cell, the host cell cytoplasm, the parasite vacuole membrane and eventually the parasite plasma membrane. Furthermore, toxic end-products of hemoglobin digestion must be removed from the food vacuole. There have been many studies to define these transport pathways (19). It has been shown that human erythrocytes infected with *P. falciparum* have a greatly increased permeability to polyols, neutral amino acids and organic ions while remaining impermeant to disaccharides (20). Additionally, a report has described an induced transport of nucleosides and other low molecular weight solutes with pharmacokinetic properties different from those of normal red blood cell transporters (21). These pathways are not inhibited by the classical inhibitors of sugar or nucleoside transport such as cytochalasin B and nitrobenzylthioinosine, respectively. This opens the possibilty that new transporters, presumably synthesized by the parasite, can be blocked by chemotherapeutic agents that will selectively act on parasite-infected cells and not normal cells. Recent pharmacological evidence suggests that much of the parasite-encoded transport of a diverse range of hydrophilic and hydrophobic solutes is via a pathway with similar properties. Where these transporters are located in the infected cell is still an open question as they have been identified only on the basis of pharmacokinetics of uptake into parasite-infected erythrocytes and not isolated membranes. Parasite synthesized proteins which have the properties of transporters have not been isolated as yet, although there is an effort to identify them. Once they have been isolated and characterized, they can be located in the infected cell by immunocytochemistry.

There is also evidence for vesicles within the erythrocyte cytoplasm that are involved in transport. However, there is no evidence that the infected red blood cell, like the normal red cell, engages in bulk endocytosis. Therefore, these vesicles are probably involved in transport of proteins from the intracellular parasite out to the plasma membrane of the red blood cell (22).

ANTI-MALARIAL DRUGS

Malaria is becoming increasingly refractory to treatment through resistance of the parasite to many anti-malarials (23).   Several classes of drugs are widely used therapeutically and prophylactically in malaria. Since the 1940s two major classes of anti-malarial drugs have been used. The antifolates inhibit DNA synthesis and include DHFR (dihydrofolate reductase inhibitors) and the sulfa drugs (23).   The cinchona alkaloids or quinoline containing drugs include quinine and its derivatives (chloroquine, amodiaquine and mefloquine) which are the more frequently used (23).   Other anti-malarials are gaining importance as resistance mounts and include the tetracyclines and the artemisinin derivatives.

Artemisinin Derivative- Artemisinin (1) is a naturally occurring sesquiterpene lactone peroxide, isolated from Artemisia annua (wormwood). It has been used by the Chinese in the treatment of fever since A.D. 341 (4,24). Structurally, artemisinin is unrelated to any other anti-malarials and the plant extracts showed efficacy against both P. falciparum and P. vivax malaria (25).   Problems associated with high rates of recrudescense, low solubility, short plasma half-life and poor oral bioavailability prompted scientists to discover derivatives with better physiochemical properties (26, 27). Several derivatives, e.g. arteether (2), artemether (3), and sodium artesunate (4) all have in vitro and in vivo activity against erythrocytic forms (blood schizonticides) of malaria (4,27- 30).

Artemisinin-related compounds are concentrated in parasite infected erythrocytes and the bridging endoperoxide group appears to be crucial for anti-malarial activity of the drug (4). Critical for parasite survival is the anti-oxidant effect of the erythrocyte. It appears that the anti-malarial activity of the artemisinin compounds are mediated by increasing the levels of activated oxygen radicals (4,31,32). Artemisinin-related compounds are quick acting and reduce parasitemia and therefore are important in treating severe cerebral malaria. However, it is uncertain whether these compounds will decrease mortality with severe malaria (4).

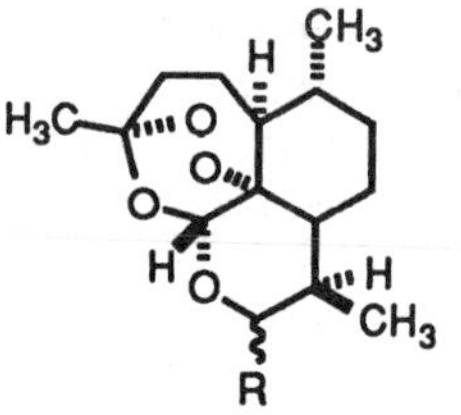

1 R = =O
   Artemisinin
2 R = OCH$_2$CH$_3$
   β-Arteether
3 R = OCH$_3$
   β-Artemether
4 R = OCOCH$_2$CH$_2$COONa
   α-Sodium Artesunate

Although artemisinin-related compounds are quick acting against drug sensitive and resistant *P. falciparum* strains, there is high recrudescent rates when administered alone. Combination of artemisinin and mefloquine (e.g., artesunate) are more effective than artesunate monotherapy by decreasing the rate of recrudescense and improving the overall cure rate (90-100%) (32). Other derivatives of artemisinin including new tricyclic trioxanes (33) and novel ring derivatives are potent *in vitro* (27,34-37) and are as effetive as arteether against drug resistant *P. falciparum* when tested in *in vivo* animal models (38).

Antimetabolites - Antimetabolites exist in two types and exert their activity on the folic acid cycle. Essential to parasite survival is their ability to carry out pyrimidine synthesis *de novo*, as they are unable to scavenge preformed pyrimidines from their host using salvage pathways (39).

The type 1 antifolates including sulphonamides and sulphones (dapsone, sulfamethoxazole, sulfisoxazole) compete with PABA (p-aminobenozoic acid) for the active site of dihydropteroate synthetase. Type 2 antifolates (pyrimethamine, cycloguanil, metabolite of proguanil) compete with dihydrofolate and inhibit DHFR (39). The sulfa drugs alone have weak anti-malarial activity and are used in combination with proguanil and pyrimethamine (23).

Aminoquinolines- Quinine (**5**) has been used for centuries in treating malaria, and it is still used to control *P. falciparum* malaria. Chloroquine (**6**) is a derivative of quinine; other modern analogs include amodiaquine (**7**), mefloquine (**8**) and halofantrine (**9**).

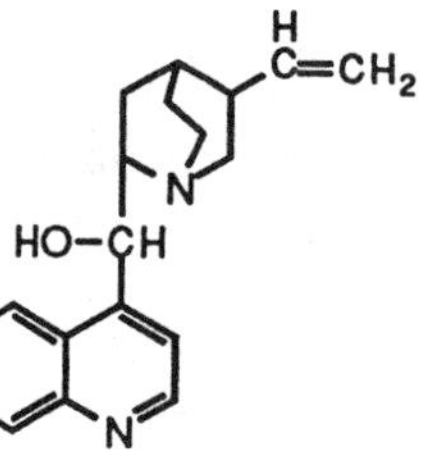

6 R = −NHCHCH$_2$CH$_2$CH$_2$N(C$_2$H$_5$)$_2$
   Chloroquine

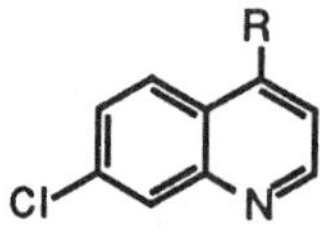

7 R = Amodiaquine

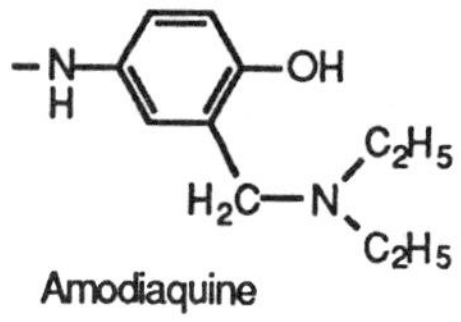

5 Quinine

These drugs are blood schizonticides and accumulate by a weak base mechanism in the acidic food vacuoles of trophozoite-infected cells, thereby preventing hemoglobin degradation from occurring in the organelle (40). The quinolines inhibit the novel heme polymerase enzyme resulting in specific toxicity of these drugs during parasite development (40).

Quinine is a cinchona alkaloid derived from the bark of the cinchona tree and remains an important anti-malarial drug due to the emergence of chloroquine resistance and multi-drug resistant (MDR) strains of malarial parasites. Quinidine is the dextra rotatory diastereomer of quinine and is more active and more toxic than quinine (17,32). Quinine monotherapy is now associated with high failure rate (40-70%) in SE Asia and Africa (41). Although a combination therapy of quinine plus tetracycline is still effective in Thailand for chloroquine-resistant *P. falciparum* malaria with cure rates of 90%, increasing recrudescent rates and MICs suggest that failure rates will rise. (42)

Despite widespread resistance, chloroquine is still used in treating *falciparum* malaria in Sub-Saharan Africa. However, reports indicate that chloroquine is no longer an effective therapy in these areas (43). The level of resistance has altered standard treatment for *P.falciparum* malaria from chloroquine to a combination therapy (fansidar or quinine plus clindamycin) (44). However with our new understanding and knowlege of chloroquine resistance, researchers should be better able to rationally design new classes of antimalarials such as heme polymerase inhibitors.

Mefloquine (**8**) is a 4-aminoquinoline methanol structurally related to quinine. It is administered orally and it possess a long half-life which may have facilitated drug resistance (41). Mefloquine appears to act by inhibiting heme polymerase, like chloroquine and it was first used clinically in 1975. It has been a very effective agent against drug-resistant *P. falciparum*. Mefloquine is selectively active against the intraerythrocytic mature forms, (trophozoites and schizonts) of malaria and has no activity against mature gametocytes (45). Both *in vitro* and *in vivo* resistance has been reported in malaria endemic regions and the mechanism of resistance may involve the *P. falciparum* MDR gene family (46). Cross resistance can be expected between mefloquine, halofantrine and maybe quinine. Therefore, the future of mefloquine monotherapy in the treatment of MDR malaria is uncertain (32).

Related to mefloquine and quinine is halofantrine (**9**), a phenanthrene methanol analog which is a recent addition of anti-malarial to treat (MDR) *P. falciparum*. *In vitro*, it is more active than mefloquine and was introduced in 1984 when clinical trials began against *P. falciparum* (47). Halofantrine is mainly used or indicated for treating mild to moderate acute malaria in sensitive strains of *P. falciparum* and *P. vivax* (48). Problems associated with this drug are mainly oral bioavailability leading to considerable inter-individual variation and treatment failures as well as cross resistance with mefloquinine (49). Cure rates varied and high recrudescent rates were observed (49). Also recent reports have raised serious questions concerning its safety (32).

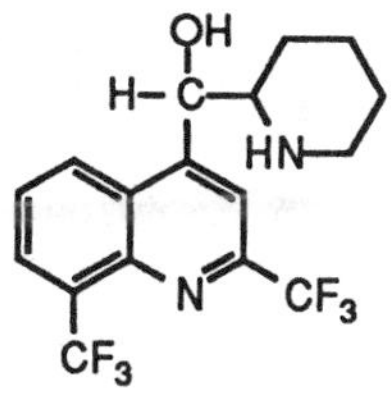

8 Mefloquine

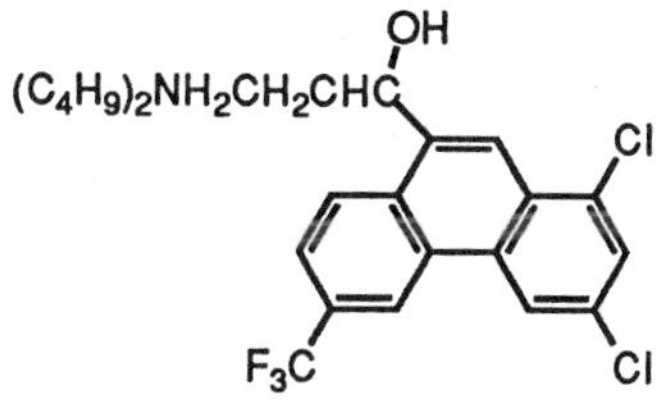

9 Halofantrine

Antibiotics -The use of antibiotics in the treatment of malaria is old; however, renewed interest in this drug class emerged with the appearance of chloroquine-resistant malaria (50).   Doxycycline and azithromycin possess anti-malarial activity. Doxycycline is a tetracycline which inhibits parasite mitochondrial protein synthesis (51). It is an effective agent when used prophylactically against MDR *P. falciparum* and is the preferred agent on Thai borders and Cambodia (32).

Azithromycin is a semisynthetic azalide antibiotic use as a prophylactic agent. It concentrates in tissues and demonstrates significant activity against exoerythrocytic stage of malaria (liver) (52). It's anti-malarial activity is probably due to inhibition of protein synthesis. Advantages over the tetracyclines include its pharmacokinetic properties possibly allowing less frequent administration plus it can be used to treat the general patient population (e.g., children) (32).

Other Drug Classes- The chalcones (1,3-diphenyl-2-propen-lone) have been identified as novel potential anti-malarials. The series screened *in vitro* showed activity against both chloroquine sensitive and resistant strains of *P. falciparum* in the nanomolar range (53). Recently, licochalcone A (**10**), isolated from Chinese licorice roots was also reported to possess *in vitro* activity against both chloroquine sensitive and resistant strains of *P. falciparum* (53).

Chalcones were designed to target the malarial cysteine protease which is believed to mediate hemoglobin degradation (54) along with aspartyl proteases. Chloroquine does not act via this mechanism which may be a promising target in treatment of chloroquine-resistant malaria (53).

Another chemical series, adenosine analogs, targets transmethylation reactions by indirectly inhibiting S-adenosyl-L-homocysteine hydrolase (SAHH), a potent inhibitor of transmethylation (55). A novel 4',5'-unsaturated 5'-fluoroadenosine, MDL-28,842 (**11**), as well as 2'-deoxyadenosine or adenine arabinoside are inhibitors of SAHH (55), and MDL-28,842 was found to inhibit *P. falciparum* and *P. berghei* intraerythrocytic development (55). As a chemotherapeutic agent, MDL-28,842 has limited use due to its indirect activity on *plasmodia* and observed toxicity. However, less toxic inhibitors of SAHH may be a target for future therapy in chloroquine-resistant malarial strains.

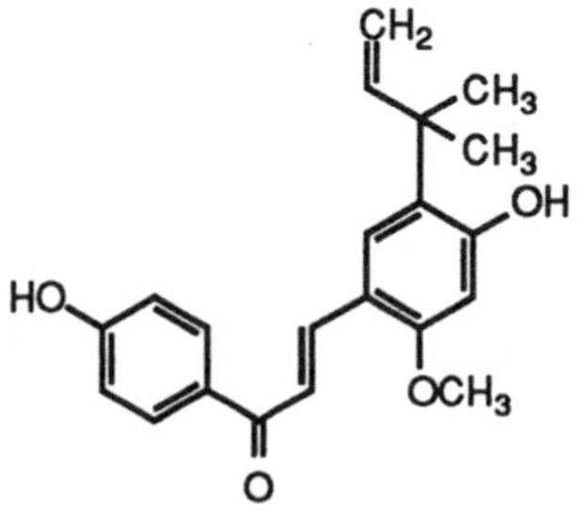

10 Licochalcone A 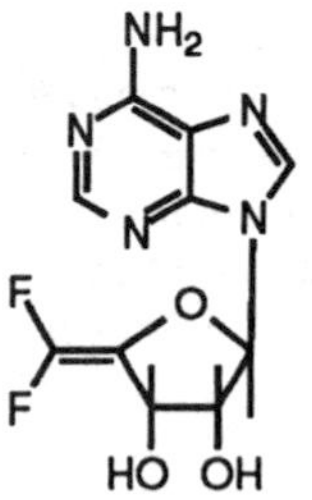   **11** MDL 28,842

Atovaquone, hydroxynaphthoquinone, is an inhibitor of protozoan electron transport (32).   Administration of atovaquone alone induces rapid resistance. However combination therapy (atovaquone plus proguanil) is quite effective with > 90% cure rates (56). Malarone (atovaquone plus proguanil) is currently under Phase III clinical trails in Thailand.

DRUG RESISTANCE IN *P. FALCIPARUM*

Resistance of *P. falciparum* to chloroquine was first reported in the late 1950s in SE Asia. At first resistance was slow to evolve but spread rapidly within SE Asia in the late 1960s and early 1970s. Resistance was first reported in East Africa in 1979. The South American focus of chloroquine resistance spread rapidly to encompass most countries of the continent by the late 1960s. Today, drug resistance has been reported from all endemic countries and there are some areas of the world, namely, SE Asia and East Africa where certain isolates are highly resistant to all known drugs (57).

The pattern of cross-resistance to other anti-malarial drugs is complex. Several studies suggest that many chloroquine resistant strains are cross-resistant to other anti-malarials such as quinine and mefloquine (58). In other studies, resistance to quinine has been associated with resistance to chloroquine. These studies suggest that parasites, although only exposed to one drug, do become resistant to multiple drugs, that is, they display a MDR phenotype (59). However, the picture is clouded by the fact that in some instances chloroquine resistant parasites are found not to be cross-resistant to other quinolines. Several factors suggest that chloroquine and quinoline resistance are not the result of a single mutational event. Firstly, the geographic spread of resistance indicates that it is not the result of a single event. Secondly, the biochemical studies on chloroquine targets within the infected cell suggest multiple targets for chloroquine. One possible mode of action is that chloroquine is a lysosomal tropic drug that increases lysosomal pH and inhibits lytic enzymes. A second mode of action has been proposed to involve chloroquine binding to heme, the toxic bi-product of hemoglobin degradation. Heme normally is sequestered in the hemozoin pigment in a membrane vacuole, but chloroquine binding of hemozoin may release it from the vacuole and render it accessible to the parasite cytoplasm. A third mode of action has recently implicated chloroquine as an inhibitor of heme polymerization (40).

In some isolates of *P. falciparum* that are resistant to mefloquine there is a correlation with increased expression of an MDR gene. MDR proteins are transporters that regulate the efflux of low-molecular weight metabolites and electrolytes and by default, drugs (60). They have been shown to be overexpressed in drug resistant mammalian cells and responsible for an increased efflux of drugs in these cells. The fact that verapamil and other inhibitors of MDR can reverse drug resistance in some isolates of *P. falciparum*, support a role for this transporter in drug resistance (61).

The geographic pattern of resistance to antifolate drugs, pyrimethamine and proguanil is similar to that in bacteria. It is rapid and multi-focal suggesting a point mutation. Dihydrofolate reductase, the target of sulfur drugs, has been sequenced in drug-resistant *P. falciparum*, and resistance has been attributed to a single amino acid change (62).

DRUG SCREENING AND EVALUATION

Guidelines for screening anti-malarial drugs were established by WHO in 1973. The guidelines can be summarized as the following steps: Stage 1 (Primary screening)- To determine whether a compound has any activity against malarial parasites. This would involve testing *in vivo* with certain animal models and in the available *in vitro* cultivation systems described below. Stage 2 (secondary screening)- To obtain information on the relative safety of the compound and to test chemical analogues for anti-malarial activity and toxicity. Stage 3 (tertiary screening)- To determine efficacy in animal models which most closely approximate human malaria, with the view to deciding whether it is safe to be tested in man. Stage 4 (Clinical Screening).

A brief description of the *in vivo* and *in vitro* models that have been useful in the screening stage are described below.

In vivo Rodent Models- A large drug screening program in the U.S. was organized during the 1960s and early 1970s and coincided with the Vietnam war, in which there was a large number of causalties from malaria, both *P. falciparum* and *P. vivax*. In this drug screening program, the animal model of choice was the rodent, principally because of the low cost of the host animals. However, there are difficulties with this species which should be kept in mind. These include parasite dependent factors, such as differences in response within a single species of *Plasmodium*, due to strain differences. There are also host dependent factors, such as innate immune resistance or tolerance to infection. For instance, infections with a Keybourg 173 (N) strain of *P. berghei* is lethal in many strains of mice including Swiss albinos, but is not in NMRI mice. There can be further differences between infection rates depending on whether it is initiated by sporozoite infections or injection of blood stages. Other factors should be taken into consideration, such as dosing regimen, diet and concomitant infections. Drugs have been tested for activity as both prophylactic and curative treatment. Prophylactic activity is defined as treatment that will protect against infection for 4-6 weeks, if administered at or prior to the time of infection. Curative treatment is one which cures an active infection within 7 days of treatment.

The four most common rodent species used are *P. berghei*, *P. yoelli*, *P. chabaudi* and *P. vinckei*. *P. berghei* and *P. yoelli* are the most commonly used species and within these species there are certain long-lived laboratory strains that have proved very reliable. Selection of drugs that can be considered schizonticidal positive can be based on several pharmokinetic criteria but a simple test has proved useful (63). In this test mosquitoes are fed on mice having male and female gametocytes. After 8 days, mature sporozoites are extracted from the insects and injected i.v. into mice. A single dose of the test drug is administered after inoculation and thin blood-films are examined. A compound is considered to be causally prophylatic if all blood films are negative on Day+14.

Avian malaria models are less attractive than the rodent models because of the availability and cost of the host animals. However, *P. gallinaceum* has proved very useful for studies to test gametocidal agents, that is, drugs that are active against gametocytes, as this species produces large quantities of the sexual stages.

In vivo Simnian Models- *P. cynamolgi* has proved a useful model system because it reproduces the relapse course of disease of *P. vivax* infection in man. *P. knowlesi* infections of rhesus monkeys are widely used because they represented an excellent system to obtain large amounts of parasites before the *in vitro* culture systems became available.   Monkeys are the only system for testing human malarias, *P. falciparum, P. vivax, P. ovale* and *P. malaria*. Aotus trivirgatus is susceptable to both *P. falciparum* and *P. vivax* and infections follow a similar course to that in humans. These monkeys have been very useful in establishing the correct prophylatic and curative doses for chloroquine, pyrimethamine and quinine (64).

In Vitro Models- The introduction of the *in vitro* system for the cultivation of *P. falciparum* by Trager and Jensen (65) in 1976 has superseded all other culture systems, most of which were only short term. Furthermore, it is the most important of the human parasites, for which new drugs are urgently needed. The *P. falciparum/* human blood system allows for the easy growth of large numbers of parasites and is perfect for testing new drugs and evaluating drug resistant isolates. Large numbers of drugs have been tested in this system. Typically, the drug is left in the culture medium for 24-48 hr, washed out and parasitemia counted in control and treated cultures at 96 hr.

NEW CHEMOTHERAPEUTIC TARGETS

The rapid increase in drug resistance has lead to an increased attention for the need to develop new anti-malarials. Unfortunately, due to the lack of economic interest in development, efforts have been minimal. All the anti-malarials used for treatment have their origin in drug discovery in the early part of this century or even earlier. Quinine was first isolated from Cinchona bark in 1820. Many potentially interesting new targets have been identified in *P. falciparum* in recent years. Several promising targets are listed below: 1- Parasite proteins are involved in invasion into host hepatocytes and erythrocytes. Drugs that mimic the docking or receptor site could block this highly specific interaction (66). 2- Transport proteins are inserted into the host membrane or parasitophorous membrane to enable rapid influx of nutrients. Inhibitors of the transporters would block the uptake of essential nutrients (19). 3- Vesicles containing parasite proteins have been identified that transport essential parasite products to the external host membrane. The movement of the vesicles could be inhibited (22). 4- The enzymes involved in hemoglobin digestion have been identified. Inhibitors of these enzymes would block the accumulation of essential nutrients (15). 5- The malarial parasite synthesizes novel lipids and require novel lipid synthetases. 6- The malarial parasite contains organelles that have some of the morphological characteristics of mitochondria including a small genome. However, sequencing of small extrachromosomal genomes revealed some homology with chloroplast genes. To what extent products of these genes are expressed is not known. This has led to the testing of herbicides as anti-malarials with some success (67).

With our ability to sequence new target proteins and advances in drug design, it should be technically straightforward to develop a range of new anti-malarials based on the knowledge of some of these products.

References

1.   N.J. White, Antimicrob. Chem., 30, 571 (1992).
2.   D.J. Wyler, Clin. Infect.. Dis., 16, 449 (1993).
3.   P.C. Garnham, Malaria Parasites and Other Haemosporidia. Oxford: Blackwell Scientific Publications, 1966.
4.   L.B. Barradell and A. Fitton, Drugs (New Zealand) 50(4), 714 (1995).
5.   World Health Organization, ed, Drug in Parasitic Diseases, WHO Prescribing Inform (1990).
6.   D.J. Wyler, *Plasmodium*, In G.L. Mandell, R.G. Douglas and J.E. Bennett (eds), Principles and Practice of Infectious Diseases, 3rd Edition, 2056, Churchhill Livinstone, NY (1990).
7.   N.J. White and S. Pukrittayakamee, The Med. J. Australia, 159, 197 (1993).
8.   C. Cerami, U. Frevert, P. Sinnis, C. Takacs, M.J. Santos and V. Nussenzweig, Cell, 70, 1021 (1992).
9.   J.A. Dvorak, L.H. Miller, W.C. Whitehouse and T. Shiroishi, Science, 187, 748 (1975).
10.   R. Ladda, M. Aikawa and H. Sprinz, J. Parasitol., 55, 633 (1969).
11.   L.H. Miller, M. Aikawa and J.A. Dvorak, J. Immunol., 114, 1237 (1975).
12.   M. Perkins, J. Cell. Biol., 90, 563 (1981).
13.   L.H. Miller, S.J. Mason and D.F. Clyde, N. Engl. J. Med., 295, 302 (1976).
14.   A.A. Divo, J. Protozoal, 32, 59 (1985).
15.   P.L. Oiliaro and D.E. Goldberg, Parasitology Today, 11(8), 294 (1995).
16.   S.E. Francis, EMBO J., 13, 306 (1994).

17.   A.F. Slater, Pharmac. Ther., 57, 203 (1993).
18.   A. Yahon and H. Ginsburg, Cell. Biol. Int. Rep., 7, 895 (1983).
19.   A. Gero and K. Kirk, Parasitology Today, 10, 395 (1994).
20.   H. Ginsburg, M. Krugliak, O. Eidelman and C.I. Cabantchik, Mol. Biochem. Parasitol., 8, 177 (1983).
21.   A.M. Gero, Mol. Biochem. Parasitol., 27, 159 (1988).
22.   Z. Etizon and M. Perkins, Eur. J. Cell. Biol., 48, 174 (1989).
23.   S.J. Foote and A.F.Cowman, Acta. Tropica., 56,157 (1994).
24.   D.L. Klayman, Science, 228, 1049 (1985).
25.   A.N. Chawira, D.C. Warhurst, B.L. Robinson and W. Peters,Trans. Roy. Soc. Trop. Med. Hyg., 81, 554 (1987).
26.   I.S. Lee and C.D. Hufford, Pharmcal. Ther., 48, 345 (1990).
27.   A.J. Lin and R.E. Miller, J. Med. Chem., 38, 764 (1995).
28.   T. Wang and R. Xu, J. Trad. Chin. Med., 5, 240 (1985).
29.   Q. Yang, W. Shi, R. Li and J. Gan, J. Trad. Chin. Med., 2, 99 (1982).
30.   Y. Yuthavong, P. Butthep, A. Bunyaratvej and S. Fucharoen, J. Clin. Invest., 83, 502 (1989).
31.   S. Meshnick, A. Thomas, A. Tanz, C. Xu and H. Pan, Mol. Biochem. Parasitol., 49, 181 (1991).
32.   K.C. Kain, Wilderness and Env. Med., 6, 307 (1995).
33.   G. Posner, C. Oh, L. Gerena and W. Milhous, J. Med. Chem., 35, 2459 (1992).
34.   Y.M. Pu and H. Ziffer, J. Med. Chem., 38, 613 (1995).
35.   B. Venugopalan, C.P. Bapat, P.J. Karnik, C.K. Chatterjee, N. Iyer and D. Lepcha, J. Med. Chem., 38, 1922 (1995).
36.   M.A. Avery, J.D. Bonk, W.K.M. Chong, S. Mehrotra, R. Miller, W. Milhous, D.K. Goins, S. Venkatesan, C. Wyandt, I. Khan and B.A. Avery, J. Med. Chem., 38, 5038 (1995).
37.   D.S. Torok, H. Ziffer, S.R. Meshnick and X.Q. Pan, J. Med. Chem., 38, 5045 (1995).
38.   G.H. Posner, C.H. Oh, H.K. Webster, A.L. Ager and R.N. Rosson, Am. J. Trop. Med. Hyg., 50, 522 (1994).
39.   D.C. Warhurst, J. Antimicrobiol. Chemother., 18 suppl B, 51 (1986).
40.   A.F. Slater and A. Cerami, Nature, 335, 167 (1992).
41.   N.J. White, J. Antimicrob. Chemother., 30, 571 (1992).
42.   S. Looareesuwan, S. Vanijanonta, D. Kyle and K. Webster, Lancet, 339, 369 (1992).
43.   L. Miller, Science, 257, 367 (1992).
44.   K.C. Kain, Current Opinion in Inf. Dis., 6, 803 (1993).
45.   K.J. Palmer, S.M. Holliday and R.N. Brogen, Drug Eval., 45(3), 430 (1993).
46.   C. Wilson, S. Volkman, S. Thaithong, R. Martin, D. Kyle, W. Milhous and D. Wirth, Mol. Biochem. Parasitol., 57, 151 (1993).
47.   Antiprotozoal Chemotherapy, Pharm. J., 243(6557), 560 (1989).
48.   USP DI 1992 Update, US Pharmacopeial Convention, Rockville, 78 (1992).
49.   N.J. White, FRCP and F. Nosten, Current Opinion Inf. Dis., 6, 323 (1993).
50.   E.I. Ferreira, Rev. Form. Bioquim., 29(1), 1 (1993).
51.   R. Kiatfuengfoo, T. Suthiphongchai, P. Prapunwohana and Y. Yuthavong, Mol. Biochem. Parasitol., 34, 109 (1989).
52.   R.A. Kuschner, D.G. Heppner, S.L. Anderson, B.T. Wellde, T. Hall, I. Schneider, W.R. Ballou, G. Foulds, J.C. Sadoff, B. Schuster and D.W. Taylor, Lancet, 343,1396 (1994).
53.   L. Rongshi, G.L. Kenyon, F.E. Cohen, X. Chen, B. Gong, J.N. Dominguez, E. Davidson, G. Kurzban, R.E. Miller, E.O. Nuzum, P. Rosenthal and J.H. McKerrow, J. Med. Chem., 38, 5031 (1995).
54.   J.H. McKerrow, E. Sun, P.J. Rosenthal and J. Bouvier, Annu. Rev. Microbiol., 47, 821 (1993).
55.   A.J. Bitonti, R.J. Baumann, E.T. Jarvi, J.R. McCarthy and P.P. McCann, Biochem. Pharm., 40(3), 601 (1990).
56.   A.T. Hudson, M. Dickens, C.D. Ginger, W.E. Gutteridge, T. Holdich, D.B. Hutchinson, M. Pudney, A.W. Randall and U.S. Latter, Drugs Exp. Clin. Res., 17, 427 (1991).
57.   W. Peters, Parasitology, 90, 705 (1985).
58.   P. Brasseur, P. Druilhe, J. Kouamouo, O. Brandicourt, M. Danis and S.R. Moyou, Am. J. Trop. Med. Hyg., 35, 711 (1986).
59.   A.F. Cowman, Parasitology Today, 7, 70 (1991).
60.   C.M. Wilson, A.E. Serrano, A. Wasley, M. Bogenschutz, A.H. Shankar and D.F. Wirth, Science, 244, 1186 (1989).
61.   S.K. Martin, A.M.J. Oduola and W.K. Milhous, Science, 235, 899 (1987).
62.   D.S. Peterson, W.K. Milhous and T.E. Wellems, P. N. A. S., 87, 3018 (1990).
63.   W. Peters, Annals of Trop. Med. Parasitol., 69, 155 (1975).
64.   W. Peters, In Chemotherapy and Drug Resistance in Malaria: Academic Press, Ch 1-6.
65.   W. Trager and J. Jensen, Science, 193, 673 (1976).
66.   K.L. Sim, Parasitol. Today, 11, 213 (1995).
67.   D.S. Roos, N.S. Morrissette, K.Keenan and J. Lippencott-Schwartz, Mol. Parasitol. Meeting, Wood's Hole, Sept 1993, Abstract 213.

Chapter 16. Recent Advances in Tyrosine Kinase Inhibitors

David W. Fry
Parke-Davis Pharmaceutical Research
Division of Warner-Lambert Company
Ann Arbor, Michigan 48105

Introduction - Protein tyrosine kinases (PTK) are enzymes that phosphorylate specific tyrosine residues within the sequence of a wide variety of functional proteins and have been found to be a common mechanism for transmitting mitogenic signals and regulating numerous cellular processes (1-4). A potential role for certain protein tyrosine kinases in tumorigenesis is evident by their ability to transform normal cells to a neoplastic phenotype when expressed in a mutated, unregulated form or to an abnormally high level. Indeed half of the protooncogenes identified to date encode for proteins having PTK activity (5,6). This potential to transform normal cells is compatible with existing data implicating tyrosine phosphorylation and dephosphorylation as events intimately involved in growth regulation and mitogenesis (3,7-10). For these reasons tyrosine kinases are considered attractive targets for cancer therapies (11-15), as well as for other proliferative diseases including psoriasis (16,17) atherosclerosis (18) and restenosis (19,20).

Within the past two years, new structural classes of tyrosine kinase inhibitors have begun to emerge which exhibit enormous improvements in potency and specificity over prior compounds. Most of these newer compounds are directed against either the epidermal growth factor (EGF) or platelet derived growth factor (PDGF) receptor tyrosine kinases and have the capacity to effectively suppress their target in cells. This chapter will highlight these recently disclosed compounds and discuss their significance in terms of the available biological data. New inhibitors that were disclosed mainly within the last two years will be included and only those where enough biological data are given to indicate their novel character or therapeutic usefulness. For earlier work, the reader is directed to several review articles on tyrosine kinase inhibitors (13, 14, 21-23).

INHIBITORS OF RECEPTOR TYROSINE KINASES

Inhibitors of the EGF Receptor Tyrosine Kinase - Inhibition of the EGF receptor tyrosine kinase by a series of anilinoquinazolines provided the first indication that considerable improvement in potency and specificity was feasible. Compounds **1**, **2** and **3** were first described in the patent literature and reported to have IC$_{50}$ values of 20, 20 and 180 nM, repectively, for inhibition of isolated EGF receptor tyrosine kinase and potencies of 0.8 to 5 µM for inhibition of EGF-dependent mitogenesis in KB naso-pharyngeal cells (24). This represented an increase in potency of at least an order of magnitude over previously described inhibitors. A subsequent patent (25) greatly extended this series and showed that potency could be further improved by substituting electron-donating substituents at the 6 and 7 positions. Within these claims compound **4** was shown to be 30 to 40 times more potent in the EGFR tyrosine kinase and KB cell assays than the unsubstituted derivative **3**. Halogen substitution at the meta position, however, was far superior to methyl and three publications (26-28) have highlighted the remarkable potency, selectivity and utility of compounds **5** and **6**. Compound **5** was was reported to have a Ki value of 5.2 pM for inhibition of an immunpurified preparation of the EGF receptor (26). This molecule was remarkably specific for the EGF receptor tyrosine kinase and inhibited other purified tyrosine kinases only at concentrations approaching 50 µM. It rapidly suppressed autophosphorylation of the EGF receptor at low nanomolar concentrations in fibroblasts or in human epidermoid carcinoma cells and selectively blocked EGF-mediated cellular processes including mitogenesis, early gene expression and morphological transformation. Compound **5** demonstrates an increase in potency over previous agents of four to five orders of magnitude for inhibition of isolated EGF receptor tyrosine kinase and two to three orders of magnitude for inhibition of cellular phosphorylation.

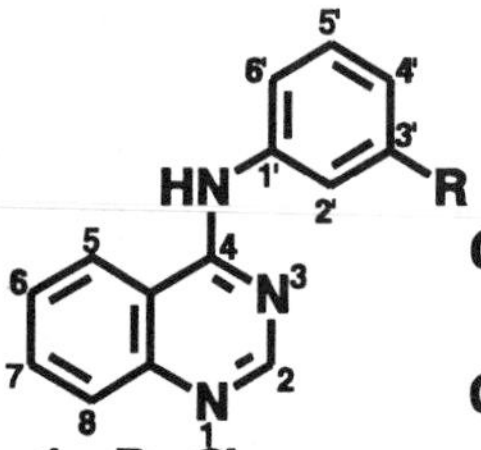

1   R=Cl
2   R=Br
3   R=CH₃

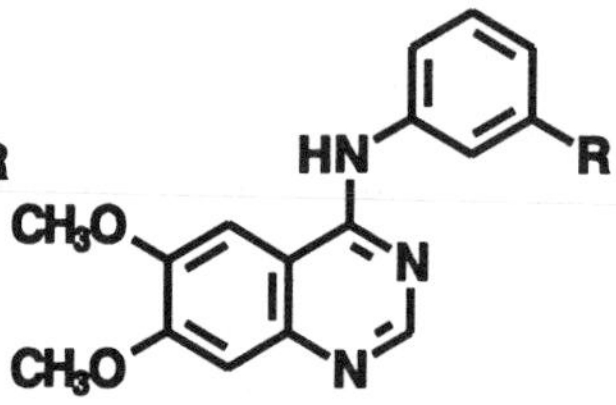

4   R=CH₃
5   R=Br
6   R=Cl

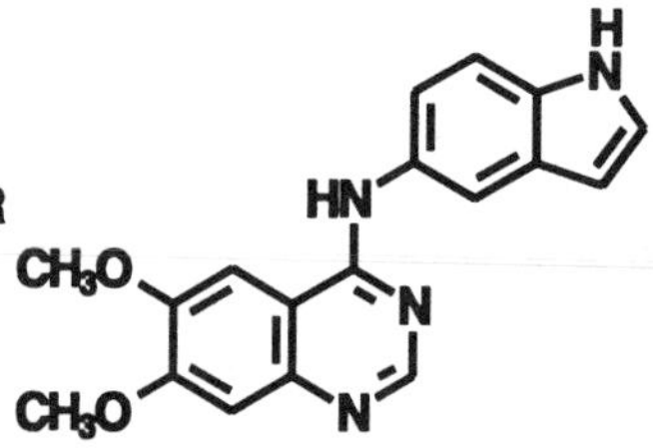

7

This compound has been shown to abolish EGF receptor associated phosphotyrosine in A431 tumor xenografts implanted in mice within 15 minutes of a single intravenous injection (29). Likewise, compound **6** was highlighted in a report describing the method for identifying this series employing database searching using a 3-dimensional structure based on the proposed mechanism of ATP binding and phosphoryl transfer [27]. The compound was competitive with ATP and exhibited a Ki of approximately 16 nM. The utility of **6** was demonstrated in a study which addressed the role of the EGF receptor in activating pp60^{src} kinase (28). Two very recent publications describing a more detailed SAR around the 4-anilinoquinazoline series have appeared (30,31). These studies further emphasized the benefit of electron donating groups at the 6 and 7 position and demonstrated a a very broad SAR resulting in compounds ranging in potency from 6 pM to >100 µM. A third publication (32) that explores benzylic methylation of 4-benzylaminoquinazolines shows that (*R*) enantiomers inhibit the EGF receptor tyrosine kinase 30- to 500-fold more potently than (*S*) and indicates that inhibitor potency is quite sensitive even to moderate out of plane bulk in certain directions. The 4-amino substituent claims were expanded to include a wide variety of bicyclic heteroaromatic side chains exemplified by **7** (33). This compound inhibited EGFR tyrosine kinase with an IC$_{50}$ of 1 nM and EGF-dependent proliferation in KB cells at 400 nM. The 6,7 position has now been expanded to include a series of 7-halo substituted 4-anilinoquinazolines (34).

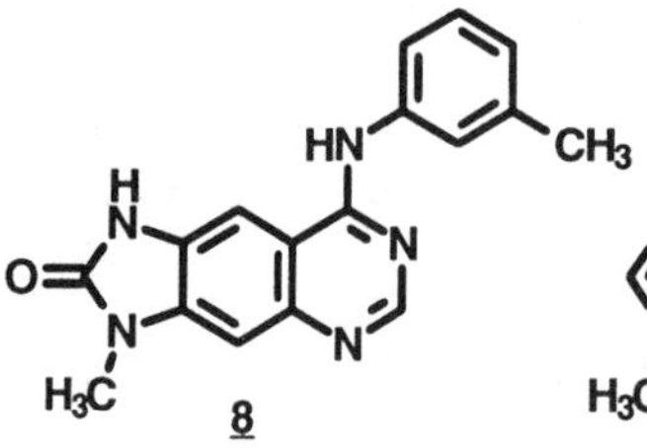

8

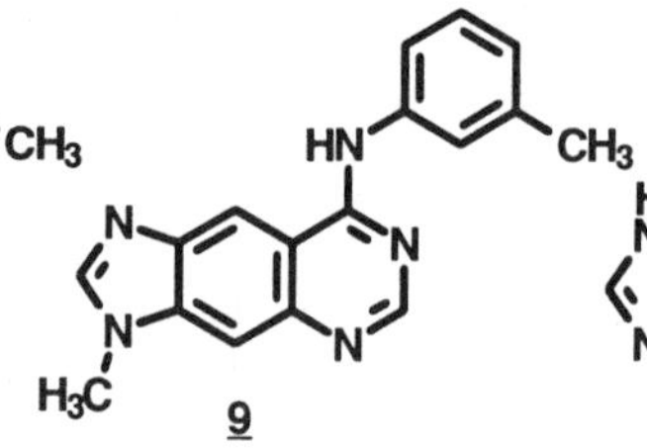

9

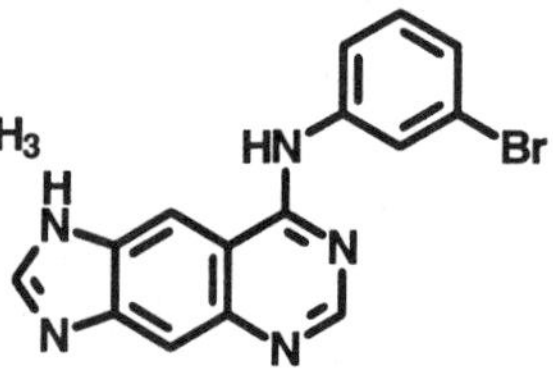

10

Finally, the quinazoline nucleus was broadened to include several different types of fused linear tricyclic derivatives (35). The 2-oxoimidazolinoquinazoline **8** and the imidazoloquinazoline **9** had IC$_{50}$ values of 35 and 16 nM, respectively, against the isolated enzyme, and both inhibited EGF-dependent proliferation of KB cells at 1 µM. Compound

10 has been reported to have an IC$_{50}$ value 8 pM against EGFR tyrosine kinase which was six orders of magnitude more potent than against a selected panel of other tyrosine kinases (36,37). The compound inhibited ligand-induced EGF receptor autophosphorylation at 43 nM but had very little effect on FGF or PDGF receptor autophosphorylation in Swiss 3T3 fibroblasts at concentrations as high as 10 µM. Another tricyclic compound recently reported was a benzothieno[3,2-d]pyrimidine, **11** (36,38). This compound was active at 0.27 nM and 86 nM, respectively, against the isolated enzyme and cellular EGFR autophosphorylation in A431 cells.

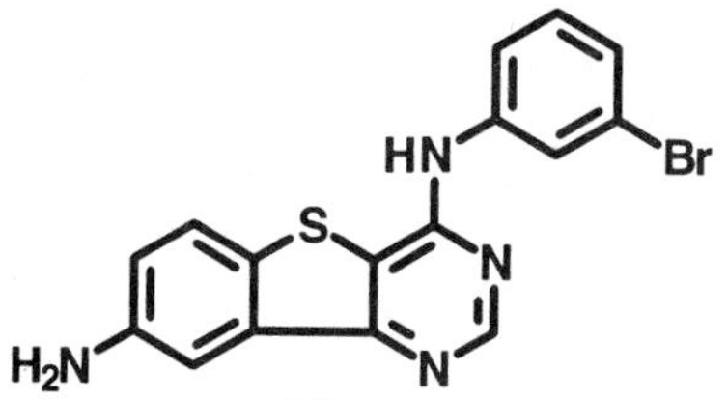

11 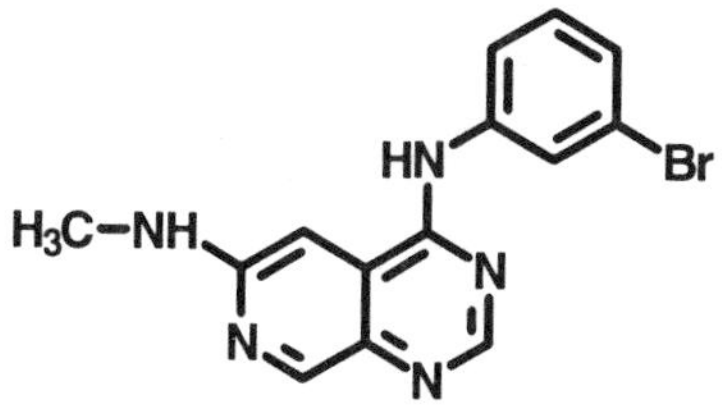   **12**

A recent patent and publication have described a series of 4-anilino pyridopyrimidines that inhibit isolated EGFR tyrosine kinase with subnanomolar potency (39,40). This series was highlighted by compound **12**,which inhibited the enzyme by 50% at 8 pM and EGF receptor autophosphorylation in A431 human epidermoid carcinoma cells at 13 nM. This compound inhibited EGF-dependent mitogenesis in Swiss 3T3 fibroblasts with an IC$_{50}$ of 0.1 µM but required greater than 5 µM to inhibit bFGF- and PDGF-dependent mitogenesis. At concentrations of 0.25 µM, **12** reversed the transformed morphology of fibroblasts transfected with high expression levels of the EGF receptor, and inhibited soft agar colony formation of EGFR, EGF and *neu* transfected fibroblasts but not *raf* or mutant *ras* transformed lines (36,40). This compound was also quite active against other members of the EGF receptor family. Partially purified heregulin from the conditioned media of MDA-MB-231 cells was used to stimulate SK-BR-3 or MCF-7 breast carcinoma cells. Tyrosine phosphorylation was inhibited with an IC$_{50}$ value of 50 nM. A recent presentation indicates that pyrrolo[2,3-]pyrimidines are also strong inhibitors of the EGFR tyrosine kinase (41). The most potent compounds were the meta-chloro and bromo derivatives, **13** and **14**, which inhibited the enzyme with IC$_{50}$ values of 27 nM and 25 nM, respectively.

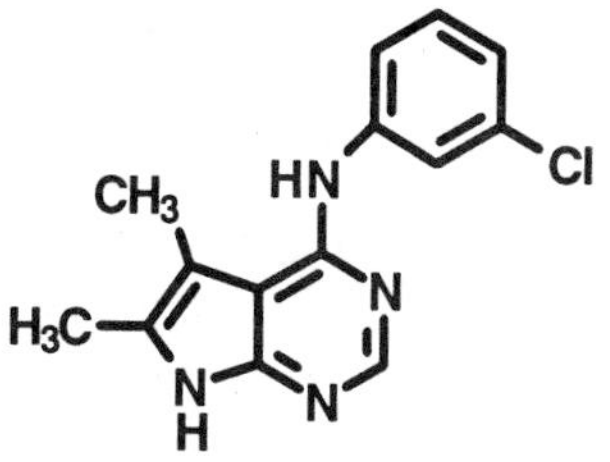

13 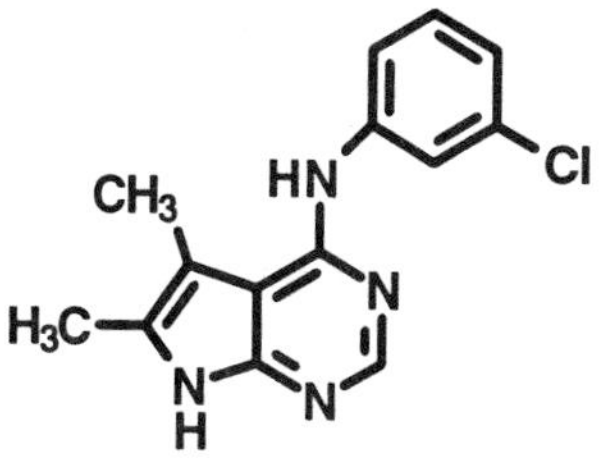   **14**

Compounds **13** and **14** inhibited EGF dependent mitogenesis at 1.7 µM and 1.4 µM and EGFR autophosphorylation in A431 cells at 0.3 µM and 1.5 µM, repectively. Likewise, both compounds inhibited EGF induced c-fos expression while not affecting PDGF or PMA induction. Compound **13** produce significant growth inhibition of the A431 xenograft model in nude mice with an ED50 of 1.6 mg/kg after once a day oral dosing for 14 days. Finally, it has been shown that the basic pharmacophore can apparently be stripped down to dianilinopyrimidine of which the most potent compound in this series was **15**, with an IC_{50} of 1 nM (42).

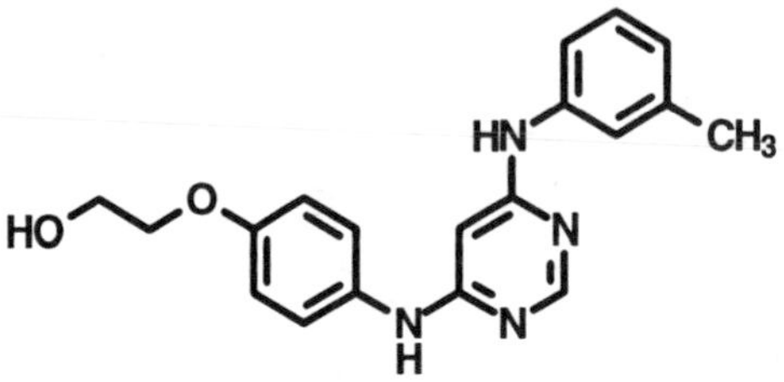

Another interesting chemical class of EGF receptor tyrosine kinase inhibitors are the dianilinophthalimides (43). Expanded biological testing of one of these derivatives, compound **16** (DAP1,CGP 52411), shows that it inhibits purified EGF receptor tyrosine kinase at 0.3-0.8 μM, and EGF and erbB-2 receptor autophosphorylation in A431 human epidermoid carcinoma or SKBR-3 human breast carcinoma respectively, in the 10 μM range with no effect on PDGF receptor autophosphorylation in balb/c 3T3 fibroblasts. This compound also selectively blocked EGF-dependent c-fos induction and cellular proliferation and produced a small response *in vivo* using the SK-OV-3 human ovarian carcinoma grown as a xenograft in nude mice [44]. Compound **16** was shown to be extensively metabolized *in vivo* and a modified structure **17** was synthesized which was resistent to metabolism (45). Compound **17** inhibited EGFR tyrosine kinase with an IC_{50} of 0.7 μM but also inhibited PKCβ-2 equally well. The compound inhibited growth of the EGFR overexpressing A431 human epidermoid carcinoma grown *in vivo* as xenografts in nude mice but not against a PDGF transformed cell line transfected with the v-*sis* oncogene or the T24 human bladder carcinoma (45).

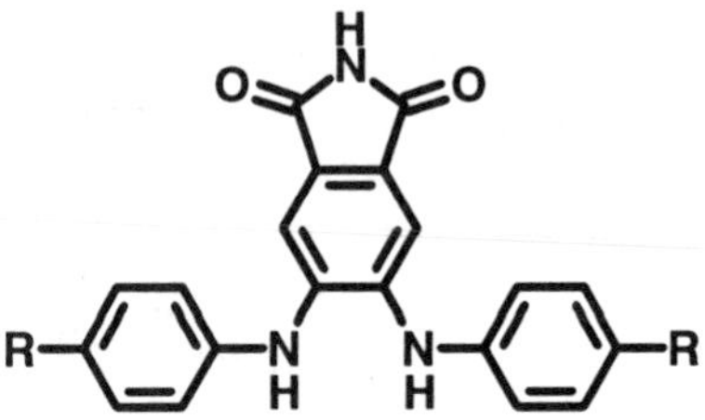

Considerable improvement in potency against the EGF receptor tyrosine kinase was attained through structural modification of a series of 5-[(2,5-dihydroxybenzyl)amino] salicylate derivatives (46).

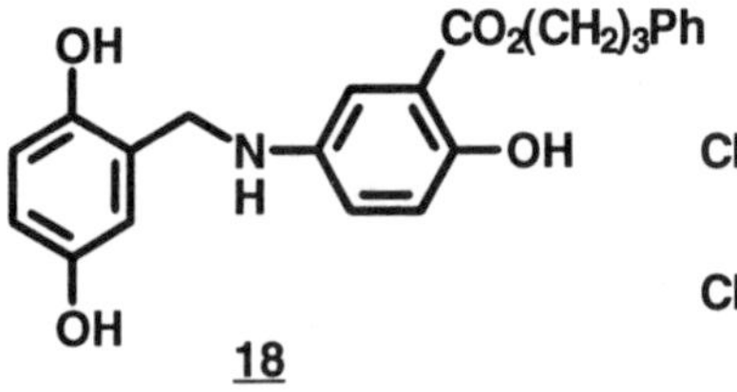

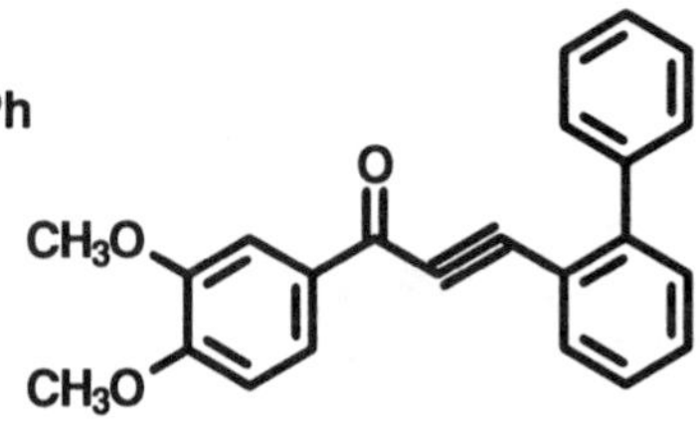

The most potent compound in this series was **18** which attained an IC_{50} value of 30 nM against cell-free enzyme preparations. Unfortunately, the potency of this series of compounds against EGF-stimulated DNA synthesis in cells did not correlate well as a whole with that of the isolated enzyme, indicating perhaps that many of the compounds had problems with cellular penetration or compartmentation (51).

Finally, a novel series of benzoylacetylene derivatives have been shown to be potent inhibitors of the EGF receptor tyrosine kinase as well as cell growth in EGF dependent cell lines. Compound **19** inhibited the tyrosine kinase activity of partially purified preparations of the EGF receptor from A431 and K13 cells with an IC_{50} value of 12 nM (47).

Inhibitors of erbB-2/HER-2 Related Phosphorylation - Few specific inhibitors of erbB-2 kinase activity have appeared in the literature. This may in part be due to the fact that a specific ligand for this receptor has yet to be conclusively identified and that cellular studies with this receptor are complicated by its tendency to heterodimerize with other members of the EGF receptor family in cells that co-express multiple constituents of this group. A recent patent has published (48) which broadly claims structures based essentially on what has been referred to as tyrphostins in past literature (28). Most were inhibitors of EGF receptor tyrosine kinase in the micromolar range; however, one specific structural class was clearly different with regard to specificity toward erbB-2 receptor. Compounds **20**, **21**, and **22** inhibited cellular erbB-2 phosphotyrosine content in BT474 human breast carcinoma with IC_{50} values of 0.17, 0.11, and 0.29 µM, respectively, with no effect on EGF receptor autophosphorylation in an EGF receptor overexpressing line of NIH-3T3 mouse fibroblasts at 100 µM. Furthermore, these compounds inhibited the growth of an erbB-2 overexpressing line of MCF-7 human breast carcinoma with IC_{50} values of 0.05, 0.06 and 0.09 µM. These results are somewhat complicated by the fact that these cell lines express other members of the erbB family and reductions in phosphotyrosine content of erbB-2 may be due not only to inhibition of autophosphorylation, but also transphosphorylation, (e.g., of erbB-2 by erbB-4). Nevertheless, these results are of significance since all four erbB family receptors have been associated with a functional role in human cancer.

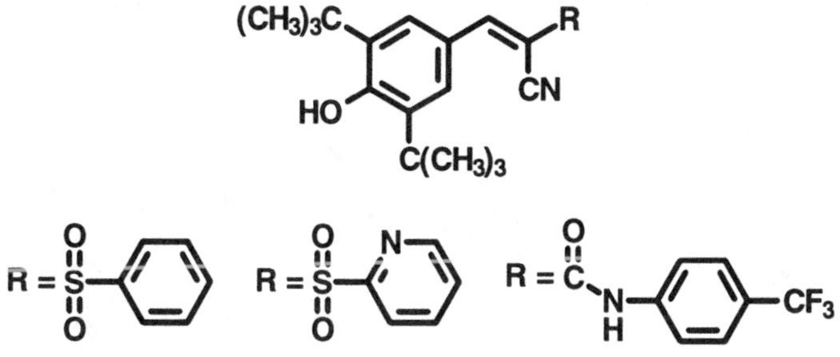

Inhibitors of the PDGF Receptor Tyrosine Kinase - Another area where considerable progress has been made is identification of inhibitors of the PDGF receptor tyrosine kinase. Recently a series of 2-phenylaminopyrimidines were synthesized and screened for inhibitory activity against a panel of protein kinases. One compound in particular, **23** (CGP53716), showed remarkable selectivity against the PDGF receptor tyrosine kinase (49). This compound inhibited immunoprecipitated PDGF receptor tyrosine kinase from BALB/c 3T3 fibroblasts with an IC_{50} of approximately 0.1 µM. IC_{50} values were greater than 100 µM for several other tyrosine kinases including the EGF receptor, TPK-IIB, a protein-tyrosine kinase purified from murine spleen, the protein products of c-*src*, c-*lyn* and *csk* as well as serine/threonine kinases such as protein kinase A and C or cdc2/cyclin B. This selectivity was also apparent in cellular studies.

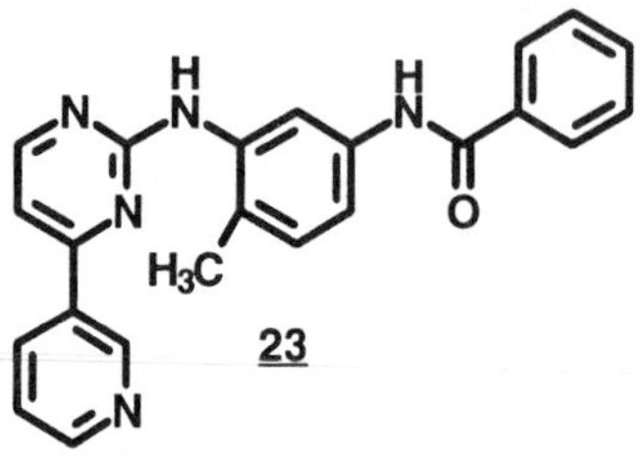

Compound **23** inhibited PDGF receptor autophosphorylation in ligand-stimulated BALB/c 3T3 cells with an IC_{50} of between 0.03 and 0.1 µM, whereas no effect was observed at concentrations as high as 100 µM on EGF, insulin or insulin-like growth factor-1 receptor autophosphorylation. Similarly, **23** inhibited PDGF-mediated induction of c-fos at 0.1-0.3 µM but not in EGF, bFGF or PMA stimulated cells. The selective effects of **23** were also apparent in its antiproliferative properties where the compound inhibited cell growth with low potency (IC_{50} values of 15 -19 µM) in EGF-dependent BALB/MK cells, interleukin 3-dependent FDC-P1 cells and the H-*ras* -transformed T24 bladder carcinoma, but was more than 20-fold more active against a v-*sis* transformed BALB/c 3T3 fibroblast. Compound **23** has demonstrated *in vivo* antitumor efficacy in tumor-bearing nude mice. When given orally, once a day for 15 or 21 consecutive days, 50 mg/kg **23** reduced subcutaneously implanted tumors of v-*sis* or c-*sis* transformed BALB/c 3T3 cells by 86% or 94%, respectively. In contrast, the compound was inactive against the A431 human epidermoid carcinoma which overexpresses the EGF receptor. Another derivative in this series, **24** (CGP 57148) demonstrated similar inhibitory properties against PDGF receptor tyrosine kinase and related cellular activities but also was active against *abl* kinase activity with an IC_{50} value of 0.1 µM. The compound had demonstrated antitumor activity against v-*sis* and v-*abl* transformed balb/c 3T3 fibroblasts grown *in vivo* but not a v-*src* transformed cell line (50).

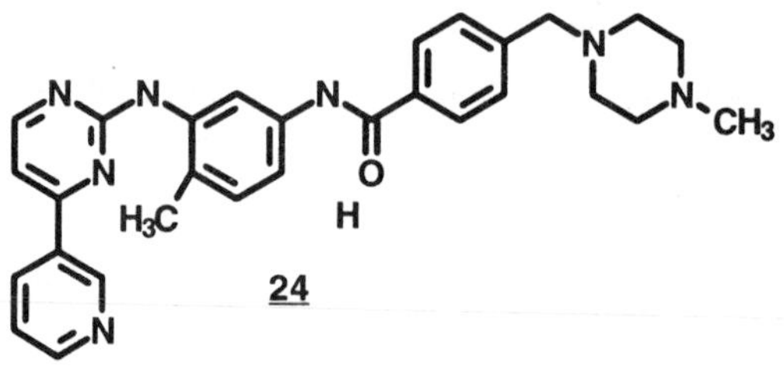

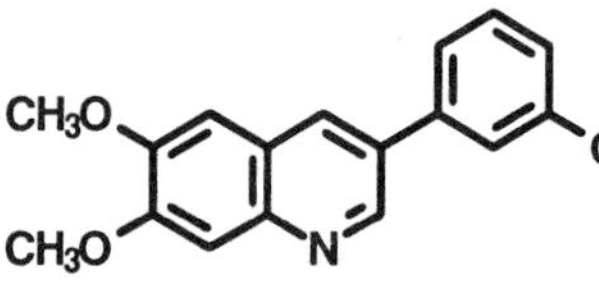

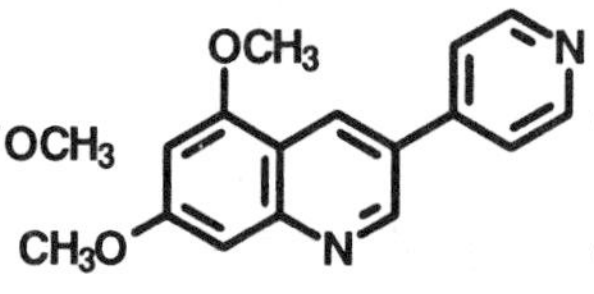

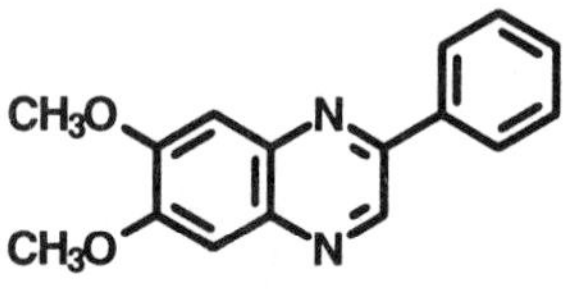

Another reported series of PDGF receptor tyrosine kinase inhibitors are the 3-substituted quinoline derivatives (51). An SAR study of 63 compounds showed that the presence of 6,7-dimethoxy was advantageous and a lipophilic group attached to the quinoline 3-position considerably enhanced activity. One of the most potent compounds was the 6,7-dimethoxyquinoline substituted in the 3-position with 4-methoxyphenyl **25**. This compound inhibited PDGF receptor autophosphorylation with IC_{50} values ranging between 1-15 nM using an immunoprecipitated PDGF receptor from NIH 3T3 murine fibroblasts. No specificity data with other enzymes or cellular data were given in this publication. Two other groups have taken this nucleus and inserted a nitrogen into the ring system to produce compounds with similar specificity for PDGF receptor tyrosine kinase. The first was 5,7-dimethoxy-3-(4-pyridinyl) quinoline **26** which was found to inhibit PDGF receptor autophosphorylation with an IC_{50} of 80 nM and exhibited a Ki value of 14 nM as a competitive inhibitor with respect to ATP (52). It was inactive at concentrations as high as 10 µM against a number of other kinases including the EGF receptor, p185[erbB2], p56[lck], PKC and PKA. This compound could specifically inhibit PDGF-induced thymidine incorporation into DNA in primary human vascular smooth muscle cells. The second chemical class is derived from quinoxalines which were also reported to selectively inhibit PDGF receptor tyrosine kinase activity with relatively good potency (53). Compound **27** inhibited PDGF receptor autophosphorylation in Swiss 3T3 fibroblast cell membranes with an IC_{50} value between 0.3 and 0.5 µM, while having no effect on EGF receptor autophosphorylation in the same membrane preparation. A similar potency and selectivity was observed when the compound was tested against PDGF and EGF receptor autophosphorylation in intact cells. Other indications for the selective nature of **27** were its inhibitory activity on PDGF-mediated DNA synthesis and the ability to reverse the transformed morphology and prevent soft agar colony formation of *sis*-transfected fibroblasts but not *src* transformed cells.

INHIBITORS OF NONRECEPTOR TYROSINE KINASES

Inhibitors of pp60[src] - Whereas the the greatest progress in the last 2 years in terms of increasing potency and specificity of inhibitors has been directed toward receptor tyrosine kinases, several reports are of significance with regard to nonreceptor kinases. Certain analogs of 5,10-dihydropyrimido[4,5-b]quinolin-4(1H)-one have been recently shown to be selective inhibitors of pp60[c-src] (54). Compound **28** inhibited this enzyme with an IC_{50} value of 0.5 µM while having a value of 75 µM for the EGF receptor tyrosine kinase and greater than 3 mM for protein kinase A. Similarly, pp60[c-src] could be inhibited in the 0.3 - 1 µM range with a series of carbazole analogues of elegaic acid with specificities relative to PKA as high as 690-fold (55). Finally, pp60[src] inhibition data were recently published on a series of 3-(N-phenyl)carbamoyl-2-iminochromene derivatives in which compound **29** produced an IC_{50} of 35 nM (56). This compound also had equal activity against p56[lck] but not against p56[lyn] or p55[fyn].

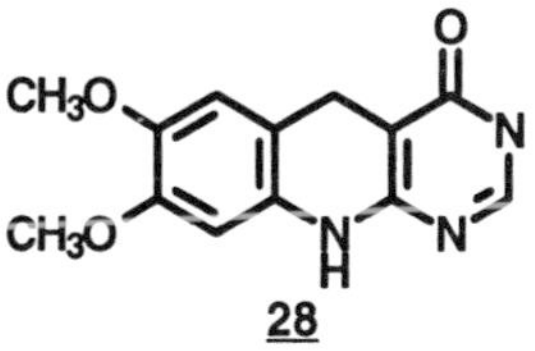

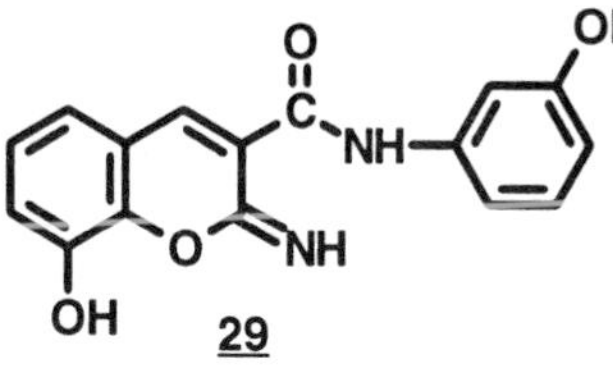

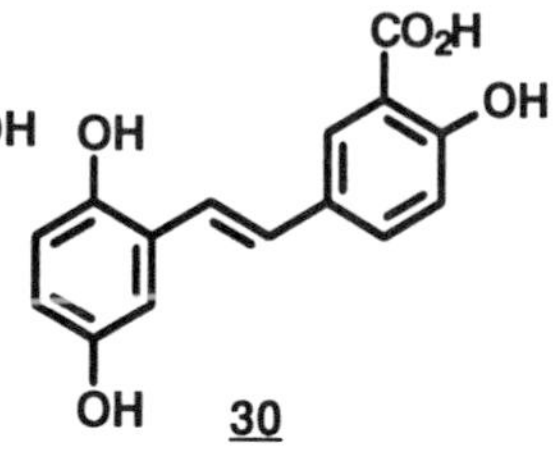

Inhibitors of p56[lck] - Inhibitors of p56[lck] tyrosine kinase have been of interest because of the involvement of this enzyme in T-cell lymphomas as well as suppression of T-lymphocyte activation. Non-amine based analogues of lavendustin A have proven to be potent and specific inhibitors of this kinase as illustrated by compound **30** which inhibited p56[lck] with an IC_{50} value of 60 nM (57). Another series of compounds that show potent activity against p56[lck] tyrosine kinase is 7,8-dihydroxyisoquinoline derivatives exemplified by the 3-carboxylate, **31**, which showed an IC_{50} of 200 nM (58). A recent paper describing the properties of compound **32** (WIN 61651) shows moderate inhibition of p56[lck] and while not particularly potent at 18 µM, it inhibited T cell activation as measured by IL-2 production in purified CD4 positive peripheral blood T lymphocytes at 10 - 15 µM (59). In addition, tyrosine phosphorylation in response to treatment with antibodies to CD3 or CD4 was inhibited at 45 µM by **32**.

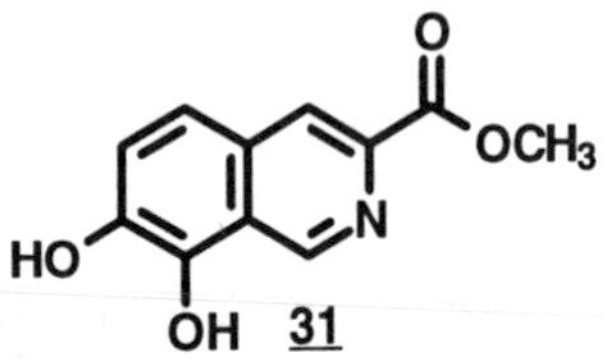

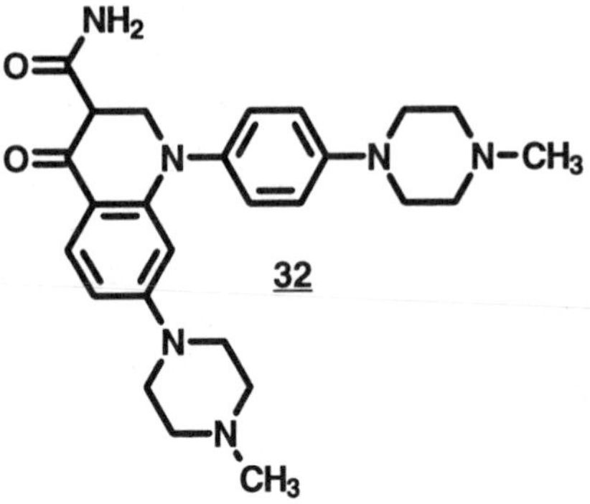

Finally, 4-amino-5-(4-methylphenyl)-7-(t-butyl)pyrazolo[3,4-d]pyrimidine **33** has recently been reported to be a potent inhibitor of p56[lck] and p59[lyn],having IC_{50} values of 5 and 6 nM, respectively (60). Specificity was demonstrated by its lack of effect on other tyrosine kinases including ZAP-70 and JAK2 at concentrations as high as 100 µM and 50-fold less inhibitory activity against the EGF receptor tyrosine kinase. Compound **33** reduced anti-CD3-stimulated tyrosine phosphorylation and thymidine incorporation in purified human T cells by 50% at 0.5 µM, and IL-2 mRNA was almost completely abolished at 1 µM.

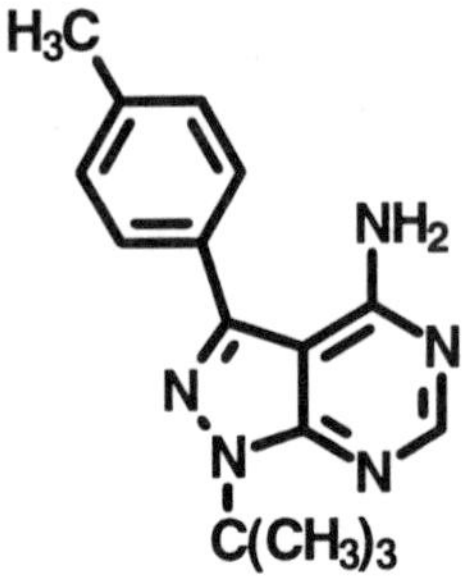

33

CONCLUSION

Many of the compounds just described appear to finally have the needed potency, specificity and cellular pharmacology to truly test the concept that suppression of specific tyrosine kinases might have therapeutic benefits in certain disease states. Some of these agents have clearly set new boundaries for defining potency and specificity for inhibitors of the EGF receptor tyrosine kinase and it is important to distinguish them from most if not all of previously reported compounds. Perhaps one of the reasons that it has taken so long to achieve this potency and specificity is the fact that these properties appear to be most readily attained with inhibitors that are competitive at the ATP site. Since homology is highest in the catalytic domains of protein kinases, there was a possibility that inhibitors affecting the ATP site might be intrinsically nonspecific. This has turned out not to be the case and indeed it appears that this site or somewhere near it imparts the most exquisite specificity and potency to date. Thus far this success has been attained mainly with the EGF and PDGF receptor tyrosine kinases; however, the lessons that have been learned through these initial accomplishments can and are being applied to the discovery of inhibitors of other kinases such as *src* family tyrosine kinases.

For many of the inhibitors described in this document, there has been a distinct convergence to certain common chemical features, regardless of the kinase. These include in many cases a substituted fused bicyclic ring system combined with another aromatic ring positioned off the main nucleus. This may be a reflection of the known high sequence homology in the catalytic domains of different kinases which allows basal interactions with a common chromophore. Substitutions which confer selectivity and dramatically increase potency for a specific kinase, however, begin to diverge and may be a function of the subtle structural differences between these kinases. Now that it has been realized that inhibitors of tyrosine kinases with inhibition constants in the nM to pM range are feasible, more sophisticated technology can be used to understand these specific chemical interactions that make this tight enzyme/inhibitor complex possible. This would certainly be greatly facilitated if co-crystallization of inhibitor-complexed enzymes could be accomplished, and it is encouraging that the first crystal structure for a tyrosine kinase has recently appeared (61,62). Unquestionably, the stage has been set for very significant progress to occur in the area of tyrosine kinase inhibitors in the next few years.

REFERENCES

1. S.K. Hanks and T. Hunter, FASEB J., 9, 576 (1995).
2. D.L. Cadena and G.N. Gill, FASEB J., 6, 2332 (1992).
3. J. Schlessinger and A. Ullrich, Neuron, 9, 383 (1992).
4. P. Vandergeer, T. Hunter and R.A. Lindberg, Ann Rev Cell Biol, 10, 251 (1994).
5. P.J. Chiao, F.Z. Bischoff, L.C. Strong and M.A. Tainsky, Cancer Metast Rev, 9, 63 (1990).
6. T. Hunter, Cell, 64, 249 (1991).
7. S.A. Aaronson, Science, 254, 1146 (1991).
8. L.C. Cantley, K.R. Auger, C. Carpenter, B. Duckworth, A. Graziani, R. Kapeller and S. Soltoff, Cell, 64, 281 (1991).
9. D.R. Kaplan, A. Perkins and D.K. Morrison in "Oncogenes and tumor suppressor genes in human malignancies," C.C. Benz and E.T. Liu, Ed., Kluwer Academic Publishers, Boston, MA, 1993, p. 265.
10. A. Ullrich and J. Schlessinger, Cell, 61, 203 (1990).
11. V.J. Brunton and P. Workman, Cancer Chem. Pharmacol., 32, 1, (1993).
12. G. Powis, Pharmacol. Therap. 62, 57 (1994).
13. A. Levitzki and A. Gazit, Science, 267, 1782 (1995).
14. D.W. Fry, Exp. Opin. Invest. Drugs, 3, 577 (1994).
15. S.P. Langdon and J.F. Smyth, Cancer Treat. Rev., 21, 65 (1995).
16. J.J. Elder, G.J. Fisher, P.B. Lindquist, G.L. Bennet, M.R. Pittelkow, R.J. Coffey, L. Ellingsworth, R. Denrynck and J.J. Voorhees, Science, 243, 811 (1989).
17. R. Vassar and E. Fuchs, Gene & Dev. 5, 714 (1991).
18. D.P. Hajjar and K.B. Pomerantz, FASEB J., 6, 2933 (1992).
19. P. Libby, D. Schwartz, E. Brogie, H. Tanaka and S.K. Clinton, Circulation 86, 47 (1992).
20. R.S. Schwartz, D.R. Holmes and E.J. Topol, J. Am. Coll. Cardiol., 20, 1284 (1992).
21. D.W. Fry and A.J. Bridges, Curr. Opin. Biotechnol., 6, 662 (1995).
22. A.J. Bridges, Exp. Opin. Ther. Patents, 5, 1245 (1995).
23. A.P. Spada and M. R. Myers, Exp. Opin. Ther. Patents, 5, 805 (1995).
24. A.J. Barker and D.H. Davies, EP patent 0,520,722,A1(1992).
25. A.J. Barker, EP patent 0,566,226,A1(1993).
26. D.W. Fry, A.K. Kraker, A. McMichael, L.A. Ambroso, J.M. Nelson, W.R. Leopold, R. W.Connors and A.J. Bridges, Science, 265, 1093 (1994).
27. W.H.F. Ward, P.N. Cook, A.M Slater, H. Davies, G.A. Holdgate and L.R. Green, Biochem. Pharmacol., 48, 659 (1994).
28. N. Osherov and A. Levitzki, Eur. J. Biochem. 225, 1047, (1994).
29. K.E. Hook, M.W. Kunkel, W.L. Elliott, C.T. Howard and W.R. Leopold, Am. Assoc. Cancer Res. 36, 434 (1995).

30. G.W. Rewcastle, W.A Denny, A.J. Bridges, H. Zhou, D.R. Cody, A. McMichael and D.W. Fry, J. Med. Chem., 38, 3482 (1995).

31. A.J. Bridges, H. Zhou, D.R. Cody, G.W. Rewcastle, A. McMichael, H.D.H. Showalter, D.W. Fry, A.J. Kraker and W.A. Denny, J. Med. Chem. 39, 267 (1996).

32. A.J. Bridges, D. R. Cody, H. Zhou, A. McMichael and D.W. Fry. Bioorg. Med. Chem., 3, 1651 (1995).

33. A.J. Barker, EP patent 602,851 (1993).

34. A.J. Barker, EP patent 635,498 (1994).

35. A.J. Barker, EP patent 635,507 (1995).

36. D.W. Fry, A.J. Bridges, A.J. Kraker, A. McMichael M. Nelson and W.A.Denny. Am. Assoc. Cancer Res. 36, 689 (1995).

37. G.W. Rewcastle, B.D. Palmer, A.J. Bridges, H.D.H. Showalter, L. Sun, J.A.Nelson, A. McMichael, A.J. Kraker, D.W. Fry and W.A. Denny. J. Med. Chem. in press (1996).

38. A.J. Bridges, W.A. Denny, D.W. Fry, A.J. Kraker, R. Meyer, G.W. Rewcastle, WO patent 95/19970 (1995).

39. A.J. Bridges, W.A. Denny, D.W. Fry, A.J. Kraker, R. Meyer, G.W. Rewcastle and A.M. Thompson, WO patent 95/19774 (1995).

40. A.M. Thompson, A.J. Bridges, D.W. Fry, A.J. Kraker and W. A. Denny, J. Med. Chem., 38, 3780 (1995)

41. P.M. Traxler, E. Buchdunger, P. Furet, T. Meyer, H. Mett and N.B. Lydon, Med. Chem. Sym., American-Japanese Chemical Society, Tokyo, September 1995.

42. A.P. Thomas, WO patent 95/15952 (1995)

43. U. Trinks, E. Buchdunger, P. Furet, W. Kump, H. Mett, T. Meyer, M. Muller, U. Regenass, G. Rihs, N. Lydon and P. Traxler, J. Med. Chem., 37, 1015 (1994).

44. E. Buchdunger, U. Trinks, H. Mett, U. Regenass, M. Muller, T. Meyer, E. Mcglynn, L.A. Pinna, P. Traxler and N.B. Lydon, Proc. Nat. Acad. Sci. USA, 91, 2334 (1994).

45. E. Buchdunger, H. Mett, U. Trinks, U. Regenass, M. Muller, T. Meyer, P. Beilstein, B. Wirz, P. Schneider, P. Traxler and N.B. Lydon. Clin. Cancer Res., 1, 813 (1995).

46. H.X. Chen, J. Boiziau, F. Parker, P. Mailliet, A. Commercon, B. Tocque, J.B. Lepecq, B.P. Roques and C. Garbay, J. Med. Chem. 37, 845 (1994).

47. H. Takayanagi, Y. Kitano, H. Inokawa and T. Suzuki WO patent 0,645,379,A2 (1995).

48. H. Chen, A. Gazit, K.P. Hirth, A. Levitzki, E. Mann, L.K. Shawver, J. Tsai, and P.C. Tang, WO patent 95/24190 (1995).

49. E. Buchdunger, J. Zimmermann, H. Mett, T. Meyer, M. Muller, U. Regenass and N.B. Lydon, Proc. Natl. Acad. Sci. USA, 92, 2558 (1995).

50. E. Buchdunger, J. Zimmermann, H. Mett, T. Meyer, M. Muller, B.J. Druker and N.B. Lydon, Cancer Res., 56, 100 (1996).

51. M.P. Maguire, K.R. Sheets, K. Mcvety, A.P. Spada and A. Zilberstein, J. Med. Chem. 37, 2129 (1994).

52. R.E. Dolle, J.A. Dunn, M. Bobko, B. Singh, J.E. Kuster, E. Baizman, A.L. Harris, D.G. Sawutz, D. Miller., S. Wang, C.R. Faltynek, W. Xie, J. Sarup, D.C. Bode, E.D. Pagani and P.J. Silver, J. Med. Chem. 37, 2627 (1994).

53. M. Kovalenko, A. Gazit, A. Bohmer, C. Rorsman, L. Ronnstrand, C.H. Heldin, J. Waltenberger, F.D. Bohmer and A. Levitzki, Cancer Res., 54, 6106 (1994).

54. R.L. Dow, B.M. Bechle, T.T. Chou, C. Goddard and E.R. Larson, Bioorg. Med. Chem. Lett., 5, 1007 (1995).

55. R.L. Dow, T.T. Chou, B.M. Bechle, C. Goddard and E.R. Larson, J. Med. Chem., 37, 2224 (1994).

56. C.-K. Huang, F.-Y. Wu and Y.-X. Ai, Bioorg. Med. Chem. Lett., 5, 2423 (1995).

57. M.S. Smyth, I. Stefanova, I.D. Horak and T.R. Burke, J. Med. Chem. 36, 3015 (1993).

58. T.R. Burke, B. Lim, V.E. Marquez, Z.-H. Li, J.B. Bolen, I. Stefanova and I.D. Horak, J. Med. Chem. 36, 425 (1993).

59. C.R. Faltynek, S. Wang, D. Miller, P. Mauvais, B. Gauvin, J. Reid, W. Xie, S. Hoekstra, P. Juniewicz, J. Sarup, R. Lehr, D.G. Sawutz and D. Murphy. J. Enzyme Inhibition, 9, 111 (1995).

60. J.H. Hanke, J.P. Gardner, R.L. Dow, P.S. Changelian, W.H. Brissette, E.J. Weringer, B.A. Pollok and P.A. Connelly, J. Biol. Chem., 271, 695 (1996).

61. S.R. Hubbard, L. Wei, L. Elis and W.A. Hendrickson, Nature, 372, 746 (1994).

62. L. Wei, S.R. Hubbard, W.A. Hendrickson, L. Ellis, J. Biol. Chem., 270 8122 (1995).

Chapter 17. Recent Advances in the Chemistry and Biology of Antimycobacterial Agents

William R. Baker,* Lester A. Mitscher,** Taraq M. Arain,* Ribhi Shawar,* C. Kendall Stover*
*PathoGenesis Corporation, Seattle, WA 98119
**Department of Medicinal Chemistry, Kansas University, Lawerence, KS 66045

Introduction - The resurgence of *M. tuberculosis* in industrialized countries in recent years, after several decades in which the disease was thought to have been conquered, has led to an increased need for improved therapeutics and diagnostics. Also alarming has been the rise of multidrug-resistant (MDR) tuberculosis. The World Health Organization (WHO) has estimated that within 10 years, 30 million people worldwide will die from tuberculosis (1). There is an equally urgent need for new research approaches to help combat the spread of tuberculosis and other mycobacterial infections, including *M. avium* and *M. leprae*. In particular, new knowledge is needed in the areas of *in vitro* and *in vivo* drug susceptibility assay methods, mycobacterial molecular biology, and antimycobacterial drug mechanism of action and drug discovery.

MYCOBACTERIAL MOLECULAR BIOLOGY AND BIOCHEMISTRY

INH Mechanism of Action - With the exception of rifampin, little is known about the mechanisms of action for the drugs used to treat tuberculosis. The principal antitubercular drug, isonicotinic acid hydrazide (INH), is highly specific for mycobacteria of the tuberculosis complex. The reasons for this specificity have yet to be clarified. However, it has become increasingly clear that there is a complex interplay between the mechanism of INH action and the means by which tubercle bacilli resist oxidative stress. Before exhibiting mycobactericidal properties, INH first requires activation by the *M. tuberculosis* KatG catalase-peroxidase. KatG is also thought to be an important determinant for detoxification of reactive oxygen intermediates in the macrophage phagolysosome, the preferred environment for *M. tuberculosis* (2). The KatG mediated oxidation of INH to a potent electrophile proceeds at the expense of hydrogen peroxide which acts as an electron sink for the reaction (3). The activated INH electrophile is then free to interact with any number of cellular nucleophiles. Activated INH appears to inhibit a step in the biosynthesis of cyclopropanated cell wall mycolic acids, which are also thought to form the primary insulating barrier to toxic oxygen species (4). Genetic evidence obtained with a faster growing saprophytic soil mycobacterium indicates that INH exacts its toll on an enoyl-coA hydralase homolog, InhA (5). Recent crystal structure data and *in vitro* studies with purified InhA also suggest that activated INH can covalently modify this enzyme (6,7). However, these data appear to conflict with biochemical findings in the pathogen *M. tuberculosis* which indicate that INH inhibits a desaturase involved in long chain fatty acid biosynthesis. The most common and potent mechanism of resistance to INH involves loss of KatG activity to prevent activation of the drug. However, in doing so, the tubercle bacilli sacrifice one important component of their intracellular survival machinery (KatG) and would be expected to be deficient in the detoxification of peroxides generated by growth in the macrophage phagolysosome. In fact, *M. tuberculosis* KatG mutants are attenuated *in vivo* (2). The resolution of this seeming paradox and the conclusive identification of the enzymatic target for INH in *M. tuberculosis* are under investigation.

METHODS FOR *IN VITRO* EVALUATION OF ANTIMYCOBACTERIAL ACTIVITY

Agar-Based - Present practices and future trends in antimycobacterial susceptibility testing have recently been reviewed by Inderlied (8). There are three agar-based methods for determining the *in vitro* antimycobacterial activity: i) proportion method; ii) absolute concentration method; and iii) resistance ratio method (9-11). The agar-based methods, though very useful for determining the susceptibility of clinical isolate, are not amenable for use in screening as they are slow, cumbersome, often hard to interpret, and require large amounts of test material.

Radiometric - Susceptibility testing of mycobacteria by radiometric (BACTEC™) broth dilution is the preferred procedure for determining MICs, especially for *M. tuberculosis*. The radiometric method measures $^{14}CO_2$ produced by the mycobacteria as they grow in liquid medium supplemented with ^{14}C-labeled palmitic acid (8,11). The BACTEC™ method has significantly reduced the time for evaluation of susceptibility. However, the method has three disadvantages: 1) radioactive waste disposal; 2) high cost; and 3) it is not suitable for high throughput screening.

Intracellular Methods - Since mycobacteria are intracellular pathogens, methods for the determination of antimycobacterial activity in cell culture is important. Cultured macrophages are used for evaluation of the intracellular activity of new compounds and several methods have been published (12-16).

Alternative Methods - As mentioned above, conventional susceptibility methods are too slow for high throughput screening of compounds and chemical libraries. As a result, newer methods for the evaluation of antimycobacterial agents have emerged and a number of innovative approaches have been reported. These include:

1. Modifications of conventional broth dilution. Several groups have reported useful methods for drug susceptibility screening which reduce the time and increase the capacity of the screen. Broth macro- and micro-dilution methods using turbidimetric end-points have been described for rapid- and slow-growing mycobacteria, including *M. tuberculosis* and *M. avium* (17-21). Others have employed similar methodology but used rapid growing mycobacteria, such as *M. smegmatis* as a marker for screening of compounds for antimycobacterial activity (22). Recently, a high throughput broth based screening assay was reported (23). The authors used the saprophyte *M. aurum* as a surrogate marker in a broth based assay in which end-points were determined by measuring uptake of radiolabelled uracil.

2. Bioluminescence-based assays. Measurement of ATP levels by bioluminescence was applied to susceptibility testing of *M. tuberculosis* (24). Strategies for susceptibility testing of *M. tuberculosis* and *M. avium* using luciferase reporter gene assays were also successfully demonstrated (25,26). The highly infectious nature of *M. tuberculosis* precludes its use in a high throughput screen. Therefore, some authors have indicated that bioluminescent *Mycobacterium smegmatis,* produced by insertion of the firefly luciferase gene into plasmids or mycobacteriophages, was an alternative organism which could be employed for screening anti-mycobacterial agents (27,28). A rapid, quantitative broth-dilution method using recombinant strains of bacille Calmette Guerin and *M. intracellulare* expressing firefly luciferase was reported. This approach was used to screen crude plant extracts for antimycobacterial activity in a high throughput format (29). These results compared favorably with data obtained using Alamar blue colorimetric assay. In this, as in other luciferase reporter gene assays, the ability of a compound to inhibit growth of a mycobacterial reporter strain was indicated by a decrease in luminescence.

3. Other.   A new method in which an oxygen-quenched fluorescent indicator is incorporated in the bottom of test tubes containing 7H9 broth has been reported to be effective in detecting drug resistance in *M. tuberculosis* (30). Other novel, but yet to be proven approaches have also been reported including the use of flow-cytometry, gel microdrop encapsulation, hybridization protection assay and paraffin slide culture for *M. avium* (31-34).

Successful propagation of *M. leprae in vitro* is impossible to achieve and growth of *M. leprae* in the mouse footpad remains the only accepted method for drug susceptibility testing (see below). However, the use of cell-free broth culture systems in which biochemical parameters, such as ATP content, uptake of tritiated thymidine, or metabolism of ^{14}C-palmitic acid, have been reported as useful assays in evaluation of drug susceptibility of *M. leprae* and results compared favorably with the classical mouse footpad model (35-38).

METHODS FOR *IN VIVO* EVALUATION OF ANTIMYCOBACTERIAL ACTIVITY

M. tuberculosis   - With the advent of the antibiotic era came the discovery of streptomycin in the 1940s, and the necessity to demonstrate the efficacy of chemotherapeutic agents against *M. tuberculosis* in an animal model.   Mice are ideally suited for efficacy studies as they are easily infected with tuberculosis by either intravenous administration or an inhalation aerosol. Recent work to rank the virulence of several mycobacterial strains in mice demonstrated a change in lung histopathology during infection with all strains, leading eventually to complete destruction of organ architecture (39).   Related work with drug-resistant strains indicated a wide range of virulence in mice, with no correlation to *in vitro* characteristics or degree of resistance (40). This proved contrary to the popular belief that the acquisition of resistance was associated with decreased virulence.

Descriptions of the *in vivo* activity of levofloxacin provides good examples of a clasical tuberculosis mouse protection study (41, 42). Since treatment of tuberculosis in humans requires a multiple antibiotic regimen spanning several months, longer term experiments examining single and combination therapies are often described for a new chemotherapeutic agent, as exemplified by recent studies with rifabutin (43).   The emergence of multidrug resistance has also prompted investigators to test new and old drugs against resistant strains isolated from humans (42, 44). A noteworthy advance which has greatly accelerated *in vivo* drug evaluation studies employs recombinant *M. bovis* BCG expressing firefly luciferase as the challenge organism (45). Measurements of luminescence in spleen homogenates after drug treatment quickly determines efficacy, thereby eliminating laborious and time-consuming procedures associated with determinations of bacterial counts on agar plates.

Guinea pigs are also used in drug studies. One of the major advantages of the guinea pig model is that the course of disease after aerogenic infection is very similar to that in humans, leading to formation of typical granulomatous lesions in the lung (46). However, compared with mice, guinea pigs are expensive and, having greater body weight, require more drug per experiment to deliver comparable doses.   Nevertheless, their use in the evaluation of chemotherapeutic agents remains important (47). Rabbits suffer the same disadvantages and reports of their use in evaluations of antitubercular therapies are infrequent. Of note, however, is the description of an isoniazid co-polymer which, when implanted on the backs of rabbits, allowed release of active drug for up to 63 days. Drug levels measured in serum and urine were fairly high, with urine samples at 6 weeks being capable of inhibiting growth of *M. tuberculosis in vitro* (48).

MAC Group - For the *in vivo* evaluation of chemotherapeutic agents against the MAC group, murine models are employed almost exclusively, with a preponderance of data relating to the beige mouse model of disseminated infection (49). This animal species has a defect in natural killer and T-helper cell functions, mimicking some of the conditions encountered in patients with AIDS. MAC strains from AIDS patients were more virulent in this model than environmental isolates and strains from non-AIDS cases (50). The general basis of the experimental protocol is similar to that used for *M. tuberculosis*: animals are inoculated intravenously with high doses of viable MAC bacteria and, after a period of treatment, organs are removed, homogenized and plated onto agar medium. Bacterial loads for treated and untreated groups are then compared. Lungs, spleens and livers all yield useful data in such studies. The length of treatment can vary from 10 days to several weeks (51-53).

Since the most probable route of infection in AIDS patients is via the gastrointestinal tract, a beige mouse model has been proposed wherein disseminated infection occurs after colonization of the intestinal tract (54). This model may offer advantages in the evaluation of prophylactic oral antibiotics. The prohibitive cost of beige mice has led some investigators to look at immunocompetent mice as a less expensive and more commonly available alternative. In one such case, treatment of normal C57BL/6 mice commenced five weeks after inoculation and continued for a further three weeks, allowing a longer term model of established infection (55).

M. leprae - An experimental model of *M. leprae* infection was not available until 1960, when it was demonstrated that small inocula of *M. leprae* could replicate in the footpads of mice (56). Although infection can also be established at alternative sites in other rodents, the mouse footpad remains the model of choice for evaluation of chemotherapeutic regimens. In immunologically competent mice, the disease process is self-limiting, and levels greater than 10^6 bacteria per footpad are seldom achieved. Immunologically deficient animals, particularly athymic nude mice, support the enhanced multiplication of *M. leprae* (up to 10^9 bacteria per footpad), with the appearance of lesions similar to those encountered in lepromatous leprosy (57). Several new antibiotics have been evaluated in the nude mouse model, including ofloxacin (58), sparfloxacin (59) and KRM-1648 (60). These experiments generally last for five months, although some studies have been described with timespans of up to one year (61).

ANTIMYCOBACTERIAL AGENTS

Rifamycins - The rifamycins and semi-synthetic rifamycins belong to a novel class of macrolide antibiotics that feature a propionate-derived chain bridging a tricyclic napthalene core. The rifamycins are active *in vitro* against Gram-positive bacteria including *M. tuberculosis* and, to a lesser extent, Gram-negative bacteria (62, 63). However, the rifamycins, in particular rifamycin S (**1**), are not orally active and clinical use of these novel macrolides was limited to intraparenteral administration (64). The discovery of rifampicin (**2**) solved the problems associated with poor oral absorption (65), and modification of the C3 position with appropriate substituents produced potent, orally active derivatives. Modification of the C3/C4 position of the rifamycin core continues to produce potent analogs. For example, KRM-1648 (**4**) a novel rifamycin derivative possessing a 3'-hydroxy-5'(piperazinyl)benzoxazino moiety at the C3/C4 position was synthesized and tested for antimycobacterial activity (66). KRM-1648 is more potent *in vitro* and *in vivo* against *M. tuberculosis* and *M. avium* complex than rifampicin. For example, the MIC_{50} and MIC_{90} for KRM-1648 when measured against 30 fresh clinical isolates of *M. tuberculosis* were 0.016 and 2 µg/ml, respectively. These values were considerably lower than those for rifampicin which were 4 and >128 µg/ml, respectively. When tested for *in vitro* activity against *M. leprae*, KRM-1648 displayed potent activity

(MIC 0.05 µg/ml), 6-fold more active than rifampicin. In combination, KRM-1648 and the fluoroquinolone, ofloxacin, were tested together against *M. leprae* at concentrations of 0.006 and 0.375 µg/ml, respectively. Complete inhibition of growth and synergy was observed (67). The effect of both drugs on *M. leprae* was bactericidal. The potent anti-mycobacterial activity of KRM-1648 was largely due to its increased ability to penetrate the mycobacterial cell walls (68). KRM-1648 demonstrated excellent *in vivo* efficacy

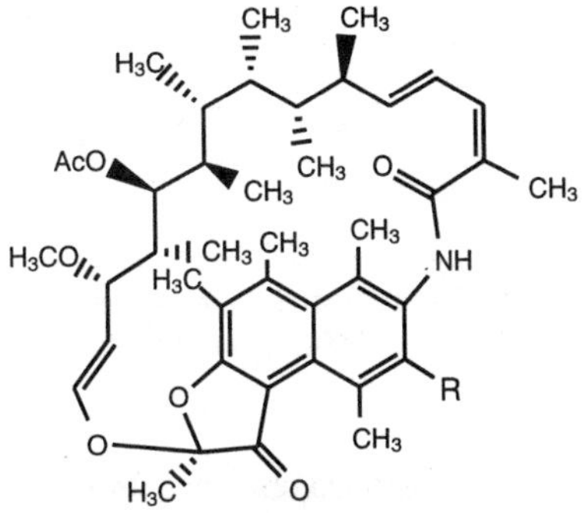

1 R = CHO, Rifamycin SV

R = C=N—N⟷N—R'

2 R' = CH$_3$, Rifampicin
3 R' = CH$_2$CH=CHPh , T9

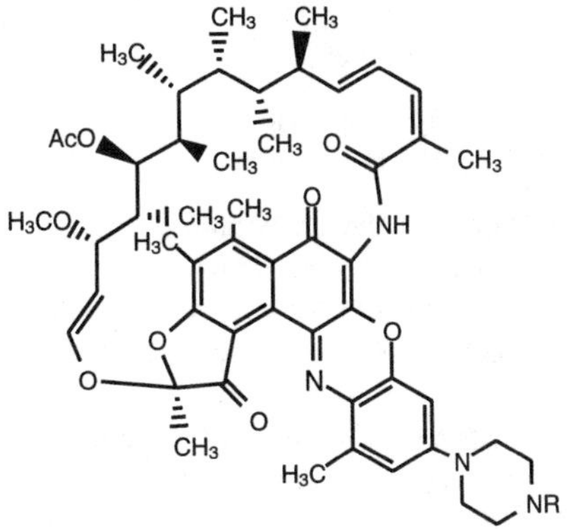

4 R = CH$_2$CH(CH$_3$)$_2$ KRM -1648

against *M. tuberculosis* infection produced by rifampicin-sensitive organisms, having activity in the murine model superior to rifampicin (69). In addition, KRM-1648 achieved high tissue-drug levels in mice (66). The high tissue drug levels and long plasma half-life of KRM-1648 significantly contributes to its excellent *in vivo* efficacy.

A modified rifampicin derivative (T9, **3**) was synthesized and evaluated for anti-mycobacterial activity (70). Like KRM-1648, T9 was more potent *in vitro* against rifampicin susceptible strains of *M. tuberculosis* and *M. avium* complex than rifampicin. MIC$_{90}$ values for T9 against *M. tuberculosis* and *M. avium* were <0.25 and <0.125 µg/ml, respectively. T9 showed high bactericidal activity with MIC values two- to four-fold lower than rifampicin. As expected, T9 displayed excellent *in vivo* efficacy in mice. The increased potency associated with T9 was most likely due to its increased lipophilicity.

Pyrazinoic Acid Derivatives - Pyrazinamide (PZA, **6**) was introduced in 1949 for the treatment of pulmonary tuberculosis (71). When used with INH, PZA was highly effective in treating *M. tuberculosis* infections in the murine model. The combination therapy of PZA, INH and rifampicin represents the first-line treatment for human tuberculosis. The mechanism of action of PZA in unknown; however, researchers have hypothesized that PZA functions as a prodrug of pyrazinoic acid (**5**), being converted to pyrazinoic acid by an intracellular amidases (72). This hypothesis suggested that pyrazinoic acid esters could also serve as prodrugs of pyrazinoic acid. The pyrazinoates would be converted to pyrazinoic acid by mycobacterial esterases rather than amidases. A large series of 5-substituted (R' = F, Cl, and CH$_3$) pyrazinoates were synthesized and evaluated for *in vitro* activity against *M. tuberculosis* and *M. avium* isolates. The SAR revealed that the lipophilic pyrazinoate esters, *n*-decyl (R = O(CH$_2$)$_9$CH$_3$), *n*-pentadecyl (R = O(CH$_2$)$_{14}$CH$_3$) and 5-chloro pyrazinoates (R' = Cl) *n*-heptyl (R = O(CH$_2$)$_6$CH$_3$ and *n*-octyl (R = O(CH$_2$)$_7$CH$_3$ were the most potent derivatives prepared with *in vitro* MIC values against *M. tuberculosis* between 0.03 - 0.5 µg/ml (73). Like PZA, the pyrazinoic esters were inactive against *M. avium.* Derivatives with the 5-chloro substitution were more potent than analogs with 5-

methyl and 5-fluoro substitutions. A related series of two pyrazinoic esters and two thiopyrazinamide derivatives was evaluated for *in vitro* antimycobacterial activity (74). All four compounds studied were bactericidal against *M. avium* and *M. tuberculosis* at 200 and 100 µg/ml, respectively. PZA was not bactericidal at 200 µg/ml.

5 R' = H, R = OH
6 R' = H, R = NH$_2$

7 U -100480

Oxazolidinones - The antibacterial effects of the oxazolidones has been extended to include antimycobacterial activity. When tested against a panel of five drug-sensitive and five drug-resistant *M. tuberculosis* isolates, U-100480 (**7**) had MIC values of 0.03-0.5 µg/ml and 0.125-0.5 µg/ml, respectively (75, 76).

Fluoroquinolones - Because of their potency and breadth of antimicrobial spectrum, the fluoroquinolones are today among the most widely employed synthetic antimicrobial agents. Two members of this class have found significant secondary use as orally active antimycobacterial agents: ciprofloxacin and ofloxacin; and, a number of others show promise in recent studies (sparfloxacin, irloxacin, clinafloxacin and PD 131628, Bay y 3118, and levofloxacin). An emphasis is being placed upon agents which are still quite potent and bactericidal in the more acidic internal environment of the macrophage. Since the present agents were developed as general purpose antimicrobials, it is not immediately obvious that they should turn out to be the optimal fluoroquinolones for any individual pathogen. Comparatively, few detailed studies have appeared in which an extensive retrospective search of existing fluoroquinolone libraries has been made in an attempt to identify an optimal antimycobacterial. In one such study, potency *in vitro* against comparatively fast growing *M. fortuitum* and *M. smegmatis* was used as surrogates for anti-*M. tuberculosis* activity (77). The N-1 substituent was aromatic (selected by analogy to lipophilic temafloxacin, which has been withdrawn from the market because of toxicity problems) and an attempt was made to correlate activity with pKa, c log P and log D. The correlations were less than average, perhaps because activity is a composite of cellular penetration as well as intrinsic ability to inhibit DNA gyrase and/or topoisomerase IV and the structural features dictating these different properties need not be the same. Of particular interest is the finding that those agents possessing best antimycobacterial activity are those with greatest activity against other, Gram-positive and -negative bacteria. These data indicate that the molecular mode of action was the same for all these microorganisms. This idea was further supported by analyses of genetic changes in resistant strains in which the resulting amino acid substitutions take place at the same loci (78). The inference was also made that c log P contributions to the molecule due to the C-7 substituent were more important than those due to the N-1 substituent. None of the agents examined was superior to ciprofloxacin or sparfloxacin although the *in vivo* efficacy of the N-1-aryl series was fairly often under predicted by *in vitro* data. In a more recent study, the contributions of the N-1 substituent were found to decrease from t-Bu > c-Pr > 2,4-Difluorophenyl >Ethyl equiv. to c-Bu and > *i*-Pr (79).

Another group searched through a library from the same source for agents active *in vitro* following 14 days incubation against *M. avium* complex (MAC) (80-82). They measured MIC$_{90}$ values of 88 fluoroquinolones, including ciprofloxacin and sparfloxacin, against 14 MAC strains. The library strongly emphasized N-1-cyclopropyl analogs, including only one t-butyl and one aryl example. While ciprofloxacin and sparfloxacin stood out, some presently undeveloped leads, notably PD 125354 (**8**), were even more potent *in vitro*. Using the MULTICASE computer program to analyze the SAR trends in the series, it was concluded that a C-6-F group and a C-7-NCH2R were optimal, N-1 should be cyclopropyl and that the C-8 substituent should be either CH or CF. Molecular features deleterious to antimycobacterial potency were replacement of C-8 by N and the possession of an N-1-ethyl group.

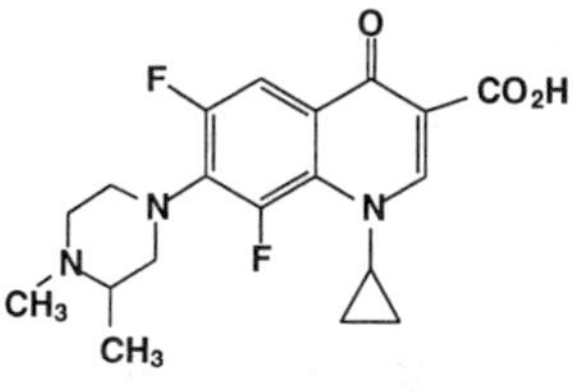

8   PD 125,354 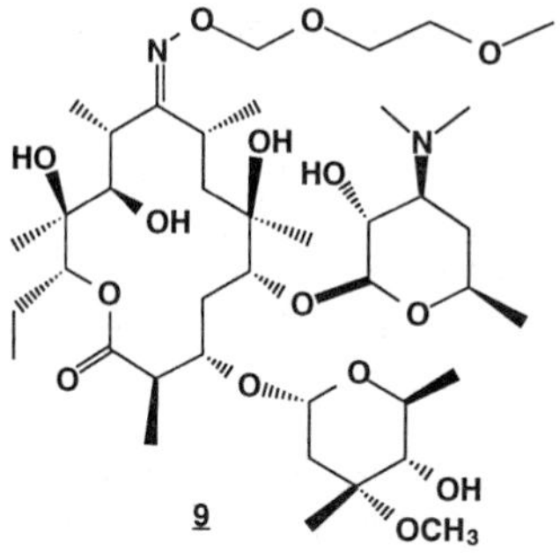   9 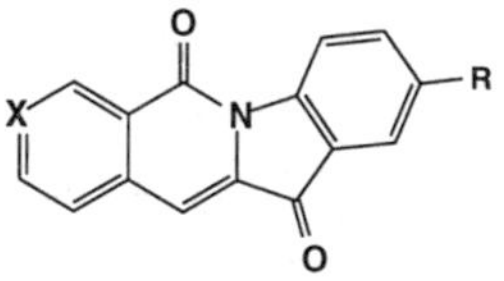   10   X = C H, R = H, Tryptanthrin
                               11   X = N, R = *n*-octyl, PA-342

The recently revealed 2-pyridone analogs show promising potency against mycobacteria, but no detailed study of antimycobacterial SAR has appeared as yet (83).

Macrolides - Classical macrolides have found no particular place in antitubercular chemotherapy. On the other hand, the ability of certain members of this class to concentrate in tissues of the respiratory tract as well as macrophages, neutrophils and T-lymphocytes, has led to the clinical use of newer macrolides (azithromycin, clarithromycin) against opportunistic mycobacteria such as MAC. These newer agents are superior to classic macrolides in producing higher oral blood levels and in causing less G.I. upset. More recent animal studies suggest that roxithromycin (**9**) was comparable to the established agents in its clinical promise (84, 85). Preliminary data indicates that ketolide RU 004 possesses roughly equivalent though inferior *in vitro* activity to clarithromycin against susceptible and resistant opportunistic mycobacteria including MAC, *M. xenopi*, and *M. marinum* (86). *In vivo* data for the ketolides are lacking. Against some strains, however, antagonism is seen with clarithromycin and amikacin as well as with clarithromycin and INH. The molecular interactions underlying these findings is unknown.

Miscellaneous Agents - Tryptanthrin (**10**), an azaindoloquinazolinedione alkaloid from Chinese *Strobilanthes cusia* and other sources, was active *in vitro* against sensitive and multiply resistant *M. tuberculosis* strains. Synthetic studies including combinatorial methods produced more than a hundred analogs with PA-342 (**11**) being the best. Limited oral activity was seen in a murine disease model (87). PS 15 (**12**) , a dihydrofolate

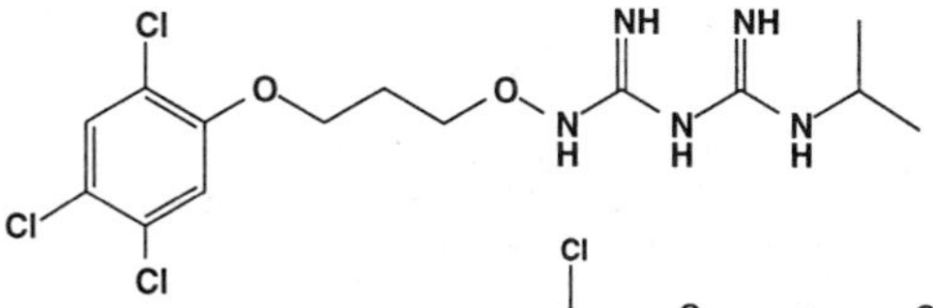

12   PS 15 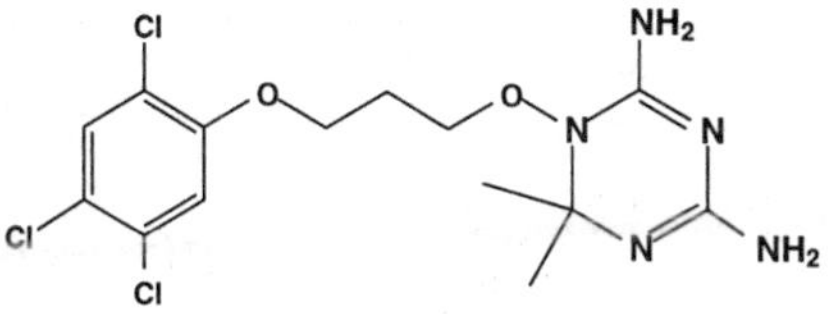   13   WR 99210 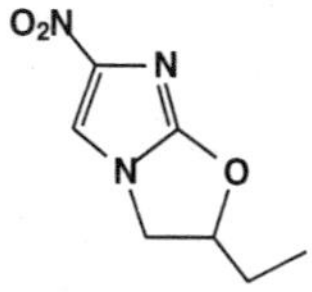   14   CGI 17341

reductase inhibitor, and its more active metabolite WR 99210 (**13**), were active *in vitro* against MAC (16-64 µg/ml) (88). A number of phenanthrolines have shown antimycobacterial activity *in vitro* (intermediate in potency between INH and streptomycin). Interestingly, activity against other microorganisms was greatly enhanced by Cu(II) ions, but cupric ion was not required for antimycobacterial action. Some related dipyridyls have also shown activity. Nitroimidazole CGI 17341 (**14**) was orally active against MDRTB in mice at 8 mg/kg, it was not cross resistant with existing antitubercular compounds, and its action was essentially non pH dependent (89, 90).

A rather long list of miscellaneous agents has been studied. The list includes new aminoglycosides (91), antihistamines (92), beta-lactams (93, 94), colistin (95), dihydromycoplanecin A (96), fusidic acid (97), imidazothiazoles (98), indolizinones (99), indoles (100), phenothiazines (101), pyridines (102), pyrimidines (103), pyrazines (104), rescorcinomycin (105), tetrazoles (106), thiazolidinones (107), thiohydrazides (108) and a variety of as yet undefined plant products (109).

The pathologies of HIV-1 and tuberculosis infections are mutually reinforcing to the patient's detriment. This synergism is not fully understood although it is clear that up regulation of HIV-1 expression in cell lines is stimulated by the mycobacterial cell wall component lipoarabinomannan. Racemic thalidomide inhibits up regulation of HIV-1 expression. Sufficient interest has been generated by this observation to suggest a thalidomide clinical trial in HIV-1/tuberculosis infected patients (110).

References

1.   N. Moran, Nature Medicine, 2, 337, 1996.
2 .   T.M. Wilson, G.W. DeLisle and D.M. Collins, Mol. Micro. 15, 1009 (1995).
3 .   K. Johnsson, P. G. Schultz, J. Am. Chem. Soc., 116, 7425 (1995).
4.   Y. Yuan, R. E. Lee, G. S. Besra, J. T. Belisle, C. E. Barry III, Proc. Natl. Acad. Sci., 92, 6630 (1995).
5.   A., Bannerjee, E. Dubnau, A. Quemard, V. Balasubramanian, K. Sun Um, T. Wilson, D. Collins, G. De Lisle, G and W.R. Jacobs, Jr., Science, 263, 227 (1994).
6 .   A. Dessen, A. Quemard, J.S. Blanchard, W.R. Jacobs, Jr. and J.C. Saccettini., Science, 267, 1638 (1995).
7 .   K. Johnsson, D.S. King, and P.G. Schultz J. Am. Chem. Soc., 117, 5009 (1995).
8.   C.B. Inderlied. Eur. J. Clin. Microbiol. Infect. Dis., 13, 980 (1994).
9.   P.T. Kent and G. P. Kubica, 1985. Public Health Mycobacteriology. A Guide to the Level III Laboratory. Centers for Disease Control, Public Health Service, U.S. Department of Health and Human Services, Atlanta, GA (1985).
10.   D. Hacek., Clinical Microbiology Procedure Handbook , p 5.13.1(1992).
11.   C.B. Inderlied and M. Salfinger, Antimicrobial Agents and Susceptibility Tests: Mycobacteria. pp. 1379, In P. R. Murray et. al (ed.). Manual of Clinical Microbiology, American Society for Microbiology, Washington, DC (1995).
12.   N. Mor, J. Vanderkolk and L. Heifets., Antimicrob. Agents. Chemother., 38, 1161 (1994).
13.   J. Luna-Herrera, M.V. Reddy and P.R.J. Gangadhara,m, Antimicrob. Agents. Chemother., 39, 440 (1995).
14.   C. O. Onyeji, C.H. Nightingale, D.P. Nicolau and R. Quitiliani., Antimicrob. Agents. Chemother., 38, 523 (1994).
15.   N. Rastogi, V. Labrousse and A. Bryskier., Antimicrob. Agents. Chemother., 39, 976 (1995).
16.   N. Osawa, Kitasato Arch. Exp. Med .,64, 213 (1991).
17.   R.J. Wallace, Jr., D. R. Nashi, L. C. Steele, and V. Steingrube, J. Clin. Microbiol., 24, 976(1986).
18.   B.A. Brown, R.J. Wallace Jr., and G.O. Onyi, Antimicrob. Agents Chemother., 36, 1987 (1992).
19.   R.T. Mehta, A. Keyhani, T. J. McQueen, B. Rosenbaum, K. V. Rolstorn, and J. J. Tarrand, Antimicrob. Agents Chemother. 37, 2584 (1993).
20.   R. Gomez-Flores, S. Gupta, R. Tamez-Guerra and R. T. Mehta, J. Clin. Microbiol., 33, 1842 (1995).
21.   D.M. Yajko, J.J. Madej, M. V. Lancaster, C.A Sanders, V.L. Cauthon, B. Ababst, and W.K. Hadley, J. Clin. Microbiol., 33, 2324 (1995).
22.   Y. Boily and L. Van Puyvelde, J Ethnopharmacol., 16,1 (1986).
23.   G.A.C. Chung, Z. Aktar, S. Jackson, and K. Duncan, Antimicrob. Agents Chemother., 39, 2235 (1995).
24.   L.E. Nilsson, S. E. Hoffner, and S. Ansehn, Antimicrob. Agents Chemother., 32, 1208 (1988).
25.   R.C. Cooksey, J.T. Crawford, W.R. Jacobs, Jr., and T.M. Shinnick, Antimicrob. Agents Chemother., 37, 1348 (1993).

26.   R.C. Cooksey, G.P. Morlock, M. Beggs, and J. T. Crawford, Antimicrob. Agents Chemother. 39, 754 (1995).
27.   P.W. Andrew and I.S. Roberts, J. Clin. Microbiol. 31, 2251 (1993).
28.   G. J. Sarkis, W. R. Jacobs, Jr., G. F. Hatfull, Mol. Microbiol. , 15,1055 (1995).
29.   R.M. Shawar, D.J. Humble, V.R. Mugford, C.K. Stover and M.J. Hickey, 95th ASM Genral Meeting,   U-52, (1995).
30.   S.E. Kodsi, S.B. Walters, D.T. Stitt, and B.A. Hanna, 94th ASM Gereral Meeting, C114,(1994).
31.   M.A. Norden, T.A. Kurzynski, S.E. Bownds, S.M. Callister and R.F. Schell., J. Clin. Microbiol., 33, 1231 (1995).
32.   C. Ryan, B.T. Nguyenand S.J. Sullivan. J. Clin. Microbiol., 33,1720 (1995).
33.   J. Miyamoto, H. Koga, S. Kohno, H. Ohno, M. Fukuda, K. Ogawa, K. Tomono, M. Kaku, K. Hara and A. Hashimoto, Kekkaku, J. 70,377 (1995).
34.   R.A. Ollar, S. Brown, J.W. Felder, I.N. Brown, F.F. Eduards and D. Armstrong, Tubercle., 72, 198 (1991).
35.   S.G. Fransblau, J. Clin. Microbiol., 26, 18 (1988).
36.   A.M. Dhople, and I. Ortega., Ind. J. Lepr., 62, 66(1990).
37.   A. M. Dhople, M.A. Ibanez, and A.A. Dhople, Antimicrob. Agents Chemother., 38, 2908 (1994)
38.   L.B. Adams, S.G. Franzblau, Z. Vaurin, J.B. Hibbs, Jr. and J.L. Krahenbuhl. J. Immunol., 147, 1642 (1991).
39.   P.L. Dunn and R.J. North. Infect. Immun. 63, 3428 (1995).
40.   D.J. Ordway, M.G. Sonnenberg, S.A. Donahue, J.T. Belisle, and I.M. Orme, Infect. Immun., 63, 741 (1995)
41.   S.P. Klemens, C.A. Sharpe, M.C. Rogge, and M.H. Cynamon, Antimicrob. Agents Chemother., 38, 1476 (1994).
42.   B. JI, N. Lounis, C. Truffot-Pernot, and J. Grosset, Antimicrob Agents Chemother., 39, 1341 (1995).
43.   D. Jabes, C. Della Bruna, R. Rossi, and P. Olliaro. Antimicrob Agents Chemother.,38, 2346 (1994).
44.   C. Jagannath, M.V. Reddy, S. Kailasam, J.F. O'Sullivan, and P.R. Gangadharam, Am. J. Respir. Crit. Care Med,151, 1083 (1995).
45   M.J. Hickey, T.M. Arain, R.M. Shawar, D.J. Humble, M.H. Langhorne, J.N. Morgenroth, and C. Kendall Stover, Antimicrob. Agents and Chemother., 40, 400-407 (1996).
46.   V. Balasubramanian, E.H. Wiegeshaus, and D.W. Smith, Immunobiology 191, 395 (1994).
47.   D.W. Smith, V. Balasubramanian, and E.H. Wiegeshaus, Tubercle., 72, 223 (1991).
48.   S. Kailasam, D. Daneluzzi, and P.R. Gangadharam, Tuber. Lung. Dis., 75, 361 (1994).
49.   M.A. Bertram, C.B. Inderlied, S. Yadegar, P. Kolanoski, J.K. Yamada, and L.S. Young, J. Infect. Dis., 154, 194 (1986).
50.   V.M. Reddy, K. Parikh, J. Luna-Herrera, J.O. Falkinham 3rd, S. Brown, and P.R. Gangadharam, Microb. Pathog., 16, 121 (1994).
51.   M.H. Cynamon, S.P. Klemens, and M.A. Grossi, Antimicrob. Agents Chemother., 38, 1452 (1994).
52.   S.P. Klemens, M.A. Grossi, and M.H. Cynamon, Antimicrob. Agents Chemother.,38, 234 (1994).
53.   N. Lounis, B. Ji, C. Truffot-Pernot, and J. Grosset. Antimicrob. Agents Chemother., 39, 608 (1995).
54.   L.E. Bermudez, M. Petrofsky, P. Kolonoski, and L.S. Young, J. Infect. Dis. , 165, 75 (1992).
55.   Y. Cohen, C. Perronne, T. Lazard, C. Truffot-Pernot, J. Grosset, J.L. Vilde, and J.J. Pocidalo, Antimicrob. Agents Chemother., 39, 735 (1995).
56.   C.C. Shepard, J. Exp. Med. , 112, 445 (1960).
57.   D.K. Banerjee and R.D. McDermott-Lancaster, Int. J. Lepr. Other Mycobact. Dis., 60, 4107 (1992).
58.   M. Gidoh and S. Tsutsumi, Lepr. Rev., 63, 108 (1992).
59.   M. Gidoh, G. Matsuki, S. Tsutsumi, T. Hidaka, and S. Nakamura, Lepr. Rev., 66, 39 (1995).
60.   M.J. Colston and G.R.F. Hilson, Nature, 262, 399 (1976).
61.   H. Tomioka and H. Saito, Int. J. Lepr. Other Mycobact. Dis., 61, 255 (1993).
62.   P. Sensi, Farmaco, Ed. Sci., 14, 146 (1959).
63.   P. Sensi, Experentia, 16, 412 (1960).
64.   Pallanza, R. Arzneim. Forschung 15, 800 (1965).
65.   N.Maggi, C.R. Pasqualucci, R. Ballotta and P. Sensi, Chemotherapia 11, 285 (1966).
66.   T. Yamane, T. Hashizume, K. Yamashita, E. Konishi, K. Hosoe, T. Hidaka, K. Watanabe, H. Kawaharada, T. Yamamoto and F. Kuze, Chem. Pharm. Bull., 41, 148 (1993).
67.   V. M. Reddy, G. Nadadhur, D. Daneluzzi, V. Dimova, P. R. J. Gangadharam, Antimicrob. Agents Chemother., 39, 2320 (1995).
68.   A.M. Dhople and M. A. Ibanez, J. of Antimicrobial Chemotherapy, 35, 463, (1995).
69.   K. Fujii, H. Saito, H. Tomioka, T. Mae, K. Hosoe, Antimicrob. Agents Chemother., 39, 1489 (1995).
70.   T. Hirata, H. Saito, H. Tomioka, K. Sato, J. Jidoi, K. Hosoe, T. Hidaka, Antimicrob. Agents Chemother., 39, 2295 (1995).
71.   R.L. Yeager, W.G.C. Munroe and F.I. Dessau Am. Rev. Tuberc., 65, 523 (1952).
72.   R.J. Speirs, J. T. Welch, M. H. Cynamon, Antimicrob. Agents Chemother., 39, 1269 (1995).

73.   M. H. Cynamon, R. Gimi, F. Gyenes, C. A. Sharpe, K. E. Bergmann, H. J. Han, L. B. Gregor, R. Rapolu, G. Luciano, J. T. Welch, J. Med. Chem., 38, 3902 (1995).

74.   S. Yamamoto, I. Toida, N. Watanabe and T. Ura, Antimicrob. Agents Chemother., 39, 2088 (1995).

75.   M.R. Barbachyn, S.J. Brickner, D.K. Hutchinson, J.O. Kilburn, G. E. Zurenko, S. Glickman, 35th ICAAC, (San Francisco), Abstr. F227, (1995).

76.   M.R. Howard, I.J. Martin, M.J. Ackland, M.L. Sedlock, R.A. Anstadt, M.R. Barbachyn, 35th ICAAC, (San Francisco), Abstr. F230, (1995).

77.   T.E. Renau, J.P. Sanchez, M.A. Shapiro, J.A. Dever, S.J. Gracheck and J.M. Domegala, J. Med. Chem., 38, 2974 (1995).

78.   G.J. Alangaden, E.K. Manavathu, S.B. Vakulenko, N.M. Zvonok, S.A. Lerner, Antimicrob. Agents Chemother., 39, 1700 (1995).

79.   T.E. Renau, J.P. Sanchez, J.W. Gage, J.A. Dever, M.A. Shapiro, S.J. Gracheck and J.M. Domegala, J. Med. Chem., 39, 729 (1996).

80.   G. Klopman, S. Wang, M.R. Jacobs, S. Bajaksouzian, K. Edmonds and J.J. Ellner, Antimicrob. Agents Chemother., 37, 1799 (1993).

81.   G. Klopman, S. Wang, M.R. Jacobs and J.J. Ellner, Antimicrob. Agents Chemother., 37, 1807 (1993).

82.   M.R. Jacobs, Drugs 49 (S2), 67 (1995).

83.   A.K.L. Fung, D.T. Chu, Y.L. Armiger, Q. Li, S.K. Tanaka, R.K. Flamm, L. Shen, J. Baranowski, K. Marsh, D. Crowell and J. Plattner, 35th ICAAC, (San Francisco), Abst. F9, 114 (1995).

84.   N. Rastogi, K.S. Goh, P. Ruiz and M. Casal, Antimicrob. Agents Chemother., 39, 1162 (1995).

85.   L. Struillou, Y. Cohen, N. Lounis, G. Bertrand, J. Grosset, J.-L. Vilde, J.-J. Pocidalo and C. Perronne, Antimicrob. Agents Chemother., 39, 878 (1995)

86.   C. Truffot-Pernot, N. Lounis, J.F. Chantot and J. Grosset, Abstr. 35th ICAAC, (San Francisco), F167, 142 (1995)

87.   W.R. Baker, L.A. Mitscher, B. Feng, S. Cai, M. Clark, T. Leung, J.A. Towell, I. Darwish, K. Stover,
B.   Kereiswirth, S. Moghazeh, T. Henriquez, A. Resconi and T. Arain, 35th ICAAC, (San Francisco), Abst. F17, 116 (1995)

88.   S.C.C. Meyer, S.K. Majumder and M.H. Cynamon, Antimicrob. Agents Chemother., 39, 1862 (1995).

89.   K. Nagrajan, R. G. Shankar, S. Rajappa, S. J. Shenoy, R. Costa-Perira Eur. J. Med. Chem., 24, 631 (1989).

90.   D.P. Ashtekar, R. Costa-Perira, K. Nagrajan, N. Vishvanathan, A.D. Bhhatt, W. Rittel, Antimicrob. Agents Chemother., 37, 183 (1993).

91.   T. Yokota, N. Kato and R. Nozawa, Chemotherapy (Tokyo), 33S5, 22 (1985).

92.   W. Meindl, Arch. Pharm. 322, 493 (1989).

93.   M.-H. Yang, J.-S. Lin, Y.-C. Lee and R.-P. Perng, Proc. Natl. Science Council ROC, B, Life Sciences, 19, 80 (1995).

94.   L.J. Utrup, T.D. Moore, P. Actor and J.A. Poupard, Antimicrob. Agents Chemother., 39, 1454 (1995).

95.   N. Rastogi, M.C. Potar and H.L. David, Ann. Inst. Pasteur/Microbiol., 137A, 45 (1986.)

96.   T. Haneishi, M. Nakajima, A. Shiraishi, T. Katayama, A. Torikata, Y. Kawahara, K. Kurihara, M. Arai and T. Arai, Antimicrob. Agents Chemother., 32, 110 (1988.)

97.   S.E. Hoffner, B. Olsson-Liljequist, K.J. Rydgard, S.B. Svenson and G. Kaellenius, Eur. J. Clin. Microbiol. Infect. Dis. 9, 295 (1990).

98.   Z. Cesur, H. Guner and G. Otuk, Eur. J. Med. Chem., 29, 981 (1994).

99.   G. Dannhardt, W. Meindl, S. Gussmann, S. Ajili and T. Kappe, Eur. J. Med. Chem., 22, 505 (1987).

100.   S. Mahboobi, T. Burgemeister and F. Kastner, Arch. Pharm., 328, 29 (1995).

101.   A.J. Crowle, G.S. Douvas and M.H. May, Exptl. Chemotherapy 38, 410 (1992).

102.   J.K. Seidel, H. Van der Goot and H. Timmerman, Chemotherapy , 40, 124 (1994).

103.   G. Hachtel, R. Haller and J.K. Seydel, Arzneim.-Forsch., 38, 1778 (1988).

104.   M. Dolezal, J. Hartl, A. Lycka, V. Buchta and Z. Odlerova, Coll. Czech. Chem. Commun., 60, 1236 (1995).

105.   S. Masaki, T. Konishi, N. Tsuji and J. Shoji, J. Antibiotics, 42, 463 (1989).

106.   K. Waisser, J. Kunes, A. Hrabalen and Z. Odlerova, Coll. Czech. Chem. Commun., 59, 234 (1994).

107.   A. Solankee and K. Kapadia, J. Inst. Chem. (India), 65, 95 (1993).

108.   K. Waisser, Folia Pharm. Un. Carolinae, 16, 15 (1992.)

109.   L. Van Puyvelde, J.D. Ntawukiliyayo, F. Portaels and E. Hakizamungu, Phytotherapy Res., 2, 65 (1994).

110.   P.K. Peterson, G. Gekker, M. Bornemann, D. Chatterjee and C.C. Chao, Antimicrob. Agents Chemother., 39, 2807 (1995).

Chapter 18. Ras Farnesyltransferase Inhibitors

Semiramis Ayral-Kaloustian and Jerauld S. Skotnicki
Wyeth-Ayerst Research, Pearl River, NY 10965-1299

<u>Introduction</u> - Mammalian H-, K-, and N-Ras proteins, encoded by H-, K-, and N-*ras* proto-oncogenes, respectively, are 21 kD GTP-binding proteins which possess intrinsic GTPase activity and play a fundamental role in cell proliferation and differentiation. The Ras proteins have been the subject of intense study over the last decade and comprehensive reviews of their structural properties and biochemical function have appeared (1,2). Specific mutations in the *ras* gene impair GTPase activity of Ras, leading to uninterrupted growth signals and to the transformation of normal cells into malignant phenotypes. It has been shown that normal cells transfected with mutant *ras* gene become cancerous and that unfarnesylated, cytosolic mutant Ras protein does not anchor in cell membranes and cannot induce this transformation (2,3). Posttranslational modification and plasma membrane association of mutant Ras is essential for this transforming activity. The first and required step in the processing of Ras is farnesylation at the cysteine residue of its carboxyl terminal motif, CAAX (C = Cys-186, A = aliphatic amino acid, X = usually methionine, serine or glutamine).

<u>Therapeutic Rationale</u> - Mutant *ras* oncogenes are found in approximately 25% of all human cancers, including 90% of pancreatic, 50% of colon, and 50% of thyroid tumors (4). Several potential strategies for interfering with Ras function, such as antisense oligonucleotides, Ras-specific antibodies, and blockers of Ras/protein interactions, have been under investigation for several years, with emphasis on inhibition of posttranslational modification of Ras (1,2,5-12). Since its identification, the enzyme protein farnesyl transferase (FTase) that catalyzes the first of these processing steps has emerged as the most promising target for therapeutic intervention (1,13-20). In the last two years, major milestones have been achieved with small molecules that show efficacy without toxicity *in vitro* as well as in mouse models bearing *ras*-dependent tumors or human xenografts with H-, N-, or K-*ras* mutations. This chapter will review the progress that has been reported with FTase inhibitors.

RATIONAL DESIGN-BASED INHIBITORS

In the absence of a published X-ray structure for FTase, rational design of inhibitors has been spurred by information on substrate specificities of the enzyme, hypothetical active-site models, and substrate conformations (*e.g.*, β-turn) derived from molecular dynamics or NMR studies of enzyme-bound substrates and their analogs (1,19).

<u>Analogs of the CAAX Sequence or β-Turn Mimics</u> - It has been shown that the CAAX tetrapeptide motif of Ras is sufficient for binding to FTase and is the smallest substrate for farnesylation. The nature of the terminal amino acid X confers specificity to the peptide, in terms of recognition by FTase *vs.* GGTase (geranylgeranyl transferase). The CAAX analogs and early peptide mimics, including the β-turn mimic <u>1</u> (BZA-2B; IC_{50} = 0.85 nM), an *in vitro* inhibitor which was designed to fit the proposed enzyme-bound conformation of the tetrapeptide, have been reviewed (1,5,19). The absolute

configuration of the active diastereomer of **1** was determined, thus shedding more light on the structural requirements of the peptide binding site of the enzyme (21). Recent enzyme studies have shown that **1** inhibits farnesylation of K-Ras at 8-fold higher concentrations than those required to inhibit farnesylation of H-Ras. These findings underscore the importance of using K-Ras-based assays in order to identify potential anti-cancer agents for human cancers that are known to contain predominantly oncogenic K-Ras4B. In H-*ras*-transformed cells, but not in untransformed cells, the ester (**2**; BZA-5B) reduces the activities of enzymes further downstream of Ras in the signaling pathway, suggesting that normal growth continues in untransformed cells through prenylation of other forms of Ras not inhibited by the compound (22,23). Furthermore, **2** blocks farnesylation of various cellular proteins, but does not interfere with their localization or function, including cell growth or viability, suggesting unknown salvage pathways or farnesylation-independent mechanisms for normal growth (24).

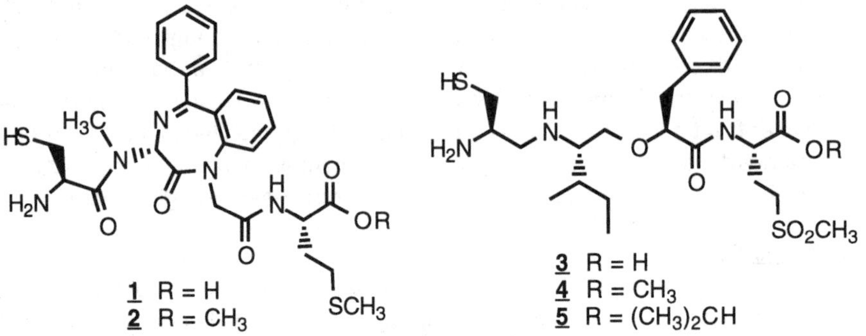

1 R = H
2 R = CH₃

3 R = H
4 R = CH₃
5 R = (CH₃)₂CH

Further progress was made in understanding the spatial arrangements for inhibitor binding by the NMR determination that the FTase-bound conformations of analogs related to the potent FTase inhibitor **3** (IC$_{50}$ =1.8 nM) approximate a type III β-turn, and that the backbone conformation is not significantly perturbed by minor changes in the methionine residue (25). Series of analogs related to **3**, bearing a flexible backbone and a variety of amide bond replacements have been reported (26). The prodrugs **4** (L-739,749) and **5** (L-744,832) have been studied extensively in transformed cell lines and in human tumor cells (27-30). Compound **4** afforded the first proof of *in vivo* activity in nude mice bearing H-, N- and K-*ras* tumors (51-66% inhibition of tumor growth), and **5** provided the first demonstration of complete regression of pre-existing, doxorubicin-insensitive tumors in H-*ras* transgenic mice, with no systemic toxicity to the animals. New studies in *ras*-transformed cells treated with **4** revealed significant suppression of the angiogenic factor, VEGF, suggesting that *ras* oncogenes may contribute indirectly to tumor angiogenesis *in vivo*, and that FTase inhibitors may be controlling tumor growth not only by direct inhibition of proliferation but also by suppression of angiogenesis (31). The role of **4** on vesicular localization of Rho proteins was investigated, and its ability to suppress malignant growth was shown to be at least in part due to its interference with Rho, which is necessary for Ras transformation (32).

The strategy of replacing the two internal aliphatic amino acids of the CAAX sequence with a hydrophobic, non-peptidic spacer has provided potent, specific (FTase *vs*. GGTase) inhibitors which are, in general, competitive with the peptide but not substrates for farnesylation. Studies with compound **6**, its prodrug **7** (FTI-244), and other related rigid analogs demonstrated that the β-turn conformation, previously presumed necessary, was not required for inhibition of FTase (33-35). Compound **8**, with the cysteine amide bond reduced, exhibited a 6-fold lower potency relative to **6**. Introduction of a phenyl group to increase the hydrophobicity of the central portion of the molecule, and to fully occupy the FTase binding pocket, as with the tetrapeptides CVFM and CVIM (the terminal sequence of K-Ras), provided **9**. This analog was found to be 400-fold more potent ($IC_{50} = 0.5$ nM) than **8**, and 100-fold selective for FTase over GGTase I (33). Substitution of leucine for methionine in **9** gave, GGTI-287, which exhibited 5-fold selectivity for GGTase I ($IC_{50} = 5$ nM) over FTase ($IC_{50} = 25$ nM) and was more effective than **9** in inhibiting oncogenic K-Ras4B processing and signaling in cells (36). A similar substitution in another peptide mimic reportedly also enhanced GGTase inhibition over FTase by 40-fold (37). In nude mice, both **9** and its ester **10** (FTI-277) blocked the growth of tumors bearing the two most prevalent genetic alterations found in human cancers, namely K-Ras mutation and p53 deletion, with no evidence of toxicity even with prolonged dosing at four times the efficacious dose (38). Tumor inhibition was shown to be Ras-dependent, with no inhibition of normal signaling. The compound induced cytoplasmic accumulation of inactive, non-farnesylated Ras/Raf complexes, hence preventing downstream signaling events in the transformed cells, but not in normal cells (39). A less potent analog, PM-061, similar to **9** with the cysteine amide bond replaced by an E-alkene moiety, was reported to have good enzymatic and *in vivo* activity, but poorer selectivity (40).

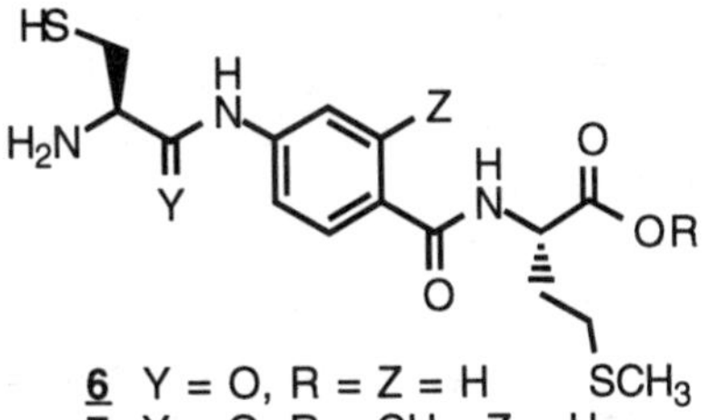

6 Y = O, R = Z = H SCH_3
7 Y = O, R = CH_3, Z = H
8 Y = H, H, R = Z = H
9 Y = H, H, R = H, Z = Ph
10 Y = H, H, R = CH_3, Z = Ph

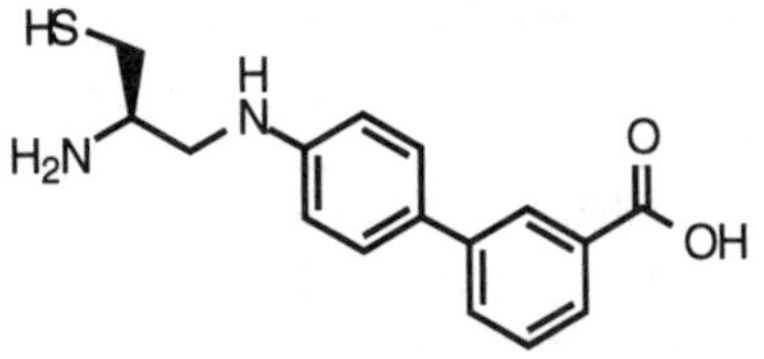

11

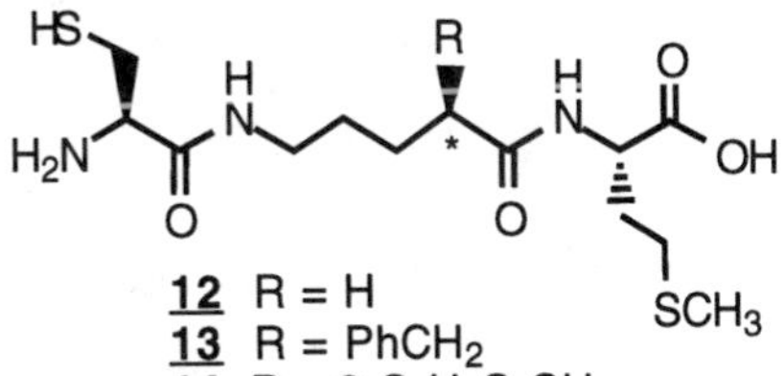

12 R = H SCH_3
13 R = $PhCH_2$
14 R = 2-C_4H_3S-CH_2

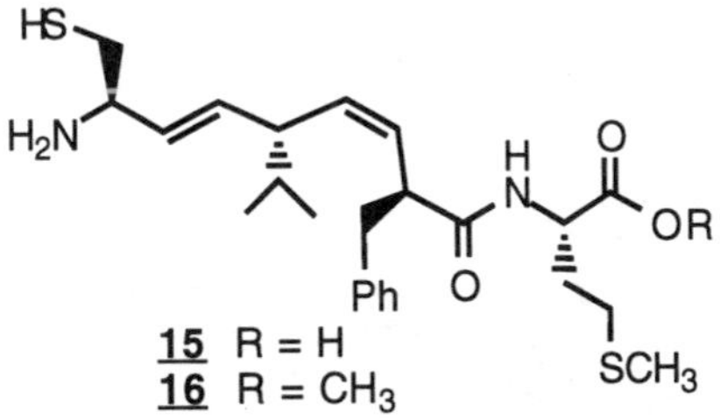

15 R = H SCH_3
16 R = CH_3

Compound **11** (FTI-265), a non-peptide mimetic of CAAX, was designed to incorporate the key pharmacophores on a scaffold with restricted rotation, positioned at a proper distance from each other, as suggested by molecular modeling of the extended conformation of CVIM. Despite the lack of methionine, **11** was more potent ($IC_{50} = 150$ nM) than **8** and highly selective for FTase (666-fold over GGTase) and, unlike **8** or other CAAX mimics, did inhibit Ras processing in cells without having to mask the carboxylate as an ester. Both the presence and the position of the carboxyl group were crucial for this activity, corroborating information derived from molecular modeling (33,41).

Non-aromatic spacers have also been used to link the cysteine and methionine residues (42, 43). The importance of the bulk and hydrophobicity of the spacer for enhanced binding was demonstrated further with compound **12** (IC_{50} = 6600 nM; R = H) vs. **13** (IC_{50} = 20 nM, with 205-fold specificity; R = $PhCH_2$, S-configuration at marked center), differing only in the absence or presence of the phenylmethyl group attached to the backbone. Substitution of thiophene for phenyl gave an even more potent (IC_{50} = 8 nM) and selective (638-fold) inhibitor, **14**. Corresponding analogs of **13** and **14** of R-configuration were also potent, but less selective. Reduction of the N-terminal amide bond of **12** resulted in diminished potency and specificity. A variety of more rigid spacers, including alkenes in the carbon backbone, were explored (26,44). The potent (IC_{50} = 12 nM, FTase) compound **15** and its ester **16** (B1086) were found to inhibit the growth of cells expressing H-, N-, and K-*ras*, and the growth of human tumor xenografts in mice. Differences were observed in the sensitivities of cell lines, and some tumors that were not *ras*-transformed were also sensitive to higher doses of **15**. These studies suggest that Ras dependency of tumor growth may need to be determined before a treatment protocol can be chosen, and that FTase inhibitors may be effective in the treatment of certain tumors that do not express oncogenic *ras*.

The replacement of the phenylalanine residue of CVFM-type mimics with unnatural amino acids, which confer rigidity and metabolic stability to the peptide, afforded highly potent inhibitors that fit the spatial requirements and the preferred extended conformation predicted by molecular modeling studies. Compounds **17-21**, were reported to inhibit FTase with IC_{50} values below 1 nM, however the specificity relative to GGTase inhibition varied dramatically depending on the nature of the R and Y groups (approximately 4000-, 200-, and 10-fold, respectively, for **18, 19** and **20**). Replacement of the methionine branch of **19** by an *n*-butyl group resulted in a potent and selective GGTase inhibitor. Various analogs in this series were active in cells. The ester prodrug of **20** exhibited activity in mice (45-48).

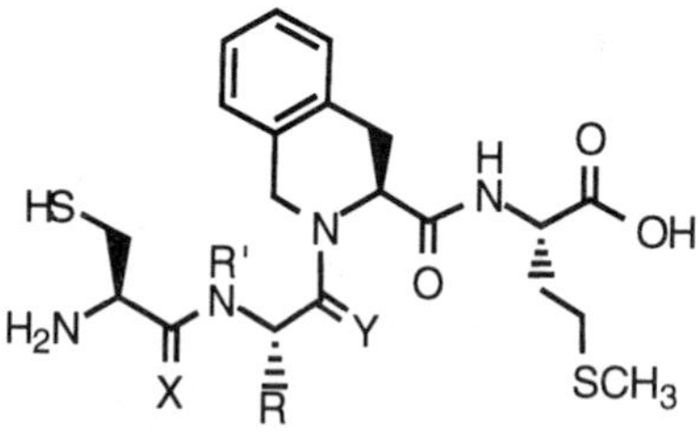

17 X = Y = O, R = $(CH_3)_2CH$, R'= CH_3
18 X = Y = H,H, R = $(CH_3)_2CH$, R'= H
19 X = H,H, Y =O, R = $(CH_3)_2CH$, R'= H
20 X = H,H, Y =O, R = $(CH_3)_3C$, R'= H
21 X = Y = H,H, R = $CH_3(C_2H_5)CH$, R'= H

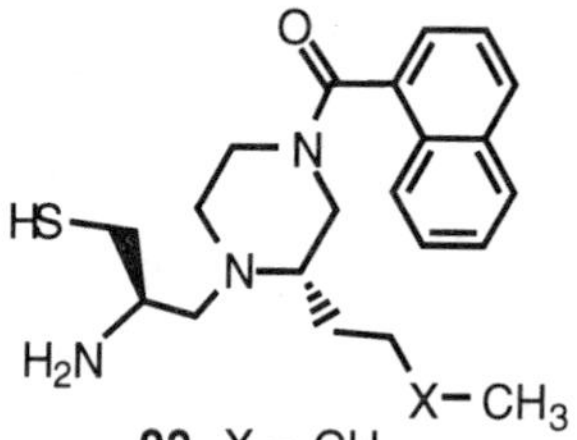

22 X = CH_2
23 X = O

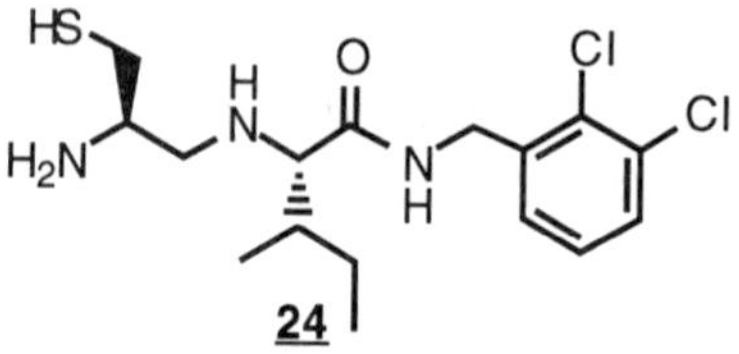

24

Rigid inhibitors utilizing a piperazine scaffold to support the key pharmacophores were reported (49). These non-peptidic compounds lack the terminal carboxyl group which was deemed necessary for binding to FTase, and have a reduced cysteine amide with only an appendage to mimic the methionine chain. Despite this lack of most of the CAAX motif, compounds such as **22** and **23** were found to be potent inhibitors of FTase (2 and 5 nM, respectively, with FTase and 0.5 µM in cells). A further example of truncation of the carboxyl terminal was provided by a series of reduced dipeptides linked to a hydrophobic group, as depicted in **24** (23 nM, FTase). These compounds were shown to be active and specific without the terminal amino acid, presumed to confer specificity, but their usefulness was limited by the observed non-mechanism-based cytotoxicity (50). Rigid analogs of **24** have been reported recently (51).

Cysteine replacements have been designed in an effort to make the CAAX mimics more stable, cell permeable, and/or bioavailable. Inhibitor **25** (IC_{50} = 0.79 nM), derived from **19** by replacing the thiol and amine of cysteine with a chemically stable imidazole ring, displayed good cellular activity (IC_{50} = 3.8 µM) without the need for prodrug formation. No cytotoxicity was observed with untransformed cells at a 40-fold higher concentration of the compound (52). Related structures with different N-heterocycles instead of the imidazole are under investigation (53). Phenol-based thiol replacements in CVLS and CVFM mimics, as depicted in structures **26** and **27**, have proven to be less effective (IC_{50} = 29 and 0.45 µM, respectively, for FTase). Steric and electronic effects of subtitutuents on the phenol ring, and the position of the phenolic hydroxyl group appeared to influence activity dramatically (54,55). Less potent inhibitors lacking the thiol group and incorporating N-terminal lactam, carbamate or thiocarbamate rings have also been reported (56). Thus, if positioned at the correct orientation, a group that can provide a site for the putative zinc coordination may act as a surrogate for the key thiol moiety. Pseudopeptides related to **19**, have been reported (46). Prodrug variations beyond the standard carboxyl, thiol or amine protective groups are also emerging (57).

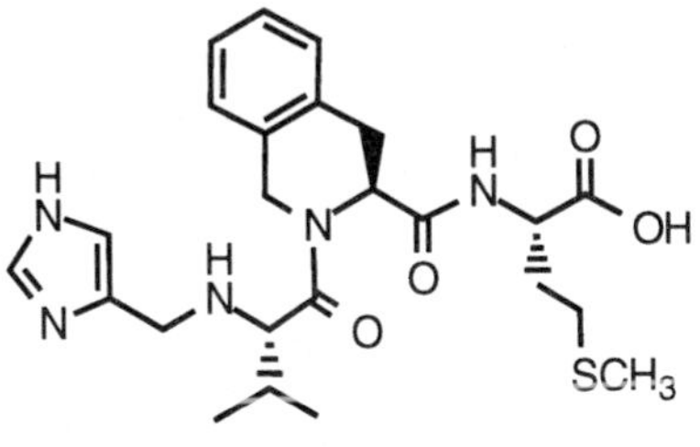

25

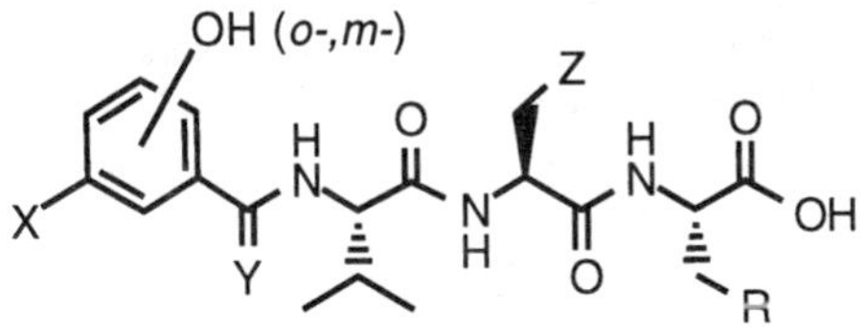

26 X = H, Y = O, Z = (CH$_3$)$_2$CH, R = OH; *meta*-OH

27 X = Br, Y = H,H, Z = Ph, R = CH$_3$SCH$_2$; *ortho*-OH

Bisubstrate Analogs - A hypothetical active site model for the farnesylation of Ras has been used for the rational design of bisubstrate inhibitors of FTase, which were expected to exhibit better affinity and specificity for the enzyme than either substrate alone. Stable analogs were provided by replacement of the labile sulfhydryl group with functional equivalents which can interact with putative metals in the active site. The intact farnesyl group and the AAX terminal portion of the peptide were used with various flexible linkers to mimic the collected substrate model or the transition state of the transformation. The carboxylic acid-based inhibitors **28** and **29** were evaluated as 1:1 diastereomeric mixtures (IC_{50} = 0.9 and 0.15 µM, respectively, FTase) and **30** was synthesized in enantiomerically pure forms (S and R, IC_{50} = 0.033 and 3.4 µM, respectively, FTase), revealing the importance of the linker configuration (58). No cell activity was reported for this series, while the more potent phosphinic acid-based analog **31** exhibited specificity for FTase (IC_{50} = 6 nM) and its prodrug **32** (BMS-186511) was active in H-*ras*-transfected cells at 100 µM concentration and less effective in K-*ras*-transfected cells (59,60). The latter was also reported to inhibit the growth properties of the malignant phenotype of neurofibromatosis type I, whose growth depends on up-regulation of wild-type Ras (61). Although stable and specific inhibitors have been synthesized and potency relative to the original tetrapeptides has been enhanced by the bisubstrate approach, bioavailabilty and potency in cells relative to the newer CAAX mimics appears poor. There is no indication as to whether some of these compounds may also inhibit endoproteases or other processing enzymes, as reported for some farnesyl-CAAX analogs (10,11).

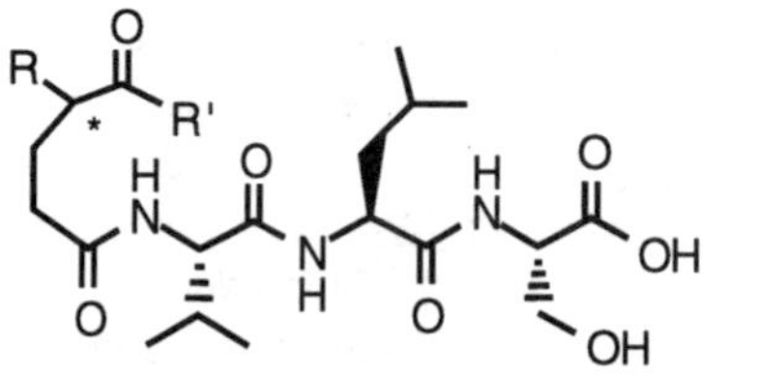

28 R = Farnesyl, R' = OH
29 R = Farnesyl, R'= NHCH$_2$PO$_3$H$_2$
30 R = Farnesyl-CONH, R' = OH

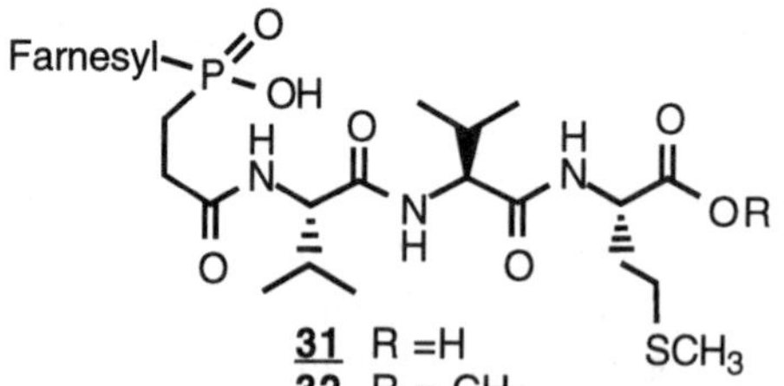

31 R =H
32 R = CH$_3$

Farnesyl =

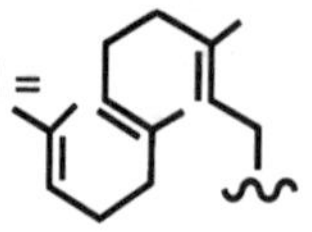

Analogs of Farnesyl Pyrophosphate - The design of non-substrate analogs of farnesyl pyrophosphate as inhibitors of the farnesylation reaction continues to be explored. Amide **33** was identified as a potent, selective inhibitor of porcine brain FTase (IC_{50} = 75 nM). Methylation of the amide or transpositioning of the amide nitrogen and carbonyl groups furnished potent congeners (IC_{50} = 530 and 50 nM, respectively). Alteration of

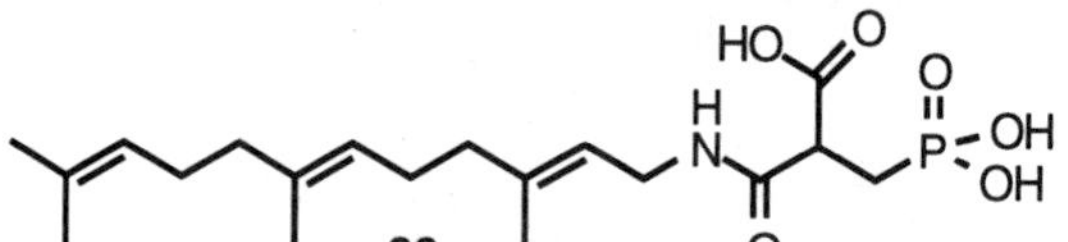

33

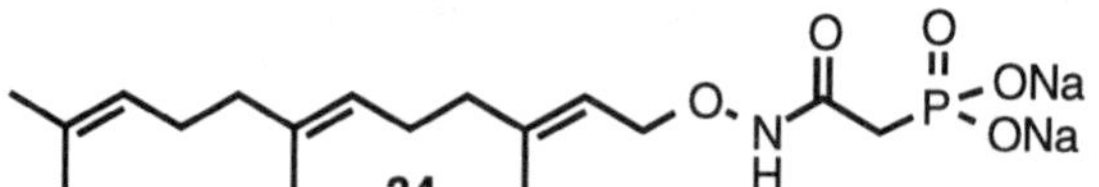

34

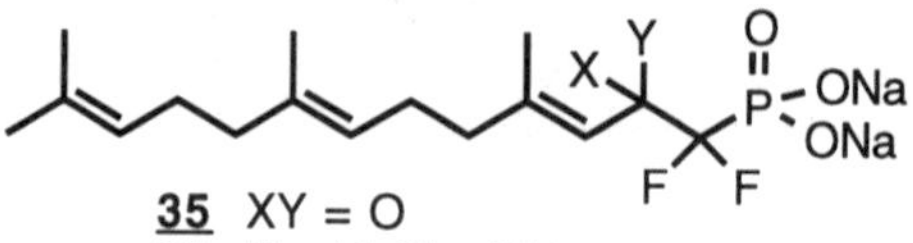

35 XY = O
36 X = H, Y = OH

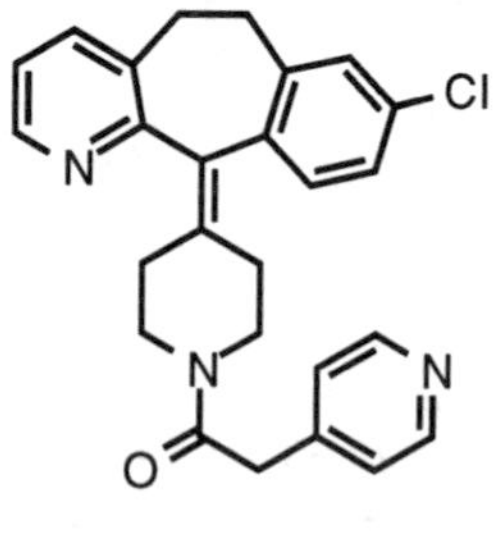

37

the farnesyl portion, removal of the carboxylic acid, or replacement of the phosphonic acid with a carboxylic acid gave compounds with significantly diminished potency (62).

The related hydroxamate **34** displayed potent and selective *in vitro* inhibition (IC_{50} = 0.075 μM and 24.5 μM, FTase and GGTase, respectively), but was ineffective in whole cells (NIH 3T3). Its mono pivaloyloxymethyl ester prodrug exhibited reduced potency *in vitro* (IC_{50} = 2.6 μM) but did inhibit Ras processing in cells at 100 μM (63). The introduction of α–gem-difluoro substitution was shown to be consistent with improved *in vitro* potency. Thus, ketone **35** (IC_{50} = 350 nM) and alcohol **36** (IC_{50} = 780 nM) were 5 and 15 times more potent than the corresponding non-fluorinated analogs (64).

SCREENING-BASED INHIBITORS

<u>Natural Products</u> - Screening of microbial extracts continues to aid in the identification of novel inhibitors. Preussomerins and deoxypreussomerins, keto-epoxides which contain spiroketal unit(s), represent two classes of fungal metabolites whose members display interesting activity (IC_{50} = 1.2-17 μM) *vs.* bovine brain FTase (65). Related keto-di-epoxides SCH 49209 and its acetyl derivative SCH 50672 demonstrated inhibition of the growth of *ras*-transformed cells (IC_{50} = 0.6 and 2.4 μM, respectively) and of murine epidermoid carcinoma M27 cells implanted in mice (2.5 mg/kg, i.p.). Both compounds, which were inactive by i.v. and p.o. administration, showed toxic effects (weight loss) at accumulated doses of 40 mg/kg (66). Cylindrol A, a bicyclic compound in which a resorcinol carboxaldehyde is tethered to a functionalized cyclohexanone by a substituted pentene, was found to be a non-competitive inhibitor (IC_{50} = 2.2 μM) of bovine brain FTase, but inactive against GGTase (67). Actinoplanic Acids A and B, 30-carbon highly functionalized acids, exhibited IC_{50}'s of 230 and 50 nm, respectively, against human recombinant FTase, and no effect against human squalene synthase or bovine brain GGTase (68-70). Di-epoxybenz[a]anthracene SCH 58450 was found to be a selective inhibitor of FTase *vs.* GGTase with IC_{50} values of 29 and 740 μM, respectively (71). Fusidienol, a tricycle composed of benzopyran fused to an oxepine ester, inhibited bovine brain and human recombinant FTase (IC_{50} = 300 and 2700 nM, respectively) and was inactive against rat liver squalene synthase and bovine brain GGTase (72). Radiciol (UCS 1006), a macrolactone which is comprised of an epoxide, dienone, and a *m*-resorcinol, suppressed the growth of *ras*- and *mos*-transformed NIH 3T3 cells at 0.025 and 0.0125 μg/ml, respectively (73). A novel vinca alkaloid III-121C was reported to be the first alkaloid that inhibits Ras functions (IC_{50} = 0.29 μg/ml) when evaluated in K-Ras-NRK cells (74). An octahydronaphthalene carboxylic acid further functionalized with an acylimide, TAN-1813, inhibited FTase activity and growth of K-ras NIH 3T3 cells by 50% (12 μg/ml and 0.4 μg/ml, respectively) and, at 50 mg/kg, reduced the size of tumors resulting from human fibrosarcoma cells grafted onto nude mice (75).

<u>Small Molecules</u> - Tricycle SCH 44342 (**37**) is reported to be a representative example of a series of novel, non-peptidic, potent and selective FTase inhibitors (76). SCH 44342 inhibited rat brain FTase and human recombinant FTase (IC_{50} = 250 and 280 nM, respectively), and the processing of Ras-CVLS in Cos cells (IC_{50} = 3 μM), but was ineffective *vs.* rat brain GGTase (IC_{50} > 114 μM).

CONCLUSIONS

Targeting mutant Ras function has emerged as an exciting, novel, and viable alternative to traditional cancer therapy. Great progress has been made towards proof-of-principle in animals, especially with the CAAX mimics. Potency, specificity, and surprising lack of toxicity have been demonstrated. Yet, clinical efficacy of these anticancer agents in humans bearing tumors with multiple genetic transformations remains to be proven. Since these agents appear to act mainly as cytostatic agents, and will presumably need to be administered for prolonged periods, problems of toxicity and drug resistance may also be major hurdles to overcome (30).

The FTase inhibitors appear to impede more processes that are essential for transformed cells than for normal cells. The lack of cytotoxicity in normal cells and in animals has been attributed to a variety of factors, but is still not well understood. Some of these contributing factors have been demonstrated with compounds such as **2**, **4** and **9** (19,24,31,32,39). Ongoing *in vitro* studies with Ras corroborate some of the results seen with the inhibitors. Unfarnesylated, oncogenic, GTP-bound Ras forms a stable complex with Raf which accumulates in the cytosol and inhibits aberrant signaling, even with only partial inhibition of farnesylation, whereas GDP-bound, unfarnesylated normal Ras does not sequester Raf and normal signaling continues (39,77). Further elucidation of protein/protein interactions, such as Ras/Raf and Ras/GAP, in the Ras-signaling pathway may provide other feasible approaches for the inhibition of mutant Ras and a clarification of its complex role in transforming cells (2,32,77-79). While results to date are quite encouraging, better understanding of the relative requirements for maintaining prenylation and signaling of normal Ras, *vs.* inhibiting farnesylation and/or geranylgeranylation of different mutant Ras forms may provide answers for effective, non-toxic long-term treatments. More information on the X-ray crystal structure of FTase, on the detailed mechanism of the farnesyl transfer reaction, and on the proteases and other enzymes involved in Ras processing can lead to new approaches to drug design (80).

References

1. G.L. Bolton, J.S. Sebolt-Leopold and J.C. Hodges, Annu.Rep.Med.Chem., 29, 165 (1994).
2. R.J.A. Grand in "New Molecular Targets in Cancer Chemotherapy" J.D. Kerr, and P. Workman, Eds., CRC Press, Boca Raton, FL, 1994, p. 97.
3. J.F. Hancock, H. Paterson, and C.J. Marshall, Cell, 63, 133 (1990).
4. J.L. Bos, Cancer Res., 49, 4682 (1989).
5. J.B. Gibbs, A. Oliff, and N.E. Kohl, Cell, 77, 175 (1994).
6. B. Jansen, H. Wadl, S.A. Inoue, B. Truelzsch, E. Selzer, M. Duchene, H.-G. Eichler, and K. Wolff, Antisense Res.Dev., 5, 271 (1996).
7. K. Aoki, T. Yoshida, T. Sugimura, and M. Terada, Cancer Res., 55, 3810 (1995).
8. G. Schwab, C. Chavany, I. Duroux, G. Goubin, J, Lebeau, C. Hélène and T. Saison-Behmoaras, Proc.Natl.Acad.Sci.USA., 91, 10460 (1994).
9. M. Marom, R. Haklai, G. Ben-Baruch, D. Marciano, Y. Egozi, and Y. Kloog, J.Biol.Chem., 270, 22263 (1995).
10. S.L. Graham, Patent GB2276618-A (1994).
11. C.C. Hall, J.D. Watkins, S.B. Ferguson, L.H. Foley, and N.H. Georgopapadakou, Biochem.Biophys.Res.Commun., 217, 728 (1995).
12. R.Wolin, D. Wang, J. Kelly, A. Afonso, L. James, P. Kirschmeier, and A.T. McPhail, Bioorg.Med. Chem.Lett., 6, 195 (1996).
13. Y. Reiss, J.L. Goldstein, M.C. Saebra, P.J. Casey, and M.S. Brown, Cell, 62, 81 (1990).
14. E. S. Furfine, J.J. Leban, A. Landavazo, J. F. Moomaw, and P. J. Casey, Biochemistry, 34, 6857 (1995).
15. P.J. Casey and M.C. Seabra, J.Biol.Chem., 271, in press (1996).
16. J.M. Dolence, P.B. Cassidy, J.R. Mathis, and C.D. Poulter, Biochem., 34, 16687(1995).
17. A. Vogt, J. Sun, Y. Qian, E. Tan-Chiu, A.D. Hamilton, and S.M. Sebti, Biochemistry, 12398 (1995).

18. S. Omura and H. Takeshima, Drugs of the Future, 19, 751 (1994).
19. B.E. Buss and J.C. Marsters, Jr., Chemistry and Biology, 2, 787 (1995).
20. S.L. Graham, Exp.Opin.Ther. Patents, 5, 1269 (1995).
21. T.E. Rawson, T.C. Somers, J.C. Marsters, Jr., D.T. Wan, M.E. Reynolds, and D.J. Burdick, Bioorg.Med.Chem.Lett., 5, 1335 (1995).
22. G.L. James, J.L. Goldstein, and M.S. Brown, J.Biol.Chem., 270, 6221 (1995).
23. G.L. James, M.S. Brown, M.H. Cobb, and J.L. Goldstein, J.Biol.Chem., 269, 27705 (1994).
24. M.B. Dalton, K.S. Fantle, H.A. Bechtold, L. DeMaio, R.M. Evans, A. Krytosek, and M. Sinensky, Cancer Res., 55, 3295 (1995).
25. K.S. Koblan, J.C. Culberson, S.J. Desolms, E.A. Giuliani, S.D. Mosser, C.A. Omer, S.M. Pitzenberger, and M.J. Bogusky, Protein Sci., 4, 681 (1995).
26. S.J. deSolms, V.M. Garsky, E..A. Giuliani, R.P. Gomez, and S.L. Graham, Int. Patent WO 95/09001 (1995).
27. N.E. Kohl, F.R. Wilson, S.D. Mosser, E. Giuliani, S.J. deSolms, M.W. Conner, N.J. Anthony, W.J. Holtz, R.P. Gomez, T-J. Lee, R.L. Smith, S.L. Graham, G.D. Hartman, J.B. Gibbs, and A. Oliff, Proc.Natl.Acad.Sci.USA., 91, 9141 (1994).
28. N.E. Kohl, M.W. Conner, J.B. Gibbs, S.L. Graham, G.D. Hartman, and A.. Oliff, J.Cell.Biochem., Suppl.22, 145 (1995).
29. L. Sepp-Lorenzino, Z. Ma, E. Rands, N.E. Kohl, J.B. Gibbs, A.. Oliff, and N. Rosen, Cancer Res., 55, 5302 (1995).
30. N.E. Kohl, C.A. Omer, M.W. Conner, N.J. Anthony, J.P. Davide, S.J. deSolms, E. Giuliani, R.P. Gomez, S.L. Graham, K. Hamilton, L.K. Handt, G.D. Hartman, K.S. Koblan, A.M. Kral, P.J. Miller, S.D. Mosser, T.J. O'Neill, E. Rands, M.D. Schaber, J.B. Gibbs, and A. Oliff, Nature Medicine, 1, 792 (1995).
31. J. Rak, Y. Mitsuhashi, L. Bayko, J. Filmus, S. Shirasawa, T. Sasazuki, and R.S. Kerbel, Cancer Res., 55, 4575 (1995).
32. P.F. Lebowitz, J.P. Davide, and G.C. Prendergast, Mol.Cell.Biol., 15, 6613 (1995).
33. A. Vogt, Y. Qian, M.A. Blaskovich, R.D. Fossum, A.D. Hamilton, and S.M. Sebti, J.Biol.Chem., 270, 660 (1995).
34. T.F. McGuire, Y. Qian, M.A. Blaskovich, R.D. Fossum, J. Sun, T. Marlowe, S.J. Corey, S.P. Wathen, A. Vogt, A.D. Hamilton, and S.M. Sebti, Biochem.Biophys.Res.Commun., 214, 295 (1995).
35. Y. Qian, M.A. Blaskovich, C.-M. Seong, A. Vogt, A.D. Hamilton, and S.M. Sebti, Bioorg.Med.Chem.Lett., 4, 2579 (1994).
36. E.C. Lerner, Y. Qian, A.D. Hamilton, and S.M. Sebti, J.Biol.Chem., 270, 26770 (1995).
37. J.B. Gibbs and S.L. Graham, Int. Patent WO95/20396 (1995).
38. J. Sun, Y. Qian, A.D. Hamilton, and S.M. Sebti, Cancer Res., 55, 4243 (1995).
39. E. Lerner, Y. Qian, M.A. Blaskovich, R.D. Fossum, A. Vogt, J. Sun, A.D. Cox, C.J. Der, A.D. Hamilton, and S.M. Sebti, J.Biol.Chem., 270, 26802 (1995).
40. M.D. Lewis, J.J. Kowalczyk, A.E. Christuk, E.M. Harrington, and X.C. Sheng, Int. Patent WO95/25086-A1 (1995).
41. Y. Qian, A. Vogt, S.M. Sebti, and A.D. Hamilton, J.Med.Chem., 39, 217 (1996).
42. Y. Qian, M.A. Blaskovich, M. Saleem, C.M. Seong, S.P. Wathen, A.D. Hamilton, and S.M. Sebti, J.Biol.Chem., 269, 12410 (1994).
43. E.M. Harrington, J.J. Kowalczyk, S.L. Pinnow, K. Ackermann, A.M. Garcia, and M.D. Lewis, Bioorg.Med.Chem.Lett., 4, 2775 (1994).
44. T. Nagasu, K. Yoshimatsu, C. Rowell, M.D. Lewis, and A.M. Garcia, Cancer Res., 55, 5310 (1995).
45. F.-F. Clerc, J.-D. Guitton, N. Fromage, Y. Lelièvre, M. Duchesne, B. Tocqué, E. James-Surcouf, A. Commerçon, and J. Becquart, Bioorg.Med.Chem.Lett., 5, 1779 (1995).
46. G. Byk, M. Duchesne, F. Parker, Y. Lelievre, J.D. Guitton, F.F. Clerc, J. Becquart, B. Tocque, and D. Scherman, Bioorg.Med.Chem.Lett., 5, 2677 (1995).
47. K. Leftheris, T. Kline, G.D. Vite, Y.H. Cho, R.S. Bhide, D.V. Patel, M.M. Patel, R.J. Schmidt, H.N. Weller, M.L. Andahazy, J.M. Carboni, J.L. Gullo-Brown, F.Y.F. Lee, C. Ricca, W.C. Rose, N. Yan, M. Barbacid, J.T. Hunt, C.A. Meyers, B.R. Seizinger, R. Zahler, and V. Manne, J.Med.Chem., 39, 224 (1996).
48. S.J. deSolms and S.L. Graham, US Patent 5,439,918 (1995).
49. S.L. Graham and T.M. Williams, Int. Patent WO 95/00497 (1995).

50. S.J. deSolms, A.A. Deana, E.A. Giuliani, S.L. Graham, N.E. Kohl, S.D. Mosser, A.I. Oliff, D.L. Pompliano, E. Rands, T.H. Scholz, J.M. Wiggins, J.B. Gibbs, and R.L. Smith, J.Med.Chem., 38, 3967 (1995).

51. S.J. deSolms and S.L. Graham, US Patent 5,491,164-A (1996).

52. J.T. Hunt, V.G. Lee, K. Leftheris, B. Seizinger, J. Carboni, J. Mabus, C. Ricca, N. Yan, and V. Manne, J.Med.Chem., 39, 353 (1996).

53. K. Leftheris, European Patent EP-696593-A2 (1996).

54. D.V. Patel, M.M. Patel, S.S. Robinson, and E.M. Gordon, Bioorg.Med.Chem.Lett., 4, 1883 (1994).

55. J.J. Kowalczyk, K. Ackermann, A.M. Garcia, and M.D. Lewis, Bioorg.Med.Chem.Lett., 5, 3073 (1995).

56. S.J. deSolms, E.A. Guiliani, and S.L. Graham, US Patent 5,468,733 (1995).

57. C.J. Springer and R. Marais, Int. Patent WO 95/03830-A (1995).

58. R.S. Bhide, D.V. Patel, M.M. Patel, S.P. Robinson, L.W. Hunihan, and E.M. Gordon, Bioorg.Med.Chem.Lett., 4, 2107 (1994).

59. D.V. Patel, E.M. Gordon, R.J. Schmidt, H.N. Weller, M.G. Young, R. Zahler, M. Barbacid, J.M. Carboni, J.L. Gullo-Brown, L. Hunihan,C. Ricca, S. Robinson, B.R. Seizinger, A.V. Tuomari, and V. Manne, J.Med.Chem., 38, 435 (1995).

60. V. Manne, N. Yan, J.M. Carboni, A.V. Tuomari, C. Ricca, J. Gullo-Brown, M.L. Andahazy, R.J. Schmidt, D. Patel, R. Zahler, R. Weinmann, C.J. Der, A.D. Cox, J.T. Hunt, E.M. Gordon, M. Barbacid, and B.R. Seizinger, Oncogene, 10, 1763 (1995).

61. N. Yan, C. Ricca, J. Fletcher, T. Glover, B.R. Seizinger, and V. Manne, Cancer Res., 55, 3569 (1995).

62. D.V. Patel, R.J. Schmidt, S.A. Biller, E.M. Gordon, S.S. Robinson, and V. Manne, J.Med.Chem., 38, 2906 (1995).

63. V. Manne, C.S. Ricca, J. Gullo-Brown, A.V. Tuomari, N. Yan, D. Patel, R. Schmidt, M.J. Lynch, C.P. Ciosek, Jr., J.M. Carboni, S. Robinson, E.M. Gordon, M. Barbacid, B.R. Seizinger, and S.A. Biller, Drug Develop.Res., 34, 121 (1995).

64. M.S. Kang, D.M. Stemerick, J.H. Zwolshen, B.S. Harry, P.S. Sunkara, and B.L. Harrison, Biochem.Biophys.Res.Commun., 217, 245 (1995).

65. S.B. Singh, D.L. Zinc, J.M. Liesch, R.G. Ball, M.A. Goetz, E.A. Bolessa, R.A. Giacobbe, K.C. Silverman, G.F. Bills, F. Pelaez, C. Cascales, J.B. Gibbs, and R.B. Lingham, J.Org.Chem., 59, 6296 (1994).

66. I. King, C. Blood, M. Chu, M. Patel, M. Liu, Z. Li, N. Robertson, E. Maxwell, and J.J. Catino, Oncology Research, 7, 1 (1995).

67. S.B. Singh, D.L. Zink, G.F. Bills, R.G. Jenkins, K.C. Silverman, and R.B. Lingham, Tetrahedron Lett., 36, 4935 (1995).

68. S.B. Singh, J.M. Liesch. R.B. Lingham, K.C. Silverman, J.M. Sigmund, and M.A. Goetz, J.Org.Chem., 60, 7896 (1995).

69. K.C. Silverman, C. Cascales, O. Genilloud, J.M. Sigmund, S.E. Gartner, G.E. Koch, M.M. Gagliardi, B.K. Heimbuch, M. Nallin-Omstead, M. Sanchez, M.T. Diez, I. Martin, G.M. Garrity, C.F. Hirsch, J.B. Gibbs, S.B. Singh, and R.B. Lingham, Appl.Microbiol.Biotechnol., 43, 610 (1995).

70. S.B. Singh, J.M. Liesch, R.B. Lingham, M.A. Goetz, and J.B. Gibbs, J.Am.Chem.Soc., 116, 11606 (1994).

71. D.W. Phife, R.W. Patton, R.L. Berrie, R.Yarborough, M. S. Puar, M. Patel, W.R. Bishop, and S.J. Coval, Tetrahedron Lett., 36, 6995 (1995).

72. S.B. Singh, E.T. Jones, M.A. Goetz, G.F. Bills, M. Nallin-Omstead, R.G. Jenkins, R.B. Lingham, K.C. Silverman, and J.B. Gibbs, Tetrahedron Lett., 35, 4693 (1994).

73. J.-F. Zhao, H. Nakano, and S. Sharma, Oncogene, 11, 161 (1995).

74. K. Umezawa, T. Ohse, T. Yamamoto, T. Koyano, and Y. Takahashi, Anticancer Res., 14, 2413 (1994).

75. T. Ishii, T. Hida, Y. Nozaki, K. Ootsu, European Patent Application 0677513 A1 (1995).

76. W.R. Bishop, R. Bond, J. Petrin, L. Wang, R. Patton, R. Doll, G. Njoroge, J. Catino, J. Schwartz, W. Windsor, R. Syto, J. Schwartz, D. Carr, L. James, and P. Kirschmeier, J.Biol.Chem., 270, 30611 (1995).

77. M. Miyake, S. Mizutani, H. Koide, and Y. Kaziro, FEBS Lett., 378, 15 (1996).

78. C.-Y. Chen, D.V. Faller, J.Biol.Chem., 271, 2376 (1996).

79. M. Spaargarten, J.R. Bischoff, F. McCormick, Gene Expression, 4, 345 (1995).

80. YQ. Mu, C.A. Omer, R.A. Gibbs, J.Amer.Chem.Soc., 118, 1817 (1996).

SECTION IV. IMMUNOLOGY, ENDOCRINOLOGY AND METABOLIC DISEASES

Editor: William K. Hagmann
Merck Research Laboratories, Rahway, NJ 07065

Chapter 19. Estrogen Receptor Modulators: Effects in Non-Traditional Target Tissues

Timothy A. Grese and Jeffrey A. Dodge
Endocrine Research, Eli Lilly & Co., Indianapolis, IN 46285

Introduction - Endogenous estrogens, such as 17β-estradiol (1) and estrone (2), have long been recognized as the primary hormones involved in the development and maintenance of the female sex organs, mammary glands, and other sexual characteristics (1). More recently their involvement in the growth and/or function of a number of other tissues, such as the skeleton, the cardiovascular system, and the central nervous system, in both males and females has been recognized (2,3). The decreased ovarian production of estrogens which occurs after the climacteric has been linked to a number of pathologies including osteoporosis (4), coronary artery disease (5), depression (6), and Alzheimer's disease (7). Estrogen replacement therapy (ERT) has demonstrated effectiveness in reducing the frequency and severity of these pathologies, however concerns relating to the increased risk of endometrial cancer have necessitated the development of therapeutic regimens in which the uterine effects of estrogen are opposed by progestin treatment (8). Side-effects of progestin treatment, such as resumption of menses, and the possibility of attenuated cardiovascular benefits, have decreased patient compliance (9). Furthermore, recent studies which confirm the increased risk of breast cancer associated with long term ERT have led to the search for treatment alternatives (10).

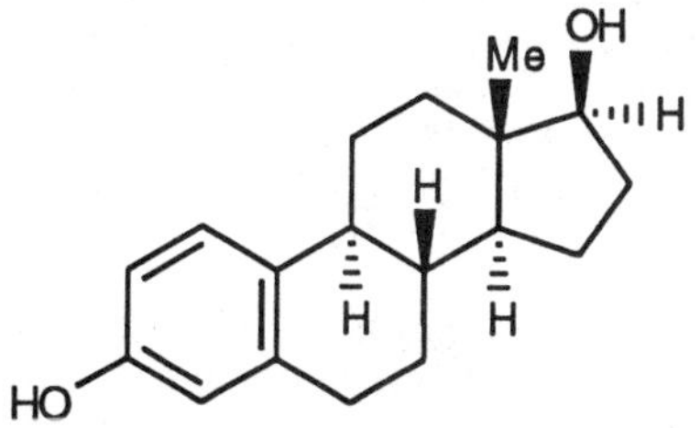

1 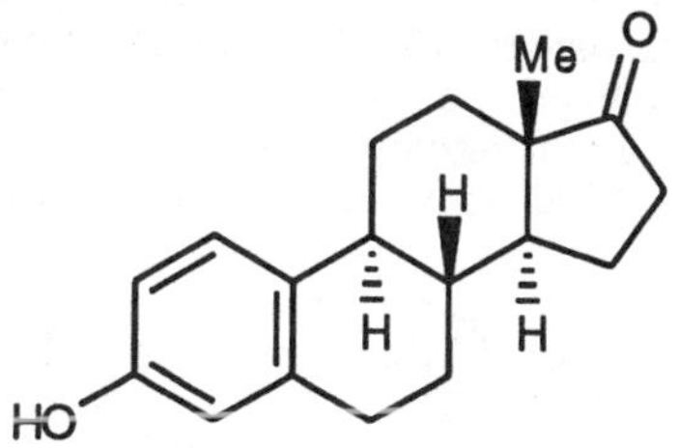   **2**

Over the years a variety of steroidal and non-steroidal compounds which interact with the estrogen receptor (ER) have been developed as contraceptives and for the treatment of breast cancer, uterine dysfunction, and other disorders of the female reproductive system and they have been recently reviewed (11-14). The utility of synthetic estrogens such as diethylstilbestrol (DES, 3) has been greatly diminished due to concerns similar to those encountered with the natural hormones. Conversely, compounds such as tamoxifen (4), which antagonize the action of estrogen in breast tissue, have found great utility in the treatment of breast cancer. Early concerns that the long-term use of "antiestrogens" would lead to increased risks of osteoporosis and cardiovascular disease have been dispelled by the surprising finding that some "antiestrogens" actually mimic the effects of estrogen in skeletal and cardiovascular tissues (15). These findings have led to a reevaluation of the mechanisms of ER activation and have resulted in a reclassification of estrogen receptor ligands (16).

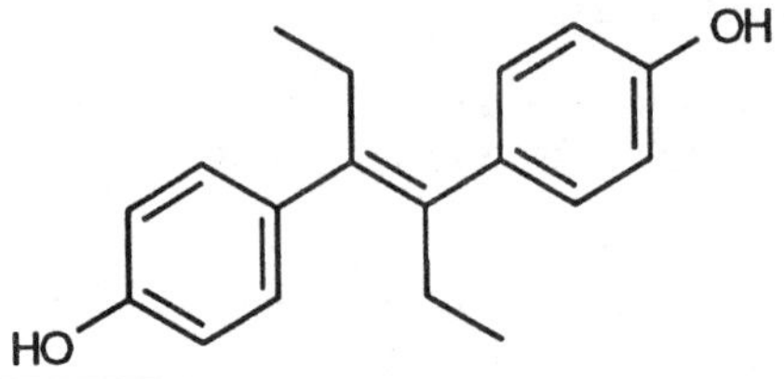

$OCH_2CH_2NMe_2$

3 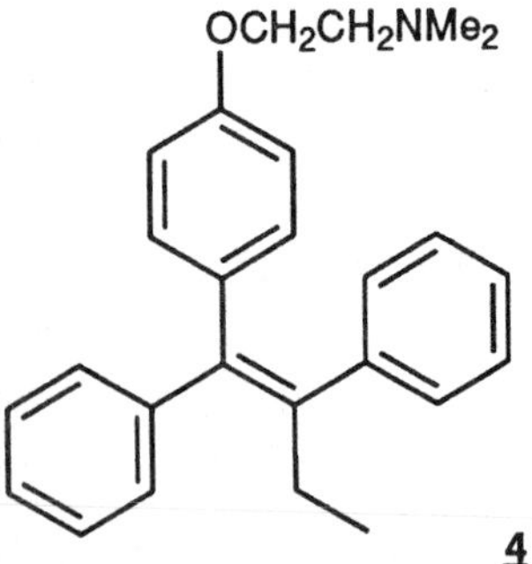   **4**

In this chapter, we will focus primarily on the effects of these ER modulators in non-mammary tissue. For organizational purposes we have divided the discussion into four segments and classified compounds on the basis of their ability to mimic the effects of estrogen in skeletal and/or cardiovascular tissue and to induce uterine stimulation (see Table 1). This classification is consistent with one recently proposed on the basis of distinct DNA transcriptional profiles observed for the different classes (*vide infra*) (16). Nevertheless, it should be noted these distinctions may be artificial, since there is likely to be a continuum of activities from full agonist to full antagonist and the relative activity of an individual compound may be different for each tissue or animal species examined.

CLASSIFICATION	Uterine Stimul.	Bone/ CV
Estrogen Agonists	yes	agonist
Partial Agonists	yes	agonist
Selective Estrogen Receptor Modulators	no	agonist
Pure Antiestrogens	no	antag./ neutral

Table 1: Classification of Estrogen Receptor Modulators

Estrogens and Estrogen Agonists – The therapeutic utility of synthetic and natural estrogens for the replacement of endogenous hormones in postmenopausal women is well established (4,5,8,9). To date, most studies have focused on the efficacy of ERT in the prevention of osteoporosis and cardiovascular disease. More recently, benefits in the central nervous system including improvements in cognitive function, and palliation of Alzheimer's disease and postmenopausal depression have been described (6,17,18).

In the cardiovascular system, ERT has been shown to result in a 40-50% reduction in the relative risk of coronary disease and to prevent atherosclerosis (19-21). The effects of estrogens on cardiovascular risk factors include raising serum levels of high-density lipoprotein (HDL) cholesterol, triglycerides, and apolipoprotein A-1, and lowering levels of low-density lipoprotein (LDL) cholesterol, lipoprotein (a), and apolipoprotein B (22,23). It is unlikely that these effects on serum markers are solely responsible for estrogen's cardioprotective effects, as direct effects on blood vessel walls have also been demonstrated (24).

In the prevention of osteoporosis, estrogens function primarily as antiresorptive agents with effects on both cortical and cancellous bone (25). This benefit is offset by a subsequent decrease in bone formation, however the overall result of ERT is a substantial increase in bone mineral density and a decrease in fracture incidence (26). Although ER has been detected in both osteoblasts and osteoclasts, it is currently unclear if the effects of estrogens on bone metabolism are direct or indirect (27).

The combination of these strong benefits of ERT in non-reproductive tissues and its significant liabilities in the breast and uterus have led to a search for compounds which regulate these activities differentially (8,10). Early efforts to identify selective estrogens focused on changes in the parent steroid to icit the desired tissue specificity. For example, the estrogen metabolites estriol (5) and 17α-estradiol (6) were found to be time-dependent mixed agonist/antagonists, stimulating early uterotrophic responses but with little effect on true uterine hypertrophy and hyperplasia (28). In fact, estriol causes significantly less uterine hyperplasia than 17β-estradiol (1) and inhibits the development of breast cancer in rodents (29). Another estrogen metabolite, 2-methoxyestradiol, has been implicated in the angiogenesis of vascular tissue (30). Improvements in tissue selectivity have been observed with a family of D-ring halogenated estrones (7) which have demonstrated potent lipid lowering yet diminished uterine hypertrophy relative to estrone (31). Other attempts to attenuate the estrogenic activity of steroids via opening of the steroid nucleus, such as 9,11-seco steroids (9), have met with limited success (32,33). More recently, the components of conjugated estrogens derived from the urine of pregnant mares have been evaluated individually for their lipid lowering effects. In this study, 17α-dihydroequilenin (8) reduced total plasma cholesterol concentrations in the ovariectomized (OVX) rat without inducing uterine growth (34). This compound also behaved as an estrogen agonist on bone although uterine hypertrophy was observed at higher doses (35,36).

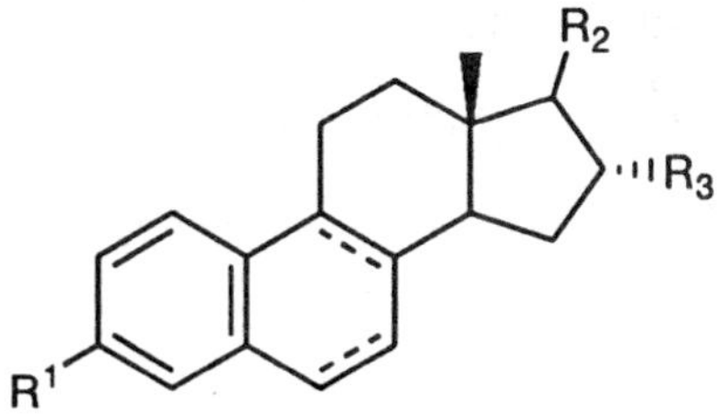

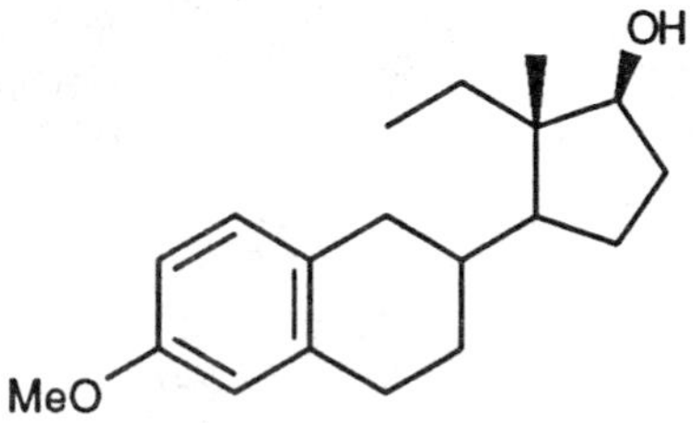

5   R_1 = OH, R_2 = β-OH, R_3 = OH
6   R_1 = OH, R_2 = α-OH, R_3 = H
7   R_1 = OMe, R_2 = O, R_3 = Cl
8   R_1 = OH, R_2 = β-OH, R_3 = H, 6,8 = diene

9

A variety of non-steroidal environmental estrogens have also been studied in relation to their effects on reproductive and non-reproductive tissue. This class of compounds consists of naturally occurring or commercially produced chemicals derived from common sources such as plants (phytoestrogens), pesticides, plastics, and animal health products. Of the representative agents studied, a variety behaved as estrogen agonists in their cholesterol and bone responses. For example, both zeranol (10) and coumestrol (11) were found to bind to the ER, lower serum lipids, and prevent bone loss in the OVX rat (37,38). In addition, zeranol has been studied clinically in women as a drug to alleviate menopausal symptoms (39).

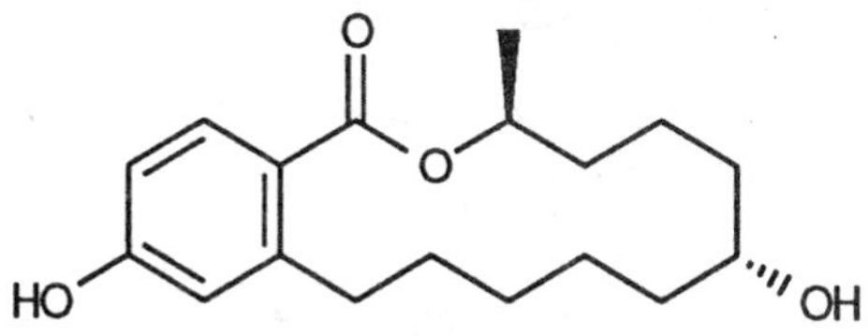

10

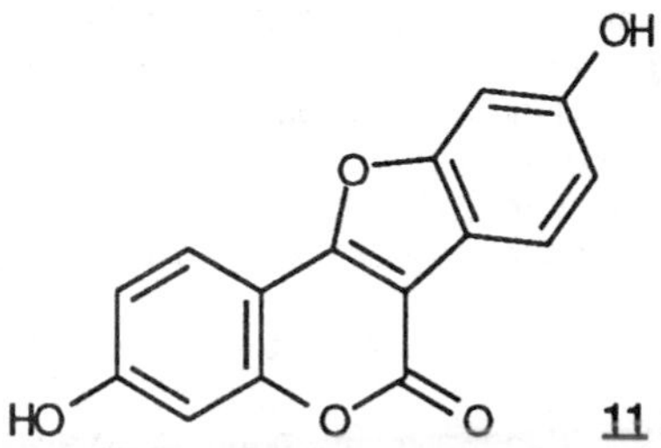

11

<u>Partial Agonists and Triphenylethylenes</u> - By far the most investigated of the non-steroidal ER modulators are the triphenylethylenes (TPEs). Originally investigated as contraceptives, compounds such as tamoxifen (4) have been developed for the treatment of breast cancer on the basis of strong antagonism of estrogen action in mammary tissue (14). Only recently have the estrogen agonist activities of these compounds in the skeletal and cardiovascular systems been described (15). Although

they partially antagonize the effects of estrogen on the uterus, evidence to date suggests that in the absence of estrogen the members of this structural class also tend to induce some level of uterine stimulation, hence their classification here as partial agonists (14,35).

One of the first TPEs to achieve clinical significance was clomiphene (12), available as a one-to-one mixture of double-bond isomers. Paradoxically, although developed as a contraceptive, it has been mainly utilized for the induction of ovulation in anovulatory women (40). Estrogen agonist effects of clomiphene in skeletal tissue have been reported in an OVX rat model of post-menopausal osteoporosis (41,42). Clomiphene has also been reported to inhibit bone resorption *in vitro* and to decrease serum markers of bone resorption in menopausal/castrated women (43,44).

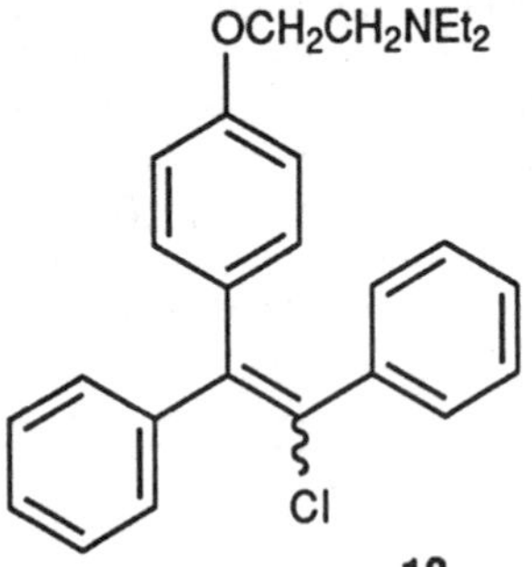

$OCH_2CH_2NEt_2$

Cl

12

Effects on the cardiovascular system have also been reported with clomiphene. In rats, an estrogen-like reduction in serum cholesterol was observed, however interpretation of this effect is complicated by clomiphene's ability to inhibit cholesterol biosynthesis by an ER-independent mechanism (45).

In the rat uterus, clomiphene has been shown to potently stimulate the epithelium, and to stimulate a variety of estrogenic effects in OVX animals (45-47). Similar results have been obtained in OVX baboons (45).

The widespread clinical use of tamoxifen (4) for the treatment of breast cancer has resulted in a large body of evidence with respect to its effects in other tissues (48). It has been demonstrated to protect against osteopenia in the OVX rat model and to inhibit bone resorption *in vitro* (43,49). In several clinical trials, tamoxifen has demonstrated effectiveness in the preservation of bone mineral density at the lumbar spine, femoral neck, and forearm in postmenopausal women (50-53). These effects on bone density are accompanied by estrogen-like reduction of serum markers of bone turnover (50-52,54).

Likewise, significant decreases in risk factors for cardiovascular disease have been observed in postmenopausal women treated with tamoxifen (44,55). In general, significant decreases in LDL cholesterol and lipoprotein(a) have been observed with modest increases in triglycerides and little or no change in HDL cholesterol (55-58). These effects have coincided with a reduction in mortality due to cardiovascular disease in patients treated with tamoxifen (59-61). Whether the cardiovascular benefits of tamoxifen are mediated via its interaction with the ER have not been fully established. Alternative mechanisms including inhibition of cholesterol synthesis (62), inhibition of lipid peroxidation (63), and decreases in membrane fluidity have been proposed (64). In a primate model, tamoxifen has also been shown to significantly inhibit the progression of coronary artery atherosclerosis (65).

Although clearly an antagonist of estrogen action in breast tissue (48), tamoxifen has been shown to induce endometrial hyperplasia in the OVX rat model (66). Notwithstanding the considerable positive effects described above, there continues to be concern about the increased risk of endometrial cancer associated with tamoxifen use (67-70). Tamoxifen has also been shown to induce DNA-adduct formation and liver cancer in rats (71,72).

Concern over this potential carcinogenicity in the uterus and liver has led to the development of tamoxifen analogs. It has been speculated that blocking the metabolic hydroxylation of tamoxifen, which occurs primarily at the 4-position, might lead to agents in which these unwanted side-effects are reduced (14). Several of these drugs, including droloxifene (K 060 E, **13**), toremifene (Fc-1157a, **14**), idoxifene (CB-7432, **15**), and TAT-59 (**16**) are being evaluated for the treatment of breast cancer (73-76).

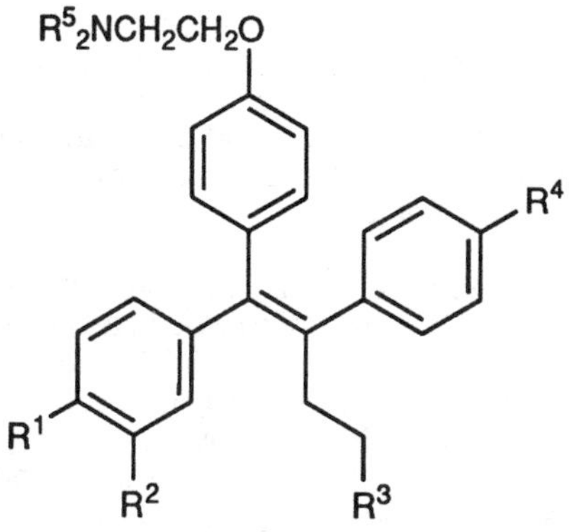

$R^5_2NCH_2CH_2O$

R^4

R^1

R^2 R^3

13: R^1= R^3= R^4= H; R^2= OH; R^5= Me
14: R^1= R^2= R^4= H; R^3 = Cl; R^5= Me
15: R^1= I; R^2= R^3= R^4= H; R^5= -(CH$_2$)$_4$-
16: R^1= OH; R^2= R^3= H; R^4= *i*-pr; R^5= Me

Reports on the effects of these compounds in non-traditional target tissues have also begun to appear (77,78,81-84). Toremifene (**14**) has been reported to reduce serum cholesterol in breast cancer patients and to be less uterotrophic than tamoxifen in the rat (77,78). In postmenopausal women, however, its estrogenic effects on the uterus have been reported to be comparable to tamoxifen (79).

In OVX rats, droloxifene (**13**) has been reported to reduce serum cholesterol by 40-46% and to protect against loss of bone mineral density, but with less uterine stimulation than tamoxifen (80-81). Estrogenic effects on the skeleton have also been observed by histomorphometry in both cortical and cancellous bone (81-84). Neither **13** nor **14** mimics tamoxifen's ability to induce DNA adduct formation or hepatic cancer in rats (84,85).

Recently, triphenylethylene **17** with a non-basic side chain has been described to have full estrogen agonist activity in bone with minimal uterine agonist activity in the OVX rat (86).

Ormeloxifene (centchroman, CDRI 67/20, **18**) is a structurally distinct ER modulator which has been reported to have tissue-selective effects on bone in OVX rats (87). Other authors, however, have reported potent uterine stimulation with **18** in OVX rats (88). Ormeloxifene has also been found to inhibit osteoclastic bone resorption *in vitro* (89).

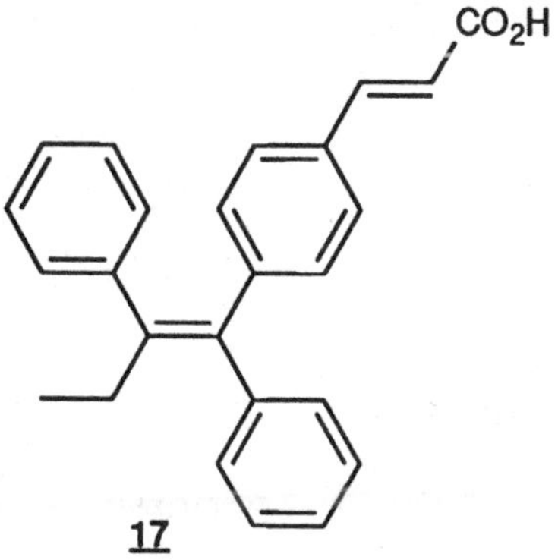

CO_2H

17

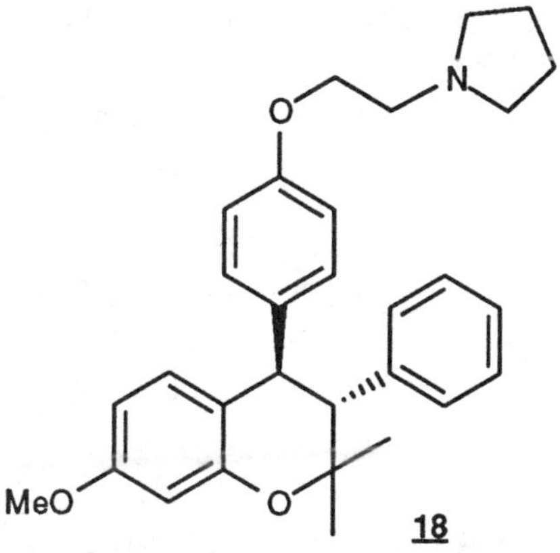

MeO   **18**

<u>Selective Estrogen Receptor Modulators (SERMs)</u> - The discovery that some compounds are able to mimic the effects of estrogen in the skeletal and cardiovascular system yet produce almost complete antagonism in the breast and uterus has led to

the coining of the term Selective Estrogen Receptor Modulator (SERM) (90). The most investigated of these compounds, raloxifene (keoxifene, LY139481 HCl, LY156758, **19**), was originally investigated for utility in the treatment of breast cancer (91,92).

The effects of raloxifene in non-traditional estrogen target tissues have recently been described. In OVX rats, raloxifene has been shown to reduce serum cholesterol by 50-75% after 1-5 weeks of daily dosing (93,94). Similar effects have been demonstrated in postmenopausal women where raloxifene has been demonstrated to reduce both total serum cholesterol and LDL cholesterol (95).

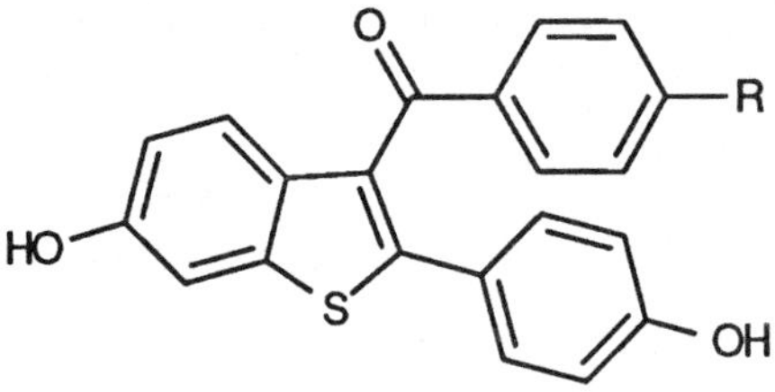

19: R = OCH$_2$CH$_2$N(CH$_2$)$_5$

Raloxifene has also been shown to protect against OVX-induced osteo-penia in several studies (93,96). Positive effects on both cortical and cancellous bone sites have been observed (96). Interestingly, although raloxifene appears to suppress bone resorption with efficacy equal to that of estrogen, bone formation may be suppressed to a lesser degree, resulting in a net gain in bone mass with raloxifene (96). Positive effects on bone strength been described and raloxifene has been demonstrated to have estrogen-like effects on serum and urinary markers of bone metabolism in postmenopausal women (95,97). *In vitro* studies have demonstrated similar effects of raloxifene and 17β-estradiol on osteoclastogenesis (98).

Uterine effects observed with raloxifene have been qualitatively different from those observed with TPEs (35). Although a modest increase in uterine weight has been observed in OVX rats treated with raloxifene, this increase was not dose-dependent and was not accompanied by similar changes in epithelial cell height or other estrogen sensitive parameters of uterine histology (93). In a direct comparison of raloxifene and TPEs, raloxifene was a significantly more effective antagonist of estrogen action in the immature female rat uterus (35). In OVX rats, the TPEs tamoxifen, droloxifene, and idoxifene were found to induce a larger maximal stimulation of uterine weight and to induce uterine eosinophilia while raloxifene did not (34). Eosinophils have previously been suggested to be important components of uterine physiology and pathophysiology (99). Furthermore, raloxifene has been shown to effectively antagonize the uterotrophic effects of tamoxifen in OVX rats (100). In postmenopausal women, raloxifene has been shown to exert a significant suppression of estrogen effects on the uterine epithelium relative to placebo-treated controls (95).

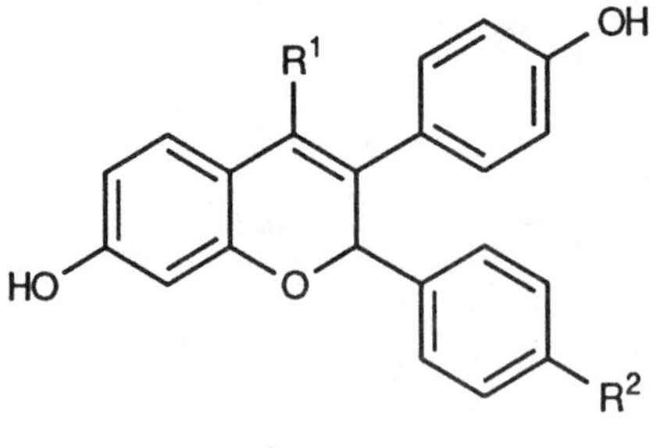

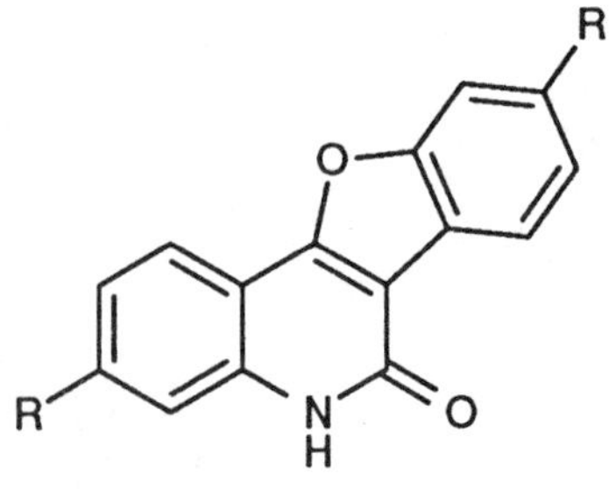

20: R^1 = Me; R^2= -OCH$_2$CH$_2$N(CH$_2$)$_5$
21: R^1 = H; R^2= -OCH$_2$CH$_2$N(CH$_2$)$_5$

22: R = OCONMe$_2$

Benzopyrans EM 343 (**20**) and **21** have been reported to have an activity profile similar to raloxifene with respect to bone protection and serum cholesterol reduction in OVX rats (101). Interestingly, neither **20** or **21** show any tendency to increase uterine weight or eosinophilia in these assays (101). Benzofuroquinoline

KCA-098 (**22**), which is related to coumestrol, has been reported to increase bone strength in OVX rats without a corresponding increase in uterine weight (102). Compound **22** has also been found to inhibit bone resorption in organ cultures of fetal rat femora (103).

Pure Antiestrogens – While estrogen agonists, partial agonists, and SERMs can mimic the pharmacology of the natural hormone, pure antiestrogens represent a class of therapeutic agents which are devoid of estrogen agonism regardless of the target tissue. These compounds exhibit no estrogenic activity in the rat uterus, vagina, and hypothalamic-pituitary axis, as well as effectively antagonizing the stimulatory effects of estrogen (104). Pure antiestrogens are estrogen antagonists in non-reproductive tissues as well. For example, ZM 189,154 (**23**) reversed the bone-sparing action of 17α-ethynyl estradiol in OVX animals while remaining neutral in intact animals (105). Similarly, ICI 164,384 (**24**) and ICI 182,790 (**25**) exhibited no capacity for lowering serum cholesterol or sparing bone loss in the OVX rat model (106-108). Because of the antagonist effects on bone, concerns remain over the long term clinical use of pure antiestrogens (109).

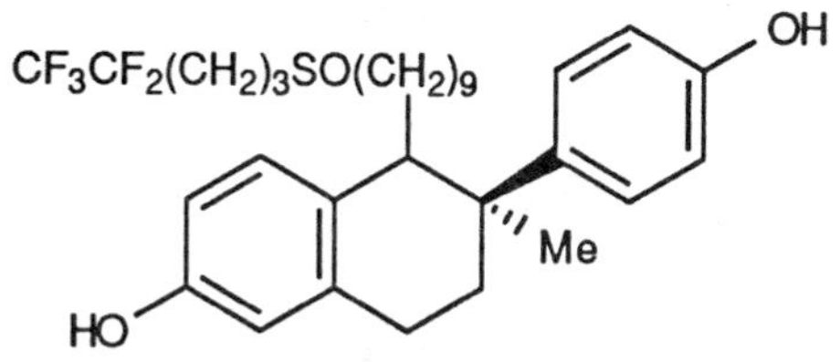

$CF_3CF_2(CH_2)_3SO(CH_2)_9$

23

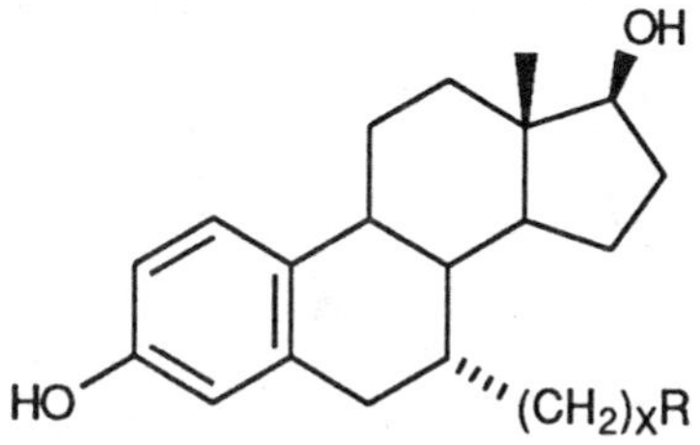

24 X = 10, R = CON(Me)n-Bu
25 X = 9, R = SO(CH$_2$)$_3$CF$_2$CF$_3$

Mechanisms of Estrogen Receptor Modulation - Although a variety of alternative mechanisms have been advanced (110-113), it is generally accepted that the actions of estrogen agonists and antagonists are primarily mediated by interaction with the ER. The details of this interaction and the subsequent events are, however, the subject of intensive debate and research (114,115). The mechanisms by which a single ligand and its interactions with a single receptor can produce agonist effects in some tissues and antagonist effects in others are fundamental to this debate.

The ER contains two transcriptional activation functions, AF-1 and AF-2 which operate upon genes containing an estrogen response element (ERE) to initiate transcription (116). Of these activation functions only AF-2 requires ligand binding. It has been demonstrated that tamoxifen binding inhibits AF-2 function and therefore gene transcription in some cell types (117). In other cell types, however, AF-1 activation is sufficient to induce gene transcription and in these tissues tamoxifen functions as an estrogen agonist (117). Whether or not AF-1 is sufficient to activate transcription is dependent upon the cell type and the individual gene promoter involved (118). This mechanistic picture has been further complicated by the discovery that AP-1 transcriptional sites within DNA can also respond to the ER-ligand complex (119). Once again, activation by a particular ligand is dependent upon cell type, as tamoxifen has shown agonist activity at AP-1 sites in uterine but not breast tissue (120). A raloxifene-inducible element (RIE) has also been characterized within the TGF-β$_3$ gene promoter which responds transcriptionally to raloxifene bound ER (121). It has been hypothesized that the agonist effects of raloxifene may be exerted via activation of genes containing this RIE promoter sequence (121).

It has been proposed that the binding of alternative ligands to the ER may induce different conformations of the receptor-ligand complex (16). These different conformations may, in turn, demonstrate unique profiles in terms of their ability to interact with various gene promoters or to interact with other pieces of the cellular machinery which are involved with gene transcription (16). Altered transcriptional profiles have been observed for 17β-estradiol, TPEs, raloxifene, and ICI 164384, thus four distinct classes of ER modulators have been identified to date (16,122). Clearly, the agonist/antagonist profile for any ER modulator must be determined for each target tissue in question. It is expected that as additional target tissues are explored and new estrogen-regulated genes are uncovered new classes of ER modulators will be developed with unique profiles of selectivity.

References

1. B.E. Henderson, R.K. Ross and L. Bernstein, Cancer Res., 48, 246 (1988).
2. E.P. Smith, J. Boyd, G.R. Frank, H. Takahashi, R.M. Cohen, B. Specker, T.C. Williams, D.B Lugahn and K.S. Korach, N.Engl.J.Med., 331, 1056 (1994).
3. D.R. Ciocca and L.M. Vargas Roig, Endocr.Rev., 16, 35 (1995).
4. R. Lindsay in "Osteoporosis: Etiology, Diagnosis, and Management;" B.L. Riggs and L.J. Melton, Eds., Raven, New York, 1988, p. 333.
5. R.K. Ross, M.C. Pike, T.M. Mack and B.E. Henderson in "HRT and Osteoporosis," J. O. Drife, J. W. W. Studd, Eds., Springer-Verlag, London, 1990, p. 209.
6. D. Gath and S. Iles in "HRT and Osteoporosis," J. O. Drife, J. W. W. Studd, Eds., Springer-Verlag, London, 1990, p. 35.
7. V.W. Henderson, A. Paganini-Hill, C.K. Emanuel, M.E. Dunn and J.G. Buckwalter, Arch.Neurol., 51, 896 (1994).
8. C.F. Holinka, Ann.N.Y.Acad.Sci., 734, 271 (1994).
9. D.R. Session, A.C. Kelly and R. Jewelewicz, Fertility Sterility, 2, 277 (1993).
10. G.A. Colditz, S.E. Hankinson, D.J. Hunter, W.C. Willett, J.E. Manson, M.J. Stampfer, C. Hennekens, B. Rosner and F.E. Speizer, N.Engl.J.Med., 332, 1589 (1995).
11. J.-F. Miquei and J. Gilbert, J.Steroid.Biochem.Molec.Biol., 31, 525 (1988).
12. R.A.. Magarian, L.B. Oveacre, S. Singh and K.L. Meyer, Curr.Med.Chem., 1, 61 (1994).
13. S.K. Chander, S.S. Sahota, T.R.J. Evans and Y.A. Luqmani, Crit.Rev.Oncol.Hematol., 15, 243 (1993).
14. V.C. Jordan, J.Cell.Biochem., 51 (1995).
15. G.L. Evans and R.T. Turner, Bone, 17, 181S (1995).
16. D.P. McDonnell, D.L. Clemm, T. Hermann, M.E. Goldman and J.W. Pike, Mol.Endocrinol., 9, 660 (1995).
17. H. Honjo, H. Tanaka, T. Kashiwagi, M. Urabe, H. Okada, M. Hayashi and K. Hayashi, Hormone Metab.Res., 27, 204 (1995).
18. B.B. Sherwin, Ann.N.Y.Acad.Sci., 749, 213 (1994).
19. E. Barrett-Connor and D. Kritz-Silverstein, J.Am.Med.Assn., 265, 1861 (1991).
20. M.J. Stampfer and G.A. Colditz, Prev.Med., 20, 47 (1991).
21. R.H. Punnonen, H.A. Jokela, P.S. Dastidar, M. Nevala and P.J. Laippala, Maturitas, 21, 179 (1995).
22. Writing Group for the PEPI Trial, J.Am.Med.Assn., 273, 199 (1995).
23. J.C. LaRosa, Fertility and Sterility, 62 (suppl. 2), 140S (1994).
24. M.E. Mendelsohn and R.H. Karas, Curr.Opin.Cardiol., 9, 619 (1994).
25. R.T. Turner, B.L. Riggs and T.C. Spelsberg, Endocrine Rev., 15, 275 (1994)
26. R. Lindsay, J.M. Aitken and J.B. Anderson, Lancet, 1, 1038 (1976).
27. M.J. Oursler, J.P. Landers, B.L. Riggs and T.C. Spelsberg, Ann.Med., 25, 361 (1993).
28. J.H. Clark and B.M. Markaverich, Pharmacol.Ther., 21, 621 (1982).
29. L.H. Lemon, Acta Endocrinol.Supp., 233, 17 (1980).
30. T. Fotsis, Y. Zhang, M.S. Pepper, H. Adlercreutz, R. Montesano, P.P. Nawroth and L. Schweigerer, Nature, 268, 237 (1994).
31. G.P. Mueller, W.F. Johns, D.L. Cook and R.A. Edgren, J.Am.Chem.Soc., 80, 1769 (1958).
32. L.J. Chinn, J.H. Dygos, S.E. Mares, R.L. Aspinall and R.E. Ranney, J.Med.Chem., 17, 351 (1974).
33. J.H. Dygos and L.J.Chinn, J.Org.Chem., 40, 685 (1975).
34. S.A. Washburn, M.R. Adams, T.B. Clarkson and S.J. Adelman, Am.J.Obstet.Gynecol., 169, 251 (1993).
35. H.U. Bryant, P.K. Wilson, M.D. Adrian, H.W. Cole, L.L. Short, J.A. Dodge, T.A. Grese, J.P. Sluka and A.L. Glasebrook, J.Soc.Gynecol.Invest. 3 (Suppl.), 152A (1996).
36. J.A. Dodge, D.E. Magee, P. Shetter, H. Cole, D. Adrian and H.U. Bryant, J.SteroidBiochem.Molec.Biol., in press.
37. N. Tsutimi, Biol.Pharm.Bull., 18, 1012 (1995).
38. J.A. Dodge, A. Glasebrook, D.L. Phillips, H.W. Cole, D.E. Magee, B. Serlin and H.U. Bryant, J. Bone Miner. Res., 9 (suppl. 1), S134 (1994).
39. W.H. Utian, Br.Med.J., 1, 579 (1973).

40.   L.C. Huppert, Fertility and Sterility, 31, 1 (1979).
41.   P.T. Beall, L.K. Misra, R.L. Yound, H.J. Spjut, H.J. Evans and A. Leblanc, Calcif.Tissue Int., 36, 123 (1984).
42.   P.K. Chakraborty, J.L. Brown, C.B. Ruff, M. . Nelson and A.S. Mitchell, J.Steroid Biochem. Molec.Biol., 40, 725 (1991).
43.   P.J. Stewart and P.H. Stern, Endocrinology, 118, 125 (1986).
44.   R.L. Young, J.W. Goldzieher, K. Elkind-Hirsch, P.G. Hickox and P.K. Chakraborty, Int.J. Fertility, 36, 167 (1991).
45.   J.H. Clark and B.M. Markaverich, Pharmac.Ther., 15, 467 (1982).
46.   J.H. Clark and S.C. Guthrie, Biol.Reprod., 25, 667 (1981).
47.   R.L. Young, J.W. Goldzieher, P.K. Chakraborty, W.B. Panko and C.N. Bridges, Int.J. Fertility, 36, 291 (1991).
48.   V.C. Jordan, Br.J.Pharmacol., 110, 507 (1993).
49.   R.T. Turner, G.K. Wakley, K.S. Hannon and N.H. Bell, Endocrinology, 122, 1146 (1988).
50.   A.B. Grey, J.P. Stapleton, M.C. Evans, M.A. Tatnell, R.W. Ames and I.R. Reid, Am.J.Med., 99, 636 (1995).
51.   R.R. Love, H.S. Barden, R.B. Mazess, S. Epstein and R.J. Chappell, Arch.Intern.Med., 154, 2585 (1994).
52.   R.L. Ward, G. Morgan, D. Dalley and P.J. Kelly, BoneMineral, 22, 87 (1993).
53.   B. Kristensen, B. Ejlertsen, P. Dalgaard, L. Larsen, S.N. Holmegaard, I. Transbøl and H.T. Mouridsen, J.Clin.Oncol., 12, 992 (1994).
54.   A.M. Kenny, K.M. Prestwood, C.C. Pilbeam and L.G. Raisz, J.Clin.Endocr.Metab., 80, 3287 (1995).
55.   P. Sismondi, N. Biglia, M. Giai, L. Sgro and C. Campagnoli, Anticancer Res., 14, 2237 (1994).
56.   A.B. Grey, J.P. Stapleton, M.C. Evans and I.R. Reid, J.Clin.Endocr.Metab., 80 (1995).
57.   D.A. Shewmon, J.L. Stock, C.J. Rosen, K.M. Heiniluoma, M.M. Hogue, A. Morrison, E.M. Doyle, T. Ukena, V. Weale and S. Baker, Arterioscler.Thromb., 14, 1586 (1994).
58.   R.R. Love, D.A. Wiebe, J.M. Feyzi, P.A. Newcomb and R.J. Chappell, J.Natl.Cancer Inst., 86, 1534 (1994).
59.   L.E. Rutqvist and A. Mattson, J.Natl.Cancer Inst., 85, 1398 (1993).
60.   B. Fisher, J.P. Constantino, C.K. Redmond, E.R. Fisher, D.L. Wickerham and W.M. Cronin, J.Natl.Cancer Inst., 86, 527 (1994).
61.   Early Breast Cancer Trialists' Collaborative Group, Lancet, 339, 1 (1992).
62.   H. Gylling, E. M äntylä and T.A. Miettinen, Atherosclerosis, 96, 245 (1992).
63.   V. Guetta and R.O. Cannon III, Circulation, 92 (suppl. 1), 164 (1995).
64.   H. Wiseman, P. Quinn and B. Halliwell, FEBS, 330, 53 (1993).
65.   J.K. Williams and M.R. Adams, Circulation, 92 (suppl. 1), 627 (1995).
66.   R. Fuchs-Young, A.L. Glasebrook, L.L. Short, M.W. Draper, M.K. Rippy, H.W. Cole, D.E. Magee, J.D. Termine and H.U. Bryant, Ann.N.Y.Acad.Sci., 761, 355 (1995).
67.   R.P. Kedar, T.H. Bourne, T.J. Powles, W.P. Collins, S.E. Ashley, D.O. Cosgrove and S. Campbell, Lancet, 343, 1318 (1994).
68.   V.J. Assikis and V.C. Jordan, Endocr.Relat.Cancer, 2, 235 (1995).
69.   D.Y.-S. Kuo and C.D. Runowicz, Med.Oncol., 12, 87 (1995).
70.   D.C. Robinson, J.D. Bloss and M.A. Schiano, Gynecol.Oncol., 59, 186 (1995).
71.   M.R. Osborne, A. Hewer, I.R. Hardcastle, P.L. Carmichael and D.H. Phillips, Cancer Research, 56, 66 (1996).
72.   G.M. Williams, M.J. Iatropoulos, M.V. Djordjevic and O. . Kaltenberg, Carcinogenisis, 14, 315 (1993).
73.   W. Rauschning and K. . Pritchard, Breast Cancer Res.Treat., 31, 83 (1994).
74.   L.E. Sterbygaard, J. Hernstedt and J. F. Thomsen, Breast Cancer Res.Treat., 25, 57 (1993).
75.   R.C. Coombes, B.P. Haynes, M. Dowsett, M. Quigley, J. English, I.R. Judson, L.J. Griggs, G.A. Potter, R. McCague and M. Jarman, Cancer Res., 55, 1070 (1995).
76.   J.-I. Koh, T. Kubota, F. Asanuma, Y. Yamada, E. Kawamura, Y. Hosoda, M. Hashimoto, O. Yamamoto, S. Sakai, K. Maeda and E. Shiina, J.Surg.Oncol., 51, 254 (1992).
77.   H. Gylling, S. Pyrhönen, E. Mäntylä, H. Mäqenpää, L. Kangas and T.A. Miettinen, J.Clin. Oncol., 13, 2900 (1995).
78.   E. di Salle, T. Zaccheo and G. Ornati, J.Steroid Biochem.Molec.Biol., 36, 203 (1990).
79.   E. Tomás, A. Kauppila, G. Blanco, M. Apaja-Sarkkinen and T. Laatikainen, Gynecol.Oncol., 59, 261, (1995).
80.   R. Löser, K. Seibel, W. Roos and U. Eppenberger, J.Cancer Clin.Oncol., 21, 985 (1985).
81.   H.Z. Ke, H.A. Simmons, C.M. Pirie, D.T. Crawford and D.D. Thompson, Endocrinology, 136, 2435, (1995).
82.   H.K. Chen, H.Z. Ke, C.H. Li, Y.F. Ma, H. Qi, D.T. Crawford, C.M. Pirie, H.A. Simmons, W.S.S. Jee and D.D. Thompson, Bone, 17 (suppl.), 175S (1995).
83.   H.Z. Ke, H.K. Chen, H. Qi, C.M. Pirie, H.A. Simmons, Y.F. Ma, W.S.S. Jee and D.D.

Thompson, Bone, 17, 491 (1995).

84. I.N.H. White, F. de Mattheis, A. Davis, L.L. Smith, C. Croften-Sleigh, S. Venitt, A. Hewer and D.H. Phillips, Carcinogenesis, 13, 2197 (1992).
85. G.C. Hard, M.J. Iatropoulos, K. Jordan, L. Radi, O.P. Kaltenberg, A.R. Imondi and G.M. Williams, Cancer Res., 53, 4534 (1993).
86. T.M. Willson, B.R. Henke, T.M. Momtahen, P.S. Charifson, K.W. Batchelor, D.B. Lubahn, L.B. Moore, B.B. Oliver, H.R. Sauls, J.A. Triantafillou, S.G. Wolfe and P.G. Baer, J.Med. Chem., 37, 1550 (1994).
87. S.D. Bain, D.L. Celino, M.C. Bailey, M.J. Strachan, J.R. Piggott and V.M. Labroo, Calcified Tiss., 55, 338 (1994).
88. R.N. Trivedi, S.C. Chauhan, A. Dwivedi, V.P. Kamboj and M.M. Singh, Contraception, 51, 367 (1995).
89. T.J. Hall, H. Nyugen, M. Schaueblin and B. Fournier, Biochem.Biophys.Res.Comm., 216, 662 (1995).
90. M. Sato, A.L. Glasebrook and H.U. Bryant, J.Bone Miner.Metab., 12 (suppl. 2), S9 (1995).
91. M.A. Anzano, C.W. Peer, J.M. Smith, L.T. Mullen, M.W. Shrader, D.L. Logsdon, C.L. Driver, C.C. Brown, A.B. Roberts and M.B. Sporn, J.Natl.Cancer inst., 88, 123 (1996).
92. C.D. Jones, M.G. Jevnikar, A.J. Pike, M.K. Peters, L.J. Black, A.R. Thompson, J.F. Falcone and J.A. Clemens, J.Med.Chem., 27, 1057 (1984).
93. L.J. Black, M. Sato, E.R. Rowley, D.E. Magee, A. Bekele, D.C. Williams, G.J. Cullinan, R. Bendele, R.F. Kauffman, W.R. Bensch, C.A. Frolik, J.D. Termine and H.U. Bryant, J.Clin. Invest., 93, 63 (1994).
94. R.F. Kauffman and H.U. Bryant, Drug News Perspec., 8, 531 (1995).
95. M.W. Draper, D.E. Flowers, W.J. Huster and J.A. Neild in "Proceedings of the Fourth International Symposium on Osteoporosis and Consensus Development Conference, 1993," C. Christiansen, B. Riis, Eds., Handelstrykkeriert Aalborg Aps, Aalborg, Denmark, 1993, p. 119.
96. G. Evans, H.U. Bryant, D. Magee, M. Sato and R.T. Turner, Endocrinology, 134, 2283 (1994).
97. C.H. Turner, M. Sato, H.U. Bryant, Endocrinology, 135, 2001 (1994).
98. G. Fiorelli, F. Gori, U. Frediani, A.M. Morelli, A. Falchetti, S. Benvenuti, L. Masi and M.L. Brandi, Biochem.Biophys.Res.Comm., 211, 857 (1995).
99. M.C. Perez, E.E. Furth, P.D. Matzumura and C.R. Lyttle, Biol.Reprod., 54, 249 (1996).
100. R. Fuchs-Young, D.E. Magee, H.W. Cole, L. Short, A.L. Glasebrook, M.K. Rippy, J.D. Termine and H.U. Bryant, Endocrinology, 136 (suppl.), 57 (1995).
101. T.A. Grese, J.P. Sluka, H.U. Bryant, H.W. Cole, J.R. Kim, D.E. Magee, E.R. Rowley and M. Sato, Biorg.Med.Chem.Lett., 6, in press (1996).
102. M. Kojima, N. Tsuttsumi, H. Nagata, F. Itoh, A. Ujiie, K. Kawashima, H. Endo and M. Okazaki, Biol.Pharm.Bull., 17, 504 (1994).
103. N. Tsutsumi, K. Kawashima, N. Arai, H. Nagata, M. Kojima, A. Ujiie and H. Endo, Bone and Mineral, 24, 201 (1994).
104. A.E. Wakeling and J. Bowler, J.Steroid Biochem.Molec.Biol., 31, 645 (1988).
105. M. Dukes, R. Chester, L. Yarwood and A. E. Wakeling, J. Endocrinology, 141, 335 (1994).
106. J.A. Dodge, M.G. Stocksdale, L.J. Black, E.R. Rowley, A. Bekele, H.W. Cole, C.E. Brown, D.E. Magee, G.A. Dehoney and H.U. Bryant, J.Bone Miner.Res., 8 (suppl. 1), S278 (1993).
107. A.E. Wakeling, Breast Cancer Res.Treat., 25, 1 (1993).
108. P.G. Baer, T.M. WIllson and D.C. Morris, Calcifed Tissue Int., 55, 338 (1994).
109. V.C. Jordan, Cancer, 70, 977 (1992).
110. D.J. Weiss and E. Gurpide, J.Steroid Biochem.Molec.Biol., 31, 671 (1988).
111. I. Nemere, L.-X. Zhou and A.W. Norman, Receptor, 3, 277 (1993).
112. A.A. Colletta, J.R. Benson and M. Baum, Breast Cancer Res.Treat., 31, 5 (1994).
113. L.B. Hendry, C.K. Chu, M.L. Rosser, J.A. Copland, J.C. Wood and V.V. Mahesh, J.Steroid Biochem.Molec.Biol., 49, 269 (1994).
114. D.M. Wolf and S.A.W. Fuqua, Cancer Treat.Rev., 21, 247 (1995).
115. B.S. Katzenellenbogen, M.M. Montano, P. Le Goff, D.J. Schodin, W.L. Kraus, B. Bhardwaj and N. Fujimoto, J.Steroid Biochem.Molec.Biol., 53, 387 (1995).
116. V. Kumar, S. Green, G. Stack, M. Berry, J. R. Jin and P. Chambon, Cell, 54, 199 (1988).
117. M. Berry, D. Metzger and P. Chambon, EMBO J., 9, 2811 (1990).
118. M.T. Tzukerman, A. Esty, D. Santiso-Mere, P. Danielian, M.G. Parker, R.B. Stein, J.W. Pike and D.P. McDonnell, Mol.Endocrinol., 8, 21 (1994).
119. Y. Umayahara, R. Kawamori, H. Watada, E. Imano, N. Iwams, T. Morishima, Y. Yamasaki, Y. Kajimoto and T. Kamada, J.Biol.Chem., 269, 16433 (1994).
120. P. Webb, G.N. Lopez, R.M. Uht and P.J. Kushner, Mol.Endocrinol., 9, 443 (1995).
121. N.N. Yang and S. Hardiker, J.Bone Miner.Res., 9 (suppl. 1), S144 (1994).
122. D.P. McDonnell, S.L. Dana, P.A. Hoener, B.A. Lieberman, M.O. Imhof and R.B. Stein, Ann. N.Y.Acad.Sci., 761, 121 (1995).

Chapter 20. Cell Adhesion Integrins as Pharmaceutical Targets

V. Wayne Engleman, Michael S. Kellogg, and Thomas E. Rogers
Searle Research and Development, Monsanto Company
700 Chesterfield Parkway North, St. Louis, MO 63198

Introduction - Rapid advances in understanding of cell adhesion and communication engendered the "golden age for adhesion" (1). Faithfully orchestrated cellular behavior observed by embryologists and developmental biologists of the past century are now appreciated in molecular terms. The complex, dynamically regulated phenotypic expression of plasma membrane linked glycoprotein and carbohydrate molecules and their ability to uniquely recognize soluble and insoluble ligands helps define the biological basis of current concepts of development, life, and death. As understanding of the pivotal role of cell adhesion and communication in normal life processes improves, we can further discern consequences of atypical adhesion and communication. The promise of more lucid molecular pathology is modification of disease states through drug treatment. We review here the integrin class of cell adhesion molecules. The platelet integrin, $\alpha_{IIb}\beta_3$, is addressed in Chapter 10. Our focus is toward pathologies for which disease modifying agents could be developed based on understanding of the specific molecular interactions.

THE INTEGRINS

The integrin superfamily is made up of structurally and functionally related glycoproteins distributed over three families: the VLA family (β_1), the Leucam family (β_2), and the Cytoadhesin family (β_3). Beta subunits 4-8 are not yet classified into families. Integrins are α,β heterodimeric, transmembrane receptor molecules found in combinations on every mammalian cell type except red blood cells. There are fifteen α subunits and eight β subunits that are noncovalently linked and expressed on the surfaces of cells in combinations (Table 1). The integrin name was derived from their role in "integrating" the extracellular matrix (ECM) with the cytoskeleton (2). The β subunits show higher sequence homology (40-48%) than the α subunits, and are 90-110 kD in MW. The β subunits have short cytoplasmic domains of 40-50 amino acids, highly conserved transmembrane domains, a large extracellular domain with four cysteine rich domains (48-56 Cys residues), and an EF-hand like divalent cation binding domain adjacent to a highly conserved N-terminus. β_4 notably differs from other β subunits due to an unusually long (1018 amino acid) cytoplasmic domain containing four fibronectin (Fn) type III repeats, which may be involved in assembly of structures forming hemidesmosomes of epithelial cells (3-5). Alternatively spliced variants of β_1 (6), β_3 (7), β_4 (8-12), α_3 (13), α_6 (14,15) and α_{IIb} (16) exist which alter cellular response to receptor ligation.

The fifteen known α subunits contain light chains with short cytoplasmic domains (13-58 amino acids) containing the sequence KXGFFKR, a transmembrane domain, and a large extracellular domain. There is absolute fidelity of the KXGFFKR sequence near the C-terminus, which appears critical to the assembly and proper function of $\alpha_1\beta_1$, suggesting similar importance in other integrins (17). Most α subunits are post-translationally cleaved resulting in a membrane spanning light chain disulfide linked to the heavy chain. The heavy chain contains seven repeating domains. Domains IV, V, VI, and VII resemble EF-hand like structures capable of binding divalent cations. The α_4 subunit has a cleavage site that yields heavy and light chains that are not disulfide bonded. α_1, α_2, α_L, α_M, and α_X lack the disulfide linkage but gain an extra domain between the second and third repeat domains (I-domain). In addition, α_1 contains an extra inserted domain between the fifth and sixth repeat domains. Subunit molecular weight ranges from 150-200 kD (18).

Of integrin ligands, the Fn receptor $\alpha_5\beta_1$ was the first for which a ligand binding structural motif was identified (19). The cell binding site on Fn was localized to the 10th type III domain and has a minimal active sequence of Arg-Gly-Asp (RGD). $\alpha_5\beta_1$ is the predominant receptor for Fn. While $\alpha_5\beta_1$ binds only RGD in Fn, $\alpha_v\beta_3$ and $\alpha_{IIb}\beta_3$, may recognize RGD and similar tri-peptide sequences in several ligands, including Fn, vitronectin (Vn), fibrinogen (Fb), von Willebrand factor (vWf) osteopontin (Opn), laminin (Ln), thrombospondin (TS), and denatured collagens (Col) (Table 1) (20).

Understanding of distribution and function is advancing. β_7 was previously called β_p due to its involvement in homing of lymphocytes to Peyer's patch. LPAM-1, ($\alpha_4\beta_7$) is a member of the emerging β_7 family of integrins found primarily on mucosal lymphocytes (21-23). The preferred ligand for $\alpha_4\beta_7$ appears to be Ig superfamily adhesion molecule MadCAM-1, a vascular addressin that directs lymphocyte traffic into Peyer's patches, although with higher states of activation, $\alpha_4\beta_7$ can bind VCAM-1 and Fn (24,25). The expression of this and related β_7 integrins (α_E, α_{IEL}) is related to T-cell ontogeny (26,27). They may exist in various activation states depending on signal (28,29). The β_7 integrins have been localized to restricted populations of B-lymphocytes and certain well differentiated cells of the monocyte lineage (30,31).

The only known member of the β_6 family is $\alpha_v\beta_6$, apparently a Fn receptor (32). Its discovery on cultured endothelial cells suggested wide distribution but localization studies would argue that it is an integrin of transition (33,34). Immunolocalization studies conclude that $\alpha_v\beta_6$ is developmentally regulated. Expression is significantly reduced in adult tissues. It is re-expressed in wounded dermal or inflamed airway epithelia, is responsive to TGF-β, and is highly expressed in oral squamous carcinoma tissue (35). An 11 amino acid sequence on the cytoplasmic domain of β_6 has been shown to be responsible for growth proliferation in a human colon carcinoma cell line (36). This suggests a role in tumor cell invasion similar to $\alpha_v\beta_3$.

Integrin β_8 is anomolous. It is only 31%-37% homologous to other betas and lacks 6 of the 56 conserved cysteines of the extracellular domain. Most striking is complete lack of homology of the cytoplasmic domain and apparent lack of cytoskeletal interaction and signal transduction (37,38).

Subunit, α_8, is not yet well characterized. It was first found in chicken where it localized primarily in the nervous system, and recently, in a mammalian source where localization is in smooth muscle or contractile cells (39,40). Homology is 78% and an association with β_1 noted initially, is consistent between the two sources.

The remaining new alpha subunit, α_9, is a β_1 family member that was purified from guinea pig airway epithelium, from human lung, small intestine, and tumor cell cDNA libraries, and from rat liver on a GRGDSPC affinity column (41-43). It is critical to note that the cysteine is essential for this interaction and the RGD sequence only cooperates in this context. Known ligands for $\alpha_9\beta_1$ are immobilized fragment E8 of EHS-laminin and CNBr fragment 8 of collagen alpha 1 (43). $\alpha_9\beta_1$ does not bind to Fn, Ln, Vn, Fb, TS, or type I or IV collagen but shows enhanced binding on transfected cells to the third type III repeat of tenascin (Tn) (44). No reports of α_9 binding sensitive to RGD peptides are known. Immunolocalization of $\alpha_9\beta_1$ has been found in airway epithelium (but not alveolar epithelium), squamous epithelial basal layer, skeletal and smooth muscle, hepatocytes, and developing cornea (42,45,46).

Signalling -The term "communication" has been appended to cell adhesion and provides an historically correct view of the transition from entity to function. Initial observations with the platelet integrin $\alpha_{IIb}\beta_3$ framed the concept of two-way signalling through integrins (47). Growth suppression, apoptosis protection ($\alpha_5\beta_1$), or cooperation with growth factors ($\alpha_v\beta_3$) has been observed (48). Vn signaling via $\alpha_v\beta_3$ depends on ligand state, soluble or substratum bound. Chemotaxis is stimulated by soluble Vn through coupling to G protein pathways, while haptotaxis, anlaogous to directional cell spreading, involves tyrosine phosphorylation of paxillin (49). More coherent understanding of the spatial and temporal regulation of other integrins is evolving. Integrins coexist with hormones, growth factors, and cytokines to help regulate overall control of growth, differentiation, and cell death (50-54).

TABLE 1.   INTEGRIN SUBUNIT ASSOCIATIONS, NATURAL/SYNTHETHETIC LIGANDS, AND CD NOMENCLATURE

SUBUNIT (VARIANT)	α1 CD49a	α2 CD49b, gpla	α3(a,b) CD49c	α4 CD49d	α5 CD49e	α6 (a,b) CD49f	α7 (a,b) αOL	α8	α9	αIIb gpIIb, CD41	αV CD51	αIel	αL CD11a	αM CD11b	αX CD11c
β1 (3v, s) VLA, CD29	Ln C-I, C-IV DGEA	C-I, Fn (Ln) RGD	Fn, Ln C-I Ep	V-1/d1,2 Fn, In RGD	Fn	Ln	Ln	? RGD	Ln, C-1 Tn GRGDSPC	Fn, Vn Opn RGD					
β2 LEUCAM, CD18													I-1/d1 I-2 I-3	I-1/d3 Fb IC3b	IC3b Fb GPRP
β3 (a, b) CYTOADHESIN CD41a, CD61 GPIIb/IIIa, VnR										Fb, Fn vWF, Vn C-1, Bb TS, RGD KQAGDV	Vn Ln, TS Opn, vWF RGD VI, Tat				
β4							Ln Ep								
β5 βx, βs											Vn, Fn Opn Tat, RGD				
β6											Fn RGD				
β7 (βp) (LPAM-1)				MAdCAM-1 V-1, Fn EILDV								?			
β8									?						

Key to Ligands: Bb = Borella burgdorieri, C = collagen-type, Ep = epiligrin, Fb = fiberogen, Fn = fiberonectin, iC3b = inactivated complement 3b, in = invasin, Ln = iaminin, Opn = osteopotin, Tat = HIV TAT protein, Tn = tenascin, Ts = thrombospondin, I-id = ICAM-type/domain, V-id = VCAM-type/domain, VI = virus (adenovirus, echovirus, FMDV) Vn = vitronectin, vWF = von Willebrand Factor

Structure - VCAM-1 binds to VLA-4 ($\alpha_4\beta_1$) via its homologous first and fourth Ig domains. Similarly, $\alpha_4\beta_7$ interacts with these domains. Different levels of cation activation were required for binding of the two integrins. These differences may be important in cellular response to integrin occupancy (55). A panel of functional Mabs to the α_4 chain were used to map aggregation and adhesion regions (56). The crystal structure of a VLA-4 binding fragment on VCAM-1 indicated a distinct conformation of the Q38IDSPL integrin recognition moiety in a loop between beta strands on domain 1 of VCAM-1. This motif may be common to integrin binding Ig superfamily molecules (57). A cyclic peptide that mimics the loop structure inhibited binding of VLA-4 to VCAM-1 (58,59). Biochemical evaluations suggest conformational dependence of $\alpha_v\beta_3$ on matrix, antibody or calcium binding (60).

THERAPEUTIC IMPLICATIONS

Thrombosis- The role of the $\alpha_{IIb}\beta_3$ integrin in platelet aggregation and arterial thrombosis has been broadly investigated resulting in clinical studies with Mabs, cyclic peptides, and peptidomimitecs. This work is reviewed in detail in Chapter 10.

<u>Atherosclerosis, Restenosis</u> - Oxidized low-density lipoprotein (ox ldl) was associated with monocyte-macrophage-like differentiation, increased CD11b expression and monocyte adhesion to human umbilical vein endothelial cells (61). Adhesion was blocked by an anti α_M (CD11b) antibody. Similarly, ox LDL stimulated CD11b/CD18 (MAC 1) expression in whole blood and leukocyte adhesion to endothelium in hamsters (62). After surgical injury (eg. balloon angioplasty) to restore patency to a stenotic artery, a neointimal hyperplasia leading to restenosis is common (63,64). Activated smooth muscle cells (SMC) migrate and proliferate in the neointima, accelerating reclosure of the artery (65). Vn receptor integrin $\alpha_v\beta_3$ is implicated in this process (66). An $\alpha_v\beta_3$ specific mAb LM609 blocked PDGF induced human SMC migration in vitro, as did a cyclic peptide Gpen*GRGDSPC*A (* indicates cyclization points) that is relatively specific for $\alpha_v\beta_3$. This peptide administered perivascularly by osmotic pump at a rate of 2 µg/hr to injured rabbit carotid artery reduced neointimal thickening relative to inactive peptide control (67). Similarly, a nonspecific cyclic peptide antagonist, G4120, inhibited neointima formation in a dose dependent fashion after catheter damage to hamster carotid arteries (68). Balloon catheter injury in the rat gave rise to Opn and β_3 integrin mRNA expression, peaking between 0.3 and 14 days after injury, and returning to undetectable levels at 42 days when regeneration was complete. The protein levels followed the course of mRNA expression. Neither Opn nor the β_3 integrin was detected in uninjured rat arterial endothelium. In vitro, Opn from rat SMC and mouse recombinant protein both stimulated adhesion and migration of bovine aortic endothelial cells (69). A humanized monoclonal antibody to β_3 (ReoPro, c7E3), which blocks both the platelet ($\alpha_{IIb}\beta_3$) and the Vn ($\alpha_v\beta_3$) receptors was effective in reducing secondary events and possibly restenosis in high risk angioplasty patients (70). Integrelin, a cyclic heptapeptide more specific to $\alpha_{IIb}\beta_3$ also reduced short term ischemic complications (71).

<u>Wound Healing</u> - Adhesion among keratinocytes is mediated by integrins localized to specific areas of the cell. At the basement membrane proximal surface of basal cells, $\alpha_6\beta_4$ is localized and codistributed with Ln and nicein/kalinin, whereas the lateral regions contain primarily the $\alpha_2\beta_1$ and $\alpha_3\beta_1$ integrins that function to maintain colony morphology through cell-cell interactions. During wound healing, keratinocytes express $\alpha_v\beta_5$ and $\alpha_5\beta_1$ to facilitate attachment, proliferation, and migration over provisional matrix in the wound (72). In hyperproliferative skin diseases, the polarized expression of integrins disappears. $\alpha_v\beta_3$ is expressed at neovascular sites in wound granulation tissue, but it is not detected in normal skin, while β_1 integrins are expressed on blood vessels in normal skin, granulation tissue, and the basal cells of the epithelium (73). On the chick chorioallanotic membrane (CAM), angiogenisis induced by basic fibroblast growth factor (bFGF), tumor necrosis factor α (TNFα) or human melanoma fragments, was blocked by $\alpha_v\beta_3$ antibody LM609. LM609 treatment indicated that $\alpha_v\beta_3$ is also involved in neovascularization in actively growing chick embryos, but it had no effect on preexisting blood vessel growth.

<u>Remodeling Bone</u> - The role of RGD-containing proteins and their corresponding integrins in the formation and turnover of bone has been reviewed (74,75). Inhibition of bone resorption in a rat model of osteoporosis by systemic administration of a monoclonal antibody to β_3 integrin was recently reported (76).

<u>Cancer</u>- Integrins are implicated in a variety of pathologies associated with malignant tumours, including adhesion to tissue, metastasis to remote tissues, angiogenesis of host vessels for growth, and immune response. The snake venom disintegrin triflavin blocked adhesion of human cervical carcinoma cells to Fn, Fb, and Vn, but was less

effective in blocking adhesion to Ln and Col (types I and IV). Triflavin did not kill the cells. The peptide GRGDS was competitive with triflavin binding to cells (77). When a variety of cancer cells were evaluated for integrin involvement in adhesion and extravastation to IL-1 treated endothelial monolayers, it was found that blocking antibodies to VLA-4 ($\alpha_4\beta_1$) inhibited adhesion of some melanomas, sarcomas, and a lung carcinoma, while anti-Vn receptor antibodies blocked adhesion of 65% of the human cell lines investigated (78). The functional involvement of VLA-4 on melanoma cell metastasis to lung was demonstrated by blocking Mab studies on IL-1 treated nude mice (79). VLA-4 was also reported to have significance in the early phase of adherence of malignant lymphoblastic cells to marrow stromal cells, and may be important to the homing process (80). Association of VLA-4 expression with clinical outcome in primary cutaneous malignant melanoma was examined through immunohistochemical screening. Detection of this integrin may be indicative of greater risk of developing metastases (81). The role of integrins $\alpha_v\beta_3$, $\alpha_3\beta_1$, $\alpha_{v4}\beta_1$, $\alpha_5\beta_1$, $\alpha_6\beta_1$ and their cognate cellular and matrix molecules in melanocytic tumor progression was reviewed with respect to melanoma invasion, metastasis, and evasion of lysis by effector cells (82). Squamous carcinoma cell lines adhered to matrix proteins and were examined for their invasive capabilities through matrigel. Invasion was blocked by RGD peptides, and adhesion to Vn could also be inhibited by anti α_v mAbs (83).

The $\alpha_v\beta_3$ integrin has been implicated as a marker of blood vessels associated with malignant human breast tumors (84). Human breast tumor cells ($\alpha_v\beta_3$ negative) present in human skin transplanted to a severe combined immunodeficiency (SCID) mouse, induced an angiogenic response. Administration (iv, 250μg/50μL twice weekly for 3 weeks) of the blocking mAb LM609 prevented tumor growth or reduced tumor cell proliferation in the graft. The treated tumors also appeared less invasive than sham treated controls. Angiogenisis depends on redundant systems to initate blood vessel cell proliferation and regulation of the invasion and differentiation of new blood vessels. While mAb LM609 blocks angiogenisis induced by bFGF and TNFα in the chick CAM model and a rabbit corneal model, it does not affect neovascularization initiated by vascular endothelial cell growth factor (VEGF) or transforming growth factor α (TGFα) (85). The latter events are, however, blocked specifically by mAb P1F6 (anti $\alpha_v\beta_5$). Cyclo Arg*-Gly-Asp-D-Phe-Val* blocks both α_v integrins and effectively blocks angiogenisis induced by bFGF, VEGF, TNFα, and TGFα (85). The role of $\alpha_v\beta_3$ in angiogenisis and apoptosis was recently reviewed (86). Angiogenisis in bone marrow of patients with active or non-active multiple myeloma (MM) was investigated by immunohistochemistry. Angiogenic activity was low or absent in patients with non-active MM, but it was markedly increased in those with active MM. Adhesion molecule expression (LFA-1, VLA-4, LAM-1 and CD-44) was also highest in the the latter patients. It is unclear what functional significance the correlation may have, but it suggests a connection between adhesive interactions and tumor dissemination and angiogensis (87).

Inflammation - Eosinophils release mediators involved in the pathogenesis of bronchial asthma. Asthmatic subjects with air flow limitation were strongly positive for LFA-1 (CD11a/CD18), Mac-1 (CD11b/CD18), and VLA-4. Expression of these integrins prior to spontaneous asthma attacks suggest that they may be important contributors to eosinophil infiltration to the lung (88). Experimental lung eosinophilia induced by iv Sephadex injection in guinea pigs could be blocked by an anti IL-5 mAb. Lung tissue eosinophil peroxidase (EPO) activity was also reduced, but the concomitant influx of neutrophils and monocytes was not affected. Blocking mAbs to VLA-4 and CD-18 caused incomplete suppression of eosinophilia in the bronchoalveolar lavage (BAL) and had no effect on lung tissue EPO activity. When simultaneously administered, however, the mAbs to VLA-4 and CD-18 completely blocked eosinophil accumulation in the BAL and inhibited the lung tissue EPO activity as well as mononuclear cell and neutrophil influx (89). IgG immobilized to tissue culture plates served as a platform for adhesion and degranulation of eosinophils. These events were inhibited by antibodies to CD18 and CD11b, but not by a mAb to CD29 (β_1) (90). Evidence was obtained for compartmentalized roles of CD11a and CD11b in the vascular and alveolar compartments, respectively, of animals challenged with intrapulmonary IgG. Antibodies administered either IV or intrathecally had much different effects on lung injury and reductions in BAL levels of TNFα, MPO, and neutrophil retrieval (91). Evidence for involvement of pericytes expressing VCAM next to endothelial cells, and regulating T-cell infiltration to the CNS, was obtained through immunohistochemical analysis of multiple sclerosis brain tissue. Interactions of LFA-1/ICAM-1 and VLA-4/VCAM-1 were equally involved in T-cell adhesion to TNFα stimulated endothelial cells, but TNFα stimulated pericytes were dominated by the latter interactions (92).

Changes in integrin expression and matrix protein distribution in chronic periodontal inflammation were investigated (93). Both Fn and Tn decreased in chronically inflamed connective tissue in this study, and basement membrane components in abnormal locations were postulated to function as new ligands for inflammatory cells. While the VCAM/α_4 integrin and ICAM/β_2 integrin interactions are relatively well known as mediators of leukocyte-endothelial cell adhesion, a report has appeared including CD-31/$\alpha_v\beta_3$ as a third immunoglobulin superfamily integrin pair of adhesion molecules with this function, and a potential target for modulation of inflammation and metastasis (94).

Infection - Integrin cognates can function in the pathogenesis of infectious disease through host and microorganism recognition and adhesion. *Candida albicans* was reported to express a protein functionally equivalent to $\alpha_5\beta_1$ which is recognized and bound by Fn. This interaction was blocked by mAb to α_5 or β_1 and by GRGDSP, but not by GRGESP (95). VCAM and ICAM also act as cellular receptors for viral and parasitic agents (96). Internalization of adenovirus was shown to be dependent on α_v integrins β_3 and β_5 via interaction of viral coat proteins displaying RGD sequences (97). Bacterial water soluble proteins induced integrin expression in granulocytes, and a unique pattern of integrin expression was found as a result of *H. pylori* water soluble protein, which may explain the observation that *H. pylori* is a powerful stimulus for granulocyte infiltration (98).

INHIBITORS

Snake Venom Disintegrins- The first integrin antagonists reported were a family of relatively low molecular weight RGD containing proteins isolated from viper venom, the 'disintegrins' (DI). These inhibit several classes of integrins and have served as the starting motif for design of peptide and peptide mimic antagonists (99,100). Inhibitory capacities of six DIs against $\alpha_{IIb}\beta_3$, $\alpha_v\beta_3$ and $\alpha_5\beta_1$ were published (100). Eristostatin exhibits an anti-metastatic effect (101), contortostatin inhibits β_1-integrin mediated human metastatic melanoma cell adhesion and *in vivo* lung colonization (102), triflavin inhibits *in vitro* human cervical carcinoma adhesion (77) and echistatin blocks integrin mediated bone resorption both in vitro and in vivo (103). Integrin collagen receptor, $\alpha_2\beta_1$, is inhibited by the snake venom proteins jararhagin and jaracetin (104). Cellular DIs, a family of membrane bound proteins, with homology to venom DIs, have been reported with an RGD sequence in the DI domain (105).

Inhibitors to $\alpha_v\beta_3$.- Development of blocking mAbs to integrins has been a pivotal tool to their classification and definition of potential utility. As noted above, the antibody c7E3 to β_3 (ReoPro) is showing promise in clinical trials (70). LM609 directed to $\alpha_v\beta_3$ has been instrumental in identifying the role of this integrin in the vasculature and on tumors (67,73,84,85) including a report   that a single intravascular injection of LM609 or cyclic peptide (R*GDfV*, **1**) promoted tumor regression by inducing apoptosis of angiogenic blood vessels (106). Similarly the mAb P1F6 (anti $\alpha_v\beta_5$) inhibited $\alpha_v\beta_5$ mediated angiogenesis (85). Antibodies F4 and F11 to rat β_3 were used to characterize the adherence of rat osteoclasts to a range of matrix proteins (107, 108). Various RGD peptides have been useful to complement mAb studies to implicate alpha v integrins in restenosis and squamous carcinoma cell invasion (67, 83). A set of cyclic and linear RGD containing peptides were examined as antagonists of $\alpha_{IIb}\beta_3$, $\alpha_v\beta_3$ and $\alpha_5\beta_1$ (109). The most potent inhibitor toward $\alpha_v\beta_3$ was **1** (50 ± 20 nM). This antagonist was at least an order of magnitude less potent toward the other two integrins (109,110). Following a progression reminiscent of clinical candiates for integrin mediated antiplatelet function (cf Chapter 10) from mAbs, to peptides (109), to cyclic peptides (111) such as G4120 (112,113), to peptidomimetics, the search for inhibitors to modify specific disease targets is progessing rapidly.   Two recent patent applications and presentations described   selective   non-peptide antagonists of $\alpha_v\beta_3$ of structure types **2** and **3**, for the treatment of osteoporosis (114, 115).   N-terminal acylation of RGDS and RLDS to provide acetyl-DRGDS and trimesyl-(DRGDS)$_3$ gave derivatives   with   potent antimetastatic activity (116).

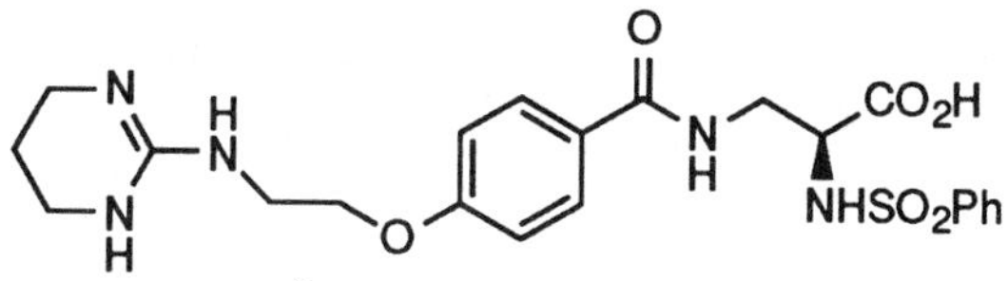

2

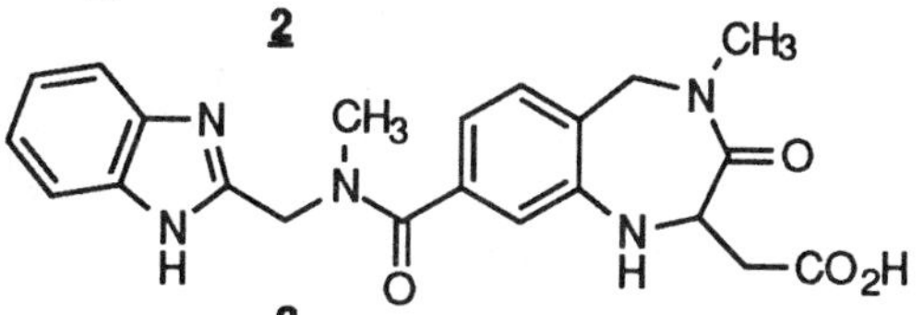

3

<u>Inhibitors of β_1</u> - Members of the β_1 integrin family recognize multiple ligands and conformational changes may effect avidity (117). Anti β_1 mAb 18D3 inhibits the induction of T-cell DNA synthesis. The effect was reversible and did not lead to cell death but appears to selectively inhibit IL-2 production (118). Another anti β_1 mAb was reported to inhibit metastatic tumor growth in lung tissue but had minimal effect on intravasion, adhesion to target organs or extravasion (119) while antibodies directed to the alpha chain of the major collagen receptor on endothelial cells, $\alpha_2\beta_1$ inhibit the attachment to and proliferation on collagen (120). A conformationally restricted cyclic peptide containing a hydrophobic amino terminus, adamantyl-acetyl-C*GAGDSPC* effectively inhibits $\alpha_4\beta_1$ mediated cell adhesion (121).

<u>Inhibitors of β_2</u> - The hook worm derived neutrophil adhesion inhibitor, NIF, targets the A-domain of β_2 integrin CR3 (CD11 / CD 18) and is a novel disintegrin sharing no obvious homology with the snake venom disintegrins (122,123) and may provide new non-RGD motifs for small molecule inhibitor design. Likewise, a novel tick protein, TAI, inhibits platelet and endothelial cell adhesion to collagen apparently through antagonism of $\alpha_1\beta_2$ integrin (124).   Recently, rubralins A-C were isolated which demonstrated moderate inhibition of β_2 integrin mediated cell adhesion (125) and mAb 7A10 preferentially binds to $\alpha_m\beta_2$ on activated leukocyte cells (126). In a study of cyclic peptides directed toward finding a potent $\alpha_{IIb}\beta_3$ antagonist   moderate antagonists of $\alpha_5\beta_1$ and $\alpha_v\beta_5$ were also reported (111).

Conclusion - Integrin research is rapidly evolving. Antiplatelet agents are in the clinic and discovery efforts are ranging across a spectrum of opportunities for treatment of disease, both metabolic and infectious. Recent progress with monoclonal antibodies peptides and peptide mimetics portend effective new therapeutics based on modulation of these cell adhesion molecules.

References

1.   G.M. Edelman, CellAdhes.Commun. 1, 1 (1993).
2   R.O. Hynes, Cell 48, 549 (1987)
3.   M.A. Stepp, S. Spurr Michaud, A. Tisdale, J. Elwell and I.K. Gipson, Proc.Natl.Acad.Sci.USA 87, 8970 (1990).
4.   A. Sonnenberg, J. Calafat, H. Janssen, H. Daams, van der Raaij L.M. Helmer, R. Falcioni, S.J. Kennel, J.D. Aplin, J. Baker, M. Loizidou and D. Garrod., J.Cell.Biol. 113, 907 (1991).
5.   M.A. Kurpakus, V. Quaranta and J.C. Jones, J.CellBiol, 115, 1737 (1991).
6.   F. Altruda, P. Cervella, G. Tarone, C. Botta, F. Balzac, G. Stefanuto and L. Silengo,Gene 95, 261 (1990).
7.   T.H. van Kuppevelt, L.R. Languino, J.O. Gailit, S. Suzuki and E. Ruoslahti, Proc.Natl.Acad.Sci.USA 86, 5415 (1989).
8.   S. Suzuki and Y. Naitoh, EMBO J. 9, 757 (1990).
9.   F. Hogervorst, I. Kuikman, A.E. von dem Borne and A. Sonnenberg, EMBO J. 9, 765 (1990).
10.   R.N. Tamura, C. Rozzo, L. Starr, J. Chambers, L.F. Reichardt, H.M. Cooper and V. Quaranta, J.Cell.Biol. 111, 1593 (1990).
11.   A.S. Clarke, M.M. Lotz and A.M. Mercurio, Cell.Adhes.Commun. 2, 1 (1994).
12   C.M. Niessen, O. Cremona, H. Daams, S. Ferraresi, A. Sonnenberg, and P.C. Marchisio, J.Cell.Sci. 107, 543 (1994).
13.   R.N. Tamura, H.M. Cooper, G. Collo and V. Quaranta, Proc.Natl.Acad.Sci. USA 88, 10183 (1991).
14.   F. Hogervorst, I. Kuikman, A.G. van Kessel and A. Sonnenberg, Eur.J.Biochem. 199, 425 (1991).
15.   H.M. Cooper, R.N. Tamura and V.J. Quaranta, Cell.Biol. 115, 843 (1991).
16.   P.F. Bray, C.S. Leung and M.A. Shuman, J.Biol.Chem. 265, 9587 (1990).
17.   R. Briesewitz, A. Kern and E.E. Marcantonio, Mol.Biol.Cell 6, 997 (1995).
18.   R. Pigott and C. Power in "Adhesion Molecule Factsbook," Academic Press, New York, N.Y., 1993, p. 9-11.
19.   M.D. Pierschbacher and E. Ruoslahti, Nature 309, 30 (1984).
20.   R.O. Hynes, Cell 69, 11 (1992).
21.   P.J. Kilshaw and S.J. Murant, Eur.J.Immunol. 21, 2591 (1991).
22   C.M. Parker, K.L. Cepek, G.J. Russell, S.K. Shaw, D.N. Posnett, R. Schwarting and M.B.Brenner, Proc.Natl.Acad.Sci. USA 89, 1924(1992).
23.   N. Cerf Bensussan, B. Begue, J. Gagnon and T. Meo, Eur.J.Immunol. 22, 885 (1992).
24.   C. Berlin, E.L. Berg, M.J. Briskin, D.P. Andrew, P.J. Kilshaw, B. Holzmann, I.L. Weissman, A. Hamann and E.C. Butcher, Cell 74, 185 (1993).
25.   U.G. Strauch, A. Lifka, U. Gosslar, P.J. Kilshaw, J. Clements and B. Holzmann, Int.Immunol. 6, 263 (1994).
26.   D.J. Erle, M.J. Briskin, E.C. Butcher, A. Garcia Pardo, A.I. Lazarovits and M. Tidswell, J.Immunol. 153, 517 (1994).
27.   L. Lefrancois, T.A. Barrett, W.L. Havran and L. Puddington, Eur.J.Immunol. 24, 635 (1994).
28.   D.T. Crowe, H. Chiu, S. Fong and I.L. Weissman, J.Biol.Chem. 269, 14411 (1994).
29.   G.J. Russell, C.M. Parker, K.L. Cepek, D.A. Mandelbrot, A. Sood, E. Mizoguchi, E.C. Ebert, M.B. Brenner and A.K. Bhan, Eur.J.Immunol. 24, 2832 (1994).
30.   A.A. Postigo, P. Sanchez Mateos, A.I. Lazarovits, F.Sanchez Madrid and M.O. de Landazuri, J.Immunol. 151, 2471 (1993).
31.   S. Tiisala, T. Paavonen, and R. Renkonen, Eur.J.Immunol. 25, 411 (1995).
32.   M. Busk, R. Pytela and D. Sheppard, J.Biol.Chem. 267, 5790 (1992).
33.   D. Sheppard, C. Rozzo, L. Starr, V. Quaranta, D.J. Erle and R. Pytela, J.Biol.Chem. 265, 11502 (1990).
34.   J.M. Breuss, N. Gillett, L. Lu, D. Sheppard and R. Pytela, J.Histochem.Cytochem. 41, 1521 (1993).
35.   J.M. Breuss, J. Gallo, H.M. DeLisser, I.V. Klimanskaya, H.G. Folkesson, J.F. Pittet, S.L. Nishimura, K. Aldape, D.V. Landers, W. Carpenter,N. Gillett, D. Sheppard, M.A. Matthay, S. M. Albelda, R. H. Kramer and R. Pytela, J.Cell.Sci. 108, 2241 (1995).
36.   M.V. Agrez and R.C. Bates, Eur.J.Cancer, 30a 2166 (1994).
37.   M. Moyle, M.A. Napier and J.W. McLean, J.Biol.Chem. 266, 19650 (1991).
38.   S.L. Nishimura, D. Sheppard and R. Pytela, J.Biol.Chem. 269, 28708 (1994).
39.   B. Bossy, E. Wetzel and L.F. Reichardt, EMBO J. 10, 2375 (1991).
40.   L.M. Schnapp, J.M. Breuss, D.M. Ramos, D. Sheppard and R. Pytela, J.Cell.Sci. 108, 537

(1995).

41. D.J. Erle, D. Sheppard, J. Breuss, C. Ruegg and R. Pytela, Am.J.Respir.Cell. Mol.Biol 5, 170 (1991).
42. E.L. Palmer, C. Ruegg, R. Ferrando, R. Pytela, and D. Sheppard, J.Cell.Biol. 123, 1289 (1993).
43. E. Forsberg, B. Ek, A. Engstrom, and S. Johansson, Exp.Cell Res. 213, 183 (1994).
44. Y. Yokosaki, E.L. Palmer, A.L. Prieto, K.L. Crossin, M.A. Bourdon, R. Pytela and D. Sheppard, J.Biol.Chem. 269, 26691 (1994).
45. A. Weinacker, R.L. Ferrando, M. Elliott, J. Hogg, J. Balmes and D. Sheppard, Am.J.Respir.Cell.Mol.Biol. 12, 547 (1995).
46. M.A. Stepp, L. Zhu, D. Sheppard, and R.L. Cranfill, J.Histochem.Cytochem. 43, 353 (1995).
47. S.J. Shattil, M.H. Ginsberg and J.S. Brugge, Curr.Opin.Cell Biol. 6, 695 (1994).
48. E. Ruoslahti, Tumor Biol. 17, 117 (1996).
49. S. Aznavoorian, M.L. Stracke, J.Parsons, J.McClanahan and L.A. Liotta, J.Biol.Chem. 271, 3247 (1996).
50. K.M. Yamada and S. Miyamoto, Curr.Opin.CellBiol. 7, 681 (1995).
51. E.A. Clark and J.S. Brugge, Science 268, 233 (1995).
52. S. Miyamoto, S.K. Akiyama and K.M. Yamada, Science 267, 883 (1995).
53. N. Boudreau, C.J. Sympson, Z. Werb and M.J. Bissell, Science 267, 891 (1995).
54. M.A. Schwartz and K. Denninghoff, J.Biol.Chem. 269, 11133 (1994).
55. G. Kilger, L.A. Needham, P.J. Nielsen, J. Clements, D. Vestweber and B. Holzmann, J.Biol.Chem. 270, 5979 (1995).
56. S.G. Schiffer, M.E. Hemler, R.R. Lobb, R. Tizard, and L. Osborn, J.Biol.Chem. 270, 14270 (1995).
57. E.Y. Jones, K. Harlos, M.J. Bottomley, R.C. Robinson, P.C. Driscoll, R.M. Edwards, J.M. Clements, T.J. Dudgeon and D.I. Stuart, Nature 373, 539 (1995).
58. J.H. Wang, R.B. Pepinsky, T. Stehle, J.H. Liu, M. Karpusas, B. Browning and L. Osborn, Proc.Natl.Acad.Sci. USA 92, 5714 (1995).
59. B.J. Graves, Nature Struct.Biol. 2, 181 (1995)
60. A.J. Pelletier, T.Kunicki and V.Quaranta, J.Biol.Chem. 271, 1364 (1996).
61. C. Weber, W. Erl and P.C. Weber, Biochem.Biophys.Res.Commun. 206, 621 (1995).
62. H.A. Lehr, F. Krombach, S. Munzing, R. Bodlaj, S.I. Glaubitt, D. Seiffge, C. Hubner, U.H. von Andrian and K. Messmer, Am.J.Pathol. 146, 218 1995).
63. J.H. Ip, V. Fuster, D. Israel, L. Badimon, J. Badimon and J.H. Chesebro, J.Am.Coll.Cardiol. 17, 77B (1991).
64. M. Bobuyoski, T. Kimura, H. Ohishi, H. Horiuchi, H. Nosaka, N. Hamasaki, H. Yokoi and K. Kim, J.Am.Coll.Cardiol. 17, 433 (1991).
65. G.A. Ferns, A.L. Stewart-Lee and E.E. Auggard, Atherosclerosis 92, 89 (1992).
66. D.I. Leavesley, M. A. Schwartz, M. Rosenfeld and D.A. Cheresh, J.Cell.Biol. 121, 163 (1993).
67. E.T. Choi, L. Engel, A.D. Callow, S. Sun, J. Trachtenberg, S. Santoro and U. Ryan, J.Vasc.Surg. 19, 125 (1994).
68. H. Matsuno, J.M. Stassen, J. Vermylen and H. Deckmyn, Circulation 90, 2203 (1994).
69. L. Liaw, V. Lindner, S. M. Schwartz, A. F. Chambers and C.M. Giachelli, Circ.Res., 77, 665 (1995).
70. E.J. Topol, Am.J.Cardiol. 75, 27b (1995).
71. D. Cox, ExpertOpin.Invest. Drugs 4, 413 (1995).
72. M. De Luca, G. Pellegrini, G. Zambruno and P.D. Marchisio, J.Dermatol. 21, 821 (1994).
73. P.C. Brooks, R.A.F. Clark and D. A. Cheresh, Science 264, 569 (1994).
74. P.G. Robey, Ann.Rep.Med.Chem. 28, 227 (1993).
75. J.M. Stadel, A.J. Nichols, D.R. Bertolini and J.M. Samanen in "Integrins" D.A. Cheresh and R.P. Mecham Ed., Academic, San Diego (1994).
76. B.A. Crippes, V.W. Engleman, S.L. Settle, J. Delarco, R.L. Ornberg, M.H. Helfrich, M.A. Horton and G.A. Nickols, Endocrinology 137, 918 (1996).
77. J.R. Sheu, C.H. Lin, H.C. Peng and T.F. Huang, Peptides 15, 1391 (1994).
78. R.M. Lafrenie, S. Gallo, T.J. Podor, M.R. Buchanan and F.W. Orr, Eur.J.Cancer 30A, 2151 (1994).
79. A. Garofalo, R.G. Chirivi, C. Foglieni, R. Pigott, R. Mortarini, I. Martin-Padura, A. Anichini, A.J. Gearing, F. Sanchez-Madrid, E. Dejana, and R. Giavazzi, Cancer Res. 55, 414 (1995).
80. C.W. Patrick Jr., H.S. Juneya, S. Lee, F.C. Schmalstieg and L.V. McIntire, Blood 85, 168 (1995).
81. D. Schadendorf, J. Heidel, C. Gawlik, L. Suter and B.M. Czarnetzki, J.Natl.Cancer Inst. 87, 366 (1995).
82. M. Edward, Curr.Opin.Oncol. 7, 185 (1995).
83. E. Kawahara, K. Imai, S. Kumagai, E. Yamamoto and I Nakanishi, J.Cancer Res.Clin.Oncol. 121, 133 (1995).
84. P.C. Brooks, S. Stromblad, R. Klemke, D. Visscher, F.H. Sarkar and D. Cheresh, J.Clin.Invest. 96, 1815 (1995).
85. M. Friedlander, P.C. Brooks, R.W. Shaffer, C.M. Kincaid, J.A. Varner, and D.A. Cheresh, Science, 270, 1500 (1995).

86.   J.A. Varner, P.C. Brooks and D.A. Cheresh, CellAdhes.Commun. 3, 367, (1995).

87.   A. Vacca, M. Di Loreto, D. Rabatti, R. DiStefano, G. Gadaleta-Caldarola, G. Iodice, D. Caloro and F. Dammacco, Am.J. Hematol. 50, 9 (1995).

88.   Y. Ohkawara, K. Yamauchi, N. Maruyama, H. Hoshi, I. Ohno, M. Honma, Y. Tanno, G. Tamura, K. Shirato and H. Ohtani, Am.J.Respir.Cell.Mol.Biol. 12, 4 (1995).

89.   A.M. Das, T.J. Williams, R. Lobb and S. Nourshargh, Immunology 84, 41 (1995).

90.   M. Kaneko, S. Horie, M. Dato, G.J. Gleich and H. Kita, J.Immunol. 155, 2631 (1995).

91.   M.S. Mulligan, A.A. Vaporciyan, R.L. Warner, M.L. Jones, K.E, Foreman, M. Miyasaka, R.F. Todd and P.A. Ward, J.Immunol. 154, 1350 (1995).

92.   M.M. Verbeek, J.R. Westphal, D.J. Ruiter and R.M. de Waal, J.Immunol. 154, 5876 (1995).

93.   K. Haapasalmi, M. Makela, O. Oksala, J. Heino, K.M. Yamada, V.J. Uitto, and H. Larjava, Am.J.Pathol. 147, 193 (1995).

94.   L. Piali, P. Hammel, C. Uherek, F. Bachmann, R.H. Gisler, D. Dunon and B.A. Imhof, J.Cell.Biol. 130, 451 (1995).

95.   G. Santoni, A. Gismondi, J.H. Liu, A. Punturieri, A. Santoni, L. Frati, M. Piccoli and J.Y. Djeu, Microbiol. 140, 2971 (1994).

96.   E.Y. Jones, K. Harlos, M.J. Bottomley, R.C. Robinson, P.C. Driscoll, R.M. Edwards, J.M. Clements, T.J.Dudgeon and D.I. Stuart, Nature 373, 539 (1995).

97.   T.J. Wickham, P. Mathias, D.A. Cheresh and G. R. Nemerow, Cell 73, 309 (1993).

98.   G. Enders, W. Brooks, N. von Jan, N. Lehn, E. Bayerdorffer and R. Hatz, Infect.Immun. 63, 2473, (1995).

99.   R.J. Gould, M.A. Polokoff, P.A. Friedman, T.F. Huang, J.C. Holt, J.J. Cook and S. Niewiarowski, Proc.Soc.Exp.Biol.Med. 195, 168, (1990).

100.   M. Pfaff, M.A. McLane, L. Beviglia, S. Niewiarowski and R. Timpl, Cell.Adhes.Commun. 2, 491 (1994).

101.   V.L. Morris, E.E. Schmidt, S. Koop, I.C. MacDonald, M. Brattan, R. Knokna, M.A. McLane, S. Niewiarowski, A.F. Chambers and A.C. Groom, Exp.Cell Res. 219, 571, (1995).

102.   M. Trikha, Y.A. De Clerck and F.S. Markland, Cancer Res. 54, 4993 (1994).

103.   R. Dresner-Pollak and M. Rosenblatt, J.Cell.Biochem. 56, 323 (1994).

104.   M. De Luca, C.M. Ward, K. Ohmori, R.K. Andrews and M.C. Berndt, Biochem.Biophys.Res.Commun. 206, 570 (1994).

105.   J. Kratzschmar, L. Lum and C.P.Blobel, J.Biol.Chem. 271, 4593, (1996)

106.   P.C. Brooks, A.M. Montgomery, R. Rosenfeld, R.A. Reisfeld, T. Hu, G. Klier and D.A. Cheresh, Cell 79, 1157 (1994).

107.   M.H. Helfrich, S.A. Nesbitt, E.L. Dorey and M.A. Horton, J.Bone Miner.Res. 7, 335 (1992).

108.   M.H. Helfrich, S.A. Nesbitt and M.A. Horton, J.Bone Miner.Res. 7, 345 (1992).

109.   M. Pfaff, K. Tangemann, B. Muller, M. Garruth, G. Muller, H. Kessler, R. Timpl and J. Engel, J.Biol.Chem. 269, 20233 (1994).

110.   P. Brooks and D.A. Cheresh, PCT WO95/25543 (1995).

111.   S. Cheng, W.S. Craig, D. Mullen, J.F. Tschopp, D. Dixon and M.S. Pierschbacher, J.Med.Chem. 37, 1 (1994).

112.   H. Matsuno, J.M. Stassen, J. Vermylen and H. Deckmyn, Circulation 90, 2203 (1994)

113.   H. Matsuno, M.F. Hoylaerts, J. Mermylen and H. Deckmyn, Nippon Yakurigaku Zasshi 106, 143 (1995).

114.   M.E. Duggan, J.E. Fisher, M.A. Gentile, G.D. Hartman, W.J.Hoffman, J.R. Huff, N.C. Ihle, A.E. Krause, T-C. Leu, R.M. Nagy, J.J, Perkins, G.A Rodan, S.B. Rodan, G.Weslolwski and D.B. Whitman, Abstract MEDI 234, 211th ACS National Meeting, 1996.

115.   R. Keenan, W. Miller, F.Ali, L.Barton, B, Bondinell, J. Burgess, J. Callahan, R. Calvo, R. Cousins, M. Gowen, W. Huffman, S. Hwang, D. Jakas, T. Ku, C. Kwon, A. Lago, V. Mombouyran, T. Nugyen, S. Ross, J. Samanen, D. Takata, I. Uzinskas, J. Venslavsky, A. Wong, T. Yellin a and C. Yuan, Abstract MEDI 236, 211th ACS National Meeting, 1996

116.   H. Fujii, H. Komazawa, H.Mori, M. Kojima, I.Itoh, J. Murata, I.Azuma and I.Saiki, Biol.Pharm.Bull. 18, 1681 (1995(.

117.   Y. Takada and W. Puson, J.Biol.Chem. 268, 17597 (1993).

118.   T.K. Teague and R.W. McIntyre, Cell.Adhes.Commun. 2, 169 (1994).

119.   B.E. Elliott, P. Ekblom, H. Pross, A. Niemann and K. Rubin, Cell.Adhes.Commun. 1, 319 (1994).

120.   J.R. Gamble, L.J. Mathias, G. Meyer, P. Kaur, G. Russ, R. Jaull, M.C. Berndt and M.A. Vadas, J.Cell.Biol. 121, 931 (1993).

121.   P.M. Cardarelli, R.R. Cobb, D.M. Nowlin, W. Scholz, F. Gorscan, M. Moscinski, Yasuhara, S.L. Chiang and T.J. Lobl, J.Biol.Chem. 269, 18668 (1994).

122.   P. Rieu, T.Ueda, I. Haruta, C.P. Sharma and M.A. Arnaout, J.Cell.Biol. 127, 2081 (1994).

123.   P.J. Muchowski, L. Zhang, E.R. Chang, H.R. Soule, E.F. Plow and M. Moyle, J.Biol.Chem. 269, 26419 (1994). Erratum, J.Biol. Chem. 270, 6420 (1995).

124.   J. Karczewski, L. Waxman, R.R. Endris and T.M. Connolly, Biochem.Biophys.Res. Commun. 208, 532 (1995).

125.   L.L. Musza, L.M. Killar, P. Speight, C.J. Barrow, A.M. Gillu and R. Cooper, Phytochemistry, 39, 621 (1995).

126.   G.S. Elemer and T.S. Edgington, J.Immunol. 152, 5836 (1994).

Chapter 21.   Treating Obesity in the 21st Century

Donald R. Gehlert, David J. Goldstein and Philip A. Hipskind
Lilly Research Laboratories
Indianapolis, IN 46285

Introduction - Clinically relevant obesity is now prevalent in a third of Americans (1,2). It is responsible for various adverse effects on health, being associated with an increase in morbidity and mortality from non-insulin dependent diabetes mellitus (NIDDM), hypertension, hypercholesterolemia, sleep apnea, and other medical conditions (3). Of particular interest is the comorbidity of NIDDM and obesity. More than 80% of the NIDDM patients in the U.S. are obese, and there may be a pathophysiological link between these two disorders (4). Insulin resistance is believed to play a role in other pathological states such as dyslipidemias, atherosclerosis, hypertension and other cardiovascular disorders. Weight loss can result in a significantly lower risk for these endocrine and cardiovascular disorders (5,6). The question is not whether obese patients should lose weight but rather how that should be accomplished. Dietary restriction with or without pharmacotherapy can be a fairly effective measure in the short term but rarely results in long term reduction in body weight (7-9). The goal of a weight loss therapy should be sufficient weight loss to improve health and function, and since obesity is a chronic disease, therapy should be continuous, lack serious side effects and not be addictive.

Understanding more about the pathophysiology of obesity may be the best pathway to novel and efficacious pharmacotherapy. Many factors may lead to the development of obesity including heredity, genetic diseases, tumors and endocrine or hypothalamic disorders (10). The regulation of adiposity is the result of a large number of neuroendocrine influences. At the center of this regulation is the hypothalamus which plays a critical role in appetite and metabolism (Figure 1). This region of the brain is under the influence of higher brain centers that can influence appetite and, therefore, appetite disorders are often comorbid with psychiatric diseases. The hypothalamus is also under the influence of a variety of humoral substances that are secreted from peripheral tissues such as the gastrointestinal tract, pancreas and adipocytes. Abnormalities in the secretion and synthesis of these hormones have been observed in many animal models of obesity. Nevertheless, in Western society the two greatest factors implicated in the etiology of obesity are diet composition and lack of exercise. An increase in the percentage of fat in the diet preferentially leads to fat storage since the body is more efficient at storing dietary fat than dietary carbohydrate. Also sedentary occupations and the advent of the automobile have contributed to diminished caloric expenditure and may lead to increased fat deposition. Thus, the ideal pharmacotherapy should increase metabolism and reduce appetite, particularly the desire for high fat foods.

THE ROLE OF CURRENT AGENTS FOR THE TREATMENT OF OBESITY

Existing pharmacologic therapies for weight loss are only modestly effective. For the average patient, maximal weight loss effect is modest and occurs by 6 months of therapy (11,12). However, the amount of weight loss is typically sufficient to result in some health benefit, particularly for improving glycemic control, blood pressure, and lipid profile (5). When treatment is discontinued, weight is rapidly regained demonstrating the need for continuing therapy (12). Because weight loss with present treatments alone is less than optimal, weight reduction therapy must be combined with regiments that include caloric restriction, behavioral modification, and increased physical activity. Also existing agents commonly display a less than optimal side effect profile. All of the available agents today are controlled substances which limits

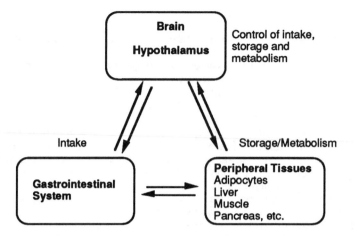

Figure 1. Regulation of food intake, storage and metabolism. Adiposity is the result of a complex regulation through a variety of neuroendocrine influences and behavioral cues. Traditional drug therapies seek to reduce food intake by reduction of appetite through CNS influences. Some newer therapies will attempt to increase metabolism or reduce absorption of nutrients.

prescriber and patient acceptance. In many states, use of conventional appetite suppressants is limited to three month period, a less than optimal treatment duration for a chronic disease. The goal of newer therapies will be to improve efficacy and safety while allowing for chronic treatment of the disorder.

STRATEGIES DESIGNED TO INCREASE METABOLISM

Thermogenic Agents - The sympathetic nervous system plays an important role in the regulation of both energy intake and energy expenditure offering a dual role in the pharmacological intervention for obese patients. Pharmacological stimulation with sympathomimetic compounds suppresses appetite and increases energy expenditure through stimulation of a variety of receptor subtypes (13). Available thermogenic agents are not very selective and thus cause cardiac effects including tachycardia and hypertension and tremor (14). Ephedrine (**1**), caffeine (**2**), and phenobarbital (**3**) have been used to increase thermogenesis (15). Ephedrine and caffeine in combination have considerably more effect than either alone; tremor, insomnia, and dizziness were reported as side effects (16-18).

β_3 adrenergic receptor agonists induce brown adipose tissue lipolysis. This topic was recently reviewed in this series (19). Based upon animal studies, these agents are also expected to have an anti-diabetic effect independent of weight loss and selective reduction in fat (20). Several compounds, which were developed based on desirable rodent receptor pharmacologies, have now been tested in man. Included in this list are CL316243 (**4**), Ro40-2148 (**5**) and ZD7114 (**6**) (21-25). It is clear from the mixed results reported by these studies that compounds should be developed based on the human receptor and not rat receptor pharmacologies (19). Since the last review, agents that claim to be developed based on such a scheme have appeared.

An isoproterenol SAR using a cloned human β_3 receptor (hβ_3) yielded two compounds, L-739,574 (**7**) and L-742,791 (**8**) (26,27). These compounds were reported to induce lipolysis in human adipocytes from obese patients. A series of benzoimidazolyl-propanolamines including **9** have appeared with good hβ_3 adenylyl cyclase activity (EC$_{50}$ 0.9 mM) and selectivity verses cloned hβ_1 and hβ_2 receptors (28).

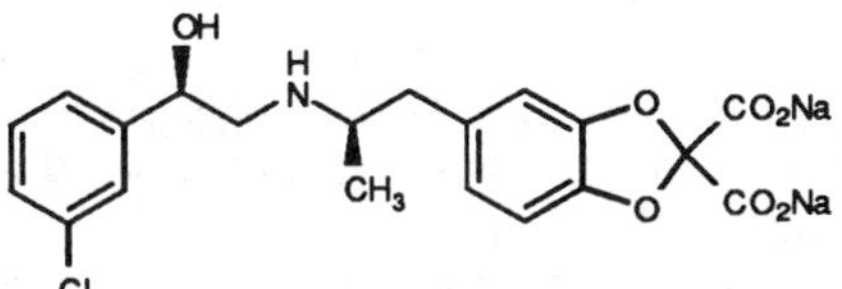

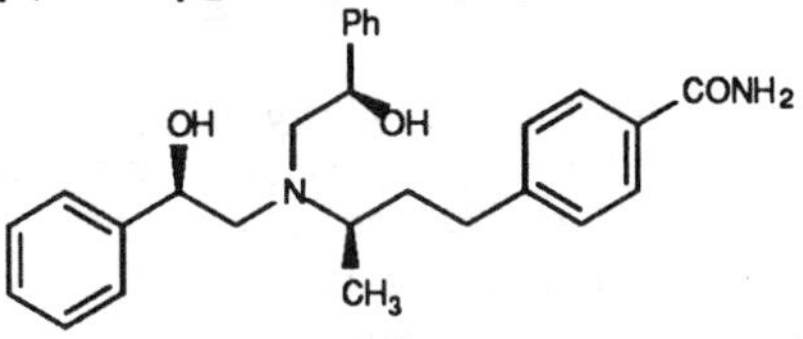

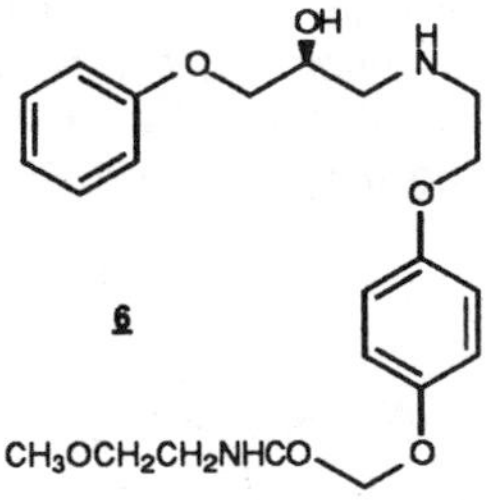

STRATEGIES DESIGNED TO REDUCE ABSORPTION

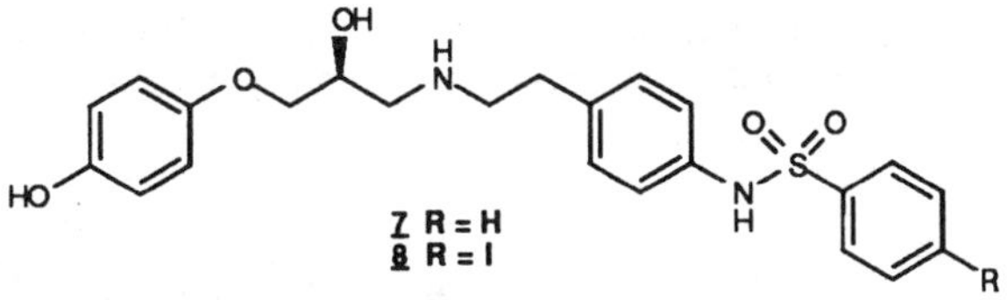

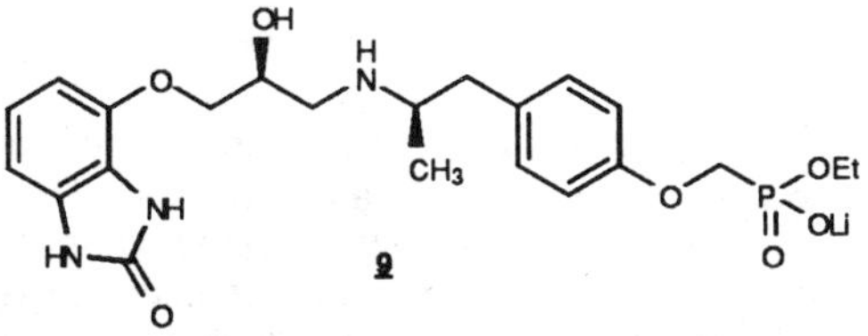

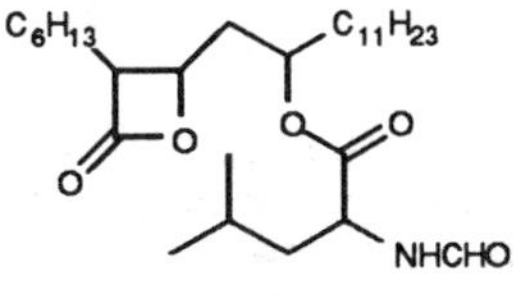

<u>Absorption Inhibitors</u> - The role of such agents is to limit overingestion of food. While the malabsorbtion of food by the intestinal tract will have a direct effect on the amount of various substances, such as fat, that enters systemic circulation, this malabsorption can also be associated with flatulence and diarrhea providing futher reinforcement towards prescribed dietary regiments. Tetrahydrolipostatin (Orlistat, **10**) is an inhibitor of gastric and pancreatic lipase which reduces dietary fat absorption (29). However, because it reduces fat absortion, patients have resulting steatorrhea and malabsorption of fat, which is more tolerable when fat intake is limited to 30% of the diet or less (30,31). Acarbose, a high molecular weight cyclitol-containing pseudotetrasaccharide, is a glucosidase inhibitor which reduces carbohydrate absorption and is also used to reduce hyperglycemia in patients with diabetes (32).

STRATEGIES DESIGNED TO REDUCE APPETITE

<u>Amphetamines and Amphetamine-like Compounds</u> - These agents act by releasing catecholamines from presynaptic nerve terminals and stimulating adrenergic receptors in the medial paraventricular nucleus in the hypothalamus. After the anorectic effect of amphetamine (**11**) was described in 1938 (33), its isomer d-amphetamine, methamphetamine (**12**), phenmetrazine (**13**) and benzphetamine (**14**) were soon noted to share this effect. Nevertheless, these agents were rarely used for weight loss because of their addiction potential and adverse cardiovascular properties including tachycardia and hypertension. Further, the amphetamines and their derivatives cause heightened nervousness, irritability, and insomnia.

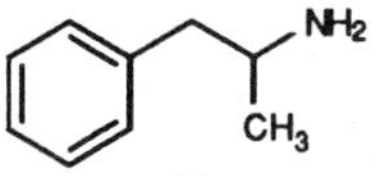

11 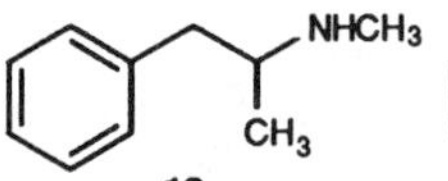   12 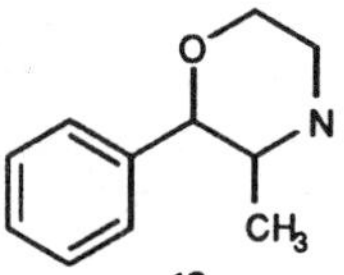   13 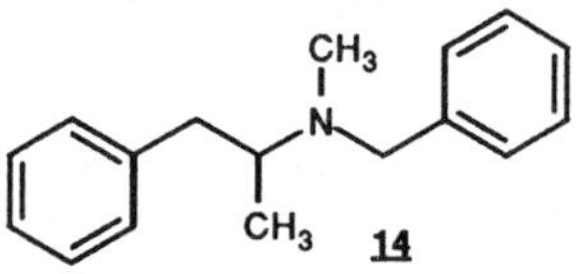   14

Phenylethylamine Derivatives - These agents are also activating towards the sympathetic nervous system, but lack significant abuse potential. Weight loss is similar to that obtained with the amphetamines. Phentermine (**15**) is associated with insomnia, nervousness, headache, and irritability. Diethylpropion (**16**) is associated with insomnia. Mazindol (**17**), an imidazoisoindole derivative, increases locomotor activity, decreases gastric acid secretion and motility, decreases glucose absorption, decreases food intake via the lateral hypothalamus, and inhibits insulin secretion (12,13). It is also associated with nervousness, irritability, insomnia, dry mouth, sweating, nausea, and constipation (34). Phenylpropanolamine (**18**), a racemic mixture of norephedrine isomers, is weakly effective in inducing weight loss (35,36).

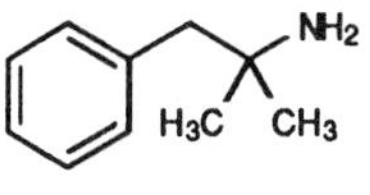

15 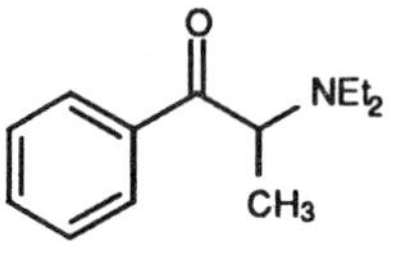   16 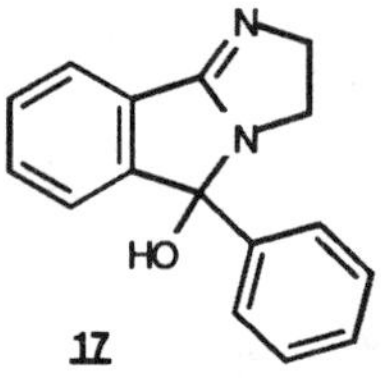   17 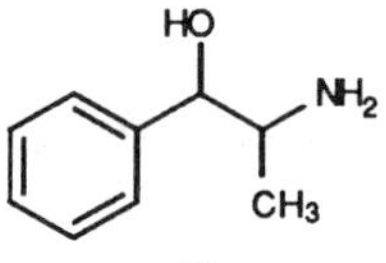   18

Serotonergic Agents - Serotonergic drugs are thought to work through promotion of the serotonergic satiety effect on carbohydrate intake in the hypothalamus. Dexfenfluramine (**19**) primarily releases serotonin from presynaptic neuron, but it is also a serotonin reuptake inhibitor. In human trials, patients treated with **19** lost more weight (3.8 vs 1.1 kg for placebo group), had lower diastolic blood pressure, lower cholesterol levels, and showed improved insulin sensitivity (37). In other human trials **19** was shown to have selective effects on carbohydrate ingestion (38). Side effects have included nausea, dry mouth, diarrhea and somnolence (39) . Of potential greater concern is the rare report of pulmonary hypertension and the animal data suggesting a potential for neurotoxicity (40-42). Nevertheless, because it can deplete presynaptic serotonin, it can induce depression both during continued therapy and after abrupt discontinuation (43,44). Recently **19** became the first new agent approved for the treatment of obesity in twenty years (45).

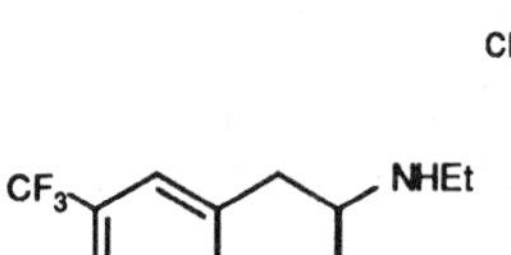

19 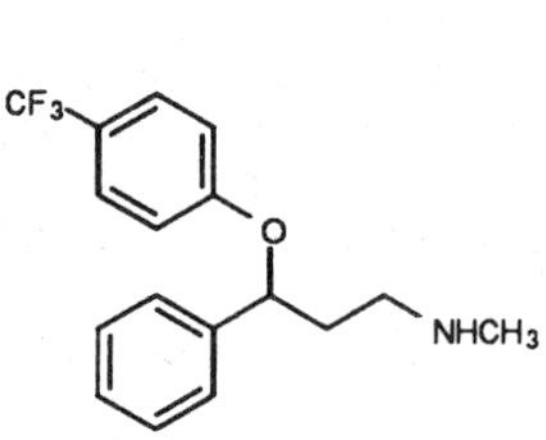   20 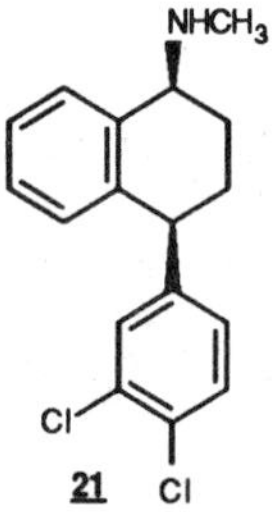   21 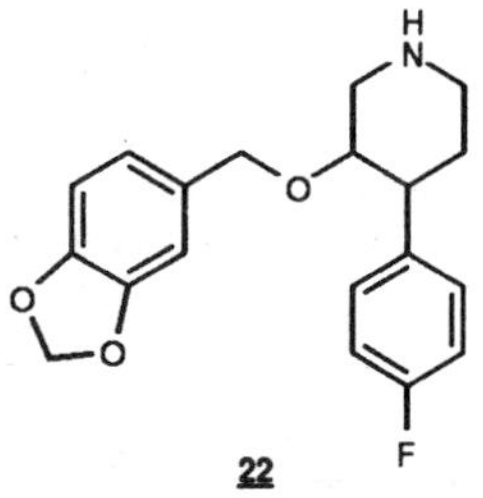   22

    The mechanism of serotonin reuptake inhibition also increases serotonin in the synapse, but does not result in depletion of serotonin from the presynaptic terminals. Fluoxetine (**20**), a serotonin uptake inhibitor, has demonstrated 12 month weight loss in some studies, but not in others (46-50). Dose-related effects in obese patients are somnolence, asthenia, sweating and tremor. Other effects include nausea, diarrhea, and insomnia. Sertraline (**21**) and paroxetine (**22**) are also serotonin uptake inhibitors with effects on food intake in animal models, but their potential in controlled human clinical trial has not been reported (51,52). Various other serotonergic targets for the treatment of obesity have recently been reviewed (53).

Other Monoamine Uptake Inhibitors - Sibutramine (**23**) is a mixed noradrenergic uptake inhibitor and a serotonergic uptake inhibitor, thus combining in one molecule the effects of the phentermine and fenfluramine combination (54,55). In human studies **23** has produced a dose-related decrease in body-weight (6.1 to 6.5 kg after 24 weeks treatment) and a low incidence of side effects (dry mouth, insomnia, constipation and headache) (56-58). Also, glucose tolerance testing in obese NIDDM patients showed a trend towards improved glucose handling with **23** but not placebo (59). Combined, these data suggest the potential therapeutic utility of **23** in NIDDM.

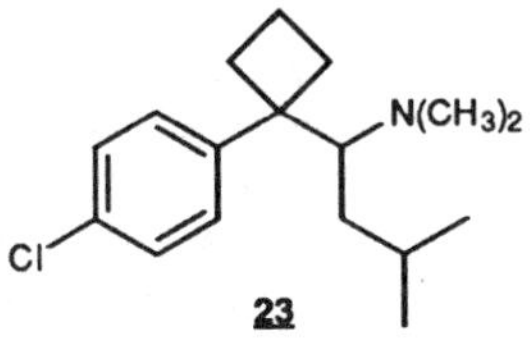

23

Combination Therapy - Because of the modest effects of various individual agents, phentermine (**15**) and dexfenfluramine (**19**) have been studied in combination (60). This combination of an adrenergic agent and a serotonergic agent had a greater effect than either of the mechanisms independently, and the activating effects of the adrenergic and the sedating effects of the serotonergic components serve to negate each other for many patients (61) .

Peptides and Obesity - A large number of peptides are thought to affect feeding behavior when administered centrally or peripherally (Table 1). In addition, several peptides alter metabolism when given centrally or peripherally and drug therapies based on these peptides may show additional benefit beyond appetite reduction. The profound effects of these substances has made them attractive targets for drug discovery. In general, medicinal chemistry efforts have concentrated on developing specific antagonists for the substances that stimulate feeding. However, notable exceptions exist such as the efforts to develop peptide-like agonists and the inhibition of proteolytic degradation of certain peptides. Recent developments in several of the peptide targets are listed below.

Table 1. Endogenous Substances That Effect Feeding.

Increase Feeding	Decrease Feeding
Galanin	Amylin
Gamma-amino butyric acid	Cholecystokinin
Growth Hormone Releasing Hormone	Corticotropin Releasing Hormone
Neuropeptide Y	Cytokines
Norepinephrine	Dopamine
Opiates: Dynorphin, β-endorphin	Enterostatin
	Glucagon
	Glucagon-like peptide-1
	Insulin
	Leptin
	Neurotensin
	Serotonin

<u>Cholecystokinin (CCK)</u> - Peripheral administration of the 33 amino acid peptide CCK reduces meal size by the inhibition of gastric emptying and by activation of the vagal afferents which exert control over the hypothalamus (62). Administration of the CCK-A receptor antagonist, devazepide (**24**), increased food consumption in the lean but not obese animals suggesting an abnormality in this pathway in obesity (63). Drug discovery efforts in this area have concentrated on the discovery of specific CCK receptor agonists such as U-67827E, an octapeptide analog of CCK that has a longer plasma half life with similar potency to the native peptide (64). Recently a small molecule, non-peptide CCK-A receptor agonist (**25**) with *in vivo* anorexia activity was reported (65). Another strategy is to reduce the degradation of CCK. Butabindide (**26**), a molecule which prevents the degradation of CCK by tripeptidyl peptidase II, has been reported to reduce feeding in starved rats (66).

<u>Enterostatin</u> - Enterostatin is the activation pentapeptide derived from procolipase, a 101 amino acid peptide that is synthesized in the pancreas and stomach (67). By either peripheral or central administration, enterostatin selectively reduces the intake of fat. Interestingly, ingestion of fat increases the secretion of pancreatic colipase indicating this may be one of the physiological pathways by which dietary preference is regulated. Structure-activity studies have identified cyclo-aspartyl-proline (cyclo-DP, **27**) as the active portion of the peptide (68). Recent studies suggest that enterostatin and cyclo-DP reduce food intake through a kappa opioid receptor (69).

<u>Galanin</u> - Human galanin is a 30 amino acid peptide found in the brain. Central administration of galanin will increase food consumption in satiated rats with most prominent increase in macronutrient consumption being fat (70). However, chronic administration of galanin does not appear to induce the obese state in rodents. To date, progress in the development of galanin antagonists has been confined to peptide antagonists. By exchanging portions of the galanin peptide sequence with that of other neuropeptides, several chimeric peptide antagonists have been discovered. Antagonists such as M35 and M40 are very potent and can block galanin induced and natural food consumption in rats (71,72).

<u>Glucagon-like Peptide-1 (GLP-1)</u> - GLP-1 is a peptide that is found in the hypothalamus and produces reductions in feeding in fasted animals (73). This effect can be blocked by the peptide GLP-1 receptor antagonist, extendin. Interestingly, extendin administration alone can enhance neuropeptide Y-induced feeding suggesting that GLP-1 works through a pathway that is distinct from neuropeptide Y. While this is an exciting new lead into the pathophysiology of obesity, the role of GLP-1 in the adiposity and metabolic abnormalities obeserved in obesity remains to be established.

<u>Leptin</u> - Animal models of obesity may also point to potential therapeutic modalities. A genetic defect in one animal model of obesity, the ob/ob mouse (74), has been identified (75). This gene encodes for a 16 kDa protein called leptin that is secreted from adipocytes and found in the plasma. Exogenous administration of this protein to ob/ob mice results in a reduction in feeding and body weight (76-78). This effect appears to be mediated by interaction of circulating leptin with receptors in the brain resulting in suppression of hypothalamic neuropeptide Y gene expression (78,79). Another mouse model of obesity, the db/db mouse, presumably has a defect in the

receptor for this substance. Administration of leptin to these animals produces no effect (77,78). The role of leptin in human obesity , however, remains unclear since expression of this protein directly correlates with body weight (80,81).

Neuropeptide Y (NPY) - NPY is a 36 amino acid peptide found in the central and peripheral nervous systems. Central administration of NPY produces a rapid and profound increase in food consumption (82,83) and chronic administration results in the physical and metabolic manifestations of obesity (84). The effects of NPY center around the hypothalamus where NPY levels are elevated in a variety of animal models of obesity (85-90). Given the profound actions of NPY on feeding and metabolism, the development of nonpeptide antagonists has been an area of intensive research (91). The receptor that mediates the feeding response to NPY appears to differ from the Y1 receptor that most drug discovery efforts have targeted (92,93) and little information is available in the scientific literature on the effects of Y1 antagonists on feeding.

Opioids - The opioid peptides β-endorphin and dynorphin stimulate food consumption in rodents through both kappa and mu receptor subtypes. Elevated levels of plasma and pituitary β-endorphin have been observed in a variety of obese animals (94). In experimental animals, administration of opioid antagonists such as naloxone (**28**), naltrexone (**29**) and nalmefene (**30**) has been shown to reduce food intake (95). Limited studies in human patients suggest opioid antagonists produce a modest reduction in food intake and can alter food preferences (95).

28	R = CH_2CHCH_2	X = O
29	= CH_2-cyclopropyl	= O
30	= CH_2CHCH_2	= CH_2

REGULATORY ASPECTS OF ANTIOBESITY DRUG DEVELOPMENT

The regulatory requirements for antiobesity agents is becoming more clear. An encouraging sign is the recent approval of dexfenfluramine. This is the first "new" agent approved by the FDA since fenfluramine was approved in 1973. Since dexfenfluramine is simply the active enantiomer of the racemic mixture, fenfluramine, this approval represents a rather small, but significant step in the development of safer and more efficacious agents. When considering the many potential breakthrough therapies that are under development, the next generation will undoubtedly present new challenges to the approval process. The essence of the regulatory concerns for these agents is found in the reality that obesity is a chronic condition that will require chronic drug therapy. Therefore the benefits realized from this chronic therapy should outweigh the potential risks associated with the drug. An FDA guidance was formulated by panels consisting of academic and industrial experts in the treatment of obesity (96). While this guidance is not binding, it indicates the direction of future requirements for the clinical development of antiobesity therapies. Since the present generation of antiobesity agents are approved only for short term therapy (< 3 mos.), future drugs will need to show longer term benefits. To that end, the FDA guidance suggests a new drug demonstrate a weight loss after 12 months that exceeds at least 5% of the baseline weight. Weight loss should consist of a reduction in body fat without muscle wasting or loss of body water. In addition, there should be significant improvements in co-morbid conditions such as glucose tolerance, blood pressure, serum lipid profile or improved quality of life. Following the 12 month placebo controlled trial, the study should continue as an open label study for an additional year to provide safety data.

Summary - The amount of research in obesity is disproportionately small compared with the societal cost due to high prevalence and serious health consequences. Negative public and medical perceptions of obesity includes the impression that obesity is due to a weakness of character and a lack of will power to control diet (97,98). For the available treatments, long term drug therapy is restricted in most states. Without long-term success in weight maintenance, public perceptions will not be challenged and reimbursement for obesity treatment will not be generally available. The environment for approval of weight reduction therapies appears to be improving as a result of our understanding of the mechanisms causing obesity is leading toward development of new improved therapies.

References

1.   R.J. Kuczmarski, K.M. Flegal, S.M. Campbell and C.L. Johnson, JAMA, 272, 205 (1994).
2.   L.F. Martin, S.M. Hunter, R.M. Lauve and J.P. O'Leary, Southern Medical J., 88, 895 (1995).
3.   N.I.o.H.C.D. Panel., Ann.Int.Med., 103, 1073, (1985).
4.   J.K. Wales, Pract. Diabetes, 10, 7 (1993).
5.   D.J. Goldstein, Int.J.Obes., 16, 1, (1992).
6.   G.L. Blackburn and J.L. Read, Postgrad.Med.J., 60, 13 (1984).
7.   N. Finer, S. Finer and R.P. Naoumova, Int.J.Obes., 13, 91 (1989).
8.   G.K. Goodrick and J.P. Foreyt, J.Am.Diet.Assoc., 91, 1243 (1991).
9.   T.A. Wadden, J.A. Sternberg, K.A. Letizia, A.J. Stunkard and G.D. Foster, Int.J.Obes., 13, 39 (1989).
10. M.L. Drent and E.A. Van Der Veen, Netherlands J.Med., 47, 127 (1995).
11. B.A. Scoville in "Obesity in Perspective DHEW Pub. No. (NIH 75-708)," G. A. Bray, Ed., U.S. Government Printing Office., 1975, p. 441.
12. G.A. Bray, Ann.Int.Med., 119, 707 (1993).
13. A. Astrup, S. Toubro, N.J. Christensen and F. Quaade, Am.J.Clin.Nutr., 55, 246S (1992).
14. A. Astrup, Int.J.Obesity, 19, S24 (1995).
15. A. Astrup, L. Breum and S. Toubro, Obesity Res., 3, S 537 (1995).
16. S. Toubro, A. Astrup, L. Breum and F. Quaade, Int.J.Obes.Relat.Metab.Disord., 17, S73 (1993).
17. L. Breum, J.K. Pedersen, F. Ahlstrom and J. Frimodt-Moller, Int.J.Obes.Relat.Metab.Disord., 18, 99 (1994).
18. S. Toubro, A.V. Astrup, L. Breum and F. Quaade, Int.J.Obes.Relat.Metab.Disord., 17, S69 (1993).
19. T.H. Claus and J.D. Bloom, Ann.Rep.Med.Chem., 30, 189 (1995).
20. T. Yen, Obesity Res., 2, 472, (1994).
21. E. Haesler, A. Golay, C. Guzelhan, Y. Schutz, D. Hartmann, E. Jequier and J.P. Felber, Int.J.Obes.Relat.Metab.Disord., 18, 313 (1994).
22. G.R. Goldberg, A.M. Prentice and P.R. Murgatroyd, Int.J.Obes., 19, 625 (1995).
23. E. Danforth, J. Himms-Hagen and P. D., Curr.Opin.Endocr.Diab., 59 (1996).
24. J.D. Bloom and T.H. Claus, Drugs Future, 19, 23 (1994).
25. R. Munger, A. Buckert, E. Jequier and J.-P. Felber, Diabetes, 39 (suppl. 1), abstr. 1100 (1990).
26. M.H. Fisher, E.M. Naylor, D. Ok, H. Ok, T. Shih and A.E. Weber, PCT WO9529159A1 (1995).
27. M.H. Fisher, R.J. Mathvink, H.O. Ok, E.R. Parmee and A.E. Weber, EP 611003 A1 (1994).
28. L.J. Beeley and J.M. Berge, PCT WO 9504047 A1 (1995).
29. J. Prous, N. Mealy and J. Castaner, Drugs Future, 19, 1003 (1994).
30. M.L. Drent, I. Larsson, T. William-Olsson, F. Quaade, F. Czubayko, K. von Bergmann, W. Strobel, L. Sjostrom and E.A. van der Veen, Int.J.Obes.Relat.Metab.Disord., 19, 221 (1995).
31. J. Zhi, A.T. Melia, H. Eggers, R. Joly and I.H. Patel, J. Clin. Pharm., 35, 1103 (1995).
32. M. Berger, Am. J. Clin. Nutr., 55, 318S, (1992).
33. M.F. Lesses and A. Myerson, NEJM, 218, 119 (1938).
34. T. Silverstone in "Obesity: Theory and Therapy.," A. J. Stunkard and T. A. Wadden, Ed., Raven Press, 1993, p. 85.
35. F. Greenway, Am.J.Clin.Nutr., 55, 203S (1992).
36. M. Weintraub, G. Ginsgerb, C. Stein, P.R. Sundaresan, B. Schuster and P. O'Connor, Clin.Pharmacol.Ther., 39, 501 (1985).
37. I.M. Holdaway, E. Wallace, L. Westbrooke and G. Gamble, Int.J.Obes.Relat.Metab.Disord. 19, 749 (1995).
38. J. Wurtman, R. Wurtman, E. Berry, R. Gleason, H. Goldberg, J. McDermott, M. Kahne and R. Tsay, Neuropsychopharmacology, 9, 201 (1993).
39. H.T. O'Connor, R.M. Richman, K.S. Steinbeck and I.D. Caterson, Int.J.Obes.Relat.Metab.Disord. 19, 181 (1995).
40. P. Cacoub, R. Dorent, P. Nataf, J.P. Houppe, J.C. Piette, P. Godeau and Gandjbakhch, Eur.J.Clin.Pharmacol., 48, 81 (1995).

41. P.d. Groote, A. Millaire, E. Decoulx, F. Passard, F. Puisieux and G. Ducloux, Therapie, 48, 373 (1993).
42. R. Voelker, J. Am. Med. Assoc., 272, 1087, (1994).
43. A.C. Toornvliet, H. Pijl and A.E. Meinders, Int.J.Obes.Relat.Metab.Disord., 18, 650 (1994).
44. C. Galletly, A. Clark and L. Tomlinson, Int.J. Eating Disord., 19, 209 (1996).
45. A. Choi, Wall Street Journal, B8, April 30, (1996).
46. M.L. Fernandez-Soto, A. Gonzalez-Jimenez, F. Barredo-Acedo, J.D. Luna del Castillo and F. Escobar-Jimenez, Ann.Nutr.Metab., 39, 159 (1995).
47. D.J. Goldstein, A.H. Rampey, Jr., G.G. Enas, J.H. Potvin, L.A. Fludzinski and L.R. Levine, Int.J.Obes.Relat.Metab.Disord., 18, 129 (1994).
48. L.L. Darga, L. Carroll-Michals and S.J. Botsford, Am.J.Clin.Nut.r, 54, 321 (1991).
49. M.D. Marcus, R.R. Wing, L. Ewing, E. Kern and McDermott, Am.J.Psych., 147, 876 (1990).
50. D.T. Wong, F.P. Bymaster and E.A. Engleman, Life Sci., 57, 411 (1995).
51. W. Meyerowitz and D.C. Jaramillo, Curr.Ther.Res., 55, 1176 (1994).
52. J.A. Nielsen, D.S. Chapin, J.L. Johnson and L.K. Torgersen, Am.J.Clin.Nutr., 55, 185S (1992).
53. C.T. Dourish, Obes.Res., 3, 449S (1995).
54. Drugs Future, 19, 806 (1994).
55. G.A. Bray, Int.J.Obes., 18, 60 (1994).
56. G.A. Bray, D.H. Ryan, D. Gordon, S. Heidingsfelder, R. Macchiavelli and K. Wilson, J.Invest.Med., 43, 244A (1995).
57. J.M. Ferguson, G.A. Bray, G.L. Blackbiurn, F.L. Greenway, A. Jain, P.E. Kaiser, J. Mendels, D. Flyan and S.L. Schwartz, Psychopharmacol.Bull., 31, 568 (1995).
58. P.E. Kaiser and J.L. Hinson, J.Clin.Pharmacol., 34, 1019 (1994).
59. R. Varqas, F.G. McMahon and A.K. Jain, Clin.Pharmacol.Ther., 55, 188 (1994).
60. D.J. Goldstein and J.H. Potvin, Am.J.Clin.Nutr., 60, 1101 (1994).
61. M. Weitraub, P.R. Sundaresan and B. Schuster, Clin.Pharmacol.Ther., 51, 581 (1992).
62. R.J. Lieverse, J.B.M.J. Jansen and C.B.H.W. Lamers, Neth.J.Med., 42, 146 (1993).
63. A.J. Strohmayer and D. Greenberg, Physiol.Behav., 56, 1037 (1994).
64. T.H. Moran, T.K. Sawyer, D.H. Seeb, P.J. Ameglio, M.A. Lombard and P.R. McHugh, Am.J.Clin.Nutr., 55, 286S (1992).
65. C.J. Aquino, D.R. Armour, J.M. Berman, L.S. Birkemo, R.A.E. Carr, D.K. Croom, M. Dezube, R.W. Dougherty, Jr., G.N. Ervin and E.E. Sugg, J.Med.Chem., 39, 562 (1996).
66. C. Rose, F. Vargas, P. Facchinetti, P. Bourgeat, P. Bambal, P.B. Bishop, S.M.T. Chan, A.N.J. Moore, C.R. Ganellin and J.-C. Schwartz, Nature, 380, 403 (1996).
67. C. Erlanson-Albertsson, Scand. J.Nutr.Naringsforsk., 38, 11 (1994).
68. L. Lin, S. Okada, D.A. York and G.A. Bray, Peptides, 15, 849 (1994).
69. D.A. York in "Pennington Symposium Series: Molecular Biology of Obesity," G. A. Bray and D. Ryan, Ed., 1996, p. In press.
70. S.F. Leibowitz, Obesity Res., 3, 573S (1995).
71. R.L. Corwin, J.K. Robinson and J.N. Crawley, Eur. J. Neurosci., 5, 1528 (1993).
72. S.F. Leibowitz and T. Kim, Brain Res., 599, 148 (1992).
73. M.D. Turton, D. O'Shea, I. Gunn, S.A. Beak, C.M.B. Edwards, K. Meeran, S.J. Choi, G.M. Taylor, M.M. Heath, P.D. Lambert, J.P.H. Wilding, D.M. Smith, M.A. Ghatei, J. Herbert and S.R. Bloom, Nature, 379, 69 (1996).
74. G.A. Bray and D.A. York, Physiol. Rev., 59, 719 (1979).
75. Y. Zhang, R. Proenca, M. Maffei, M. Barone, L. Leopold and J.M. Friedman, Nature, 372, 425 (1994).
76. M.A. Pelleymounter, M.J. Cullen, M.B. Baker, R. Hecht, D. Winters, T. Boone and F. Collins, Science, 269, 540 (1995).
77. J.L. Halaas, K.S. Gajiwala, M. Maffei, S.L. Cohen, B.T. Chait, D. Rabinowitz, R.L. Lallone, S.K. Burley and J.M. Friedman, Science, 269, 543 (1995).
78. T.W. Stephens, M. Basinski, P.K. Bristow, J.M. Bue-Valleskey, S.G. Burgett, L. Craft, J. Hale, J. Hoffmann, H.M. Hsiung, A. Kriauciunas, W. Mackellar, P. Rosteck, B. Schoner, D. Smith, F.C. Tinsley, X.Y. Zhang and M. Heiman, Nature, 377(6549), 530 (1995).
79. L.A. Campfield, F.J. Smith, Y. Guisez, R. Devos and P. Burn, Science, 269, 546 (1995).
80. M. Maffei, J. Halaas, E. Ravussin, R.E. Pratley, G.H. Lee, Y. Zhang, H. Fei, S. Kim, R. Lallone, S. Ranganathan, P.A. Kern and J.M. Friedman, Nature Med., 1, 1155 (1995).
81. F. Lonnqvist, P. Amer, L. Nordfors and M. Shalling, Nature Med., 1, 950 (1995).
82. B.G. Stanley and S.F. Leibowitz, Proc.Natn.Acad.Sci. USA, 82, 3940 (1985).
83. B.G. Stanley and S.F. Leibowitz, Life Sci., 35, 2635 (1984).
84. N. Zarjevski, I. Cusin, R. Vettor, F. Rohner-Jeanrenaud and B. Jeanrenaud, Endocrinology, 133, 1753 (1993).
85. A. Sahu, J.D. White, P.S. Kalra and S.P. Kalra, Brain Res. Mol. Brain Res., 15, 15, (1992).
86. G. Williams, H.M. Cardoso, Y.C. Lee, J.M. Ball, M.A. Ghatei, M.J. Stock and S.R. Bloom, Clin.Sci., 80, 419 (1991).
87. B. Beck, A. Burlet, J.P. Nicolas and C. Burlet, J.Nutr., 120, 806 (1990).
88. B. Beck, A. Burlet, J.P. Nicolas and C. Burlet, Physiol.Behav., 47, 449 (1990).

89.   P.E. McKibbin, S.J. Cotton, S. McMillan, B. Holloway, R. Mayers, H.D. McCarthy and G. Williams, Diabetes, 40, 1423 (1991).
90.   G. Williams, L. Shellard, D.E. Lewis, P.E. McKibbin, H.D. McCarthy, D.G. Koeslag and J.C. Russell, Peptides, 13, 537 (1992).
91.   P.A. Hipskind and D.R. Gehlert, Ann.Rep.Med.Chem., 31, 1 (1996).
92.   B.G. Stanley, W. Magdalin, A. Seirafi, M.M. Nguyen and S.F. Leibowitz, Peptides, 13, 581 (1992).
93.   S.P. Kalra, M.G. Dube, A. Fournier and P.S. Kalra, Physiol. Behav., 50, 5 (1991).
94.   D.L. Margules, B. Moisset, M.J. Lewis, H. Shibuya and C.B. Pert, Science, 202, 988 (1978).
95.   M. de Zwaan and J.E. Mitchell, J.Clin.Pharmacol., 32, 1060 (1992).
96.   L. Lutwak in "Obesity: Advances in Understanding and Treatment; Section IV.," D. J. Goldstein, Ed., International Business Communications, 1996, p. 1.
97.   G.A. Bray, Ann.Int.Med., 115, 152 (1991).
98.   A. Frank, JAMA, 269, 2132 (1993).

Chapter 22.   Anti-Osteoporosis Agents

Paul Da Silva Jardine and David Thompson
Pfizer Central Research
Eastern Point Road, Groton, CT 06340

Introduction - Osteoporosis is a disease characterized by low bone mass and enhanced bone fragility resulting in an increased risk of fractures. It results from a deficit in new bone formation versus resorption during the ongoing remodeling process. Bone mass increases during the first two decades of life reaching a "peak bone mass" then steadily declines (0.5%/year) thereafter. Immediately after menopause women, who already have a lower peak bone mass than men, experience a sharp increase in bone mass decline (2-4%/year), putting them at greater risk for fractures. Osteoporosis is a major public health concern of increasing magnitude as the population ages. Today, osteoporosis affects 25 million Americans who either already have an osteoporotic fracture or are at high risk of one because of dangerously low bone mass. One out of every two women or one of five men will suffer from an osteoporotic fracture. Currently, estrogen replacement therapy (ERT), calcitonin and the bisphosphonate, alendronate, are the only approved therapies for osteoporosis in the US, but in the next few years several important new classes of agents should become available. This chapter will review both the current and emerging therapies with an emphasis, where possible, on the structure activity relationships in the various classes. Reviews on anti-osteoporosis therapy have recently been published(1,2).

There are two types of bone: cortical bone and trabecular or cancellous bone. Cortical bone is the dense outer shell enclosing the medullary cavity, which contains marrow interlaced with a supporting network of thin plates of bone called cancellous or trabecular bone. Both types of bone are lost with age and estrogen deficiency following menopause, but since cancellous bone has a higher turnover rate, it is lost more rapidly. Osteoclasts are large multinucleated cells that are responsible for resorbing the bone while osteoblasts are responsible for depositing new bone.

The objective of anti-osteoporosis therapy is to maintain or increase bone mass and bone strength thereby reducing the risk of fractures. Anti-osteoporosis therapy falls into two classes: anti-resorptive and bone restorative (or anabolic) agents. Anti-resorptive therapy targets bone resorption reducing bone turnover and thereby preventing bone loss. It has been the major focus and current therapies fall into this class. More recently the focus has shifted onto bone restoration. Bone restoration therapy when approved would seek to restore bone to osteopenic skeletons thereby reducing the likelyhood of skeletal fracture. Anti-resorptive therapy is preventative while bone restoration is treatment of osteoporosis.

ANTIRESORPTIVE AGENTS

Bisphosphonates - All bisphosphonates are of the general structure shown in Table 1. The P-C-P [P=P(O)(OH)$_2$] unit, a non-hydrolyzable isostere of pyrophosphate, binds avidly to hydroxyapatite and targets the bisphosphonates to the resorbing surfaces of bone, resulting in inhibition of resorption as well as mineralization through an unknown mechanism. The first generation agents, etidronate and clodronate, 1 and 2, inhibited both resorption and bone mineralization at similar concentrations which precluded their chronic use in the treatment of osteoporosis. The second generation agents exemplified by pamidronate (3), alendronate(4), tiludronate(5) and risedronate(6) are

100-1000 times more potent inhibitors of bone resorption than of bone mineralization (Table1). Alendronate (<u>4</u>) has recently been approved in the US for the prevention of osteoporosis. In postmenopausal women, alendronate at 10 mg daily dosing for three years increased vertebral, femoral neck and trochanter bone mineral density by 8.8%, 5.9% and 7.8%, respectively, versus placebo (3). The proportion of women with new vertebral fractures was reduced by 48% and the average number of vertebral fractures was reduced by 63%. There was also a trend towards a reduction in the incidence of non-vertebral fractures. No adverse effects on bone mineralization were seen. Several other bisphosphonates are currently undergoing clinical evaluation (4).

<u>Table 1</u>: **Bisphosphonates in development**

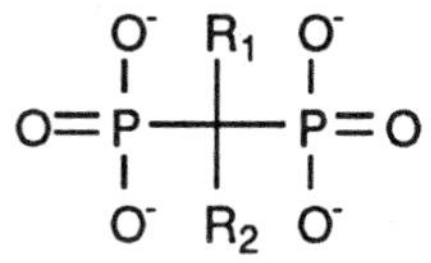

	Compound	R1	R2
<u>1</u>	Etidronate	OH	CH_3
<u>2</u>	Clodronate	Cl	Cl
<u>3</u>	Pamidronate	OH	$-(CH_2)_2NH_2$
<u>4</u>	Alendronate	OH	$-(CH_2)_3NH_2$
<u>5</u>	Tiludronate	H	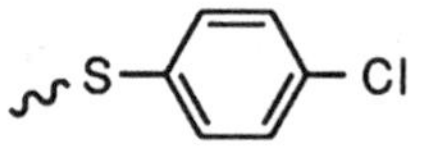
<u>6</u>	Risedronate	OH	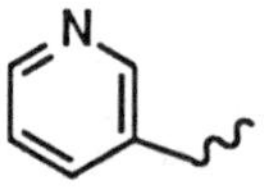
<u>7</u>	EB-1053	OH	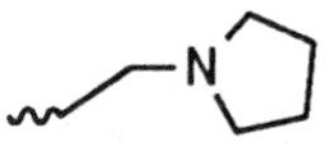
<u>8</u>	CGP-42446	OH	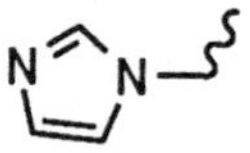
<u>9</u>	YM-529	OH	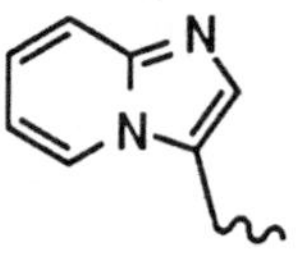

Some important SAR trends of this class of agents have emerged (5, 6). A key finding from the first generation of bisphosphonates is that incorporation of a hydroxyl group at C1 maximizes affinity for hydroxyapatite and increases the anti-resorption potencies. In second generation compounds, incorporation of amino alkyl groups at C1 increased anti-resorptive potency at least 10 fold. The length of the aliphatic chain is important with lengths of 2-5 carbon atoms being optimum (4). Incorporation of this nitrogen into a heterocyclic ring as in risedronate (<u>6</u>), EB-1053 (<u>7</u>), CGP-42446 (<u>8</u>), YM-529 (<u>9</u>) has resulted in very potent third generation bisphosphonates (7,8,9). The methylene chain connecting the heterocycle to C1 is imperative, as analogs with longer chains or other heteroatom linkers are >50 fold less potent. The basicity of the nitrogen is less critical albeit still important (10).

Oral absorption of the bisphosphonates is very poor (1-5%) and is highly dependent on the presence of food and calcium in the stomach both of which can reduce absorption to negligible amounts. The bisphosphonates have very long half-lives in bone (several years) which has raised questions about the long term safety profile and effects on bone turnover and mechanical strength (11). However, it is expected that these agents will still have significant benefits in the prevention of osteoporosis

Estrogens - At menopause estrogen (10) levels in women decline to very low levels resulting in increased bone turnover rates and rapid bone loss. Estrogen replacement therapy (ERT) is efficacious at slowing bone turnover and preventing bone loss (12). Recent studies have shown that long term estrogen usage reduces the risk of vertebral fractures by 50-60%. The effect on hip fracture risk was harder to determine, as the incidence of hip fractures was low, but it was clear that estrogen usage does not reduce hip fracture risk by half, as was widely believed, but by about 25% (13,14). Women with intact uteri are recommended to have progesterone added to protect the endometrium: hormonal replacement therapy (HRT). Several classes of estrogens and conjugated estrogens, either human or equine, including 17β-estradiol, estrone sodium sulfate, 17-ethinyl estradiol, and equilin sodium sulfate are used clinically. Various dosing regimens with or without progesterone and various delivery systems are in use (15,16).

In addition to its beneficial effects on bone, ERT/HRT has several other benefits. Most importantly, it reduces the risk of coronary heart disease by up to 50% (17). A portion (25-50%) of estrogens cardioprotective effect is believed to be due to its favorable impact on lipid profiles, reducing LDL and raising HDL levels by 10-15%(18). However, several adverse side effects, most notably the increase in risk of endometrial and mammary tumors, have been associated with estrogen usage. Progesterone add-back virtually eliminates the risk of endometrial cancer. The effects of HRT on breast cancer risk is controversial, but the findings from the latest Nurses' Health Study clearly indicated a significant increase in the risk of breast cancer among women who had used estrogen, with or without progestin, for more than five years (relative risk = 1.4) versus postmenopausal women who had never used hormones (19).   Other objectionable side effects of estrogen usage are hot flashes, breakthrough bleeding and breast tenderness. Ongoing multicenter studies of Postmenopausal Estrogen/Progestin Intervention (PEPI trial) and the Hormone/Estrogen Replacement Study (HERS) will cast further light on the balance between the risks and balances of estrogen usage. Current compliance with HRT is poor. Only 13.7% of eligible women in the US are on HRT, of those, 20% stop taking medication within 9 months (20).

Much progress has been made in the last 2-3 years in determining the mechanism of the bone and cardiovascular protective effects of estrogen, but much is still unknown (21). One possible mechanism of estrogen's bone protective effect is that it inhibits osteoclastogenesis through several potential mechanisms; for instance, estrogen is a potent inhibitor of cytokine-induced osteoclastogenesis (eg., IL-1, TNF-α, and IL-6) both in vitro and in vivo, (22). Beyond its beneficial effects on lipid levels, even less is known about the mechanism of its cardioprotective effects (21).

Estrogen Agonists/Antagonists - An ideal estrogenic agent would be one that was an agonist in bone, but an antagonist in reproductive tissue, breast and uterus, while maintaining estrogen's cardioprotective properties. The first clue that a tissue selective estrogen agonist was achievable came with tamoxifen (11), an antiestrogen in mammary tissue, used in breast cancer therapy. Instead of inducing bone loss in postmenopausal women, as initially predicted, tamoxifen protected these women against bone loss while apparently maintaining much of estrogen's cardioprotective effects (23,24,25). The mechanism through which tamoxifen achieves this tissue selectivity, especially in view of the fact that the estrogen receptor is identical in the different tissues, is under active investigation.

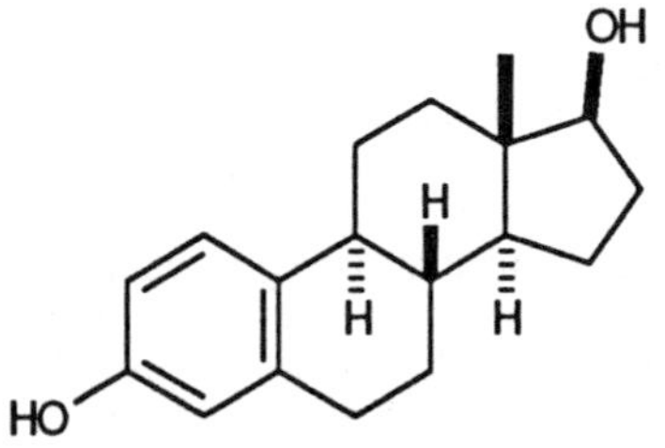

10

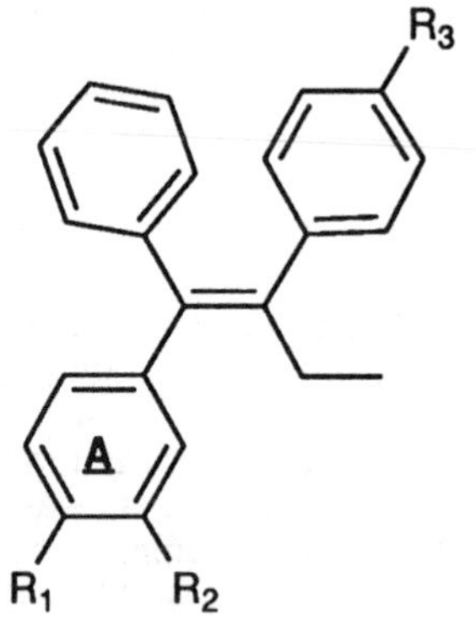

11 R_1= R_2= H, R_3=O$(CH_2)_2$NMe$_2$
13 R_1= H, R_2= OH, R_3=O$(CH_2)_2$NMe$_2$
15 R_1= OH, R_2= H, R_3=O$(CH_2)_2$NMe$_2$
16 R_1= R_2= H, R_3=*trans*-CH=CHCOOH

Tamoxifen, albeit less uterotrophic than estrogen, still has significant proliferative effects on uterine tissue. There are several other concerns about the safety profile of long tern usage of tamoxifen which would likely limit its use in osteoporosis therapy (26). Two more potent and tissue selective estrogen agonists/antagonists with better safety profiles, raloxifene (**12**) (Phase III) and droloxifene (**13**) are currently being evaluated in clinical trials for their use in osteoporosis. In a 12-week Phase II study, raloxifene (50-200 mg/day) reduced bone turnover, as measured by serum markers, equivalent to Premarin (27). This reduction in turnover would be expected to translate to prevention of bone loss. Additionally, as with estrogen, raloxifene lowered plasma LDL levels but, unlike estrogen, HDL levels were unaffected at these doses. No significant uterine hypertrophy was observed. In animal models, droloxifene also potently protected against ovariectomy induced bone loss and lowered serum LDL levels, without causing significant uterine hypertrophy (28). Since both these agents are potent antiestrogens in mammary tissue, the risk of breast cancer should be significantly reduced versus estrogen (29). In fact, droloxifene has been reported to be efficacious in the treatment of advanced breast cancer and thus, it does appear to have the expected antiestrogenic properties (30).

Very few SAR conclusions with respect to tissue specific functional activity of these agents can be drawn at this point, but some general features of the SAR, especially with respect to estrogen receptor binding activity, are listed below. The only pure antiestrogens appear to be the 7α-substituted estrogens (eg., ICI 182,780, (**14**)) (31). Tamoxifen, raloxifene and droloxifene are all mixed agonists/antagonists and were derived from the same structural class of triarylethylenes. In this class, a phenolic group in the A-ring is critical for receptor binding and functional activity as estrogens or antiestrogens (compare tamoxifen and its active metabolite 4-hydroxy tamoxifen (**15**) which binds 100 fold more tightly than tamoxifen). It is likely that this hydroxyl group mimics the critical 3-hydroxyl group of estrogen (32,33,34). The relative orientation of the three aryl rings in a propeller arrangement is important for tight receptor binding and functional activity (note the wide range of activity seen with several conformational constrained tamoxifen analogs) (35,36). The ethanolamine sidechain, present in almost all estrogen agonists/antagonists was thought to confer the antagonist activity to these compounds. Recently, this sidechain has been replaced by an acrylic acid group (16), which still maintains similar functional tissue specific activity as the ethanolamine analogs (37). A comprehensive review of these and other estrogen receptor modulators has been published (38).

Calcitonin - This normally circulating 32-amino acid peptide protects against postmenopausal bone loss when administered parenterally or nasally (39). Salmon calcitonin is more potent than the human variant and is more commonly used. Additionally calcitonin, a powerful analgesic, significantly ameliorates the pain that can be associated with osteoporosis. Calcitonin therapy reduces the risk of fracture but reduced efficacy, high cost and tachyphylaxis will limit its use in osteoporosis therapy. Much is known about the SAR of this peptide but is beyond the scope of this review (40).

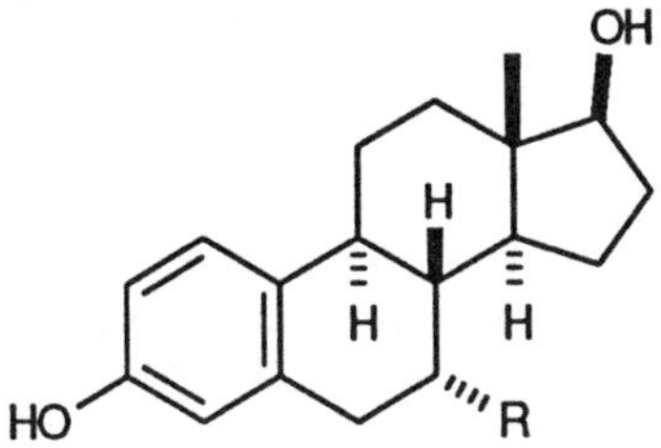

14 R=$(CH_2)_{10}CON(Me)(CH_2)_3Me$ 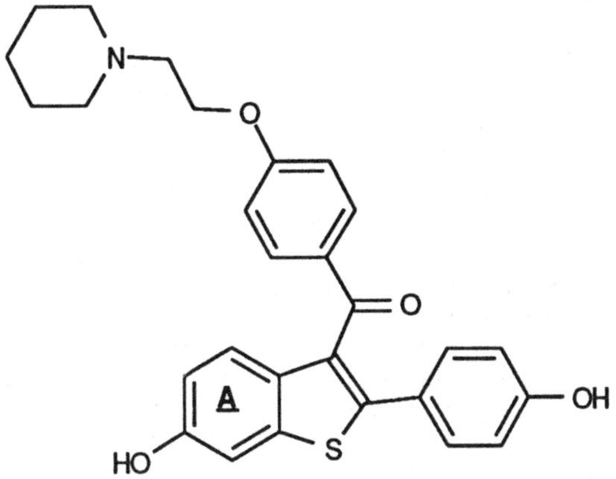   **12**

Other Agents - Numerous Vitamin D analogs have been synthesized, most notably $1\alpha,25$-dihydroxyvitamin D_3 (calcitriol) and 1α–vitamin D_3 which are used in Japan for osteoporosis, and have been reviewed (41). The major goal of research in the use of Vitamin D analogs in the prevention of osteoporosis is to separate their bone protection properties from their hypocalcemic effects. Since their use for osteoporosis remains experimental they will not be discussed here (42).

Osteoclasts attach to bone during resorption through osteoclast integrin receptors ($\alpha_v\beta_3$ for example) and RGD-containing bone matrix proteins. Inhibitors of these interactions are being developed as anti-resorptive therapy. A detailed discussion of the progress in this area of research has recently been published (43).

The major phenotype of the src-kinase knockout mouse was discovered to be osteopetrosis, resulting from impaired bone resorption, thus prompting a search for src-kinase inhibitors for osteoporosis therapy (44).

BONE RESTORATIVE APPROACHES

Parathyroid Hormone - Parathyroid hormone (PTH) is the major circulating hormone, secreted by the parathyroid gland, that functions to maintain calcium homeostasis. In osteopenic rats, single daily injection of (8-25 nmoles/kg) of hPTH (1-84) or hPTH (1-34) can strongly induce formation of normal, biomechanically sound new trabecular bone (45). Paradoxically, continuous administration of PTH will result in the further loss of trabecular bone (46). In humans the picture is less clear, but intermittent administration of PTH(1-34) for 6-24 months in osteoporotic men and postmenopausal women significantly increased vertebral trabecular mineral density (32-98%) with only a slight decrease or no effect in cortical bone (47). Human PTH (1-34) administered subcutaneously at 25 µg/day to osteoporotic patients on estrogen increased lumbar vertebral bone mass 11% within two years, with little further increase in the third year (48). This increase in vertebral bone mass can be maintained after cessation of PTH treatment by continued treatment with estrogen (48).

Progress has been made in understanding the SAR of the osteogenic effects of PTH. The PTH/PTH receptor complex can utilize either of two signal transduction mechanisms: the adenylate cyclase-cAMP path or the phospholipase C-protein kinase C path. Fragment 1-34 retains the full hormone's bone anabolic activity and can activate either signaling mechanism. Human PTH (1-31)NH$_2$, which can stimulate only the adenylate cyclase path, is anabolic in trabecular bone. In contrast, hPTH(8-84) which only activates the phospholipase C-protein kinase C path, is not bone anabolic (47).

Due to the high cost and the inconvenience of parenteral administration of PTH, efforts have been undertaken to downsize it to a minimum effective peptide or non-peptide mimic that stimulates bone formation. One such analog is RS-66271 which was derived from PTHrP (1-34) by replacing the 20-34 region, which is postulated to form an amphiphilic α-helix critical for receptor binding of PTHrP to the PTH receptor, by a model α-helical peptide (MAP) sequence (49,50,51).

Another strategy to restore bone to osteopenic skeletons may be the targetting of the calcium receptor or sensor on parathyroid cells (52). This receptor regulates the secretion of PTH in response to changes in serum Ca^{2+} levels. Activation of this receptor by elevated serum Ca^{2+} levels will trigger a decrease in PTH secretion. Conversely, antagonists of this receptor may stimulate the secretion of PTH and mimic the bone anabolic effects seen with PTH administration.

Growth Hormone Secretagogues - With aging, growth hormone levels decrease precipitously and there is much interest in restoring these levels in the aged to those of a young adult. Growth hormone directly and indirectly via IGF-1 is a potent osteogenic agent (53). Twelve months of growth hormone therapy in GH-deficient adults increased bone mineral density in the spine and cortical bone by 5% and 4%, respectively (54). But proof that growth hormone administration will restore bone to human osteopenic skeletons is still needed. Growth hormone therapy in the elderly have a multitude of benefits partially reversing some of the effects of aging (55). Administration of pharmacological doses of growth hormone as a single daily bolus have resulted in several objectionable side effects, Carpal Tunnel syndrome, for example. A more desirable agent would restore growth hormone levels in the aged to young adult levels by augmentation of the endogenous pulsatile secretion. Several years ago a series of hexapeptides (17), derived from methionine enkephalin and named Growth Hormone Releasing Peptides (GHRP's), were discovered by Bowers (56). They act directly on the hypothalamus and pituitary to stimulate the release of growth hormone. The SAR of these peptides has been reviewed; one finding that a D-amino acid like D-Trp or D-napAla at position 2 is important for activity is especially significant (57).

A series of non-peptide, GHRP mimetic, benzolactam growth hormone secretagogues, exemplified by L-692,429 (18), was reported to elevate growth hormone levels in young males and healthy older subjects (58). Some key aspects of the SAR of this important class of compounds are summarized below (59). First, the seven member lactam ring size was important; the corresponding six and eight member lactams had 10-50 fold less activity. Second, at C3 of the benzazepinone

ring, 3R was the active configuration. Both these points support the hypothesis that
the benzolactam ring of L-692,429 overlaps with the crucial D-Trp in the peptidyl
GHRP-6 secretagogue. Third, the amino group of the α-Me Ala sidechain at C3 of the
benzazepinone is critical for growth hormone release activity although it can be
substituted with various hydroxyalkyl groups (the 1, 2-dihydroxypropyl group in
L-700,653 (**19**) is especially active) (60). Fourth, the acidic tetrazole in L-692,429
could be replaced by neutral groups such as carboxamides (61).

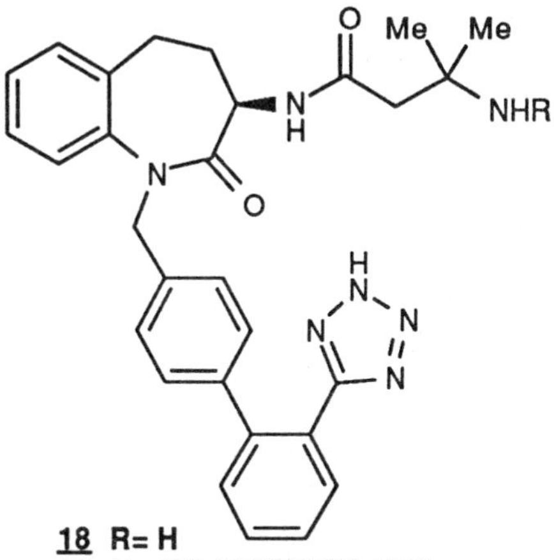

18 R= H
19 R= CH$_2$CH(OH)CH$_2$(OH)

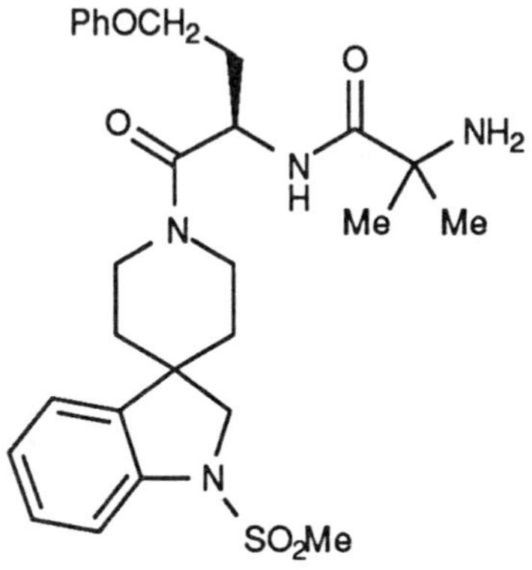

20

These secretagogues had limited oral bioavailabilty, but a related class of potent,
orally active (60% oral bioavailabilty in dogs), dipeptide growth hormone
secretagogues, represented by L-163,191 (MK-677, **20**) has been disclosed (62). The
SAR of this class is similar to that of the first series: the amino group in the α-MeAla
residue and the D-configuration at the O-benzylserine residue are again critical; the
SAR in the spiroindane piperidine unit is less sensitive, except that incorporation of
polar groups at the 3 position significantly improved the potency of these compounds.
As with growth hormone itself, proof that these agents will be efficacious at restoring
lost bone to osteopenic skeletons is required, but judging from the wide-ranging
beneficial effects seen with administration of rhGH in human studies, MK-677 may find
important utility beyond the treatment of osteoporosis. Dipeptide **21** is one member of
the newest series of very   potent growth hormone secretagogues obtained by
downsizing the hexapeptide GHRP's to di- and tripeptides (63).

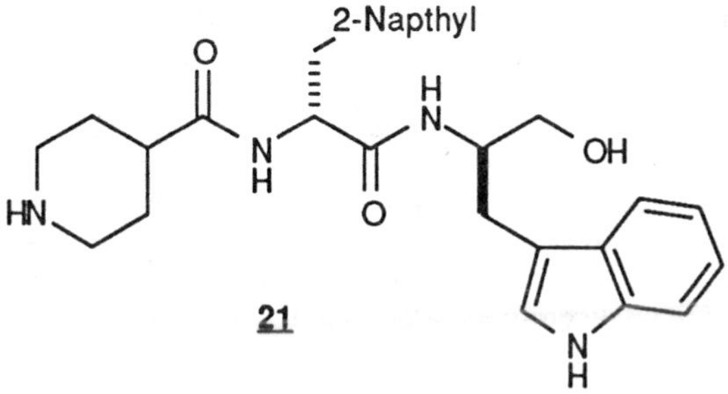

2-Napthyl

21

<u>Sodium Fluoride</u> - Initial clinical studies in the US with high doses (75 mg/day for 4
years) of sodium fluoride showed that it did not decrease vertebral fracture rates
although it markedly increased the bone mineral density at that site (64). Also an
increase in the rate of appendicular fractures, hip and wrist was observed. In an

earlier study with a lower dose (50 mg/day for 2 years) of sodium fluoride, the fluoride group showed a significantly lower rate of new vertebral fractures and the number of femoral neck fractures was not significantly different from the non-fluoride group (65). In recent studies with a new slow release formulation at lower doses (25 mg twice daily), sodium fluoride substantially increased vertebral bone mass 4.8% per year for 4 years and reduced the rate of new vertebral fractures by 66%, while no evidence of increased appendicular fracture rates and microfractures was seen (66). Additionally the gastric side effects seen with plain sodium fluoride was not seen to the same extent with the slow release formulation. Some limitations of slow release fluoride are: it has no apparent effect on the rate of recurrent vertebral fractures; in patients with severe osteopenia only marginal effects on the fracture rate were seen; finally although the incidence of overall appendicualr fractures was not increased, no decrease in the rate of such fractures was seen. Data on the hip fracture rates are not available yet.

Miscellaneous - Studies in several animal species and humans have demonstrated that Prostaglandin E_2 and $F_{1\alpha}$ are potent and powerful bone anabolic agents (67). Remarkably PGE_2, unlike PTH and sodium fluoride, is able to restore bone even in severely osteopenic animal skeletons. PGE_2 can restore connectivity in perforated trabeculae and also build bone on endocortical surfaces which dramatically increases the biomechanical strength of the new bone. After the termination of PGE_2 administration, the new bone formed is rapidly lost, but combination of PGE_2 with the bisphosphonate risedronate, an anti resorptive agent, was successful in maintaining the new bone added (68). Severe adverse side effects limit the use of endogenous prostaglandins, like PGE_2 in osteoporosis therapy. The recent cloning of four PGE_2 receptor subtypes may shed some light on the mechanism of its powerful bone anabolic effects (69).

The bone morphogenic proteins (BMP's) are members of the TGF-β superfamily that were originally identified by their remarkable ability to induce ectopic bone formation when implanted subcutaneously *in vivo* (70). To date nine BMP's have been cloned and sequenced (71). Like other members of the TGF-β superfamily, they exert their effects by forming heteromeric complexes of Type I and Type II serine/threonine kinase receptors (72). The BMP's play critical roles in the development of the embryonic skeleton (73). In mature skeletons, they are involved in new bone formation and bone repair (74). The availability of recombinant BMP's have resulted in their successful application in craniofacial regeneration and fracture healing (75). Recombinant hBMP-2, for example, promotes rapid and complete union of large segmental defects in adult rat femurs (76). Their use to restore bone to osteopenic skeletons has not yet been demonstrated but is under active investigation. Recently, the three-dimensional structure of recombinant human osteogenic protein 1 (BMP-7) has been determined by x-ray crystallography, which may help in the design of small molecule mimetics of the BMP's (77).

Conclusion - Although significant progress has been made in both its prevention and treatment, osteoporosis remains a major health concern of increasing magnitude for several reasons: an aging population, improved diagnostic procedures and increasing public awareness. As more emphasis is placed on the treatment of osteoporosis, restoration of lost bone, it will not be sufficient just to increase bone mass; to reduce the rate of fractures the new bone must be biomechanically sound.

References

1. M. Zaidi and M. Pazianas in "Therapy of Osteoporosis," T. Dyson,Ed., Derwent Information, London, UK, 1995.
2. G. Rodan, Ann.Rep.Med.Chem., 29, 275 (1994).
3. U. Liberman, S. Weiss, J. Broll, H. Minne, H. Quan, N. Bell, J. Rodriguez-Portales, R. Downs, J. Dequeker, M. Favus, E. Seeman, R. Recker, T. Capizzi, A. Santora, A. Lombardi, R. Shah, L. Hirsch and D. Karpf, N.Engl.J.Med., 333, 1437 (1995).
4. W. Sietsema and F. Ebetino, Expert Opin.Invest.Drugs, 3, 1255 (1994).
5. W. Sietsema, F. Ebetino, A. Salvagno and J. Bevan, Drugs Exp.Clin.Res., 15, 389　　(1989).
6. A. Geddes, S. D'Souza, F. Ebetino and K. Ibbotson, Bone Miner.Res., 8, 265 (1994).
7. J. Green, K. Muller and K. Jaeggi, J.Bone Miner.Res., 9, 745 (1994) for CGP-42446.
8. Y. Isomura, M. Takeuchi and A. Tetsushi,European Patent 354806 (1990) for YM-529.
9. G. Van der Pluijm, L. Binderup, E. Bramm, L. Van der Wee-Pals, H. DeGroot, E. Binderup, C. Lowik and S. Papapoulos, J. Bone Miner.Res., 7, 981 (1992) for EB-1053.
10. M. Rogers, X. Xiong, R. Brown, D. Watts, R. Russell, G. Graham, A. Bayless and F. Ebetino, Mol.Pharmacol., 47, 398 (1995).
11. P. Sambrook, N.Engl.J.Med., 333, 1495 (1995).
12. E. Lufkin, H. Wahner, W. O'Fallon, S. Hodgson, M. Kotowitcz, A. Lane, A. Judd, R. Caplan,and B. Riggs, Ann.Intern.Med., 117, 1 (1992).
13. J. Cauley, D. Seeley, K. Ensrud, B. Ettinger, D. Black and S. Cummings, for the Study of Osteoporotic Fractures Research Group, Ann.Int.Med., 122, 9 (1995).
14. P. Maxim, B. Ettinger and G. Spitalny, Osteoporosis Int., 5, 23 (1995).
15. B. von Schoultz in "The Menopause," J. Studd and M. Whitehead, Eds., Blackwell Scientific, Oxford, England, 1988, p.130.
16. C. Maschchak, R. Lobo, R. Dozono, R. Takano, P. Eggena, R. Nakamura, P. Brenner and D. Mishell, am.J.Obstet.Gynecol., 144, 511 (1982).
17. R. Lobo, Am.J.Obstet.Gynecol., 173, 982 (1995).
18. The Writing Group for the PEPI Trial, JAMA, 273, 199 (1995).
19. G. Colditz, S. Hankinson, D. Hunter, W. Willett, J. Manson, M. Stampfer, C. Hennekens, B. Rosner and F. Speizer, N.Eng.J.Med., 332, 1589 (1995).
20. J. Cauley, S. Cummings and D. Black, Am.J.Obstet.Gynecol., 163, 1438 (1990).
21. R. Turner, L. Riggs and T. Spellberg, Endocrine Reviews, 15, 275 (1994).
22. S. Manolagas and R. Jilka, N.Eng.J.Med., 332, 305 (1995).
23. V. Jordan, E. Phelps and J. Lindgreen, Breast Cancer Res.Treat., 10, 31 (1987).
24. R. Love, R. Mazess, H. Barden, S. Epstein, P. Newcomb, V. Jordan, P. Carbone and D. De mets, N.Engl.J.Med., 326, 852 (1992).
25. R. Love, D. Weibe, P. Newcomb, L. Cameron, H. Leventhal, V. Jordan, J. Feyzi and D. De Mets, Ann.Int.Med., 115, 860 (1991).
26. R. Nease and J. Ross, Am.J.Med., 99, 180 (1995).
27. M. Draper, D. Flowers, W. Huster and J. Neid, 4th Int.Symp.Osteoporosis, Hong Kong, 1993, p119.
28. H. Ke, H. Simmons, C. Pirie, D. Crawford, and D. Thompson, Endocrinology, 136, 2435　　(1995).
29. M. Hasmann, B. Rattel and R. Loser, Cancer Lett., 84, 101 (1994).
30. B. Chevalier, C. Brown, P. Bruning, L. Deschenes, R. Hegg and A. Malzyner, Eur.J.Cancer, 27 Suppl.2, 555 (1991).
31. A. Wakeling and J. Bowler, J.Endocrinol., 112, R7 (1987).
32. R. Hahnel, E. Twaddle and T. Ratajczak, J.Steroid.Biochem., 4, 21 (1973).
33. N. Fanchenko, S. Sturchak, R. Shchedrina, K. Pivnitsky, E. Novikov and V. Ishkov, Acta.Endocrinol., 90, 167 (1979).
34. P. Kym, G. Anstead, K. Pinney, S. Wilson and J. Katzenellenbogen, J.Med.Chem., 26, 3910 (1993).
35. R. Kuroda, S. Cutbush, S. Neidle and O-J. Leung, J.Med.Chem., 28, 1497 (1985).
36. D. Acton, G. Hill and B. Tait, J.Med.Chem., 26, 1131 (1983).
37. T. Willson, B. Henke, T. Momtahen, P. Charifson, K. Batchelor, D. Lubahn, L. Moore, B. Oliver, H. Sauls, J. Triantafillou, S. Wolfe and P. Baer, J.Med.Chem., 37,　　1550 (1994).
38. R. Magarian, L. Overcare, S. Singh and K. Meyer, Curr.Med.Chem., 1, 61 (1994).
39. C. Gennari and D. Agnusdei, Br.J.Clin.Pharmac., 48, 196 (1994).
40. M. Zaidi and M. Pazianas in "Therapy of Osteoporosis", T. Dyson,Ed., Derwent Information, London, UK, 1995, Chp 8.
41. M. Calverley and G. Jones in "Antitumor Steroids", R. Blickenstaff, Ed., Academic Press, 1992, p193.
42. D. Bikle, Rhematic Diseases Clinics of North America, 20, 759 (1994).
43. M. Chorev, R. Dresner-Pollack, Y. Eshel and M. Rosenblatt, Biopolymers (Peptide Science), 37, 367 (1995).
44. P. Soriano, C. Montgomery, R. Geske and A. Bradley, Cell, 64, 693 (1991).

45. D. Dempster, F. Cosman, M. Parisien, V. Shen and R. Lindsay, Endocrin.Rev., 14, 690   (1993).
46. C. Tam, J. Heersche, T. Murray and J. Parsons, Endocrinology, 110, 506 (1982).
47. J. Whitfield and P. Morley, TiPS, 16, 382 (1995).
48. R. Lindsay, F. Cosman, V. Shen, J. Nieves and D. Dempster, J.Bone MineralRes., 10, S1, PS356 (1995).
49. J.L. Krstenansky, T.J. Owen, K.A. Hagaman and L.R. McLean, FEBS Lett., 242, 409 (1989).
50. B. Vickery, Z. Avnur, T. Ho and J. Krstenansky, J.Bone Min.Res., 8, Suppl 1, S141,#100 (1993).
51. V. Wray, T. Federau, W. Gronwald, H. Mayer, D. Schomburg, W. Tegge and E. Wingender, Biochem., 33, 1684 (1994).
52. E. Brown, G. Gamba, D. Riccardi, M. Lombardi, R. Butters, O. Kifor, A. Sun, M. Hediger, J. Lytton and S. Herbert, Nature, 336, 575 (1993).
53. M. Slootweg, Horm.Metab.Res., 25, 335 (1993).
54. D. O'Halloran, A. Tsatsoulis and R. Whitehouse, J.Clin.Endocrin.Met., 76, 1344 (1993).
55. E. Corpas, S. M. Harman and M. Blackman, Endocrine Rev., 14, 20 (1993).
56. A. Felix, E. Heisner and T. Mowles, Ann.Rep.Med.Chem., 20, 185 (1986).
57. W. Schoen, M. Wyvratt and R. Smith, Ann.Rep.Med.Chem., 28, 177(1993).
58. R. Smith, K. Chen, W. Schoen, S. Pong, G. Hickey, T. Jacks, B. Butler, W.W.-S. Chan, L.-Y. Chuang, F. Judith, J. Taylor, M. Wyvratt and M. Fisher, Science, 260, 1640 (1993).
59. W. Schoen, J. Pisano, K. Prendegast, M. Wyvratt, M. Fisher, K. Cheng, W. W.-S. Chan, B. Butler, R. Smith and R. Ball, J.Med.Chem., 37, 897 (1994).
60. R. DeVita, W. Schoen, M. Fisher, A. Frontier, J. Pisano, M. Wyvratt, K. Cheng, W. W.-S. Chan, B. Butler, G. Hickey, T. Jacks and R. Smith, Bioorg.Med.Chem.Lett., 4, 2249 (1994).
61. R. DeVita, W. Schoen, D. Ok, L. Barash, J. Brown, M. Fisher, M. Hodges, M. Wyvratt, K. Cheng, W. W.-S. Chan, B. Butler and R. Smith, Bioorg.Med.Chem.Lett., 4, 1807 (1994).
62. A. Patchett, R. Nargund, J. Tata, M.-H. Chen, K. Barakat, D. Johnston, K. Cheng, W. W.-S. Chan, B. Butler, G. Hickey, T. Jacks, K. Schleim, S.-S. Pong, L.-Y.P. Chuang, H. Chen, E. Frazier, K. Leung, S.-H. Chiu and R. Smith, Proc.Natl.Acad.Sci. USA, 92, 7001 (1995).
63. R. McDowell, K. Elias, M. Stanley, D. Burdick, J. Burnier, K. Chan, W. Fairbrother, R.G. Hammonds, G. Ingle, N. Jacobsen, D. Mortensen, T. Rawson, W. Won, R. Clark and T. Sommers, Proc.Natl.Acad.Sci. USA, 92, 11165 (1995).
64. B. Riggs, S. Hodgson, W. O'Fallon, E. Chiu, H. Wahner and J. Muhs, N.Eng.J.Med., 322, 802 (1990).
65. N. Mamelle, R. Dusan, J. Martin, A. Prost, P. Meunier, M. Guillaume, A. Gaucher and G. Zeigler, Lancet, 2, 361 (1988).
66. C. Pak, K. Sakhaee, B. Adams-Huet, V. Piziak, R. Peterson and J. Poindexter, Ann.Internal.Med., 123, 401 (1995).
67. W. Jee, Y. Ma, M, Li, X., Liang, B. Lin, X. Li, H-Z. Ke, S. Mori, R. Setterberg and D. Kimmel, Ernst Schering Res.Found. Workshop , 9, 119 (1994).
68. L. Tang, W. Jee, H-Z. Ke and D. Kimmel, J.Bone Miner.Res., 7, 1093 (1992).
69. K. Pierce, D. Gil, D. Woodward and J. Regan, TiPS, 16, 253 (1995).
70. H. Reddi, Curr.Opin.Biol., 4, 850 (1992).
71. J. Wozney and V. Rosen, Handbook Exp.Pharm., 107, 723 (1992).
72. P. ten Dijke, H. Yamashita, T.K. Sampath, H, Reddi, M. Estevez, D. Riddle, H. Ichijo, C-H. Heldin and K. Miyazono, J.Biol.Chem., 269, 16985, (1994).
73. D. Kingsley, Genes and Develop., 8, 133 (1994).
74. S. Vukicevic, A. Stavljenic and M. Pecina, Eur. J. Clin.Chem.Clin.Biochem., 33, 661   (1995).
75.   H.C. Anderson, Curr.Opin.Ther.Patents, 4, 17 (1994).
76. A. Yasko, J. Lane, E. Fellingen, V. Rosen and J. Wozney, J.Bone.Jt.Surg., 74A, 659 (1992).
77. D. Griffith, P. Keck, K. Sampath, D. Rueger and W. Carlson, Proc.Natl.Acad.Sci. USA, 93, 878 (1996).

Chapter 23. Nitric Oxide Synthase Inhibitors

James E. Macdonald
Astra Arcus USA
Rochester, NY

Several reviews are available on the activity and role of the isozymes of nitric oxide synthase (NOS) and the nitric oxide (NO) they produce (1,2). The medicinal chemistry of inhibitors of NOS has also been examined (3,4,5). Reviews of the role of NO and NOS in specific conditions are also available (6,7).

Nitric oxide synthase is a family of three heme enzymes of mammalian systems that produce NO by the oxygen-dependent oxidation of a terminal nitrogen on the guanidine of the amino acid L-arginine (L-Arg) (1,2). This oxidation reaction requires calmodulin, NADPH, tetrahydrobiopterin, FMN and FAD as cofactors. The three isozymes are differentially expressed and play different physiological roles (5). The neuronal expressed NOS (nNOS, NOS-I) shows a dependence on calcium and calmodulin, and is also expressed in skeletal muscle, pancreatic islet cells and kidney macula densa cells. The endothelial NOS (eNOS, NOS-III) also shows a dependence on calcium and calmodulin, and plays a major role in the control of blood pressure. The third form of NOS is the inducible NOS (iNOS, NOS-II), and it is unique in that its expression is induced by inflammatory stimuli, and its catalytic activity does not require elevation of intracellular calcium levels.

The activity of the NOS inhibitors has been reported by various authors as Ki and /or IC_{50}'s. The IC_{50}'s are dependent on the concentration of L-Arg in the assay media, and range from a low of 0.3 µM to a high of 120 µM. Whenever possible, comparative data for standard inhibitors such as nitroarginine 1 are included for comparison. Furthermore, the species source of the NOS isozyme is important, and comparisons between species have shown differences, particularly between the mouse and human iNOS (m-iNOS and h-iNOS).

AMINO ACID INHIBITORS

Modification of an enzyme's substrate is a well tested strategy to discover inhibitors of the enzyme, and has been successful in the discovery of new inhibitors of NOS. The first inhibitors of NOS were L-Arg derivatives in which the terminal guanidine of the arginine was modified to provide nitroarginine 1 (L-NNA) and methylarginine 2 (L-NMA). Although these compounds continue to be useful in the study of the function of NOS, their lack of selectivity among the isozymes has driven the search for more selective and potent agents including compounds in which L-Arg has been modified to provide inhibitors with some selective advantage.

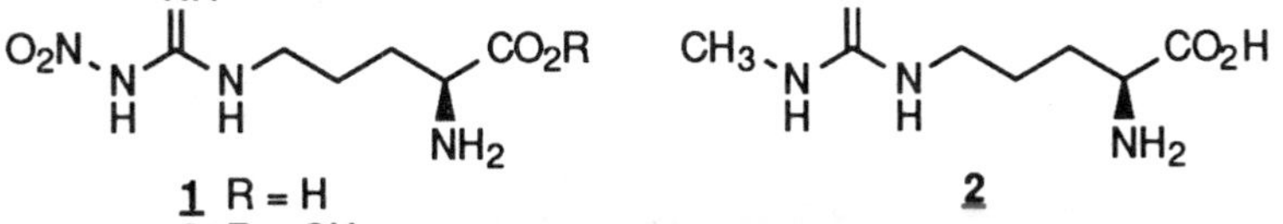

1 R = H
3 R = CH₃

2

A derivative of 1 that has found wide use as an inhibitor of NOS is N^{G}-nitro L-arginine methyl ester 3 (L-NAME). L-NAME, however, binds muscarinic receptors as well as inhibiting all isozymes of NOS (8). Evidence for an NO independent action of L-NAME was a antagonistic rightward shift in the response to acetylcholine in denuded endothelium with L-NAME that was not reversed by L-Arg. L-NMA did not produce a similar rightward shift of this acetylcholine response. As L-NAME has often been employed at up to 100 µM, adverse and confounding effects from muscarinic antagonist activity may be expected in tissues expressing muscarinic receptors.

An L-Arg derivative that modified the guanidine to an acetamidine and lengthened the chain to that of Lys provided L-N^6-(1-iminoethyl)lysine **4** (L-NIL) which is a selective inhibitor of iNOS (9). It is clear that **4** has a significant advantage of potency for the m-iNOS over the rat nNOS (r-nNOS) in comparison with **1** and **2** and the other amidine derivatives, L-N^5-(1-iminoethyl)ornithine **5** (n = 1, L-NIO) and D,L-N^7-(1-iminoethyl)homolysine **6** (n = 3) with shorter and longer amino acid chains.

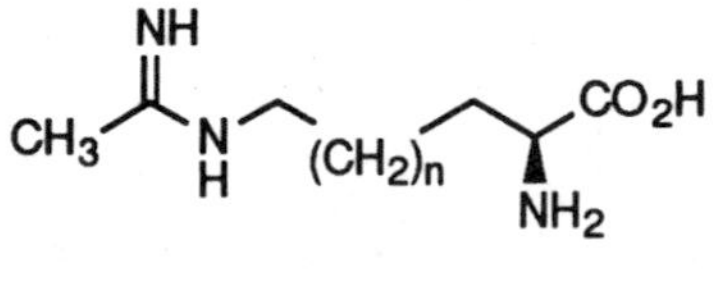

	n	r-nNOS IC$_{50}$ µM	m-iNOS IC$_{50}$ µM	r-nNOS/ m-iNOS
4	2	92	3.3	28
5	1	3.9	2.2	1.8
6	3	808	73	11
2		8.3	18	0.5
1		0.5	20	0.025

(30 µM L-Arg)

Other modifications of the guanidine of L-Arg have also successfully provided new inhibitors of NOS. A stereospecific, heme-binding inhibitor of NOSs has been obtained by replacing one of the terminal nitrogens of the guanidine of L-Arg with sulfur to provide the thiourea, L-thiocitrulline, **7** (10). L-Thiocitrulline is a potent and reversible inhibitor of both r-nNOS, Ki = 0.06 µM and r-iNOS, Ki = 3.6 µM, and is competitive with L-Arg in both forms of NOS. L-Thiocitrulline gives a "type II" difference spectrum, suggesting that **7** contributes a sixth ligand to the heme iron. Further modification of **7** was achieved by its alkylation with iodoalkanes to form the S-alkyl L-thiocitrullines (11). These isothiourea derivatives are potent inhibitors of r-nNOS and r-iNOS, with the S-methyl L-thiocitrulline **8** inhibiting rat nNOS with an IC$_{50}$ of 1.05 µM in an assay media containing 20 µM arginine. The inhibition by S-alkyl L-thiocitrullines is not very selective between rat nNOS and rat iNOS. This potency was comparable to the non-amino acid S-methylisothiourea **12**.

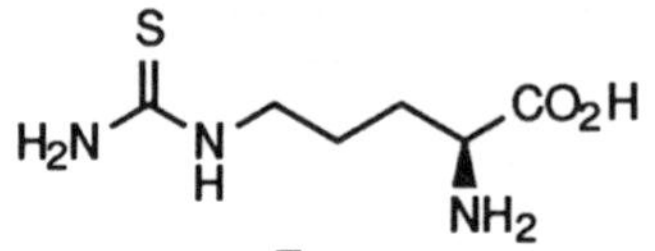

The Ki for S-alkyl L-thiocitrullines were determined for both r-NOS forms (11). S-methyl L-thiocitrulline **8** is significantly selective for r-iNOS. These compounds are reversible inhibitors, competitive with L-Arg, and are stereoselective. They differ from thiocitrulline, in that their binding to NOS does not involve direct interaction with the iron of the heme, as determined by spectral studies of the interaction. The activity of both **8** and **9** has also been examined for the human isoforms of NOS and found to be time-dependent, slowly reversible, and competitive with L-Arg (12).

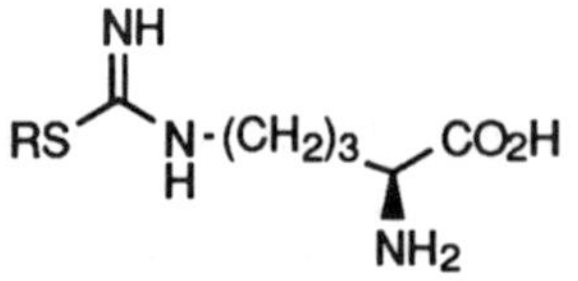

		r-nNOS IC$_{50}$ µM	r-iNOS IC$_{50}$ µM
8	R = CH$_3$	1.05	2.2
9	R = C$_2$H$_5$	1.56	1.56
10	R = C$_3$H$_7$	3.7	2.5
11	R = C$_4$H$_9$	43	54
12	CH$_3$SC(=NH)NH$_2$	1.1	0.63
1	L-NNA	0.18	6

(20 µM L-Arg)

	r-iNOS Ki μM	r-nNOS Ki μM	h-nNOS Ki μM	h-iNOS Ki μM	h-eNOS Ki μM
8	0.05	0.84	0.0012	0.040	0.011
9	0.17	0.41	0.0005	0.020	0.024
10	0.4	0.66			
11	4.6	14			

Although no enzymatic measurements were presented for the endothelial form of rat NOS, the treatment of anesthetized rats with **8**-**10** at 20 mg/kg iv resulted in a dramatic rise in blood pressure up to 75 mm Hg (11). Furthermore, **8** at 20 mg/kg iv was effective in restoring blood pressure in a canine model of LPS-mediated cytokine shock. The activity **8** and **9** were compared in rat brain slices, and rat aortic ring relaxation, and the results were compared with **1** and **2** (12). In these whole cell preparations, **8** was found to have the best selectivity based on the IC_{50}'s for nNOS versus eNOS in rat tissues.

	Rat Brain Slice IC_{50} μM	Rat aortic ring IC_{50} μM
8	0.31	5.4
9	1.2	7
1	0.75	1.2
2	4.1	11

Analogs of **5** with an intervening heteroatom with good enzymatic selectivity for iNOS and nNOS and weaker activity at eNOS have been reported (13). The potent and selective compounds appeared to be the 2-amino-6-(1-imino-2-fluoroethylamino)-4,4-dioxo-4-thiahexanoic acid, **13**, and its non fluorinated analog, **14**. **13** was strongly time dependent, with a Kd of 0.1 μM. The activity of **13** and **14** were examined in rat aorta, with intact endothelium to examine eNOS, or LPS induced endothelium-denuded rings to examine iNOS. In this tissue preparation, **13** showed spectacular selectivity, lacking <8% effect on the normal eNOS mediated relaxation at 300 μM, but blocking the iNOS response at 1.7 μM.. The arginine derivative Nε-iminoethyl-aminomethylselenocysteine **15** has been reported to have moderate activity and selectivity for the mouse iNOS, and good activity in a whole cell assay, but poor activity versus human iNOS (14).

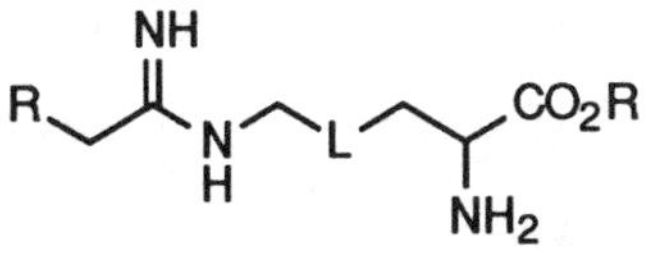

			h-iNOS Ki μM	h-eNOS Ki μM	h-nNOS Ki μM	Rat aorta iNOS IC_{50} μM	Rat aorta eNOS IC_{50} μM
13	R = F	L = SO$_2$	1.9	23	2.6	1.7	>>300
14	R = H	L = SO$_2$	2.9	79	18	1.5	101
			m-iNOS IC_{50} μM	h-iNOS IC_{50} μM	r-nNOS IC_{50} μM	Raw Cell IC_{50} μM	
15	R = H	L = Se	5.0	36	229	0.37	
	(30 μM L-Arg)						

Derivatives of nitroarginine **16** and **17** in which a cis olefin has been introduced and the carboxylate reduced to an alcohol show reasonable activity for the rat nNOS and some selectivity over the other NOS isoforms (15). Other patents report derivatives of nitroarginine that are instructive in that they are weakly active or show poor whole cell activity (16).

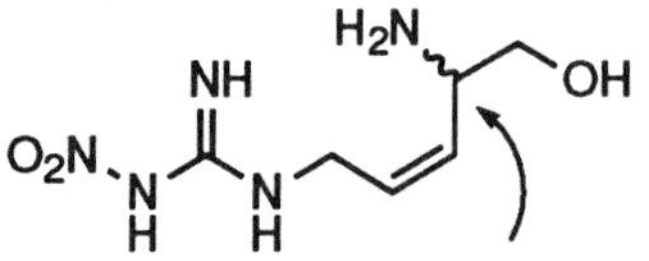

	r-nNOS $IC_{50} \mu M$	m-iNOS $IC_{50} \mu M$	b-eNOS $IC_{50} \mu M$
16	3	75	72
17	5	60	
10 µM L-Arg			

16 (R)
17 (S)

NON-AMINO ACID INHIBITORS

Isothioureas - Non-amino acid isothioureas are potent and in some cases selective inhibitors of human NOS isozymes (17). Simple lower alkyl isothioureas **18** and **19** are potent and competitive with L-Arg, yet not greatly selective. Larger (**20**) or longer (**21**) groups lead to a large loss in the potency against all NOS isozymes. The S-(2-phenylethyl)isothiourea **23** was much more effective than the benzyl (**22**) or phenylpropyl (**24**) analogs. Substitution on nitrogen of the isothiourea with one or two methyl groups resulted in a dramatic reduction of the activity.

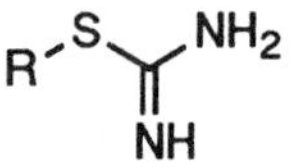

		h-iNOS $Ki \mu M$	h-eNOS $Ki \mu M$	h-nNOS $Ki \mu M$
12	R = CH_3	0.12	0.2	0.16
18	C_2H_5	0.019	0.039	0.029
19	$CH(CH_3)_2$	0.0098	0.022	0.037
20	$C(CH_3)_3$	0.24	1.2	0.62
21	$CH_2CH_2CH_3$	0.24	0.67	0.63
22	CH_2Ph	5.6	23	14
23	CH_2CH_2Ph	0.50	1.9	0.80
24	$CH_2CH_2CH_2Ph$	9.7	17	14

Bis-isothioureas are very selective and potent inhibitors of the human iNOS (17). The compounds were competitive with L-Arg, and as in the monoisothioureas, the bis-isothioureas showed a strong preference for an ethyl spacer from the central phenyl ring to the isothiourea. The one and three carbon spacers gave compounds with greatly reduced activity. The activity of the most potent and selective compounds, **18, 25,** and **26** were examined in a cultured cell assay of cytokine-induced NO production in human cells (17). It was found that the compounds were significantly less potent than the enzymatic Ki's, suggesting that they penetrated cells poorly.

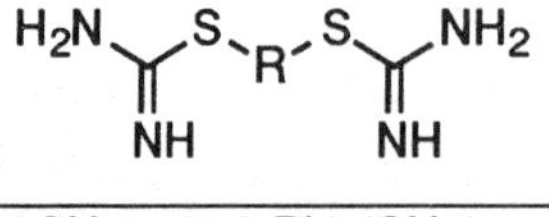

		h-iNOS $Ki \mu M$	h-eNOS $Ki \mu M$	h-nNOS $Ki \mu M$
25	-$(CH_2)_2$-(1,3-Ph)-$(CH_2)_2$-	0.047	9.0	0.25
26	-$(CH_2)_2$-(1,4-Ph)-$(CH_2)_2$-	0.0074	0.36	0.016
27	-(CH_2)-(1,4-Ph)-(CH_2)-	0.2	2.9	0.1
28	-$(CH_2)_3$-(1,3-Ph)-$(CH_2)_3$-	3.8	39	4.8

The NOS inhibitory activity of a series of N-substituted phenyl S-alkyl isothioureas has been described (18). The most potent inhibitors of nNOS are S-ethyl-N-(2-methoxyphenyl)isothiourea, **29**, and S-ethyl-N-(2-chlorophenyl)isothiourea, **30**. The compounds show reasonable potency and selectivity for the nNOS.

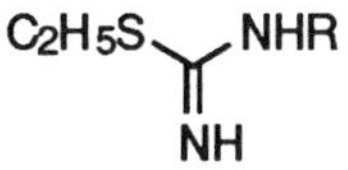

C_2H_5S NHR	h-iNOS Ki μM	h-eNOS Ki μM	h-nNOS Ki μM
29 R = N-(2-methoxyphenyl)	1.4	1.6	0.17
30 R = N-(2-chlorophenyl)	2.9	1.8	0.17
31 R = N-(4-phenoxyphenyl)	3.1	4.0	0.19
(1.0 μM L-Arg)			

The potent and selective inhibition of iNOS by 2-amino-5,6-dihydro-6-methyl-4H-1,3-thiazine (AMT) **32**, and ethyl isothiourea **18** have been reported (19). The inhibition by both compounds was competitive with L-Arg and reversible. These compounds were effective and reversible in a whole cell assay of iNOS inhibition. Interestingly , **32** had been previously reported to be an analgesic (20).

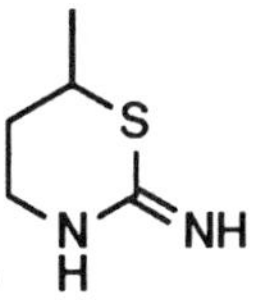

	r-nNOS IC$_{50}$ μM	m-iNOS IC$_{50}$ μM	b-eNOS Ki μM	J774 whole cell Ki μM
1	0.55	7.71	3.78	27
18	0.25	0.013	0.37	0.082
32	0.011	0.0036	0.15	0.015
(10 μM L-Arg)				

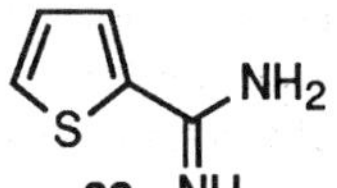

Amidines-The activity of simple amidines as inhibitors of NOS has been described (21). The compounds in general showed little discrimination between the r-nNOS and RAW 264.7 celll iNOS. The most potent compound was the 2-thiopheneamidine, **33**, with an iNOS IC$_{50}$ of 1.0 μM and a nNOS IC$_{50}$ of 1.3 μM. in an assay media containing 30 μM arginine.

Cyclic amidines have also been disclosed, with some potent compounds showing moderate to over 100 fold selectivity for iNOS versus nNOS and have demonstrated ability to penetrate cells in a whole cell assay (22). The iNOS was from LPS treated RAW 264.7 mouse cells, and the nNOS was from rat brain. The small cyclic amidines **34** and **35** were the most potent enzyme inhibitors, but the amidines **36, 37,** and **38** with longer side chains and greater lipophilicity showed better activity in the whole cell assay. The introduction of sulfur or oxygen into the cyclic amidine in the place of (CH$_2$)$_n$ provided **39, 40** and **41** with similar potency and selectivity (23).

The inhibition of nitrite production from induced mouse j774 cells and rat aortic smooth muscle cells in culture by a variety of amidines, guanidines and pyridines was examined (24). The most effective amidine was 2-iminopiperidine, **39**, which showed a EC$_{50}$ of 10 μM for the J774 cells, and 19 μM in the rat induced smooth muscle cell culture. Several other compounds showed weaker but measurable activity in the assays. When the arginine levels were increased from 0.3 μM to 30 μM, the EC$_{50}$ value of **39** showed a substantial drop in potency to 480 μM, suggesting that the amidines were competitive inhibitors at the arginine binding site. They also reported that **39** increased blood pressure in anesthetized rats (0.1, 1 and 10 mg/kg iv) and that the increase was equal or greater than that seen with L-NNA at the same dose levels.

The potential for selective nNOS activity of N-aryl amidines similar in structure to the isothioureas **29-31** but replacing the ethylthio function with a small heterocycle is suggested by several other publications, but biological data is lacking (25,26).

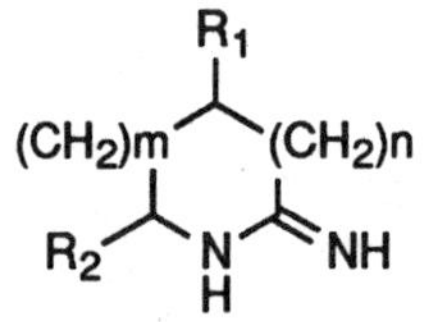

$$R_1$$
$$(CH_2)m \qquad (CH_2)n$$
$$R_2 \qquad N \qquad NH$$
$$H$$

		iNOS IC_{50} μM	nNOS IC_{50} μM	RAW Cell IC_{50} μM
34	n = m=1, R_1=CH$_3$, R_2= H	0.055	0.285	
35	n=m=1, R_1=CH$_3$, R_2= CH$_3$	0.043	0.127	
36	n=2, m=1, R_1=H, R_2= (CH$_2$)$_2$CH$_3$	0.081	7.8	0.55
37	n=2, m=1, R_1=H, R_2= (CH$_2$)$_3$CH$_3$	0.55	106	0.65
38	n=1, m=0, R_1= CH$_3$, R_2= CH$_2$CH$_3$	0.17	1.6	0.25
		h-iNOS μM	h-eNOS μM	h-nNOS μM
39	n = m = 1, R_1=H, R_2= H	1.0	4.7	1.1
40	n = O, m = 1, R_1=H, R_2= H	2.9	7.1	3.2
41	n = S, m = 1, R_1=H, R_2= H	1.8	8.2	3.7

30 μM L-Arg

Guanidines - Guanidines substituted with small alkyl groups are weak inhibitors of NOS, but still define a minimum structure required for recognition at the binding site for amidines and guanidines of the NOS isoforms. Ethylguanidine **42** has been shown to be weak inhibitor of iNOS with a EC_{50} of 120 μM for the inhibition of iNOS in the mouse and rat whole cell (21). Methylguanidine **43** inhibits mouse iNOS with an IC_{50} of 200 μM, and nNOS with an IC_{50} of 500 μM (27).

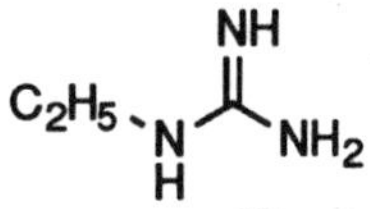

NH   NH   HO$_2$C   NH

C_2H_5–N–NH$_2$ CH_3–N–NH$_2$ HO$_2$C   N–NH$_2$

H   H   H

42 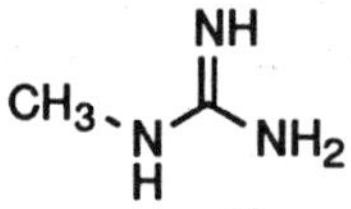   **43** 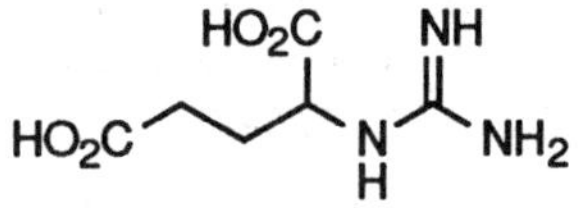   **44**

α-Guanidinoglutaric acid (GGA, **44**) has been shown to be a NOS inhibitor (28). GGA is normally present in cerebral cortex at low levels, but increases in the cobalt induced epileptic focus. GGA has been shown to be an inhibitor of nNOS with a Ki of 2.7 μM, similar in potency to L-NMA. The inhibition by GGA was reversible by increasing concentrations of L-Arg. ICV injection of 10 μL of a 100 μM GGA caused sporadic spike discharges in rats, and the response was blocked by addition of 100 μM L-Arg to the injection solution. This result suggests that nNOS inhibition may play a role in development of seizure activity.

Other inhibitors- A number of nitrogen containing heterocycles have been reported as NOS inhibitors. Imidazole **45** is an L-Arg competitive inhibitor of porcine b-NOS (29). The effects on formation of L-citrulline from L-Arg were examined along with the formation of H_2O_2 in absence of L-Arg and the reduction of cytochrome C. Imidazole had an IC_{50} of 6.2 mM for the inhibition of citrulline formation with 100μM L-Arg. Imidazole was more effective in preventing the formation of H_2O_2, with an IC_{50} of 83 μM. The arginine dependence of imidazole inhibition of nNOS showed a purely competitive relationship and a Ki for imidazole of 390 μM. Imidazole also inhibited the binding of tritiated L-NNA to NOS with a Ki of 198 μM. Imidazole showed weak effects on [³H] H$_4$biopterin binding, with 100 μM imidazole reducing H$_4$biopterin binding by 50%.

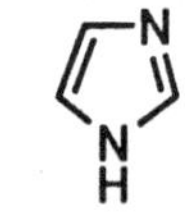

45

Biopterin-site directed NOS inhibitors are also selective nNOS inhibitors (30). The selective nNOS inhibitor activity of 5-aryl substituted 2,4-diaminopyrimidines is presented in an assay with brain cytosol NOS. Whole cell NOS inhibitory activity in brain slices stimulated with veratrine was also measured. The most potent compounds were the 2,4-diamino-5-(3,4-cholrophenyl)pyrimidine **46** and the 2,4-diamino-5-(4-methoxyphenyl)pyrimidine **47**, both with an IC_{50} of 0.09 μM. Claims to the use of 7-nitroindazole **48** (7-NI) as a selective inhibitor of nNOS are also reported. Pteridine derivatives have also been reported to inhibit NOS, but with little supporting data (31).

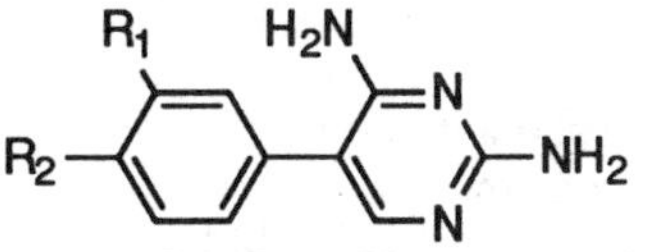

46 R_1, R_2 = Cl
47 R_1 = H; R_2 = OCH$_3$

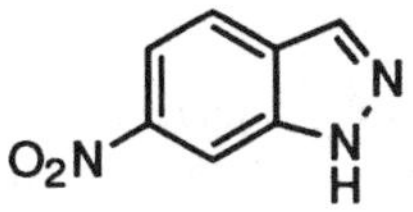

48

7-NI has antinociceptive effects as well as its activity as a NOS inhibitor (32). Mice treated with 10-50 mg/kg ip of 7-NI rapidly developed an antinociceptive effect in a formalin hind paw lick model, and an acetic acid induced abdominal constriction model. At 25 mg/kg of 7-NI, the mice had reduced nNOS activity in cerebella homogenates that correlated with the observed antinociception. 7-NI given at 25 or 80 mg/kg ip had no effect on mean arterial pressure in urethane anaesthetized mice.

The molecular mechanisms of inhibition of porcine nNOS by 7-NI were investigated and compared with imidazole (33). 7-NI was competitive with L-Arg, had a Ki of 2.8 μM, and also blocked NOS mediated H_2O_2 formation with and IC_{50} of 0.28 μM. This is similar but much more potent than the results for imidazole. 7-NI antagonized the binding of tritiated L-NNA with a Ki of 0.09 μM, and the binding of radiolabeled H_4biopterin was also similarly antagonized with a Ki of 0.12μM. suggesting that the arginine site and the biopterin site are functionally linked.

The monosodium salt of 7-NI (7-NINA) was prepared in an attempt to improve on the poorly soluble 7-NI (34). The 7-NINA was then employed in microdialysis studies. Microdialysis perfusion with 1 mM 7-NINA rapidly increased striatal free dopamine to 203 to 229% of baseline without raising the levels of the dopamine metabolites 3,4-dihydroxyphenylacetic acid or homovanillic acid. This increase in striatal free dopamine was diminished by coperfusion with 1 mM L-arginine.

The effect of 1-(2-trifluoromethylphenyl)imidazole (TRIM), **49**, on the isoforms of NOS has also been examined (35). TRIM was very weak versus eNOS, and equipotent for iNOS and nNOS, but weaker than L-NNA. TRIM was examined in late phase formalin hind paw licking and showed an ED_{50} of 16.7mg/kg. This effect of TRIM was partially reversed by L-Arg.

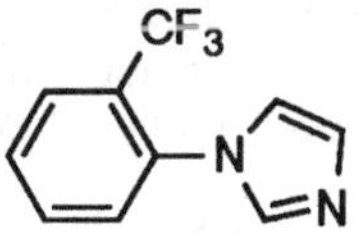

49

		mouse iNOS IC_{50} μM	Bovine eNOS IC_{50} μM	Mouse nNOS IC_{50} μM
49	TRIM	27	1057	28
1	L-nitroarginine	10.6	6.5	0.66

The calmodulin inhibitors calmidazolium, **50**, and W-7, **51**, can block the effect of acetylcholine on endothelium dependent relaxation in rat aorta.(36). Calmidazolium inhibited at 3 μM and blocked totally at 10 μM. The effect of W-7 was similar but weaker, with inhibition at 30 μM, and blocked totally at 100 μM. This activity at the eNOS was selective as **50** had no effect on the IL-1β induced iNOS relaxation at the doses effective for inhibiting eNOS. In normal aorta **50** at 10 μM blocked the formation of c-GMP, and that **50** did not block the formation of c-GMP in endothelium free IL-1β induced arteries.

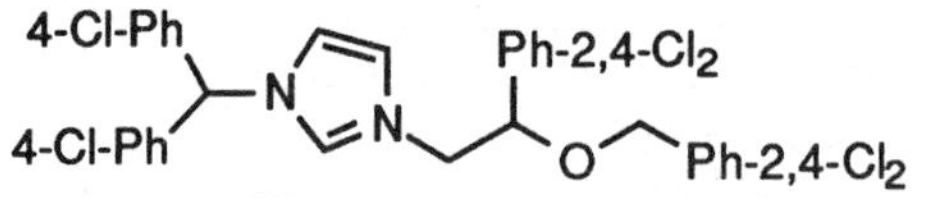

4-Cl-Ph
4-Cl-Ph   **50**   Ph-2,4-Cl₂
                                 Ph-2,4-Cl₂

$SO_2NH(CH_2)_6NH_2$

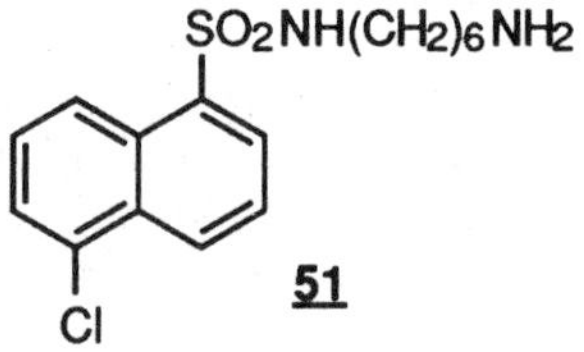

51

Cl

Phencyclidine **52** a psychotomimetic agent and drug of abuse, is a suicide inhibitor of brain NOS (37). It gives a weak initial reversible complex with a Ki of 4.9 µM, and irreversible inhibition with a k_{inact} of 0.3 min^{-1} in rat brain cytosol. The irreversible inhibition was dependent on the presence of NADPH, and the NOS enzyme was protected by L-Arg but not by D-Arg. The high concentration required for the formation of the initial weak complex, and the protection by L-Arg suggest that NOS inhibition plays little role in the effects of **52**.

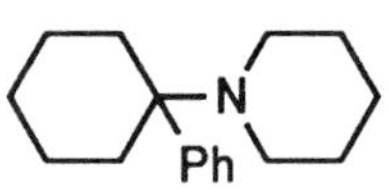

52

Chlorpromazine **53** has been described as effective in preventing endotoxic shock, which prompted investigation into **53** for effects on NOS (38). **53** at 100 µM gave a 54% reduction of citrulline production from r-nNOS, and at 12 mg/kg **53** reduced the ex vivo production of citrulline from nNOS by 25%. **53** was not effective against iNOS at 100 µM. LPS-induced iNOS production in the rat lung was reduced 90% by **53** at 12 mg/kg 6 hours after LPS instillation. In mice, 4 mg/kg of **53** prevented death from LPS, reduced citrulline production by 88% in the lung, and reduced plasma nitrite levels as well. Blockade of the induction of iNOS would be an attractive an alternate means of treating iNOS mediated disease states, but **53** is certainly not an ideal agent.

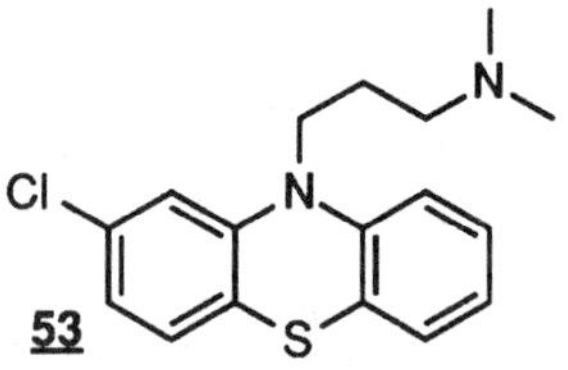

53

1,2-Diaminobenzimidazoles were found to be selective inhibitors of NOS (39). The most selective compound was the 2-amino-3-methylaminobenzimidazole **54** (r-nNOS IC$_{50}$ = 5.8 µM) which had no significant activity at 100 µM against mouse or human iNOS or human eNOS. The structure and neuronal selectivity suggest a relationship to the biopterin site inhibitors.

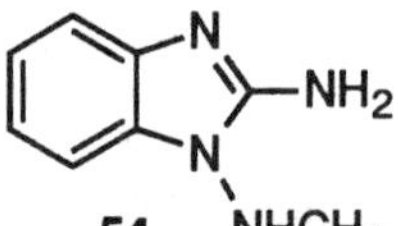

54   NHCH₃

Neuronal Ischemia- Several reviews of the effects of NOS inhibition on cerebral blood flow and in cerebral ischemia models have appeared (6,7). The results are suggestive that partial NOS inhibition after ischemia will be effective, but predosing with high doses of NOS inhibitors will be ineffective. This has been shown with **1**, which has a narrow therapeutic window in a model of transient ischemic injury (40).

As nNOS selective inhibitors have been discovered, they have been examined in models of stroke. The NOS inhibitor, 7-NI, dosed after ischemia decreased focal infarct volume (41). The dose (25 mg/kg ip) provides a 50% inhibition of nNOS by an ex vivo assay, suggestive that total block of nNOS is not needed to give protection. The nNOS selective **8** was also protective but a narrow dose window was observed (42). The level of nNOS activity in brain microvessels was found to rise almost 6 fold within 2 hours of focal ischemia and 2 hours of reperfusion, and has been suggested to play a role in the developing injury (6). Finally, mice deficient nNOS had smaller infarcts than the controls (43). These results suggest that nNOS inhibitors will be effective in the treatment of stroke, but the narrow dose windows may lead to difficult clinical development.

Septic Shock - iNOS inhibitors in septic shock to control hypootension has been the subject of intense investigation (44,45). The reports to date suggest a moderate doses of iNOS selective inhibitors can reverse the fall in blood pressure, but high doses appear to have an adverse effect on mortality. For example, the isothioureas **18** and **32** were examined in a rat model of LPS mediated cytokine shock (46). It was shown that in rats induced with 5 mg/kg LPS, both **18** and **32** reduced nitrite production with IC_{50} doses of 0.4 and 0.2 µM, respectively. Unfortunately, this treatment also resulted in a significant rise in mortality in the treated rats at higher doses with a 75% mortality in the rats given **32** (4 mg/kg) and 1/8 at 0.2 mg/kg, and 75% for **18** at 20 mg/kg. **32** and **18** were not lethal in normal rats. The observed mortality in LPS-treated rats administered **18** (4 mg/kg) was reduced to 1/8 by pretreatment with 300 mg/kg L-Arg but not by D-Arg. This pretreatment with L-Arg did not increase plasma nitrate levels. Concurrent administration of LPS and dexamethasone at 1 mg/kg to block iNOS induction, followed by **18** did not significantly reduce the toxicity seen with this compound in LPS treated animals. The toxicity may be due to nonselective inhibition of other isozymes of NOS.

In contrast to the above results, related work showed improved survival in rodent models of septic shock with the closely related **12** HSO_4^- (47). The studies differ in that the species is changed to mouse, the LPS dose is raised to 60 mg/kg, and **12** was administered at 1 mg/kg. Higher doses were not reported. With these conditions survival is improved at 24 hours. **12** showed a greater presser response in LPS treated animals than in controls, suggesting it is selective for iNOS in vivo as well as in vitro.

The contribution of iNOS to shock has also been investigated through the development of iNOS knock outs, in which the gene for iNOS has been eliminated. The initial report suggested that the iNOS knock out had improved survival in a LPS shock model (48). Further investigation has shown that in general the iNOS knock outs are not protected in septic shock models and show poorer survival (49,50). These results suggest that complete inhibition of iNOS may be counterproductive in septic shock.

In summary, the scope of inhibitors for iNOS and nNOS are rapidly increasing, with many new structural types being reported over the last few years. The need for isoform selectivity is still critical, and safe inhibitors that show higher selectivity and good cell penetration are still needed to delineate the role of the isoforms of NOS in disease models and in the clinic.

REFERENCES

1.   S.H. Snyder and D.S. Bredt, Sci.Am. **266**, 68, 74 (1992).
2   P.L. Feldman, O.W. Griffith and D.J. Stuehr, Chem.Eng.News. Dec. 20 (1993).
3.   J.F. Kerwin and M. Heller, Med.Res.Rev. **14**, 23 (1994).
4.   M.A. Marletta, J.Med.Chem. **37**, 1899, (1994).
5.   J.F. Kerwin, J.R. Lancaster and P.L. Feldman, J.Med.Chem. **38**, 4343 (1995).
6.   T. Nagafuji, M. Sugiyama, T. Matsui, A. Muto and S. Naito, Molec.Chem.Neuropath. **26**, 107 (1995).
7.   C. Iadecola, D.A. Pelligrino, M.A. Moskowitz and N.A. Lassen, J. Cereb.Blood Flow Metab. **14**, 175 (1994)
8.   I.L. Buxton, D.J. Cheek, D. Eckman, D.P. Westfall, K.M. Sanders and K.D. Keef, Circ.Res. **72**, 387 (1993).
9.   W.M. Moore, R. K. Webber, G. M. Jerome, F. S. Tjoeng, T. P. Misko and M.G. Currie, J.Med.Chem. **37**, 3886 (1994).
10. C. Frey, K. Narayanan, K. McMillan, L. Spack, S.S Gross, B.S. Masters and O.W. Griffith, J.Biol.Chem. **269**, 26083 (1994)

11. K. Narayanan, L. Spack, K. McMillan, R.G. Kilbourn, H.A. Hayward, B.S. Masters, O.W. Griffith, J.Biol.Chem. 270, 11103 (1995).
12. E.S. Furfine, M.F. Harmon, J.E. Paith, G.R. Knowles, M. Salter, R.J. Kiff, C. Duffy, R. Hazelwood, J.A. Oplinger and E.P. Garvey, J.Biol.Chem. 269, 26677 (1994).
13. H. F. Hodson, R.M.J. Palmer, D.A. Sawyer,R.G. Knowles,K.W. Franzmann, M.J. Drysdale, S. Smith, P.I.Davies, H.A.R.Clark and B.G.Shearer, PCT Pat.Applic. WO95/34534 (1995).
14. M.G.Currie, K. Webber, F.S. Tjoeng and K.F. Fok, PCT Pat.Applic. WO 95/25717 (1995).
15. F. Murad, J.F. Kerwin and L.D. Gorsky, PCT Pat.Applic. WO 95/28377 and US 5,380, 945 (1995).
16. EA. Hallinan, F.O. Tjoeng, K.F. Fok, T.J. Hagen, M.V. Toth, S. Tsymbalov and B.S. Pitzele, PCT Pat.Applic. WO 95/24382, (1995).
17. E.P. Garvey, J.A. Oplinger, G.J. Tanoury, P.A. Sherman. M. Fowler, S. Marshall, M.F. Harmon, J.E. Paith and E.S. Furfine, J.Biol.Chem. 269, 26669 (1994).
18. J.A. Oplinger, B.S. Shearer, E. Bigham, E.S. Furfine and E. P.Garvey, PCT Pat.Applic. WO95/09619 (1995).
19. M. Nakane, V. Klinghofer, J.E. Kuk, J.L. Donnelly, G.P. Budzik, J.S. Polloc, F. Basha and G.W. Carter, Molec.Pharm. 47, 831 (1995).
20. L.L. Skaletzky and R.J. Collins, US 3,169,090(1965).
21. F.S. Tjoeng, K.F. Fok and R.K. Webber, PCT Pat. Applic. WO95/11014 (1995).
22. D.W. Hansen Jr, M.G.Currie, E.A. Hallinan, K.F. Fok, J. Timothy, A.A. Bergmanis, S.W. Kramer, L.F. Lee, S. Metz, W.M. Moore, K.B. Peterson, B.S. Pitzele, D.P. Spangler, R.K. Webber, M.V. Toth, M. Trivedi and F.S.Tjoeng, PCT Pat.Applic. WO95/11231 (1995).
23. W.M. Moore, R.K. Webber, K. F. Fok, G.M. Jerome, J.R.Conner, P.T. Manning, P.S. Wyatt, T.P.Misko, F.s. Tjoeng and M.G.Currie, J.Med.Chem. 39, 669 (1996).
24. G.J. Southan, C. Szabo, M.P. O'Connor, A.L. Salzman and C. Thiemermann, Eur.J.Pharm.Molec.Pharm., Sect. 291, 311 (1995).
25. R.J.Gentile, R.J. Murray, J.E.Macdonald and W.C. Shakespeare, PCT Pat.Applic. WO95/05363 (1995).
26. J. E. Macdonald, W.C.Shakespeare, R.J.Murray and J.R.Matz, PCT Pat.Applic.WO 96/01817 (1996).
27. R.G. Tilton, K. Chang, K.S. Hasan, S.R. Smith, J.M. Petrash, T.P. Misko, W.M. Moore, M.G. Currie, J.A. Corbett and M.L. McDaniel, Diabetes 42, 221 (1993).
28. I. Yokoi, H. Kabuto, H. Habu and A. Mori, J.Neurochem. 63, 1565 (1994).
29. B. Mayer, P. Klatt, E.R. Werner and K. Schmidt, FEBSLett. 350, 199 (1994).
30. E.C. Bigham, J.F. Reinhard Jr, P.K. Moore, R. C. Babbedge, R. G. Knowles, M.S. Nobbs and D. Bull, PCT Pat. Applic. WO94/14780 (1994).
31. W. Pfleiderer, H. Schmidt and R. Henning, Pat.Applic. DE4418096-A1; DE 4418097-A1.
32. P.K. Moore, P. Wallace, Z. Gaffen, S.L. Hart and R.C. Babbedge, Brit.J.Pharm. 110, 219 (1993).
33. B. Mayer, P. Klatt, E.R. Werner and K. Schmidt, Neuropharm. 33, 1253 (1994).
34. M.T. Silva, S. Rose, J.G. Hindmarsh, G. Aislaitner, J.W. Gorrod, P.K. Moore, P. Jenner and C.D. Marsden, Brit.J.Pharm. 114, 257 (1995).
35. R.L.C. Handy, P. Wallace, Z. Gaffen and P.K. Moore, Brit.J.Pharm. 116 (Proc.Suppl), 446 (1995).
36. V.B. Schini and P.M. Vanhoutte, J.Pharm.Exp.Ther. 261, 553 (1992).
37. Y. Osawa and J.C. Davila, Biochem.Biophys.Res.Comm. 194, 1435 (1993).
38. M. Palacios, J. Padron, L. Glaria, A. Rojas, R. Delgado, R. Knowles and S. Moncada, Biochem.Biophys.Res.Comms. 196, 280 (1993).
39. P. Hamley and A.C. Tinker, Bioorg.Med.Chem.Lett. 5, 1573 (1995).
40. T. Nagafuji, M. Sugiyama, T. Matsui and T. Koide, Eur.J.Pharm.-EnvironTox.Pharm.Sect. 248, 325 (1993).
41. T. Yoshida, V. Limmroth, K. Irikura and M.A. Moskowitz, J.Cereb.BloodFlowMetab. 14, 924 (1994).
42. T.Nagafuji, M. Sugiyama, A. Muto, T. Makino, T. Miyauchi and H. Nabata, Neuroreport. 6, 1541 (1995).
43. Z.H. Huang, P.L. Huang, N. Panahian, T. Dalkara, M.C. Fishman and M.A. Moskowitz, Science 265,1883 (1994).
44. T.A. Wolfe and J.F. Dasta, Ann.Pharmacotherapy 29, 36 (1995)
45. A. Petros, G. Lamb, A. Leone, S. Moncada, D. Bennett and P. Vallance, Cardiovasc.Res. 28, 34 (1994).
46. W.R. Tracey, M. Nakane, F. Basha and G. Carter, Can.J.Phys.Pharm. 73, 665 (1995).
47. C. Szabo, G.J. Southan and C. Thiemermann, Proc.Nat.Acad.Sci.-USA 91, 12472 (1994).
48. X.Q. Wei, I.G. Charles, A. Smith, J. Ure, C.J. Feng, F.P. Huang, D.M. Xu, W. Muller, S. Moncada and F.Y. Liew, Nature 375, 408 (1995).
49. J.D Macmicking,C. Nathan, G. Horn, N. Chartrain, D.S. Fletcher, M. Trumbauer, K. Stevens, Q.W. Xie, K. Sokol, N. Hutchinson, H. Chen and J.S Mudgett, Cell 81, 641(1995);
50. V.E. Laubach, E.G. Shesely, O. Smithies and P.A. Sherman, Proc.Nat.Acad.Sci.-USA 92, 10688 (1995).

Chapter 24. Inhibition of Matrix Metalloproteinases.

William K. Hagmann[1], Michael W. Lark[2] and Joseph W. Becker[1]
Merck Research Laboratories[1], Rahway, NJ 07065
SmithKline Beecham Pharmaceuticals[2], King of Prussia, PA19406

Introduction - Extracellular matrix (ECM) remodeling is essential to maintain tissue homeostasis in virtually every organ in the body. Both physiological and pathological ECM remodeling have been proposed to be mediated by members of the matrix metalloproteinase (MMP) family (1). Many physiological processes require rapid ECM degradation including connective tissue remodeling seen in development and uterine involution (1,2). In contrast, homeostatic turnover of cartilage and proteoglycan may require hundreds of days (3). When considering pathology, there are diseases in which there is rapid ECM degradation including deterioration of the heart during congestive heart failure and ECM degradation observed in extravasation of highly metastatic tumor cells (4,5) as well as conditions in which there is slow ECM degradation including atherosclerotic lesion formation and rupture (6), cartilage matrix loss in osteoarthritis (OA) and rheumatoid arthritis (RA) (7,8), bone matrix degradation in osteoporosis (9), gingival degradation in periodontal disease (10), and matrix remodeling and deposition in Alzheimer plaque formation (11). In this chapter, we will briefly describe the MMP family of enzymes and the putative role it plays in both physiology and pathology, the structures of these enzymes and the current status of MMP inhibitor development for the treatment of this wide range of diseases.

MMP FAMILY MEMBERS

The MMP family currently includes fourteen members encoded by unique genes, ten of which are secreted from cells in a soluble form and four new members which are bound to the cell membrane (Table 1). The MMPs are zinc dependent, calcium requiring enzymes, which are expressed as inactive zymogens. These enzymes are also inhibited by one or more members of the tissue inhibitor of metalloproteinase (TIMP) family, of which three have been cloned and sequenced (12,13). The physiological activators of many of these enzymes remain unknown; however, some of the members of the MMP family have the capacity to activate other family members (14). Therefore, inhibition of the enzyme(s) ultimately responsible for this activation cascade could result in blocking the activity of multiple enzymes without directly inhibiting their activity.

Table 1. The MMP Family

MMP Number	Enzyme name	MMP Number	Enzyme name
MMP-1	collagenase-1	MMP-11	stromelysin-3
MMP-2	gelatinase A	MMP-12	metalloelastase
MMP-3	stromelysin-1	MMP-13	collagenase-3
MMP-7	matrilysin	MMP-14	membrane-type 1 (MT1-MMP)
MMP-8	neutrophil collagenase	MMP-15	membrane-type 2 (MT2-MMP)
MMP-9	gelatinase B	MMP-16	membrane-type 3 (MT3-MMP)
MMP-10	stromelysin-2	MMP-17	membrane-type 4 (MT4-MMP)

A role for these enzymes has been implicated in both physiology and pathology, since they are found, often at elevated levels, within the various target tissues at the time that matrix remodeling is taking place. For example, MMPs -1, -2, -3 and -9 have all been shown to be elevated in OA (7), RA (8), periodontal disease (10), atherosclerosis (6), and congestive heart failure (4). MMP-2 appears to play an important role in matrix degradation that occurs upon extravasation of highly metastatic tumor cells (5). MMP-7 was initially cloned from a tumor cell library and shown to be identical to the short MMP isolated and characterized from involuting uterus (15). This somewhat circumstantial evidence of involvement in pathology has resulted in a significant effort to develop of inhibitors for these enzymes.

Recently four membrane type MMPs (MT-MMPs) have been cloned and sequenced (16-19). Two of these enzymes have the capacity to activate MMP-2 suggesting that thay may be a target to consider for intervention of tumor metastasis (16,18). These enzymes are unique among the MMPs since they appear to be associated with the cell membrane and not released into the extracellular space. Using MMP inhibitors, shedding of a number of cell associated proteins, including TNFα (20), IL-6 and TNF receptors (21), FAS ligand (22), ACE (23), TSH receptor ectodomain (24) and CD-23 (25), may be mediated by membrane associated MMPs.

STRUCTURAL CHARACTERIZATION OF THE MMPs

MMPs are synthesized as inactive precursors and contain four distinct structural domains: an N-terminal propeptide of about 80 residues, a catalytic domain of about 180 amino acid residues, a linker of 17-63 residues, and a C-terminal domain of approximately 180 residues that appears to play a role in recognition of macromolecular substrates and interaction with macromolecular inhibitors. The C-terminal domain is homologous to hemopexin and contains an internal 4-fold sequence repeat. MMP-7 and MMP-12 do not contain the hemopexin domain. MMPs -2 and -9 contain domains homologous to the type II gelatin-binding domain of fibronectin. Active enzyme is produced by cleavage of the propeptide (26-30). Most structural studies on MMPs have been performed on isolated catalytic domains because the full-length proteins are frequently unstable and structurally heterogeneous (26,28,31,32). Use of truncated MMPs in the structure-based design of inhibitors of the full-length protein has been validated by the observations that the enzymatic activity, specificity, and sensitivity to inhibitors of the catalytic domains of MMP-3 are similar to those of the full-length proteins (33,34).

Three-dimensional structures of the complexes between the catalytic domains of MMPs and various inhibitors have been published (35-46). One structure of an MMP proenzyme has been published (35) as well as four structures containing hemopexin domains (47-50). The structures of the catalytic domains show considerable similarity to one another and belong to the "metzincin" family of proteins characterized by an HExxHxxGxxH sequence motif and includes digestive enzymes, snake venom metalloproteases, and bacterial proteases as well as the MMP's (51-53). This family of proteins is also structurally related to bacterial metalloenzymes such as thermolysin that contain a related sequence motif, HExxH(~20x)NExSD (54).

The catalytic domain is folded into a single globular unit approximately 35 Å in diameter, and the structure is dominated by a single five-stranded β-sheet, with one antiparallel and four parallel strands, and three α-helices. The propeptide of MMP-3 makes up a separate smaller domain, approximately 20 Å in diameter that contains three α-helices and an extended peptide that occupies the active site. The catalytic domain contains two tetrahedrally-coordinated Zn^{2+} ions: a "structural" zinc ion and a "catalytic" zinc ion whose ligands include the side chains of the three histidyl residues in the signature HExxHxxGxxH sequence. In the inhibited complexes, the fourth ligand of the catalytic zinc is a group of the inhibitor such as an hydroxamate or carboxylate; in the proMMP-3 propeptide, it is the sulfhydryl group of Cys^{75}. The catalytic domains also contain 1-3 Ca^{2+} ions. C-terminal to the final His residue in the catalytic zinc site is another conserved sequence containing a Met residue. This residue is in a tight turn just below the catalytic zinc ion, and the conservation of these features has led to the designation of this class of protein as "metzincins" (51,53).

The active site consists of two distinct regions: a groove in the protein surface, centered on the catalytic zinc ion and an S_1' specificity site that varies considerably among members of the family. Bound inhibitors adopt extended conformations witin the groove, make several β-structure-like hydrogen bonds with the enzyme and provide the fourth ligand for the catalytic zinc ion. In proMMP-3, the propeptide makes interactions with the same groups of the enzyme as the inhibitors mentioned above, but the direction of the polypeptide chain is opposite to that seen in those structures (35). The S_1' subsite apparently plays a significant role in determining the substrate specificity in the active enzymes. The volume of this subsite varies widely, with a relatively small hydrophobic site in MMP-7 (43) and MMP-1 (38,39) as compared with a very large site in MMP-8 (41,42,44,45), and a site that extends all the way through the MMP-3 molecule, open to solution at both ends (35).

LOW MOLECULAR WEIGHT INHIBITORS

The discovery of low molecular weight inhibitors of the MMPs has largely evolved from substrate-based designs that incorporate structural features located on either side of the scissile bond of a peptide substrate. This design strategy has resulted in many different structural classes and has been reviewed (55-64). This section will focus on more recent results that have expanded on this strategy and those that have capitalized on structural features obtained from recent X-ray and NMR structures of the MMPs as well as non-peptidyl inhibitors. The inhibitors are generally grouped together according to the functional group that serves as the ligand to the catalytic zinc in the active site with separate sections on conformationally restricted and nonpeptidyl inhibitors.

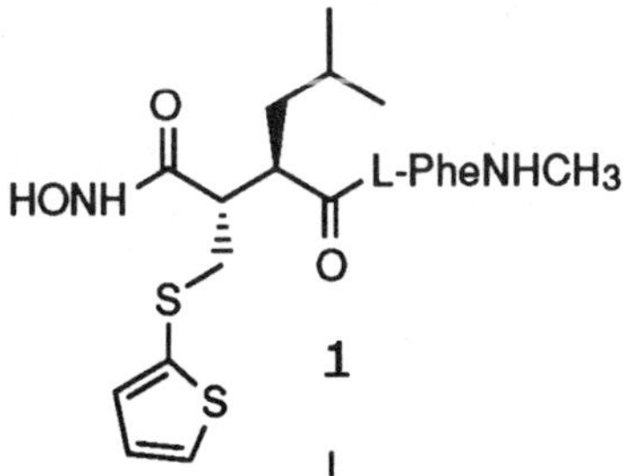

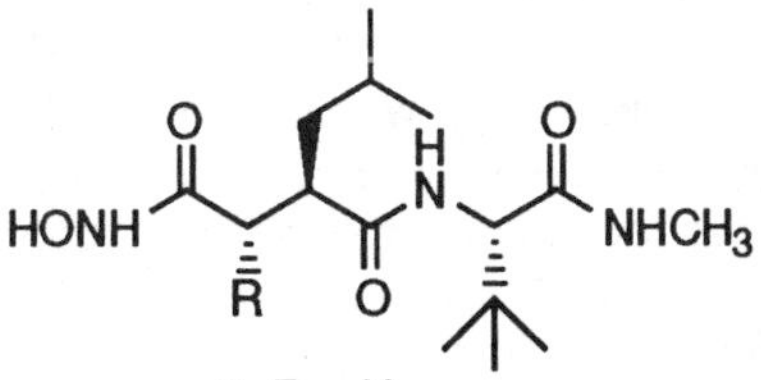

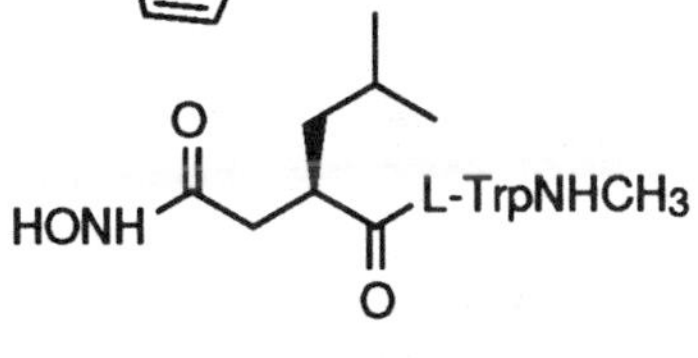

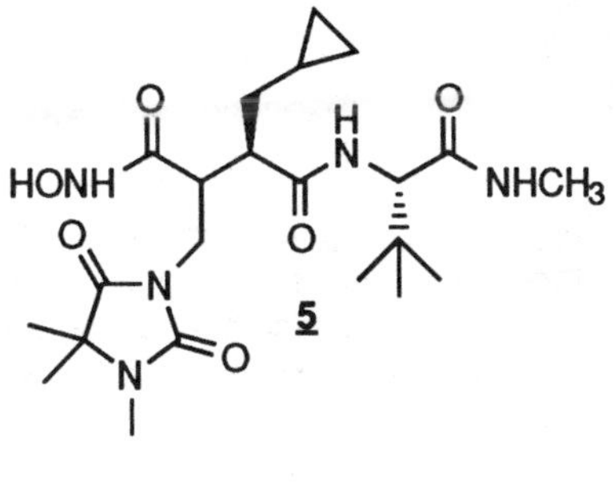

Hydroxamic Acids - Incorporation of the strong zinc binding offered by the bidentate ligation of the hydroxamic acid group onto an optimized peptidyl backbone has resulted in the discovery of extremely potent and, in some cases, selective MMP inhibitors. At this point the leading clinical candidates are BB-94 **1** (batimastat), BB-2516 **3** (marimastat) and GM-6001 **4** (galardin). Batimastat **1** was shown to be effective in several animal models of human metastatic cancers (65,66) and is in Phase III clinical studies for the treatment of malignant pleural effusion (67). Marimastat **3** was reported to be effective at 25, 50 or 75 mg oral doses/bid in Phase II trials as measured by drops or no rises in cancer antigen levels in four types of metastatic cancers (68). Galardin **4** has completed Phase II/III studies for the treatment of corneal ulcers (69). RO31-9790 **2** had been shown to be effective in experimental allergic encephalomyelitis (61,70). Recently disclosed variants of **2** have cycloalkyl groups at P_1' and more elaborate functionalities α (P_1) to the hydroxamic acid (71). Hydroxamate **5** is reported to have an IC_{50} of 1.0 nM vs MMP-1.

Carboxylic Acids - The weaker binding of carboxylates to zinc compared to the hydroxamates seems to necessitate the addition of more, mostly hydrophobic, binding interactions between inhibitor and enzyme. The phenethyl group at P_1' in **6** was needed in a series of N-carboxyalkyl dipeptide inhibitors to yield inhibitors with submicromolar potencies (MMP-1 Ki = 0.76 μM; MMP-2 Ki = 0.2 μM; MMP-3 Ki = 0.47 μM (72). Greater selectivity with respect to MMP-1 was obtained with long chain alkyl or ω-aminoalkyl groups at P_1' (**7**: MMP-1 Ki > 10 μM; MMP-2 Ki = 0.50 μM; MMP-3 Ki = 0.24 μM) (73). Potency approaching the hydroxamates was further enhanced by the addition of small alkyl groups onto the 4-position of the P_1' phenethyl group in (eg., **8** (MMP-1 Ki = 5.9 μM; MMP-2 Ki = 3.5 nM; MMP-3 Ki = 18 nM) (74). Large hydrophobic groups at the P_1 position also enhance potency over simple P_1 substitution (eg., the phthalimidobutyl group in **9** (MMP-1 Ki = 720 nM; MMP-2 Ki = 86 nM; MMP-3 Ki = 8 nM), the tosylamidoethyl group in **10** (MMP-3 IC_{50} = 50 nM) and the 2,3-naphthalimidoethyl group in **11** (MMP-1 Ki = 20 nM; MMP-3 Ki = 91 nM; MMP-9 Ki = 5 nM)) (75-77).

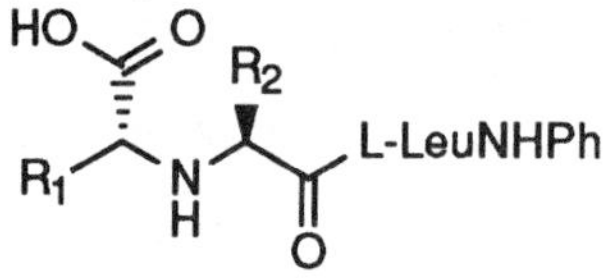

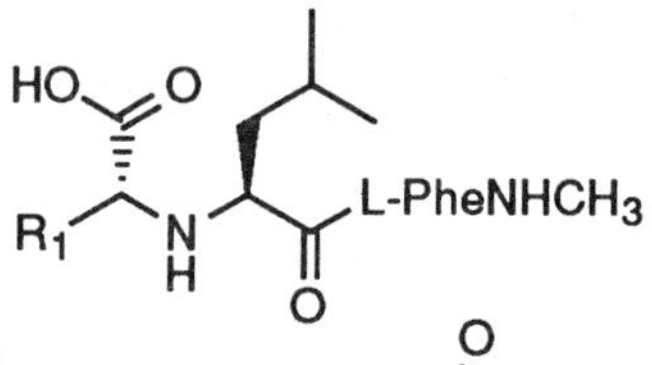

6 R_1 = CH$_3$; R_2 = (CH$_2$)$_2$Ph
7 R_1 = CH$_3$; R_2 = (CH$_2$)$_9$NH$_2$
8 R_1 = CH$_3$; R_2 = (CH$_2$)$_2$Ph-4-C$_3$H$_7$

9 R_1 = [phthalimide]·(CH$_2$)$_4$- ; R_2 = (CH$_2$)$_2$Ph

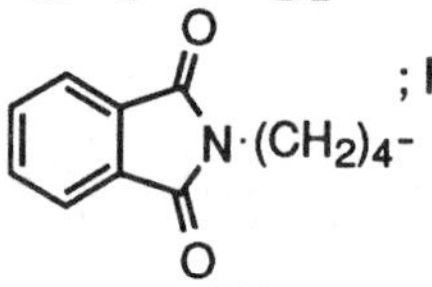

11 R_1 = [2,3-naphthalimide]·(CH$_2$)$_2$-

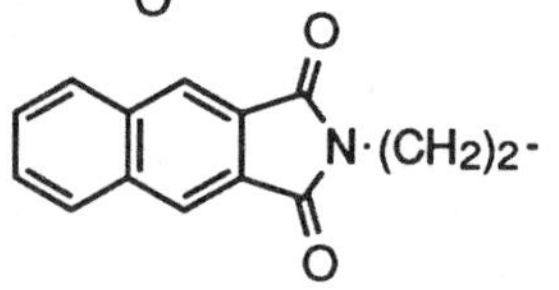

10 R_1 = CH$_3$-[phenyl]-SO$_2$NH(CH$_2$)$_2$- ; R_2 = (CH$_2$)$_2$Ph

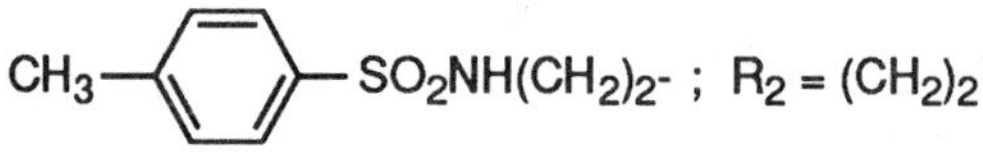

Substituted glutaric acid amides have also been developed as MMP-3 inhibitors (78). Glutarate **12** (MMP-1 Ki > 10 μM; MMP-2 Ki = 310 nM; MMP-3 Ki = 68 nM) was more active than **8** when dosed orally in a model of MMP-3-induced degradation of transferrin in the mouse pleural cavity. P_1' Cα-disubstitution in glutaric acid amides has has also been reported to yield potent MMP-1 inhibitors (79). Succinate **13** is reported to be a very selective inhibitor of MMP-2 (MMP-2 IC_{50} = 30 nM; MMP-1, -3 and -7 IC_{50}'s > 100 μM) (80).

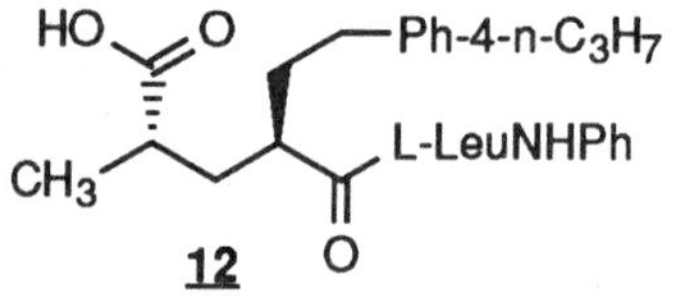

12 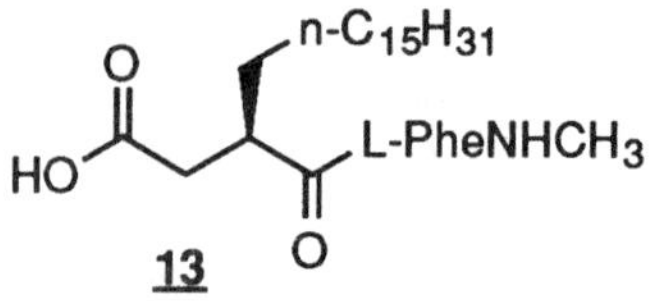   **13**

Phosphorous-containing Inhibitors - Previous reports of phosphorous-containing inhibitors of the MMPs have been reviewed (63). More recently, phosphinic acid **16** was found to be more potent against MMP-1 (IC_{50} = 4.1 μM) and MMP-3 (IC_{50} = 0.43 μM) than its analogous phosphonamidate **14** and phosphonate **15** (81). This order of potency (X = CH$_2$> NH > O) is different from other metalloproteinases, such as thermolysin, CPase-A and ACE. The most potent and selective inhibitor of MMP-3 in this study was phosphinic acid **17** (MMP-1 Ki >10 μM; MMP-2 Ki = 20 nM; MMP-3 Ki = 1.4 nM).

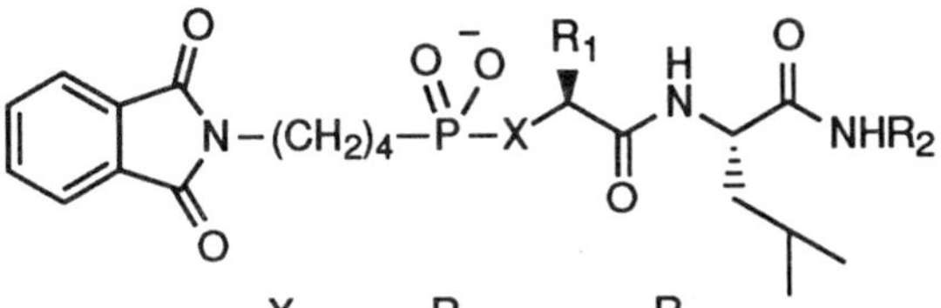

	X	R$_1$	R$_2$
13	NH	CH$_2$Ph	CH$_2$Ph
14	O	CH$_2$Ph	CH$_2$Ph
15	CH$_2$	CH$_2$Ph	CH$_2$Ph
16	CH$_2$	(CH$_2$)$_2$Ph	Ph

Sulfur-containing Inhibitors - Modeling studies suggested thiophenol-containing inhibitors of MMP-1 would place the mercapto group in spatial proximity to a linear inhibitor having the thiol in its usual aliphatic position (α to the P$_1$' substituent) (82). Thiophenol **18** inhibited MMP-1 (IC$_{50}$ = 55 nM) with potency comparable to the acyclic analog **19** (MMP-1 IC$_{50}$ = 17 nM). Interestingly, the preferred configuration at P$_1$' in the thiophenol inhibitors is (S) which corresponds to the unnatural (R)-Leu configuration in MMP-1 substrates and **19**. α–Mercaptoacetamides (**20**) are claimed as inhibitors of MMPs -1, -2 and -3 as well as TNFα release (83). If the isobutyl group is binding into the S$_1$' subsite, then the amide carbonyl and the α-thiol group may be forming the zinc ligand as shown. While not formally containing a thiol ligand, sulfoximines (**21**) have been claimed as MMP inhibitors (84).

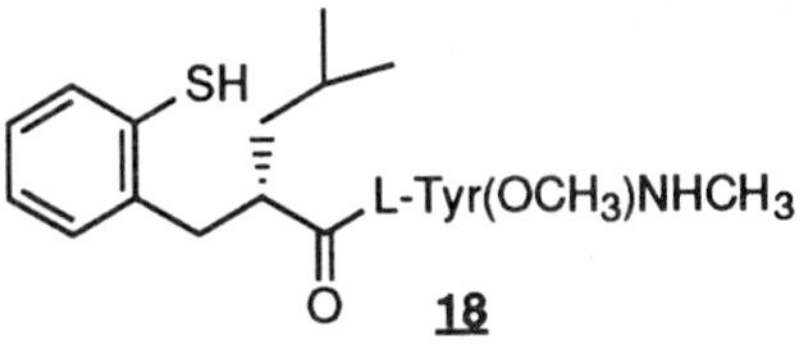

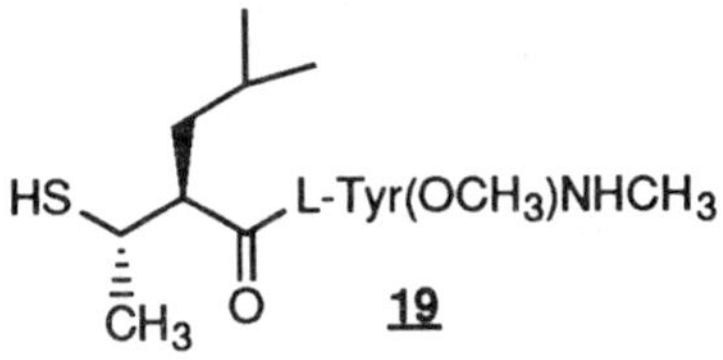

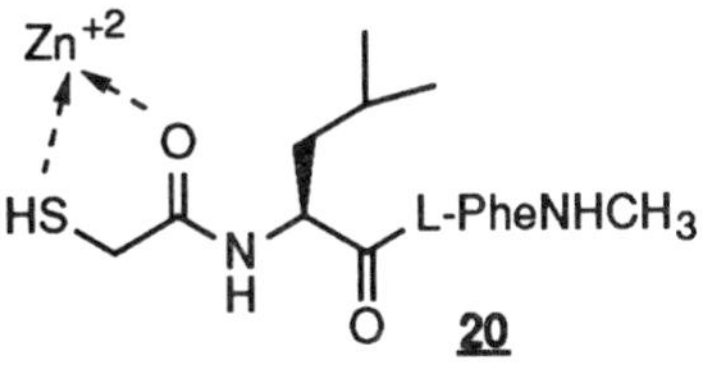

Conformationally Constricted Inhibitors - Cyclization of P$_2$'-P$_3$' substituents to yield potent inhibitors of MMP-1 had been reported (55). Extension of this work to include an azalactam that may improve aqueous solubility gave compounds such as **22** (MMP-1 IC$_{50}$ = 12 nM) (85). Similar cyclization incorporating a more rigid indole ring gave indololactams (eg., **23**, MMP-1 Ki = 0.1 nM; MMP-2 Ki = 0.2 nM; MMP-3 Ki = 9 nM; MMP-7 Ki = 3 nM; MMP-8 Ki = 0.4 nM) which were more potent than the analogous acyclic inhibitors (86). The large ring size (13-membered) for these inhibitors was needed to maintain the trans- conformation of the P$_2$'-P$_3$' amide bond; a feature observed in the X-ray and NMR structures and deemed to be necessary to establish the hydrogen bond network between the inhibitor backbone and the enzymes. However, a trans-amide bond between P$_2$' and P$_3$' does not seem to be an absolute requirement for potent inhibitors. The P$_2$'-P$_3$' carbostyril and benzazepine derivatives **24** and **25** were still very potent inhibitors (**24**: MMP-1 IC50 = 2.5 pM; MMP-2 IC$_{50}$ = 18 nM; MMP-3 IC$_{50}$ = 23 nM; **25**: MMP-1 IC$_{50}$ = 0.04-4 μM)) (87-89) as was the imidazole derivative **26** (90).

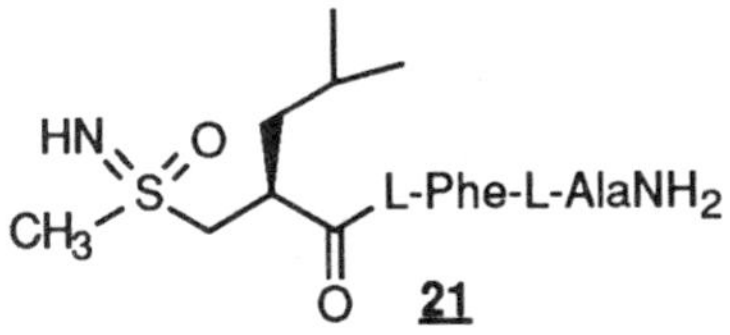

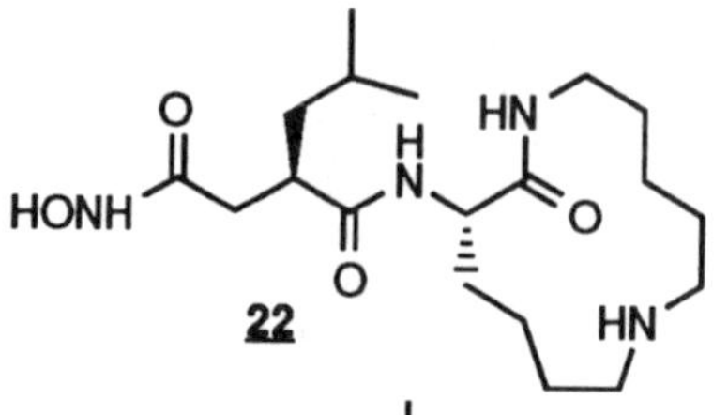

A different constriction cyclized the P_2' substituent onto its amide bond nitrogen (91,92). Inspired by the matlystatins (93,94), piperazic acid derivatives as well as other heterocycles were incorporated into the P_2' position containing a variety of side chains with hydroxamate and carboxylate zinc ligands (91). Hydroxamate **27** (MMP-3 Ki < 50 nM) was demonstrated to have oral bioavailability and compounds of this invention reportedly inhibited cartilage degradation and TNF release *in vivo*. The P_3' pocket of MMP-3 was also probed with this structure type (95). The long chain alkyl at P_1' in **28** enhanced potencies (MMP-1 IC_{50} = 5.6 nM; MMP-2 IC_{50} = 0.78 nM; MMP-3 IC_{50} = 10 nM) whereas the P_1' cycloalkyl group in **29** (MMP-1 IC_{50} = 2.4 nM) provides a replacement for the usual isobutyl group (92,96).

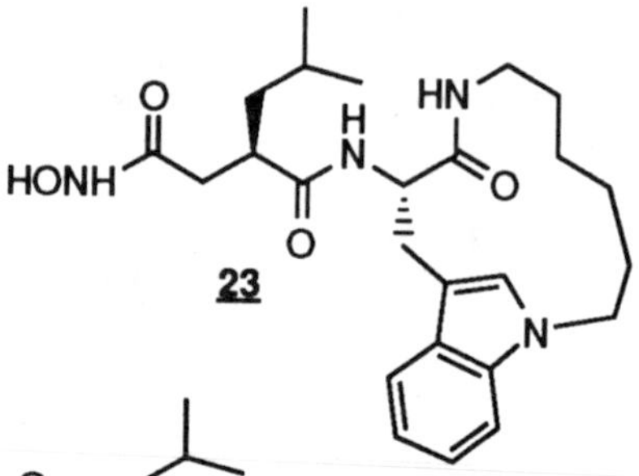

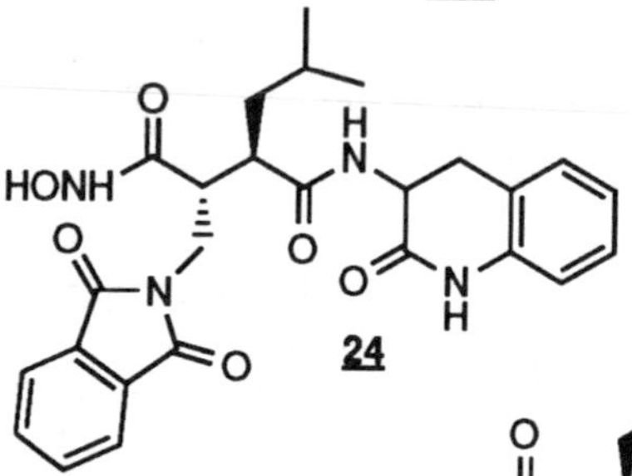

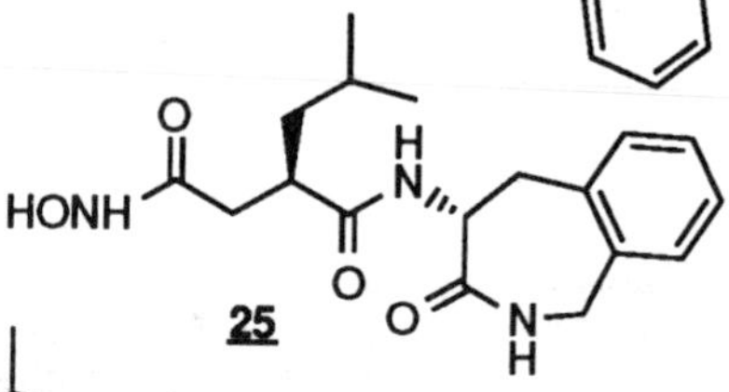

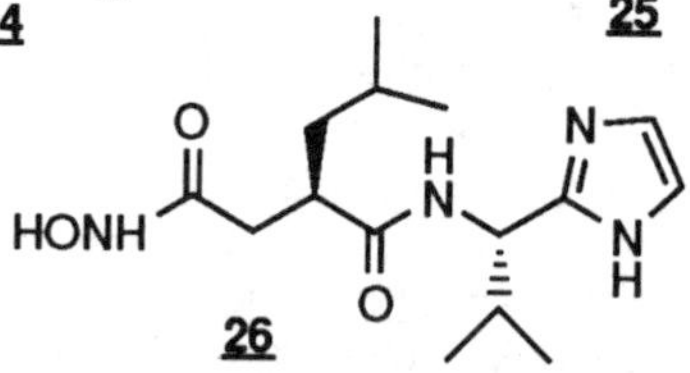

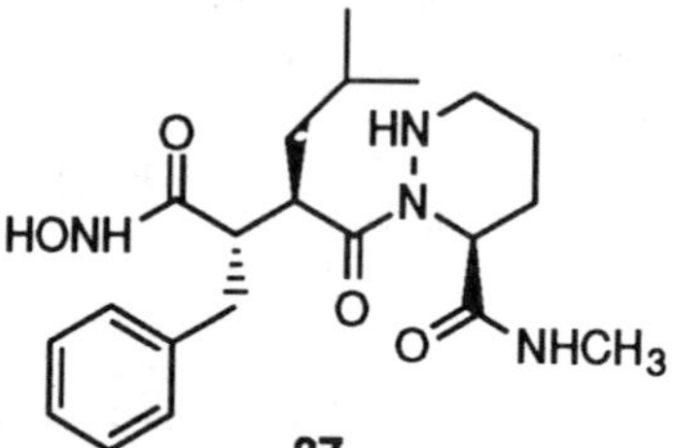

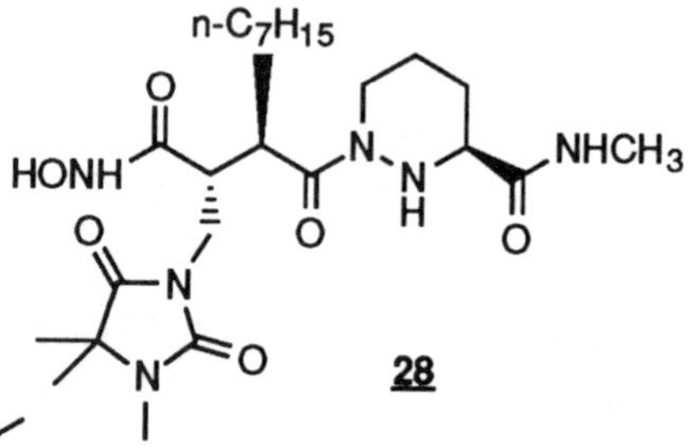

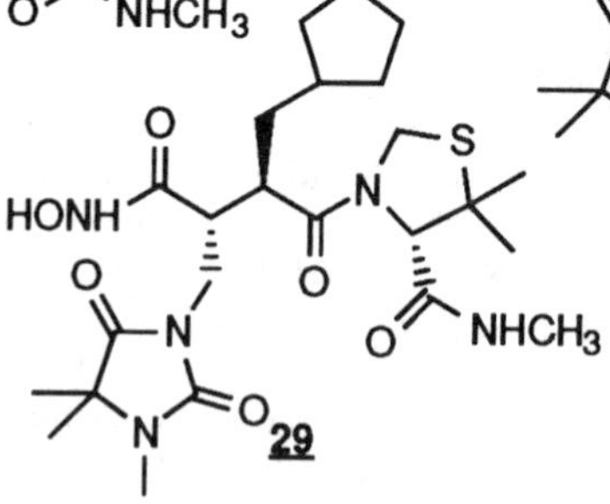

Non-peptidyl Inhibitors - Screening of compounds with potential zinc ligating functionalities led to the discovery of CGS 23161 **30** (MMP-3 Ki = 70 nM) (63,97). Further elaboration of this lead resulted in the discovery of CGS 27023A **31** (MMP-1 Ki = 380 nM; MMP-3 Ki = 17 nM). CGS 27023A (100 mpk/day po) was efficacious in inhibiting loss of cartilage proteoglycan in a partial meniscectomy model of OA in rabbits (98). It did not prevent proteoglycan loss from IL-1-induced bovine nasal cartilage cultures but did block collagen loss (99). Related compounds **32** have been claimed to inhibit MMP-12 (100). Long chain alkyl groups at P_2' and P_1' have also been incorporated into this design strategy (101,102).

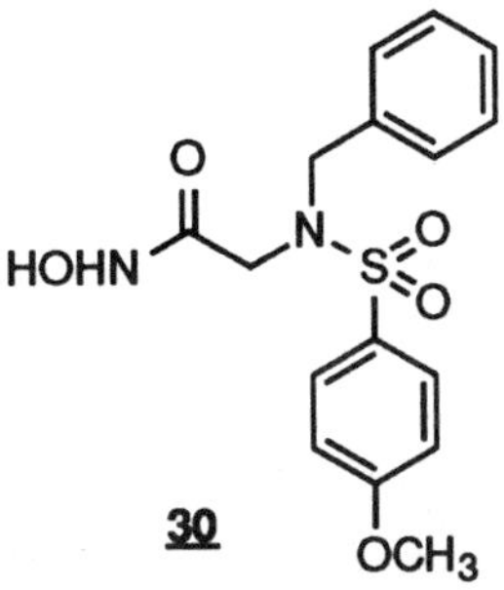

Metalloproteinase Inhibitors of TNFα Release - While not formally a member of the MMP family of metalloproteinases, the recently described inhibition of TNFα release by metalloproteinase inhibitors and the subsequent involvement in models of inflammation presents a potentially new therapeutic target for many of the inhibitors in this area. The metalloproteinase purported to be essential for the cleavage of the active 17K TNFα from its membrane-bound presursor protein (Mr = 26K) has been termed TNFα converting enzyme or TACE. Several hydroxamic acid inhibitors have been reported to block TNFα release from a variety of cell types (20,103,104). They have also been effective inhibitors of TNFα release from LPS treated rodents and, in one case, increased survivability from a lethal dose of LPS was demonstrated (103). However, inhibition of the shedding of various receptors, ligands and enzymes by metalloproteinase inhibitors has also been observed and may indicate that TACE (or members of a group of like enzymes) may have other roles (21-25).

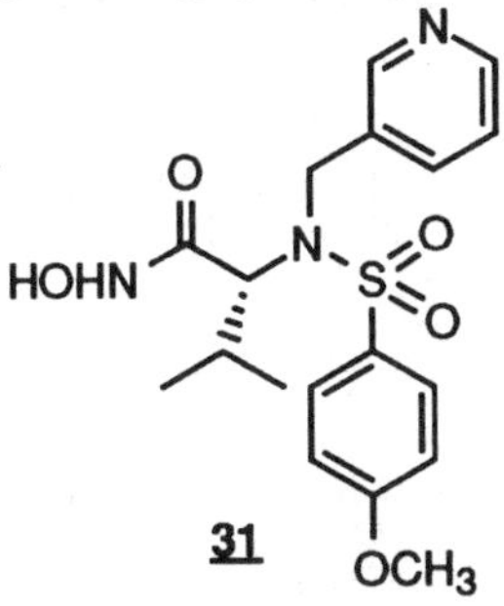

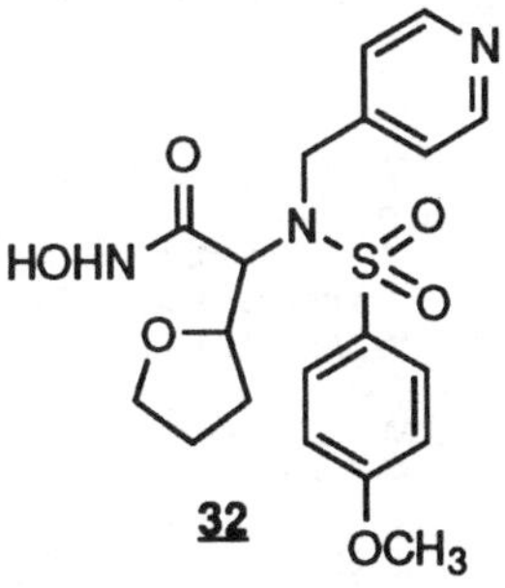

INHIBITION OF MATRIX DEGRADATION

One of the limitations of studies in this field is the limited evidence for matrix cleavage by these enzymes. To determine if MMPs have degraded any of the components of the ECM, specific degradation products are being identified and characterized. For example, MMP-1 generated cleavage products of type II collagen have been identified in aging and OA cartilage (106). Further characterization of these cleavage products in combination with enzyme localization studies are underway to help identify which of the collagenases (MMP-1, MMP-8 or MMP-13) may be primarily responsible for the observed cartilage collagen degradation. Antibodies against aggrecan fragments have also been generated and used to localize MMP-cleaved aggrecan fragments in OA, RA and control articular cartilage, indicating that these enzymes may be involved in these arthritic diseases, but they also appear to be involved in normal cartilage remodeling (107). In these studies, an additional aggrecan degrading enzyme which cleaves at a site C-terminal to the classical MMP site has been identified (108). The enzyme has yet to be purified and therefore its identity and relationship to known MMPs is yet unknown.

There is an extreme amount of overlap in enzyme expression and substrate cleavage among the various members of the MMP family and, thus, it is likely that multiple members of the family may be involved in any of the pathological and physiological processes described above. If this is indeed the case, it may be difficult to develop a specific MMP inhibitor that would alone block matrix degradation. The other complicating factor when considering targeting this family of enzymes for development of therapeutics, is that many of these enzymes appear to be involved in normal ECM homeostasis and in physiologically normal ECM turnover. Having said this, potent gelatinase inhibitors have shown activity in animal models of metastasis

(109) and are generating encouraging results in human clinical trials (64). Other inhibitors have been efficacious in the rat adjuvant arthritis model (110) and in the rabbit meniscectomy and spontaneous guinea pig models of OA (111). As additional selective MMP inhibitors with appropriate pharmacokinetics become available, a more defined role for the MMP family members may become apparent.

REFERENCES

1. C.W. Alexander and Z. Werb, in Cell Biology of the Extracellular Matrix (E.D. Hay, ed.) 2nd Edn., Plenum Press, New York, p. 255 (1991).
2. L.M. Matrisian, BioEssays 14, 455 (1992).
3. S. Lohmander, Arch.Biochem.Biophys. 186, 93 (1987).
4. P.W. Armstrong, G.W. Moe, R.J. Howard, E.A.. Grima and T.F. Cruz, Exp.Cardiol. 10, 214 (1994).
5. J.M. Ray and W.G. Stetler-Stevenson, EMBO J. 14, 908 (1995).
6. Z.S. Galis, G.K. Sukhova, M.W. Lark and P. Libby, J.Clin.Invest. 94, 2493 (1994).
7. L.S. Lohmander, L.A. Hoerrner and M.W. Lark, Arthrit.Rheum. 36, 181 (1993).
8. L.A. Walakovits, V.L. Moore, N. Bhardwaj, G.S. Gallick and M.W. Lark, Arthrit.Rheum. 35, 35 (1992).
9. K. Tezuka, K. Nemato, Y. Tezuka, T. Sato, Y. Ikeda, M. Kobori, H. Kawashima, H. Eguchi, Y. Hakeda and M. Kumegawa, J.Biol.Chem. 269, 15006 (1994).
10. A. Haerian, E. Adonogianaki, J. Mooney, J.P. Docherty and D.F. Kinane, J.Clin.Periodontol. 22, 505 (1995).
11. G. Perides, R.A. Asher, M.W. Lark and A. Bignami, Biochem.J. 312, 377 (1995).
12. G. Murphy, in AAS 35: Progress in Inflammation Research and Therapy 69, Verlag Birkhauser, Basel (1991).
13. S.S. Apte, B.R. Olsen and G. Murphy, J.Biol.Chem. 270, 14313 (1995).
14. G.M. Murphy, M.I. Cockett, P.E. Stephens, B.J. Smith and A.J.P. Docherty, Biochem.J. 248, 265 (1987).
15. B. Quantin, G. Murphy and R. Breathnach, Biochem. 28, 5327 (1989).
16. H. Sato, T. Takino, Y. Okada, J. Cao, A. Shinagawa, E. Yamamoto and M. Seiki, Nature 370, 61 (1994).
17. B. Will and B. Hinzmann, Eur.J.Biochem. 231, 602 (1995).
18. T. Takino, H. Sato, A. Shinagawa and M. Seiki, J.Biol.Chem. 270, 23013 (1995).
19. X.S. Puente, A.M. Pondas, E. Llano, G. Velasco and C. Lopez-Olin, Canc.Res. 56, 944 (1996).
20. G.M. McGeehan, J.D. Becherer, R.C. Blast Jr., C.M. Boyer, B. Champion, K.M. Connolly, J.G. Conway, P. Furdon, S. Karp, S. Kidao, A.B. McElroy, J. Nichols, K.M. Pryzwansky, F. Schoenen, L. Sekut, A. Truesdale, M. Verghese, J. Warner and J.P. Ways, Nature 370, 558 (1994).
21. J. Mullberg, F.H. Durie, C. Otten-Evans, M.R. Alderson, S. Rose-John, D. Cosman, R.A. Black and K.M. Mohler, J.Immunol. 155, 5198 (1995).
22. N. Kayagaki, A. Kawasaki, T. Ebata, H. Ohmoto, S. Ikeda, S. Inoue, K. Yoshino, K. Okumura and H. Yagita, J.Exp.Med. 182, 1777 (1995).
23. R. Ramachandran and I. Sen, Biochem. 34, 12645 (1995).
24. J. Couet, S. Sar, A. Jolivet, M.-T.V. Hai, E. Milgrom and M. Misrah, J.Biol.Chem. 271, 4545 (1996).
25. G. Christie and B.J. Weston PCT WO 96/02240 (1996).
26. Y. Okada, E.D. Harris and H. Nagase, Biochem.J. 254, 731 (1988).
27. Y. Okada and I. Nakanishi, FEBS.Lett. 249, 353 (1989).
28. H. Nagase, J.J. Enghild,K. Suzuki and G. Salvesen, Biochem. 29, 5783 (1990).
29. P.A. Koklitis, G. Murphy, C. Sutton and S. Angal, Biochem.J. 276, 217 (1991).
30. G.A.C. Murrell, D. Jang and R.J. Williams, Biochem.Biophys.Res.Comm. 206, 15 (1995).
31. Y. Okada, H. Konomi, T. Yada, K. Kimata and H. Nagase, FEBS.Lett. 244, 473 (1989).
32. Y. Okada, H. Nagase and E.D. Harris, J.Biol.Chem. 261, 14245 (1986).
33. A.I. Marcy, L.L. Eiberger, R. Harrison, H.K. Chan, N.I. Hutchinson, W.K. Hagmann, P.M. Cameron, D.A. Boulton and J.D. Hermes, Biochem. 30, 6476 (1991).
34. Q.-Z. Ye, L.L. Johnson, I. Nordan, D. Hupe and L. Hupe, J.Med.Chem. 37, 206 (1994).
35. J.W. Becker, A.I. Marcy, L.L. Rokosz, M.G. Axel, J.J. Burbaum, P.M.D. Fitzgerald, P.M. Cameron, C.K. Esser, W.K. Hagmann, J.D. Hermes and J.P. Springer, ProteinSci. 4, 1966 (1995).
36. P.R. Gooley, J.F. O'Connell, A.I. Marcy, G.C. Cuca, S.P. Salowe, B.L. Bush, J.D. Hermes, C.K. Esser, W.K. Hagmann, J.P. Springer and B.A. Johnson, NatureStruct.Biol. 1, 111 (1994).
37. N. Borkakoti, F.K. Winkler, D.H. Williams, A. Darcy, M.J. Broadhurst, P.A. Brown, W.H. Johnson and E.J. Murray, NatureStruct.Biol. 1, 106 (1994).

38. B. Lovejoy, A. Cleasby, A.M. Hassell, K. Longley, M.A. Luther, D. Weigl, G. McGeehan, A.B. McElroy, D. Drewry, M.H. Lambert and S.R. Jordan, Science 263, 375 (1994).
39. B. Lovejoy, A.M. Hassell, M.A. Luther, D. Weigl and S.R. Jordan, Biochem. 33, 8207 (1994).
40. J.C. Spurlino, A.M. Smallwood, D.D. Carlton, T.M. Banks, K.J. Vavra, J.S. Johnson, E.R. Cook, J. Falvo, R.C. Wahl, T.A. Pulvino, J.J. Wendoloski and D.L. Smith, Proteins: Struct.Func.Genet. 19, 98 (1994)
41. W. Bode, P. Reinemer, R. Huber, T. Kleine, S. Schnierer and H. Tschesche, EMBO J. 13, 1263 (1994).
42. T. Stams, J.C. Spurlino, D.L. Smith, R.C. Wahl, T.F. Ho, M.W. Qoronfleh, T.M. Banks and B. Rubin, NatureStruct.Biol. 1, 119 (1994).
43. M.F. Browner, W.W. Smith and A.L. Castelhano, Biochem. 34, 6602 (1995).
44. F. Grams, M. Crimmin, L. Hinnes, P. Huxley, M. Pieper, H. Tschesche and W. Bode, Biochem. 34, 14012 (1995).
45. F. Grams, P. Reinemer, J.C. Powers, T. Kleine, M. Pieper, H. Tschesche, R. Huber and W. Bode, Eur.J.Biochem. 228, 830 (1995).
46. S.R. Van Doren, A.V. Kurochkin, W.D. Hu, Q.Z. Ye, L.L. Johnson, D.J. Hupe and E.R.P. Zuiderweg, ProteinSci. 4, 2487 (1995).
47. H.R. Faber, C.R. Groom, H.M. Baker, W.T. Morgan, A. Smith and E.N. Baker, Structure 3, 551 (1995).
48. J. Li, P. Brick, M.C. O'Hare, T. Skarzynski, L.F. Lloyd, V.A. Curry, I.M. Clark, H.F. Bigg, B.L. Hazleman, T.E. Cawston and D.M. Blow, Structure 3, 541 (1995).
49. A.M. Libson, A.G. Gittis, I.E. Collier, B.L. Marmer, G.I. Goldberg and E.E. Lattman, NatureStruct.Biol. 2, 938 (1995).
50. U. Gohlke, F.-X. Gomis-Rüth, T. Crabbe, G. Murphy, A.J.P. Docherty and W. Bode, FEBS Lett. 378, 126 (1996).
51. W. Bode, F.X. Gomis-Rüth and W. Stockler, FEBSLett. 331, 134 (1993).
52. F.X. Gomis-Rüth, L.F. Kress, J. Kellermann, I. Mayr, X. Lee, R. Huber and W. Bode, J.Molec.Biol. 239, 513 (1994).
53. W. Stöcker, F. Grams, U. Baumann, P. Reinemer, F.X. Gomis-Rüth, D.B. McKay and W. Bode, ProteinSci. 4, 823 (1995).
54. B.H. Matthews, Acct.Chem.Res. 21, 333 (1988).
55. W.H. Johnson, N.A. Roberts and N. Borkakoti, J.Enz.Inhib., 2, 1 (1987).
56. B. Henderson, A.J.P. Docherty and N.R.A. Beeley, Drugs Future, 15, 495 (1990).
57. W.H. Johnson, Drug News Perspective, 3, 453 (1990).
58. R.C. Wahl, R.P. Dunlap and B.A.Morgan, Ann.Rep.Med.Chem., 25, 177 (1990).
59. M.A. Schwartz and H.E. Van Wart in "Progress in Medicinal Chemistry," Vol. 29, G.P. Ellis and D.K. Luscombe, Eds., Elsevier Sci. Publ., London, 1992, Ch. 8, p. 271.
60. N.R.A. Beeley, P.R.J. Ansell and A.J.P. Docherty, Curr.Opin.Ther.Patents, 4, 7 (1994).
61. P.A. Brown, in "Osteoarthritis: Advances in Pathology, Diagnosis and Treatment", Internal.Bus.Commun., London, 1994, p. 1.
62. P.D. Brown, Adv.Enz.Regul. 35, 293 (1995).
63. J.R. Morphy, T.A. Millican and J.R. Porter, Curr.Med.Chem., 2, 743 (1995).
64. P.D. Beckett, A.H. Davidson, A.H. Drummond, P. Huxley and M. Whittaker, DrugDesignToday 1, 16 (1996).
65. S.A. Watson, T.M. Morris, G. Robinson, M.J. Crimmin, P.D. Brown and J.D. Hardcastle, Canc.Res. 55, 3629 (1995).
66. G.W. Sledge, M. Qulali, R. Goulet, E.A. Bone and R. Fife, J.Natl.Canc.Inst. 87, 1546 (1995).
67. Scrip 2041, 7 (1995).
68. Scrip 2084, 23 (1995).
69. Scrip 1775, 24 (1992).
70. A.K. Hewson, T. Smith, J.P. Leonard and M.L. Cuzner, Inflamm.Res. 44, 345 (1995).
71. M. J. Broadhurst, P.A. Brown and W.H. Johnson, PCT WO 95/33709 (1995).
72. K.T. Chapman, I.E. Kopka, P.L. Durette, C.K. Esser, T.J. Lanza, M. Izquierdo-Martin, L. Niedzwiecki, B. Chang, R.K. Harrison, D.W. Kuo, T.-Y. Lin, R.L. Stein and W.K. Hagmann, J.Med.Chem., 36, 4293 (1993).
73. C.K. Esser, I.E. Kopka, P.L. Durette, R.K. Harrison, L.M. Niedzwiecki, M. Izquierdo-Martin, R.L. Stein and W.K Hagmann, Bioorg.Med.Chem.Lett., 5, 539 (1995).
74. S.P. Sahoo, C.G. Caldwell, K.T.Chapman, P.L. Durette, C.K. Esser, S.A. Polo, K.M. Sperow, L.M. Niedzwiecki, M. Izquierdo-Martin, B.C. Chang, R.K. Harrison, R.L. Stein, M. MacCoss and W.K. Hagmann, Bioorg.Med.Chem.Lett., 5, 2441 (1995).
75. K.T. Chapman, J. Wales, S.P. Sahoo, L.M. Niedzwiecki, M. Izquierdo-Martin, B.C. Chang, R.K. Harrison, R.L. Stein and W.K. Hagmann, Bioorg.Med.Chem.Lett., 6, 329 (1996).
76. C.-B. Xue, X. He, J. Roderick, W.F. Degrado, C. Decicco and R.A. Copeland, Bioorg.Med.Chem.Lett., 6, 379 (1996).
77. F.K. Brown, P.J. Brown, D.M. Bickett, C.L. Chambers, H.G. Davies, D.N. Deaton, D. Drewry, M. Foley, A.B. McElroy, M. Gregson, G.M. McGeehan, P.L. Myers, D. Norton, J. Salovich, F.J. Schoenen and P. Ward, J.Med.Chem., 37, 674 (1994).

78. K.T.Chapman, P.L. Durette, C.G. Caldwell, K.M. Sperow, L.M. Niedzwiecki, R.K. Harrison, C. Saphos, A.J. Christen, J.M. Olszewski, V.L. Moore, M. MacCoss and W.K. Hagmann, Bioorg.Med.Chem.Lett., $\underline{6}$, 803 (1996).
79. K.M. Donahue, B.J. Cronin, B.P. Jones, P.G. Mitchell, L. Lopresti-Morrow, J.A. Ragan, L.M. Reeves, J.P Rizzi, R.P. Robinson and S.A. Yocum, Abstracts of Papers, 211th ACS Nat'l.Mtg., New Orleans, LA, March 1996, Abs.# MEDI 212..
80. A. Miller, P.R. Beckett, F.M. Martin and M. Whittaker, PCT WO 95/32944 (1995).
81. C.G. Caldwell, S.P. Sahoo, S.A. Polo, R.R. Eversole, T.J. Lanza, S.G. Mills, L.M. Niedzwiecki, M. Izquierdo-Martin, B.C. Chang, R.K. Harrison, D.W. Kuo, T.-Y. Lin, R.L. Stein, P.L. Durette and W.K. Hagmann, Bioorg.Med.Chem.Lett., $\underline{6}$, 323 (1996).
82. I. Hughes, G.P. Harper, E.H. Karran, R.E. Markwell and A.J. Miles-Williams, Bioorg.Med.Chem.Lett., 5, 3039 (1995).
83. J. Montana, J.Dickens, D.A. Owen and A.D. Baxter, PCT WO 95/13289 (1995).
84. M.A. Schwartz and H. Van Wart, PCT WO 95/9620 (1995).
85. J. Bird, G.P. Harper, I. Hughes, D.J. Hunter, E.H. Karran, P.E. Markwell, A.J. Miles-Williams, S.S. Rahman and R.W. Ward, Bioorg.Med.Chem.Lett., $\underline{5}$, 2593 (1995).
86. A.L. Castelhano, R. Billedeau, N. Dewdney, S. Donnelly, S. Horne, L.J. Kurz, T.J. Liak, R. Martin, R. Uppington, Z. Yuan and A. Krantz, Bioorg.Med.Chem.Lett., $\underline{5}$, 1415 (1995).
87. M. Sakamoto, T. Imaoka, M. Motoyama, Y. Yamamoto and H. Takasu PCT WO 94/21612 (1994).
88. M. Nitta, R. Hirayama and M. Yamamoto, Jap. Pat. Applic. JP7304770 (1995).
89. M. Nitta and M. Yamamoto, Jap. Pat. Applic. JP7304746 (1995).
90. J.S. Frazee, J.G. Gleason and B.W. Metcalf, PCT WO95/23790 (1995).
91. C.P DeCicco, I.C. Jacobson, R.L. Magolda, D.J. Nelson and R.J. Cherney, PCT WO95/29892 (1995).
92. M.J. Broadhurst, P.A. Brown and W.H. Johnson, PCT WO 95/33731 (1995).
93. K. Tanzawa, M. Ishii, T. Ogita and K. Shimada, J.Antibiotics, $\underline{45}$, 1733 (1992).
94. K. Tamaki, K. Tanzawa, S. Kurihara, T. Oikawa, S. Monma, K. Shimada and Y. Sugimura, Chem.Pharm.Bull., $\underline{43}$, 1883 (1995).
95. D.A. Nugiel, K. Jacobs, C.P. DeCicco, D.J. Nelson, R.A. Copeland and K.D. Hardman, Bioorg.Med.Chem.Lett., $\underline{5}$, 3053 (1995).
96. M.J. Broadhurst, P.A. Brown and W.H. Johnson, EPA 0684240A1 (1995)
97. L.J. MacPherson and D.T. Parker, U.S. Patent 5,455,258 (1995).
98. E.M. O'Byrne, D.T. Parker, E.D. Roberts, R.L. Goldberg, L.J. MacPherson, V. Blancuzzi, D. Wilson, H.N. Singh, R. Ludewig and V.S. Ganu, Inflamm. Res., $\underline{44}$ (suppl. 2), S177 (1995).
99. S. Spirito, J. Doughty, E. O'Byrne, V. Ganu and R.L. Goldberg, Inflamm. Res., $\underline{44}$ (suppl. 2), S131 (1995).
100. L.J. MacPherson, D.T. Parker and A.Y. Jeng, PCT WO 96/00214 (1996).
101. A. Miller, M. Whittaker and R.P. Beckett, PCT WO 95/35275(1995).
102. A. Miller, M. Whittaker and R.P. Beckett, PCT WO 95/35276(1995).
103. K.M. Mohler, P.R. Sleath, J.N. Fitzner, D.P. Cerretti, M. Alderson, S.S. Kerwar, D.S. Torrance, C. Otten-Evans, T. Greenstreet, K. Weerawarna, S.R. Kronheim, M. Peteren, M. Gerhart, C.J. Kozlosky, C.J. March and R.A. Black, Nature, $\underline{370}$, 218 (1994).
104. A.J.H. Gearing, P. Beckett, M. Christodoulou, M. Churchill, J. Clements, A.H. Davidson, A.H. Drummond, W.A. Galloway, R. Gilbert, J.L. Gordon, T.M. Leber, M. Mangan, K. Miller, P. Nayee, K. Owen, S. Patel, W. Thomas, G. Wells, L.M. Wood and K. Wooley, Nature, $\underline{370}$, 555 (1994).
105. P.D. Crowe, B.N. Walter, K.M. Mohler, C. Otten-Evans, R.A. Black and C.F. Ware, J.Exp.Med., $\underline{181}$, 1205 (1995)
106. A.P. Hollander, I. Pidoux, A. Reiner, C. Rorabeck, R. Bourne and A. R. Poole, J.Clin.Invest. $\underline{96}$, 2859, 1995.
107. M.W. Lark, E.K. Bayne, and L.S. Lohmander, Acta Orthp. Scand. 66, 92 (1995).
108. M.W. Lark, J.T. Gordy, J.R. Weidner, J. Ayala, J.H. Kimura, H.R. Williams, R.A. Mumford, C.R. Flannery, S.S. Carlson, M. Iwata, and J.D. Sandy, J. Biol. Chem. $\underline{270}$, 2550 (1995).
109. X. Wang, X. Fu, P.D. Brown, M.J. Crimmin and R.M. Hoffman, Can. Res. $\underline{54}$, 4726 (1994).
110. J.G. Conway, J.A. Wakefield, R.H. Brown, B.E. Marron, L. Sekut, S.A. Stimpson, A. McElroy, J.A. Menius, J.J. Jeffreys, R.L. Clark, G.M. McGeehan and K.M. Connolly, J.Exp.Med. $\underline{182}$, 449 (1995).
111. M.A. Prata, Inflam. Res. $\underline{44}$, 458, 1995.

SECTION V. TOPICS IN BIOLOGY

Editor: Winnie W. Wong, BASF Bioresearch Corporation, Worcester, Massachusetts

Chapter 25. Cell Cycle Control and Cancer

Giulio Draetta[1] and Michele Pagano[2]

[1]Department of Experimental Oncology, European Institute of Oncology, Via Ripamonti 435, 20141 Milan, Italy and [2]Department of Pathology, New York University 560 First Avenue, New York, NY 10016, USA

Introduction - In recent years, there has been an exciting convergence in studies of the cell cycle and studies of oncogenesis. Indeed, accumulating evidence indicates that a derangement in the cell cycle machinery contributes to both uncontrolled cell growth and genetic instability, two characteristics of tumor cells (1).

The eukaryotic cell division cycle is a coordinated and highly regulated series of events during which cells replicate their DNA content, and then divide. Formally, the cell cycle has been divided in four phases: G1 (gap1), S (DNA synthesis), G2 (gap2) and M (mitosis). During G1 the cell monitors its environment and growth before committing to DNA replication, while during G2 it ensures that DNA replication is completed, while preparing for mitosis. Mitosis is the phase of nuclear division leading up to the division of a cell into two daughter cells (2). It is now known that the sequential activation of different cyclin-dependent kinases plays a central role in controlling the passage through the different phases of the cell cycle (3). There are several distinct molecular mechanisms for controlling the activity of the different cyclin-dependent kinases: regulated synthesis and destruction of the cyclin regulatory subunit (4); post-translational modification of the kinase subunit by highly specific kinases and phosphatases (5), and association/dissociation with a variety of inhibitory proteins, that are either able to inhibit all cyclin-dependent kinases such as p21 (also called Cip1, Pic1, Sdi1 and Waf1; (6-10), p27 (also called Ick, Kip1 and Pic2; (11, 12) and p57 (also called Kip2; (13, 14); or that are specific for the Cdk4-Cdk6 cyclin kinases. The latter include p16 (also called Ink4A, Mts1, Cdkn2 and Cdk4i (15, 16), p15 (also called Ink4B, Mts2; (16-19) p18 (also called Ink4C and Ink6A (17, 20) and p19/p20 (also called Ink4D and Ink6B; (20-22). These inhibitors regulate the cell cycle through the inhibition of Cdk4-cyclin D1, Cdk2-cyclin E and Cdk2-cyclin A complexes. The importance of these molecules in controlling the cell cycle and in oncogenic events is demonstrated by the finding that they can be the direct or indirect targets of mutational events in a wide variety of human tumors and tumor cell lines (1, 23).

The G1 Phase of the Cell Cycle and the G1/S Transition - The cell cycle can be represented also as the cycle of the retinoblastoma gene product, pRb (24, 25). This protein is in a hypophosphorylated state during the early G1 phase and is subsequently phosphorylated by cyclin-dependent kinases in mid-G1 and kept in a hyperphosphorylated state until mitosis, at which time it is dephosphorylated again. In early G1, the pRb protein is bound to transcription factors of the E2F family and as such it inhibits the activation of transcription of genes required for DNA synthesis. Upon phosphorylation, pRb releases E2F allowing cells to enter S-phase. The phosphorylation of pRb is initiated by cyclin D/cdk complexes, which have a rate-limiting function during G1. Indeed, it has been shown that microinjection of antibodies or antisense plasmids to cyclin D1 prevents entry into S-phase (26, 27), while overexpression of D-cyclins accelerates entry into S (27-29). Furthermore, overexpression of the p16 inhibitor prevents entry into S-phase (17, 30, 31). Both p16 expression and inhibition of cyclin D1 function by antibody/antisense injection are completely ineffective when performed on cells that lack the pRb function, demonstrating that the primary function of the cyclin D1 kinase complex is the phosphorylation and inactivation of pRb (17, 31-34). This is not the case for the Cdk2/cyclin E complex. In fact, microinjection of cyclin E antibodies is able to inhibit the entry into S-phase of both Rb$^-$ and Rb$^+$ cells (35), demonstrating that substrates other than pRb are involved in regulating the G1/S transition in response to cyclin E activation.

p53 and Cell Cycle Control - The existence of feedback controls allows cells to monitor progression though the cell cycle and to react to any perturbation by arresting the cycle in order to respond to any possible damage (36). One key checkpoint function is exerted by the tumor suppressor protein p53. p53 is a transcription factor that becomes activated in response to DNA damage caused by ionizing radiation or chemical agents (37). The increase in p53 abundance in response to DNA damage is primarily the result of an increase in its half-life. If one compares the half life of p53 in growing cells as opposed to the one in irradiated cells, it is possible to measure an increase of at least five to ten fold. In response to DNA damage, p53 activates the transcription of a number of genes which affect both cell cycle progression and apoptosis (38, 39). p53 could induce cell cycle arrest to allow cells to repair their damaged DNA prior to restarting the division process. The p53 mediated arrest is caused by the accumulation of p21, which binds and inactivates cyclin-dependent kinases. (7-9, 40-45). It is thought that in the presence of severe stress conditions, and in tumor cells that carry alteration in other growth control pathways, p53 can induce the entrance of cells in apoptosis (46). p53 is altered in almost fifty percent of all human cancers (47). In addition to genetic alterations in the p53 gene, other molecular alterations can be present in human tumors that affect its function, either through direct physical interaction with p53, or through interaction with its downstream targets. This is particularly important since it has been demonstrated that alterations in p53 function are associated with an increased resistance of tumors to the action of anti-proliferative drugs (48).

Ubiquitin-Mediated Degradation and Cell Cycle Control - The degradation of most intracellular proteins is thought to involve the activation of a signaling cascade that results in the covalent attachment of a poly-ubiquitin chain to a target molecule (49, 50). This covalent modification occurs through the formation of isopeptide bonds between the carboxy terminus of ubiquitin and ε-amino groups of lysine residues on the target protein. The formation of a poly-ubiquitin chain, through the addition of multiple ubiquitin molecules to the target protein, acts as a signal for degradation by the proteasome, a large 26S multimeric protein complex (51). Recent publications have involved the ubiquitin system in the regulated degradation of several proteins including the tumor suppressor p53 (52-57), the c-jun and c-fos transcription factors (58), the cyclin A and B proteins (4, 59, 60), the NFκB transcription factor and its inhibitor IκB (61), the p27 cell cycle inhibitor (62). Furthermore, it has been shown that the generation of peptides for antigen presentation through the MHC 1 system requires conjugation to ubiquitin (63). Ubiquitin-conjugation to a target protein requires the presence of three key enzymes that have been called E1, E2 and E3 (64). In the presence of ATP, the E1 or ubiquitin-activating enzyme binds ubiquitin forming a thioester bond between a cysteine in its active site and the carboxy-terminus of ubiquitin. The ubiquitin is then transferred to an E2 (ubiquitin conjugating enzyme, UBC) which can either directly transfer ubiquitin to a lysine on the target substrate or donate its ubiquitin first to an E3 enzyme (ubiquitin ligase) which then specifically recognizes the substrate and generates the isopeptide link with ubiquitin. The multiplicity of E2 and E3 enzymes is thought to be responsible for the specific targeting of protein substrates for degradation.

<u>Involvement of Cell Cycle Proteins in Cancer</u> Deregulation of cell cycle controls in the G1 phase of the cell cycle appears to be one of the most frequent molecular alterations in cancer (65). To date the following have been described (see also Fig. 1):

- Mutations in the retinoblastoma tumor suppressor gene (24, 25)

- Overexpression of cyclin D1, caused in some cases by the presence of gene amplification or gene translocation (1, 66-74). It has also been shown that transgenic mice carrying cyclin D1 constructs targeted to the breas,t or the lymphoid tissue, develop mammary adenocarcinomas and lymphomas, respectively (71, 75)

- Cyclin E overexpression in breast cancer (76, 77)

- Altered expression of the p21 protein in cells that lack p53 function. p21 is transcriptionally induced by the tumor suppressor p53 (8, 41). In p53 minus cells, p21 is poorly expressed and is not associated with cyclin-dependent kinases (78).

- Inactivation of the p16/p15 tumor suppressors by gene deletion/ mutations or DNA methylation(reviewed by (79)). Indeed, it has been shown that homozygous mice carrying a disruption of the p16 locus develop tumors at high frequency (80).

- Altered expression of the Cdc25 A and B genes in human breast cancer (81) (see below)

- Lack of expression of the p27 protein due to an acceleration of its degradation (82) (see below), in the absence of apparent structural alterations or point mutations in the p27 gene (83-85).

- Mutations in Cdk4 that render it insensitive to inhibition by p16 (86) (see below).

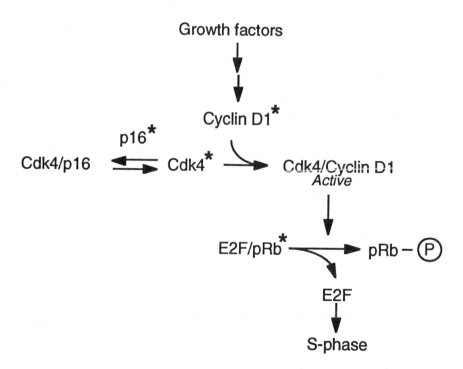

Fig. 1: Signaling cascade linking growth factor stimulation with S-phase entry (asterisks indicate proteins whose function is often altered in tumors, see text for details)

In the following paragraphs we will describe in more detail some of the most recent findings on this topic.

Cdk4 Mutations in Melanoma. A mutation in Cdk4, Arg24Cys, has been identified in a human melanoma (86). The original mutation was due to a somatic mutation occurred in the patient's tumor cells and has thereafter been found in other melanoma cases, although with low incidence. The mutant Arg24Cys Cdk4 protein can form stable complexes with cyclin D1, p27 and p21, but not with the p16 inhibitor. The protein is also enzymatically active in phosphorylating pRb with kinetics comparable to the ones of non-mutant Cdk4. These findings indicate that Cdk4 Arg24Cys is selectively deficient in its interaction with p16, and this could have a role in the generation of the malignant phenotype of the affected melanoma cells.

Oncogenic Properties of Cdc25 Phosphatases. Cdc25 was identified as the product of a yeast (S. pombe) gene required for entry into mitosis (87). Further to that it was shown that homologues of Cdc25 are present in all eukaryotes and that Cdc25 is a mixed-specificity protein phosphatases able to dephosphorylate and activate Cdc2 (88-92). In subsequent years it was shown that in human cells at least two other proteins with sequence homology to Cdc25 exist, Cdc25A and B (93). Cdc25A is required for the G1/S transition in mammalian cells (94, 95) and most likely act by dephosphorylating Cdk4 or Cdk2. Interestingly, it has been shown that Cdc25A and B bind to, and are substrates of the Raf1 protein kinase, a downstream target of Ras (96). Cdc25 A or B immunoprecipitations from human cells, showed a strong association of these proteins with Raf1. Furthermore Cdc25 A and B can bind Raf1 in vitro, and Raf1 can phosphorylate and activate both protein phosphatases. Furthermore, a fraction of Cdc25 A and B colocalize with Raf1 to the cell membrane, the site at which activation of Raf1 occurs (96). These findings demonstrate the existence of a direct molecular link between mitogenic signals and cell cycle control mechanisms. Indeed, Cdc25 A or B can cooperate with Raf1 in inducing the formation of foci of transformed cells upon transfection in mouse embryo fibroblasts. Cells cotransfected with Cdc25 A or B and H-Ras^{G12V} can grow in soft agar. Furthermore, the transfected cells can induce tumor formation in nude mice (81). As for the possible involvement of Cdc25 in human tumors, there is to date no published evidence on the occurrence of mutations or genetic rearrangements. In a recent study though, Cdc25 B was found to be frequently overexpressed in human breast tumors and that this correlated with the progression of the disease. The published study involved a series of 124 human primary breast cancers, that had been characterized for tumor size, hystological and nuclear grade, mitotic rate, and survival after mastectomy over a period of eleven years. Cdc25 B was overexpressed in approx. thirty percent of all tumors, with no detectable expression in normal tissue. Tumor-specific expression of Cdc25 B correlated with a less favorable prognosis and survival (81).

Degradation of the p53 Tumor Suppressor - It has been shown that cells lacking p53 function are more resistant to drug treatment compared to normal cells, through a relative resistance to apoptosis. p53 is a transcription factor which becomes activated in response to cellular damage, e.g. radiation. The protein functions as a tetramer and activates the transcription of genes involved in the regulation of the cell cycle and apoptosis (38, 97-99). The p53 gene is frequently altered in cancer, due to mutations or deletions. p53 can be inactivated though other mechanisms, which include the overexpression of mdm-2 a protein which binds to p53 and prevents its function, or the binding of p53 to HPV-E6. The E6 protein encoded by the human papilloma viruses (HPV) of the 16 and 18 serotypes, binds to and induces the degradation of p53 through the ubiquitin pathway (53, 54). Ubiquintaion of p53 can be reconstructed in vitro in the presence of E6, the ubiquitin conjugating enzyme UBC4, ubiquitin and a protein called E6-associated protein (E6AP) (56, 57, 100). Furthermore, microinjection of anti-UBC4 antibodies or antisense DNA to UBC4 or E6AP prevents the disappearance of p53 in HPV-infected cells (57). In cells carrying an inactivation of the retinoblastoma protein pathway it is possible to induce apoptosis by increasing the intracellular abundance of p53. In various experimental systems it has been demonstrated that cells that overexpress one of the pRb targets, E2F-1, or that lack pRb, are exquisitely sensitive to the induction of apoptosis in the presence of p53 (101-103). It should be possible therefore to induce apoptosis in HPV-infected cells that have both inactivated pRb and p53, by restoring the accumulation of p53 in those cells through inhibition of its degradation.

Regulated Degradation of the p27 Cdk Inhibitor - p27 is a specific inhibitor of cyclin-dependent kinases that is present in high abundance in quiescent cells (104, 105). Its level slowly decreases as cells are stimulated to enter the cell cycle, primarily as a result of an accelerated degradation of the protein (62) (106). Indeed, the intracellular abundance of p27 when added to asynchronously growing cells increases upon addition of inhibitors of the proteasome. Furthermore, it has been shown that p27 is ubiquitinated in intact cells and in cell lysates. Two human ubiquitin conjugating enzymes, UBC2 and UBC3 were found to be involved in p27 ubiquitination since their addition to an ubiquitination reaction accelerated p27 ubiquitination, while addition of mutated proteins carrying a substitution in the active site cysteine, inhibited the process In addition, there is a decrease in the ubiquitination rate for p27 in quiescent cells compared to proliferating cells, accounting for the longer half life of p27 in these cells.

An important finding has come with the demonstration that the regulation of p27 degradation might be altered in cancer. Although a potential role for cyclin kinase inhibitors as tumor suppressors has been suggested, such as p15 and p16, because of the inactivating of mutations in tumors, no similiar molecular alteration in the p27 gene has been demonstrated in tumors. However, in a study conducted on colon cancer cases, p27 was found to be expressed at some levels in ninety percent of the tumors. Interestingly, patients whose tumors lacked p27 expression had a median survival of 69 moths compared to 151 months for patients whose tumors expressed the protein. Furthermore, there was no direct correlation between mRNA and protein expression in the tumors, with twelve percent of the positive tumors showing expression of the protein. Interestingly, a biochemical study in a subset of tumors showed direct correlation between lack of p27 protein in a tumor and amount of p27 degradation activity via the proteasome in the same tumor. No correlation was found with the presence/absence of other cell cycle regulators such as p21 or cyclin A and the ability of tumor extracts to degraded these proteins, suggesting that the inactivation of p27 via increased degradation was a highly specific effect. p27 could be, by analogy the role of p53 in HPV-positive cervical cancer, a selective target for accelerated degradation in a subset of human tumors.

Therapeutic Perspectives - It has become increasingly clear that cancer develops as a disease with multiple underlying genetic defects (107). While the transformed phenotype is described as the ability of a cell to grow in an uncoordinated fashion compared to the other cells of an organism, the spectrum of molecular alterations present make each cancer distinct from the other. Therefore the need emerges for treatments that are targeted to the particular pattern of molecular defects identified in a patient's tumor. There will be a further need for the development of corresponding molecular diagnostics and of novel therapeutics that affect specifically a given molecular reaction. Should the prospect of having to develop hundreds of different therapeutics, each with a relatively small patient basis, be seen as a deterrent for molecular drug discovery in cancer? We would like to think that each of the protein molecules chosen for anti-cancer screening could on its own become a legitimate target for other diseases involving cell proliferation, and that each treatment most likely could be utilized in cancers of different tissue origin. Thus, this approach should constitute an incentive for continued research efforts in the cancer field. The cell cycle regulatory proteins, protein kinases, phosphatases, proteases etc. are very interesting potential targets given that they serve very specific functions, which are often non-redundant. Although many still think about the need for a magic bullet as a cure for all cancers, our knowledge of the molecular mechanisms underlying this disease make the prospect of developing such a universal cure very unlikely.

References

1.   T. Hunter, and J. Pines, Cell, 79, 573 (1994)
2.   D. C. Anderson, L. J. Wible, B. J. Hughes, C. W. Smith, and B. R. Brinkley, Cell, 31, 719 (1982)
3.   C. Sherr, Cell, 79, 551 (1994)
4.   R. King, J. Peters, and M. Kirschner, Cell, 79, 563 (1994)
5.   T. Coleman, and W. Dunphy, Curr. Op. in Cell Biol., 6, 877 (1994)
6.   Y. Gu, C. Turck, and D. Morgan, Nature, 366, 707 (1993)
7.   J. W. Harper, G. R. Adami, N. Wei, K. Keyomarsi, and S. J. Elledge, Cell, 75, 805 (1993)
8.   W. S. El-Deiry, T. Tokino, V. E. Velculescu, D. B. Levy, R. Parsons, J. M. Trent, D. Lin, W. E. Mercer, K. W. Kinzler, and B. Vogelstein, Cell, 75, 817 (1993)
9.   Y. Xiong, G. Hannon, H. Zhang, D. Casso, R. Kobayashi, and D. Beach, Nature, 366, 701 (1993)
10.   A. Noda, Y. Ning, S. Venable, O. Pereira-Smith, and J. R. Smith, Exp. Cell. Res., 211, 90 (1994)
11.   K. Polyak, M. Lee, H. Erdjement-Bromage, A. Koff, J. Roberts, P. Tempst, and J. Massague, Cell, 79, 59 (1994)
12.   Toyoshima, and T. Hunter, Cell, 78, 67 (1994)
13.   M. Lee, I. Reynisdottir, and J. Massague, Genes & Dev., 9, 639 (1995)
14.   S. Matsuoka, M. Edwards, C. Bai, S. Parker, P. Zhang, A. Baldini, W. Harper, and S. Elledge, Genes & Dev., 9, 650 (1995)
15.   M. Serrano, G. J. Hannon, and D. Beach, Nature, 366, 704 (1993)
16.   D. Quelle, R. Ashmun, G. Hannon, P. Rehberger, D. Trono, H. Richter, C. Walker, D. Beach, C. Sherr, and M. Serrano, Oncogene, 11, 635 (1995)
17.   K. Guan, C. Jenkins, M. Nichols, X. Wu, C. O'Keefe, G. Matera, and Y. Xiong, Genes & Dev., , 2939 (1994)
18.   G. J. Hannon, and D. Beach, Nature, 371, 257 (1994)
19.   J. Jen, W. Harper, S. Bigner, D. Bigner, N. Papadopoulos, S. Markowitz, J. Willson, K. Kinzler, and B. Vogelstein, Cancer Research, 54, 6353 (1994)
20.   H. Hirai, M. Roussel, J. Kato, R. Ashmun, and C. Sherr, Mol Cell Biol, 15, 2672 (1995)
21.   F. Chan, J. Zhang, L. Cheng, D. Shapiro, and A. Winoto, Mol Cell Biol, 15, 2682 (1995)
22.   K. Guan, C. Jenkins, Y. Li, M. Zariwala, S. Noh, X. Wu, and Y. Xiong, Mol Biol Cell, 7, 57 (1996)
23.   E. Lees, and E. Harlow. 1995. Cell cycle control. IRL Press, Oxford, NewYork, Tokyo.
24.   M. Ewen, Cancer and Metastasis Reviews, 13, 45 (1994)
25.   R. A. Weinberg, Cell, 81, 323 (1995)
26.   V. Baldin, J. Lukas, M. J. Marcote, M. Pagano, and G. Draetta, Genes & Dev., 7, 812 (1993)
27.   D. Quelle, R. Ashmun, S. Shurtleff, J. Kato, D. Bar-Sagi, M. Roussel, and C. Sherr, Genes & Dev., 7, 1559 (1993)
28.   W. Jiang, S. M. Kahn, P. Zhou, Y. Zhang, A. M. Cacace, A. S. Infante, S. Doi, R. M. Santella, and I. B. Weinstein, Oncogene, 8, 3447 (1993)
29.   D. Resnitzky, M. Gossen, H. Bujard, and S. Reed, Mol. Cell. Biol., 14, 1669 (1994)
30.   R. Medema, R. E. Herrera, F. Lam, and R. A. Weinberg, Proc. Natl. Acad. Sci USA, , (1995)
31.   J. Lukas, D. Parry, L. Aagard, D. J. Mann, J. Bartkova, M. Strauss, G. Peters, and J. Lukas, Nature, , (1995)
32.   J. Lukas, J. Bartkova, M. Rohde, M. Strauss, and J. bartek, Mol. Cell. Biol., 15, 2600 (1995)
33.   J. Koh, G. H. Enders, B. D. Dynlacht, and E. Harlow, Nature, 375, 506 (1995)
34.   S. W. Tam, J. W. Shay, and M. Pagano, Cancer Research, 54, 5816 (1994)
35.   M. Ohtsubo, A. M. Theodoras, J. Schumacher, R. J. M., and M. Pagano, Mol. Cell. Biol., 15, 2612 (1995)
36.   L. H. Hartwell, and T. A. Weinert, Science, 246, 629 (1989)
37.   R. Haffner, and M. Oren, Curr Opin Genet Dev, 5, 84 (1995)

38. D. P. Lane, C. A. Midgley, T. R. Hupp, X. Lu, B. Vojtesek, and S. M. Picksley, Philos Trans R Soc Lond B Biol Sci, 347, 83 (1995)

39. C. J. Leonard, C. E. Canman, and M. B. Kastan, Important Adv Oncol, , 33 (1995)

40. S. Parker, G. Eichele, P. Zhang, A. Rawls, A. Sands, A. Bradley, E. Olson, W. Harper, and S. Elledge, Science, 267, 1024 (1995)

41. V. Dulic, W. Kaufmann, S. Wilson, T. Tlsty, E. Lees, W. Harper, S. Elledge, and S. Reed, Cell, 76, 1013 (1994)

42. H. Zhang, G. Hannon, and D. Beach, Genes & Dev., 8, 1750 (1994)

43. S. Waga, G. Hannon, D. Beach, and B. Stillman, Nature, 369, 574 (1994)

44. R. Li, S. Waga, G. Hannon, D. Beach, and B. Stillman, Nature, 371, 534 (1994)

45. H. Zhang, Y. Xiong, and D. Beach, Mol. Biol. Cell, 4, 897 (1993)

46. D. E. Fisher, Cell, 78, 539 (1994)

47. M. S. Greenblatt, Bennett, W.P., Hollstein, M., Harris, C.C., Cancer Research, , 4855 (1994)

48. M. B. Kastan, C. E. Canman, and C. J. Leonard, Cancer Metastasis Rev, 14, 3 (1995)

49. M. Hochstrasser, Curr. Op. Cell Biol., 7, 215 (1995)

50. A. Ciechanover, Cell, 79, 13 (1994)

51. A. L. Goldberg, Science, 268, 522 (1995)

52. J. Huibregtse, M. Scheffner, and P. Howley, EMBO J., 10, 4129 (1991)

53. M. Scheffner, B. Werness, J. Huibregtse, A. Levine, and P. Howley, Cell, 63, 1129 (1990)

54. M. Scheffner, M. Munger, J. Huibregtse, and P. Howley, EMBO J., 11, 2425 (1992)

55. M. Scheffner, J. Huibregtse, R. Vierstra, and P. Howley, Cell, 75, 495 (1993)

56. M. Scheffner, U. Nuber, and J. Huibregtse, Nature, 373, 81 (1995)

57. M. Rolfe, P. Romero, S. Glass, J. Eckstein, I. Berdo, A. Theodoras, M. Pagano, and G. Draetta, Proc. Natl. Acad. Sci. USA, 92, 3264 (1995)

58. M. Treier, L. Staszewski, and D. Bohmann, Cell, 78, 787 (1994)

59. A. Hershko, D. Ganoth, V. Sudakin, A. Dahan, L. Cohen, F. Luca, J. Ruderman, and E. Eytan, J. Biol. Chem., 269, 4940 (1994)

60. M. Glotzer, A. Murray, and M. Kirschner, Nature, 349, 132 (1991)

61. V. Palombella, O. Rando, A. Goldberg, and T. Maniatis, Cell, 78, 773 (1994)

62. M. Pagano, S. W. Tam, A. M. Theodoras, P. Romero-Beer, G. Del Sal, V. Chau, R. Yew, G. Draetta, and M. Rolfe, Science, 269, 682 (1995)

63. K. Rock, C. Gramm, L. Rothstein, K. Clark, R. Stein, L. Dick, D. Hwang, and A. Goldberg, Cell, 78, 761 (1994)

64. A. Hershko, and Ciechanover, Annu. Rev. Biochem., 61, 761 (1992)

65. J. E. Karp, and S. Broder, Nature Medicine, 1, 309 (1995)

66. G. A. Lammie, R. Smith, J. Silver, S. Brookes, C. Dickson, and G. Peters, Oncogene, , 2381 (1992)

67. W. Jiang, Y. Zhang, S. M. Kahn, M. Hollstein, M. Santella, S. Lu, C. C. Harris, R. Montesano, and I. Weinstein, Proc. Natl. Acad. Sci. USA, 90, 9026 (1993)

68. T. Motokura, and A. Arnold, Curr. Opini. in Genet. and Dev., 3, 5 (1993)

69. J. Lukas, H. Müller, J. Bartkova, D. Spitkovsky, A. Kjerulff, P. Jansen-Durr, M. Strauss, and J. Bartek, J. of Cell Biol., 125, 625 (1994)

70. J. Lukas, D. Jadayel, J. Bartkova, E. Nakeva, M. Dyer, M. Strauss, and J. Bartek, Oncogene, , in press. (1994)

71. H. Lovec, A. Grzeschiczek, M. Kowalski, and T. Moroy, EMBO J., 13, 3487 (1994)

72. E. A. Musgrove, C. S. L. Lee, M. F. Buckley, and R. L. Sutherland, Proceedings of the National Academy of Sciences of the United States of America, 91, 8022 (1994)

73. J. Bartkova, J. Lukas, M. Strauss, and J. Bartek, International Journal of Cancer, 58, 568 (1994)

74. S. Bodrug, B. Warner, M. Bath, G. Lindeman, A. Harris, and J. Adams, EMBO J., 13, 2124 (1994)

75. T. C. Wang, R. D. Cardiff, L. Zukerberg, E. Lees, A. Arnold, and E. V. Schmidt, Nature, 369, 669 (1994)

76. A. Dutta, R. Chandra, L. Leiter, and S. Lester, Proc. Natl. Acad. Sci. USA, 92, 5386 (1995)

77. K. Keyomarsi, and A. Pardee, Proc. Natl. Acad. Sci, USA, 90, 1112 (1993)

78. Y. Xiong, H. Zhang, and D. Beach, Genes & Dev., 7, 1572 (1993)

79. R. J. Sheaff, and J. M. Roberts, Current Biol., 5, 28 (1995)

80. M. Serrano, Cell, , (1996)

81. K. Galaktionov, A. Lee, J. Eckstain, G. Draetta, J. Meckler, M. Loda, and D. Beach, Science, in press, (1995)

82. M. Loda, B. Cukon, S. Tam, P. Lavin, M. Fiorentino, G. Draetta, J. Jessup, and M. Pagano, submitted, , (1996)

83. V. Ponce-Castañeda, M. Lee, E. Latres, K. Polyak, L. Lacombe, K. Montgomery, S. Mathew, K. Krauter, J. Sheinfeld, J. Massague, and C. Cordon-Cardo, Cancer Research, 55, 1211 (1995)

84.   A. N. Kawamat, R. Morosetti, C. Miller, D. Park, K. Spirin, T. Nakamaki, S. Takeuchi, Y. Hatta, J. Simpson, S. Wilczynski, Y. Lee, C. Bartram, and H. Koeffler, Cancer Res., 55, 2266 (1995)

85.   J. Pietenpol, S. Bohlander, Y. Sato, N. Papadopoulos, B. Liu, C. Friedman, B. Trask, J. Roberts, K. Kinzler, J. Rowley, and B. Vogelstein, Cancer Research, 55, 1206 (1995)

86.   T. Wölfel, M. Hauer, J. Schneider, M. Serrano, C. Wölfel, E. Klehmann-Hieb, E. De Plaen, T. Hankeln, K. Meyer zum Büschenfelde, and D. Beach, Science, 269, 1281 (1995)

87.   P. Russell, and P. Nurse, Cell, 45, 145 (1986)

88.   J. Gautier, M. J. Solomon, R. N. Booher, J. F. Bazan, and M. W. Kirschner, Cell, 67, 197 (1991)

89.   J. Millar, C. H. McGowan, G. Lenaers, R. Jones, and P. Russell, EMBO J., 10, 4301 (1991)

90.   U. Strausfeld, J. C. Labbe, D. Fesquet, J. C. Cavadore, A. Picard, K. Sadhu, P. Russell, and M. Doree, Nature, 351, 242 (1991)

91.   W. G. Dunphy, and A. Kumagai, Cell, 67, 189 (1991)

92.   M. S. Lee, S. Ogg, M. Xu, L. Parker, D. Donoghue, J. Maller, and H. Piwnica-Worms, Mol. Biol. Cell, 3, 73 (1992)

93.   K. Galaktionov, and D. Beach, Cell, 67, 1181 (1991)

94.   I. Hoffmann, G. Draetta, and E. Karsenti, EMBO J., 13, 4302 (1994)

95.   S. Jinno, K. Suto, A. Nagata, M. Igarashi, Y. Kanaoka, H. Nojima, and H. Okayama, EMBO J., 13, 1549 (1994)

96.   K. Galaktionov, C. Jessus, and D. Beach, Genes & Develop., 9, 1046 (1995)

97.   P. Hainaut, Curr Opin Oncol, 7, 76 (1995)

98.   D. P. Lane, X. Lu, T. Hupp, and P. A. Hall, Philos Trans R Soc Lond B Biol Sci, 345, 277 (1994)

99.   M. L. Hooper, J Cell Sci Suppl, 18, 13 (1994)

100. M. Scheffner, J. Huibregtse, and P. Howley, Proc. Natl. Acad. Sci. USA, 91, 8797 (1994)

101. X. Q. Qin, D. M. Livingston, W. G. Kaelin, Jr., and P. D. Adams, Proc Natl Acad Sci U S A, 91, 10918 (1994)

102. A. Almasan, S. P. Linke, T. G. Paulson, L. C. Huang, and G. M. Wahl, Cancer Metastasis Rev, 14, 59 (1995)

103. X. Wu, and A. J. Levine, Proc Natl Acad Sci U S A, 91, 3602 (1994)

104. C. Sherr, and J. Roberts, Genes& dev., 9, 1149 (1995)

105. S. Elledge, and W. Harper, Curr. Opin. Cell Biol., 6, 847 (1994)

106. L. Hengst, and S. I. Reed, Science, 271, 1861 (1996)

107. J. E. Karp, and S. Broder, Nature Medicine, 1, 309 (1995)

Chapter 26. Regulation of Apoptosis by Members of the ICE Family and the Bcl-2 Family

Douglas K. Miller
Merck Research Laboratories
Rahway, NJ 07065

Introduction - Members of an expanding family of cellular cysteine proteases, the ICE family, have been found to play key roles in the initiation of programmed cell death caused by a diverse assortment of stimulators. ICE family proteases initiate cell death by functional inactivation of key nuclear and membrane proteins involved in the maintenance of cellular structure and DNA repair. Activation of the ICE family of proteases is partially controlled by the Bcl-2 family. These proteins appear to function downstream from an emerging third diverse group of proteins that contain death domains which enable the signal transduction of the pro-apoptotic receptors to the ICE proteases. It is the purpose of this review to elaborate on the structure, activity, and interaction of members of the Bcl-2 and ICE protein families following apoptotic stimulation.

THE APOPTOTIC PROCESS

Cells can be induced to die by a series of carefully controlled, physiologically regulated cellular events or more suddenly in an uncontrolled fashion. The first is an active, enzymatic process termed either 'programmed cell death' (1) or 'apoptosis' (2). While 'programmed cell death' is associated with a developmental process, the term 'apoptosis' refers to the actual morphological changes associated with the dying cells (3). Derived from the Greek meaning "falling away" as a leaf in autumn, apoptosis is characterized by a sequence of distinct cellular changes that occur in a time-dependent fashion. The term 'necrosis' is associated with sudden pathological cell death induced by physical trauma, ischemia, hyperthermia, or oxidative or chemical injury. The morphological changes that distinguish apoptosis from necrosis are summarized in Table 1 (4-11). Typically, necrotic cells die quickly en masse as clusters whereas apoptotic cells die individually surrounded by healthy cells. In apoptotic cells, The increased exposure of 3'-OH ends of DNA has enabled the development of the popular TUNEL labeling technique, in which terminal deoxynucleotidyltransferase adds a tagged dUTP to the nicked ends (8).

Table 1 Comparison of morphological changes in apoptosis and necrosis.

Characteristic	Apoptosis	Necrosis
Cause of Death	Internal	External
Cell Size	Shrinks 30-50%	Swells
Cell Density	Heavy	Light
Plasma Membrane	Intact Extensive Blebs	Ruptured Blebs
Nucleus	Fragmented	Little Change
Chromatin	Condensed	Little Change
DNA	Nucleosome Ladder (180 bp)	Random Cleavage
Reversibility	No	Variable

Apoptosis can be induced in cells by removal of necessary growth factors (12) or the addition of hormones such as glucocorticoids, anticancer agents, ionizing or UV-irradiation (13-16). Cross-linking of antigen receptors also stimulates cell death of B and T cells. In thymic maturation, the high affinity interaction between the T cell receptor and autoantigens induces the death of autoreactive T cells (17). DNA damage or viral infection stimulates p53 which induces apoptosis through upregulation of genes such as Bax and p21/WAF1 (18-21). Overexpression of proteins involved in regulating cell cycle progression e.g. c-Myc and cyclin D1 also induce apoptosis (22,23).

ACTIVATION AND TRANSMISSION OF THE CELL DEATH SIGNAL

Receptors with Cell Death Domains - Several members of the TNF receptor superfamily induce apoptosis in cells when stimulated by ligands or antibodies. The best characterized members of this family include the p55 TNF receptor (TNFR1) and the Fas or APO-1 molecule and the 75kDa neurotrophin receptor (Fig. 1). These receptors contain in their cytoplasmic tail an 80 amino acid long "death domain" of five tandem alpha helices that mediate protein-protein interactions (24-27). The fifth alpha helix is highly homologous to the amphipathic helix formed by the mastoparans, and may act similarly as a coil to helix transition switch to activate the proteins to which it is bound (28). Partial deletion or point mutations of conserved amino acids in the death domain disrupt the proapoptotic signalling of Fas and TNFR1 (25,26,29). Aggregation of the receptors normally occurs upon binding of TNFα or Fas (26, 30). Increased expression of the receptor death domain enhances cytotoxicity by promoting self association within the domains (31). The inducible TNF receptor (p75, TNFR2) has little sequence homology with the TNFR1 in the cytoplasmic domain and is largely

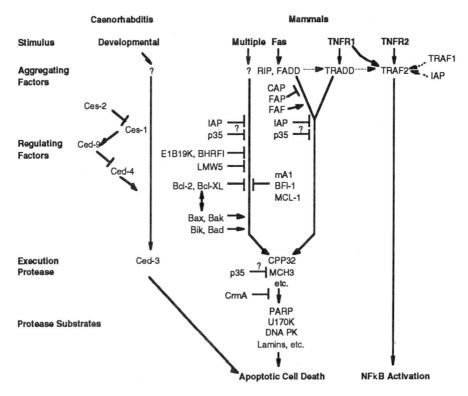

Figure 1. Factors involved in the activation, regulation, and execution of apoptosis in *Caenorhabditis*, and mammals.

associated with the activation downstream of the transcription factor NFKB (32). Recently, a number of different proapoptotic proteins have been identified by the yeast two-hybrid system which aggregate to the death domain and propagate the apoptotic signal.

<u>Proteins that Interact with Fas</u> - The 23-kDa protein FADD (**Fas-a**ssociated **death** domain) (33) or MORT1 (mediator of receptor-induced toxicity) (34) contained a homologous death aggregation domain and interacted with the death domain of Fas, but had little interaction with the TNFR1. A point mutation in the Fas death domain that inactivated Fas (as found in the *lpr*, autoimmune mouse)(26) prevented its association with FADD. Furthermore, the removal of the C-terminal 15 amino acids of Fas enhanced cell killing activity (24) by increasing FADD association (33, 34). Transfection of a FADD mutant lacking the N-terminal domain necessary for apoptosis could inhibit apoptosis induced by Fas. A 74-kDa protein, termed RIP (receptor interacting protein) (35), was also identified with a homologous death domain as well as a larger N terminal region containing a Tyr kinase homologous domain. RIP was found to interact with Fas via its death domain region, and could induce apoptosis when expressed in cells. RIP was expressed both cytoplasmically and in a perinuclear punctate pattern.

<u>Proteins that Interact with the TNFR</u> - A 34-kDa protein called TRADD (**TNRF1-a**ssociated **death domain**) (36) contained a homologous death domain and directly interacted with the TNFR1 and dimerizes with itself and with Fadd, but did not interact with Fas (37). Overexpression of TRADD in cells induced apoptosis and the activation of NF-ĸB. The simultaneous transfection of the antiapoptotic Bcl-2 or E1B 19K proteins could not prevent this cell death (36). The proapoptotic effects of TRADD may be controlled by regulatory proteins such as A20, a Zn RING finger protein (38,39). TRAF2 (TNF receptor-associated factor) is a 56-kDa protein which bound directly to the cytoplasmic domain of TNFR2, and TRAF1 is a 45-kDa protein that could bind to TRAF2 on the TNFR2 (43). None of the TRAF molecules could interact with the TNFR1 or Fas, and only TRAF2 was able to bind directly to TNFR2 and induce NF-ĸB activation. Hence, TRAF2 for the TNFR2 is the functional counterpart of TRADD for the TNFR1. The TRAF molecules contain both coiled-coil protein sequences that enable heterodimerization by packing of amphipathic helices, as well as RING or Zn-finger sequence motifs that enable binding to nucleic acids. Associated with the activated TNRF2/TRAF2 complex and bound to a distinct TRAF domain, are the inhibitor of apoptosis proteins (hc-IAP1 and hcIAP2) which are recruited following binding of TRAF1 (42). All the IAPs contain RING finger motifs, as well as BIR (baculovirus iap repeat) regions, the latter of which is required for c-IAP binding to TRAF. The c-IAPs have, however, been shown to suppress apoptosis induced by withdrawal of growth factors (43), but they do not inhibit NF-ĸB activation, and they do not block apoptosis induced by TNFRI, Fas, actinomycin D or staurosporin. These observations suggest that mammalian IAPs differentially interact with and modulate the stimulatory proteins that form aggregated complexes early during the transduction of certain receptor-mediated signals.

<u>Other Death Domain Associated Proteins</u> - Other proteins found associated with the Fas cytosolic domain by coimmunoprecipitation include CAP3 and CAP4 (cytotoxicity-dependent APO-1-associated proteins) (44). The yeast two-hybrid system was used to identify murine FAP-1 (Fas-associated phosphatase) that inhibited Fas-induced cell death (45). By contrast to the inhibitory effects of FAP-1, the FAF1 protein (Fas-associated protein factor) potentiates Fas-induced apoptosis by an unknown mechanism (46). The TRAP-1 and TRAP-2 (TNF receptor-associated protein) members of the 90-kDa family of heat shock proteins associated with TNFR1 are homologous to members of TNFR1 (47).

Downstream of the death-promoting, receptor-mediated signalling events lie a series of proteins that both regulate and initiate programmed cell death. In the small roundworm *Caenorhabditis elegans* , three genes directly control the onset of apoptosis during development: Ced-9, Ced-4, and Ced-3 (48). Ced-9 protects cells from apoptosis and is functionally and structurally similar to and replaceable by the antiapoptotic regulatory protein Bcl-2 (49-51). The Ced-3 protein is necessary to induce cell death and is highly homologous to ICE (55, 56). The novel protein Ced-4 (54) interacts with Ced-9 in a fashion comparable to the interaction of prodeath effectors such as bax (see below) with Bcl-2 (55). Two other uncharacterized gene products function upstream of Ced-9 to control its activity: Ces-1 (cell death specification) acts positively to regulate Ced-9, while Ces-2 inhibits Ced-9 by inhibiting Ces-1 (56, 57).

THE BCL-2 FAMILY OF APOPTOSIS REGULATORS

The Bcl-2 gene (B-cell lymphoma/leukemia-2) was identified when it translocated to t(14:18), adjacent to the immunoglobulin heavy-chain locus, where it was shown to be involved in B cell malignancies (58). This gene increased lymphoma cell numbers by blocking cellular apoptosis and prolonging cell survival (59-61). Bcl-2 expression inhibited the cell death induced by the loss of cytokines, heat shock, alcohol treatment, UV and γ-irradiation, chemotherapeutic drugs, and glutamate (62-64). Bcl-2 cannot, however, prevent neuronal apoptosis induced by the beta amyloid protein, TNF-induced death, or negative selection in thymocytes via Fas. In Bcl-2 sensitive systems, the extent to which Bcl-2 protein provides protection varies from 10,000 fold for protection against dexamethasone to 5 fold for the anticancer drug 2-chlorodeoxyadenosine (65). Bcl-2 is highly expressed in some long lived cells, such as neurons, stem cell populations in epithelial cells, and medullary immature thymic T cells and long-lived memory B and peripheral T cells. Bcl-2 is low or absent in terminally differentiated cells that apoptose and slough off, in the short lived B cells in lymph nodes, or the immature thymocytes, many of which die by apoptosis (66-69).

The 26-kDa Bcl-2 protein is present in a variety of membranes including the plasma membrane, the junction regions of inner and outer membranes of mitochondria and around the nuclear pore complexes (73). Bcl-2 contains a hydrophobic domain near its C-terminus (Fig. 2) that is not essential for activity but may act to concentrate the protein at membrane sites (74-77). The mitochondrial localization of Bcl-2 was originally thought to be important for protection from the oxidative species generated in mitochondria (75), but recent evidence suggests that reactive oxygen species are not required for apoptosis, and Bcl-2 can prevent apoptosis induced by other means (78- 80).

Members of the Bcl-2 family have either pro-apoptotic or anti-apoptotic functions. Three conserved regions within the Bcl-2 family members are essential for protein-protein interactions and enable dimerization with other Bcl-2 homologs. The first Bcl-2 homology region (BH1) contains FRDG and NWGR sequences which are essential for activity, and the second homologous region (BH2) contains conserved WI and GGW sequences (81). A third homologous region BH3 containing conserved LR residues was identified (82). The BH3 region is the most important region of the proapoptotic Bcl-2 family members in enabling its interaction with the antiapoptotic family members in their BH1, BH2, and BH3 regions(83, 84). A fourth N-terminal region was essential for bcl-2 homodimerization but not heterodimerization (81,85,86).

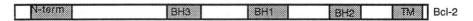

Figure 2. Structure of Bcl-2 family members containing the conserved N-terminal, BH1, BH2, BH3, and C-terminal hydrophobic domains.

Bcl-2 Family Members with Anti-Apoptotic Activity - In addition to Bcl-2, mammalian homologs of this group include Bcl-x$_L$, Bfl-1 (A-1), and Mcl-1.   Bcl-x$_L$, highly homologous to Bcl-2 in the conserved BH1 and BH2 domains, prevented apoptosis caused by growth factor withdrawal (84).   Bcl-x$_L$ also contained the C-terminal hydrophobic domain and the BH3 domain which enabled its binding to the proapoptotic homologs Bak and perhaps to Bax as well (80,85). Different residues in BH1 and BH2 of Bcl-x$_L$ and Bcl-2 were necessary for protection against cell death (86).   An alternatively spliced form Bcl-x$_S$, containing a 63 amino acid deletion comprising both the BH1 and BH2 domains did not prevent apoptotic death but could antagonize the protective effect of Bcl-2 or Bcl-x$_L$ (82,84,85). Bcl-x$_L$ and Bcl-2 were complementarily expressed in different tissues (97). A search of novel human fetal liver genes using expressed sequence tags yielded the Bfl-1 gene (Bcl-2-related gene expressed in fetal liver) (88). Bfl-1 was highly homologous to Bcl-2 in the conserved regions and was expressed abundantly in hematopoetic cells. Overexpression of Bfl-1 in stomach cancers, but not in hepatoma cancers, indicated that it may be involved with the progression of certain cancers by permitting cell survival (89). Mcl-1 (myeloid cell leukemia) was discovered by its increased expression in phorbol ester-induced differentiation of the myeloblastic cell line ML-1 (90,91).   It was mapped to a chromosomal location that is often duplicated or rearranged in a variety of cancers (92). Like Bcl-2, Mcl-1 levels were high in differentiated epithelial tissues, such as skin and intestine, but low in the self-renewing stem cells. In addition, Mcl-1 contains a hydrophobic C-terminal region that permits it to be associated with mitochondria and other light membrane organelles (93). Cell death induced by overexpression of c-Myc was prevented by the coexpression of Mcl-1 but it was less efficient than the effects of Bcl-2 (94).

Bcl-2 Family Members with Proapoptotic Properties - These family members include Bax, Bak, Bad, and Bik. Bax was the first of a series of proapoptotic homologs of Bcl-2 identified: overexpression of Bax in cells resulted in their accelerated apoptosis, but only after a signal for apoptosis such as growth factor withdrawal was initiated. Bax could form homodimers as well as heterodimers with Bcl-2; cell death was increased with large numbers of Bax homodimers, while cell survival was associated with Bcl-2 heterodimers.   Overexpression of Bcl-2 in cells reduced the amount of Bax homodimerization, although the majority of Bcl-2 was not associated with Bax. Bcl-2 could only inhibit cell death if it first bound to Bax, and a disruption of Bcl-2/Bax heterodimer formation by mutation in the BH1 and BH2 regions of Bcl-2 promoted cell death. Hence, Bcl-2 can be viewed as a regulator that opposes the death-inducing activity of Bax (78,95,96). While BH1 and BH2 were necessary in Bcl-2 to enable it to bind to Bax, they were not necessary in Bax to enable its binding to Bcl-2. Instead, only the BH3 domain of Bax was essential for its proapoptotic activity and binding to Bcl-2, Bcl-x$_L$, or the E1B 19K viral homolog (80,81). Conversion of the BH3 domain of Bcl-2 to that of Bax converts Bcl-2 into a proapoptotic molecule (97). The exact molecular interaction of Bax with other homologs such as Bcl-x$_L$ is unclear (86). Reciprocal expression of Bax and Bcl-2 is consistent with their opposing effects in apoptosis (98).

Bak (Bcl-2 homologous antagonist/killer), contains BH1 and BH2 and was identified by its ability to interact with the antiapoptotic adenovirus protein E1B 19K protein (99-101). Bak interacted directly with and antagonized the antiapoptotic effects of Bcl-2 and Bcl-x$_L$. Bak could induce cell death unlike Bax, even in the presence of serum (100). Only the BH3 domain of Bak is essential for its cell death promotion activity and its ability to bind Bcl-x$_L$ and Bcl-2 (86). Bad, a proapoptotic 22-kDa protein Bcl-2-interacting protein (Bcl-2 associated death promoter) showed limited homology with the Bcl-2 family in the BH1 and BH2 regions, and did not contain a putative C-terminal transmembrane region (102). In cells, Bad was associated with both Bcl-x$_L$ and Bcl-2, but not with Bax. The association of Bad with Bcl-x$_L$ was the strongest, because Bad expression was able to overcome the inhibition of cell death by Bcl-x$_L$. Bik (Bcl-2 interacting killer) is a 26.5-kDa proapoptotic protein localized in the nuclear envelope and perinuclear area similarly to that of Bcl-2. Bik lacked the BH1 and BH2 homologous regions found in many of the Bcl-2 family members but contained the BH3 homologous region (103). Bik can interact with both Bcl-2 and Bcl-x$_L$, which contain the BH1 and BH2 domains, as well as Bcl-x$_S$, which lacks the BH1 and BH2 domains, by means of the BH3 domain (103).

Bcl-2 Family Transgenics - In contrast to prolonged life of lymphocytes in the Bcl-2 transgenics (58), Bcl-2 knockout mice exhibited a shortened lifespan of mature T and B cells. While embryonic life developed normally in these knockouts, loss of Bcl-2 led to apoptosis of kidney cells producing a polycystic kidney disease, apoptosis of hair follicle melanocytes producing a loss of hair color, and increased death of crypt and differentiating epithelial cells in the intestine producing shortened, deformed villi (104-106). Bcl-x$_L$, in contrast, is required for normal embryonic development: Bcl-x$_L$ knockout mice died early in gestation with massive apoptotic neuronal and hematopoietic liver cell degeneration. Unlike the Bcl-2 knockout, the Bcl-x$_L$ knockout showed a reduction in the developing T and B cells, while the differentiation and lifespan of the mature lymphocytes was uneffected (107). Knockout of the proapoptotic Bax homolog produced more modest effects: deficient mice exhibited increased number of B and T cells, an increase in ovarian granulosa cells, and an increase in premeiotic sperm cells together with abnormal seminiferous tubules producing male infertility (108). These results indicate that the importance of the various Bcl-2 homologs varies from cell to cell.

Other Bcl-2 Associated Proteins - Using the yeast two-hybrid system, a number of unique proteins that associate with Bcl-2 have been found that are not structurally similar to Bcl-2. Nip-1, -2, and -3 (Nineteen-kDa-interacting proteins) interacted with the E1B 19K protein as well as Bcl-2 and BHRF1 (109). Nip-1 and -2 were proteins expressed at the nuclear envelope while Nip-3 was associated with mitochondria. Nip-1 shared some sequence homology the catalytic domain of mammalian cyclic nucleotide phosphodiesterases, while Nip-2 shared sequence homology with the GTPase-activating protein RhoGAP. BAG-1 (Bcl-2-associated athanogene (Gr, "antideath")) is an acidic, amphipathic 219 residue protein with homology to ubiquitin-like proteins but no sequence homology to Bcl-2 or its other associated proteins. Because BAG-1 protects cells from the induction of apoptotic death when both Bcl-2 and BAG-1 are cotransfected, it was speculated that BAG-1 may also play an aggregation role (110).

Viral Homologs of Bcl-2 - Oncogenic DNA viruses have evolved numerous strategies to prevent the apoptotic death of cells that they infect. While they stimulate cell activation to promote increased virion production, such induction normally would induce cellular apoptosis. Adenovirus contains a viral homolog to Bcl-2, an early gene product termed the E1B 19K protein that inhibits cellular apoptosis (111,112). While the 19K protein is evolutionarily very distant from Bcl-2, it contains modest homology to the BH1 and BH2 regions, and Bcl-2 is able to complement the antiapoptotic requirement for the 19K protein (113). Despite its sequence divergence from Bcl-2, the 19K protein reacts with a common group of antiapoptotic cellular proteins, Nip-1, 2, 3 (109) as well as the proapoptotic protein Bak and Bax (81,101). The E1B 19K protein displays a wider range of antiapoptotic activity than does Bcl-2 itself. For example, cell death induced by TNFα or anti-Fas treatment is only partially inhibited by Bcl-2, whereas it is largely protected by the 19K protein (113,114). Furthermore, overexpression of E1B 19K protein but not Bcl-2 can protect against cell death caused by the expression of Bcl-x$_S$. This suggests that E1B 19K interacts with more effectors of apoptosis in cells than do Bcl-2 or Bcl-x$_L$ alone (115).

Epstein-Barr virus infects B cells and produces a series of latent proteins, such as LMP-1 (latent membrane protein), that enables their long term survival by upregulating Bcl-2 expression in those cells (116). The virus also produces a 17-kDa protein early during the lytic phase of the infection cycle, BHRF1 (*Bam*H1 **H RF**1 open reading frame), that itself bears a distant homology to Bcl-2. Besides containing homology to the BH1 and BH2 domains, BHRF1 also contains the C-terminal hydrophobic domain for membrane association (117,118). Cellular expression of BHRF1 protects B cells from apoptosis due to serum withdrawal or calcium ionophore addition (119) and protects CHO cells from apoptosis due to treatment with DNA damaging agents such as cisplatin and etoposide (120). Other viral proteins with homologous BH1 and BH2 sequences to Bcl-2 and similar antiapoptotic functions include the 20-kDa LMW5-HL protein of the African Swine Fever virus (121), a 16-kDa protein of Herpes saimiri virus (122) and the $\gamma_1$34.5 gene of Herpes simplex virus (123).

Baculovirus p35 and IAP Proteins–Two classes of cytoplasmic baculoviral proteins inhibited the cellular apoptotic pathway induced by the viral infection. The first was a 35-kDa baculoviral late expressing gene product, p35, identified by loss of function mutants that increased the apoptotic cell death of the infected host insect cells (124-127). By complementing p35-deleted viruses with DNA from other baculoviruses, a second 30-kDa group of antiapoptotic gene products were discovered, termed IAPs (inhibitor of apoptosis) (128,129). The IAPs contained two 80 amino acid BIR (baculovirus iap repeat) putative metal binding sequences as well as a Zn-finger binding motif, and both are critical for the antiapoptotic activity (127,129). Similarly the N-terminal region of p35 was necessary for its anti-apoptotic activity (130). Expression of p35 in mammalian cells could, like Bcl-2, prevent apoptosis by a variety of stimuli (131,132), and p35 could rescue the programmed cell death found in *C. elegans* mutant containing a defective ced-9 (133). Mammalian homologs for the IAPs include NIAP, the 140-kDa gene product for the neurodegenerative disorder spinal muscular atrophy (SMA), in which loss of a BIR domain was characterized by the apoptosis of motor neurons (134). Recently, three more human IAPs (hIAP-1, hIAP-2, xIAP) have been identified (42,43). Overexpression of IAP in CHO cells protected them from apoptosis induced by TNF or free radicals, and expression of the other three human IAPs could suppress apoptosis of CHO cells following serum withdrawal. Since NIAP is expressed only in the motor neurons that become apoptotic in the SMA disease and not in sensory nerves, these various IAPs may be necessary for inhibition of apoptosis in different cells (43).

The Bcl-2 family members and the baculoviral and mammalian p35 and IAP members represent nonoverlapping regulatory proteins of apoptosis. The antiapoptotic effects of p35 were not inhibited by the overexpression of the proapoptotic Bcl-x$_S$, and the antiapoptotic Bcl-2 and E1B 19K could not substitute for p35 or IAP to reduce baculoviral apoptosis, indicating that p35 and IAP had different mechanisms of action from the Bcl-2 family (115,127,130). While their molecular mechanism is not known, Bcl-2 family members do not affect all apoptotic signals and generally they only modulate the intensity of apoptotic signals that are already underway. In contrast, the inhibitory effects of p35 and the IAPs are more general than those of mammalian or viral Bcl-2 homologs, and they appear early in conjunction with an apoptotic stimulus. This suggests that p35 and the IAPs inhibit directly the initial activation pathway of apoptosis, a role that is becoming more evident from the observed associations of IAPs with the active receptor death domain complexes (42,43). p35 itself may have a separate or additional role as an inhibitor of activities of the ICE protease family. Recent evidence has indicated that p35 can serve as a pseudosubstrate, providing a competitive ICE-like cleavage site for subsequent protease inhibition (135,136). This would imply that p35 functions identically to the protease inhibitory activity of CrmA. Whether or not this represents the physiological antiapoptotic activity of p35 during virus infection remains to be demonstrated.

THE ICE FAMILY OF CYSTEINE PROTEASES

Interleukin-1β converting enzyme (ICE), was originally identified as the monocytic cell enzyme responsible for activation of IL-1β (137,138). IL-1β is a "master cytokine" that stimulates synthesis and secretion of matrix matalloproteinases, other cytokines such as IL-6, the inducible form of nitric oxide synthetase, cyclooxygenase 2, and a number of leukocyte adhesion proteins (139,140). The importance of IL-1β in disease is highlighted by the observation that i) antiIL-1β antibodies prevent and reverse collagen-induced arthritis in animal models (141,142); ii) mice with the IL-1β gene knocked out had an impairment in fever and acute-phase responses following the proinflammatory stimulus lipopolysaccharide (LPS) treatment (143); and iii) IL-1β is specifically controlled during inflammation by high levels of soluble Type II IL-1 receptor (144,145). To diminish host response mediated by IL-1β, pox viruses have targeted both the activation of IL-1β by producing an inhibitor of ICE, CrmA, as well as the binding of IL-1β to its receptor by producing B15R, a protein homologous to the Type II IL-1 receptor (146,147). The importance of IL-1β in defense against viral infection is underscored by the observation that the IL-1β knockout mouse is much more sensitive to influenza infection (148).

Following proinflammatory stimulus such as LPS, IL-1β is synthesized in monocytes as a soluble cytosolic 31-kDa precursor (pIL-1β) which is subsequently released extracellularly as a mature 17-kDa C-terminal fragment (mIL-1β) (149). The 31-kDa pIL-1β is inactive on the Type I IL-1 signalling receptor unless it is first processed by ICE at a site following Asp^{116} to yield the mIL-1β (150). ICE was originally identified as an atypical cysteine protease with an absolute specificity for Asp in P_1 (153,154). The minimal substrate for this enzyme was a tetrapeptide with a large hydrophobic group such as Tyr, Phe, or Leu present in P_4, similar to what was normally observed at the $YVHD^{116}$ cleavage site of pIL-1β (151,155,156). AcYVAD-aminomethylcoumarin is used routinely as a substrate ($K_m = 14$ μM). The aldehyde inhibitor, AcYVAD-H (Fig. 3) was a potent, reversible inhibitor ($K_i = 0.8$ nM). By substituting Lys for Ala, a useful affinity reagent was created that allowed purification of ICE from a crude cytosolic mixture (151,156)

Purified, active ICE contained two subunits, p10 and p20, that were processed from a common 45-kDa (p45) precursor (151,152,157). The crystal structure of ICE bound to either AcYVAD-H or AcYVAD-CMK inhibitors (Fig. 3) showed that the enzyme was a homodimer of two p20/p10 heterodimers, in which the homodimer interface was largely between the two p10s (158,159). Two inhibitors were bound/tetramer with the P_1 Asp in a deep pocket stabilized by Arg^{179} and Arg^{341}. The catalytic His^{237} and the Cys^{285} residues were provided by the p20 subunit, whereas the p10 subunit provided the residues for binding the P_2-P_4 residues. The P_2 and P_3 side chains of the substrate were surface exposed, consistent with their tolerance for substitution. The binding of the tetrapeptide occurs in an extended β-sheet conformation where the P_1 and P_3 amido nitrogens are critically involved in binding to the enzyme. The P_4 binding pocket was a broad shallow pocket with His^{342} and Pro^{343} closely opposed to the N-terminal acetyl group and the Tyr ring (158,159).

ICE Inhibitors - Both reversible and irreversible peptide-based ICE inhibitors block pIL-1β- cleavage in vitro and in vivo. In addition to the aldehyde described above (163-165), peptide ketones are also potent reversible inhibitors (164,165). Derivatives of acyloxymethyl ketones (165, 166) were potent, time dependent irreversible inhibitors of ICE. WIN 67694 (Fig. 3) inhibited ICE with a second-order inactivation rate of about 400,000 (167-169). Inactivation occured by displacement of the carboxylate leaving group to form a thiomethyl ketone with the active site Cys^{285} as determined by mass spectral analysis (168). ICE showed a surprising tolerance for D-stereochemistry in the P_1 Asp position unlike other cysteine proteases (171).

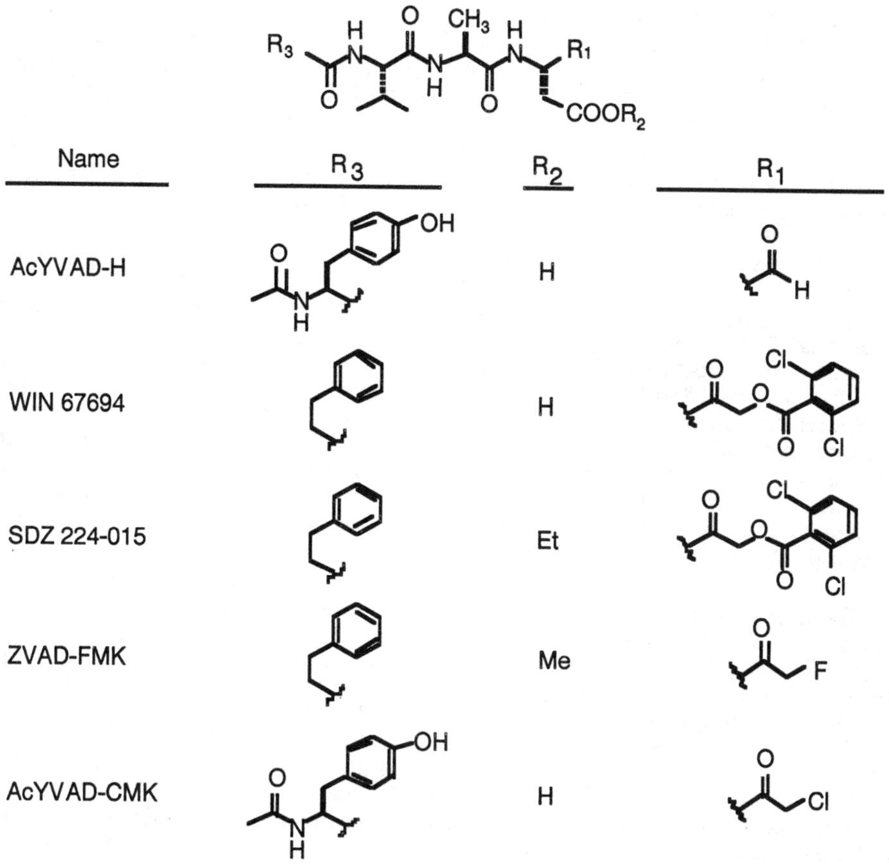

Figure 3. Structures of ICE family protease inhibitors.

CrmA protein (cytokine response modifier gene), a 38-kDa early gene product produced by cowpox virus with sequence similarity to serpins (serine protease inhibitors) promoted severe hemorrhagic lesions in host animals. Expression of CrmA prevented accumulation of inflammatory cells in the lesion, reducing the host inflammatory response(172-175). Extracts of the lesions containing the CrmA were found to inhibit ICE processing of pIL-1β. Purified CrmA with a Leu-Val-Ala-Asp sequence in the reactive site loop, was a pseudosubstrate, (176) and a specific inhibitor of ICE with (K_i of 4-7 pM)(180). When CrmA was coexpressed in cells with ICE and pIL-1β, the processing and secretion of mIL-1β was totally inhibited (178).

<u>Effects of ICE inhibitors in cells and animal models of inflammation</u> - Ac-YVAD-H inhibited the release of mIL-1β from monocytes in blood with an ED_{50} of about 0.7-4 mM without blocking IL-1α, IL-6, or TNFα production (151,155, 179). Likewise, the irreversible ICE inhibitors typified by the ZVAD-acyloxymethyketone (WIN 67694) gave similar inhibition (0.2-1.8 μM) in LPS-activated THP.1 cells and in macrophages, without effects on the secretion of other cytokines (180,181). In a study of LPS-induced production of cytokines in mice, 10 mg/kg of Ac-YVAD-H produced a transitory 80% inhibition of IL-1β in blood without affecting IL-1α or IL-6 production (179). In a murine subcutaneous tissue chamber implantation model of inflammation,

IL-1β production stimulated by zymosan was reduced 55% at 100 mg/kg of WIN 67694 without effects on IL-1α , IL-6, or TNFα (180). SDZ 224-015, dosed p.o., gave 50% inhibition of swelling at 25 mg/kg in a carrageenin-induced paw edema model and similarly inhibited LPS-induced pyrexia at 11 mg/kg, results that were significantly more effective than the cyclooxygenase inhibitor diclofenac (ED$_{50}$ of 100 mg/kg) (181). In a murine collagen-induced arthritis model, SDZ 224-015 (a.k.a. VE-13,045) given prophylactically or therapeutically 10 days after a collagen boost at 50 mg/kg significantly reduced joint destruction (182,183). These results indicate that ICE inhibition effectively reduces both mIL-1β production and inflammation in various animal models.

Role of ICE in apoptosis - Transfection of cDNA encoding active ICE induced apoptosis. Both the p20 and p10 subunits of ICE were required, and active site minus mutants were ineffective in causing apoptosis (53,184). Coexpression of CrmA or Bcl-2 in the cells inhibited the apoptosis induced by ICE and to a lesser effect the apoptosis induced by Ced-3 (184,185). Athough peptidic ICE inhibitors blocked Fas-induced apoptosis (186,187), the selectivity of such inhibitors for ICE was questioned. In cells that secrete mIL-1β and undergo apoptosis following IL-2 withdrawal or ATP treatment, inhibition of IL-1β activity had no effect on apoptosis (189,190). Furthermore, cells without ICE protein could undergo apoptosis, showing that apoptosis was independent of ICE (191-194).

Transgenic mice lacking ICE were generated to determine the role of ICE in mIL-1β production and apoptosis (196,197). The absence of ICE did not affect the generation of cytoplasmic pIL-1β or pIL-1α, but the amount of mIL-1β released from monocytic cells following LPS stimulation was inhibited by >99%. Of interest was the observation that mIL-1α production was also inhibited by about 80% in the same cells, despite the inability of the YVAD-H inhibitor to affect mIL-1α release. Similar results were observed in vivo in mice treated with LPS, and these mice were resistant to LPS-induced lethality (195). No effects on apoptosis induced by high ATP treatment or dexamethasone or γ-irradiation were observed in macrophages and thymocytes from ICE knockout animals, and there were no abnormalities in the number of B cells or T cell subsets (195,196). There was, however, partial resistance to apoptosis induced by anti-Fas treatment, even though these animals showed no evidence for any autoimmune pathology as in the Fas deficient lpr/lpr mouse (26).

The identification of apoptotic ICE-like cleavage activities and substrates - Cell-free apoptotic extracts may contain Asp-specific cysteine protease activities distinct from ICE. An avian protease resembling ICE (prICE) cleaved the DNA repair enzyme poly(ADP-ribose) polymerase (PARP) at an internal DEVD sequence (192). Extracts from mitotically blocked cells contained a separate ICE-like cleavage activity that cleaved lamins A and B (197,198) at a site following the sequence VEID^{230} (199). Extracts of irradiated apoptotic cells cleavaged and inactivatied the 70-kDa U1 small nuclear ribonucleoprotein (U170K) and the DNA-dependent protein kinase (DNA-PK) (191,200,201) and protein kinase C (202) at discreet internal sites following DXXD motifs. The GDP dissociation inhibitor protein for the Rho GTPase family D4-GDI was cleaved at a DELD^{19} site during Fas-activated apoptosis (203). A microfilament associated protein, the growth arrest-specific 2 (Gas2) protein, associated with the maintenance of cells in a noncycling state, was specifically cleaved during apoptosis at a SRVD sequence in a manner analagous to that of PARP cleavage (204). Actin itself was cleaved during apoptosis following a ELPD sequence by an ICE-like protease. Nonerythroid spectrin (fodrin) was specifically cleaved during apoptosis, although it has not been determined if these occurred following Asp (205). ICE-like proteases have also been implicated in the apoptotic cleavage of Topoisomerase I (206), the hypophosphorylated retinoblastoma protein RB (207), and the nuclear matrix associated protein NuMA (206-208). In each case the cleavages were either not made by ICE itself, or required nonphysiologically high levels of enzyme for cleavage.

Furthermore, they were blocked only by micromolar rather than nanomolar concentrations of YVAD-H or YVAD-CMK inhibitors.

These endoproteolytic cleavages after Asp could be attributed to various members of an expanding family of ICE homologs that bore close structural similarity to each other. Phylogenetic analysis indicates that there are at least two subfamilies: one containing ICE, and a second containing Ced-3 and other closely related mammalian homologs (Fig. 4A). These proteases contained two subunits processed from an inactive precursor with conserved catalytic Cys and His and Arg179, Arg341, Ser347, and Gln283 which were necessary for the binding of P_1 Asp (Fig. 4B) (158,159). The residues involved in binding the P_2, P_3, and particularly P_4 were less well conserved and these probably define substrate specificity for each homolog.

Nedd-2/ICH-1 - The first ICE homolog identified was mouse Nedd-2, which was expressed in embryonic tissue undergoing high rates of programmed cell death (211). Overexpression of Nedd-2 induced apoptosis which was inhibitable by Bcl-2 (212). Its human counterpart, ICE Ced-3 homolog (ICH-1), was cloned and two transcripts were found: ICH-1$_L$, which encoded a 435 amino acid protein comparable to ICE, and ICH-1$_S$, a truncated version encoding 312 amino acids (213). Transfection of cDNA encoding ICH-1$_S$ prevented apoptosis in serum-deprived cells in a fashion similar to that for Bcl-2. Cotransfection of the cDNA for CrmA had little effect in preventing ICH-1$_L$-induced apoptosis.

ICE$_{rel}$II (TX, ICH-2) and ICE$_{rel}$III (Ty) - The closely related ICE homolog ICE$_{rel}$II (ICE2) (52% identity to ICE) induced apoptosis when overexpressed in mammalian cells (214-217). ICE2 was much less efficient than ICE at generating mature IL-1β at high concentrations. It also cleaved the precursors of ICE, PARP and its own precursor. AcYVAD-H was 20-fold less effective as an inhibitor of ICE2 relative to ICE (217). Using the ICE2 cDNA as probe, a second related homolog was discovered termed ICE$_{rel}$III (214) or TY (219) (denoted as ICE3). This protease induced apoptosis following transfection and did not cleave pIL-1β.

CPP32/Yama/apopain - CPP32 (cysteine protease protein 32-kDa) or Yama (after the Hindu god of death) was cloned from published Expressed sequence tag database (EST)(220,221). CPP32 also induced apoptosis when expressed in cells (220) . CPP32 is the only homolog that has the same Asn-Ser residues as Ced-3 in the P_4 peptide binding region (Fig. 4) (158,159). Isolation of the active proapoptotic enzyme that cleaved PARP (192) from THP.1 cells yielded a protease identical to CPP32 (193). The protease (with p17 and p12 subunits) was elevated during apoptosis, and its removal from apoptotic cell extracts reduced apoptosis in an in vitro assay. Because this purified protease, but not ICE, could restore apoptosis, it was termed apopain (193). Subnanomolar concentrations of an AcDEVD-H inhibitor blocked apopain activity and cell free nuclear fragmentation (193). In addition to PARP, apopain cleaved purified U1-70K ribonuclear protein, DNA-dependent protein kinase (201), and D4-GDI (203) at DXXD sites.   All the cleavages were inhibitable by nanomolar concentrations of the AcDEVD-H inhibitor, but not by the Ac-YVAD-H ICE inhibitor, indicating that the probable cause of the apoptotic cleavages was due to active apopain in the dying cells (201). A major role for apopain in apoptosis was further established when this enzyme was found to be activated in dying Jurkat cells after anti-Fas treatment and that apoptosis was inhibited by AcDEVD-H but not by AcYVAD-H (194,222). A protease related to apoppain also cleaved a membrane-bound sterol-regulatory transcription factor at a DEPD site in hamster cells (224,225).

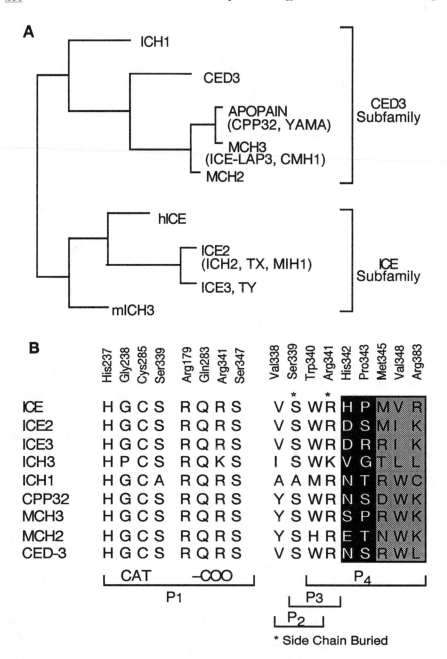

Fig. 4.   Homology of the ICE family members. **A.** Phylogenetic tree as determined by the GCG Pileup, EGCG Jukes Cantor Distances and UPGMA Growtree programs. Human ICE homologs are shown with the *C. elegans* Ced-3 and murine ICH-3 proteins (210) . **B.** Comparison of residues potentially involved in the binding of the tetrapeptide inhibitors. **Cat,** residues involved in catalysis; **-COO,** residues involved in the formation of the P_1 Asp binding pocket; P_2, P_3, P_4 refer to residues on the enzyme that associated with the respective positions on the inhibitor. The residues surrounding the P_4 pocket are shaded, with the black box denoting those residues that are closest to the P_4 side chain (158,159)

Mch-2 - In a search for cDNA sequences containing the conserved QACRG motif, a new homolog termed Mch2 was discovered (**M**ammalian-**c**ed **h**omolog 2)(226). Mch-2 was the protease responsible for cleavage of lamins during apoptosis following a VEID sequence(199). While Mch-2 could induce apoptosis in cells following its overexpression and could cleave PARP, the efficiency of cleavage was poor relative to that of CPP32. CPP32 did not cleave lamins and the two proteases showed differential inhibition by Asp-containing peptide inhibitors (199).

Mch-3/ICE-LAP3/CMH1 - A closely related homolog to CPP32 was separately cloned and called Mch-3 (**M**ammalian-**c**ed **h**omolog 3; (227)), ICE-LAP3 (**ICE**-**L**ike **A**poptotic **P**rotease 3; (223)), or CMH1; (**C**ed **M**ch2 **H**omolog 1) (228)). This enzyme could also cleave PARP and induce apoptosis when expressed in cells, but could not cleave pIL-1β. The close similarity between Mch-3 and CPP32 was born out by the demonstration that chimeric molecules utilizing p17 from one enzyme and p12 from the other were active. In functional comparisons, Mch-3 showed a 4 fold higher K_m for a DEVD-AMC substrate and a 3 fold higher K_i for the DEVD-H inhibitor (227,228). Like CPP32, Mch-3 is activated from an inactive precursor following anti-Fas or TNFα activation (223).

The activation and inhibition of multiple ICE family members in apoptosis - Apoptosis may result from the activation of several ICE family members simultaneously or in succession. This process induced by cytotoxic lymphocytes is in part mediated by the Fas system and in part by the introduction of granzyme B, an Asp specific serine protease, into cells (229). Fas stimulation activates multiple ICE homologs such as CPP32, Mch-3 and another CrmA inhibitable enzyme (230). Similarly, granzyme B can activate CPP32 and potentially other ICE homologs by cleaving at the conserved Asp residue at the C-terminus of the catalytic subunit (231,232). The proapoptotic role of these proteases is substantiated by the use of various ICE family specific compounds. Peptide based inhibitors of both ICE-like activities (YVAD-H, YVAD-CMK, or ZVAD-FMK) and CPP32-like activities (DEVD-H) have been shown to prevent apoptosis in a variety of systems (193,199,233-240). Likewise, the transfection of the ICE-specific viral serpin CrmA has been used to show the role of ICE family members in apoptosis. It is unlikely that CrmA evolved to be antiapoptotic since it confers no selective advantage in viral production during *in vitro* cell growth (174,175). Nonetheless, CrmA inhibited CPP32/apopain and Mch-3 at relatively high concentrations (193,227). The homolog ICH1, in contrast, was almost totally resistant to CrmA inhibition (241). Cellular apoptosis induced by Fas or TNFα activation can be inhibited by CrmA (242), but apoptosis induced by staurosporine, irradiation or serum deprivation were not (230,243) despite their inhibition by the generic ICE family inhibitor ZVAD-FMK (238,244). These events are hence dependent on selective and sequential activation of CrmA sensitive or insensitive members (239).

ICE has been shown to be essential for the cleavage of pIL-1ß, and specific peptide-based ICE inhibitors are useful therapeutically in animal models to reduce IL-1ß-mediated inflammation. The ICE homologs have different but overlapping substrate specificities. Each protease is structurally unique in its peptide-binding region, suggesting that selective inhibitors can be developed for each member. Because more than one ICE family protease is likely to be activated following an apoptotic stimulus, it remains to be determined whether or not apoptosis in certain cells or via certain stimuli can be blocked in a therapuetically meaningful manner.

SUMMARY

A recognition of the various intracellular proteins involved in the induction and execution of apoptotis has occurred recently. The aggregating factors utilize the death domain to transfer a signal from a stimulated receptor. It seems likely that the death domains can induce conformational changes as described for the role of mastoparan in the activation of G proteins (245). Associatied with death domain proteins are regulatory factors such as the IAPs and other functional counterparts. The IAPs seem to impart a general inhibition of apoptotic activation in the systems in which they have been transfected. Members of the Bcl-2 family also aggregate by means of their BH1, BH2, and BH3 domains, but it is unclear how these proteins directly control the activation of the ICE proteases. It is clear that only certain pathways of apoptosis are affected by Bcl-2 members: Fas-induced apoptosis, for example, is not inhibited by Bcl-2 members, although such inducers as serum deprivation, irradiation, and staurosporine treatment are inhibited (230,243,246). Of the Bcl-2 members, the viral homologs, such as E1B19K, generally appear to have a wider spectrum of inhibition than do the mammalian members in transfection systems. Other viral inhibitors of apoptosis such as CrmA and p35 can be direct inhibitors of the executioner proteases. The ICE family proteases are restrictive in their substrate specificity, recognizing peptides with a P_1 Asp. Despite their sequence similarities in structural regions of the subunits, the homologs appear to contain unique residues within their peptide binding pockets, particularly for P_4. This specificity should enable the development of inhibitors to individual homologs. The current challenge is the determination of whether unique proteases in this family are associated with certain types of cell death. Undesirable apoptotic death such as that associated with various neurodegenerative disorders might be attributable to the activation of one family member whereas the more desirable apoptosis of autoreactive lymphocytes and aberrantly growing cancer cells might be controlled by a different family member. Better knowledge of such protease specificity could lead to the development of therapeutically useful inhibitors.

References

1.   R. A. Lockshin and C. M. Williams, J. Insect Physiol. 11, 123 (1965).
2.   J. F. R. Kerr, A. H. Wyllie, and A. R. Currie, Br. J. Cancer 26, 239 (1972).
3.   L. M. Schwartz, Cell Death and Differentiation 2, 83 (1995).
4.   J. J. Arends and A. H. Wyllie, in Int. Rev. Exp. Path., G.W. Richter and M.A. Epstein, Ed, Academic Press, New York, 223 (1991).
5.   S. Sen, Biol. Rev. 67, 287 (1992).
6.   D. Wilcock and J. A. Hichman, Biochim. Biophys. Acta 946, 359 (1988).
7.   F. Oberhammer, J. W. Wilson, C. Dive, I. D. Morriss, J. A. hickman, A. E. Wakeline, P. R. Walker, and M. Sikorska, EMBO J 12, 3679 (1993).
8   Y. Gavrieli, Y. Sherman, and S. A. Ben-Sasson, J. Cell Biol. 119, 493 (1992).
9.   V. A. Fadok, D. R. Voelker, P. A. Campbell, J. J. Cohen, D. L. Bratton, and P. M. Henson, J. Immunol. 148, 2207 (1992).
10.   J. Savill, V. Fadok, P. Henson, and C. Haslett, Immunology Today 14, 131 (1993).
11.   G. Majno and I. Joris, Am. J. Pathol. 146, 3 (1995).
12.   M. C. Raff, Nature 356, 397 (1992).
13.   A. H. Wyllie, Nature 284, 555 (1980).
14.   K. S. Sellins and J. J. Cohen, J. Immunol. 139, 3199 (1987).
15.   A. Eastman, in Cancer Cells, Cold Spring Harbor Press, 275 (1990).
16.   C. L. Sentman, J. R. Shutter, D. Hockenbery, O. Kanagawa, and S. J. Korsmeyer, Cell 67, 879 (1991).
17.   B. A. Osborne, Seminars Cancer Biol. 6, 27 (1995).
18.   R. M. Elledge and W.-H. Lee, BioEssays 17, 923 (1995).
19.   T. Enoch and C. Norbury, Trends Biochem. Sci. 20, 426 (1995).
20   E. White, Curr. Top. Microbiol. Immunol. 199, 34 (1995).
21   S. Rowan, R. L. Ludwig, H. Y., S. Bates, X. Lu, M. Oren, and K. H. Vousden, EMBO J. 15, 827 (1996).
22.   G. Packham and J. L. Cleveland, Biochim. Biophys. Acta 1242, 11 (1995).
23.   O. Kranenburg, A. J. van der Eb, and A. Zantema, EMBO J. 15, 46 (1996).
24.   N. Itoh and S. Nagata, J. Biol. Chem. 268, 10932 (1993).
25.   L. A. Tartaglia, T. M. Ayres, G. H. W. Wong, and D. V. Goeddel, Cell 74, 845 (1993).
26.   S. Nagata and P. Golstein, Science 267, 1449 (1995).
27.   E. Feinstein, A. Kimchi, D. Wallach, M. Boldin, and E. Varfolomeev, Trends Biochem. Sci. 20, 342 (1995).
28.   B. S. Chapman, FEBS Letters 374, 216 (1995).

29.   I. Cascino, G. Papoff, R. De Maria, R. Testi, and G. Ruberti, J. Immunol. 156, 13 (1996).
30.   M. P. Boldin, I. L. Mett, V. E.E., I. Chumakov, Y. Shemer-Avni, J. H. Camonis, and D. Wallach, J. Biol. Chem. 270, 387 (1995).
31.   C. A. Smith, T. Farrah, and R. G. Goodwin, Cell 76, 959 (1994).
32   P. Vandenabeele, W. Declercq, R. Beyaert, and W. Fiers, Trends Cell Biol. 5, 392 (1995).
33.   A. M. Chinnaiyan, K. O'Rourke, M. Tewari, and V. M. Dixit, Cell 81, 505 (1995).
34.   M. P. Boldin, E. E. Varfolomeev, Z. Pancer, I. L. Mett, J. H. Camonis, and D. Wallach, J. Biol. Chem. 270, 7795 (1995).
35.   B. Z. Stanger, P. Leder, T.-H. Lee, E. Kim, and B. Seed, Cell 81, 513 (1995).
36.   H. Hsu, J. Xiong, and D. V. Goeddel, Cell 81, 495 (1995).
37.   H. Hsu, H.-B. Shu, M.-G. Pan, and D. V. Goeddel, Cell 84, 299 (1996).
38..   M. Jaattela, H. Mouritzen, F. Elling, and L. Bastholm, J. Immunol. 156, 1166 (1996).
39.   M. Tewari, F. W. Wolf, M. F. Seldin, K. S. O'Shea, V. M. Dixit, and L. A. Turka, J. Immunol. 154, 1699 (1995).
40.   M. Rothe, S. C. Wong, W. J. Henzel, and D. V. Goeddel, Cell 78, 681 (1994).
41.   M. Rothe, V. Sarma, V. M. Dixit, and D. V. Goeddel, Science 269, 1424 (1995).
42   M. Rothe, M.-G. Pan, W. J. Henzel, T. M. Ayres, and D. V. Goeddel, Cell 83, 1243 (1995).
43.   P. Liston, N. Roy, K. Tamai, C. Lefebvre, S. Baird, G. Cherton-Horvat, R. Farahani, M. McLean, J.-E. Ikeda, A. MacKenzie, and R. G. Korneluk, Nature 379, 349 (1996).
44.   F. C. Kischkel, S. Hellbardt, I. Behrmann, M. Germer, M. Pawlita, P. H. Krammer, and M. E. Peter, EMBO J. 14, 5579 (1995).
45   T. Sato, S. Irie, S. Kitada, and J. C. Reed, Science 268, 411 (1995).
46.   K. Chu, N. X., and L. T. Williams, Proc. Natl. Acad. Scie. USA 92, 11894 (1995).
47.   H. Y. Song, J. D. Dunbar, Y. X. Zhang, D. Guo, and D. B. Donner, J. Biol. Chem. 270, 3574 (1995).
48.   R. E. Ellis, J. Yuan, and H. R. Horvitz, Annu. Rev. Cell Biol. 7, 663 (1991).
49.   M. O. Hengartner, R. E. Ellis, and H. R. Horvitz, Nature 356, 494 (1992).
50.   D. L. Vaux, I. L. Weissman, and S. K. Kim, Science 258, 1955 (1992).
51.   M. O. Hengartner and H. R. Horvitz, Cell 76, 665 (1994).
52.   H. M. Ellis and H. R. Horvitz, Cell 44, 817 (1986).
53.   J. Yuan, S. Shaham, S. Ledoux, H. M. Ellis, and H. R. Horvitz, Cell 75, 641 (1993).
54.   J. Yuan and H. R. Horvitz, Development 116, 309 (1992).
55.   S. Shaham and H. R. Horvitz, Genes Dev. 10, 578 (1996).
56.   R. E. Ellis and H. H.R., Development 112, 591 (1991).
57.   J. Yuan, J. Cell. Biochem. 60, 4 (1996).
58.   Y. Tsujimoto, J. Cossman, E. Jaffe, and C. Croce, Science 228, 1440 (1985).
59.   D. Vaux, S. Cory, and J. Adams, Nature 335, 440 (1988).
60.   T. J. McDonnell, N. Deane, F. M. Platt, G. Nunez, U. Jaeger, J. P. McKearn, and S. J. Korsmeyer, Cell 57, 79 (1989).
61.   D. Hockenbery, G. Nunez, C. Milliman, R. D. Schreiber, and S. J. Korsmeyer, Nature 348, 334 (1990).
62.   Y. Tsujimoto, Oncogene 4, 1331 (1989).
63.   C. Behl, L. I. Hovey, S. Krajewski, D. Schubert, and J. C. Reed, Biochem. Biophys. Res. Comm. 197, 949 (1993).
64.   J. C. Reed, J. Cell Biol. 124, 1 (1994).
65.   J. C. Reed, Hematol. Oncol. Clin. N. 9, 451 (1995).
66.   D. P. LeBrun, R. A. Warnke, and M. L. Cleary, Am. J. Pathol. 142, 743 (1993).
67.   Q.-L. Lu, R. Poulsom, L. Wong, and A. M. Hanby, J. Pathol. 169, 431 (1993).
68.   J. Gratiot-Deans, L. Ding, T. A. Turka, and G. Nunez, J. Immunol. 151, 83 (1993).
69.   D. J. Veis, C. L. Sentman, E. A. Bach, and S. J. Korsmeyer, J. Immunol. 151, 2546 (1993).
70.   S. Krajewski, S. Tanaka, S. Takayama, and e. al, Cancer Res. 53, 4701 (1993).
71.   Z. Chen-Levy and M. L. Cleary, J. Biol. Chem. 265, 4929 (1990).
72.   D. M. Hockenbery, Z. N. Oltvai, X.-M. Yin, C. L. Milliman, and S. J. Korsmeyer, Cell 75, 241 (1993).
73.   C. Borner, I. Martinou, C. Mattmann, M. Irmler, E. Schaerer, J.-C. Martinou, and J. Tschopp, J. Cell Biol. 126, 1059 (1994).
74.   M. Nguyen, P. E. Branton, P. A. Walton, A. N. Oltvai, S. J. Korsmeyer, and G. C. Shore, J. Biol. Chem. 269, 16521 (1994).
75.   M. D. Jacobson, J. F. Burnett, M. D. King, T. Miyashita, J. C. Reed, and M. C. Raff, Nature 361, 365 (1993).
76.   M. D. Jacobson and M. C. Raff, Nature 374, 814 (1995).
77.   S. Shigeomi, Y. Eguchi, H. Kosaka, W. Kamlike, H. Matsuda, and Y. Tsujimoto, Nature 374, 811 (1995).
78.   X.-M. Yin, Z. N. Oltvai, and S. J. Korsmeyer, Nature 369, 321 (1994).
79   J. M. Boyd, G. J. Gallo, B. Elangovan, A. B. Houghton, B. Malstrom, B. J. Avery, R. G. Ebb, T. Subramanian, T. Chittenden, R. J. Lutz, and G. Chinnadurai, Oncogene 11, 1921 (1995).

80.   T. Chittenden, C. Fl;emington, A. B. Houghton, R. G. Ebb, G. J. Gallo, B. Elangovan, G. Chinnadurai, and R. J. Lutz, EMBO J. <u>14</u>, 5589 (1995).
81.   J. Han, P. Sabbatini, D. Perez, L. Rao, D. Modha, and E. White, Genes Develop. <u>10</u>, 461 (1996).
82.   T. Sato, M. Hanada, S. Bodrug, S. Irie, N. Iwama, L. H. Boise, C. B. Thompson, E. Golemis, F. L., H.-G. Wang, and J. C. Reed, Proc. Natl. Acad. Sci. USA <u>91</u>, 9238 (1994).
83.   M. Hanada, C. Aime-Sempe, T. Sato, and J. C. Reed, J. Biol. Chem. <u>270</u>, 11962 (1995).
84.   L. H. Boise, M. Gonzalez-Garcia, C. E. Postema, L. Ding, T. Lindsten, L. A. Turka, X. Mao, G. Nunez, and C. B. Thompson, Cell <u>74</u>, 597 (1993).
85.   A. J. Minn, L. H. Boise, and C. B. Thompson, J. Biol. Chem. <u>271</u>, 6306 (1996).
86.   E. H.-Y. Cheng, B. Levine, L. H. Boise, C. B. Thompson, and J. M. Hardwick, Nature <u>379</u>, 554 (1996).
87.   G. Nunez, R. Merino, D. Grillot, and M. Gonzalez-Garcia, Immunol. Today <u>15</u>, 582 (1994).
88.   L. D. Moscinski and M. B. Prystowsky, Oncogene <u>5</u>, 31 (1990).
89.   S. S. Choi, I.-C. Park, J. W. Yun, Y. C. Sung, S.-I. Hong, and H.-S. Shin, Oncogene <u>11</u>, 1693 (1995).
90.   K. M. Kozopas, T. Yang, H. L. Buchan, P. Zhou, and R. W. Craig, Proc. Natl. Acad. Sci. USA <u>90</u>, 3516 (1993).
91.   T. Yang, K. M. Kozopas, and R. W. Craig, J. Cell Biol. <u>128</u>, 1173 (1994).
92.   R. W. Craig, E. W. Jabs, P. Zhou, K. M. Kozopas, A. L. Hawkins, J. M. Rochelle, M. F. Seldin, and C. A. Griffin, Genomics <u>23</u>, 457 (1994).
93.   S. Krajewski, S. Bodrug, M. Krajewska, A. Shabaik, R. Gascoyne, K. Berean, and J. C. Reed, Am. J. Pathol. <u>146</u>, 1309 (1995).
94   J. E. Reynolds, T. Yang, L. Qian, J. D. Jenkinson, P. Zhou, A. Eastman, and R. W. Craig, Cancer Res. <u>54</u>, 6348 (1994).
95.   Z. N. Oltvai, C. L. Milliman, and S. J. Korsmeyer, Cell <u>74</u>, 609 (1993).
96.   Z. N. Oltvai and S. J. Korsmeyer, Cell <u>79</u>, 189 (1994).
97   J. J. Hunter and T. G. Parslow, J. Biol. Chem. <u>271</u>, 8521 (1996).
98.   S. Krajewski, M. Krajewska, A. Shabaik, T. Miyashita, H. G. Wang, and J. C. Reed, Am. J. Path. <u>145</u>, 1323 (1994).
99.   M. C. Kiefer, M. J. Brauer, V. C. Powers, J. J. Wu, S. R. Umansky, L. D. Tomei, and P. J. Barr, Nature <u>374</u>, 736 (1995).
100.   T. Chittenden, E. A. Harrington, R. O'Connor, C. Flemington, R. J. Lutz, G. I. Evan, and B. C. Guild, Nature <u>374</u>, 733 (1995).
101.   S. N. Farrow, J. H. M. White, I. Martinou, T. Raven, K.-T. Pun, C. J. Grinham, J.-C. Martinou, and R. Brown, Nature <u>374</u>, 731 (1995).
102.   E. Yang, J. Zha, J. Jockel, L. H. Boise, C. B. Thompson, and S. J. Korsmeyer, Cell <u>80</u>, 285 (1995).
103.   J. M. Boyd, G. J. Gallo, B. Elangovan, A. B. Houghton, S. Malstrom, B. J. Avery, R. G. Ebb, T. Subramanian, T. Chittenden, R. J. Lutz, and G. Chinnadurai, Oncogene <u>11</u>, 1921 (1995).
104.   D. J. Veis, C. M. Sorenson, S. R. Shutter, and S. J. Korsmeyer, Cell <u>75</u>, 229 (1993).
105.   K.-I. Nakayama, K. Nakayama, I. Negishi, K. Kuida, Y. Shinkai, M. C. Louie, L. E. Fields, P. J. Lucas, V. Stewart, F. W. Alt, and D. Y. Loh, Science <u>261</u>, 1584 (1993).
106.   S. Kamada, A. Shimono, Y. Shinto, T. Tsujimura, T. Takahashi, T. Noda, Y. Kitamura, H. Kondoh, and Y. Tsujimoto, Cancer Res. <u>55</u>, 354 (1995).
107.   N. Motoyama, F. Wang, K. A. Roth, H. Sawa, K. i. Nakayama, K. Nakayama, I. Negishi, S. Senju, Q. Zhang, S. Fujii, and D. Y. Loh, Science <u>267</u>, 1506 (1995).
108   C. M. Knudson, K. S. K. Tung, W. G. Tourtellotte, G. A. J. Brown, and S. J. Korsmeyer, Science <u>270</u>, 96 (1995).
109.   J. M. Boyd, S. Malstrom, T. Subramanian, L. K. Venkatesh, U. Schaeper, C. D'Sa-Eipper, B. Elangovan, and G. Chinnadurai, Cell <u>79</u>, 341 (1994).
110.   S. Takayama, T. Sato, S. Krajewski, K. Kochel, S. Irie, J. A. Millan, and J. C. Reed, Cell <u>80</u>, 279 (1995).
111.   L. Rao, M. Debbas, P. Sabbatini, D. Hockenbery, S. Korsmeyer, and E. . White, Proc. Natl. Aca. Sci. USA <u>89</u>, 7742 (1992).
112.   E. White, P. Sabbatini, M. Debbas, W. S. M. Wold, D. I. Kusher, and L. Gooding, Mol. Cell. Biol. <u>12</u>, 2570 (1992).
113.   S.-K. Chiou, C.-C. Tseng, L. Rao, and E. White, J. Virol. <u>68</u>, 6553 (1994).
114.   S. A. Hashimoto, A. Ishii, and S. Yonehara, Int. Immunol. <u>3</u>, 343 (1991).
115.   I. Martinou, P.-A. Fernandez, M. Missotten, E. White, B. Allet, R. Sadoul, and J.-C. Martinou, J. Cell Biol. <u>128</u>, 201 (1995).
116.   S. Henderson, M. Rowe, C. Gregory, D. Croom-Carter, F. Wang, R. Longnecker, E. Kieff, and A. Rickinson, Cell <u>65</u>, (1991).
117.   M. L. Cleary, S. D. Smith, and J. Sklar, Cell <u>47</u>, 19 (1986).
118.   G. R. Pearson, J. Luka, L. Petti, J. Sample, M. Birkenbach, D. Braun, and E. Kieff, Virology <u>160</u>, 151 (1987).
119.   S. Henderson, D. Huen, M. Rowe, C. Dawson, G. Johnson, and A. Rickinson, Proc. Natl. Acad. Sci. USA <u>90</u>, 8479 (1993).

120. B. Tarodi, T. Subramanian, and G. Chinnadurai, Virology 201, 404 (1994).
121. J. G. Neilan, Z. Lu, C. L. Afonso, G. F. Kutish, M. D. Sussman, and D. L. Rock, J. Virol. 67, 4391 (1993).
122. C. A. Smith, Trends Cell Biol. 5, 344 (1995).
123   J. Chou and B. Roizman, Proc. Natl. Acad. Sci. USA 89, 3266 (1992).
124. P. D. Friesen and L. K. Miller, J. Virol. 61, 2264 (1987).
125. R. J. Clem, M. Fechheimer, and L. K. Miller, Science 254, 1388 (1991).
126. S. G. Kamita, K. Majima, and S. Maeda, J. Virol. 67, 455 (1993).
127. R. J. Clem and L. K. Miller, Mol. Cell. Biol. 14, 5212 (1994).
128. N. E. Crook, R. J. Clem, and L. K. Miller, J. Virol. 67, 2168 (1993).
129. M. J. Birnbaum, R. J. Clem, and L. K. Miller, J. Virol. 68, 2521 (1994).
130. P. L. Cartier, P. A. Hershberger, and P. D. Friesen, J. Virol. 68, 7728 (1994).
131. S. Rabizadeh, D. J. LaCount, P. D. Friesen, and D. E. Bredesen, J. Neurochem. 61, 2318 (1993).
132. D. R. Beidler, M. Tewari, P. D. Friesen, G. Poirier, and V. M. Dixit, J. Biol. Chem. 270, 1 6526 (1995).
133. A. Sugimoto, P. D. Friesen, and J. H. Rothman, EMBO J. 13, 2023 (1994).
134. N. Roy, M. S. Mahadevan, M. McLean, G. Shutler, Z. Yaraghi, R. Farahani, S. Baird, A. Besner-Johnston, C. Lefebvre, X. Kang, M. Salih, H. Aubry, K. Tamai, X. Guan, P. Ioannou, T. O. Crawford, P. J. de Jong, L. Surh, J.-E. Ikeda, R. G. Korneluk, and A. MacKenzie, 80, 1 67 (1995).
135. N. J. Bump, M. Hackett, M. Hugunin, S. Seshagiri, K. Brady, P. Chen, C. Ferenz, S. Franklin, T. Ghayur, P. Li, P. Licari, J. Mankovich, L. Shi, A. Greenberg, L. K. Miller, and W. W. Wong, Science 269, 1885 (1995).
136. D. Xue and H. R. Horvitz, Nature 377, 248 (1995).
137. M. J. Kostura, M. J. Tocci, G. Limjuco, J. Chin, P. Cameron, A. G. Hillman, N. A. Chartrain, and J. A. Schmidt, Proc Natl Acad Sci U S A 86, 5227 (1989).
138. R. A. Black, S. R. Kronheim, J. E. Merriam, C. J. March, and T. P. Hopp, J Biol Chem 264, 5323 (1989).
139. C. A. Dinarello, Blood 77, 1627 (1991).
140. C. A. Dinarello, Eur. Cytokine Netw. 5, 517 (1994).
141. T. Geiger, H. Towbin, A. Cosenti-Vargas, Z. O., J. Arnold, C. Rordorf, M. Glatt, and K. Vosbeck, Clin. Exp. Rheumatol. 11, 515 (1993).
142. W. B. Van den Berg, L. A. B. Joosten, M. Helsen, and F. A. J. van de Loo, Clin. Exp. I mmunol. 95, 237 (1994).
143. H. Zheng, D. Fletcher, W. Kozak, M. Jiang, K. J. Hofmann, C. A. Conn, E. Soszynski, C. Grabiec, M. E. Trumbauer, A. Shaw, M. J. Kostura, K. Stevens, H. Rosen, R. J. North, H. Y. Chen, M. J. Tocci, M. J. Kluger, and L. H. T. Van der Ploeg, Immunity 3, 9 (1995).
144. J. G. Giri, J. Wells, S. K. Dower, C. E. McCall, R. N. Guzman, J. Slack, T. A. Bird, K. Shanebeck, K. H. Grabstein, and J. E. Sims, J. Immunol. 153, 5802 (1994).
145. S. K. Dower, W. Fanslow, C. Jacobs, S. Waugh, J. E. Sims, and M. B. Widmer, Ther. I mmunol. 1, 113 (1994).
146. A. Alcami and G. L. Smith, Cell 71, 153 (1992).
147   M. K. Spriggs, D. E. Hruby, C. R. Maliszewski, D. J. Pickup, J. E. Sims, R. M. L. Buller, and J. VanSlyke, Cell 71, 145 (1992).
148. W. Kozak, H. Zheng, C. A. Conn, D. Soszynski, L. H. van der Ploeg, and M. J. Kluger, Am. J. Physiol. 269, 969 (1995).
149. J. A. Schmidt and M. J. Tocci, in The Handbook of Experimental Pharmacology I. Peptide Growth Factors and their Receptors, M. Sporn and A. Roberts, Ed, Springer Verlag, Berlin, 473 (1990).
150. B. Mosley, D. L. Urdal, K. S. Prickett, A. Larsen, D. Cosman, P. J. Conlon, S. Gillis, and S. K. Dower, J Biol Chem 262, 2941 (1987).
151. N. A. Thornberry, H. G. Bull, J. R. Calaycay, K. T. Chapman, A. D. Howard, M. J. Kostura, D. K. Miller, S. M. Molineaux, J. R. Weidner, J. Aunins, K. O. Elliston, J. M. Ayala, R. J. Casano, J. Chin, G. J.-F. Ding, L. A. Egger, E. P. Gaffney, G. Limjuco, O. C. Palyha, S. M. Raju, A. M. Rolando, J. P. Salley, T. T. Yamin, T. D. Lee, J. E. Shively, M. MacCoss, R. A. Mumford, J. A. Schmidt, and M. J. Tocci, Nature 356, 768 (1992).
152. D. P. Cerretti, C. J. Kozlosky, B. Mosley, N. Nelson, K. Van Ness, T. A. Greenstreet, C. J. March, S. R. Kronheim, T. Druck, L. A. Cannizzaro, K. Huebner, and R. A. Black, Science 256, 97 (1992).
153. P. R. Sleath, R. C. Hendrickson, S. R. Kronheim, C. J. March, and R. A. Black, J Biol Chem 265, 14526 (1990).
154. A. D. Howard, M. J. Kostura, N. Thornberry, G. J. F. Ding, G. Limjuco, J. Weidner, J. P. Salley, K. A. Hogquist, D. D. Chaplin, R. A. Mumford, J. A. Schmidt, and M. J. Tocci, J. Immunol. 147, 2964 (1991).
155. D. K. Miller, J. R. Calaycay, K. T. Chapman, A. D. Howard, M. J. Kostura, S. M. Molineaux, and N. A. Thornberry, in Immunosuppressive and Antiinflammatory Drugs, a.C. Allison, K.J. Lafferty, andH. Fliri, Ed, New York Academy of Sciences, New York, 133 (1993).

156.   N. Thornberry, Methods in Enzymology 244, 615 (1994).
157.   T.-T. Yamin, J. M. Ayalan, and D. K. Miller, J. Biol. Chem. In press, (1996).
158.   N. P. C. Walker, R. V. Talanian, K. D. Brady, L. C. Dang, N. J. Bump, C. R. Ferenz, S. Franklin, T. Ghayur, M. C. Hackett, L. D. Hammill, L. Herzog, M. Hugunin, W. Houy, J. A. Mankovich, L. McGuiness, E. Orlewica, M. Paskind, C. A. Pratt, P. Reis, A. Summani, M. Terranova, J. P. Welch, L. Xiong, A. Moller, D. E. Tracey, R. Kamen, and W. W. Wong, Cell 78, 343 (1994).
159.   K. P. Wilson, J. F. Black, J. A. Thomson, E. E. Kim, J. P. Griffith, M. A. Navia, M. A. Murcko, S. P. Chambers, R. A. Aldape, S. A. Raybuck, and D. J. Livingston, Nature 370, 270 (1994).
160.   K. T. Chapman, Biorg. med. Chem. Lett. 2, 613 (1992).
161.   T. L. Graybill, R. E. Dolle, C. T. Helaszek, R. E. Miller, and M. A. Ator, Int. J. Peptide Protein Res. 44, 173 (1994).
162   M. D. Mullican, D. J. Lauffer, R. J. Gillespie, S. S. Matharu, D. Kay, M. Porritt, P. L. Evans, J. M. C. Golec, M. A. Murcko, Y.-P. Luong, S. A. Raybuck, and D. J. Livingston, Bioorg. Med. Chem. Lett. 4, 2359 (1994).
163.   A. M. M. Mjalli, K. T. Chapman, M. MacCoss, and N. A. Thornberry, Bioorg. Med. Chem. L ett. 3, 2689 (1993).
164.   A. M. M. Mjalli, K. T. Chapman, J. J. Zhao, N. A. Thornberry, E. P. Peterson, and M. MacCoss, Bioorg. Med. Chem. Let. 5, 1405 (1995).
165.   A. Krantz, L. J. Copp, P. J. Coles, R. A. Smith, and S. B. Heard, Biochemistry 30, (1991).
166.   D. H. Pliura, B. J. Bonaventura, R. A. Smith, P. J. Coles, and A. Krantz, Biochem. J. 288, 759 (1992).
167.   R. E. Dolle, D. Hoyer, C. V. C. Prasad, S. J. Schmidt, C. T. Helaszek, R. E. Miller, and M. A. Ator, J. Med. Chem. 37, 563 (1994).
168.   N. A. Thornberry, E. P. Peterson, J. J. Zhao, A. D. Howard, P. R. Griffin, and K. T. Chapman, Biochemistry 33, 3934 (1994).
169.   L. Revesz, C. Briswalter, R. Heng, A. Leutwiler, R. Mueller, and W. Hangs-Juerg, Tetrahedron Lett. 35, 9693 (1994).
170.   R. E. Dolle, J. Singh, J. Rinker, D. Hoyer, C. V. C. Prasad, T. L. Graybill, J. M. Salvino, C. T. Helaszek, R. E. Miller, and M. A. Ator, J. Med. Chem. 37, 3863 (1994).
171.   C. V. C. Prasad, C. P. Prouty, D. Hoyer, T. M. Ross, J. M. Salvino, M. Awad, T. L. Graybill, S. J. Schmidt, I. K. Osifo, R. E. Dolle, C. T. Helaszek, R. E. Miller, and M. A. Ator, Bioorg. Med. Chem. Let. 4, 315 (1995).
172.   D. J. Pickup, B. S. Ink, W. Hu, C. A. Ray, and W. K. Joklik, Proc. Natl. Aca. Sci. USA 83, 7698 (1986).
173.   G. J. Palumbo, D. J. Pickup, T. N. Fredrickson, L. J. McIntyre, and R. M. Buller, Virology 171, 262 (1989).
174.   G. J. Palumbo, R. M. Buller, and W. C. Glasgow, J. Virol. 68, 1737 (1994).
175.   C. A. Ray and D. J. Pickup, Virology 217, 384 (1996).
176.   C. A. Ray, R. A. Black, S. R. Kronheim, T. A. Greenstreet, P. R. Sleath, G. S. Salvesen, and D. J. Pickup, Cell 69, 597 (1992).
177.   T. K. Komiyama, C. A. Ray, D. J. Pickup, A. D. Howard, N. A. Thornberry, E. P. Peterson, and G. Salvesen, J. Biol. Chem. 19331 (1994).
178.   A. D. Howard, O. C. Palyha, P. R. Griffin, E. P. Peterson, A. B. Lenny, G. J.-F. Ding, D. J. Pickup, N. A. Thornberry, J. A. Schmidt, and M. J. Tocci, J. Immunol. 154, 2321 (1995).
179.   D. Fletcher, L. Agarwal, K. Chapman, J. Chin, L. Egger, G. Limjuco, S. Luell, D. MacIntyre, E. Peterson, N. Thornberry, and M. Kostura, J. Interferon Cytokine Res. 15, 243 (1995).
180.   B. E. Miller, P. A. Krasney, D. M. Gauvin, K. B. Holbrook, D. J. Koonz, R. V. Abruzzese, R. E. Miller, K. A. Pagani, R. E. Dolle, M. A. Ator, and S. C. Gilman, J. Immunol. 154, 1331 (1995).
181.   P. R. Elford, R. Heng, L. Revesz, and A. R. MacKenzie, Brit. J. Pharmacol. 115, 601 (1995).
182.   M. W. Harding, G. Ku, T. Faust, L. L. Lauffer, and D. J. Livingston, Arthritis Rheumatism 38, P1482 (1995).
183.   D. J. Livingston, J. Cell. Biochem. in press, (1996).
184.   M. Miura, H. Zhu, R. Rotello, E. A. Hartweig, and J. Yuan, Cell 75, 653 (1993).
185.   V. Gagliardini, P. A. Fernandez, R. K. Lee, H. C. Drexler, R. J. Rotello, M. C. Fishman, and J. Yuan, Science 263, 826 (1994).
186.   M. Enari, H. Hug, and S. Nagata, Nature 375, 78 (1995).
187.   M. Los, M. Van de Craen, L. C. Penning, H. Schenk, M. Westendorp, P. A. Baeuerle, W. Droge, P. H. Krammer, W. Fiers, and K. Schulze-Osthoff, Nature 375, 81 (1995).
188.   M. S. Williams and P. A. Henkart, J. Immunol. 153, 4247 (1994).
189.   J. P. Vasilakos, T. Ghayur, R. T. Carroll, D. A. Giegel, J. M. Saunders, L. Quintal, K. M. Keane, and B. D. Shivers, J. Immunol. 155, 3433 (1995).
190.   M. Nett-Fiordalisi, K. Tomaselli, J. H. Russell, and D. D. Chaplin, J. Leukoc. Biol. 58, 717 (1995).
191.   L. A. Casciola-Rosen, D. K. Miller, G. J. Anhalt, and A. Rosen, J. Biol. Chem. 269, 30757 (1994).

192. Y. A. Lazebnik, S. H. Kaufmann, S. Desnoyers, G. G. Poirier, and W. C. Earnshaw, Nature <u>371</u>, 346 (1994).
193. D. W. Nicholson, A. Ali, N. A. Thornberry, J. P. Vaillancourt, C. K. Ding, M. Gallant, Y. Gareau, P. R. Griffin, M. Labelle, Y. A. Lazebnik, N. A. Munday, S. M. Raju, M. E. Smulson, T.-T. Yamin, V. L. Yu, and D. K. Miller, Nature <u>376</u>, 37 (1995).
194. J. Schlegel, I. Peters, S. Orrenius, D. K. Miller, N. A. Thornberry, T.-T. Yamin, and D. W. Nicholson, J. Biol. Chem. <u>271</u>, 1841 (1996).
195. P. Li, H. Allen, S. Banerjee, S. Franklin, L. Herzog, C. Johnston, J. McDowell, M. Paskind, L. Rodman, J. Salfeld, E. Towne, D. Tracey, S. Wardwell, F.-Y. Wei, W. Wong, R. Kamen, and T. Seshadri, Cell <u>80</u>, 401 (1995).
196. K. Kuida, J. A. Lippke, G. Ku, M. W. Harding, D. J. Livingston, M. S.-S. Su, and R. A. Flavell, Science <u>267</u>, 2000 (1995).
197. Y. A. Lazebnik, A. Takahashi, R. D. Moir, R. D. Golman, G. G. Poirier, S. H. Kaufmann, and W. C. Earnshaw, Proc. Natl. Acad. Sci. USA <u>92</u>, 9042 (1995).
198. F. A. Oberhammer, K. Hochegger, G. Froschl, R. Tiefenbacher, and M. Pavelka, J. Cell Biol. <u>126</u>, 827 (1994).
199. A. Takahashi, E. S. Alnemri, Y. A. Lazebnik, T. Fernandes-Alnemri, G. Litwack, R. D. Moir, R. D. Goldman, G. G. Poirier, S. H. Kaufmann, and W. C. Earnshaw, Proc. Natl. Acad. Sci. USA <u>in press,</u> (1996).
200. L. A. Casciola-Rosen, G. J. Anhalt, and A. Rosen, J. Exp. Med. <u>182</u>, 1625 (1995).
201. L. Casciola-Rosen, D. W. Nicholson, T. Chong, K. R. Rowan, N. A. Thornberry, D. K. Miller, and A. Rosen, J. Exp. Med. <u>in press,</u> (1996).
202. Y. Emoto, Y. Manome, G. Meinhardt, H. Kisaki, S. Kharbanda, M. Robertson, T. Ghayur, W. W. Wong, R. Kamen, R. Weichselbaum, and D. Kufe, EMBO J. <u>14</u>, 6148 (1995).
203. S. Na, T.-H. Chuang, A. Cunningham, T. G. Turi, J. H. Hanke, G. M. Bokoch, and D. E. Danley, J. Biol. Chem. <u>271</u>, 11209 (1996).
204. C. Brancolini, M. Benedetti, and C. Schneider, EMBO J. <u>14</u>, 5179 (1995).
205. S. J. Martin, G. A. O'Brien, W. K. Nishioka, A. J. McGahon, A. Mahboubi, T. C. Saido, and D. R. Green, J. Biol. Chem. <u>270</u>, 6425 (1995).
206. C. Voelkel-Johnson, A. J. Entingh, W. S. Wold, L. R. Gooding, and S. M. Laster, J. Immunol. <u>154</u>, 1707 (1995).
207. B. An and Q. P. Dou, Cancer Res. <u>56</u>, 438 (1996).
208. H.-L. Hsu and N.-H. Yeh, J. Cell Sci. <u>109</u>, 277 (1996).
209. D. K. Miller, J. Myerson, and J. W. Becker, J. Cell Biochem. <u>in press,</u> (1996).
210. J. Yuan and M. Miura, World Patent 95/00160 (1995).
211. S. Kumar, Y. Tomooka, and M. Noda, Biochem. Biophys. Res. Commun. <u>185</u>, 1155 (1992).
212. S. Kumar, M. Kinoshita, M. Noda, N. G. Copeland, and N. A. Jenkins, Genes Develop. <u>8</u>, 1613 (1994).
213. L. Wang, M. Miura, L. Bergeron, H. Zhu, and J. Yuan, Cell <u>78</u>, 739 (1994).
214. N. A. Munday, J. P. Vaillancourt, A. Ali, F. J. Casano, D. K. Miller, S. M. Molineaux, T.-T. Yamin, V. L. Yu, and D. W. Nicholson, J. Biol. Chem. <u>270</u>, 15870 (1995).
215. L. Hillier, N. Clark, T. Dubuque, K. Elliston, M. Hawkins, M. Holman, M. Hultman, T. Kucaba, M. Le, G. Lennon, M. Marra, J. Parsons, L. Rifkin, T. Rohlfing, M. Soares, F. Tan, E. Trevaskis, R. Waterston, A. Williamson, P. Wohldmann, and R. Wilson, Unpublished (1995).
216. C. Faucheu, A. Diu, A. W. E. Chan, A.-M. Blanchet, D. Miossec, F. Herve, V. Collard-Dutilleul, Y. Gu, R. A. Aldape, J. A. Lippke, C. Rocher, M. S.-S. Su, D. J. Livingston, T. Hercend, and J.-L. Lalanne, EMBO J. <u>14</u>, 1914 (1995).
217. J. Kamens, M. Paskind, M. Hugunin, R. V. Talanian, H. Allen, D. Banach, N. Bump, M. Hackett, C. G. Johnston, P. Li, J. A. Mankovich, M. Terranova, and T. Ghayur, J. Biol. Chem. <u>270</u>, 15250 (1995).
218. Y. Gu, C. Sarnecki, R. A. Aldape, D. J. Livingston, and M. S.-S. Su, J. Biol. Chem. <u>270</u>, 18715 (1995).
219. C. Faucheu, A.-M. Blanchet, V. Collard-Dutilleul, J.-L. Lalanne, and A. Diu-Hercend, Eur. J. Biochem. <u>236</u>, 207 (1996).
220. T. Fernandes-Alnemri, G. Litwack, and E. S. Alnemri, J. Biol. Chem. <u>269</u>, 30761 (1994).
221. M. Tewari, L. T. Quan, K. O'Rourke, S. Desnoyers, Z. Zeng, D. R. Beidler, G. G. Poirier, G. S. Salvesen, and V. M. Dixit, Cell <u>81</u>, 801 (1995).
222. J. Schlegel, I. Peters, and S. Orrenius, FEBS Letters <u>364</u>, 139 (1995).
223. H. Duan, A. M. Chinnaiyan, P. L. Hudson, J. P. Wing, W.-W. He, and V. M. Dixit, J. Biol. Chem. <u>271</u>, 1621 (1996).
224. X. Wang, J.-T. Pai, E. A. Wiedenfeld, J. C. Medina, C. A. Slaughter, J. L. Goldstein, and M. S. Brown, J. Biol. Chem. <u>270</u>, 18044 (1995).
225. X. Wang, N. G. Zelenski, J. Yang, J. Sakai, M. S. Brown, and J. L. Goldstein, EMBO J. <u>15</u>, 1012 (1996).
226. T. Fernandes-Alnemri, G. Litwack, and E. S. Alnemri, Cancer Res. <u>55</u>, 2737 (1995).
227. T. Fernandes-Alnemri, A. Takahashi, R. Armstrong, J. Krebs, L. Fritz, K. J. Tomaselli, L. Wang, Z. Yu, C. M. Croce, G. Salveson, W. C. Earnshaw, G. Litwack, and E. S. Alnemri,

Cancer Res. 55, 6045 (1995).

228. J. A. Lippke, Y. Gu, C. Sarnecki, P. R. Caron, and M. S.-S. Su, J. Biol. Chem. 271, 1825 (1996).

229. M. J. Smyth and J. A. Trapani, Immunol. Today 16, 202 (1995).

230. A. M. Chinnaiyan, K. Orth, K. O'Rourke, H. Duan, G. G. Poirier, and V. M. Dixit, J. Biol. Chem. 271, 4573 (1996).

231. A. J. Darmon, D. W. Nicholson, and R. C. Bleackley, Nature 377, 446 (1995).

232 L. T. Quan, M. Tewari, K. O'Rourke, V. Dixit, S. J. Snipas, G. G. Poirier, C. Ray, D. J. Pickup, and G. S. Salvesen, Proc. Natl. Acad. Sci. USA 93, 1972 (1996).

233. S. C. Chow, W. M., G. E. N. Kass, T. H. Holmstrom, J. E. Eriksson, and S. Orrenius, FEBS Letters 364, 134 (1995).

234. H. O. Fearnhead, D. Dinsdale, and G. M. Cohen, FEBS Letters 375, 283 (1995).

235. T. Mashima, M. Naito, S. Kataoka, H. Kawai, and T. Tsuruo, Biochem. Biophys. Res. Comm. 209, 907 (1995).

236. C. E. Milligan, D. Prevette, H. Yaginuma, S. Homma, C. Cardwell, L. C. Fritz, K. J. Tomaselli, R. W. Oppenheim, and L. M. Schwartz, 15, 385 (1995).

237. G. J. Pronk, K. Ramer, P. Amiri, and L. T. Williams, Science 271, 808 (1996).

238 E. A. Slee, H. Zhu, S. C. Chow, M. MacFarlane, D. W. Nicholson, and G. M. Cohen, Biochem. J. 315, 21 (1996).

239. M. Enari, R. V. Talanian, W. W. Wong, and S. Nagata, Nature 380, 723 (1996).

240. J.-i. Hasegawa, S. Kamada, W. Kamiike, S. Shimizu, T. Imazu, H. Matsuda, and Y. Tsujimoto, Cancer Res. 56, 1713 (1996).

241. M. Miura, R. M. Friedlander, and J. Yuan, Proc. Natl. Aca. Sci. USA 92, 8318 (1995).

242. M. Tewari and V. M. Dixit, J. Biol. Chem. 270, 3255 (1995).

243. A. Strasser, A. W. Harris, D. C. S. Huang, P. H. Krammer, and S. Cory, EMBO J. 14, 6136 (1995).

244. H. Zhu, H. O. Fearnhead, and G. M. Cohen, FEBS Lett. 374, 303 (1995).

245. E. M. Ross and T. Higashijima, Methods Enzymol. 237, 26 (1994).

246. S. A. Memon, M. B. Moreno, D. Petrak, and C. M. Zacharchuk, J. Immunol. 155, 4644 (1995).

Chapter 27. The Role of JAKs and STATs in Transcriptional Regulation by Cytokines

Peter Lamb, H. Martin Seidel, Robert B. Stein and Jon Rosen
Ligand Pharmaceuticals
San Diego, CA 92121

Introduction - The growth, differentiation and function of a wide variety of cells is modulated by the action of a family of soluble proteins known as cytokines (1). Cytokines exert their effects on target cells by binding to specific cell surface receptors, triggering various intracellular events, including rapid changes in gene expression (2-5). Ultimately, these changes in gene expression determine the response of a cell to a particular cytokine. Recent advances have clarified the process by which binding of a cytokine to its receptor at the cell surface results in rapid alterations in gene expression in the nucleus (3-6). In contrast to some other receptor-ligand systems, these changes are not mediated by alterations in the concentration of small diffusible second messenger molecules, such as cAMP, Ca^{2+} or diacylglycerol. Rather, a series of ordered protein-protein interactions, regulated by tyrosyl residue phosphorylation, couples occupation of the cytokine receptor to transcriptional regulation in the nucleus (Figure 1).

Binding of a cytokine to its receptor results in the activation of particular members of a family of tyrosine kinases, known as JAKs (for Janus kinases, or Just Another Kinase), that associate non-covalently with the cytoplasmic domain of the receptor.

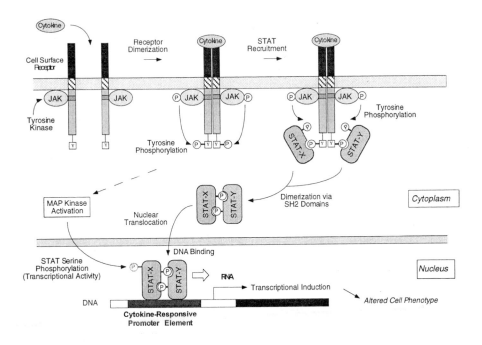

Figure 1 The JAK/STAT signal transduction pathway

Activated JAKs then phosphorylate several substrates, including the receptor itself and members of a family of latent cytoplasmic transcription factors known as STATs (Signal Transducers and Activators of Transcription). Upon phosphorylation of a specific tyrosyl residue, STAT proteins become activated, and form homo- or heterodimeric complexes, that translocate to the nucleus and bind to specific DNA sequences, termed STAT binding elements (SBEs). Analyses of SBEs in the promoters of several cytokine-responsive genes have shown that they are required for transcriptional induction following cytokine treatment. Thus, activation of STATs by JAKs at a cytokine receptor provides a rapid link that transfers information from the cell surface to the nucleus *via* a coordinated set of protein-protein interactions. A large number of studies, summarized in Table 1, have shown that cytokines that bind to either type I or type II cytokine receptors activate both JAKs and STATs (4). In addition, some growth factors that bind to receptors that possess intrinsic tyrosine kinase activity, such as EGF and PDGF, can also activate STATs. Different cytokines activate different members of the JAK and STAT families, though cytokines that share common signaling chains in their receptors (*e.g.* IL-6, OSM, LIF) often activate the same JAKs and STATs (6).

Table 1   Activation of JAKs and STATs by cytokines

Cytokine	JAK Activation	STAT Activation
IFN α/β	JAK1, Tyk2	STAT1, STAT2
IFN γ	JAK1, JAK2	STAT1
IL-10	JAK1, Tyk2	STAT1, STAT3
IL-2	JAK1, JAK3	STAT3, STAT5
IL-7	JAK1, JAK3	STAT3, STAT5
IL-9	JAK1, JAK3	STAT3, STAT5
IL-15	JAK1, JAK3	STAT3, STAT5
IL-4	JAK1, JAK3	STAT6
IL-13	JAK1	STAT6
IL-3	JAK2	STAT5
IL-5	JAK2	STAT1, STAT3
GM-CSF	JAK2	STAT5
IL-6	JAK1, JAK2, Tyk2	STAT1, STAT3
IL-11	JAK1, JAK2, Tyk2	STAT3
LIF	JAK1, JAK2, Tyk2	STAT3
OSM	JAK1, JAK2, Tyk2	STAT1, STAT3
CNTF	JAK1, JAK2, Tyk2	STAT3
IL-12	Tyk2, JAK2	STAT3, STAT4
G-CSF	JAK1, JAK2	STAT3, STAT5
Epo	JAK2	STAT5
Tpo	JAK2	STAT5
Prolactin	JAK2	STAT5
Growth Hormone	JAK2	STAT1, STAT3, STAT5
EGF	JAK1	STAT1, STAT3, STAT5
PDGF	?	STAT1, STAT3
M-CSF	?	STAT1, STAT3

The JAK Family - Four members of the JAK family have been identified, JAK1, JAK2, JAK3 and TYK2 (7-13). JAK1, JAK2 and TYK2 are widely expressed; however, JAK3 is expressed primarily in lymphoid and hematopoietic cells (8-12). JAKs share a novel domain structure depicted in Figure 2A, which distinguishes them as a subfamily within the larger family of cytoplasmic tyrosine kinases. The most C-terminal domain (JH1) is

A. JAKs

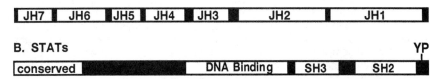

B. STATs

Figure 2 Schematic structure of JAK and STAT family members

a typical tyrosine kinase catalytic domain that is responsible for the activity of the enzyme (9). Immediately upstream is a kinase-like domain (JH2) of unknown function (9). The N-terminal portions of the JAKs share several domains that are homologous among members of the family (JH3-7), which may interact with cytokine receptors (5).

The central role of JAKs in cytokine signaling is demonstrated by studies using either somatic cells or knockout mice that lack the ability to express active JAKs. Cells lacking TYK2 are unable to induce interferon (IFN) α-responsive genes in response to IFN α (which activates TYK2 and JAK1), although they are still competent to induce IFN γ–responsive genes in response to IFN γ (which activates JAK1 and JAK2) (13,14). However, cells lacking JAK1 are unable to regulate genes in response to either IFN α, IFN γ, or IL-6, all of which activate JAK1 (15,16). JAK activation is therefore required for regulation of gene expression by these cytokines. *In vivo*, mutations resulting in a lack of functional JAK3 expression in either mice or humans lead to gross defects in lymphoid development and severe combined immune deficiency (17-19). These phenotypes result from the abrogation of signaling in response to the cytokines that normally activate JAK3 and a consequent failure of certain lymphoid lineages to proliferate and differentiate appropriately.

The STAT Family - Seven genes encoding STAT proteins have been described; STATs1, 2, 3, 4, 5a, 5b and 6 (20-29). Splice variants of STATs 1 and 3 have also been described that have altered activities (20,30). The STAT genes are widely expressed (22,23,27-29), with the exception of STAT4, which is expressed at high levels in the testes, and at lower levels in cells of the immune system (25,26). The STAT proteins constitute a novel family of transcription factors that share several structural features, illustrated in Figure 2B. The critical tyrosyl residue that becomes phosphorylated and triggers STAT dimerization is located near the C-terminus of the proteins, immediately downstream of an SH2 domain (31-33). This domain is required for STAT dimerization, and, in some cases, for association of the STAT with the corresponding cytokine receptor (see below). N-terminal to the SH2 domain is a putative SH3 domain of unknown function. The DNA binding domain, which appears to be novel, has recently been localized to a central region of the protein (34,35). The N-terminal 40 amino acids form another conserved domain that is apparently involved in STAT activation (36).

The role of STAT proteins is best understood in the context of IFN signaling. Cells lacking STAT1 show no transcriptional responses to either IFN α or γ, whereas cells lacking STAT2 are unable to respond to IFN α (37,38). Mice that lack STAT1 develop normally but are unable to respond to either IFN α or γ by induction of regulated genes (39,40). As a result, these mice are extremely susceptible to both viral and microbial infections and must be reared in a sterile environment. Although several cytokines other than IFNs activate STAT1 (see Table 1), biological responses regulated by these cytokines appear to be normal in the STAT1-deficient mice, suggesting that the role of STAT1 is restricted primarily to regulating responses to IFNs (39,40). Mice deficient in expression of the other known STATs are being generated and will be very informative.

INTERACTION OF JAKS AND STATS WITH CYTOKINE RECEPTORS

Mounting data strongly support a view that cytoplasmic domains of cytokine receptors nucleate the assembly of large complexes of signaling proteins in response to receptor occupancy (41). The constellation of proteins present in these complexes is thought to determine the type, strength and duration of signals emanating from the receptor, and ultimately the response of a cell to a particular cytokine. The key early events seem to be oligomerization of the receptor chains, resulting in activation of the JAKs and subsequent phosphorylation of the receptor cytoplasmic domains (42), which enables the receptor to recruit signaling molecules, some of which become phosphorylated themselves.

JAKs - It is now evident that members of the JAK family interact non-covalently with cytoplasmic domains of cytokine receptor subunits. Studies employing deletion mutants of receptors have localized the region required for activation of JAKs to a membrane proximal domain of about 50 amino acids, which contains a region of homology shared by all cytokine receptors termed box 1 (43-51). Many cytokines activate multiple JAKs (Table 1), and in these instances different JAKs may associate preferentially with different receptor subunits. For example, JAK1 appears to associate with the β chain of the IL-2 receptor, whereas JAK3 associates with the γ chain (52,53). This probably explains why activation of JAK3 is limited to cytokines that bind to receptors that utilize the common γ chain (IL-2, IL-4, IL-7, IL-9 and IL-15). People who inherit a mutated γ chain have severe combined immune deficiency, presumably as a consequence of defective JAK activation in response to the cytokines that utilize this receptor chain (52). In many receptor systems, mutations preventing JAK activation also prevent all functions of the receptor, including receptor phosphorylation, STAT activation and initiation of other signal transduction pathways (6). This suggests that JAKs are required for initiation of all downstream signaling events.

STATs - Like the JAKs, STAT proteins also form a complex with the receptor, but only after binding of cytokine. In the best characterized case, phosphorylation of a single tyrosyl residue on the α chain of the IFN γ receptor appears to be sufficient for binding of STAT1 to the receptor (54). A similar model has been proposed for the interaction of STAT6 with the IL-4 receptor (28). Short phosphotyrosyl residue-containing peptides derived from the cytoplasmic domain of the receptors seem to be sufficient to recruit specific STAT proteins. For example, a short peptide derived from gp130 (a component of the LIF, IL-6 and OSM receptors) directs activation of STAT3 in response to Epo when appended to the Epo receptor (which normally does not activate STAT3) (41). The SH2 domain of the STAT proteins plays a major role in recognizing these phosphopeptides; swapping the SH2 domain of one STAT protein for that of another redirects the chimeric STAT protein to the receptor recognized by the new SH2 domain (55). This shows that the STAT SH2 domains direct the binding of STATs to cytokine receptors, and suggests that the specificity of STAT activation exhibited by different cytokines can be explained, in part, by the specificity of the different STAT SH2 domains for different phosphotyrosyl residue-containing peptides embedded in the cytoplasmic domains of the cytokine receptor chains.

STATs may also associate with cytokine receptors via an indirect mechanism. In the IFN α system, a model has been proposed in which STAT1 is brought to the IFN α receptor via an interaction with receptor bound STAT2 (38). Also, mutant Epo or growth hormone receptors that lack tyrosyl residues can still activate STATs, possibly through the interaction of STATs with other proteins that bind to these receptors in a phospho-tyrosine independent fashion (51,56,57).

DNA SEQUENCES THAT BIND STAT PROTEINS: DETERMINANTS OF CYTOKINE SELECTIVITY

The STAT1 and STAT2 proteins were first identified as constituents of a heterotrimeric protein complex termed ISGF3, that bound to specific promoter sequences called IFN α/β-stimulated response elements (ISREs) (3, 58, 59). The consensus sequence of the ISRE, AGTTTCNNTTTCNC/T, is necessary and sufficient for IFN α/β-mediated transcriptional responses (3). Concurrently, it was found that in response to IFN γ, a STAT1 homodimer binds to specific promoter sequences termed gamma activation sites (GAS elements). GAS elements, which mediate IFN γ transcriptional responses, are characterized by the palindromic consensus sequence TTNCNNNAA (3, 60). IFN γ and the IFN α/β family activate different sets of genes, and the activation of different STAT complexes by IFN γ and IFN α/β coupled with the divergent nature of their response elements provide a satisfyingly simple answer to the question of signaling specificity.

Work on sequence elements that mediate transcriptional induction in response to IL-6 during the acute phase response (acute phase response elements or APREs), to prolactin in the induction of milk proteins (mammary gland factor/MGF binding sites), and to PDGF (the sis-inducible element or SIE), led to the recognition that these seemingly disparate response elements were in fact relatives of the canonical GAS element (61-63). This implied that all of the members of the STAT family share the ability to bind to similar, GAS-related STAT-binding elements (SBEs), a suggestion reinforced by *in vitro* binding site selection experiments performed for STAT1 and STAT3, which yielded the same optimal binding site for both proteins (34). The one exception to this pattern is the STAT-containing ISGF3 complex, which binds to the ISRE as described above. The fact that a diverse list of cytokines, which activate non-overlapping panels of genes, activate STATs that bind to similar, GAS-related sequences raises questions regarding how cytokine signaling specificity is achieved.

Several resolutions of this apparent conundrum can be envisaged. First, specificity may reside in the precise sequence, binding and transcriptional activities of a given SBE, and recent work supports this notion. For example, a systematic analysis by point mutation of a known GAS element led to the identification of sequences that bound selectively either STAT1 or STAT3, and in some cases this selectivity could be correlated with selective transcriptional regulation (64,65). The most dramatic effects were observed when the spacing between the half sites that make up the canonical STAT binding element were altered (66). The general palindromic structure of the GAS element consists of 'TT' half sites separated by five nucleotides, $TT(N)_5AA$; however, the promoters of several cytokine-responsive genes contain potential STAT binding elements that differ from this consensus in the number of nucleotides between the palindromic half sites. The spacing of the palindromic half sites can dramatically affect *in vitro* STAT binding (Table 2). For example, sites with a contracted spacing of 4 bp, $TT(N)_4AA$ bind selectively to STAT3, and sites with an expanded half site spacing of 6

Response Element	Core Sequence	Half-Site Spacing	*in vitro* Binding/Transcriptional Activation			
			STAT1	STAT3	STAT5	STAT6
4spC	TTCCCGAA	4	-/-	+/+	-/-	-/-
junB	GTCAGGAA	4	-/-	+/+	-/-	-/-
5spB	TTCCCGGAA	5	+/+	+/+	+/+	+/-
IRF-1	TTCCCCGAA	5	+/+	+/+	+/+	+/-
6spB	TTCCCCGGAA	6	-/-	-/-	-/-	+/-
hlε	TTCCCAAGAA	6	-/-	-/-	-/-	+/+

Table 2 STAT binding and transcriptional activation by SBEs differing in half-site spacing

bp, $TT(N)_6AA$ bind extremely selectively to STAT6. The preference of STAT6 for binding to an expanded 6 bp spacing element was later confirmed using random binding site optimization experiments (35). The canonical 5 bp spacing, $TT(N)_5AA$, in fact displays general STAT binding properties. These studies and others have demonstrated that the precise nature of the core and flanking sequences also plays a role in STAT binding specificity (65,66). In general, selective binding correlates with selective transcriptional activation; the 4 bp spacing element directs a selective response to cytokines that activate STAT3, e.g. IL-6, while the 6 bp spacing element mediates a selective response to cytokines that activate STAT6, e.g. IL-4 (Table 2). Interestingly, however, in vitro binding is not perfectly correlated with transcriptional activation. Sequences have been identified that bind well to STAT6 in vitro but that are transcriptionally inactive. The functional activity of a given element is further determined by specific sequence requirements in both the spacing and flanking nucleotides (66). The differences in selectivity exhibited by SBEs may explain the observation that cytokines can activate shared sets of genes (genes with promoters that contain SBEs with general STAT-binding properties) and also distinct sets of genes (genes with promoters that contain SBEs with restricted STAT-binding properties).

In addition to selectivity achieved through the sequence of the STAT binding element itself, a second contributing factor is the promoter context of the SBE, a notion for which there is considerable experimental support. Thus, the required presence of binding sites for additional inducible or constitutive cell type-specific transcription factors in the promoters of inducible genes helps define a selective pattern of cytokine responsiveness. For example, the induction of various milk protein genes by the hormone prolactin requires the simultaneous presence of glucocorticoids, and the promoters of some of these genes contain both an obligate SBE as well as an obligate glucocorticoid-responsive element (63,67). Another well-characterized system involves the IFN γ-induction of the FcγRI gene, which occurs specifically in myeloid cells, due to the presence in the promoter of a classical GAS element adjacent to a binding site for the tissue-restricted transcription factor PU.1 (68-70).

Still another explanation for selective cytokine responsiveness is suggested by the IFN α/β system. The exception to the 'universal' binding of STAT complexes to GAS variants is the IFN α/β-activated complex ISGF3, which binds to a sequence (the ISRE) distinct from the GAS. In addition to STAT1 and STAT2, the ISGF3 complex contains the accessory DNA binding protein p48, a distant homologue of c-myb unrelated to STATs which confers distinct DNA binding properties on the complex (71). It is therefore possible that counterparts to p48 exist for other cytokines, and such accessory proteins could redirect activated STAT complexes to novel, cytokine-selective sequences.

MODULATION OF SIGNALING VIA THE JAK/STAT PATHWAY

MAP Kinase - Recent studies show that STAT1 is phosphorylated on a serine located near the C-terminus of the protein, within a MAP kinase consensus sequence which is also present in a similar location in several other (but not all) STAT proteins (72,73). A mutant STAT1 protein incapable of being phosphorylated at this position is a less efficient transcriptional activator than wild type, serine-phosphorylated STAT1, an effect attributed to a direct influence on transcriptional activation rather than DNA binding (72). Similar results were obtained with STAT3 (72,74), but other studies have suggested that serine phosphorylation of STAT3 influences DNA binding directly (73). Although the reason for these differences is unclear, it is evident that serine phosphorylation of STATs modulates their ability to activate transcription.

Tyrosine Phosphatases - The phosphorylation of STATs, JAKs and cytokine receptors is rapid and transient, usually decaying in minutes to hours after cytokine stimulation. This decay is due, at least in part, to the action of protein tyrosine phosphatases. Treatment of cells with the general tyrosine phosphatase inhibitor pervanadate causes activation of JAKs and STATs and induction of immediate early gene expression even in the absence of cytokine, which suggests that phosphatases keep the JAK/STAT pathway inactive in untreated cells (75,76). In addition, treatment of cells with pervanadate after the addition of IFN prolongs the activation of STAT1, suggesting that phosphatases also act to terminate signaling (77). Thus, phosphatases may act in the cytoplasm, to dampen and terminate signaling from the receptor, and also in the nucleus to inactivate STAT proteins. One of the phosphatases involved in regulating JAK/STAT signaling from the Epo receptor is hematopoeitic cell phosphatase (HCP, also known as PTP1C, SH-PTP1). HCP binds to specific phospho-tyrosyl residues on the activated Epo receptor, and is then thought to dephosphorylate signaling molecules associated with the receptor, including JAK2 (78,79). Individuals who carry a mutation in the Epo receptor that prevents the association of HCP exhibit a benign erythrocytosis, and mice that lack HCP activity (moth-eaten mice) also show increased erythropoiesis, consistent with a negative-regulatory role for HCP in Epo signaling (80-82). Similar paradigms may apply to other cytokine receptors, involving either HCP or related phosphatases (83, 84).

DRUG DISCOVERY

Many cytokines that signal *via* the JAK/STAT pathway, including IFN α, IFN β, IFN γ, IL-2, growth hormone, and the hematopoietic growth factors, Epo, G-CSF, and GM-CSF, have proven clinical utility, and other cytokines have great potential (Table 3). In addition, the putative roles of many of the cytokines in the pathophysiology of various diseases suggest possible uses for cytokine antagonists (see Table 3). Examples include possible use of interleukin or interferon antagonists in the treatment of inflammation or CSF inhibitors in the treatment of leukemias (85,86). Many current or proposed therapies based on modulating cytokine action utilize recombinant proteins. Orally available small molecule cytokine mimics or antagonists would have several advantages over such proteins, being less expensive to manufacture and eliminating the need for injections.

Assay Systems - Based on the recent elucidation of the JAK/STAT signal transduction pathway, a variety of cell-based or biochemical high throughput screens can be developed for the discovery of small molecule cytokine mimics or antagonists. A typical cell-based assay system (87) includes a reporter plasmid with a promoter

Table 3 Approved and potential therapeutic uses of selected cytokine agonists and antagonists

Cytokine	Agonist	Antagonist
IFN α/β	Cancer, Viral infections, Autoimmunity	Inflammation
IFN γ	Chronic granulomatous disease, Infections	Inflammation, Autoimmunity
IL-2	Cancer	Histoincompatability
IL-3	Leukopenia, Myeloid reconstitution	Leukemia
IL-4	Inflammation, Cancer	Allergy
IL-6	Thrombocytopenia	Cancer, Osteoporosis, Inflammation
IL-11	Thrombocytopenia	
IL-12	Cancer, Infections	Histoincompatability, Autoimmunity
G-CSF	Neutropenia, Myeloid reconstitution	Leukemia
GM-CSF	Leukopenia, Myeloid reconstitution	Leukemia
Epo	Anemias, Elective blood donation	
Tpo	Thrombocytopenia	
EGF	Wound healing	Cancer
PDGF	Wound healing	Cancer, Atherosclerosis

containing SBEs, linked to a reporter gene such as firefly luciferase, which is introduced into a cell line containing all of the components of the cytokine signaling system of interest (*e.g.*, cytokine receptor, JAKs, STATs and tyrosine phosphatases). Addition of the cytokine results in a robust increase in reporter levels, and the presence of compounds that antagonize JAK/STAT signaling is indicated by a reduction in reporter activity. Such compounds are potential antagonists of the JAK/STAT pathway, potentially acting at one of many steps in the signaling cascade. Addition of compounds in the absence of cytokine, or in the presence of low levels of cytokine, allows the detection of compounds that increase the levels of reporter activity. Such compounds are potential cytokine mimics. The major advantage of these cell-based screens is that they can identify compounds that act at any of the numerous steps of the pathway.

The JAK-STAT signal transduction cascade is also amenble to the development of biochemical assays, since it consists of a series of protein-protein interactions modulated by various enzymatic activities. Screens can be established to identify compounds that inhibit the catalytic activity of the specific kinases and phosphatases in the JAK/STAT pathway, or that interfere with the ability of two components of the pathway to interact. For example, compounds that inhibit interaction between JAKs and their substrates, such as the receptor or STATs, are likely to be useful as inhibitors of the pathway, while compounds that inhibit the interaction between phosphatases and their substrates are potential cytokine mimics. The advantages of this type of assay are very high throughput, no requirement for compounds to be able to enter cells and immediate knowledge of the mechanism of action of hits.

Conclusion - Current evidence suggests that activation of JAKs is the key early event in the initiation of signal transduction in response to cytokines, and leads to the activation not only of the STAT proteins, but also of all other known signaling pathways. STAT proteins play a direct role in the regulation of immediate early gene expression, and, via the induction of other transcription factors, they are the first link in a chain leading to longer term induction of genes that regulate proliferation, differentiation and the activity of fully differentiated cells. The precise role of the STATs in eliciting these changes is an area of evolving research, and the role that STATs play in determining the specificity apparent in the biological action of cytokines remains an important question. Nevertheless, it is clear that the ordered series of steps mediated by JAKs and STATs that constitute a signaling pathway from the cell surface to the nucleus, present excellent opportunities for the discovery of novel drugs that modulate cytokine action.

References

1. K. Arai, F. Lee, A. Miyajima, S. Miyatake, N. Arai, T. Yokota, Ann. Rev. Biochem., 59, 783 (1990)
2. A. Miyajima, T. Kitamura, N. Harada, T. Yokota, K. Arai, Ann. Rev. Immunol. 10, 295, (1992)
3. J.E. Darnell, I.M. Kerr, G.R. Stark, Science, 264, 1415 (1994).
4. C. Schindler and J.E. Darnell, Ann. Rev. Biochem., 64, 621 (1995)
5. J.N. Ihle and I.M. Kerr, TIG, 11, 69 (1995).
6. J.N. Ihle, Nature, 377, 591 (1995).
7. O. Silvennoinen, B.A. Witthuhn, F.W. Quelle, J.L. Cleveland, T. Yi, J.N. Ihle, Proc. Natl. Acad. Sci. USA, 90, 8429 (1993).
8. I. Firmbach-Kraft, M. Byers, T. Shows, R. Dalla-Favera, J.J. Krolewski, Oncogene 5, 1330 (1990).
9. A.F. Wilks, A.G. Harpur, R.R. Kurban, S.J. Ralph, G. Zurcher, A. Ziemiecki, Molec. Cell. Biol., 11, 2057 (1991).
10. A.G. Harpur, A.C. Andres, A. Ziemiecki, R.R. Aston, A.F. Wilks, Oncogene, 7, 1347 (1992)
11. M. Kawamura, D.W. McVicar, J.A. Johnston, T.B. Blake, Y.-Q. Chen, B.K. Lal, A.R. Lloyd, D.J. Kelvin, J.E.Staples, J.R. Ortaldo, J.J. O'Shea, Proc. Natl. Acad. Sci. USA, 91, 6374 (1994).
12. B.A. Witthuhn, O. Silvennoinen, O. Miura, K.S. Lai, C. Cwik, E.T. Liu, J.N. Ihle, Nature, 370, 153 (1994).
13. S. Pellegrini, J. John, M. Shearer, I.M. Kerr, G.R. Stark, Molec. Cell. Biol., 9, 4605, (1989).
14. L. Velazquez, M. Fellous, G.R. Stark, S. Pellegrini, Cell, 70, 313 (1992).
15. M. Muller, J. Briscoe, C. Laxton, D. Guschin, A. Ziemiecki, O. Silvennoinen, A.G. Harpur, G. Barbieri, B.A. Witthuhn, C. Schindler, S. Pellegrini, A.F. Wilks, J.N. Ihle, G.R. Stark, I.M. Kerr, Nature, 366, 129 (1993).
16. D. Guschin, N. Rogers, J. Briscoe, B. Witthuhn, D. Watling, F. Horn, S. Pellegrini, K. Yasukawa, P. Heinrich, G.R. Stark, J.N. Ihle, I.M. Kerr, EMBO J., 14, 1421 (1995).
17. D.C. Thomis, C.B. Gurniak, E. Tivol, A.H. Sharpe, L.J. Berg, Science, 270, 794 (1995).
18. T. Nosaka, J.M.A. van Deursen, R.A. Tripp, W.E. Thierfelder, B.A. Witthuhn, A.P. McMickle, P.C. Doherty, G.C. Grosveld, J.N. Ihle, Science, 270, 800 (1995).

19. S.M. Russell, N. Tayebi, H. Nakajima, M.C. Riedy, J.L. Roberts, M.J. Aman, T.-S. Migone, M. Noguchi, M.L. Markert, R.H. Buckley, J.J. O'Shea, W.J. Leonard, Science, 270, 797 (1995).

20. C. Schindler, X.-Y. Fu, T. Improta, R. Aebersold, J.E. Darnell, Proc. Natl. Acad. Sci. USA, 89, 7836 (1992).

21. X.-Y. Fu, C. Schindler, T. Improta, R. Aebersold, J.E. Darnell, Proc. Natl. Acad. Sci. USA, 89, 7840 (1992).

22. Z. Zhong, Z. Wen, J.E. Darnell, Science, 264, 95 (1994).

23. S. Akira, Y. Nishio, M. Inoue, X.-J. Wang, S. Wei, T. Matsusaka, K. Yoshida, T. Sudo, M. Naruto, T. Kishimoto, Cell, 77, 63 (1994).

24. R. Raz, J.E. Durbin, D.E. Levy, J. Biol. Chem., 269, 24391 (1994).

25. Z. Zhong, Z. Wen, J.E. Darnell, Proc. Natl. Acad. Sci. USA, 91, 4806 (1994).

26. K. Yamamoto, F.W. Quelle, W.E. Thierfelder, B.L. Kreider, D.J. Gilbert, N.A. Jenkins, N.G. Copeland, O. Silvennoinen, J.N. Ihle, Molec. Cell. Biol., 14, 4342 (1994).

27. H. Wakao, F. Gouilleux, B. Groner, EMBO J., 13, 2182 (1994).

28. J. Hou, U. Schindler, W.J. Henzel, T.C. Ho, M. Brasseur, S.L. McKnight, Science, 265, 1701 (1994).

29. F.W. Quelle, K. Shimoda, W. Thierfelder, C. Fischer, A. Kim. S.M. Ruben, J.L. Cleveland, J.H. Pierce, A.D, Keegan, K. Nelms, W.E. Paul, J.N. Ihle, Molec. Cell. Biol., 15, 3336 (1995).

30. T.S. Schaefer, L.K. Sanders, D. Nathans, Proc. Natl. Acad. Sci. USA, 92, 9097 (1995).

31. K. Shuai, G.R. Stark, I.M. Kerr, J.E. Darnell, Science, 261, 1744 (1993)

32. K. Shuai, C.M. Horvath, L.H. Tsai Huang, S.A. Qureshi, D. Cowburn, J.E. Darnell, Cell, 76, 1 (1994).

33. F. Gouilleux, H. Wakao, M. Mundt, B. Groner, EMBO J. 13, 4361 (1994)

34. C.M. Horvath, Z. Wen, J.E. Darnell, Genes Develop., 9, 984, (1995).

35. U. Schindler, P. Wu, M. Rothe, M. Brasseur, S.L. McKnight, 2, 689 (1995)

36. S.A. Qureshi, S. Leung, I.M. Kerr, G.R. Stark, J.E. Darnell, Molec. Cell. Biol., 16, 288 (1996)

37. M. Muller, C. Laxton, J. Briscoe, C. Schindler, T. Improta, J.E. Darnell, G.R. Stark, I.M. Kerr, EMBO J., 12, 4221 (1993).

38. S. Leung. S.A. Qureshi, I.M. Kerr, J.E. Darnell, G.R. Stark, Molec. Cell. Biol., 15, 1312 (1995).

39. J.E. Durbin, R. Hackenmiller, M.C. Simon, D.E. Levy, Cell, 84, 443 (1996).

40. M.A. Meraz, J.M. White, K.C.F. Sheehan, E.A. Bach, S.J. Rodig, A.S. Dighe. D.H. Kaplan, J.K. Riley, A.C. Greenlund, D. Campbell, K. Carver-Moore, R.N. DuBois, R. Clark, M. Aguet, R.D. Schreiber, Cell, 84, 431 (1996).

41. N. Stahl, T.J. Farruggella, T.F. Boulton, Z. Zhong, J.E. Darnell, G.D. Yancopoulos, Science, 267, 1349 (1995).

42. C-H. Heldin, Cell, 80, 213-223 (1995)

43. F.W Quelle, N. Sato, B.A. Witthuhn, R.C. Inhorn, M. Eder, A. Miyajima, J.D. Griffin, J.N. Ihle, Molec. Cell. Biol., 14, 4335 (1994).

44. S.E. Nicholson, U. Novak, S.F. Ziegler, J.E. Layton, Blood, 86, 3698 (1995).

45. D.J. Tweardy, T.M. Wright, S.F. Ziegler, H. Baumann, A. Chakraborty, S.M. White, K.F. Dyer, K.A. Rubin, Blood, 86, 4409 (1995).

46. T.-C. He, N. Jiang, H. Zhuang, D.E. Quelle, D.M. Wojchowski, J. Biol. Chem., 269, 18291 (1994).

47. Y.-D. Wang and W.I. Wood, Molec. Endocrinol., 9, 303 (1995).

48. A. Sotiropoulos, M. Perrot-Applanat, H. Dinerstein, A. Pallier, M.-C. Postel-Vinay, J. Finidori, P.A. Kelly, Endocrinology, 135, 1292 (1994).

49. S.J. Frank, G. Gilliland, A.S. Kraft, C.S. Arnold, Endocrinology, 135, 2228 (1994).

50. J.A. VanderKuur, X. Wang, L. Zhang, G.S. Campbell, G. Allevato, N. Billestrup, G. Norsteddt, C. Carter-Su, J. Biol. Chem., 269, 21709 (1994).

51. R.H. Hackett, Y. Wang, A.C. Larner, J. Biol. Chem., 270, 21326 (1995)

52. S.M. Russell, J.A. Johnston, M. Noguchi, M. Kawamura, C.M. Bacon, M. Friedmann, M. Berg, D.W. McVicar, B.A. Witthuhn, O. Silvennoinen, A.S. Goldman, F.C. Schmalstieg, J.N. Ihle, J.J. O'Shea, W.J. Leonard, Science, 266, 1042 (1994).

53. T. Miyazaki, A. Kawahara, H. Fujii, Y. Nakagawa, Y. Minami, Z.-J. Liu, I. Oishi, O. Silvennoinen, B.A. Witthuhn, J.N. Ihle, T. Taniguchi, Science, 266, 1045 (1994).

54. A.C. Greenlund, M.A. Farrar, B.L. Viviano, R.D. Schreiber, EMBO J., 13, 1591 (1994).

55. M.H. Heim, I.M. Kerr, G.R. Stark, J.E. Darnell, Science, 267, 1347 (1995).

56. J.E. Damen, H. Wakao, A. Miyajima, J. Krosl, R.K. Humphries, R.L. Cutler, G. Grystal, EMBO J., 14, 5557 (1995).

57. Y. Wang, K. Wong, W.I. Wood, J. Biol. Chem., 270, 7021 (1995)

58. D.E. Levy, D.S. Kessler, R. Pine, N. Reich, J.E. Darnell, Genes Develop., 2, 383 (1988).

59. T.C. Dale, A.M.A. Imam, I.M. Kerr, G.R. Stark, Proc. Natl. Acad. Sci. USA, 86, 1203 (1989).

60. D.L. Lew, T. Decker, I. Strehlow, J.E. Darnell, Molec. Cell. Biol., 11, 182 (1991).

61. J. Yuan, U.M. Wegenka, C. Lutticken, J. Buschmann, T. Decker, C. Schindler, P.C. Heinrich, F. Horn, Molec. Cell. Biol., 14, 1657 (1994).

62. B.J. Wagner, T.E. Hayes, C.J. Hoban, B.H. Cochran, EMBO J., 9, 927 (1990).

63. G.J.R. Standke, V.S. Meier, B. Groner, Molec. Endocrinol., 8, 469 (1994).

64. P. Lamb, L.V. Kessler, C. Suto, D.E. Levy, H.M. Seidel, R.B. Stein, J. Rosen, Blood, 83, 2063

(1994).

65. P. Lamb, H.M. Seidel, J. Haslam, L. Milocco, L.V. Kessler, R.B. Stein, J. Rosen, Nucleic Acids Res., 23, 3283 (1995).

66. H.M. Seidel, L.H. Milocco, P. Lamb, J.E. Darnell, R.B. Stein, J. Rosen, Proc. Natl. Acad. Sci. USA, 92, 3041 (1995).

67. M. Schmitt Ney, W. Doppler, R.K. Ball, B. Groner, Molec. Cell. Biol., 11, 3745 (1991)

68. R.N. Pearse, R. Feinman, K. Shuai, J.E. Darnell, J.V. Ravetch, Proc. Natl. Acad. Sci. USA, 90, 4314 (1993).

69. Q.G. Eichbaum, R. Iyer, D.P. Raveh, C. Mathieu, R.A.B. Ezekowitz, J. Exp. Med. 179, 1985 (1994).

70. C. Perez, J. Wietzerbin, P.D. Bernech, Molec. Cell. Biol., 13, 2182 (1993).

71. S.A. Veals, C. Schindler, D. Leonard, X.-Y. Fu, R. Aebersold, J.E. Darnell, D.E. Levy, Molec. Cell. Biol., 12, 3315 (1992).

72. Z. Wen, Z. Zhong, J.E. Darnell, Cell, 82, 1 (1995).

73. X. Zhang, J. Blenis, H.-C. Lin, C. Schindler, S. Chen-Kiang, Science, 267, 1990 (1995).

74. T.G. Boulton, Z. Zhong, Z. Wen, J.E. Darnell, N. Stahl, Proc. Natl. Acad. Sci. USA, 92, 6915 (1995).

75. K.-I. Igarashi, M. David, A.C. Larner, D.S. Finbloom, Molec. Cell. Biol., 13, 3984 (1993).

76. P. Lamb, J. Haslam, L. Kessler, H.M. Seidel, R.B. Stein, J. Rosen, J. Interfer. Res., 14, 365 (1994).

77. M. David, P.M. Grimley, D.S. Finbloom, A.C. Larner, Molec. Cell. Biol., 13, 7515 (1993).

78. T. Yi, J. Zhang. O. Miura, J.N. Ihle, Blood, 85, 87 (1995).

79. R. Klingmuller, U. Lorenz, L.C. Cantley, B.G. Neel, H.F. Lodish, Cell, 80, 729 (1995).

80. G. Van Zant and L. Schultz, Exp. Hematol., 17, 81 (1989).

81. J.S. Bignon and K.A. Siminovitch, Clin. Immunol. Immunopath., 73, 168 (1994).

82. A. de la Chapelle, A.-L Traskelin, E. Juvonen, Proc. Natl. Acad. Sci. USA, 90, 4495 (1993).

83. T. Yi, A. L.-F. Mui, G. Krystal, J.N. Ihle, Molec. Cell. Biol., 13, 7577 (1993).

84. T. Yi and J.N. Ihle, Molec. Cell. Biol., 13, 3350 (1993).

85. H. Heremans and A. Billiau, Drugs, 38, 957 (1989).

86. N. Meydan, T. Grunberger, H. Dadi, M. Shahar, E. Arpaia, Z. Lapidot, J.S. Leeder, M. Freedman, A. Cohen, A. Gazit, A. Levitzki, C.H. Roifman, Nature, 379, 645 (1996)

87. J. Rosen, A. Day, T.K. Jones, E.T.T. Jones, A.M. Nazdan, R.B. Stein, J. Med. Chem., 38, 4855 (1995)

Chapter 28. Novel Inhibitors of the Proteasome and Their Therapeutic Use in Inflammation

Julian Adams and Ross Stein
ProScript, Inc.
Cambridge, MA 02139

The proteasome is a large, multimeric protease that catalyzes the final step of the ubiquitin-proteasome pathway for intracellular protein degradation (1, 2, 3). Once thought to be responsible only for catabolism of abnormal proteins, it is now known that the pathway plays a key role in the turnover and processing of many regulatory proteins (1, 4). Several of these proteolytic processes represents key biochemical events that contribute to the pathogenesis of human disease (1). Thus, the enzymes of the ubiquitin-proteasome pathway (UPP) are now being recognized as molecular targets for drug discovery and design. In this chapter, we summarize what is known of inhibitors of the proteasome and how they might be used to treat inflammatory disease.

BIOCHEMISTRY OF THE UBIQUITIN-PROTEASOME PATHWAY

Ubiquitin-Proteasome Pathway - Ubiquitin (Ub) was discovered in 1978 as a heat-stable polypeptide component of a then poorly characterized ATP-dependent proteolytic system in rabbit reticulocytes (5). Subsequent studies revealed that Ub is an evolutionarily conserved protein found in all eukaryotic cells and is composed of 76 residues (MW 8.6 kDa) The role that Ub plays in this new proteolytic system is that of a molecular marker. In a series of enzyme-catalyzed reactions (6), the C-terminal Gly of Ub is ligated onto the ε-amine group of Lys residues of the targeted protein to form a covalent, isopeptide bond. Lys^{48} of this Ub molecule then donates its ε-amine for isopeptide bond formation with another molecule of Ub. Subsequent rounds of ubiquitinylation ultimately build a long polyUb chain on the substrate. Ubiquitylated proteins can either be deubiquitinylated through the action of Ub C-terminal hydrolases or degraded to peptides through the action of the proteasome. Both of these reactions regenerate intact Ub.

The first step in the ubiquitinylation of a protein is the activation of Ub by Ub activating enzyme, also referred to as E1. In an ATP-dependent reaction proceeding through a Ub adenylate intermediate, the carboxylate of the C-terminal Gly of Ub forms a thioester with the active site thiol of E1. In the second step, the Ub is transferred from E1 to the active site thiol of a Ub carrier protein, or E2, to form yet another thioester bond. Finally, the Ub is transferred from E2 to the protein substrate. This reaction is often catalyzed by a Ub protein ligase, or E3. While there is a single E1 in cells, both E2 and E3 enzymes are large families with at least a dozen members in each. E2 and E3 enzymes are thought to work in combinatorial fashion.

Structure and Catalytic Mechanism of the Proteasome - The proteasome exists in multiple forms within eukaryotic cell and at the heart of all these forms is the catalytic core known as the 20S proteasome (a.k.a. multicatalytic protease). The structure of this large particle (~700 kDa) is best described as a stack of four oligomeric rings, each composed of seven subunits (see Figure I).

The unique structure of the 20S proteasome was first revealed by electron microscopic studies (7, 8) and more recently refined through X-ray diffraction studies (9). In the 20S proteasome from the archaebacterium *Thermoplasma acidophylum*, the outer and inner rings are composed of identical α-subunits and identical β-subunits, respectively. In marked contrast, eukaryotic proteasomes have seven different but similar α-subunits and seven different but similar β-subunits.

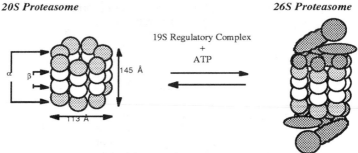

<u>Figure I</u> Structures of the 20S and 26S Proteasomes

The structures of the α- and β-subunits are remarkably similar, each being a sandwich of two five-stranded antiparallel β sheets. The β sandwich of each subunit is flanked by α helices on its top and bottom. The greatest structural difference between the α- and β-subunits lies in the N-terminal region. Here the β-subunit lacks residues 1 to 35 of the α-subunit. These residues form an α-helix that may play a critical role in the interaction of α-subunits with regulatory complexes such as the 19S regulatory complex or PA28. These helices also maintain tight contact between the α- and β-subunits, making the particle impenetrable form the outer walls of the cylinder.

To enter the cylinder, a polypeptide must negotiate a narrow passage guarded by an α-subunit loop. The seven loops of the α -subunits form a hydrophobic ring that leads to one of two symmetrically disposed minor chambers. Passage from these chambers into the major chamber of the proteasome is blocked by a bottleneck formed from β-subunit loops analogous to the α -subunits loops. The major chamber is bounded by the inner walls of the β-subunits which contain the active sites of the proteasome. For a protein to be hydrolyzed by the 20S proteasome, it must first "unwind" and then be threaded through one of the ends of the proteasome. Structural elements of the protein substrate are then recognized by β-subunit active sites and amide bonds of the substrate are hydrolyzed to liberate peptides that are roughly 8 amino acids in length (10).

The proteasome's natural substrates are poly-ubiquitinated proteins and the form of the proteasome that degrades these proteins is the 26S proteasome. This species forms in an ATP-dependent manner from the combination of one copy of the 20S proteasome with two copies of the 19S regulatory complex (11). The 19S regulatory complex is composed of approximately 16 different subunits of sizes that range from 30 kDa to 120 kDa and contains a Ub-chain receptor, ATPase activity, and isopeptidase activity.   Given these activities, a general mechanism has been proposed to describe the hydrolysis of a Ub-protein by the 26S proteasome (1): The Ub-protein conjugate substrate is first recognized and bound by the 26S proteasome through interaction with the Ub-chain receptor.   Next, chemical free energy derived from ATP hydrolysis is transduced into mechanical energy and used to pay the energetic price of unwinding the otherwise stable structure of the substrate protein. Finally, the unwound protein is threaded into the inner channel of the 20S proteasome for proteolysis, but only after Ub chains are cleaved off through the action of isopeptide activity.

The broad P_1 specificity of the proteasome is a topic of continuing research. Early investigators noted that this protease was able to cleave peptide substrates with hydrophobic, basic and acidic amino residues at the P_1 position. These experimental observations led to the definition of the chymotryptic, tryptic, and peptidylglutamyl peptide hydrolyzing (PGPH) activities of the proteasome.   More recently, two additional activities have been defined that specify cleavage of peptides with branched chain P_1 residues (e.g., Leu) and small aliphatic P_1 residues (e.g., Ala). In total, at least 9 different hydrolytic activities have been described for the proteasome. To explain how these different hydrolytic activities are mapped onto only four β-subunit active sites (12), we have proposed a new mechanism (13) in which each active site has a broad specificity that overlaps, at least to some degree, with the specificities of the other active sites. Therefore, each proteasome substrate is bound and hydrolyzed, at least to some degree, at each active site.

The identity of the catalytic nucleophile was recently determined by a combination of x-ray crystallographic and mutagenesis studies (9, 12). One of the proteasome structures that was solved has the aldehyde inhibitor, Ac-Leu-Leu-Nle-H, bound to the active site. This structure reveals that the hydroxyl of the N-terminal threonine is within bonding distance of the aldehydic carbonyl carbon of the inhibitor. Although this structure is not sufficiently well refined to access the hybridization of this carbon, it is consistent with hemiacetal formation between $Thr^1 O\gamma$ and the carbonyl carbon of the aldehyde. That Thr^1 may be the catalytic nucleophile is strongly supported by mutagenesis and kinetic studies (12).

INHIBITORS OF THE PROTEASOME

Irreversible Inhibitors - DCI - 3,4-dichloroisocoumarin (DCI) was originally described as an irreversible, mechanism-based inactivator of serine proteases, such as leukocyte elastase ($k_{inact}/[I]$ = 10,000 M^{-1} sec^{-1})(14). Recently, it was shown that DCI also inhibited several peptidase activities of the proteasome ($k_{inact}/[I]$ = 130 M^{-1} sec^{-1}, chymotryptic activity)(15, 16). However, DCI can probably not serve as a lead structure for proteasome inhibitor design due to its hydrolytic instability ($t_{1/2}$ = 15 min, pH 7.4, 37°C) and lack of enzyme selectivity (for example, DCI inactivates glycogen phosphorylase b; (17)). Furthermore, while DCI clearly inactivates the chymotryptic activity of the proteasome, it activates the caseinolytic and other peptidase activities of the 20S proteasome (15,18).

Irreversible Inhibitors - Peptidyl α',β'-Epoxyketones - Compound 1 was reported to be a potent, irreversible inactivator of the proteasome ($k_{inact}/[I]$ ~ 40,000 M^{-1} sec^{-1})(19). Significantly, the compound with 2[R] configuration was 50-fold more potent than the compound with 2[S] configuration. The mechanism of inactivation has not been reported.

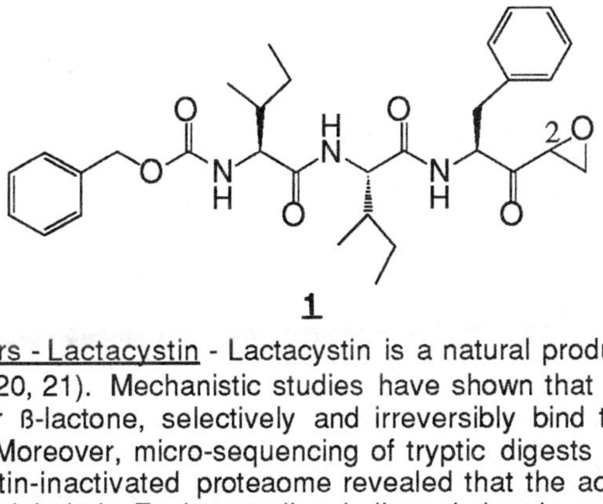

1

Irreversible Inhibitors - Lactacystin - Lactacystin is a natural product that inhibits cell cycle progression (20, 21). Mechanistic studies have shown that lactacystin, and its synthetic precursor ß-lactone, selectively and irreversibly bind to subunit X of the proteasome (22). Moreover, micro-sequencing of tryptic digests and mass spectral studies of lactacystin-inactivated proteaome revealed that the active site N-terminal threonine had been labeled. Further studies indicated that lactacystin and analogs inhibited various peptidase activities of the proteasome. The chymotrypsin-like activity is most rapidly inhibited (k_{assoc} = 200 M^{-1} sec^{-1}) while the trypsin-like and PGPH activities are more slowly inhibited (10 and 4 M^{-1} sec^{-1}, respectively). Lactacystin is a selective inhibitor of the proteasome; no inhibition was detected for common serine or cysteine proteases implying that the binding of lactacystin to the X subunit is highly discriminating. Limited SAR by Corey and Schreiber has shown that the stereochemistry of the methyl group on the lactam ring and the secondary hydroxyl group of the isobutyl side chain are required for activity (23).

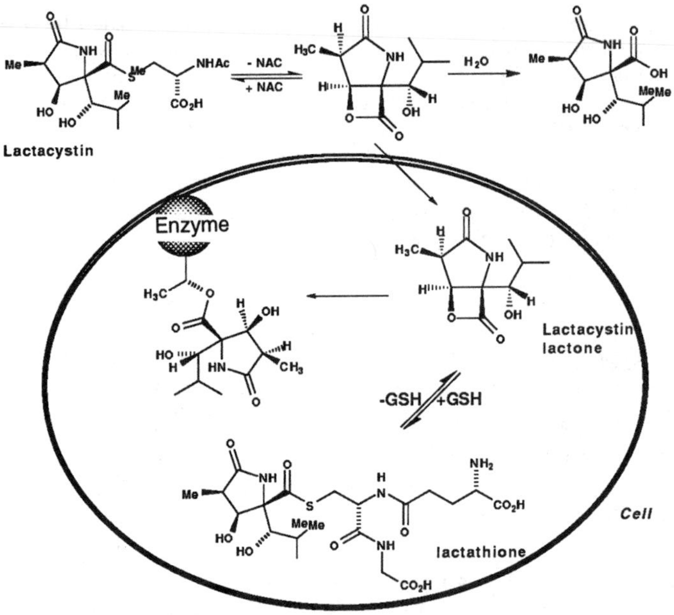

<u>Figure 2</u> Mechanism of Proteasome Inhibition by Lactacystin (24)

  Recent mechanistic studies reveal that lactacystin hydrolyzes to the inactive dihydroxy acid through a β-lactone intermediate and that it is this latter species <u>alone</u> that interacts with the proteasome and leads to its inactivation (Figure 2)(24). The addition of N-acetyl-cysteine (NAC) or glutathione (GSH) to reaction solutions of lactacystin and proteasome slow the rate of proteasome inactivation (25). A similar mechanism can occur in cells after selective uptake of the β-lactone. The presence of high intracellular concentrations of GSH converts β-lactone to lactathione which then acts as a "lactone reservoir" (25).

<u>Reversible Inhibitors - Peptide Aldehydes</u> - The best characterized inhibitors of the proteasome are peptide aldehydes. Most of the reported peptide aldehyde inhibitors of the proteasome have been directed against the chymotryptic activity of the 20S proteasome. Z-Gly-Gly-Leu-H was the first in this class and inhibits the chymotryptic activity of the 20S proteasome with a K_i of 250 µM (26). Subsequent studies in a number of laboratories established the requirement for hydrophobic residues at P_2 and P_3 of the inhibitor (18, 27, 28). For example, Ac-Leu-Leu-Nle-H and Z-Leu-Leu-Phe-H have K_i values of 6 and 0.5 µM, respectively (28). Interestingly, these compounds are also slow-binding inhibitors with similar k_{on} values of roughly 200 M^{-1} sec^{-1}. A systematic study of P_1 in the series Z-Leu-Leu-Xaa-H further established the requirement for a hydrophobic residue at this position with Leu or Phe giving the most potent inhibition (27). In a more recent mechanistic study, these compounds were

found to have identical K_i values of 4 nM (29). Finally, examination of "dipeptide" aldehydes led to a series of compounds with highly functionalized N-terminal blocking groups (30). The most potent member of this series, **2**, and has IC_{50} = 2 nM. For comparison, compounds with two other N-terminal electrophilic centers are also shown (see section below on Boronic Acids).

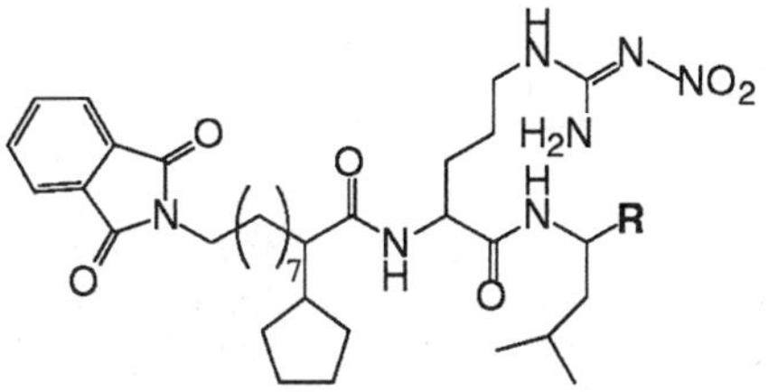

	R	K_i (nM)
2	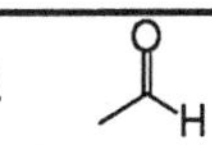 (aldehyde, CHO)	2
3	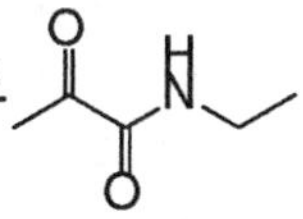 (N-ethyl oxamide)	13
4	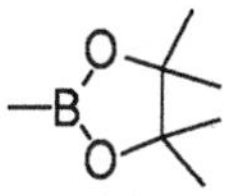 (pinacol boronate)	8

Another inhibitor of the chymotryptic activity of the proteasome is Z-Ile-Glu(OtBu)-Ala-Leu-H (31). Interestingly, this compound has an IC_{50} of 10 μM at pH 7.5, but increases in potency to IC_{50} of 0.1 μM at pH 5.0. The authors attribute this dramatic pH-dependence to inhibition of a new "acidic-chymotryptic activity" of the proteasome.

More recently, inhibition of the BrAAP acitivity of the proteasome (i.e., hydrolysis of R-Gly-Pro-Ala-<u>Leu</u>-<u>Ala</u>-R')has been investigated (32). The most potent and BrAAP-selective compound in this series is Z-Gly-Pro-Phe-Leu-H with K_i vs. the BrAAP and chymotryptic activities of 1.5 and 40 μM, respectively. Certain tripeptide aldehydes are actually more potent but not as BrAAP-selective. For example, Z-Leu-Ala-Leu-H has K_i values vs. the BrAAP and chymotryptic activities of 0.064 and 0.007 μM, respectively (29). Curiously, Z-Pro-Ala-Leu-H, which more closely mimics substrate requirements for BrAAP activity, loses potency, but gains selectivity: $K_{i,BrAAP}$ = 1,900 μM and $K_{i,Ct}$ = 11,000 μM.

<u>Reversible Inhibitors - Boronic acids</u> - Boronic acid peptides have been used extensively to inhibit serine proteases (e.g. thrombin, elastase, dipeptidyl peptidase) IV, (33-38) and have recently been examined as inhibitors of the proteasome. Boronate inhibitors of the proteasome are typically much more potent than their structurally analogous aldehydes. This enhancement in potency is thought to be due to the greater stability of the boron-Thr^{1}O$_\gamma$ dative bond that forms at the active site of the proteasome relative to carbon-Thr^{1}O$_\gamma$ bond found in the hemiacetal formed upon reaction of the proteasome with aldehyde inhibitors. A direct comparison is shown below for proteasome inhibitors(39).

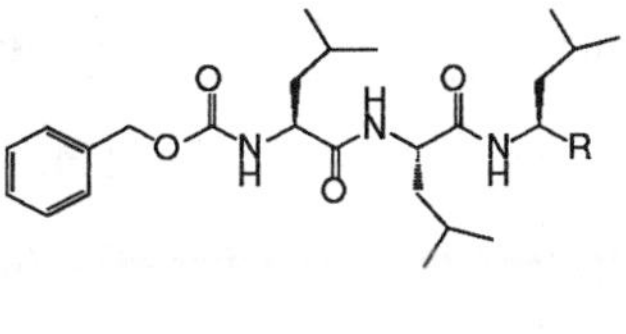

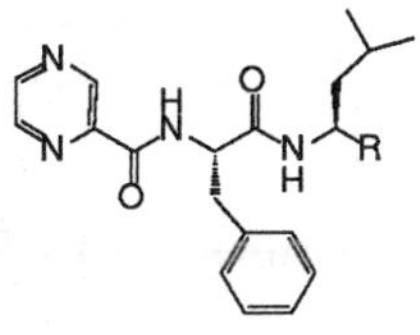

R = CHO	**5**	MG 132	K_i = 4 nM	**6**	MG 402	K_i = 1,600 nM
R = B(OH)$_2$	**7**	MG 262	K_i = 0.03 nM	**8**	MG 341	K_i = 0.6 nM

Dipeptide boronate derivatives are potent proteasome inhibitors. As a class, these compounds are active in cell-base assays (see below), orally bioavailable and active in animal studies (39). The dipeptide boronates also demonstrate a high degree of enzyme selectivity and are inactive against many common proteases (40). Recently, another group reported boronate analogs of aldehyde inhibitors, but surprisingly no potency enhancement was noted (41) .

CELL BIOLOGY OF PROTEASOME INHIBITORS

Inhibition of Intracellular Proteolysis - The ubiquitin proteasome pathway plays critical roles in cellular physiology (1, 42). Originally it was thought that the proteasome's main function was to rid the cell of misfolded, mistranslated, or damaged proteins, thereby acting as the intracellular "garbage disposal", in recent years it has become evident that the proteasome has vital regulatory functions. Approximately 70% of the cytoplasmic and nuclear proteins are potential substrates for the proteasome comprising both the long lived cyto-skeletal structural proteins as well as short-lived regulatory proteins. The regulation of the pathway for many of the cellular substrates as well as the identity of intracellular substrates is an area of active research. For example, the proteins which make up skeletal muscle (i.e. actin and myosin) are potential substrates for ubiquitination and subsequent proteolysis by the 26S proteasome, and this process is up-regulated in pathophysiologic conditions of muscle wasting. The cellular signaling which is responsible for this response is only now being unraveled. In the activation of NF-κB, the regulation of the turnover of a given protein substrate has been elucidated and the intermediate steps of activation, ubiquitination and subsequent proteolysis appear to be well defined (59).

Accumulation of Ubiquitin-Protein Conjugates - The use of proteasome inhibitors has enabled the uncoupling of ubiquitination from proteolysis in the ubiquitin-proteasome pathway. Proteasome inhibitors have been used both in cell-free and in whole cell preparations to raise steady-state levels of ubiquitin-protein conjugates that are the intermediates of this pathway. For example, Z-Ile-Glu(OtBu)-Ala-Leu-H causes the accumulation of ubiquitin-protein conjugates in a mouse neuronal cell line (43) and Z-Gly-Pro-Phe-Leu-H causes conjugate accumulation in MCF-7 breast cancer cell line (32).

Stabilization of Regulatory Proteins - Numerous examples of regulatory proteins which control cell cycle progression, transcriptional activity, and class I antigen presentation have been found to undergo ubiquitin-dependent proteolysis. The table below lists a number of mammalian regulatory proteins that are known to be substrates for the UPP. In certain cases, regulation of ubiquitination depends on the state of phosphorylation of the substrate protein.

Protein	Function	Reference
p53	tumor suppressor	(44, 45)
c-Mos	cell cycle (metaphase)	(44, 45, 46)
c-jun	transcription factor	(47)
p105(NFkB)	transcription factor	(48)
IκBα	transcription factor	(48)
Cyclin A	cell cycle progression	(49)
Cyclin B	cell cycle progression	(50, 51, 52)
p27	inhibits cyclin dep. kinase	(53)
IRP1/2	RNA binding protein	(54)

Activation of NF-κBα and NF-κBα-Dependent Gene Transcription - In this section, we will discuss the role of the UPP in the activation of NF-κBα, a key transcriptional regulator of many pro-inflammatory proteins, and describe how proteasome inhibitors suppress activation of NF-κBα and subsequent gene activation.   NF-κBα, first described to be involved in the transcription of the kappa light chain for immunoglobulin, is a member of the *rel* superfamily of dimeric transcription factors (55). Its regulation via the UPP has recently been characterized and is understood both at the molecular and functional level (see Figure 3).

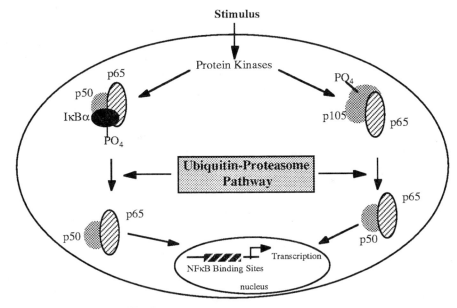

Figure 3 Activation of NF- κB

In quiescent cells, NF-κBα (p50/p65 heterodimer) is held in the cytoplasm of cells through binding to an inhibitor protein IκBα, which masks the nuclear localization sequences of p50/65 (56-58). Following signal induced phosphorylation of IκBα at Ser32 and Ser36, ubiquitination is triggered and Lys21 and Lys22 are covalently tagged with ubiquitin chains (59). The phosphorylated, ubiquitinated IκBα remains bound to p50/65 and this complex is presented to the 26S proteasome. IκBα is degraded, the active heterodimer is released and migrates to the nucleus to bind cognate κB sites in the promoter regions of various genes.

p105, the precursor protein of p50, resides in the cytoplasm of cells where it forms a heterodimer with p65. The p105/p65 heterodimer is also a substrate for the UPP and represents a second pathway for the generation of NF-κBα (see Scheme 3) (48).   The carboxy terminal of p105 is homologous to IκBα and acts as an internal inhibitor of nuclear translocation.   Remarkably, the proteasome only partially degrades the C-terminus of p105 to release free p50/65 which can then translocate to the nucleus. The mechanism for this limited proteolysis is not fully understood. In a manner similar to IκBα, p105 is also subject to phosphorylation which may regulate substrate availability for the UPP (60).

Proteasome inhibitors of diverse structural class can penetrate cells and, in a dose-dependent manner, inhibit activation of NF-κBα (39). For example, IC_{50} values for inhibition of NF-κBα activation in TNFα-treated HeLa cells are 0.5 µM, 3 µM and 10 µM for MG 341, MG 132 (Z-Leu-Leu-Leu-H), and lactacystin, respectively. The relative potencies for inhibition in this assay correlates generally with K_i values for proteasome inhibition. This is evidence that the intracellular mode of action of the inhibitors is indeed through proteasome inhibition.

NF-κBα drives the transcription of numerous pro-inflammatory cytokines including TNFα, IL-1, IL-2, IL-6, and IL-8. Proteasome inhibitors have been shown to inhibit NF-κBα-dependent gene transcription of this cytokine network (40). The close examination of the dose response curves revealed a transcriptional "amplifier" effect. For example, the IC_{50} for MG 341 inhibition of activation of NF-κBα is 0.5 μM but at this concentration of MG 341 near total inhibition of IL-2 gene expression is observed. Also, the IC_{50} for inhibition of IL-2 expression is an order of magnitude lower at 0.05 μM underscoring the magnified effect at the transcriptional level. Similar findings were obtained for the inhibitory effect of MG 341 on the cell surface adhesion molecule expression of E-selectin, ICAM and VCAM. Of note, the VCAM gene expression was blocked by >50% at doses of drug as low as 10 nM. The inducible enzymes iNOS (nitric oxide synthase and COX-2 (cyclooxygenase) are also promoted by NF-κBα binding and are also transcriptionally suppressed by proteasome inhibitors. Finally, NF-κBα is known to promote transcription of p105 and the inhibitor IκBα thereby providing additional substrate for propagating the transcriptional response. Taken together, the data suggest that inhibition of pro-inflammatory mediators by blocking the proteasome and suppressing NF-κBα gene expression provides a powerful approach to blocking the inflammatory cascade.

ACTIVITIES OF PROTEASOME INHIBITORS IN ANIMAL MODELS

Delayed Type Hypersensitivity in Mice - A classical model of T-cell dependent acute inflammation is the delayed type hypersensitivity (DTH) or contact hypersensitivity skin reaction. The phenomenon of applying a sensitizing dose of a chemical antigen to the skin followed by a re-challenge of the hapten results in a cutaneous swelling and erythema reaction at the site of challenge. The T-cell and neutrophil mediated reaction is dependent on IL-2 synthesis. Corticosteroids are active in this model and remain the only clinically effective treatment for the treatment of allergic skin reactions (i.e. poison ivy). MG 341 dosed orally at 3.0 mg/kg is efficacious in Balb-c mice sensitized and challenged with dinitrofluorobenzene (40).

Arthritis in Rats - Chronic models of inflammation of various diseases have been surveyed. The good oral potency and long half life (24h in rat) of MG 341 made it an ideal candidate for testing in a model of rheumatoid arthritis. Female Lewis rats are highly susceptible to the peptidoglycan polysaccharide antigen from *Streptococus*. A single intraperitoneal injection of antigen produces a systemic polyarthritis which manifests itself by an acute inflammation in the joints of the animals followed by a brief recovery and a second more chronic response which leads to joint destruction with cartilage and bone erosion. MG 341 dosed orally once a day at 0.3 mg/kg blunted the acute inflammation and was active in the chronic phase with almost complete inhibition of the clinical response over a 30 day period (40). Tissue histology of the joints at day 28 indicated that the high cellularity in the pannus formed in the synovial space was markedly inhibited in the drug treated animals (40). Furthermore, liver tissue samples were harvested and mRNA levels were measured for VCAM and iNOS. Both genes were overexpressed in the arthritic control rats compared with normal animals, and were attenuated in the drug treated arthritis group (40).

SUMMARY

The proteasome remains a fascinating target for enzymologists and medicinal chemists alike. The detailed mechanism of processive protein hydrolysis leaves much room for additional research. Though much has been learned from the structure of the prokaryotic 20S proteasome, the eukaryotic 26S proteasome is substantially more complex. Likewise, inhibitors thus far are derived from peptide mimicry directed towards the catalytic site and the specialized irreversible action of the natural product, lactacystin. This field is still immature and one might imagine that the future will witness additional strategies to inhibit the proteasome including targeting the active sites, allosteric sites (eg Ub binding site), or other enzyme activities (eg ATPases, isopeptidase)

Finally the therapeutic potential for proteasome inhibitors remains largely unknown. It is clear that the proteasome is vitally important for the cellular physiology of protein turnover. Only persistent research, a detailed knowledge of the substrate target, and careful dosing to achieve partial or limited and transient inhibition will allow further exploration of the potential utility of proteasome inhibitors in human disease.

REFERENCES

1.   A. L. Goldberg, R. L. Stein, J. Adams, *Chemistry & Biology* **2**, 503 (1995).
2.   M. Rechsteiner, L. Hoffman, W. Dubiel, *J. Biol. Chem.* **268**, 6065 (1993).
3.   J.-M. Peters, *TIBS* **19**, 377 (1994).
4.   M. Hochstrasser, *Current Opinion Biol.* **7**, 215 (1995).
5.   A. Ciechanover, Y. Hod, A. Hershko, *Biochem. Biophys. Res. Commun.* **81**, 1100 (1978).
6.   A. Ciechanover, *Cell* **79**, 13-21 (1994).
7.   W. Baumeister, B. Dahlmann, R. Hegerl, F. Kopp, L. Luehn, G. Pfeifer, *FEBS Lett.* **241**, 239 (1988).
8.   T. M. Schauer, M. Nesper, M. Kehl, A. Muller-Taubenberger, G. Gerisch, W. Baumeister, *J. Struct. Biol.* **111**, 135 (1993).
9.   J. Lowe, D. Stock, B. Jap, P. Swickl, W. Baumeister, R. Huber, *Science* **268**, 533 (1995).
10. T. Wenzel, C. Eckerskorn, F. Lottspeich, W. Baumeister, *FEBS Lett.* **349**, 201 (1994).
11. M. Chu-Ping, J. H. Vu, R. J. Proske, C. A. Slaughter, G. N. DeMartino, *J. Biol. Chem.* **269**, 3539 (1994).
12. E. Seemuller, A. Lupas, D. Stock, J. Lowe, R. Huber, W. Baumeister, *Science* **368**, 579 (1995).
13. R. L. Stein, F. Melandri, L. Dick, *Biochemistry* **35**, 3899(1996).
14. J. W. Harper, K. Hemmi, J. C. Powers, *Biochemistry* **24**, 1831 (1985).
15. M. E. Pereira, T. Nguyen, B. J. Wagner, J. W. Margolis, B. Yu, S. Wilk, *J. Biol. Chem.* **267**, 7949 (1992).
16. C. Cardoza, A. Vinitsky, M. C. Hidalgo, C. Michaud, M. Orlowski, *Biochemistry* **31**, 7373 (1992).
17. N. M. Rusbridge, R. J. Beynon, *FEBS Lett.* **268**, 133 (1990).
18. M. Orlowski, C. Cardozo, C. Michaud, *Biochemistry* **32**, 1563 (1993).
19. A. Spaltenstein, J. J. Leban, J. J. Huang, K. R. Reinhardt, O. H. Viveros, J. Sigafoos, R. Crouch, *Tet. Lett.* **37**, 1343 (1996).
20. S. Omura, K. Matsuzaki, T. Fujimoto, T., K. Kosuge, T. Furuya, A. Nakagawa, *J. Antibiot.* **44**, 113 (1991).
21. S. Omura, K. Matsuzaki, T. Fujimoto, T., K. Kosuge, T. Furuya, A. Nakagawa, *J. Antibiot.* **44**, 117 (1991).
22. G. Fenteany, R. F. Standaert, W. S. Lane, S. Choi, E. J. Corey, S. L. Schreiber, *Science* **268**, 726 (1995).
23. G. Fenteany, R.F. Standaert, G.A. Reichard, E.J. Corey, S.L. Schreiber, *Proc. Natl. Acad. Sci. USA* **91**, 3358 (1994).
24. L. R. Dick, A. A. Cruikshank, L. Grenier, F. Melandri, S. L. Nunes, R. L. Stein, *J. Biol. Chem.* **271**, 7273 (1996).
25. L. R. Dick, *Proteolytic Enzymes and Inhibitors in Biology and Nature, Keystone Symposia on Molecular and Cellular Biology, Keystone, Colorado* , (1996).
26. S. Wilk, M. Orlowski, *J. Neurochem.* **35**, 1172 (1980).
27. S. Tsubi, H. Kawasaki, Y. Saito, N. Miyashita, M. Inomata, S. Kawashima, *Biochem. Biophys. Res. Commun.* **196**, 1195 (1993).
28. A. Vinitsky, C. Michaud, J. C. Powers, M. Orlowski, *Biochemistry* **31**, 9421 (1992).
29. L. Dick, A. Cruikshank, L. Grenier, F. Melandri, L. Plamondon, R. Stein, *Biochemistry* **35**, 0000 (1996).
30. M. Iqbal, S. Chatterjee, J. C. Kauer, M. Das, P. Messina, B. Freed, W. Biazzo, R. Siman, *J. Med. Chem.* **38**, 2276 (1995).
31. M. E. Figueiredo-Pereira, C. W.-E., H.-M. Yuan, S. Wilk, *Arch. Biochem. Biophys.* **317**, 69 (1995).
32. A. Vinitsky, C. Cardozo, L. Sepp-Lorenzino, C. Michaud, M. Orlowski, *J. Biol. Chem.* **269**, 29860 (1994).
33. A. S. Shenvi, *Biochemistry* **25**, 1286 (1986).
34. C. Kettner, L. Mersinger, R. Knabb, *J. Biol. Chem.* **259**, 18289 (1990).
35. N. T. Soskel, S. Watanabe, R. Hardie, A.B. Shenvi, J.A. Punt, C. Kettner, *Am. Rev. Respir. Dis.* , 635 (1986).

36.   R.R. Snow, W. W. Bachovchin, R. W. Barton, S.J. Campbel, S.J. Coutts, D. M. Freeman, W. G. Gutheil, T. A. Kelly, C. A. Kennedy, D. A. Krolikowski, S. F. Leonard, C. A. Pargellis, L. Tong, J. Adams, *J. Amer. Chem. Soc.* **116**, 10860 (1994).
37.   D. H. Kinder, C.A. Elstad, G.G. Meadows, M.M. Ames, *Inv. Metastasis* **12**, 309 (1992).
38.   T. A. Kelly, J.A. Adams, W.W. Bachovchin, R.W> Barton, S.J. Campbell, S.J. Coutts, C.A. Kennedy, R.J. Snow, *J. Amer. Chem. Soc.* **115**, 12637 (1993).
39.   J. Adams, Y. T. Ma, R. Stein, M. Baevsky, L. Grenier, L. Plamondon, *World Patent Appl. WO9600000,00* , (1996).
40.   Stein, *1996 Anti-Inflammatory Drug Discovery Summit, Princeton, New Jersey* , (1996).
41.   M. Iqbal, S. Chatterjee, J. C. Kauer, P. A. Mallamo, A. Messina, A. Reiboldt, R. Siman, *Bioorg. Med. Chem. Lett.* **6**, 287 (1996).
42.   A. L. Goldberg, W. Mitch, *N. Eng. J. Med.* , (1996).
43.   M. E. Figueiredo-Pereira, K. A. Berg, S. Wilk, *J. Neurochem.* **63**, 1578 (1994).
44.   M. Scheffner, J. M. Huibregste, R. D. Vierstra, P. M. Howley, *Cell* **75**, 495 (1993).
45.   M. Rolfe, P. Beer-Romero, S. Glass, J. Eckstein, I. Berdo, A. Theodoras, M. Pagano, G. Draetta, *Proc. Natl. Acad. Sci. USA* **92**, 3264 (1995).
46.   R. J. Deshaies, *Trends Cell. Biol.* **5**, 428 (1995).
47.   M. Treir, L. M. Staszewski, D. Bohmann, *Cell* **78**, 787 (1994).
48.   V. J. Palombella, A. L. Rando, A. L. Goldberg, T. Maniatis, *Cell* **78**, 773 (1994).
49.   V. Sudakin, D. Ganoth, A. Dahan, H. Heller, J. Hershko, F. C. Luca, J. V. Ruderman, A. Hershko, *Mol. Biol. Cell* **6**, 185 (1995).
50.   A. Hersko, D. Ganoth, V. Sudakin, A. Dahan, L.H. Cohen, F.C. Luca, J.V. Ruderman, E. Eytan, *J. Biol. Chem.* **269**, 4940 (1994).
51.   W. Seufert, B. Futcher, S. Jentsch, *Nature* **373**, 78 (1995).
52.   S. Iringer, S. Piatti, C. Michaelis, K. Nasmyth, *Cell* **81**, 261 (1995).
53.   M. Pagano, S.W. Tam, A. M. Theodoras, P. Beer-Romero. G. D. Sal, V. Chau, P.R. Yew, G.F. Draetta, M. Rolfe, *Science* **269**, 682 (1995).
54.   K. Iwai, D. Klausner, A. Rouault, *EMBO J.* **14**, 5350 (1995).
55.   R. Sen, D. Baltimore, *Cell* **46**, 705 (1986).
56.   D. Thanos, T. Maniatis, *Cell* **80**, 1 (1995).
57.   U. G. Siebenist, *Annu. Rev. Cell. Biol.* **10**, 405 (1994).
58.   P. A. Bauerle, T. Henkel, *Annu. Rev. Immunol.* **12**, 141 (1994).
59.   Z. Chen, J. Hagler, V. J. Palombella, F. Melandri, D. Scherer, D. Ballard, T. Maniatis, *Genes & Develop.* **9**, 1586 (1995).
60.   K. Fujimoto, H. Yasuda, Y. Sato, K. Yamamoto, *Gene* **165**, 183 (1995).

Chapter 29. The MAP Kinase Family: New "MAPs" for Signal Transduction Pathways and Novel Targets for Drug Discovery

Bernd Stein and David Anderson
Signal Pharmaceuticals
San Diego, CA 92121

Introduction – Protein phosphorylation plays a major role in many signal transduction pathways. Protein kinases as well as protein phosphatases are involved in various cellular responses to extracellular signals. More than 400 distinct protein kinases have been identified and sequenced. In this chapter we describe the recently discovered family of mitogen-activated protein kinases (MAPK). Members of this family are pro-line-directed Ser/Thr kinases that activate their substrates by dual-phosphorylation. MAPK are activated by a variety of signals including growth factors, cytokines, ultraviolet light (UV) and stress inducing agents. Activated MAPK undergo nuclear transloca-tion, which is important for phosphorylation of nuclear transcription factors. MAPK impact many cellular processes such as proliferation, oncogenesis, development and differentiation, and cell cycle.

Several MAPK cascades have been identified over the last few years in yeast and vertebrates (1-8). Each cascade consists of a module involving protein-protein interactions that help regulate distinct cellular outputs. We are following the nomenclature of Seger and Krebs (7) and name the individual components regardless of their pathway MAPK, MAPKK and MAPKKK (Fig. 1). This chapter summarizes the mammalian MAP kinase cascades and discusses the biological role of MAPK, their involvement in disease states and their potential as drug targets.

The ERK Cascade – The extracellular-signal-regulated kinase (ERK) subgroup of MAPK has been examined in detail (9-11). ERK1 and ERK2 were the first members of this subfamily to be cloned (12). Since then other related kinases have been identified including: ERK3 (12) and ERK4 (9). ERK1/2 are activated by dual phosphorylation on Tyr and Thr by the MAPKK MEK1 and MEK2 (1, 13), which, in turn, are activated by the MAPKKK Raf-1, B-Raf and c-Mos. The ERK cascade is also activated by the recently described MEKK3 and Tpl-2 (Cot) (14, 15). A proline-rich sequence unique to MEK1 and MEK2 is required for Raf-1 binding and MEK activation (16). Upstream of the MAPKKK level a GTP-dependent physical association of Ras with Raf has been demonstrated in mammalian cells (17, 18). These and other studies suggest that Ras might function as a switch between membrane-bound receptor tyrosine kinases activated by growth factors, and soluble kinases of the ERK pathway. This alone is not sufficient for the full activation of Raf and it appears that additional Ras-independent signals are required (19, 20).

The JNK Cascade – The c-Jun N-terminal kinase (JNK) subgroup (also known as stress-activated protein kinases; SAPK) was first described as a kinase cascade lead-ing to the phosphorylation of c-Jun within the N-terminal activation domain at Ser[63] and Ser[73] in response to pro-inflammatory cytokines and exposure of cells to various forms of environmental stress (21-24). Three JNK family members (JNK1/2/3, also known as SAPK γ/α/β) with at least 10 splice variants have been identified so far. JNK is selectively phosphorylated and activated by JNKK (MKK4, SEK) (25-28), which, in turn, is activated by the MAPKKK, MEKK1/2 (14, 29, 30) and Tpl-2 (15). Upstream of

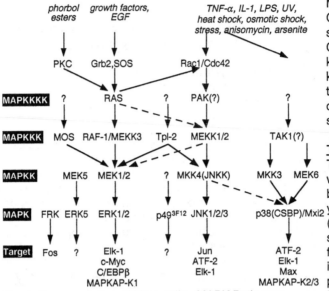

Fig. 1. Representation of Mammalian MAPK Pathways.

MEKKs a critical role of GTP-binding proteins, such as Rac-1 and Cdc42, and a Raf-like kinase (p21-activated kinase; PAK), for controlling the JNK cascade has been demonstrated (31-33).

The p38 Cascade – The p38 MAPK pathway has been identified by homology to the yeast HOG1 pathway (34-36). p38 (cytokine-suppressive anti-inflammatory drug binding protein; CSBP1/2) phosphorylates and activates ATF2 and Elk-1. A truncated splice variant of p38 with a unique C-terminus, Mxi2, efficiently phosphorylates Max (37). p38 is activated by MKK3 (27), JNKK (27, 28) and the recently discovered MEK6 (38-40). Upstream kinases for this cascade have not been described although the protein kinase TAK1 is a good candidate (41). Depending on the stimulus, Rac-1, Cdc42 and PAK couple to and regulate the p38 signaling pathway (33). p38, like JNK, is activated by pro-inflammatory cytokines and environmental stress (42).

Other MAPK Cascades – FRK is a novel Ras- and mitogen-responsive proline directed MAPK that is different from JNK, p38 and ERK (43). FRK activates c-Fos through phosphorylation at Thr[232]. MEK5, distantly related to MEK1/2, specifically interacts with ERK5 but not with members of the JNK or p38 cascade (44, 45). Interestingly, phosphorylation of ERK5 by MEK5 has not been demonstrated and no substrate has yet been identified for ERK5 (44).

CELLULAR INFORMATION TRANSMITTED VIA PROTEIN KINASE CASCADES

Activators of MAP Kinases – The activation of MAPK via specific site-directed phosphorylation events leads to conformational changes and binding to downstream kinases, phosphatases and other regulatory proteins as well as transcription factors. MAPK are highly regulated enzymes, which may be modulated naturally or with pharmacological agents.

ERK2 has been widely studied and was the first MAPK to be analyzed by crystallography (46). Dual phosphorylation events on residue Tyr[185] and on Thr[183] are required for ERK2 activation. Crystal structure analysis demonstrated that these two amino acids lie in a phosphorylation lip at the mouth of the catalytic site. Crystallographic analysis of mutant ERK2 suggested that the lip is refolded in the active enzyme structure (47). It is likely that these conformational changes are associated with activation.

The MAPK cascades are differentially activated by a wide variety of stimuli includ-

ing growth factors, mitogens, neurotransmitters, pro-inflammatory cytokines, endotox-
ins and physical and chemical stresses. Nerve growth factor (NGF) is a member of the
neurotrophin family of proteins. NGF plays an important role in the differentiation of
neurons but also influences cells of the immune system such as B- and T-cells. An
example of receptor-mediated activation of MAPK is the ERK induction by NGF in B-
lymphocytes (48). Lipopolysaccharide (LPS) stimulation of macrophages results in the
phosphorylation and activation of Raf-1, which then stimulates the MEK-ERK cascade
(49). However, simultaneous activation of p38 and ERK is seen when monocytes are
exposed to LPS or osmotic shock. These results support Raf-1-dependent and independent pathways in LPS-induced cytokine production through ERK and p38 (34, 50).

The JNK cascade is potently activated by pro-inflammatory cytokines, environmental stress such as heat shock and osmotic shock, and UV (21, 22, 51-54). The p38
cascade is activated by similar signals, including LPS (34, 35, 55). In T-cells, JNK but
not ERK is synergistically activated by costimulation with phorbol esters (PMA) and
Ca^{2+} ionophore or simultaneous activation with anti-CD3 and anti-CD28 antibodies
(56). Furthermore, Ca^{2+} acts as a second messenger for the activation of JNK but not
ERK in fibroblasts expressing muscarinic acetylcholine m1 receptor stimulated with
carbachol (57). However, the non-TPA tumor promoter, palytoxin, activates JNK but
not ERK in fibroblasts via a sodium-dependent pathway (58). In GN4 rat liver epithelial
cells, angiotensin II stimulates JNK via a novel Ca^{2+}-dependent pathway (44). Further,
vascular smooth muscle cells respond to angiotensin II by stimulation of ERK1/2 (59).

Multiple Connections – The presence of three MAPK cascades raises questions concerning their redundancy and specificity. Crosstalk between the three major MAP kinase cascades has been described, although some of these observations need to be
interpreted with caution. Protein kinases may phosphorylate proteins that are not their
physiological substrates under conditions where they are overexpressed and/or out of
their natural context. Further, studies in yeast have shown the existence of a scaffolding protein (STE5) that independently associates with MAPK family members in the
STE4, STE7 and FUS3/KSS1 pathway of the MAPKKK, MAPKK and MAPK families
(60, 61). Although the precise function of STE5 is still undefined, researchers specu-
late that mammalian cells have a protein(s) that act(s) as a scaffold and sequester(s)
the MAP kinase modules. The existence of distinct MAP kinase modules held together
by scaffolding protein(s) may contribute to signal and tissue specificity and may restrict
the cross-talk between MAP kinase cascades, as illustrated with the following example. It has been reported that MEKK1 activates MEK (29). Other reports showed that
MEKK can phosphorylate MEK1 and MEK2 but does not cause activation of ERK (62).
We and others have shown that MEKK1 can cross-talk to the p38 cascade but only at
very high expression levels. In summary, these and other studies suggest that in vivo
MEKK1 activates JNKK only. MEKK2 and Tpl-2 equally well activate the ERK and JNK
cascades (14, 15). Other examples of cross-talk are the activation of JNK as well as
p38 by JNKK (28). Further, Ras can activate c-Jun through a pathway that does not involve ERK1 and ERK2 (63). Supporting these observations, it has been shown that
Ras interacts not only with Raf-1 but also with MEKK1 (64). In addition, adapter proteins such as MEF (MAPKK-enhancing factor) may regulate MAPK enzymes in a cell-
or signal-specific fashion (65).

SUBSTRATES FOR MAP KINASES

Transcription Factors and Enzymes – MAPK are proline-directed Ser/Thr kinases.
Substrate selection by MAPK requires binding of the kinase to the target through a
domain that is independent of the phosphorylation site. The serum response element

(SRE) of the c-*fos* promoter has been found to be the target for many signal transduction cascades. The SRE binds the serum response factor (SRF) and its accessory proteins, ternary complex factor (TCF). TCF belongs to the Ets protein family. A recent review summarizes the intracellular signaling pathways that converge on the c-*fos* SRE (66). Among them are the ERK and JNK cascades that lead to the phosphorylation of TCF/Elk-1 resulting in an increase in DNA binding, transcription factor complex formation and transcriptional activation. Therefore, MAPK plays an important role in regulation of c-*fos* transcription. The c-*jun* promoter is activated by binding of c-Jun/ATF-2 heterodimers to its two response elements (67). c-Jun is activated by JNK while ATF-2 is activated by JNK and p38 (21-24, 51, 68, 69). Therefore, c-*jun* is another immediate early gene that is at least in part regulated by MAPK family members. The p38 cascade also seems to be responsible for the activation of mitogen-activated protein kinase-activated protein (MAP-KAP) kinase-2 and MAP-KAP kinase-3 *in vivo* (70). MAP-KAP kinase-2/3 phosphorylate the heat shock proteins Hsp25/Hsp27 (35). Activation of the p53 tumor suppressor protein by JNK1 is thought to be important in the cellular response to DNA-damaging agents and may be involved in cell cycle arrest or apoptosis (71).

C/EBPβ, a member of the bZIP family of transcription factors, is implicated in the regulation of several acute phase genes, cytokine promoters, and genes encoding c-Fos, albumin and adipocyte-specific proteins. The N-terminal activation domain of C/EBPβ is activated by ERK through a Ras-dependent pathway (72). The chicken homologue of C/EBPβ (NF-M) activates myelomonocyte-specific genes following phosphorylation by MAPK, which acts to derepress rather than enhance the transactivation domain (73). ERK activation also appears to be necessary for the induction of nitric oxide synthase by interleukin-1β (IL-1β) and interferon (IFN)-γ in cardiac myocytes and microvascular endothelial cells (74). Another ERK substrate is MAP-KAP kinase-1, also named p90 ribosomal S6 kinase, p90^{rsk} (48, 75). Activated ERK2 has also been shown to phosphorylate cytosolic phospholipase A2 (cPLA$_2$) at Ser^{505}, causing increased enzymatic activity of cPLA$_2$ (76). This links MAPK to arachidonic acid metabolism. Serum stimulation of quiescent mouse fibroblasts results in an increase in the amount of phosphorylated RNA polymerase II. This increase is most likely caused by a kinase that is a member of the MAP kinase family (77). *In vitro* analysis identified several additional ERK substrates including the epidermal growth factor (EGF) receptor, cytoskeletal proteins, c-Jun, c-Myc and Tal1 (78). It will be important to comprehend more fully the various substrates of the MAPK family members and their role in physiologic or pathologic responses of cells. This will result in better understanding of the biology of MAPK and how potential drugs can modulate these cascades.

Integration with other Pathways – The estrogen receptor (ER) has an N-terminal and a C-terminal transcriptional activation domain that get phosphorylated upon activation. It has been shown that Ser^{118} in the N-terminus is phosphorylated by ERK *in vitro* and in cells treated with EGF and insulin-like growth factor (IGF) (79). This implies that the Ras-ERK cascade is involved in modulation of ER activity in response to growth factor signals. Further, it has been shown that the MAPK cascade is activated by ER in MCF-7 cells (80).

MAPK (specifically ERK2) also appear to regulate IFN-α and IFN-β activation of early response genes by modifying the JAK-STAT signaling cascade through STAT phosphorylation and increased transactivation (81). Also, interleukin-6 (IL-6) signaling may involve the activation of the ERK cascade via IL-6 receptor-associated JAK kinase (82). A link between the tumor necrosis factor-α (TNF-α) induced sphingomyelin pathway and the MAPK pathway has been described in HL60 cells where the ceramide-activated protein (CAP) kinase phosphorylates Raf-1 (83). An antagonism between the cAMP-stimulated protein kinase A (PKA) and the MAPK pathway has been shown in fibroblasts and muscle cells. Increasing concentrations of cAMP blocked the activation of the ERK pathway and were accompanied by an increase in Raf-1 phosphorylation on Ser^{43} (84). In T-cells, the activity of JNK was inhibited by cAMP, which correlated with a suppression of interleukin-2 (IL-2) (85). Vitamin D3 activates the ERK pathway via a protein kinase C (PKC)-dependent pathway in 3T3 cells leading to fibroblast proliferation independent of nuclear receptor binding (86). Phorbol esters activate the ERK cascade through PKC. The general belief is that PKC has multiple sites of action within the ERK cascade, which may serve to modulate the signal transduction pathway in certain cell types (87).

REGULATION OF MAP KINASE ACTIVITY

Phosphatases – Relatively little is known about the enzymes that inactivate MAP kinases: dual-specificity phosphatases that dephosphorylate serine/threonine and tyrosine residues. Many protein phosphatases inactivate MAPK *in vitro* but their role *in vivo* is less clear. MAP kinase phosphatase 1 (MKP-1 or CL-100; 88,89) dephosphorylates ERK and thereby inhibits Ras-induced DNA synthesis, as well as the effects of angiotensin II on MAPK (90). JNK is also a likely target for MKP-1 (91). The JNK cascade is additionally responsible for transcriptional stimulation of MKP-1 in fibroblasts (92). UV and genotoxic agents that activate JNK also induce mRNA for MKP-1, which correlates with a decline in JNK activity. Constitutive expression of MKP-1 in cells inhibited JNK activity and reduced AP-1-dependent gene activation (91). Protein phosphatase 2A and other unidentified tyrosine phosphatases are also important for inactivation of ERK (93). The SV40 small T antigen is capable of inactivating protein phosphatase 2A, which results in deregulation of the ERK pathway and induction of cell growth (94, 95). The dual-specific Thr/Tyr phosphatase PAC1 shows stringent substrate specificity for ERK (96). Another report described the purification of a dual-specific protein tyrosine phosphatase that dephosphorylates ERK as well as JNK (97). A recent report showed that MKP-1 dephosphorylated *in vivo* ERK2, JNK2, and p38 equally. PAC1 did not inactivate JNK2, nor did MKP-2 inactivate p38 (98). The ERK cascade is also involved in activation of the ATP-Mg-dependent protein phosphatase in response to insulin and other growth factors (99).

Feedback Loops – The activation of MAPK-regulated signal transduction pathways by guanosine nucleotide exchange factors such as Son of Sevenless (SOS) through Ras may also be a control point for inhibition. The phosphorylation of SOS by activated MAPK may act as a negative feedback loop by altering SOS intracellular location and ability to activate Ras (100, 101). In addition, activated ERK may phosphorylate the upstream activators (MEK1 but not MEK2) acting as a negative feedback mechanism (5) and activated p38 may phosphorylate MEK6 (40). The relevance of this system for the ERK and the other MAPK cascades has not been fully established.

BIOLOGICAL ROLE OF MAP KINASES AND INVOLVEMENT IN DISEASE STATES

Various members of the MAPK family play critical roles in cell proliferation, cell cycle progression, cellular differentiation, and signal transduction. When defective, these processes may contribute to a transformed phenotype. Genes regulated by AP-1, such as stromelysin, collagenase type I and vascular endothelial growth factor, are particularly associated with oncogenic properties including transformation and the metastatic state (102, 103).

MAPK cascades respond to growth factors and upstream activators such as Raf-1 and Mos that are known to have oncogenic potential. Interestingly, no naturally occurring oncogenic forms of MAPK have been found. A constitutively active form of MEK1 induced cellular transformation (104) and generated tumors in nude mice (105). Transformation by MEK requires Ras and presumably works via an autocrine loop. Such results suggest that cells do not tolerate constitutively active MAPK and that perhaps an increase in MAPK activity is compensated by an increase in phosphatase activity. The oncogene Bcr-Abl, associated with t(9,22) Philadelphia chromosome translocation of chronic myelogenous leukemia, activates JNK in fibroblasts and hematopoietic cells. This event is associated with the transformed phenotype. Dominant negative Ras, MEKK, JNK and c-Jun but not Raf or ERK block the transformation mediated by Bcr-Abl (106). Because the regulation of MAPK and their control of many biological processes leading to cell growth and proliferation may be overactive in cancer, inhibitors of the critical control points and of selective substrate interactions for MAPK may prove to be attractive targets for anti-cancer drugs.

The stress-activated kinases (SAPK: JNK, p38) may be involved in a variety of inflammatory and immune diseases. Numerous studies have demonstrated that the production of pro-inflammatory cytokines such as IL-1 and TNF-α are regulated in part by SAPK (34, 36, 42). In addition the response of many cell types to cytokines and stress, including neutrophils (107), lymphocytes (56, 108), monocytes (24, 42), chondrocytes, synoviocytes, fibroblasts (54), neuronal cells (109), endothelial cells (74), and smooth muscle cells (110, 111) is regulated by MAPK. A block in the activation of ERK and JNK causes defective IL-2 production and anergy in T-lymphocytes (112, 113) and the ERK cascade functions in the stimulation of IL-2 gene transcription in activated T-lymphocytes (114). Understanding this effect at the molecular level will not only aid our knowledge in the biological mechanisms of tolerance but provide a novel approach for selective immunosuppressive agents. The immunodeficiency disorder, Wiskott-Aldrich syndrome, in which lymphocytes have defective signaling such as anti-CD3 stimulation for T-cells, is caused by a defect in the Rac/Cdc42–JNK pathway (115).

A large amount of cellular MAPK activated by mitogenic stimulation is physically associated with the microtubule cytoskeleton (116). This suggests that MAPK play a critical role in cytoskeletal and cell cycle regulation. It has been suggested that the *in vivo* association between p56^{lck} and ERK could be involved in IL-2 mediated S phase progression (117). An increase of JNK activity has been found after partial hepatectomy (118, 119). The unique role of c-Jun in hepatic proliferation is reflected in the failure of liver development in a c-*jun* deficient mouse (119). Ischemia and reperfusion activate JNK in both heart and kidney and this may play a role in the pathology of transplantation rejection or response to injury (120-122). The p38 cascade seems to be involved in platelet aggregation caused by thrombin (123, 124).

The balance between MAPK pathways ERK and JNK/p38 may play a role in determining cell fate — survival or death. Growth factor activation of ERK protects neuronal cells from apoptosis, which is mediated by JNK/p38 following NGF withdrawal (109). A possible splice variant of JNK3, p49^{3F12}, is expressed exclusively in the nervous system and maps to a chromosomal region that may be important in the pathogenesis of Alzheimer's disease and Down's syndrome (125). Activation of JNK but not ERK is associated with Fas-mediated apoptosis in human T-cells (126). Ceramide initiated apoptosis through the JNK cascade in monocytes and endothelial cells (127) and MEKK1 enhanced UV-induced apoptosis in fibroblasts (128). Angiotensin II is described to cause proliferation, hypertrophy and remodeling as well as protection from apoptosis in smooth muscle cells via a MAPK-mediated pathway (59, 129).

Together these findings suggest that MAPK may play a role in a wide variety of pathogenic conditions involving cell death or hyperplasia. These include reperfusion/ischemia in stroke, heart attack and organ hypoxia; autoimmune diseases due to inappropriate lymphocyte activation or selection; and vascular hyperplasia or cardiac hypertrophy. The regulation and activities of MAPK are potential targets for developing new therapeutics.

MAP KINASES AS DRUG TARGETS

Eukaryotic protein kinases and phosphatases are inhibited by a variety of natural and synthetic compounds (130, 131). Most of these compounds have a rather broad specificity. With a better understanding of protein kinase cascades and their regulation it will be possible to develop drugs with high specificity for a single protein kinase or phosphatase. A compound (PD 098059) that selec tively blocked MEK also inhibited stimulation of cell growth and reversed the phenotype of Ras-transformed BALB 3T3 mouse fibroblasts and rat kidney cells (132). Further, PD 098059 inhibited NGF-induced differentiation in PC12 cells. This suggests that the ERK pathway is required for NGF-induced neuronal differentiation in PC12 cells (133). A series of compounds (SK&F 86002) was identified that inhibited production of IL-1 and TNF-α from stimulated human monocytes (36). The inhibition was primarily at the translational level and was caused by binding of the compound to p38. Our studies with this compound showed that *in vitro* it also inhibits JNKK but not other family members of the JNK, p38 and ERK cascade. However, analogs (SB 203580) that are more selective for p38 have been synthesized. The compound lisofylline, which blocks the induction of cytokines (TNF-α), may inhibit JNK. This may explain its anti-inflammatory and immune modulating activity in preclinical and clinical studies. The Rac and Rho pathways are also attractive as drug targets for Ras-mediated malignancies (134). The anti-tumor and anti-proliferative effects of taxol may be due in part to its inhibition of MAPK (135). The downstream MAPK cascades may also serve as potential drug targets.

The broad role that MAP Kinases play in cell physiology and pathology provides a potential panoply of opportunities for drug discovery in areas such as cancer, inflammation, allergies, autoimmune diseases and neurodegeneration.

References:

1. M.H. Cobb, T.G. Boulton, and D.J. Robbins, Cell Regul., 2. 965 (1991).
2. T. Hunter, and M. Karin, Cell, 70. 375 (1992).
3. R.J. Davis, Trends Biochem., 19. 470 (1994).
4. E. Cano, and L.C. Mahadevan, Trends Biochem., 20. 117 (1995).
5. T. Hunter, Cell, 80. 225 (1995).
6. I. Herskowitz, Cell, 80. 187 (1995).
7. R. Seger, and E.G. Krebs, FASEB J., 9. 726 (1995).
8. A.J. Waskiewicz, and J.A. Cooper, Curr. Opin. Cell. Biol., 7. 798 (1995).
9. T.G. Boulton, and M.H. Cobb, Cell Regul., 2. 357 (1991).
10. M.H. Cobb, D.J. Robbins, and T.G. Boulton, Curr. Opin. Cell. Biol., 3. 1025 (1991).
11. R. Seger, N.G. Ahn, T.G. Boulton, G.D. Yancopoulos, N. Panayotatos, E. Radziejewska, L. Ericsson, R.L. Bratlien, M.H. Cobb, and E.G. Krebs, Proc. Natl. Acad. Sci. USA, 88. 6142 (1991).
12. T.G. Boulton, S.H. Nye, D.J. Robbins, N.Y. Ip, E. Radziejewska, S.D. Morgenbesser, R.A. DePinho, N. Panayotatos, M.H. Cobb, and G.D. Yancopoulos, Cell, 65. 663 (1991).
13. R. Seger, N.G. Ahn, J. Posada, E.S. Munar, A.M. Jensen, J.A. Cooper, M.H. Cobb, and E.G. Krebs, J. Biol. Chem., 267. 14373 (1992).
14. J.L. Blank, P. Gerwins, E.M. Elliott, S. Sather, and G.L. Johnson, J. Biol. Chem., 271. 5361 (1996).
15. A. Salmerón, T.B. Ahmad, G.W. Carlile, D. Pappin, R.P. Narsimhan, and S.C. Ley, EMBO J., 15. 817 (1996).

16. A.D. Catling, H.-J. Schaeffer, C.W.M. Reuter, G.R. Reddy, and M.J. Weber, Mol. Cell. Biol., 15, 5214 (1995).

17. H. Koide, T. Satoh, M. Nakafuku, and Y. Kaziro, Proc. Natl. Acad. Sci. USA, 90, 8683 (1993).

18. S.A. Moodie, B.M. Willumsen, M.J. Weber, and A. Wolfman, Science, 260, 1658 (1993).

19. D. Büscher, R.A. Hipskind, S. Krautwald, T. Reimann, and M. Baccarini, Mol. Cell. Biol., 15, 466 (1995).

20. B.W. Winston, C.A. Lange-Carter, A.M. Gardner, G.L. Johnson, and D.W.H. Riches, Proc. Natl. Acad. Sci. USA, 92, 1614 (1995).

21. B. Dérijard, M. Hibi, I.-H. Wu, T. Barrett, B. Su, T. Deng, M. Karin, and R.J. Davis, Cell, 76, 1025 (1994).

22. J.M. Kyriakis, P. Banerjee, E. Nikolakaki, T. Dai, E.A. Rubie, M.F. Ahmad, J. Avruch, and J.R. Woodgett, Nature, 369, 156 (1994).

23. T. Kallunki, B. Su, I. Tsigelny, H.K. Sluss, B. Dérijard, G. Moore, R. Davis, and M. Karin, Genes Dev., 8, 2996 (1994).

24. H.K. Sluss, T. Barrett, B. Dérijard, and R.J. Davis, Mol. Cell. Biol., 14, 8376 (1994).

25. I. Sánchez, R.T. Hughes, B.J. Mayer, K. Yee, J.R. Woodgett, J. Avruch, J.M. Kyriakis, and L.I. Zon, Nature, 372, 794 (1994).

26. B.M. Yashar, C. Kelley, K. Yee, B. Errede, and L.I. Zon, Mol. Cell. Biol., 13, 5738 (1993).

27. B. Dérijard, J. Raingeaud, T. Barrett, I.-H. Wu, J. Han, R.J. Ulevitch, and R.J. Davis, Science, 267, 682 (1995).

28. A. Lin, A. Minden, H. Martinetto, F.-X. Claret, C. Lange-Carter, F. Mercurio, G.L. Johnson, and M. Karin, Science, 268, 286 (1995).

29. C.A. Lange-Carter, C.M. Pleiman, A.M. Gardner, K.J. Blumer, and G.L. Johnson, Science, 260, 315 (1993).

30. M. Yan, T. Dai, J.C. Deak, J.M. Kyriakis, L.I. Zon, J.R. Woodgett, and D.J. Templeton, Nature, 372, 798 (1994).

31. O.A. Coso, M. Chiariello, J.-C. Yu, H. Teramoto, P. Crespo, N. Xu, T. Miki, and J.S. Gutkind, Cell, 81, 1137 (1995).

32. A. Minden, A. Lin, F.-X. Claret, A. Abo, and M. Karin, Cell, 81, 1147 (1995).

33. S. Zhang, J. Han, M.A. Sells, J. Chernoff, U.G. Knaus, R.J. Ulevitch, and G.M. Bokoch, J. Biol. Chem., 270, 23934 (1995).

34. J. Han, J.-D. Lee, L. Bibbs, and R.J. Ulevitch, Science, 265, 808 (1994).

35. J. Rouse, P. Cohen, S. Trigon, M. Morange, A. Alonso-Llamazares, D. Zamanillo, T. Hunt, and A.R. Nebreda, Cell, 78, 1027 (1994).

36. J.C. Lee, J.T. Laydon, P.C. McDonnell, T.F. Gallagher, S. Kumar, D. Green, D. McNulty, M.J. Blumenthal, J.R. Heys, S.W. Landvatter, J.E. Strickler, M.M. McLaughlin, I.R. Siemens, S.M. Fisher, G.P. Livi, J.R. White, J.L. Adams, and P.R. Young, Nature, 372, 739 (1994).

37. A.S. Zervos, L. Faccio, J.P. Gatto, J.M. Kyriakis, and R. Brent, Proc. Natl. Acad. Sci. USA, 92, 10531 (1995).

38. J. Han, J.-D. Lee, Y. Jiang, Z. Li, L. Feng, and R.J. Ulevitch, J. Biol. Chem., 271, 2886 (1996).

39. B. Stein, H. Brady, M.X. Yang, D.B. Young, and M.S. Barbosa, J. Biol. Chem., 271, 11427 (1996).

40. J. Raingeaud, A.J. Whitmarsh, T. Barrett, B. Dérijard, and R.J. Davis, Mol. Cell. Biol., 16, 1247 (1996).

41. K. Yamaguchi, K. Shirakabe, H. Shibuya, K. Irie, I. Oishi, N. Ueno, T. Taniguchi, E. Nishida, and K. Matsumoto, Science, 270, 2008 (1995).

42. J. Raingeaud, S. Guptas, J.S. Rogers, M. Dickens, J. Han, R.J. Ulevitch, and R.J. Davis, J. Biol. Chem., 270, 7420 (1995).

43. T. Deng, and M. Karin, Nature, 371, 171 (1994).

44. G. Zhou, Z.Q. Bao, and J.E. Dixon, J. Biol. Chem., 270, 12665 (1995).

45. J.M. English, C.A. Vanderbilt, S. Xu, S. Marcus, and M.H. Cobb, J. Biol. Chem., 270, 28897 (1995).

46. F. Zhang, A. Strand, D. Robbins, M.H. Cobb, and E.J. Goldsmith, Nature, 367, 704 (1994).

47. J. Zhang, F. Zhang, D. Ebert, M.H. Cobb, and E.J. Goldsmith, Curr. Biology, 3, 299 (1995).

48. R.A. Franklin, C. Brodie, I. Melamed, N. Terada, J.J. Lucas, and E.W. Gelfand, J. Immunol., 154, 4965 (1995).

49. T. Reimann, D. Büscher, R.A. Hipskind, S. Krautwald, M.L. Lohmann-Matthes, and M. Baccarini, J. Immunol., 153, 5740 (1994).

50. S.L. Weinstein, J.S. Sanghera, K. Lemke, A.L. DeFranco, and S.L. Pelech, J. Biol. Chem., 267, 14955 (1992).

51. A. Minden, A. Lin, T. Smeal, B. Dérijard, M. Cobb, R. Davis, and M. Karin, Mol. Cell. Biol., 14, 6683 (1994).

52. V. Adler, A. Schaffer, J. Kim, L. Dolan, and Z. Ronai, J. Biol. Chem., 270, 26071 (1995).

53. E. Cano, C.A. Hazzalin, and L.C. Mahadevan, Mol. Cell. Biol., 14, 7352 (1994).

54.   J.K. Westwick, C. Weitzel, A. Minden, M. Karin, and D.A. Brenner, J. Biol. Chem., _269,_
      26396 (1994).
55.   N.W. Freshney, L. Rawlinson, F. Guesdon, E. Jones, S. Cowley, J. Hsuan, and J.
      Saklatvala, Cell, _78,_ 1039 (1994).
56.   B. Su, E. Jacinto, M. Hibi, T. Kallunki, M. Karin, and Y. Ben-Neriah, Cell, _77,_ 727 (1994).
57.   F.M. Mitchell, M. Russel, and G.L. Johnson, Biochem. J., _309,_ 381 (1995).
58.   D.W. Kuroki, G.S. Bignami, and E.V. Wattenberg, Cancer Res., _56,_ 637 (1996).
59.   J.L. Duff, B.C. Berk, and M.A. Corson, Biochem. Biophys. Res. Commun., _188,_ 257 (1992).
60.   S. Marcus, A. Polverino, M. Barr, and M. Wigler, Proc. Natl. Acad. Sci. USA, _91,_ 7762
      (1994).
61.   K.-Y. Choi, B. Satterberg, D.M. Lyons, and E.A. Elion, Cell, _78,_ 499 (1994).
62.   S. Xu, D. Robbins, J. Frost, A. Dang, C. Lange-Carter, and M.H. Cobb, Proc. Natl. Acad.
      Sci. USA, _92,_ 6808 (1995).
63.   J.K. Westwick, A.D. Cox, C.J. Der, M.H. Cobb, M. Hibi, M. Karin, and D.A. Brenner, Proc.
      Natl. Acad. Sci. USA, _91,_ 6030 (1994).
64.   M. Russell, C.A. Lange-Carter, and G.L. Johnson, J. Biol. Chem., _270,_ 11757 (1995).
65.   A. Scott, C.M.M. Haystead, and T.A.J. Haystead, J. Biol. Chem., _270,_ 24540 (1995).
66.   M.A. Cahill, R. Janknecht, and A. Nordheim, Curr. Biology, _6,_ 16 (1996).
67.   H. van Dam, M. Duyndam, R. Rottier, A. Bosch, L. de Vries-Smits, P. Herrlich, A. Zantema,
      P. Angel, and A.J. van der Eb, EMBO J., _12,_ 479 (1993).
68.   M. Hibi, A. Lin, T. Smeal, A. Minden, and M. Karin, Genes Dev., _7,_ 2135 (1993).
69.   S. Gupta, D. Campbell, B. Dérijard, and R.J. Davis, Science, _267,_ 389 (1995).
70.   M.M. McLaughlin, S. Kumar, P.C. McDonnell, S.V. Horn, J.C. Lee, G.P. Livi, and P.R.
      Young, J. Biol. Chem., _271,_ 8488 (1996).
71.   D.M. Milne, L.E. Campbell, D.G. Campbell, and D.W. Meek, J. Biol. Chem., _270,_ 5511
      (1995).
72.   T. Nakajima, S. Kinoshita, T. Sasagawa, K. Sasaki, M. Naruto, T. Kishimoto, and S. Akira,
      Proc. Natl. Acad. Sci. USA, _90,_ 2207 (1993).
73.   E. Kowenz-Leutz, G. Twamley, S. Ansieau, and A. Leutz, Genes Dev., _8,_ 2781 (1994).
74.   K. Singh, J.-L. Balligand, T.A. Fischer, T.W. Smith, and R.A. Kelly, J. Biol. Chem., _271,_
      1111 (1996).
75.   K.-M. Hsiao, S.-Y. Chou, S.-J. Shih, and J.E. Ferrell, Jr., Proc. Natl. Acad. Sci. USA, _91,_
      5480 (1994).
76.   L.-L. Lin, M. Wartmann, A.Y. Lin, J.L. Knopf, A. Seth, and R.J. Davis, Cell, _72,_ 269 (1993).
77.   M.F. Dubois, V.T. Nguyen, M.E. Dahmus, G. Pagès, J. Pouysségur, and O. Bensaude,
      EMBO J., _13,_ 4787 (1994).
78.   R.J. Davis, J. Biol. Chem., _268,_ 14553 (1993).
79.   S. Kato, H. Endoh, Y. Masuhiro, T. Kitamoto, S. Uchiyama, H. Sasaki, S. Masushige, Y.
      Gotoh, E. Nishida, H. Kawashima, D. Metzger, and P. Chambon, Science, _270,_ 1491
      (1995).
80.   A. Migliaccio, M. Di-Domenico, G. Castoria, A. De-Falco, P. Bontempo, E. Nola, and F.
      Auricchio, EMBO J., _15,_ 1292 (1996).
81.   M. David, E. Petricoin III, C. Benjamin, R. Pine, M.J. Weber, and A.C. Larner, Science, _269,_
      1721 (1995).
82.   G. Kumar, S. Gupta, S. Wang, and A.E. Nel, J. Immunol., _153,_ 4436 (1994).
83.   B. Yao, Y. Zhang, S. Delikat, S. Mathias, S. Basu, and R. Kolesnick, Nature, _378,_ 307
      (1995).
84.   J. Wu, P. Dent, T. Jelinek, A. Wolfman, M.J. Weber, and T.W. Sturgill, Science, _262,_ 1065
      (1993).
85.   Y.P. Hsueh, and M.Z. Lai, J. Biol. Chem., _270,_ 18094 (1995).
86.   T.W. Lissoos, D.W.A. Beno, and B.H. Davis, J. Biol. Chem., _268,_ 25132 (1993).
87.   M.H. Cobb, and E.J. Goldsmith, J. Biol. Chem., _270,_ 14843 (1995).
88.   H. Sun, N.K. Tonks, and D. Bar-Sagi, Science, _266,_ 285 (1994).
89.   S.M. Keyse, and E.A. Emslie, Nature, _359,_ 644 (1992).
90.   J.L. Duff, B.P. Monia, and B.C. Berk, J. Biol. Chem., _270,_ 7161 (1995).
91.   Y. Liu, M. Gorospe, C. Yang, and N.J. Holbrook, J. Biol. Chem., _270,_ 8377 (1995).
92.   D. Bokemeyer, A. Sorokin, M. Yan, N.G. Ahn, D.J. Templeton, and M.J. Dunn, J. Biol.
      Chem., _271,_ 639 (1996).
93.   D.R. Alessi, N. Gomez, G. Moorhead, T. Lewis, S.M. Keyse, and P. Cohen, Curr. Biology, _5,_
      283 (1995).
94.   J. Frost, A.S. Alberts, E. Sontag, K. Guan, M.C. Mumby, and J.R. Feramisco, Mol. Cell.
      Biol., _14,_ 6244 (1994).
95.   E. Sontag, S. Fedorov, C. Kamibayashi, D. Robbins, M. Cobb, and M. Mumby, Cell, _75,_ 887
      (1993).
96.   Y. Ward, S. Gupta, P. Jensen, M. Wartmann, R.J. Davis, and K. Kelly, Nature, _367,_ 651
      (1994).

97. J.M. Denu, G. Zhou, L. Wu, R. Zhao, J. Yuvaniyama, M.A. Saper, and J.E. Dixon, J. Biol. Chem., 270, 3796 (1995).
98. Y. Chu, P.A. Solski, R. Khosravi-Far, C.J. Der, and K. Kelly, J. Biol. Chem., 271, 6497 (1996).
99. Q.M. Wang, K.-L. Guan, P.J. Roach, and A.A. DePaoli-Roach, J. Biol. Chem., 270, 18352 (1995).
100. A.D. Cherniack, J.K. Klarlund, and M.P. Czech, J. Biol. Chem., 269, 4717 (1994).
101. A. Aronheim, D. Engelberg, N. Li, N. Al-Alawi, J. Schlessinger, and M. Karin, Cell, 78, 949 (1994).
102. E. Saez, S.E. Rutberg, E. Mueller, H. Oppenheim, J. Smoluk, S.H. Yuspa, and B.M. Spiegelman, Cell, 82, 721 (1995).
103. E. Hu, E. Mueller, S. Oliviero, V.E. Papaioannou, R. Johnson, and B.M. Spiegelman, EMBO J., 13, 3094 (1994).
104. S. Cowley, H. Paterson, P. Kemp, and C.J. Marshall, Cell, 77, (1994).
105. S.J. Mansour, W.T. Matten, A.S. Hermann, J.M. Candia, S. Rong, K. Fukasawa, G.F. Vande Woude, and N.G. Ahn, Science, 265, 966 (1994).
106. A.B. Raitano, J.R. Halpern, T.M. Hambuch, and C.L. Sawyers, Proc. Natl. Acad. Sci. USA, 92, 11746 (1995).
107. C. Knall, S. Young, J.A. Nick, A.M. Buhl, G.S. Worthen, and G.L. Johnson, J. Biol. Chem., 271, 2832 (1996).
108. M. Karin, and T. Hunter, Curr. Biology, 5, 747 (1995).
109. Z. Xia, M. Dickens, J. Raingeaud, R.J. Davis, and M.E. Greenberg, Science, 270, 1326 (1995).
110. S.-i. Takewaki, M. Kuro-o, Y. Hiroi, T. Yamazaki, T. Noguchi, A. Miyagishi, K.-i. Nakahara, M. Aikawa, I. Manabe, Y. Yazaki, and R. Nagai, J. Mol. Cell. Cardiol., 27, 729 (1994).
111. A. Yao, T. Takahashi, T. Aoyagi, K.-i. Kinugawa, O. Kohmoto, S. Sugiura, and T. Serizawa, J. Clin. Invest., 96, 69 (1995).
112. W. Li, C.D. Whaley, A. Mondino, and D.L. Mueller, Science, 271, 1272 (1996).
113. P.E. Fields, T.F. Gajewski, and F.W. Fitch, Science, 271, 1276 (1996).
114. C.E. Whitehurst, and T.D. Geppert, J. Immunol., 156, 1020 (1996).
115. P. Aspenström, U. Lindberg, and A. Hall, Curr. Biology, 6, 70 (1996).
116. A.A. Reszka, R. Seger, C.D. Diltz, E.G. Krebs, and E.H. Fischer, Proc. Natl. Acad. Sci. USA, 92, 8881 (1995).
117. J. Taieb, D.A. Blanchard, M.T. Auffredou, N. Chaouchi, and A. Vazquez, J. Immunol., 155, 5623 (1995).
118. J.K. Westwick, C. Weitzel, H.L. Leffert, and D.A. Brenner, J. Clin. Invest., 95, 803 (1995).
119. F. Hilberg, A. Aguzzi, N. Howells, and E.F. Wagner, Nature, 365, 179 (1993).
120. H. Morooka, J.V. Bonventre, C.M. Pombo, J.M. Kyriakis, and T. Force, J. Biol. Chem., 270, 30084 (1995).
121. C.M. Pombo, J.V. Bonventre, J. Avruch, J.R. Woodgett, J.M. Kyriakis, and T. Force, J. Biol. Chem., 269, 26546 (1994).
122. R.J. Knight, and D.B. Buxton, Biochem. Biophys. Res. Comm., 218, 83 (1995).
123. R.M. Kramer, E.F. Roberts, B.A. Strifler, and E.M. Johnstone, J. Biol. Chem., 270, 27395 (1995).
124. J. Saklatvala, L. Rawlinson, R.J. Waller, S. Sarsfield, J.C. Lee, L.F. Morton, M.J. Barnes, and R.W. Farndale, J. Biol. Chem., 271, 6586 (1996).
125. A.A. Mohit, J.H. Martin, and C.A. Miller, Neuron, 14, 67 (1995).
126. K.M. Latinis, and G.A. Koretzky, Blood, 87, 871 (1996).
127. M. Verheij, R. Bose, X.H. Lin, B. Yao, W.D. Jarvis, S. Grant, M.J. Birrer, E. Szabo, L.I. Zon, J.M. Kyriakis, A. Haimovitz-Friedman, Z. Fuks, and R.N. Kolesnick, Nature, 380, 75 (1996).
128. N.L. Johnson, A.M. Gardner, K.M. Diener, C.A. Lange-Carter, J. Gleavy, M.B. Jarpe, A. Minden, M. Karin, L.I. Zon, and G.L. Johnson, J. Biol. Chem., 271, 3229 (1996).
129. T. Tsuda, Y. Kawahara, Y. Ishida, M. Koide, K. Shii, and M. Yokoyama, Circ. Res., 71, 620 (1992).
130. C. MacKintosh, and R.W. MacKintosh, J. Biol. Chem., 19, (1994).
131. J.C. Lee, and J.L. Adams, Curr. Opin. Biotech., 6, 657 (1995).
132. D.T. Dudley, L. Pang, S.J. Decker, A.J. Bridges, and A.R. Saltiel, Proc. Natl. Acad. Sci. USA, 92, 7686 (1995).
133. L. Pang, T. Sawada, S.J. Decker, and A.R. Saltiel, J. Biol. Chem., 270, 13585 (1995).
134. M. Symons, Curr. Opin. Biotech., 6, 668 (1995).
135. K. Nishio, H. Arioka, T. Ishida, H. Fukumoto, H. Kurokawa, M. Sata, M. Ohata, and N. Saijo, Int. J. Cancer, 63, 688 (1995).

George L. Trainor
DuPont Merck Pharmaceutical Company
Wilmington, DE 19880-0500

Chapter 30. New NMR Methods for Structural Studies of Proteins to Aid in Drug Design

Sharon J. Archer[#], Peter J. Domaille[#] and Ernest D. Laue[*]
DuPont Merck Pharmaceutical Company[#], Wilmington DE 19880 and Department of
Biochemistry, University of Cambridge[*], CB21QW, United Kingdom

Introduction - Over the last decade there has been an explosion in the use of nuclear magnetic resonance (NMR) spectroscopy to study the three-dimensional structures and dynamics of proteins in solution. This heightened activity is due to concurrent developments in NMR methods and in molecular biology and, in many cases, a desire to provide a structural and mechanistic basis for drug (ligand) design. Wüthrich and others in the 1980's showed that it was possible to determine the solution structure of proteins and peptides with molecular weight less than ca. 10 kD using homonuclear ^{1}H NMR (1). Subsequently, the incorporation of ^{13}C and ^{15}N isotopes, usually into recombinant and overexpressed proteins, has allowed researchers to extend the size range to ca. 20 kD using heteronuclear NMR techniques (2-4). More recently, partial random deuterium (^{2}H) labeling has extended resonance assignments to larger proteins (ca. 30kD) (5, 6). In addition to making full structure determinations of proteins and protein complexes feasible, isotopic enrichment has also allowed development of NMR methods to identify protein-drug interactions (7), locate bound water molecules (8), titrate sidechain groups (9), and dissect motion in these proteins on several disparate time scales (10). This chapter will focus on the use of isotopically enriched (^{13}C, ^{15}N, ^{2}H) proteins in new heteronuclear NMR experiments for detailed studies of proteins and protein complexes, and the use of this information in drug design. A collection of topics which also addresses the use of NMR in drug design involving protein-DNA, DNA-drug, and small peptides and mimetics has recently been published (11).

Isotopic Enrichment - Isotopic enrichment has made NMR amenable to larger proteins and has proven particularly powerful for studying proteins, ligands and protein-ligand complexes in solution. By controlling the extent of isotopic enrichment (e.g. by expressing the protein in isotopically varied media, or by synthesizing ligands with or without specific labels) one can selectively study just the protein, the ligand, or the interactions between them. Either the ligand, the protein, or both can be labeled. These studies are usually done with isotopically enriched protein and natural abundance ligand because the heteronuclei are required for resonance assignments and structure determination of the protein, and because the ligand is smaller and has fewer proton resonances. By using experiments that either select or filter protons attached to the heteronuclei, signals from the protein or the ligand can be observed exclusively. Experiments employing a half-filter can be used to identify interactions between the protein and its ligand.

Isotopic enrichment of proteins with ^{13}C, ^{15}N and ^{2}H is necessary because the natural abundance of these NMR-active isotopes is low. Typically this is achieved using overexpression of the protein in E. coli cells harboring a recombinant DNA plasmid, and grown in minimal media enriched with unique sources of ^{15}N and ^{13}C (12). Isotopic enrichment of non exchangeable hydrogens with deuterium is accomplished in cells grown in ^{2}H$_2$O (50% - 75%), or in an enriched medium of algal hydrolysate (13). Uniform incorporation of isotopes into proteins expressed in mammalian systems has also been achieved by adding labeled amino acids to the media (14). Although this approach was originally quite costly for producing uniformly enriched proteins, suitable media for mammalian cells is now commercially available at realistic cost. An alternative approach is to selectively enrich the proteins at specific amino acids by adding specific amino acids to the growth medium (15). Expression of proteins for use in NMR spectroscopy can also be accomplished using other high level expression systems such as yeast, e.g. *Pichia Pastoris*, or via baculoviral expression in insect cells (16, 17).

CONFORMATION OF THE LIGAND

The conformation of a bound peptide or drug ligand is required for structure-based drug design because the bound conformation can differ substantially from that of the free ligand. If the ligand (drug) has a relatively low affinity for the target protein (Kd $< 10^{-6}$) and the exchange rate of the ligand is faster than its spin-lattice relaxation rate, then transferred NOE experiments can be used to obtain information about the conformation; these experiments only require observations of the free ligand (18). The experimental conditions for transferred NOE often cannot be arranged and so we will not discuss the transferred NOE further but rather refer the reader to a recent review (19). For the more favorable and frequent situation in drug design where the ligand has a high affinity for the protein, there are several techniques that can be applied to determine the bound ligand conformation. In general, there are three approaches used to obtain information about the bound conformation: (I) natural abundance protein and selectively labeled ligand, (II) perdeuterated protein and natural abundance ligand, and (III) isotopically enriched protein and natural abundance ligand.

In method (I), the use of isotope filtered experiments allows only the signals of the isotopically labeled ligand to be observed in the presence of the numerous protein signals. One benefit of this method is that it is applicable to protein complexes in which the protein is difficult or expensive to isotopically enrich, e.g. mammalian proteins for which there is no suitable system for overexpression. If the ligand can be expressed biosynthetically, isotopically labeled ligand can be obtained in an analogous fashion to isotopic enrichment of recombinant proteins. For example, for NMR studies of the immunosuppressant cyclosporin A, ^{13}C or ^{15}N-labeled ligand was obtained by growing the cyclosporin-producing fungus on ^{13}C or ^{15}N-enriched medium (20, 21). In ^{13}C- or ^{15}N-selected experiments, signals from the labeled ligand allowed the conformation of cyclosporin A bound to cyclophilin to be determined. These experiments showed that the conformation of the bound cyclosporin was very different from that previously determined in crystalline and solution studies of uncomplexed molecule. Intermolecular NOEs between the ligand and the protein were then used to identify which portion of cyclosporin made contact with the target protein, thus identifying interesting sites for the design of drugs with better or different activities (20, 21). If the ligand is chemically synthesized, isotopic labels must be introduced during synthesis of the ligand. The amount of information available will be related to the number of sites at which labels are introduced. This will be dictated by the synthetic pathway and cost of isotopically enriched precursors, but often this type of method provides information which is hard to obtain in other ways. For example, the conformations of folate and "anti-folate" drugs bound to dihydrofolate reductase (DHFR) have been examined extensively using various selectively labeled substrates bound to the enzyme (22-24). DHFR is of interest because it is the target of the "anti-folate" drugs trimethoprim and methotrexate. NMR studies of selectively labeled folate have shown that the bound folate exists in three interconverting conformational states in the ternary complex DHFR-folate-NADP$^+$ (22, 24), while selectively labeled methotrexate exhibits only one bound conformation (24). The conformations adopted by folate are pH dependent; at low pH, form I predominates, and at high pH, forms IIa and IIb predominate. NOEs between the labeled folate and the protein indicated that the orientation of the pteridine ring in forms I and IIa resembles that of methotrexate while the orientation in form IIb is flipped by 180° and is in a catalytically active conformation (24). The ionization and tautomeric states of the substrates, which are of interest for drug design, were characterized using selectively ^{13}C or ^{15}N labeled substrates (23, 24).

The second method (II) for determining the structure of the bound ligand requires a ^{1}H NMR-silent perdeuterated protein such that the NMR signal is observed exclusively from the ligand ^{1}Hs (7, 25). This method was used to determine the bound conformations of cyclosporin A and an inactive cyclosporin analog complexed with perdeuterated cyclophilin (26).   For compact ligands or cyclic ligands, there is a wealth of information from intraligand NOEs.   For ligands that bind in an extended conformation, however, there is little information about the conformation because of the lack of long range NOEs. This method is attractive for ligands that are difficult to synthesize because it does not require special synthesis of labeled ligand. However, the method does require perdeuterated protein. For larger protein complexes where perdeuteration is already required to optimize relaxation properties of the protein, the method is readily applicable.

The remaining method (III) for determining the structure of the bound ligand is to use an unlabeled ligand complexed to a $^{13}C,^{15}N$-enriched protein. Most of the protein resonances can be eliminated by filtering all protons attached to either ^{13}C or ^{15}N using doubly isotope-filtered experiments leaving only signals from the ligand (27, 28). Although these experiments require $^{13}C,^{15}N$-enriched protein, the protein is usually already isotopically enriched for use in protein structure determination. If the ligand is small enough, doubly isotope-filtered NMR experiments can be used to determine assignments of the bound ligand and to define the conformation of the bound ligand, e.g. of an M13 peptide complexed to $^{13}C,^{15}N$-enriched calmodulin (27). The intra-ligand NOEs indicated that the conformation of the M13 peptide was α-helical when bound to calmodulin. This method is complementary to half-filtered methods which are used to identify protein-ligand close contacts ; the intra-ligand NOEs used in conjunction with those between the protein and ligand can provide enough information to define the structure of the ligand (and the protein) in full structure calculations. In subsequent work on the calmodulin/M13 complex, the intra-ligand NOEs were combined with protein-ligand NOEs and intra-protein NOEs to determine the full three-dimensional structure of the calmodulin/M13 complex (29).

CONFORMATION OF THE PROTEIN AND PROTEIN COMPLEX

NMR spectroscopy of proteins and protein-ligand complexes for use in drug design has received much attention because it can be used to determine three-dimensional structures of proteins and protein complexes in solution. The following section will outline the most recent techniques used to determine the structure of the protein in a complex with particular emphasis on larger proteins and the use of deuterium enrichment.

Sequential Resonance Assignments - In order to fully interpret NMR data, one must assign all of the NMR signals to specific atoms in the protein sequence.   A comprehensive overview of contemporary methods appeared recently (30). Homonuclear two-dimensional (2D) NMR techniques are capable of providing ^1H assignments for small proteins (< 10 kD) (1). The resonance overlap and lack of sensitivity encountered in spectra of larger proteins (up to 20kD) can be overcome by using heteronuclear (^{13}C and ^{15}N) separated or filtered methods and by increasing the dimensionality of the experiment (3D and 4D NMR) (2-4). The recent application of pulsed field gradients for coherence selection and artifact and solvent suppression has considerably improved the sensitivity of many of these experiments (31, 32), while the use of linear prediction and maximum entropy in data processing has improved resolution (33-35). However, even with these improvements, larger proteins and protein complexes (> 20kD) present a formidable challenge because their spectra have lower sensitivity and resolution. The transverse relaxation rates (T_2, particularly of ^1H and ^{13}C) become prohibitively short due to the slower tumbling rate (described by a rotational correlation time τ_c > 12ns) and limit the time available for spin manipulations in standard heteronuclear NMR experiments. ^2H isotopic enrichment (13) (in addition to ^{13}C and ^{15}N) increases the T_2 values, and thus the sensitivity and available resolution, so that resonance assignments for proteins in the 30-40 kD range are feasible. Although complete deuteration is optimal for backbone experiments, random fractional deuteration (50 - 75% ^2H) allows for assignments of both backbone and sidechain resonances in a single sample (5, 6, 36).

For proteins in which aliphatic and aromatic ^1H are replaced with ^2H, only the exchangeable ^1H_N protons are left for detection making the ^1H_N-^{15}N correlation the cornerstone for experiments on deuterated proteins. Backbone directed assignment techniques, which relate all chemical shifts to the backbone ^1H_N-^{15}N correlation, make it relatively straightforward to obtain complete heteronuclear resonance assignments for uniformly $^{13}C,^{15}N$-enriched proteins (4). Where complete or fractional deuteration is used to optimize ^{13}C relaxation properties, variations of the standard backbone directed assignment experiments are necessary due to both the size of these proteins and the extent of deuterium labels (5, 6). The "out and back" type experiments, where the magnetization both originates and is detected on the ^1H_N spin, do not involve aliphatic and aromatic protons and are the most sensitive for obtaining backbone and sidechain ^{13}C assignments of deuterated proteins (5). Clearly, full deuteration of sidechains is inappropriate when sidechain ^1H assignments are required. However, the extent of deuteration can be optimized to balance the sensitivity improvement, due to better relaxation properties, against the loss in absolute ^1H signal. The optimal level of ^2H incorporation will depend on the size of the protein, but studies of a model system indicate that ca. 50% ^2H should provide a good compromise (6). Recently, a 75% deuterated 20 kD phosphotyrosine binding domain of Shc complexed with a twelve-residue phosphopeptide was completely assigned (36). Sensitivity for larger proteins can also be improved by optimizing the experiments for detection of particular correlations or by minimizing the number of magnetization transfer steps. For example, the overall sensitivity for a 37 kD protein-DNA complex (70% ^2H) was improved by correlating the $^{13}C\alpha$ and $^{13}C\beta$ atoms with ^1H_N-^{15}N pairs in two separate experiments (5).

NOE Resonance Assignments - The basis of the distance measurement used in NMR structure determinations is the intensity of the cross peak between two ^1H nuclei in nuclear Overhauser effect spectroscopy (NOESY). To a first approximation, the intensity is proportional to the inverse sixth power of the distance between the nuclei. A number of 3D and 4D NOESY experiments are available for measuring interacting pairs (e.g. (37-40)). Generally, the 4D NOESY experiments identify $^1H_N/^{15}N$ or $^1H_C/^{13}C$ pairs in close proximity (<5A); the heteronuclei serve to resolve cross peaks and to authenticate the identity of the directly attached proton. 3D NOESY spectra, while of higher sensitivity, are more likely to produce ambiguous identification of the interacting ^1H nuclei since heteronuclear separation is restricted to only one of the protons. Both the ^{13}C and ^{15}N-separated NOESY experiments exploit the inherently large chemical shift range of the heteronuclei; the ^{13}C directed experiments are especially powerful because the majority of structurally important contacts in a protein are those involving sidechain protons that are directly attached to carbon.

For deuterated proteins in H_2O, where nonexchangeable ^1H's have been replaced with ^2H's, the NOESY experiments involving amide protons ^1H_N are the most sensitive. In 100% deuterated proteins, the H_N-H_N NOEs have been used to identify α-helical intra-strand NOEs (41) and β-sheet intra- and inter-strand NOEs (42, 43). Recently, 4D H_N-H_N NOESY experiments on large 100% deuterated proteins (29 kD, τ_C=16ns) yielded high quality spectra (42, 43) and calculations indicate that these experiments will be applicable to proteins with correlation times up to τ_C=25ns (43). For partially deuterated proteins, one can obtain high quality NOEs between the amide protons and other protons (6) with fewer complications from spin diffusion.

Many of the above experiments can also be recorded as ROESY (rotating frame NOESY) experiments to ameliorate the spin diffusion problem encountered with long NOESY mixing times (44). However, because of the short ^1H T_2 values, the ROE is less sensitive than the NOE; it is generally run as a 3D experiment and is likely to be less effective for large proteins.

Dihedral Angle Measurements - Three bond scalar coupling constants are strongly influenced by the dihedral angle between the four atoms involved and can therefore be used to measure this angle (45). Careful measurements of coupling constants can be used to obtain precise values of dihedral angles comparable to the accuracy of high resolution X-ray structures (46). Various types of experiments have been designed to measure different three-bond couplings to determine dihedral angles (47). Intensity-modulated correlation experiments appear most applicable to larger proteins (45).

These quantitative experiments can be used to measure a large number of two- and three-bond homo- and heteronuclear couplings and provide local conformational information. If there is a singe rotamer, the large number of J couplings can over determine the conformer. On the other hand, if there is averaging, the large number of J couplings can be used to better characterize this dynamic process (45). More recently developed double/zero quantum methods also appear very promising (48). Conformations on the surface of a protein, where interactions with ligands occurs and where there are few structural NOEs, are most important for designing selective ligands/drugs.

Hydrogen Exchange - Hydrogen exchange rates reflect the solvent accessibility of exchangeable protons and can be used, with structural information, to locate hydrogen bonds and to qualitatively determine the structural stability of proteins (49). Macroscopic exchange rates can be determined by using a 2D ^{1}H-^{15}N correlation experiment to monitor the disappearance ^{1}H$_N$ resonances after dissolution in ^{2}H$_2$O. Hydrogen bonds buried in the hydrophobic core of the protein exchange more slowly than those readily accessible to solvent. Alternative NMR techniques have been developed to measure more rapid exchange rates ($t_{1/2} < 5$min) (50-52). Because hydrogen exchange experiments monitor the exchange of amide protons, they are applicable to both protonated and deuterated proteins.

Secondary Structure - Because NMR spectroscopic parameters such as the chemical shift, NOE intensity, and coupling constants reflect the local environment, short range structure is readily determined. Regular elements of secondary structure are easily identified from characteristic patterns of sequential and short-range NOEs (1). These qualitative NOE patterns can be used to identify α-helices, β-strands, and turns prior to carrying out full structure calculations. For a fully deuterated protein, only ^{1}H$_N$-^{1}H$_N$ NOEs will be present, while for a partially deuterated protein, the other short range NOEs will also be accessible.

Chemical shifts of nuclei are sensitive to secondary and tertiary structural effects in proteins and can be used to identify elements of secondary structure. The chemical shifts of ^{1}H$_\alpha$, ^{13}C$_\alpha$, ^{13}C$_\beta$ and ^{13}CO have all been shown empirically to correlate with secondary structure (53-55). Recent advances in *ab initio* calculations of ^{13}C chemical shifts explain the origin of these shifts in terms of ϕ, ψ, ω angles characteristic of secondary structure (56). Thus once the backbone assignments are established the secondary structure can be predicted without detailed NOE data analysis. This will be especially useful for larger proteins where ^{13}C$_\alpha$, ^{13}C$_\beta$ and ^{13}CO chemical shifts are readily obtainable whilst proton NOEs can be scarce.

If the protein is not too large, the elements of secondary structure can also be verified using backbone ϕ angles obtained from the 3-bond coupling (^{3}J(H$_N$H$_\alpha$)); small couplings (< 6 Hz) are indicative of ϕ angles found in α-helices (-65^0) while large couplings (> 8Hz) are representative of $\phi \sim -120^0$ found in β-strands (1). Hydrogen bonding patterns from slowly exchanging ^{1}H$_N$ atoms can also confirm the presence of α-helices and β-sheets.

Active Site Identification - For a labeled protein, the simplest technique for locating the ligand binding site is to identify changes in chemical shifts of the protein in the absence and presence of ligand. Because chemical shifts are sensitive to changes in the local environment, there are often significant changes localized to the ligand binding site. The ^{1}H$_N$-^{15}N correlation experiment is often used to monitor the H$_N$ and N chemical shifts of the protein (for examples see (57, 58)); studies as a function of ligand or protein concentration allow assignments of the protein resonances in the complex. These spectra can also be used to estimate the kinetics and equilibria of the association; the linewidth and number of peaks indicates whether the complex is in fast, medium or slow exchange relative to the NMR time scale. If the protein and ligand are in fast exchange and some exchange broadening is observed, one can calculate the dissociation rate constant for the complex (57). Monitoring of either chemical shifts or other NMR properties (e.g. NOEs) can accompany the titration.

Changes in hydrogen exchange rates (quantified as protection factors) can also be used to identify surface binding sites (e.g. (59)). When a protein binds a ligand, exchange rates of protons at the active site can be substantially altered by the bulky presence of the ligand which excludes solvent and/or hydrogen bonds that are formed between the protein and the ligand.

As discussed previously, half-filtered NOE experiments can be used to readily identify protein-ligand interactions. The half-filtered experiment can be acquired as a 3D experiment so that the protein resonances can be identified by the both the proton and heteroatom chemical shifts (60). These protein-ligand NOEs are often crucial for determining the location and relative conformation of the bound ligand.

Structure Calculations - NMR structure calculations have historically been based on a large number of loose distance restraints derived from NOESY data, more precise distances based on hydrogen-bonded pairs and covalent distances, dihedral angle restraints based on coupling constants, and empirical energy terms. Several protocols involving distance geometry and dynamical simulated annealing (SA) are available to search conformational space to find an ensemble of structures consistent with the experimental data (61). Implicit in these methods is the assumption that the experimental distance restraint is unambiguously assigned to a pair of atoms. In larger proteins a NOE cross peak can arise from several protons with similar chemical shifts and the cross peak cannot always be used to provide a unique distance restraint. This situation was usually handled in an iterative manner, using only unambiguous cross peaks at an early stage, and adding additional restraints once the structure eliminated all but one possibility. Recently, as an extension of methods developed to deal with the ambiguity associated with multimeric proteins (62), automatic assignment of ambiguous NOE cross peaks was incorporated as part of the SA protocol (63). This algorithm automates the assignment of spin pairs in NOESY spectra which is the most error-prone and time-consuming component of structure determination via NMR. A distinct algorithm has also been proposed in which uniquely identified (but not assigned) NOEs, involving $^{1}H_N/^{15}N$ or $^{1}H_C/^{13}C$ pairs, are transformed into "real space" to allow chain tracing, sequence specific assignment and structure refinement in an automated procedure (64, 65)

Presently, there are a large number (> 50) of new structures of proteins and protein complexes determined by NMR reported each year; we refer the reader to the annual compilation of all newly reported atomic structures, *Macromolecular Structures* (66), for the most recently determined structures.

CHEMICAL INFORMATION AT THE ACTIVE SITE

In addition to structural information, NMR spectroscopy can provide a wealth of chemical information which is useful in drug design. Because chemical shifts are very sensitive to the local chemical environment, chemical shifts can be used to monitor and determine the ionization and tautomeric states of protein sidechains and ligands. Water-NOE and -ROE experiments can be used to identify the presence of bound water molecules for which inhibitors could be designed to displace.

Ionization and Tautomeric States of Sidechain Groups - The tautomeric and ionization states of histidines can be important for understanding detailed chemistry at the active site and elsewhere in a protein. For medium to large proteins with several histidine residues, the individual histidine resonances are readily distinguishable in 2D heteronuclear experiments. NMR spectroscopy has been used for decades for titration of histidine imidazoles (67). The chemical shifts of the nonexchangeable ring protons and carbons are monitored as a function of pH. Correlation via the two bond coupling between the nonexchangeable ring protons and the ring nitrogens in conjunction with sidechain proton assignments has been used to determine the tautomeric state of the imidazole group of histidines in larger proteins (68). In studies of a matrix metalloprotease complexed with various inhibitors, the tautomeric states of all eight hisitidines (three complexed to an active site zinc, three complexed to a second zinc, and two others) have been determined (69).

Determination of the ionization state of other titratable sidechain residues such as glutamates and aspartates is also possible using heteronuclear NMR techniques. Analogous to the histidine sidechain resonances, the ^{13}CO and $^{13}C\gamma$ chemical shifts of glutamic and aspartic acid sidechains are sensitive to the protonation state of the carboxylic acid and a heteronuclear experiment which correlates CO, $C\beta/C\gamma$ and $H\beta/H\gamma$ chemical shifts can be used to monitor the pH titration of aspartic/glutamic acids (9). This method has been used to determine the ionization state of the active site aspartate in HIV protease (9) and the active site aspartate in thioredoxin in the absence and presence of its target peptide from NFκB (70, 71). For all of these proteins, the active site acidic group had a pKa considerably higher (pKa > 7) than is typically seen for glutamic and aspartic acids. Information regarding the ionization state of acidic or basic groups at the active site is important both for designing ligands with high specificity and for understanding the activity of an enzyme where a titratable group is postulated to donate or accept a proton.

Bound Water Molecules -   Water molecules bound to proteins may help stabilize proteins and protein complexes through hydrogen bonding and those waters located in the active site of proteins may be involved in catalysis or ligand recognition. Bound water molecules can be detected via NOEs between protein resonances and water (8). For larger proteins where heteronuclear experiments are necessary to assign the protein resonances, NOEs and ROEs can be detected between bound water and protein protons attached to ^{13}C or to ^{15}N (52, 72, 73). The ^{15}N water NOE experiments involve only the amide protons and are therefore applicable to both deuterated and nondeuterated proteins while the ^{13}C water NOE experiment requires the presence of aliphatic protons.

The design of selective inhibitors of HIV protease made use of a unique water at the active site that is not found in other aspartyl proteases. A cyclic urea inhibitor was designed to selectively displace this bound water, and studies of bound waters in HIV protease were then carried out the presence of the cyclic urea inhibitor and alternative inhibitors (74). The water NOE experiments clearly indicated that the bound water molecule was displaced by the cyclic urea inhibitor.

PROTEIN DYNAMICS

NMR relaxation studies in solution can provide important information about the dynamic behavior of the protein backbone and sidechains (10). ^{15}N relaxation studies are used to characterize the overall dynamic behavior of the protein as well as to locate regions within it that are flexible on the picosecond to nanosecond time scale. Because the ^{15}N relaxation studies detect amide resonances, these methods are applicable to both protonated and deuterated proteins, and have proven useful for detection of the presence or absence of interactions between domains in larger proteins (75). ^{15}N relaxation measurements have been used to characterize the backbone dynamics of HIV protease-drug complexes (76). These relaxation measurements showed that the flaps that fold over the protein active site undergo large amplitude motions on the picosecond to nanosecond time scale and that the tips of the flaps undergo a conformational exchange on the microsecond time scale. Such functionally important mobility has also been observed in other proteins, e.g. the c-H-ras p21 protein (77).

Measurements of backbone protein dynamics also allow studies of anisotropy in rotational diffusion from which structural information about the orientation of the different ^1H_N-^{15}N bond vectors can be derived (75, 78, 79). This type of structural information promises to be very useful in larger proteins where many NOEs are often weak or absent. Recently it has been shown that studies of anisotropic magnetic susceptibility in proteins can also provide similar information (80).

The dynamics of sidechain methyl groups are of interest for drug design because aliphatic sidechains are often involved in hydrophobic interactions with ligands. For larger proteins which contain $^{13}C,^{15}N,^2H$ enrichment, the deuterium relaxation rates of sidechain methyl groups can be measured and used to characterize sidechain dynamics

(81). Deuterium relaxation studies on the methyl containing sidechains of the C-terminal SH2 domain of phospholipase C-γ1 with a bound phosphotyrosine (pTyr) containing peptide were used to examine the correlation between protein dynamics and ligand binding affinity and specificity (82). These relaxation measurements showed that the pTyr binding region, which is responsible for the high affinity of the ligand, displays substantial restriction of motion. In contrast, the hydrophobic binding site, which is responsible for recognition of sequences C-terminal to the pTyr, exhibits a high degree of mobility in both the free and bound peptide forms and this may be important for the non-specific recognition of different pTyr containing peptides (82).

Summary - Heteronuclear NMR spectroscopy of ^{13}C,^{15}N-labeled proteins has made it possible for researchers to determine the structures of proteins up to *ca.* 20 kD. More recently, partial random deuterium labeling has extended resonance assignments to larger proteins (*ca* 30 kD). Many of the heteronuclear experiments required for full structure calculations are readily applicable to partially deuterated proteins. The combination of heteronuclear triple resonance NMR experiments with partial deuterium labeling promises to make NMR structural studies tractable for protein complexes of up to *ca.* 30-40 kD.

References

1.   Wüthrich, K., Science, 243, 45 (1989).
2.   Fesik, S.W. and Zuiderweg, E., Quart. Rev. Biophys., 23, 97 (1990).
3.   Clore, G.M. and Gronenborn, A.M., Science, 252, 1390 (1991).
4.   Bax, A. and Grzesiek, S., Acc. Chem. Res., 26, 131 (1993).
5.   Yamazaki, T., Lee, W., Arrowsmith, C.H., Muhandiram, D.R. and Kay, L.E., J. Am. Chem. Soc., 116, 11655 (1994).
6.   Nietlispach, D., Clowes, R.T., Broadhurst, R.W., Ito, Y., Keeler, J., Kelly, M., Ashurst, J., Oschkinat, H., Domaille, P.J. and Laue, E.D., J. Am. Chem. Soc., 118, 407 (1995).
7.   Fesik, S.W., J. Med. Chem., 34, 2937 (1991).
8.   Otting, G., Liepinsh, E. and Wuthrich, K., Science, 254, 974 (1991).
9.   Yamazaki, T., Nicholson, L.K., Torchia, D.A., Wingfield, P., Stahl, S.J., Kaufman, J.D., Eyermann, C.J., Hodge, C.N., Lam, P.Y.S., Ru, Y., Jadhav, P.K., Chang, C.-H. and Weber, P.C., J. Am. Chem. Soc., 116, 10791 (1994).
10.   Wagner, G., Curr. Opin. Struct. Biol., 3, 748 (1993).
11.   Craik, D.J. *NMR in Drug Design* (CRC Press, New York, 1996).
12.   Muchmore, D.C., McIntosh, L.P., Russell, C.B., Anderson, D.E. and Dahlquist, F.W., Methods Enzymol., 177, 44 (1989).
13.   LeMaster, D.M., Methods Enzymol., 177, 23 (1989).
14.   Hansen, A.P., Petros, A.M., Mazar, A.P., Pederson, T.M., Rueter, A. and Fesik, S.W., Biochemistry, 31, 12713 (1992).
15.   Archer, S.J., Bax, A., Roberts, A.B., Sporn, M.B., Ogawa, Y., Piez, K.A., Weatherbee, J.A., Monica, L.-S.T., Lucas, R., Zheng, B.-L., Wenker, J. and Torchia, D.A., Biochemistry, 32, 1152 (1993).
16.   Sreekrishna, K. in *Industrial Microorganisms: Basic and Applied Molecular Genetics* (eds. Baltz, R.H., Hegaman, G.D. & Skatrud, P.L.) (American Society for Microbiology, Washington, D.C., 1993).
17.   Luckow, V.A. in *Principles and Practice of Protein Engineering* (eds. Cleland, J.L. & Craik, C.S.) (John Wiley and Sons, New York, 1995).
18.   Clore, G.M. and Gronenborn, A.M., J. Magn. Reson., 48, 402 (1982).
19.   Lian, L.Y., Barsukov, I.L., Sutcliffe, M.J., Sze, K.H. and Roberts, G.C.K., Methods Enzymol., 239, 657 (1994).
20.   Weber, C., Wider, G., Freyberg, B.v., Traber, R., Braun, W., Widmer, H. and Wuthrich, K., Biochemistry, 30, 6563 (1991).
21.   Fesik, S.W., R. T. Gampe, J., Eaton, H.L., Gemmecker, G., Olejniczak, E.T., Neri, P., Holzman, T.F., Egan, D.A., Edalji, R., Simmer, R., Helfrich, R., Hochlowski, J. and Jackson, M., Biochemistry, 30, 6574 (1991).
22.   Birdsall, B., Feeney, J., Tendler, S.J.B., Hammond, S.J. and Roberts, G.C.K., Biochemistry, 28, 2297 (1989).
23.   Selinsky, B.S., Perlman, M.E., London, R.E., Unkefer, C.J., Mitchell, J. and Blakley, R.L., Biochemistry, 29, 1290 (1990).
24.   Cheung, H.T.A., Birdsall, B., Frenkiel, T.A., Chau, D.D. and Feeney, J., Biochemistry, 32, 6846 (1993).
25.   Seeholzer, S.H., Cohn, M., Putkey, J.A., Means, A.R. and Crespi, H.L., Proc. Natl. Acad. Sci. U.S.A., 83, 3634 (1986).
26.   Hsu, V.L. and Armitage, I.M., Biochemistry, 31, 12778 (1992).
27.   Ikura, M. and Bax, A., J. Am. Chem. Soc., 114, 2433 (1992).
28.   Wider, G., Weber, C., Traber, R., Widmer, H. and Wüthrich, K., J. Am. Chem. Soc., 112, 9015 (1990).
29.   Ikura, M., Clore, G.M., Gronenborn, A.M., Zhu, G., Klee, C.B. and Bax, A., Science, 256, 632 (1992).
30.   Cavanagh, J., Fairbrother, W.J., Palmer, A.G. and Skelton, N.J. in *Protein NMR Spectroscopy Principles and Practice* 1-587 (Academic Press, New York, 1996).
31.   Kay, L.E., Curr. Opin. Struct. Biol., 5, 674 (1995).
32.   Keeler, J., Clowes, R.T., Davis, A.L. and Laue, E.D., Methods Enzymol., 239, 145 (1994).

33.   Zhu, G. and Bax, A., J. Magn. Reson., 90, 405 (1990).
34.   Zhu, G. and Bax, A., J. Magn. Reson., 98, 192 (1992).
35.   Laue, E.D., Mayger, M.R., Skilling, J. and Staunton, J., J. Magn. Reson., 68, 14 (1986).
36.   Zhou and Fesik, S., Nature, 378, 584 (1995).
37.   Kay, L., Marion, D. and Bax, A., J. Magn. Reson., 84, 72 (1989).
38.   Kay, L.E., Clore, G.M., Bax, A. and Gronenborn, A.M., Science, 249, 411 (1990).
39.   Zuiderweg, E.R.P., Petros, A.M., Fesik, S.W. and Olejniczak, E.T., J. Am. Chem. Soc., 113, 370 (1991).
40.   Vuister, G.W., Clore, G.M., Gronenborn, A.M., Powers, R., Garrett, D.S., Tschudin, R. and Bax, A., J.
      Magn. Reson., 101, 210 (1993).
41.   Torchia, D.A., Sparks, S.W. and Bax, A., Biochemistry, 27, 5135 (1988).
42.   Venters, R.A., Metzler, W.J., Spicer, L.D., Mueller, L. and II, B.T.F., J. Am. Chem. Soc., 117, 9592
      (1995).
43.   Grzesiek, S., Wingfield, P., Stahl, S., Kaufman, J.D. and Bax, A., J. Am. Chem. Soc., 117, 9594 (1995).
44.   Bax, A. and Davis, D.G., J. Magn. Reson., 63, 207 (1985).
45.   Bax, A., Vuister, G.W., Grzesiek, S., Delaglio, F., Wang, A.C., Tschudin, R. and Zhu, G., Methods
      Enzymol., 239, 79 (1994).
46.   Wang, A.C. and Bax, A., J. Am. Chem. Soc., 118, 2483 (1996).
47.   Biamonti, C., Rios, C., Lyons, B. and Montelione, G., Adv. Biophys. Chem., 4, 51 (1994).
48.   Rexroth, A., Schmidt, P., Szalma, S., Geppert, T., Schwalbe, H. and Griesinger, C., J. Am. Chem. Soc.,
      117, 10389 (1995).
49.   Englander, S.W. and Mayne, L., Ann. Rev. Biophys. Biomol. Struct., 21, 243 (1992).
50.   Spera, S., Ikura, M. and Bax, A., J. Biomol. NMR, 1, 155 (1991).
51.   Gemmecker, G., Jahnke, W. and Kessler, H., J. Am. Chem. Soc., 115, 11620 (1993).
52.   Grzesiek, S. and Bax, A., J. Biomol. NMR, 3, 627 (1993).
53.   Wishart, D.S., Sykes, B.D. and Richards, F.M., Biochemistry, 31, 1647 (1992).
54.   Spera, S. and Bax, A., J. Am. Chem. Soc., 113, 5490 (1991).
55.   Wishart, D.S. and Sykes, B.D., Methods Enzymol., 239, 363 (1994).
56.   de Dios, A.C., Pearson, J.G. and Oldfield, E., Science, 260, 1491 (1993).
57.   Archer, S.J., Vinson, V.K., Pollard, T.D. and Torchia, D.A., FEBS Lett., 337, 145 (1994).
58.   Chen, Y., Reizer, J., Saier, M.H., Fairbrother, W.J. and Wright, P.E., Biochemistry, 32, 32 (1993).
59.   Mayne, L., Paterson, Y., Cerasoli, D. and Englander, S.W., Biochemistry, 31, 10678 (1992).
60.   Vuister, G.W., Kim, S.-J., Wu, C. and Bax, A., J. Am. Chem. Soc., 116, 9206 (1994).
61.   Nilges, M., Clore, G.M. and Gronenborn, A.M., FEBS Lett., 229, 317 (1988).
62.   Nilges, M., Proteins, 17, 297 (1993).
63.   Nilges, M., J. Mol. Biol., 245, 645 (1995).
64.   Malliavin, T. E., Rouh, A., Delsuc, M. A. and Lallemand, J., Y. Compt. Rend. Acad. Sci. series II, 315,
      653 (1992).
65.   Kraulis, P.J., J. Mol. Biol., 243, 696 (1995).
66.   Hendrickson, W.A. and Wuthrich, K. Macromolecular Structures 1994 (Current Biology, Ltd., 1994).
67.   Markley, J.L., Acc. Chem. Res., 8, 70 (1975).
68.   Pelton, J.G., Torchia, D.A., Meadow, N.D. and Roseman, S., Protein Sci., 2, 543 (1993).
69.   Gooley, P.R., Johnson, B.A., Marcy, A.I., Cuca, G.C., Salowe, S.P., Hagmann, W.K., Esser, C.K. and
      Springer, J.P., Biochemistry, 32, 13098 (1993).
70.   Qin, J., Clore, G.M. and Gronenborn, A.M., Biochemistry, 35, 7 (1996).
71.   Jeng, M.-F. and Dyson, H.J., Biochemistry, 35, 1 (1996).
72.   Kriwacki, R.W., Hill, R.B., Flanagan, J.M., Caradonna, J.P. and Prestegard, J.H., J. Am. Chem. Soc.,
      115, 8907 (1993).
73.   Clore, G.M., Bax, A., Omichinski, J.G. and Gronenborn, A.M., Structure, 2, 89 (1994).
74.   Grzesiek, S., J. Am. Chem. Soc., 116, 1581 (1994).
75.   Barbato, G., Ikura, M., Kay, L.E., Pastor, R. and Bax, A., Biochemistry, 31, 5269 (1992).
76.   Nicholson, L.K., Yamazaki, T., Torchia, D.A., Grzesiek, S., Bax, A., Stahl, S.J., Kaufman, J.D., Wingfield,
      P.T., Lam, P.Y., Jadhav, P.K., Hodge, C.N., Domaille, P.J. and Chang, C., Nature Structural Biology, 2,
      274 (1995).
77.   Kraulis, P.J., Domaille, P.J., Campbell-Burk, S.L., Vanaken, T. and Laue, E.D., Biochemistry, 33, 3515 (
      1994).
78.   Hardman, C.H., Braodhurst, R.W., Raine, A.R.C., Grasser, K.D., Thomas, J.O. and Laue, E.D.,
      Biochemistry, 34, 16596 (1995).
79.   Tjandra, N., Feller, S.E., Pastor, R.W. and Bax, A., J. Am. Chem. Soc., 117, 12562 (1995).
80.   Tolman, J.R., Flanagan, J.M., Kennedy, M.A. and Prestegard, J.H., Proc. Natl. Acad. Sci., 92, 9279
      (1995).
81.   Muhandiram, D.R., Yamazaki, T., Sykes, B.D. and Kay, L.E., J. Am. Chem. Soc., 117, 11536 (1995).
82.   Kay, L.E., Muhandiram, D.R., Farrow, N.A., Aubin, Y. and Forman-Kay, J.D., Biochemistry, 35, 361
      (1996).

Chapter 31. Solid-Phase Synthesis: Applications to Combinatorial Libraries

Ingrid C. Choong and Jonathan A. Ellman
Department of Chemistry
University of California
Berkeley, California 94078

Introduction - The synthesis and evaluation of libraries of compounds has become a powerful method for the identification and optimization of lead structures. Since the last chapter in this series on the generation of moleculer diversity (1), the emphasis of library synthesis and evaluation strategies has shifted from peptides (2-6) and oligonucleotides (7) to small molecules, due to the generally unfavorable pharmacokinetic properties of these biopolymers. Small-molecule libraries have been prepared both in solution (8-13) and by employing solid-phase synthesis (14-17). Although interesting structures have been accessed by solution synthesis and lead compounds have been identified and optimized, solid-phase synthesis continues to be employed in the majority of the reported small-molecule library efforts. The focus of this chapter is to provide a comprehensive survey of templates and skeletons for which a solid-phase synthesis has been reported for library synthesis. A survey of reactions that have been performed on solid support is also provided. In addition, some issues specific to solid-phase synthesis are discussed, such as new linkage strategies and analytical techniques.

Summary of Methods for Small-Molecule Library Synthesis - The strategy that is employed to prepare and screen a library often will place constraints upon the types of molecules that can be synthesized and on the types of linkers that can be employed to attach the compound to the solid support. For this reason, the major strategies for library synthesis will be briefly summarized. Libraries of discrete compounds have been prepared by parallel synthesis in a spatially addressable fashion, where compound location indicates the compound structure (18-21). Libraries have also been prepared as pools of compounds, where pool sizes range from 10 to greater than 1000 compounds. The identification of the compound(s) responsible for activity in the most active pool(s) typically requires the iterative resynthesis and reevaluation of progressively smaller pools (3,22), although pooling strategies have been developed that may provide direct compound identification (23-27). Employing the split synthesis method for library synthesis, only one compound is produced per resin bead (28,29). If screening is performed such that the compound is localized with the resin bead (support-bound assays (30,31), assays of individual compounds in solution (32), and assays of compound mixtures (33,34) have all been reported), then structural determination provides a means of identifying the active compound. For peptides and oligonucleotides rapid and sensitive sequencing strategies are available for determining the compound structure. In contrast, sequencing methods for small molecules are currently not available, although a mass spectrometry-based method for sequencing nonbiological oligomers has been reported (35). A number of methods for introducing an encoding tag upon the resin bead in parallel with compound synthesis has been developed, with the encoding steps providing a history of the synthesis sequence. Peptide- (36,37) and oligonucleotide- (38,39) based encoding strategies were the first to be reported. In order to expand the types of chemistry that could be employed with encoding strategies, chemically inert haloaromatic compounds have been developed as the encoding tags, which are identified by gas chromatography with electron-capture detection (40,41). Two groups have also recently reported a radiofrequency method for encoding the synthesis sequence by employing microtransmitters that are encased by or are localized with the polymer support (42,43). Mass spectrometry (35,44-49) and NMR (50) can also be performed upon compounds on single beads (vide infra).

BACKBONES AND TEMPLATES

The backbones or templates for which solid-phase synthesis has been reported are categorized as non-peptide oligomers, privileged structures (51), novel templates, designed templates targeted toward a particular receptor or enzyme class, and structures for lead optimization.

Nonpeptide Oligomers - The solid-phase synthesis of a number of oligomers has been reported. The oligomer backbones were in general designed toward providing improved pharmacokinetic profiles, metabolic stability, and cell and membrane permeability relative to the natural biopolymers. The synthesis and evaluation of libraries of oligo(N-substituted) glycines, or "peptoids" 1, were the first to be reported. In the initial report, peptoids were synthesized from Fmoc-protected N-alkyl glycines as the monomer units (52). In order to expand the range of commercially available sidechains, a submonomer approach was developed whereby α-bromo acetic acid is employed to extend the backbone, and bromide displacement by primary amines serves to introduce the sidechains (53,54). A number of strategies for the post-synthesis modification of peptoids have also been employed to further expand the display of functionality (*vide infra*) (55-57). An oligocarbamate library **2** was synthesized using preformed *p*-nitrophenyl mixed carbonate monomers, which are prepared from *N*-protected α-amino alcohols and *p*-nitrophenyl chloroformate (58,59). The solid-phase synthesis of oligoureas **3** was reported using 1,2-diamines which are monoprotected as phthalimides. Activation and coupling is accomplished by *in situ* conversion to the isocyanates (60). The solid-phase synthesis of oligo(vinylogous-sulfonyl) peptides **4** using *N*-Boc γ-amino vinylsulfonyl chloride monomers, which are prepared from the corresponding *N*-Boc α-amino aldehydes in two solution steps (61), has also been reported. Initial studies toward the synthesis of oligosulfones and oligosulfoxides have also been described (59).

A number of researchers have reported the synthesis of libraries of oligomers based upon an amide backbone where monomers contain protected amine and carboxylic acid functionalities, and where the amine terminus is capped with a variety of different acylating agents (62-64). The post-synthesis modification of support-bound peptide libraries by exhaustive amide *N*-alkylation to provide *N*-alkyl peptide libraries has also been reported (65).

Privileged Structures - Privileged structures are templates that have previously provided potent therapeutic agents against a number of different receptor or enzyme targets. For example, the solid-phase synthesis of a library of 1,4-benzodiazepin-2-ones **5** has been reported using 2-aminobenzophenones, *N*-Fmoc-α-amino acids, and alkylating agents as the building blocks (20,66). This benzodiazepine synthesis sequence was later modified to introduce more diversity by using a support-bound *N*-Bpoc-2-aminoaryl stannane as the core upon which acid chlorides, *N*-Fmoc α-amino acids and alkylating agents are employed to display functionality (67,68). A different approach toward the synthesis of 1,4-benzodiazepin-2-ones was reported where α-amino acids attached to the support as α-amino esters and 2-aminobenzophenone imines were employed as the building blocks (18,69). The solid-phase synthesis of another class of benzodiazepines, the 1,4-benzodiazepine-2,5-diones **6**, has been demonstrated using anthranilic acids, α-amino esters and alkylating agents (70). An alternative route to 1,4-benzodiazepine-2,5-diones **7** from the *N*-terminus of a peptoid employs α-amino esters and *o*-azidobenzoyl chlorides (56). The solid-phase synthesis of isoquinolinones **8** employs a modification of the peptoid submonomer approach. 4-Bromo-*trans*-2-butenoic acid is attached to the support through an amide linkage. Bromide displacement with a primary amine followed by acylation with a 2-iodobenzoyl chloride with final Heck cyclization provides the support-bound isoquinolinone product (57).

A library of diketopiperazines **9** has been prepared by employing amino acids and aldehydes to introduce diversity (71). Recently the preparation of 1,4-dihydropyridines **10** where diversity can be introduced at five different positions employing β-keto esters and aromatic aldehydes (72) has been reported. Another privileged class of heterocycles, the Biginelli dihydropyrimidines **11**, has been constructed on solid support from β-keto esters, aldehydes, and ureas in a one-pot reaction (40). The solid-phase synthesis of dihydro- and tetrahydroisoquinolines **12** has also been demonstrated (73).

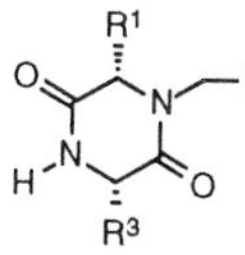

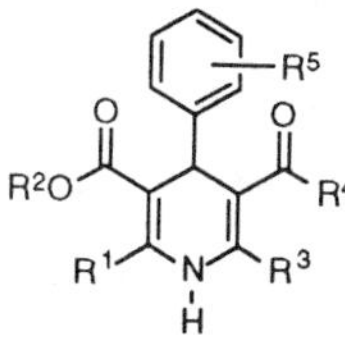

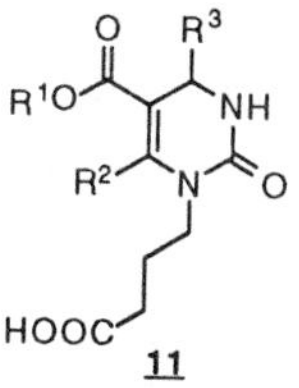

Five- and four-membered heterocycles have been prepared on solid support. Libraries of hydantoins **13** have been constructed from support-bound α-amino esters and isocyanates (18) and from primary amines and N-alkyl-amino acids which are attached to the support through a carbamate linkage (74). The synthesis of a library of N-acyl pyrrolidines **14** employing the azomethine ylide cycloaddition reaction has been reported. Support-bound α-amino esters are condensed with aromatic aldehydes followed by Lewis acid-mediated cycloaddition with acrylates. The support-bound pyrrolidine products are then acylated with mercaptoacyl chlorides (75). Imidazoles **15** have been synthesized employing ammonium acetate, diketones, primary amines, and aldehydes (76). Methods were reported whereby each of the different building blocks could be attached to the solid support. Most recently, highly functionalized pyrazoles **16** and isoxazoles **17** have been synthesized on solid support by condensing an ester with support-bound acetyl functionality. α-Alkylation followed by ring closure upon treatment with hydrazine or hydroxylamine provides the heterocyclic products (77). The preparation of β-lactams **18** using support-bound α-amino esters, aldehydes, and ketenes generated *in situ* by treatment of monosubstituted acyl chlorides with triethylamine has also been demonstrated (78).

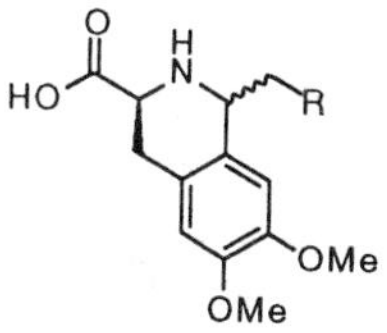

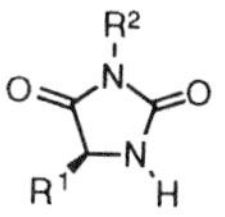

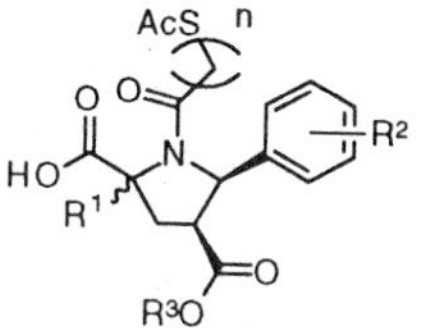

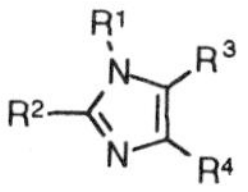

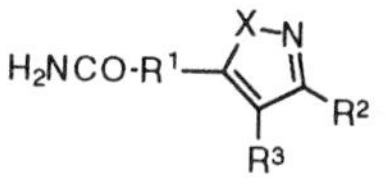

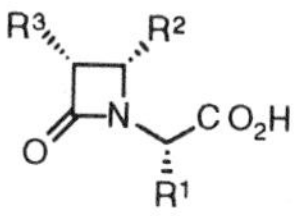

Novel Templates - Novel templates are structures upon which relatively few therapeutic agents have been based, but which are straightforward to synthesize and have the potential for the display of functionality at multiple sites in the molecule.

Libraries have been generated based upon a dihydrobenzopyran core prepared by condensation of support-bound dihydroxyacetophenones and ketones. The final library members **19** and **20** are then prepared by functionalization of the dihydrobenzopyran core and/or amine functionality introduced with the ketone component (79). The 1,2,3,4-tetrahydro-β-carbolines derivatives **21** were prepared through a support-bound diamine (only piperazine was shown) which was coupled with a tryptamine derivative either by acylation or reductive amination. The final product was then obtained by condensation with an aldehyde followed by Pictet-Spengler cyclization (80). The solid-phase synthesis of bicyclo[2.2.2]octane derivatives **22** has been demonstrated by tandem Michael additions between support-bound acrylate esters and α,β-unsaturated cyclohexenones (81). Reduction or reductive amination of the initial ketone product is followed by cleavage from the support by treatment with acid, aminolysis, or reduction to provide acid, amide, or alcohol products, respectively.

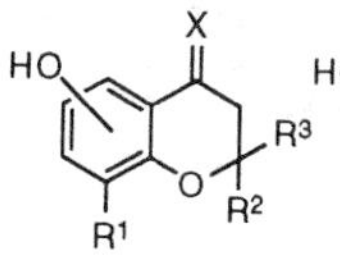

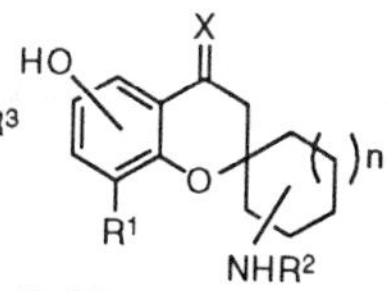

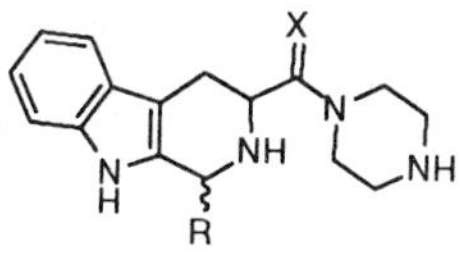

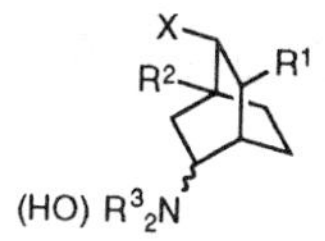

Novel five-membered ring templates have also been reported. 4-Thiazolidinones **23** have been prepared by condensing support-bound α-amino esters with aromatic aldehydes and mercaptoacetic acids (82). The synthesis of a number of N-acyl thiazolidines **24** has also been accomplished. Cysteine, which is linked to the support as an ester, is first condensed with an aldehyde (83). The thiazolidine product is then acylated and can also be oxidized to the sulfoxide. Amide bond formation has been employed to display functionality off of a rigid cyclopentane scaffold **25** (84) as well as off of Kemp's triacid **26** (85).

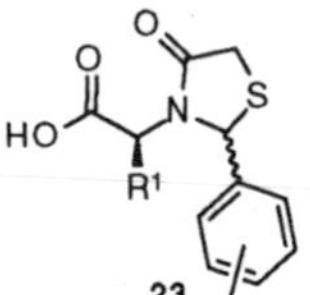

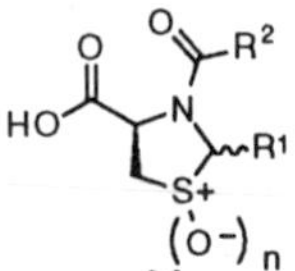

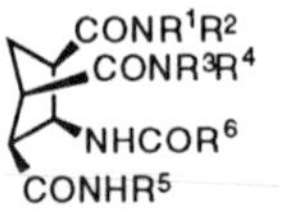

Designed Templates - These designed templates are based upon key recognition motifs for specific receptor or enzyme classes. A number of enzymes have been targeted by the display of functionality off isosteres that correspond to the transition state or intermediate of an enzyme catalyzed reaction. In an early study, statine **27** was incorporated in a library of tetrapeptides (86) for the purpose of identifying inhibitors of HIV-1 protease. A solid-phase method for incorporating a hydroxyethylamine isostere **28** into peptide derivatives has also been demonstrated (87). There are two reports of methods for displaying nonpeptide functionality about isosteres corresponding to the intermediate of peptide hydrolysis as catalyzed by the aspartic acid protease class. In one method, a core element is attached to the solid support such that different functionality could be displayed by amine displacement and two acylation steps **29** (88). Similarly, two C_2-symmetric isosteres have been employed to

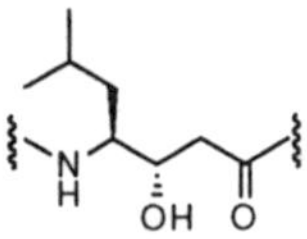

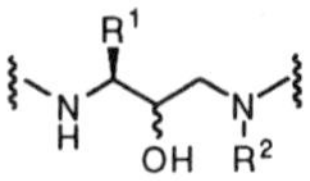

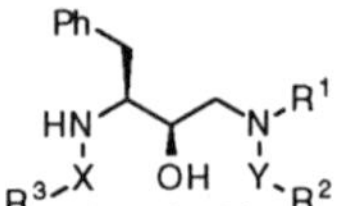

display nonpeptide functionality (**30** and **31**) (89). A peptide library incorporating the phosphonic acid transition state isostere **32** has been constructed (90). This library, which employs α-amino acids, α-hydroxy acids, and α-aminoalkylphosphonic acids as its building blocks, was synthesized for the purpose of targeting metalloproteases. A library of tetrapeptides incorporating a phosphinic acid transition state isostere **33** to target zinc metalloproteases has also been reported (91).

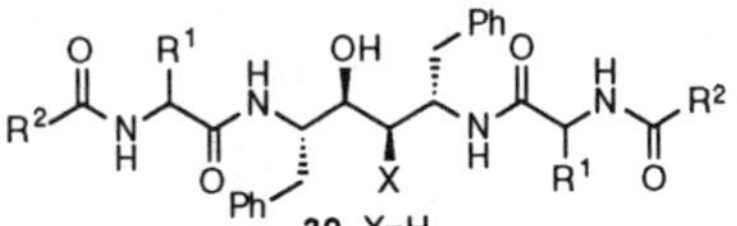

A library of β-turn mimetics **34** has been synthesized (19) using α-amino acids, aminoalkyl thiols, and α-halo acids. Potential antioxidants have also been synthesized employing carboxylic acids and aldehydes to construct a 27-membered 1,3-propanediol library **35** (44). A library of 1.1 billion amide-based oligomers directed against the SH3 domain of the tyrosine kinase Src has been constructed. Each of the monomers contained protected amine and carboxylic acid functionality, and the amine terminus was capped with a variety of different acylating agents (63). Oligomer **36** showed the highest affinity with a K_d of 3.4 μM.

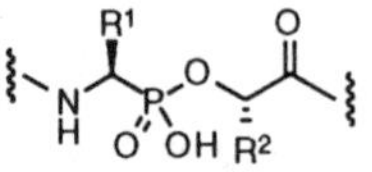

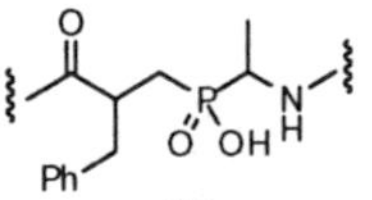

34   **35**   **36**

Lead Optimization Structures - Small-molecule libraries have also been synthesized for the purpose of lead optimization. A library of 60 compounds based on the structure of lavendustin A **37**, a known inhibitor of tyrosine kinases, employing carboxyl substituted anilines, aromatic aldehydes, and aromatic alkylating agents has been synthesized (92). The design of a library based on lavendustin A and balanol **38**, a serine/threonine kinase inhibitor has been reported (21). Using acid chlorides, isocyanates, and sulfonyl chlorides as the building blocks, a phenol library of 384 compounds was synthesized. The solid-phase synthesis of 23 hydroxystilbene derivatives **39** by coupling hydroxybenzaldehydes and benzylphosphonate anions to target the estrogen receptor has also been reported (93). Based upon the structure of captopril, the azomethine chemistry described previously (75) was employed to synthesize a library of captopril analogs **14**. This resulted in the identification of a novel inhibitor of angiotensin converting enzyme with a K_i of 160 pM.

37   **38**   **39**

REACTIONS ON SUPPORT

Many reactions have been demonstrated on the solid-support and are surveyed below. These reactions are categorized into carbon-heteroatom bond forming reactions, carbon-carbon bond forming reactions, cycloadditions, and multicomponent reactions. Literature prior to 1990 is not included, since much of this work has been reviewed (94,95).

Carbon-Heteroatom Bond Forming Reactions - The Mitsunobu coupling reaction has been employed to prepare ethers from phenols and alcohols (96,97). Either the alcohol or the phenol may be attached to the solid support. A modified Mitsunobu reaction was also employed to couple support-bound alcohols to phosphonic acids in the synthesis of peptidylphosphonates **32** (90). The conjugate addition of thiophenol derivatives to α,β-unsaturated ketones has been employed to prepare thioether bonds (44). Thioalkylation of alkyl halides has also been reported by a number of researchers (59), including for macrocyclizations (19,98). Reductive amination has been used extensively to couple amines and aldehydes or ketones. In recent work, a study to determine optimal reductive amination conditions has been reported (99). N-Alkylation of amides (20,66,67,70) and amines (19,53,54,70,88) have also been employed in a number of small-molecule synthesis efforts. $S_N Ar$ substitutions upon aromatic halides with phenylpiperazine have been demonstrated (100). Urea formation has also been reported (101) through activation of a support-bound amine as the nitrophenyl carbonate followed by amine addition, and three groups have coupled support-bound amines with isocyanates (18,21,88). Intramolecular aza-Wittig reactions were employed in the synthesis of 1,4-benzodiazepine-2,5-diones (56). A novel method for support-mediated amide formation has been reported (102). Polymer-bound 1-ethyl-3-(3-dimethylaminopropyl)-carbodiimide (P-EDC) has been employed to couple an acid with a limiting amount of an amine. Because unreacted acid remains on the polymer, pure product can be isolated by filtration and evaporation. Polymer-bound reagents, which have been extensively developed (103), will likely be further exploited in library synthesis.

Several elegant approaches toward the solid-phase synthesis of oligosaccharides have been reported (104-107), including methods for the solid-phase synthesis of peptide-oligosaccharide hybrids (105), and enzyme-mediated glycosylations (107).

Carbon-Carbon Bond Forming Reactions - Many carbon-carbon bond forming reactions have also been reported in the literature. Palladium-mediated reactions are appealing for generating diversity due to the mild reaction conditions, compatability with a range of functionality, and high yields. The Suzuki cross-coupling reaction (108,109), the Stille cross-coupling reaction (67,110,111), and the Heck reaction (57,112,113) have all been employed, including Heck-mediated macrocyclizations (114). Wittig and Horner-Wadsworth-Emmons reactions have been used extensively (44,93,115). Diastereoselective enolate alkylations have been reported (116,117), and acylsulfonamide "safety-catch" linker-based enolate alkylations have been demonstrated (109). Tandem Michael additions have been employed to prepare bicyclo[2.2.2]octane derivatives **22** (75). Aldol reactions employing zinc enolates have also been reported (44). Pictet-Spengler (80) and Bischler-Napieralski cyclization reactions (73) have been employed to prepare 1,2,3,4-tetrahydro-β-carbolines **21** and isoquinoline derivatives **12**, respectively. Grignard and organolithium additions to Weinreb-type amides have been reported in the solid-phase synthesis of ketones. Aldehydes have similarly been prepared through $LiAlH_4$ and DIBAL additions (118).

Cycloadditions - There have been several reports of cycloadditions performed on solid support. Azomethine cycloadditions and [2+2] ketene-imine cycloadditions have been employed to prepare pyrrolidines **14** (75) and β-lactams **18** (78), respectively, as described previously. Nitrile oxide-based [3+2] cycloadditions of alkenes or alkynes have also been reported to provide isoxazoles and isoxazoline, respectively (55,119,120). The Pauson-Khand cycloaddition reaction involving norbornene derivatives and support-bound alkynes has also been demonstrated (121).

Multicomponent Reactions - Multicomponent reactions provide a fast and efficient route to libraries of structurally diverse compounds (122). The previously described syntheses of 4-thiazolidinones **23** (82) and Bignelli dihydropyrimidines **11** (123) utilize three-component condensation reactions, and the synthesis of imidazoles **15** (76) involves a four-component coupling step. In the synthesis of 1,4-dihydropyridines **10**, two- or three-component cyclocondensations of support-bound eneamino esters with 2-arylidene β-keto esters, or β-keto esters and aldehydes, respectively, are performed (72). The Ugi reaction has also been demonstrated on solid support using support-bound amines, isocyanides, carboxylic acids and aldehydes (124). The use of 1-isocyanocyclohexene as a convertible isocyanide for the generation of more varied Ugi reaction products has recently been reported as well (125). Unsymmetrical phthalocyanines have also been prepared by converting support-bound phthalonitriles to diiminoisoindolines followed by condensation with three equivalents of diiminoisoindoline free in solution (126).

LINKER STRATEGIES

The linkage element for covalently attaching a compound to the solid support deserves special consideration. The linker should be stable to the reaction conditions employed in the synthesis sequence, but should be cleavable under conditions that do not degrade the product. In many small-molecule solid-phase synthesis efforts, carboxylic acid-based linkers developed for peptide synthesis have been used to provide acid or amide products. These linkers have been extensively reviewed in the peptide literature (127) and will not be summarized here. Alcohols have been linked to the solid support through ether linkages, such as tetrahydropyranyl (88,128,129), trityl (44,130), and silyl linkers (131,132), as well as through ester linkages (21). Aldehydes have been linked to the solid support as acetals (89), and several researchers have linked amines to the solid support employing carbamate functionality (74,80,133). Most recently, sulfonyl chlorides have been linked to the support through a sulfonamide linkage. Treatment with acid affords functionalized sulfonamides (134).

Linkage to the solid support is generally accomplished using a functional group that is part of the targeted molecule or that is appended onto the structure. Residual functionality which remains on the molecule after cleavage from the support can have deleterious effects upon biological activity toward certain receptor or enzyme targets. Several groups have therefore developed strategies to attach compounds to the solid support that leave no trace after cleavage from the support. There are two independent reports of silicon-based linkers for aromatic compounds which are cleaved by protodesilylation (131,132). A photochemical method for the cleavage of benzyl thioether linkages has also been described that results in a methyl substituent on the final molecule (135).

The use of linkage agents often results in additional chemical steps and can complicate small-molecule synthesis on solid support relative to the corresponding synthesis in solution. However, for many applications the linkage agent can be used advantageously. For example, when screens are performed upon compound mixtures resulting from cleavage of the compound from a collection of beads prepared by split synthesis, iterative evaluation of progressively smaller compound mixtures can be accomplished without resynthesis if sequential partial release of compounds from the individual beads is employed. Two strategies have been developed for this purpose. In the first method, multiple linkers are used which are cleaved under different reaction conditions (34). In the second method, photolabile linkers are employed where partial release of the compounds is controlled photochemically (62). Photolabile linkers that provide amides (136,137), carboxylic acids (62), and alcohols (138) have been reported.

Linkers have also been developed that are stable through a given synthesis sequence, but which can be activated for nucleophilic cleavage at the end of the synthesis to introduce additional diversity into the final molecule. Kenner's "safety-catch" linker (139) has been applied to the synthesis of aryl acetic acids and amides. Treatment with diazomethane provides the activated N-methyl acylsulfonamide which is subsequently treated with hydroxide or amine nucleophiles (109). This method was later modified to greatly enhance the level of reactivity of the activated linker by treatment of the acylsulfonamide with iodoacetonitrile to provide the N-cyanomethyl acylsulfonamide (140). Addition of limiting amounts of amine nucleophiles relative to the support-bound material results in complete consumption of the amine to provide pure amide products. The previously described addition of Grignard reagents to support-bound Weinreb amides releases ketone products into solution that also contain an added site of diveristy (118).

Cyclative cleavage strategies have also been reported in the solid-phase synthesis of hydantoins (18,74) and 1,4-benzodiazepines (18). In these strategies, the penultimate support-bound intermediate possesses functionality which, when unmasked or activated, cyclizes to release the product into solution. Truncated intermediates and/or side-products generally cannot undergo cyclization and remain attached to the resin, thus providing pure final product. Kurth has also employed cyclative release in his synthesis of lactones (116,117) and oxazoles (119,120).

ANALYTICAL TECHNIQUES

Many different analytical techniques are currently being used to evaluate solid-phase reactions or compounds bound to the solid support. Functional group titration is often used to monitor reactions. This includes ninhydrin (141), picric acid (142), trinitrobenzenesulfonic acid (143), and bromophenol blue test (144) for free amines, Ellman's test (145) for free thiols, and nitrophenylisothiocyanate-O-trityl (146) for sterically hindered or nonbasic amines. Another method includes quantitation of UV-active byproducts resulting from protecting group cleavage, such as Fmoc group cleavage with piperidine (127,147), dimethoxytrityl ether cleavage with acid (148), and nitrophenylethyloxy group cleavage with base (90). Elemental analysis has been used to quantitate resin-loading levels for a variety of atoms, and Volhard titration (149) has been used to quantitate chloride substitution levels in particular.

Mass spectroscopy has been used to characterize reaction products from single beads (44-49) and can be used in conjunction with other analytical techniques (HPLC-MS, GC-MS, MS-MS) to characterize a library (44,71,150) (see Chapter **32** in this Reports issue). NMR techniques have also been used to characterize support-bound compounds, including ^{13}C spectra of compounds attached to tentagel resin (151,152), magic angle spinning (MAS) solid-state ^1H NMR (153), MAS ^{13}C-^1H correlation experiments (154), and MAS HMQC and TOCSY experiments (155,156). Notably, NMR (50) and IR (157) have been used to characterize support-bound compounds on single resin beads.

Conclusion - Many advances continue to be made in the synthesis of small-molecule libraries on solid supports, including the development of new support materials (158,159). Equally important efforts are being made in computational methods for library design and database management, analytical evaluation, automation and instrumentation for synthesis and purification, and methods for high-throughput biological evaluation.

References

1. W.H. Moos, G.D. Green, and M.R. Pavia, Annu.Rep.Med.Chem., 28, 315 (1993).
2. M.A. Gallop, R.W. Barrett, W.J. Dower, S.P.A. Fodor, and E.M. Gordon, J.Med.Chem., 37, 1233 (1994).
3. C. Pinilla, A. Appel, S. Blondelle, C. Dooley, B. Dörner, J. Eichler, J. Ostresh, and R.A. Houghten, Biopolymers, 37, 221 (1995).
4. M.R. Pavia, T.K. Sawyer, and W.H. Moos, BioMed.Chem.Lett., 3, 387 (1993).
5. G. Jung and A.G. Becksickinger, BioMed.Chem.Lett., 31, 387 (1992).
6. W.J. Dower and S.P.A. Fodor, Annu.Rep.Med.Chem., 26, 271 (1991).
7. D.J. Ecker, T.A. Vickers, R. Hanecak, V. Driver, and K. Anderson, Nucleic Acids Res., 21, 1853 (1993).
8. D.L. Boger, C.M. Tarby, P.L. Meyers, and L.H. Caporale, J.Am.Chem.Soc., 118, 2109 (1996).

9.   T.A. Keating and R.W. Armstrong, J.Am.Chem.Soc., 117, 7842 (1995).
10. P.W. Smith, J.Y.Q. Lai, A.R. Whittington, B. Cox, and J.G. Houston, BioMed.Chem.Lett., 4, 2821 (1994).
11. M.C. Pirrung and J. Chen, J.Am.Chem.Soc., 117, 1240 (1995).
12. T. Carell, E.A. Wintner, and J. Rebek, Angew.Chem.,Int.Ed.Engl., 33, 2061 (1994).
13. Y. Ding, O. Kanie, J. Labbe, M.M. Palcic, B. Ernst, and O. Hindsgaul, Synthesis and Biological Activity of Oligosaccharide Libraries, in: Glycoimmunology (A. Alaviad, and J. S. Axford, eds.), 1995, Plenum Press, In Press.
14. L.A. Thompson and J.A. Ellman, Chem.Rev., 96, 555 (1996).
15. J.S. Früchtel and G. Jung, Angew.Chem.,Int.Ed.Engl., 35, 17 (1996).
16. N.K. Terrett, M. Gardner, D.W. Gordon, R.J. Kobylecki, and J. Steele, Tetrahedron, 51, 8135 (1995).
17. E.M. Gordon, R.W. Barrett, W.J. Dower, S.P.A. Fodor, and M.A. Gallop, J.Med.Chem., 37, 1385 (1994).
18. S.H. DeWitt, J.S. Kiely, C.J. Stankovic, M.C. Schroeder, D.M.R. Cody, and M.R. Pavia, Proc.Natl.Acad.Sci.U.S.A., 90, 6909 (1993).
19. A.A. Virgilio and J.A. Ellman, J.Am.Chem.Soc., 116, 11580 (1994).
20. B.A. Bunin, M.J. Plunkett, and J.A. Ellman, Proc.Natl.Acad.Sci.U.S.A., 6, 4708 (1994).
21. H.V. Meyers, G.J. Dilley, T.L. Durgin, T.S. Powers, N.A. Winssinger, H. Zhu, and M.R. Pavia, Molecular Diversity, 1, 13 (1995).
22. R.A. Houghten, C. Pinilla, S.E. Blondelle, J.R. Appel, C.T. Dooley, and J.H. Cuervo, Nature, 354, 84 (1991).
23. C. Pinilla, J.R. Appel, P. Blanc, and R.A. Houghten, Biotechniques, 13, 901 (1992).
24. R.A. Houghten, Gene, 137, 7 (1993).
25. C. Pinilla, J.R. Appel, S.E. Blondelle, C.T. Dooley, J. Eichler, J.M. Ostresh, and R.A. Houghten, Drug Development Research, 33, 133 (1994).
26. C.T. Dooley and R.A. Houghten, Life Sciences, 52, (1993).
27. B. Déprez, X. Williard, L. Bourel, H. Coste, F. Hyafil, and A. Tartar, J.Am.Chem.Soc., 117, 5405 (1995).
28. K.S. Lam, S.E. Salmon, E.M. Hersh, V.J. Hruby, W.M. Kazmierski, and R.J. Knapp, Nature, 354, 82 (1991).
29. M. Lebl, V. Krchnak, N.F. Sepetov, B. Seligmann, P. Strop, S. Felder, and K.S. Lam, Biopolymers, 37, 177 (1995).
30. W.C. Still, Acc.Chem.Res., 29, 155 (1996).
31. K.S. Lam, S. Wade, F. Abdullatif, and M. Lebl, J.Immun.Methods, 5, 219 (1995).
32. J.M. Quillan, C.K. Jayawickreme, and M.R. Lerner, Proc.Natl.Acad.Sci.U.S.A., 92, 2894 (1995).
33. J.J. Burbaum, M.H.J. Ohlmeyer, J.C. Reader, I. Henderson, L.W. Dillard, G. Li, T.L. Randle, N.H. Sigal, D. Chelsky, and J.J. Baldwin, Proc.Natl.Acad.Sci.U.S.A., 92, 6027 (1995).
34. S.E. Salmon, K.S. Lam, M. Lebl, A. Kandola, P.S. Khattri, S. Wade, M. Pátek, P. Kocis, V. Krchnák, D. Thorpe, and S. Felder, Proc.Natl.Acad.Sci.U.S.A., 90, 11708 (1993).
35. R.S. Youngquist, G.R. Fuentes, M.P. Lacey, and T. Keough, J.Am.Chem.Soc., 117, 3900 (1995).
36. J.M. Kerr, S.C. Banville, and R.N. Zuckermann, J.Am.Chem.Soc., 115, 2529 (1993).
37. V. Nikolaiev, A. Stierandova, V. Krchnák, B. Seligmann, K.S. Lam, S.E. Salmon, and M. Lebl, Peptide Res., 6, 161 (1993).
38. S. Brenner and R.A. Lerner, Proc.Natl.Acad.Sci.U.S.A., 89, 5381 (1992).
39. M.C. Needels, D.G. Jones, E.H. Tate, G.L. Heinkel, L.M. Kochersperger, W.J. Dower, R.W. Barrett, and M.A. Gallop, Proc.Natl.Acad.Sci.U.S.A., 90, 10700 (1993).
40. M.H.J. Ohlmeyer, R.N. Swanson, L.W. Dillard, J.C. Reader, G. Asouline, R. Kobayashi, M. Wigler, and W.C. Still, Proc.Natl.Acad.Sci.U.S.A., 90, 10922 (1993).
41. H.P. Nestler, P.A. Bartlett, and W.C. Still, J.Org.Chem., 59, 4723 (1994).
42. K.C. Nicolaou, X. Xiao, Z. Parandoosh, A. Senyei, and M.P. Nova, Angew.Chem.,Int.Ed.Engl., 34, 2289 (1995).
43. E.J. Moran, S. Sarshar, J.F. Cargill, M.M. Shahbaz, A. Lio, A.M.M. Mjalli, and R.W. Armstrong, J.Am.Chem.Soc., 117, 10787 (1995).
44. C.X. Chen, L.A.A. Randall, R.B. Miller, A.D. Jones, and M.J. Kurth, J.Am.Chem.Soc., 116, 2661 (1994).
45. C.L. Brummel, I.N.W. Lee, Y. Zhou, S.J. Benkovic, and N. Winograd, Science, 264, 399 (1994).
46. R.S. Youngquist, G.R. Fuentes, M.P. Lacey, and T. Keough, Rap.Commun.Mass.Spec., 8, 77 (1994).
47. R.A. Zambias, D.A. Boulton, and P.R. Griffin, Tetrahedron Lett., 35, 4283 (1994).
48. B.B. Brown, D.S. Wagner, and H.M. Geysen, Molecular Diversity, 1, 4 (1995).
49. B.J. Egner, G.J. Langley, and M. Bradley, J.Org.Chem., 60, 2652 (1995).
50. S.K. Sarkar, R.S. Garigipati, J.L. Adams, and P.A. Keifer, J.Am.Chem.Soc., 118, 2305 (1996).
51. E.J. Ariens, A.J. Beld, J.F. Rodrigues de Miranda, and A.M. Simonis, in: The Receptors: A Comprehensive Treaty (R. D. O'Brien, ed.) 1, 1979, Plenum, New York, pp. 33.
52. R.J. Simon, R.S. Kania, R.N. Zuckermann, V.D. Huebner, D.A. Jewell, S. Banville, S. Ng, L. Wang, S. Rosenberg, C.K. Marlowe, D.C. Spellmeyer, A.D. Frankel, D.V. Santi, F.E. Cohen, and P.A. Bartlett, Proc.Natl.Acad.Sci.U.S.A., 89, 9367 (1992).
53. R.N. Zuckermann, J.M. Kerr, S.B.H. Kent, and W.H. Moos, J.Am.Chem.Soc., 11, 10646 (1992).
54. R.N. Zuckermann, E.J. Martin, D.C. Spellmeyer, G.B. Stauber, K.R. Shoemaker, J.M. Kerr, G.M. Figliozzi, D.A. Goff, M.A. Siani, R.J. Simon, S.C. Banville, E.G. Brown, L. Wang, L.S. Richter, and W.H. Moos, J.Med.Chem., 37, 2678 (1994).
55. Y.H. Pei and W.H. Moos, Tetrahedron Lett., 35, 5825 (1994).

56. D.A. Goff and R.N. Zuckermann, J.Org.Chem., 60, 5744 (1995).
57. D.A. Goff and R.N. Zuckermann, J.Org.Chem., 60, 5748 (1995).
58. C.Y. Cho, E.J. Moran, S.R. Cherry, J.C. Stephans, S.P.A. Fodor, C.L. Adams, A. Sundaram, J.W. Jacobs, and P.G. Schultz, Science, 261, 1303 (1993).
59. E.J. Moran, T.E. Wilson, C.Y. Cho, S.R. Cherry, and P.G. Schultz, Biopolymers, 37, 213 (1995).
60. K. Burgess, D.S. Linthicum, and H.W. Shin, Angew.Chem.,Int.Ed.Engl., 34, 907 (1995).
61. C. Gennari, H.P. Nestler, B. Salom, and W.C. Still, Angew.Chem.,Int.Ed.Engl., 34, 1763 (1995).
62. J.J. Baldwin, J.J. Burbaum, I. Henderson, and M.H.J. Ohlmeyer, J.Am.Chem.Soc., 117, 5588 (1995).
63. A.P. Combs, T.M. Kapoor, S. Feng, J.K. Chen, L.F. Daude-Snow, and S.L. Schreiber, J.Am.Chem.Soc., 118, 287 (1996).
64. N.K. Terrett, D. Bojanic, D. Brown, P.J. Bungay, M. Gardner, D.W. Gordon, C.J. Mayers, and J. Steele, BioMed.Chem.Lett., 5, 917 (1995).
65. J.M. Ostresh, G.M. Husar, S.E. Blondelle, B. Dörner, P.A. Weber, and R.A. Houghten, Proc.Natl.Acad.Sci.U.S.A., 91, 11138 (1994).
66. B.A. Bunin and J.A. Ellman, J.Am.Chem.Soc., 114, 10997 (1992).
67. M.J. Plunkett and J.A. Ellman, J.Am.Chem.Soc., 117, 3306 (1995).
68. B.A. Bunin, M.J. Plunkett, and J.A. Ellman, Synthesis and Evaluation of 1,4-Benzodiazepine Libraries, in: Methods in Enzymology 267, 1995, In press
69. D.R. Cody, S.H.H. DeWitt, J.C. Hodges, J.S. Kiely, W.H. Moos, M.R. Pavia, B.D. Roth, M.C. Schroeder, and C.J. Stankovic, 5,324,483, June 28, 1994, United States.
70. C.G. Boojamra, K.M. Burow, and J.A. Ellman, J.Org.Chem., 60, 5742 (1995).
71. D.W. Gordon and J. Steele, BioMed.Chem.Lett., 5, 47 (1995).
72. M.F. Gordeev, D.V. Patel, and E.M. Gordon, J.Org.Chem., (1995).
73. W.D.F. Meutermans and P.F. Alewood, Tetrahedron Lett., 36, 7709 (1995).
74. B.A. Dressman, L.A. Spangle, and S.W. Kaldor, Tetrahedron Lett., 37, 937 (1996).
75. M.M. Murphy, J.R. Schullek, E.M. Gordon, and M.A. Gallop, J.Am.Chem.Soc., 117, 7029 (1995).
76. S. Sarshar, D. Diev, and A.M.M. Mjalli, Tetrahedron Lett., 37, 835 (1996).
77. A.L. Marzinzik and E.R. Felder, Tetrahedron Lett., 37, 1003 (1996).
78. B. Ruhland, A. Bhandari, E.M. Gordon, and M.A. Gallop, J.Am.Chem.Soc., 118, 253 (1996).
79. J.C. Chabala, J.J. Baldwin, J.J. Burbaum, D. Chelsky, L.W. Dillard, I. Henderson, G. Li, H.J. Ohlmeyer, T.L. Randle, J.C. Reader, L. Rokosz, and N.H. Sigal, Drug Discovery and Optimization using Binary-Encoded Small Molecule Combinatorial Libraries, in: Genomes, Molecular Biology and Drug Discovery 1995, Academic Press, In Press
80. K. Kaljuste and A. Undén, Tetrahedron Lett., 36, 9211 (1995).
81. S.V. Ley, D.M. Mynett, and W. Koot, Synlett, 1017 (1995).
82. C.P. Holmes, J.P. Chinn, G.C. Look, E.M. Gordon, and M.A. Gallop, J.Org.Chem., 60, 7328 (1995).
83. M. Patek, B. Drake, and M. Lebl, Tetrahedron Lett., 36, 2227 (1995).
84. M. Patek, B. Drake, and M. Lebl, Tetrahedron Lett., 35, 9169 (1994).
85. P. Kocis, V. Krchnák, and M. Lebl, Tetrahedron Lett., 34, 7251 (1993).
86. R.A. Owens, P.D. Gesellchen, B.J. Houchins, and R.D. Dimarchi, Biochem.Biophys.Res.Commun., 181, 402 (1991).
87. P.F. Alewood, R.I. Dancer, B. Garnham, A. Jones, and S.B.H. Kent, Tetrahedron Lett., 33, 977 (1992).
88. E.K. Kick and J.A. Ellman, J.Med.Chem., 38, 1427 (1995).
89. G.T. Wang, S. Li, N. Wideburg, G.A. Krafft, and D.J. Kempf, J.Med.Chem., 38, 2995 (1995).
90. D.A. Campbell and J.C. Bermak, J.Am.Chem.Soc., 116, 6039 (1994).
91. J. Jirácek, A. Yiotakis, B. Vincent, A. Lecoq, A. Nicolaou, F. Checler, and V. Dive, J. Biol. Chem., 270, 21701 (1995).
92. J. Green, J.Org.Chem., 60, 4287 (1995).
93. R. Williard, V. Jammalamadaka, D. Zava, C.C. Benz, C.A. Hunt, P.J. Kushner, and T.S. Scanlan, Chemistry and Biology, 2, 45 (1995).
94. C.C. Leznoff, Acc.Chem.Res., 11, 327 (1978).
95. J.M.J. Fréchet, Tetrahedron, 37, 663 (1981).
96. T.A. Rano and K.T. Chapman, Tetrahedron Lett., 36, 3789 (1995).
97. V. Krchnák, Z. Flegelová, A.S. Weichsel, and M. Lebl, Tetrahedron Lett., 36, 6193 (1995).
98. P.L. Barker, S. Bullens, S. Bunting, D.J. Burdick, K.S. Chan, T. Deisher, C. Eigenbrot, T.R. Gadek, R. Gantzos, M.T. Lipari, C.D. Muir, M.A. Napier, R.M. Pitti, A. Padua, C. Quan, M. Stanley, M. Struble, J.Y.K. Tom, and J. Burnier, J.Med.Chem., 35, 2040 (1992).
99. A.M. Bray, D.S. Chiefari, R.M. Valerio, and N.J. Maeji, Tetrahedron Lett., 36, 5081 (1995).
100. S.M. Dankwardt, S.R. Newman, and J.L. Krstenansky, Tetrahedron Lett., 36, 4923 (1995).
101. S.M. Hutchins and K.T. Chapman, Tetrahedron Lett., 36, 2583 (1995).
102. M.C. Desai and L.M.S. Stramiello, Tetrahedron Lett., 34, 7685 (1993).
103. P. Hodge, 1980, Synthesis and Separations Using Functional Polymers (D. C. Sherrington, ed.), John Wiley and Sons, New York.
104. S.J. Danishefsky, K.F. McClure, J.T. Randolph, and R.B. Ruggeri, Science, 260, 1307 (1993).
105. J.Y. Roberge, X. Beebe, and S.J. Danishefsky, Science, 269, 202 (1995).
106. L. Yan, C.M. Taylor, R. Goodnow, and D. Kahne, J.Am.Chem.Soc., 116, 6953 (1994).

107.   R.L. Halcomb, H. Huang, and C. Wong, J.Am.Chem.Soc., 116, 11315 (1994).
108.   R. Frenette and R.W. Friesen, Tetrahedron Lett., 35, 9177 (1994).
109.   B.J. Backes and J.A. Ellman, J.Am.Chem.Soc., 116, 11171 (1994).
110.   M.S. Deshpande, Tetrahedron Lett., 35, 5613 (1994).
111.   F.W. Forman and I. Sucholeiki, J.Org.Chem., 60, 523 (1995).
112.   K.L. Yu, M.S. Deshpande, and D.M. Vyas, Tetrahedron Lett., 35, 8919 (1994).
113.   M. Hiroshige, J.R. Hauske, and P. Zhou, Tetrahedron Lett., 36, 4567 (1995).
114.   M. Hiroshige, J.R. Hauske, and P. Zhou, J.Am.Chem.Soc., 117, 11590 (1995).
115.   C.R. Johnson and B. Zhang, Tetrahedron Lett., 36, 9253 (1995).
116.   H. Moon, N.E. Schore, and M.J. Kurth, J.Org.Chem., 57, 6088 (1992).
117.   H. Moon, N.E. Schore, and M.J. Kurth, Tetrahedron Lett., 35, 8915 (1994).
118.   T.Q. Dinh and R.W. Armstrong, Tetrahedron Lett., 37, 1161 (1996).
119.   X. Beebe, N.E. Schore, and M.J. Kurth, J.Org.Chem., 60, 4196 (1995).
120.   X. Beebe, N.E. Schore, and M.J. Kurth, J.Org.Chem., 60, 4204 (1995).
121.   N.E. Shore and S.D. Najdi, J.Am.Chem.Soc., 112, 441 (1990).
122.   I. Ugi, A. Domling, and W. Horl, Endeavour, 18, 115 (1994).
123.   P. Wipf and A. Cunningham, Tetrahedron Lett., 36, 7819 (1995).
124.   X. Cao, E.J. Moran, D. Siev, A. Lio, C. Ohashi, and A.M.M. Mjalli, BioMed.Chem.Lett., 5, 2953 (1995).
125.   A.M. Strocker, T.A. Keating, P.A. Tempest, and R.W. Armstrong, Tetrahedron Lett., 37, 1149 (1996).
126.   C.C. Leznoff, P.L. Svirskaya, B. Khouw, R.L. Cerny, P. Seymour, and A.B.P. Lever, J.Org.Chem., 56, 82 (1991).
127.   G.B. Fields and R.L. Noble, Int.J.Pept.Protein.Res., 35, 161 (1990).
128.   L.A. Thompson and J.A. Ellman, Tetrahedron Lett., 35, 9333 (1994).
129.   G.C. Liu and J.A. Ellman, J.Org.Chem., 60, 7712 (1995).
130.   C.C. Leznoff and T.M. Fyles, J.Chem.Soc.,Chem.Commun., 251 (1976).
131.   M.J. Plunkett and J.A. Ellman, J.Org.Chem., 60, 5742 (1995).
132.   B. Chenera, J.A. Finkelstein, and D.F. Veber, J.Am.Chem.Soc., 117, 11999 (1995).
133.   J.R. Hauske and P. Dorff, Tetrahedron Lett., 36, 1589 (1995).
134.   K.A. Beaver, A.C. Siegmund, and K.L. Spear, Tetrahedron Lett., 37, 1145 (1996).
135.   I. Sucholeiki, Tetrahedron Lett., 35, 7307 (1994).
136.   C.P. Holmes and D.G. Jones, J.Org.Chem., 60, 2318 (1995).
137.   B.B. Brown, D.S. Wagner, and H.M. Geysen, Molecular Diversity, 1, 4 (1995).
138.   H. Venkatesan and M.M. Greenberg, J.Org.Chem., 61, 525 (1996).
139.   G.W. Kenner, J.R. McDermott, and R.C. Sheppard, J.Chem.Soc.,Chem.Commun., 636 (1971).
140.   B.J. Backes, A.A. Virgilio, and J.A. Ellman, J.Am.Chem.Soc., (1995).
141.   V.K. Sarin, S.B.H. Kent, J.P. Tam, and R.B. Merrifield, Anal.Biochem., 117, 147 (1981).
142.   B. Gisin, Analytical Chimie, 58, 248 (1972).
143.   W.S. Hancock and J.E. Battersby, Anal.Biochem., 71, 260 (1976).
144.   V. Krchnák, J. Vágner, P. Safár, and M. Lebl, Czech.Chem.Commun., 53, 2542 (1988).
145.   G.L. Ellman, Arch.Biochem.Biophys., 82, 70 (1959).
146.   S.S. Chu and S.H. Reich, BioMed.Chem.Lett., 5, 1053 (1995).
147.   E.P. Heimer, C.-D. Chang, T.L. Lambros, and J. Meienhofer, Int.J.Pept.Protein.Res., 18, 237 (1981).
148.   M.H. Caruthers, A.D. Barone, S.L. Beaucage, D.R. Dodds, E.F. Fisher, L.J. McBride, M. Matteucci, Z. Stabinsky, and J.-Y. Tang, Chemical Synthesis of Oligodeoxyribonucleotides by the Phosphoramidite Method, in: Methods in Enzymology (R. Wu, and L. Grossman, eds.), 154, 1987, Academic Press, San Diego, pp. 287.
149.   G.S. Lu, S. Mojsov, J.P. Tam, and R.B. Merrifield, J.Org.Chem., 46, 3433 (1981).
150.   M. Stankova, S. Wade, K.S. Lam, and M. Lebl, Peptide Res., 7, 292 (1994).
151.   E. Bayer, H. Albert, H. Willisch, W. Rapp, and B. Hemmasi, Macromolecules, 23, 1937 (1990).
152.   G.C. Look, C.P. Holmes, J.P. Chinn, and M.A. Gallop, J.Org.Chem., 59, 7588 (1994).
153.   W.L. Fitch, G. Detre, C.P. Holmes, J.N. Shooley, and P.A. Keifer, J.Org.Chem., 59, 7955 (1994).
154.   R.C. Anderson, M.A. Jarema, M.J. Shapiro, J.P. Stokes, and M. Ziliox, J.Org.Chem., 60, 2650 (1995).
155.   K.D. Moeller, C.E. Hanau, and A. Davignon, Tetrahedron Lett., 35, 825 (1994).
156.   R.C. Anderson, J.P. Stokes, and M.J. Shapiro, Tetrahedron Lett., 36, 5311 (1995).
157.   B. Yan, G. Kumaravel, H. Anjaria, A. Wu, R.C. Petter, C.F. Jewell, Jr., and J.R. Wareing, J.Org.Chem., 60, 5736 (1995).
158.   H. Han, M.M. Wolfe, S. Brenner, and K.D. Janda, Proc.Natl.Acad.Sci.U.S.A., 92, 6419 (1995).
159.   M.J. Szymonifka and K.T. Chapman, Tetrahedron Lett., 36, 1597 (1995).

Chapter 32. Application of Mass Spectrometry for Characterizing and Identifying Ligands from Combinatorial Libraries

Joseph A. Loo and Dana E. DeJohn
Parke-Davis Pharmaceutical Research, Division of Warner-Lambert Company,
Ann Arbor, MI 48105

Rachel R. Ogorzalek Loo and Philip C. Andrews
University of Michigan, Department of Biological Chemistry, Ann Arbor, MI 48109

Introduction - The use of combinatorial chemistry to synthesize simultaneously larger numbers of diverse compounds for screening has become very popular (1). Traditionally, novel compounds for drug discovery in the pharmaceutical industry have been identified in receptor-based assays, evaluating large numbers of compounds by high volume screening methodologies. Combinatorial chemistry, by greatly increasing the molecular diversity available, has broadened the scope of molecular structures being surveyed for biological activity. However, characterizing these complex chemical libraries and identifying active compounds from these mixtures are challenging tasks. Chemical libraries exceeding 10^6 unique components are commonly synthesized, with improvements being made to increase the efficiency for producing libraries several orders of magnitude larger.

Approaches involving mass spectrometric identification have been developed for the affinity selection and identification of novel protein-ligand interactions. Two methods for volatilizing and ionizing thermally-labile molecules for mass spectrometric analysis developed in 1988 have revolutionized biochemical characterization: electrospray ionization (ESI) (2,3) and matrix-assisted laser desorption/ionization (MALDI) (4,5). Because of the high sensitivity and mass range of MS analysis with ESI and MALDI, the utility of MS-detection methods has been extended to several novel applications, including combinatorial chemistry.

This chapter provides a summary of reports of the use of mass spectrometry to characterize synthetic library mixtures from combinatorial chemistry as well as the direct analysis of compounds from solid-support surfaces (e.g., resin beads). In addition, methodologies integrating mass spectrometry are proving to be useful and rapid tools to screen and identify active compounds from library mixtures.

MASS SPECTROMETRIC CHARACTERIZATION OF LIBRARIES

The need for characterizing chemical libraries may seem limited; it may seem that only an examination of the active compounds is necessary. Indeed, such an approach could be successful, but it could also mislead researchers into overestimating the diversity of molecules they explore. It is important to confirm the desired randomness and diversity present in a library. As library approaches are applied toward the understanding of important receptor and enzyme interactions, it becomes more essential that the composition of the library be verified. One could imagine employing a library to explore the sequence-specificity of a phosphatase. But what would be proven if the phosphorylation step had failed during synthesis of the library and none of the peptides were actually phosphorylated? The investigator might mistakenly conclude that none of the sequences planned to explore interacted with the phosphatase, when in fact, they were not were present in the mixture.

However, the analytical characterization of these complex mixtures is a challenging task. Electrospray ionization MS is a useful method to identify potential problems in the synthesis of peptide libraries (6-9), and in combination with tandem mass spectrometry (MS/MS) (10), provides a rapid and reliable method to determine the composition. For relatively small libraries (less than 500 unique components), the resulting ESI mass spectrum shows a distribution of peaks across the m/z scale, and in many cases, each individual component can be resolved. Figure 1 shows a typical ESI mass spectrum of a tripeptide library (Ac-Ile-Ala-Xaa), where ion signals for each expected peptide are observed. Some molecules have optimum ESI-MS sensitivities with one polarity over the other, depending on the functional groups present. It can be beneficial to examine both a positive ion and negative ion spectrum to obtain a more realistic picture of the components. For larger libraries where mass resolution of each individual component is prohibitive, the shape and position of the unresolved distribution provides useful information regarding the success of the synthesis. One can compare a calculated theoretical mass distribution with the measured spectrum to gain insight (see, for example, Figure 2 for the Trp-Cys-Xaa-Xaa-Xaa-Xaa peptide library). For example, a low resolution spectrum can be used to validate a peptide library containing a chemical modification (e.g., phosphorylation) (7). However, one should be aware of one limitation of this approach: the most probable mass difference between amino acids is approximately 15 amu. Consequently, modifications of some components, e.g., oxidation, may be difficult to uncover from examination of the shape and position of an unresolved mass spectrum.

IDENTIFICATION OF COMPOUNDS ON SOLID PHASE SUPPORTS

Combinatorial libraries are most commonly synthesized using solid phase chemistry. The solid phase support usually takes the form of polystyrene beads, although other types of supports have also been utilized. An efficient and simple means of monitoring and analyzing the reaction chemistries at the molecular level is needed as the importance of combinatorial synthetic methods increases. Moreover, active compounds generated in bead-bound libraries are often identified by direct analysis of an accompanying tag molecule from a positive bead. A large library may contain several different biologically active compounds, and thus many beads requiring structural analysis of the bound compounds. Direct analysis by mass spectrometry does not depend on the synthesis of a tagging molecule or the attachment of sets of tagging molecules after each chemical separation.

Before the advent of the newer ESI and MALDI ionization methods, researchers had limited success in analyzing bead-bound materials. Ion bombardment methods such as fast atom bombardment (FAB) (11) and Californium-252 plasma desorption mass spectrometry (PDMS) (12) were used to measure molecular weights of resin-bound peptides. FAB-MS can be useful for characterizing combinatorial mixtures cleaved from the solid support (13). However, with the improvements in sensitivity offered by ESI and MALDI, analysis of single resin beads is possible and can be a routine methodology. For a typical polystyrene bead with a diameter of 50 μm, approximately 50-200 pmol of compound is bound to each bead, easily within the sensitivity limits of ESI- and MALDI-MS.

A simple MS method takes advantage of the acid lability of the Rink amide linker. Single beads are placed on a MALDI sample stage and subjected to trifluoroacetic acid

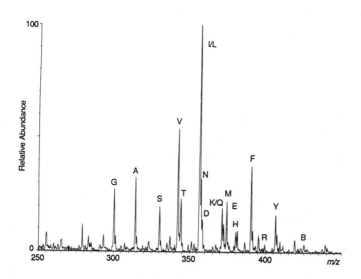

Figure 1. Positive ion electrospray ionization mass spectrum of the 18-component peptide library Ac-Ile-Ala-Xaa. Labels designate the identity of Xaa using the single letter code for the amino acids (where B = p-chloro-phenylalanine and Xaa = P, W, and C were not synthesized).

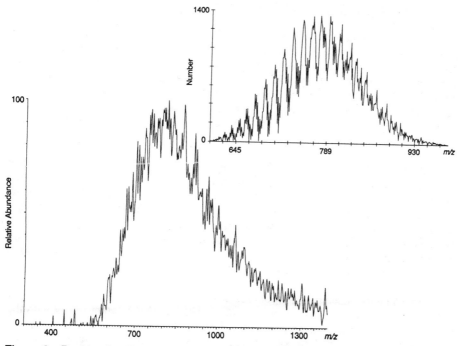

Figure 2. Positive ion ESI mass spectrum of the 160,000-component peptide library Trp-Cys-Xaa-Xaa-Xaa-Xaa. The inset shows the theoretical profile.

(TFA) vapor, thereby cleaving the compound from the linker. Subsequent addition of a suitable MALDI sample matrix (e.g., 2,5-dihydroxybenzoic acid) allows molecular weight measurements for peptides (14), peptoids (15), and other small compounds (16). In a similar fashion, ESI-MS can be applied for the direct analysis of bead-bound materials. Single beads containing angiotensin II receptor antagonists have been exposed to a TFA solution and the resulting solution injected into an ESI mass spectrometer for molecular weight measurement (17).

An advantage of MALDI-MS and ESI-MS is the capability for further structural characterization with subsequent stages of mass spectrometry. Fragment ions generated by "post-source decay" methods for MALDI-MS (15,17) or ESI interfaced with tandem mass spectrometry (17) can provide important structural information. It may also be possible to adapt MALDI-MS and ESI-MS for automated direct analysis of beads with current instrumentation.

A potentially interesting method for direct analysis of beads involves the use of imaging time-of-flight secondary ion mass spectrometry (TOF-SIMS) (17,18). SIMS uses a high energy ion beam to desorb and ionize compounds for MS measurement. The nature of the imaging TOF mass spectrometer allows not only mass resolution, but also spatial resolution of materials on a sample stage. Therefore, in addition to molecular weight and structural information (from fragment ions produced in the desorption step), it is possible to spatially resolve the composition of a single bead in the presence of other beads coated with other library members. This may be an important feature where screening procedures require a parallel assay of large numbers of active beads.

IDENTIFICATION OF ACTIVE LIGANDS BY MS-BASED METHODS

The identification of active components from these complex mixtures is the ultimate goal. Techniques that take advantage of the speed, sensitivity, and selectivity of mass spectrometry-based methods are being developed for this purpose. For example, affinity column methods can be used to selectively isolate the active materials, which are then eluted off the column and identified by mass spectrometry. An immobilized-Src SH2 domain protein column with on-line detection with HPLC/ESI mass spectrometry has been described to identify high affinity binding phosphopeptides (19).

Similar methods utilizing immunoaffinity extraction (IAE) have been described (20,21). Affinity selection of peptides from a library of 361 components with two variable sites for a 10-mer epitope known to bind to anti-HIV gp120 monoclonal antibody was verified by mass spectrometry and amino acid analysis (20). Using a benzodiazepine monoclonal antibody-protein G column interfaced to an LC/MS system, components from a small benzodiazepine combinatorial library were selectively mass analyzed. Structure identification is accomplished by tandem mass spectrometry with on-line IAE/LC/MS (21).

With affinity capillary electrophoresis/MS (ACE/MS), candidate peptides from combinatorial libraries can be identified (22). Ligands that interact tightly with the receptor, present in the electrophoresis buffer, will be retained as unbound ligands pass through the column. The bound ligands will have slower electrophoretic mobilities relative to their mobilities in the absence of the receptor. On-line ESI-MS is then used to measure the molecular weight and thus identify the interacting ligand.

    A potentially simple and efficient method for identification of specifically-bound ligands in combinatorial mixtures involves the use of a size exclusion pre-separation step (23,24). A combinatorial mixture is incubated with the receptor of interest. Afterwards, size exclusion chromatography is used to rapidly fractionate the excess unbound ligands away from the mixture. An aliquot of bound ligands is introduced into an ESI mass spectrometer via an on-line reversed-phase trap cartridge. The solution conditions used to elute the materials off the cartridge dissociates the ligand from the receptor. MS and MS/MS can then be used for identification. Peptoids that have specific affinity to human urokinase plasminogen activator receptor have been identified by this technique (23,24).

    ESI-MS has the capability of detecting biochemical complexes that are held together by noncovalent forces (25). The detection of noncovalent complexes may be a viable method to study protein-ligand binding from combinatorial mixtures. For example, the ESI mass spectrum of Src SH2 domain protein and a 6-component peptide mixture showed the expected protein-peptide complexes (compared to solution-phase binding measurements) as the dominant species in the mixture (26). A method that involves the detection of noncovalent complexes with subsequent tandem mass spectrometry to identify high binding affinity ligands was described recently (27,28). Complexes formed between a mixture of benzenesulfonamide-based inhibitors and carbonic anhydrase (29 kDa) were detected by ESI with Fourier transform MS (FTMS) (28). The relative abundances of each protein-inhibitor complex reflected the solution phase binding constants. Further confirmation and identification of the inhibitors were accomplished by subsequent MS/MS experiments. Dissociation of the gas phase complex released the ligand from the protein, providing the ligand mass. An additional stage of MS/MS of the released inhibitor ion was used to obtain structural information.

    The simultaneous identification and sequencing of an active peptide from resin-bound libraries with MALDI-MS can be achieved by incorporating mass spectrometry into the synthesis analysis (29,30). A capping reagent is used to terminate the synthesis of a fraction of the material at each coupling step during the synthesis of the peptide library. Resin beads that carry not only the full length peptide, but also smaller amounts of the sequence specific termination products are made. The peptides from the active beads are chemically cleaved (by cyanogen bromide in this case) and analyzed by MALDI-MS. The mass spectrum contains an ion for the full-length active peptide, identifying the active compound, and ions for each shortened peptide. The mass difference between adjacent peaks provides the sequence of the active peptide.

    Mass spectrometry can be used to monitor enzyme reaction products from a mixture of starting materials to determine substrate specificity. An application of ESI-MS for determining the relative k_{cat}/K_m for individual peptides in substrate mixtures has been reported (31). Turnover rates for a 20-component mixture from the substrate Ser-Gln-Asn-Tyr-Xaa-Ile-Val (where Xaa is one of the 20 naturally occurring amino acids) for HIV-1 protease were determined directly by mass spectrometry. Differences in the ion intensities of the individual components before and after incubation with HIV-1 protease was used to determine the preferred substrates.

    Mass spectrometry has been used to selectively-detect phosphorylated peptides from a synthetic peptide library pool (32). Peptide libraries incubated with the kinase of interest (e.g., cAMP-dependent protein kinase) were characterized by LC/MS with electrospray ionization. By increasing the collision energy in the ESI-MS interface, diagnostic fragment ions (m/z 63 for PO_2^- and m/z 79 for PO_3^-) can be used to identify phosphopeptides, thus identifying preferred substrates from a peptide mixture.

Substrate specificities can be determined using peptide libraries and tandem mass spectrometry (33). Peptide substrates for flavoprotein EpiD, a lantibiotic-synthesizing enzyme which catalyzes the C-terminal oxidative decarboxylation of the lantibiotic precursor peptide, was studied by ESI-MS and MS/MS methodologies. Preferred substrates from heptapeptide libraries were identified by selectively detecting decarboxylation reaction products resulting from the loss of 46 Da (i.e., neutral loss mass spectrometry).

CONCLUSIONS

At the present time, mass spectrometry is clearly useful for characterizing the fidelity of a combinatorial library. Even for large libraries, low resolution mass measurements can provide useful information on the bulk properties of a library. Higher resolution mass spectrometers, e.g., FTMS, should provide a tool to confirm the synthesis of nearly every component in a large library. In the future, mass spectrometry will be integrated more fully into the drug screening phase as the scientists involved in the use of combinatorial libraries and the mass spectrometry community increase their interactions.

Acknowledgments - We wish to thank John He and Wayne Cody (Parke-Davis Pharmaceutical Research) for the synthesis of the peptide libraries shown in the figures.

References

1. M. A. Gallop, R. W. Barrett, W. J. Dower, S. P. A. Fodor and E. M. Gordon, J. Med. Chem., 37, 1233 (1994).
2. C. K. Meng, M. Mann and J. B. Fenn, Z. Phys. D - Atoms, Molecules and Clusters, 10, 361 (1988).
3. J. B. Fenn, M. Mann, C. K. Meng, S. F. Wong and C. M. Whitehouse, Science, 246, 64 (1989).
4. M. Karas and F. Hillenkamp, Anal. Chem., 60, 2299 (1988).
5. M. Karas, U. Bahr, A. Ingendoh and F. Hillenkamp, Angew. Chem. Int. Ed. Engl., 28, 760 (1989).
6. P. C. Andrews, J. Boyd, R. Ogorzalek Loo, R. Zhao, C.-Q. Zhu, K. Grant and S. Williams in "Techniques in Protein Chemistry V," J. W. Crabb, Ed., Academic Press, San Diego, 1994, p. 485.
7. P. C. Andrews, D. M. Leonard, W. L. Cody and T. K. Sawyer in "Methods in Molecular Biology: Peptide Analysis Protocols," Vol. 36, B. M. Dunn, M. W. Pennington, Eds., Humana Press, Totowa, NJ, 1994, p. 305.
8. J. W. Metzger, K. H. Wiesmuller, V. Gnau, J. Brunjes and G. Jung, Angew. Chem. Int. Ed., 32, 894 (1993).
9. J. W. Metzger, S. Stenanovic, J. Brunjes, K.-H. Wiesmuller and G. Jung, Methods (San Diego), 6, 425 (1994).
10. J. W. Metzger, C. Kempter, K. H. Wiesmuller and G. Jung, Anal. Biochem., 219, 261 (1994).
11. J.-L. Aubagnac, M. Calmes, J. Daunis, B. El Amrani and R. Jacquier in "Peptides: Structure and Function (Proceedings of the Ninth American Peptide Symposium)," C. M. Deber, V. J. Hruby, K. D. Kopple, Eds., Pierce Chemical Co., Rockford, IL, 1985, p. 277.
12. P. A. van Veelen, U. R. Tjaden and J. van der Greef, Rapid Commun. Mass Spectrom., 5, 565 (1991).
13. J. A. Boutin, P. Hennig, P.-H. Lambert, S. Bertin, L. Petit, J.-P. Mahieu, B. Serkiz, J.-P. Volland and J.-L. Fauchere, Anal. Biochem., 234, 126 (1996).
14. B. J. Egner, G. J. Langley and M. Bradley, J. Org. Chem., 60, 2652 (1995).
15. R. A. Zambias, D. A. Boulton and P. R. Griffin, Tetrahedron Lett., 35, 4283 (1994).
16. N. J. Haskins, D. J. Hunter, A. J. Organ, S. S. Rahman and C. Thom, Rapid Commun. Mass Spectrom., 9, 1437 (1995).
17. C. L. Brummel, J. C. Vickerman, S. A. Carr, M. E. Hemling, G. D. Roberts, W. Johnson, J. Weinstock, D. Gaitanopoulos, S. J. Benkovic and N. Winograd, Anal. Chem., 68, 237 (1996).
18. C. L. Brummel, I. N. W. Lee, Y. Zhou, S. J. Benkovic and N. Winograd, Science, 264, 399 (1994).

19.   D. B. Kassel, T. G. Consler, M. Shalaby, P. Sekhri, N. Gordon and T. Nadler in "Techniques in Protein Chemistry VI," J. W. Crabb, Ed., Academic Press, San Diego, CA, 1995, p. 39.
20.   R. N. Zuckermann, J. M. Kerr, M. A. Siani, S. C. Banville and D. V. Santi, Proc. Natl. Acad. Sci. U. S. A., 89, 4505 (1992).
21.   M. Nedved, B. Ganem and J. Henion, 43rd ASMS Conference on Mass Spectrometry and Allied Topics, Atlanta, GA, 1995, p. 491.
22.   Y. H. Chu, D. P. Kirby and B. L. Karger, J. Am. Chem. Soc., 117, 5419 (1995).
23.   V. Huebner, G. Dollinger, S. Kaur, L. McGuire, D. Tang, M. Siani, R. Drummond and S. Rosenburg, Ninth Symposium of the Protein Society, Boston, MA, 1995, p. 22.
24.   S. Kaur, V. Huebner, D. Tang, L. McGuire, R. Drummond, J. Csetjey, J. Stratton-Thomas, S. Rosenberg, G. Figliozzi, S. Banville, R. Zuckermann and G. Dollinger, 43rd ASMS Conference on Mass Spectrometry and Allied Topics, Atlanta, GA, 1995, p. 30.
25.   J. A. Loo, Bioconj. Chem., 6, 644 (1995).
26.   J. A. Loo, P. Hu and V. Thanabal, 43rd ASMS Conference on Mass Spectrometry and Allied Topics, Atlanta, GA, 1995, p. 35.
27.   J. E. Bruce, G. A. Anderson, R. Chen, X. Cheng, D. C. Gale, S. A. Hofstadler, B. L. Schwartz and R. D. Smith, Rapid Commun. Mass Spectrom., 9, 644 (1995).
28.   X. H. Cheng, R. D. Chen, J. E. Bruce, B. L. Schwartz, G. A. Anderson, S. A. Hofstadler, D. C. Gale, R. D. Smith, J. M. Gao, G. B. Sigal, M. Mammen and G. M. Whitesides, J. Am. Chem. Soc., 117, 8859 (1995).
29.   R. S. Youngquist, G. R. Fuentes, M. P. Lacey and T. Keough, Rapid Commun. Mass Spectrom., 8, 77 (1994).
30.   R. S. Youngquist, G. R. Fuentes, M. P. Lacey and T. Keough, J. Am. Chem. Soc., 117, 3900 (1995).
31.   D. B. Kassel, M. D. Green, R. S. Wehbie, R. Swanstrom and J. Berman, Anal. Biochem., 228, 259 (1995).
32.   J. H. Till, R. S. Annan, S. A. Carr and W. T. Miller, J. Biol. Chem., 269, 7423 (1994).
33.   T. Kupke, C. Kempter, G. Jung and F. Gotz, J. Biol. Chem., 270, 11282 (1995).

Chapter 33.   Plasma Protein Binding of Drugs

Richard E. Olson and David D. Christ
The DuPont Merck Pharmaceutical Company
Wilmington, DE 19880-0500

<u>Introduction</u> - Current approaches to the design of small molecule therapeutics generally begin with the targeting of macromolecules apart from the *in vivo* realities of absorption, distribution and disposition. Following the identification of candidates with the desired potency and specificity, these latter hurdles must be addressed. The role of plasma protein binding is recognized as an important factor in drug disposition and efficacy (1).   This review will focus on the structures of drug-binding proteins, particularly human serum albumin (HSA) and alpha-1-acid glycoprotein (AAG), the relationship between ligand structure and protein binding, and the effect of plasma protein binding on drug disposition and pharmacological activity.   The interactions of peptides and proteins with HSA, AAG and specific transport proteins will not be discussed, but have been reviewed (2). While the discussion will focus on binding to plasma proteins, it should also be recognized that binding to tissue proteins is conceptually similar. In fact, 60% of total HSA is found in the extravascular space (3).

PLASMA BINDING PROTEINS: STRUCTURE AND LIGAND BINDING

<u>Human Serum Albumin</u> - Human serum albumin (HSA) binds a wide variety of endogenous and exogenous ligands with association constants typically in the range of 10^4 to 10^6 M^{-1} (3,4). HSA is a highly soluble 66.5 kD protein comprising a single chain of 585 amino acids containing a single tryptophan (Trp-214), low (2%) glycine content, high cystine content and a large number of charged amino acids, which provide a calculated net charge of -15 at pH 7 (3). The primary sequences of bovine, rat, mouse and a number of other serum albumins are also known (4).   HSA has high sequence homology with other mammalian serum albumins, being, for example, ~80% homologous to bovine (BSA) and rat (RSA) serum albumins. As the most abundant protein in plasma (35-50 mg/ml), it plays an important role in the maintenance of blood pH and colloidal osmotic pressure and accounts for most of the thiol content of plasma (Cys-34) (3).

<u>HSA Structure</u> - HSA is organized in a series of three repeating domains (I, II & III), each of which is composed of two subdomains (A & B) (4,5). One free cysteine (Cys-34) and 17 cystines are present, 16 of which occur in the motif shown in Figure 1, which is conserved across the serum albumin family (6). The disulfides enforce on the protein a series of 9 loops, 3 in each domain, and contribute to the stability of HSA to heat and other denaturing conditions (3,4). Changes in the structure of HSA as a function of pH have been described (4). This phenomenon was assessed recently using small-angle X-ray scattering (7). The results suggested that the subdomains within the three major domains of HSA remained invariant while the flexible links between subdomains permitted significant extension of the molecule at low pH.

Early studies and concepts of the three-dimensional structure of HSA have been recently reviewed (4). In 1989, an X-ray crystallographic study of HSA at a resolution of 6.0Å was published, revealing considerable helical content and the tri-domain homology predicted by the primary sequence (8).   Higher resolution structures revealed the protein to be heart-shaped, reminiscent of electron micrographs of the homologous α-fetoproteins (4,5,9,10).   The topology is maintained by hydrophobic and salt bridge interactions (4). Crystallographic studies of the serum albumins from other species have been reported (6,11). HSA is 67% helical, with each of the three major domains containing 10 helices. Two extended helices connect the three domains, making the total number of helices 28. The two subdomains (A & B) contain a major loop with similar helical secondary structure, resulting in additional homology (4).

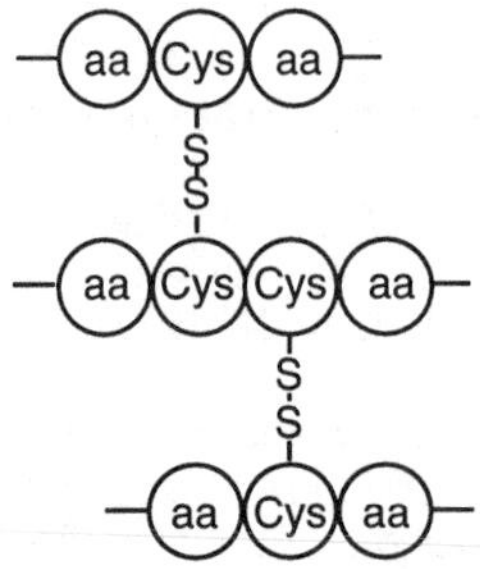

Figure 1. Loop stabilizing double cystines of serum albumins

Ligand Binding by HSA - Numerous competition and spectroscopic studies using various ligands have identified two sites (I and II) most relevant for drug binding (3,4). The availability of crystallographic data has allowed the three-dimensional localization of sites I and II, and the observed ligand selectivity is generally consistent with the results of earlier studies (4).   Site I, classically associated with warfarin (**1**) and phenylbutazone (**2**) binding, has been localized to subdomain IIA. Site II, which binds diazepam (**3**) and ibuprofen (**4**), is in subdomain IIIA. The binding of the non-selective ligand 2,3,5-triiodobenzoic acid to site I and site II has been described in some detail (4,5). Both sites feature distinct hydrophobic and basic regions, consistent with their binding of organic anions, yet differ in their respective locations within subdomains IIA and IIIA. Site II has been described as a "sock-shaped" pocket in which the foot region is primarily hydrophobic and the leg is hydrophilic (4,5). A tyrosine (Tyr-411) associated with ligand binding and esterase activity is located within site II (4). Site I associated residues include Lys-199, Trp-214 and His-242 (4).

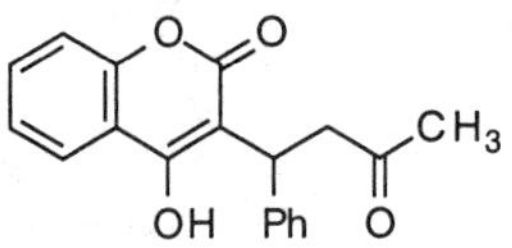

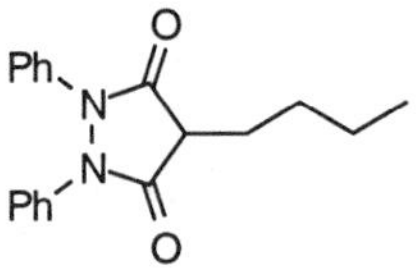

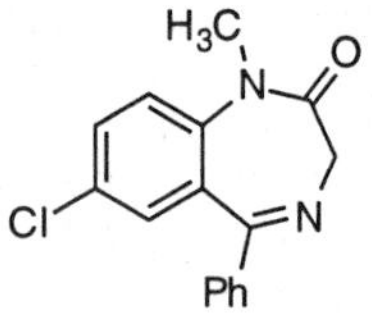

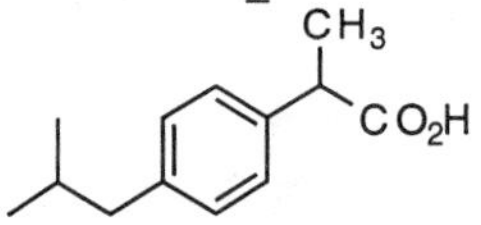

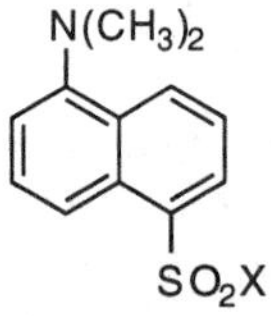

Recent solution studies of HSA in the presence of ligands complement the solid state information. Site-specific fluorescent probes such as dansylamide (**5**) (site I) and dansylsarcosine (**6**) (site II) have been used extensively (12). Evidence for the existence of pre-formed binding sites in solution was obtained from a study of the optical rotary dispersion of HSA in the presence of 28 ligands (13). While a number of ligands did induce ORD changes, 16 ligands, including phenylbutazone (**2**) (site I) and flufenamic acid (**7**) (site II) did not induce changes in the ORD, supporting the existence of pre-existing binding sites.   Microcalorimetry (14-16) has revealed differences in the thermodynamics of binding to sites I and II (15,16). Among the ligands studied were site I ligands such as **2** and site II ligands such as ibuprofen (**4**) (15,16). While the free energy of binding was similar for both classes of ligands, the site I interactions were reported to have minimal entropic contributions, while the formation of the site II complexes involved large, negative entropies. The asymmetry of sites I and II gives rise to induced circular dichroism for achiral ligands such as **2** (site I) and diazepam (**3**) (site II). The degree of induced CD in the presence of competing site-specific ligands has been used to measure dissociation constants (17).   Ultraviolet resonance Raman spectroscopy has been used to study HSA-ligand interactions. Upon binding of warfarin (**1**) (site I) to HSA, the Raman intensity of Trp-214 and **1** increased, suggesting direct hydrophobic interactions. Complexation of **4** (site II) increased the Raman intensity of

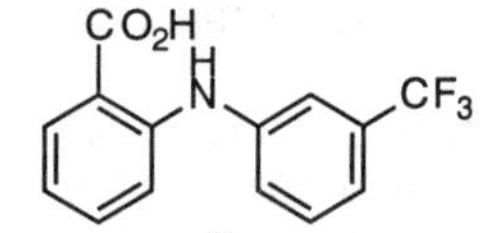

Tyr, but not of **4** which suggested an interaction of the Tyr-411 phenolic hydroxyl with the drug carboxylate (18).

The interaction of various non-steroidal antiinflammatory drugs (NSAIDs) with HSA has been studied extensively (19). In studies of the binding of nine NSAIDs to HSA using equilibrium dialysis and spectroscopic methods, compounds bearing a carboxylic acid, such as ibuprofen (**4**), selectively displaced site II fluorescent probes, whereas phenylbutazone (**2**) and related NSAIDs which lack carboxyl functionality bound to site I (20). Fluorescent probe displacement was used to show that the high affinity site for the binding of the NSAID suprofen (**8**) is site II, whereas the high affinity site for suprofen methyl ester coincides with the warfarin binding site (21). (S)-(+)-Pranoprofen (**9**) showed higher binding (by equilibrium dialysis) than (R)-(-)-**9**, but probe displacement showed that both bind HSA site II, whereas the corresponding glucuronide and methyl ester bound to site I (22). A similar stereochemical preference was shown for the enantiomers of carprofen (**10**), which has been reported to bind to both site I and II, with higher affinity for site II (23,24). Stopped-flow fluorescence measurements were used to assess the effect of the presence of various NSAIDs on the kinetics of the site II-specific dansylsarcosine (**6**) interaction with HSA (25). The enantiomers of **4** affected the association rate of **6** similarly, and only small differences were seen between fenbufen (**11**) and ketoprofen (**12**). However, **4** decreased the **6**-association constant more than **11** or **12**. Bioaffinity chromatography on immobilized HSA has been used to study the binding sites of NSAIDs (26,27). It was found that the (R)-enantiomer of **4** has only one binding site on HSA, but (S)-**4** has two, with similar association constants (26).

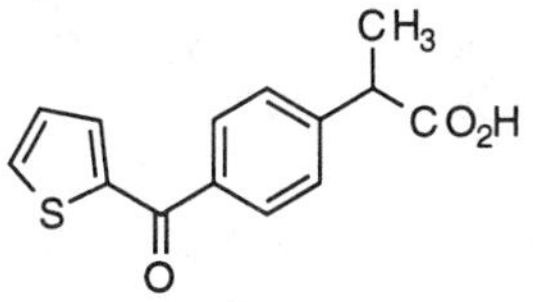

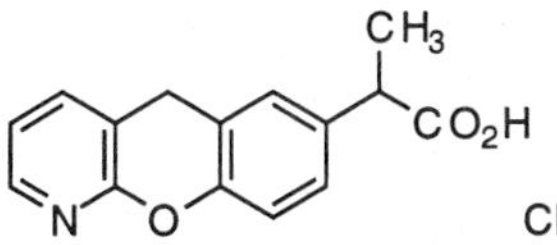

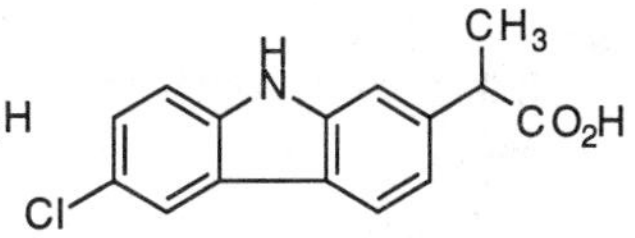

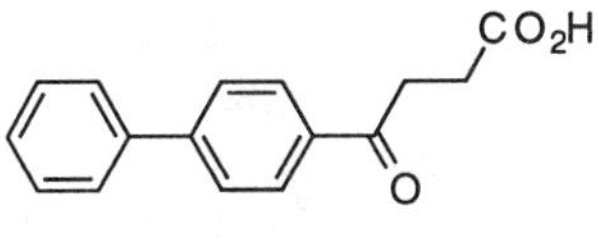

The interaction of other drug classes with HSA has been investigated. Spectroscopic methods were employed to study the binding of barbiturates such as quinalbarbitone (**13**) to HSA (28). It was concluded that these compounds interacted with subdomain IIIA (site II) of HSA, because no significant change in the microenvironment of the Trp-214 residue could be seen in the fluorescence emission spectrum. Three binding sites, I, II, and a third site, not localized, were implicated in the HSA binding of the antidiabetic tolbutamide (**14**), as evidenced by microcalorimetry and heteronuclear 2D NMR (29). The affinity of **14** for sites I and II was attributed to the presence of factors which favor site I binding (centrally localized charge) and site II binding (extended structure) (12,29). In a case of stereoselective HSA binding, the binding of folinic acid (**15**) was found by equilibrium dialysis to be selective for the (R)-enantiomer (30). Known site I and site II ligands were unable to displace **15** from HSA. Benzothiadiazides were shown by fluorescent probe displacement, induced CD and differential affinity to modified HSA to preferentially bind to site II. Hydrophobic and electrostatic factors were reported to contribute to HSA affinity (31). Chlorothiazide (**16**) showed higher affinity than hydrochlorothiazide (**17**). Hydrophobicity was also reported to contribute to the HSA affinity of a series of diazepines, including fludiazepam (**18**), which were shown to form 1:1 complexes (32). Quantitative relationships describing benzodiazepine binding to immobilized HSA have been reported, and included steric, electronic and physicochemical parameters (33).

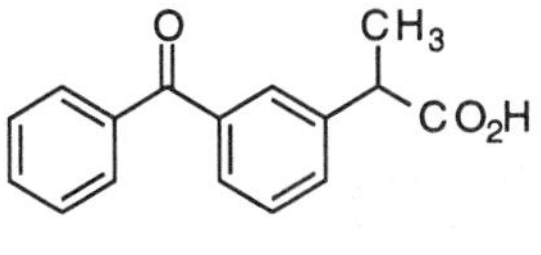

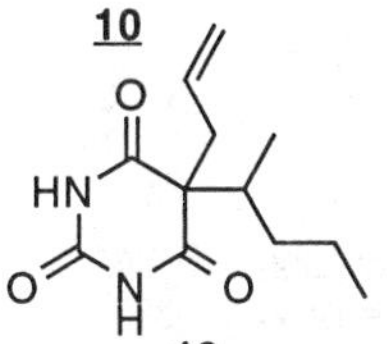

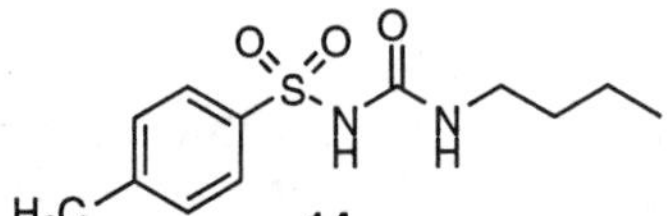

14 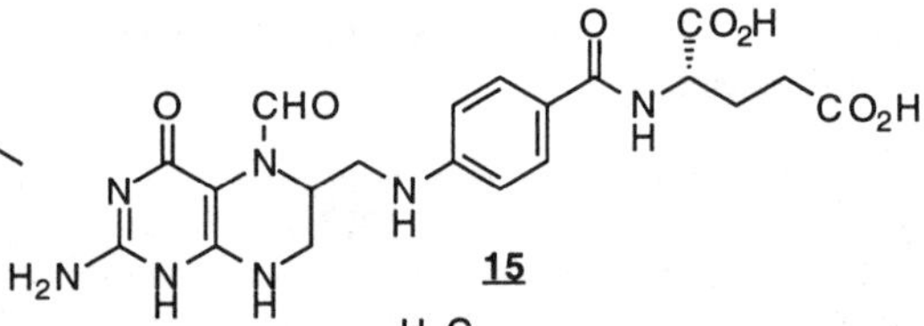   15

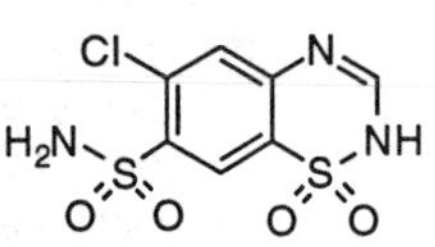

16 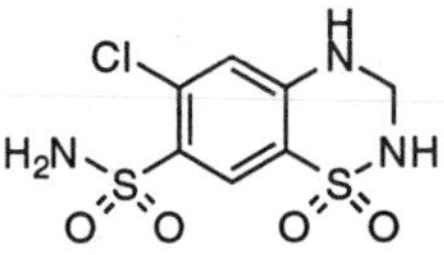   17 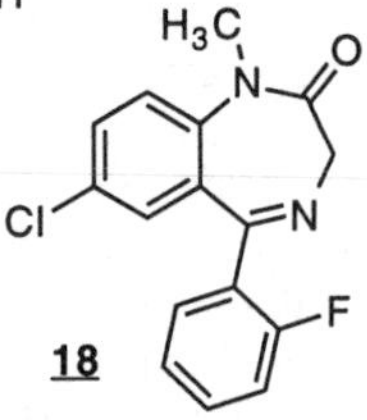   18 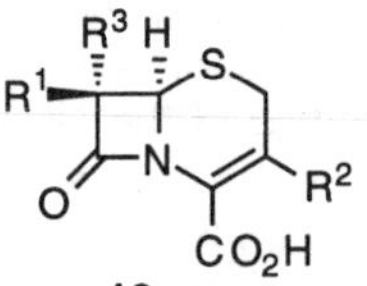   19

In a series of cephalosporins (**19**), binding to HSA as measured by equilibrium dialysis was highest for compounds with low or high lipophilicity, while those of intermediate lipophilicity showed reduced affinity (34).   Ionization was suggested as a factor complicating the usually linear relationship between lipophilicity and HSA affinity (34).   In a separate study, warfarin, but not phenylbutazone, diazepam or L-tryptophan, was found to displace cephalosporins from HSA, suggesting that these drugs bind in the vicinity of the warfarin site (35).   Three quinolone antibacterials, nalidixic acid (**20**), cinoxacin (**21**) and pipemidic acid (**22**), differed significantly in the degree of binding to HSA, with **20** showing the greatest extent (92-97%) and **22** the lowest (no significant binding). The higher lipophilicity of **20** and the positive charge (pH 7.4) associated with **22** were suggested as contributing factors (36). A species difference was observed in the binding of **21**, which showed higher binding to RSA (90%) than HSA (68%) (36).

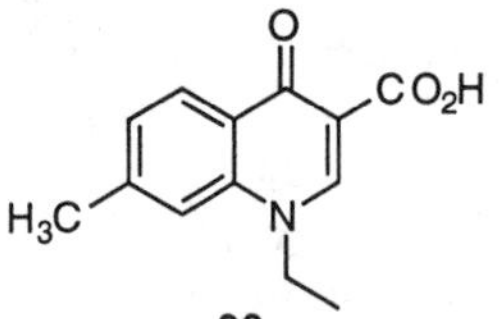

20 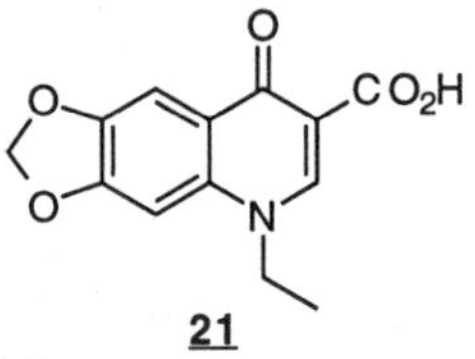   21 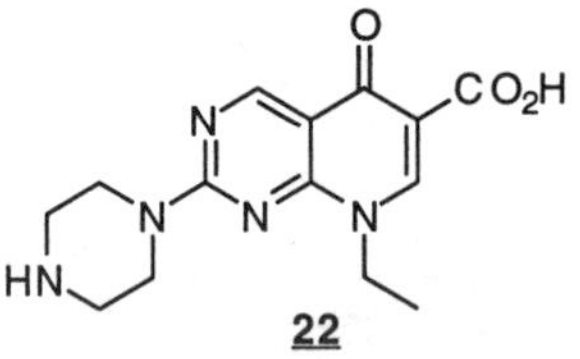   22

Alpha-1-Acid Glycoprotein - Alpha-1-acid glycoprotein (AAG; orosomucoid), is an acute phase protein (38) often associated with the binding of basic and neutral drugs, but it can also bind certain acidic compounds, and is known to function as a carrier for steroids (37). Association constants for drug ligands are similar to those for HSA complexes (4,37). The concentration of AAG ranges from 0.5-1.0 mg/ml of plasma under normal circumstances, but can vary considerably across physiological and pathological conditions, exceeding 3.0 mg/ml in certain disease states (37). AAG is a 181 [183 (37)] amino acid single polypeptide chain with a molecular weight of approximately 40,000 (37,39). Multiple isoforms of AAG exist within individuals (37,39). The carbohydrate composition of AAG, accounting for about 42% of the molecular weight, is also heterogeneous, consisting of five asparagine-linked branched glycan chains containing approximately 11% sialic acid (37,39).   The protein structure contains 21% α-helix and 21% β-sheet (37). The conformation of the protein component of AAG is reported to be unaffected by the carbohydrate groups (39). Crystals of AAG have been prepared, but the three-dimensional structure has not been reported (37,39).

Ligand Binding by AAG - AAG is believed to have one wide and flexible binding site, or two overlapping sites (37,40,41), a concept supported by the displacement of various fluorescent probes by acidic and basic ligands (41,42). The involvement of Lys-162 and several glutamic acid residues in ligand binding has been suggested by photoaffinity labeling studies, but the binding site is described as largely hydrophobic (37). Various phenothiazine ligands, including chlorpromazine (**23**), have been found to form ternary complexes with dicumarol (**24**) complexed to AAG, or to displace this acidic ligand, supporting the single acidic/basic ligand binding site hypothesis (43). The affinity of phenothiazines for AAG corresponded positively with hydrophobicity (44).

Desialylation led to a decrease in binding constants (44). The effect of pH on the binding of warfarin (**1**) and basic ligands was consistent with a higher affinity of AAG for neutral ligands (45,46). Thermodynamic analysis of the binding of phenothiazines to AAG at varying temperatures showed a positive entropic contribution, which was attributed to hydrophobic interactions (44). A similar study of the binding of basic, neutral and acidic ligands to AAG revealed a greater entropic component for basic ligands (47). A positive correlation between ligand polarity and the contribution of enthalpy to binding was attributed to hydrogen bonding (47). AAG variants were found to differ in their ligand affinity (47). A comparison of the interaction of a variety of antihistamines and adrenolytics with immobilized AAG led to a predictive equation which included terms for hydrophobicity and electron density on nitrogen, and a steric parameter (48-50). A description of the binding site as an asymmetric hydrophobic cleft containing an anionic region at the base of the cleft was suggested by molecular modeling (49,50).

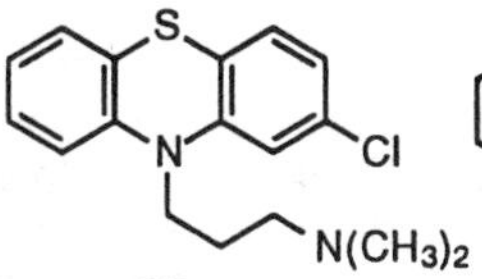

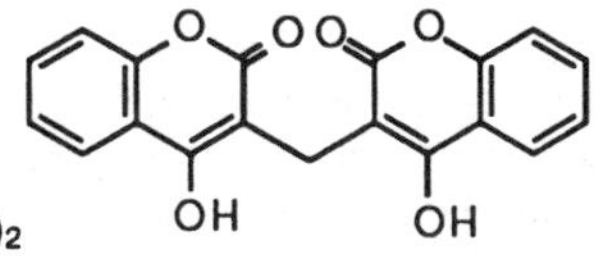

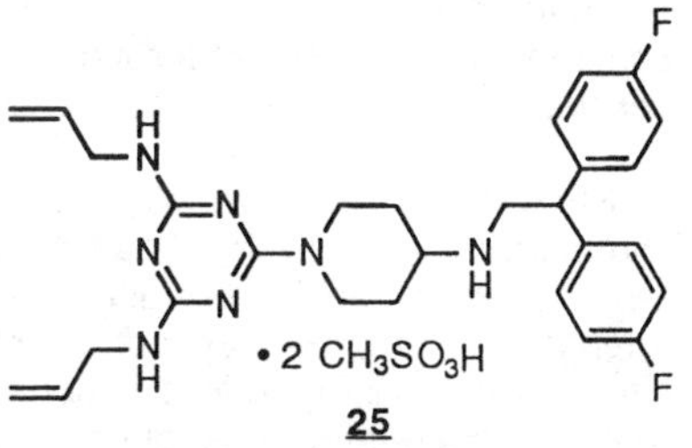

Lipoproteins: - High (HDL), low (LDL) and very low (VLDL) density lipoproteins can contribute to the plasma binding of lipophilic drugs, which dissolve in the lipid core of the lipoproteins (1). The binding of the lipophilic multidrug resistance modulator S9788 (**25**) to plasma components was assessed by the erythrocyte partitioning method. The data suggested lipoproteins were the main carriers of S9788 in blood (51).

Covalent Binding of Drugs and Metabolites - Covalent modification of plasma proteins can be produced by direct attachment of an electrophilic moiety of the parent molecule to nucleophilic protein residues, or by metabolic activation of the parent to a more reactive species. These reactions are important because of their potential to form immunogenic adducts (haptens), with the possibility of producing idiosyncratic hypersensitivity reactions (52). Direct attachment may involve nucleophilic attack by amine or thiol residues on electrophilic groups such as esters and ß-lactams. The second category is less predictable, but is exemplified by the covalent binding of acyl glucuronide metabolites of NSAIDs such as tolmetin (**26**) to albumin (53,54,55). These acid derivatives are first conjugated with glucuronic acid, bound reversibly to albumin (56), and during this transport react with nucleophilic residues on the protein to produce covalent adducts. The mechanism for the covalent binding of tolmetin glucuronide (**27**) is believed to involve migration of the drug acyl moiety from the anomeric oxygen to a secondary hydroxyl and subsequent formation of a potentially reducible Lys-imine at the anomeric center (54). The lysine residues forming the adduct, including Lys-199, were identified by mass spectrometry (54,57). Direct acylation of the protein by the acyl glucuronide can also occur (57).

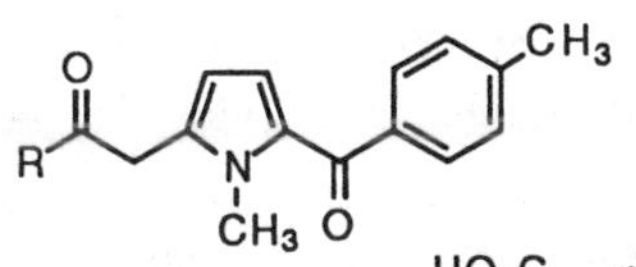

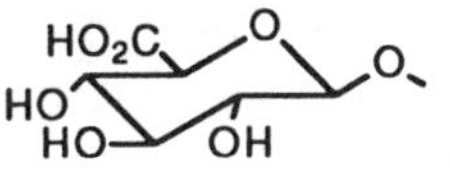

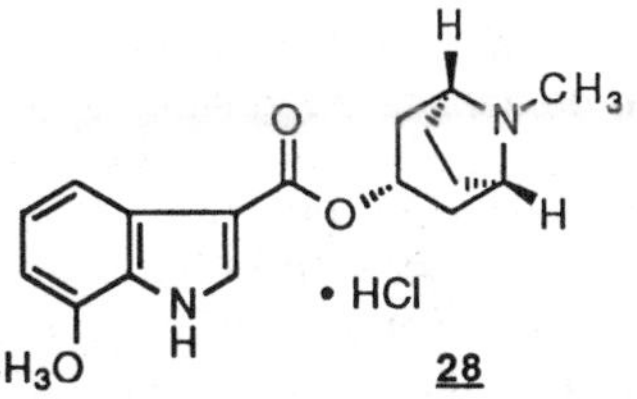

EXPERIMENTAL DETERMINATION OF PLASMA PROTEIN BINDING

This topic has been extensively reviewed recently (58). Techniques utilizing the physical separation of bound and free drug, such as equilibrium dialysis, ultrafiltration, affinity and size exclusion chromatography (HPLC), capillary electrophoresis, and microdialysis, are most often used to quantitate the extent of binding (58). Of these techniques, only microdialysis offers the potential to measure the extent of binding *in vivo* or *in vitro* in plasma, and the unbound extracellular tissue concentration (59). This approach has been utilized in the comparison of the unbound concentrations in plasma and frontal cortex of SDZ ICM 567 (**28**), a novel 5-HT$_3$ receptor antagonist (60). Analytical sensitivity is a major limitation because of the small volumes of dialysate recovered. Radiolabeled drugs are commonly used with all techniques, often with nonspecific detection. The importance of chemical and radiochemical purity has recently been discussed (61). The adsorption of lipophilic molecules to analytical components and the chemical stability of the ligands and proteins must also be considered in the selection of the appropriate technique (58).

HPLC methods utilizing protein stationary phases offer the potential for rapid screening, especially for those molecules bound to albumin (58). These methods are rapid, precise, and well suited to automation. Excellent correlation between the retention on an HSA-derived column and the fraction unbound (fu) determined by ultrafiltration was found for two series of highly bound benzodiazepines and coumarins, while little correlation was found for a series of poorly bound triazole derivatives (62).

Spectroscopic methods employing purified proteins and ligands are used to address mechanistic questions. Fluorescence markers have been developed to probe drug binding sites on albumin through competitive displacement interactions (12,63). These techniques, together with the more classical methods described previously, can also be used to determine the affinity and capacity constants for binding, typically through Scatchard plot analysis (58). Careful data analysis is required, however, for accurate quantitation (64). High field NMR spectroscopy can also be used to study the effects of molecular modifications and protein residue interactions, although the high concentrations of ligand often required may not be pharmacologically relevant (58).

THE EFFECTS OF PLASMA BINDING ON DRUG DISPOSITION

Plasma Binding, Renal and Metabolic Clearance - Systemic clearance (Cl$_s$) is defined as the irreversible removal of drug from the body, most often using blood or plasma as the sampling compartment. Clearance is related to the half-life (t$_{1/2}$) and volume of distribution (V$_d$) of a drug by the following equation;

$$t_{1/2} = \frac{0.693\ V_d}{Cl_s}$$

Lengthening t$_{1/2}$ can only be accomplished by modifying the structure to decrease Cl$_s$, or to increase V$_d$. Both the extent of plasma protein binding and the affinity of the ligand for different proteins can affect independently Cl$_s$ and V$_d$ (65).

A decrease in tubular secretion in connection with increased plasma binding has been demonstrated in studies of the highly bound cephalosporin cefonicid (**29**) using the isolated perfused rat kidney, where the rate of secretion was dependent on the fraction unbound (fu) (66). While secretion is generally determined predominantly by the fu, the secretion of the diuretic acetazolamide (**30**) by the isolated perfused rat kidney was not entirely accounted for by consideration of free drug, although no molecular mechanisms were discussed (67).

The clearance (extraction) of a drug by an organ such as liver may be less than, equal to or greater than the fu in the perfusing blood. This phenomenon has been termed "restrictive" or "nonrestrictive" (permissive) elimination, respectively (68).

Increasing the fu will increase the clearance of restrictively cleared drugs. Nonrestrictive elimination occurs when hepatic extraction exceeds the fu available, and can be envisioned by considering the extraction processes to be so avid (high affinity) and the dissociation from plasma binding sites so rapid that removal of free drug in the capillary causes dissociation and removal of the "bound" fraction of drug (68).

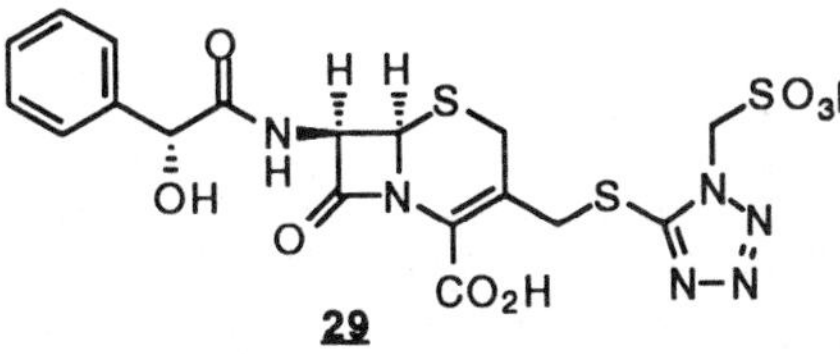

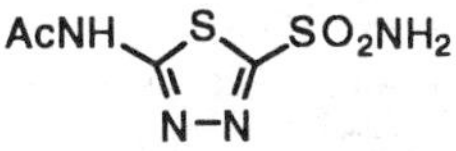

Plasma Binding, Tissue Distribution, and the Volume of Distribution - The apparent volume of distribution (V_d) relates the amount of drug in the body to the plasma concentration, and is not a physiological volume *per se*. Plasma and tissue binding will, however, influence the V_d according to;

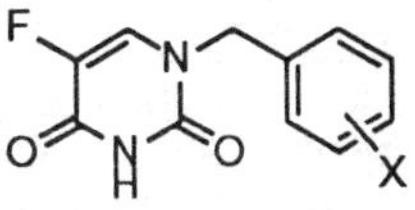

$$V_d = V_{plasma} + V_{tissue} (fu_{plasma} / fu_{tissue})$$

where V_{plasma} and V_{tissue} represent the volumes of plasma (0.07 L/kg) and total body water (minus plasma, 0.6 L/kg), respectively (1,69). Manipulating the degree of plasma binding and hence V_d by modifying lipophilicity has been demonstrated for a series of substituted 5-fluorouracil derivatives (**31**), where a correlation was shown between the plasma fu and calculated V_d after intravenous administration (70).

THE EFFECT OF PLASMA BINDING ON PHARMACOLOGICAL ACTIVITY

Enzyme Inhibition and Activity; HIV-1 Protease Inhibitors - Recent *in vitro* studies have shown the ablation of anti-HIV activity of A-77003 (**32**) by the addition of AAG, and a high degree of plasma and AAG binding by KNI-272 (**33**) (71,72). The antiviral potency of SC-52151 (**34**), which is >90% bound in plasma, is affected by AAG, but not HSA (73). The plasma binding (93%) of VX-478 (**35**) has been attributed to AAG (74). A two-fold reduction in antiviral activity of **35** was observed in the presence of 45% human plasma (75). The activity of a series of coumarin protease inhibitors including U-99499 (**36**) has been reported to be reduced by albumin; affinity for HSA site I was demonstrated (63). DMP 323 (**37**) was found not to be highly bound in plasma, and the inhibition of HIV protease was not altered in the presence of human serum or plasma (76,77).

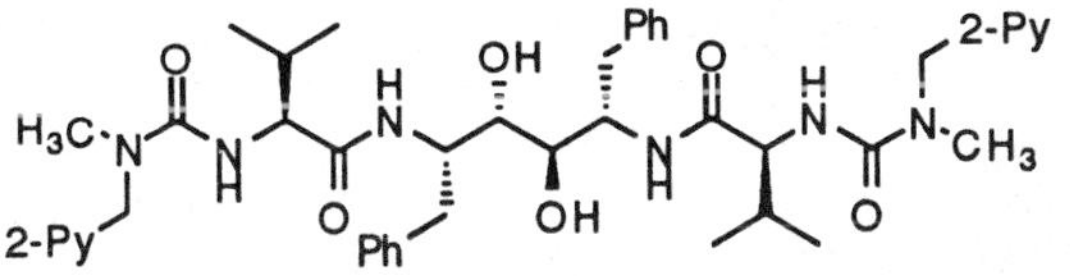

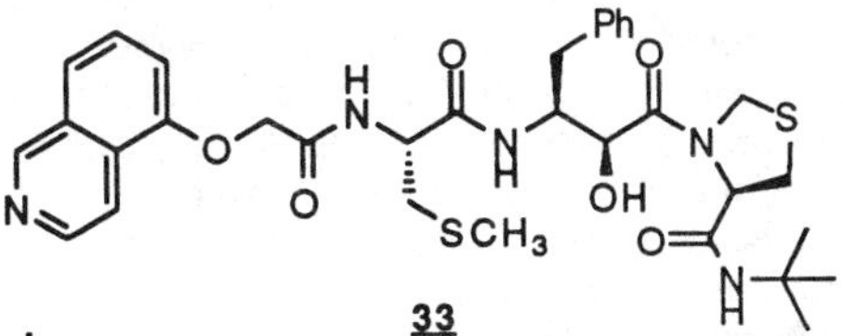

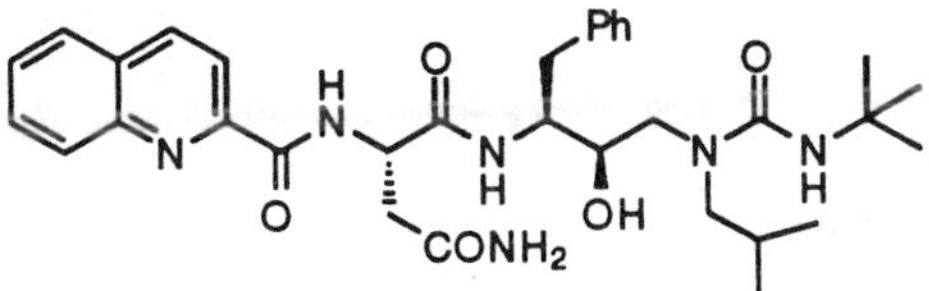

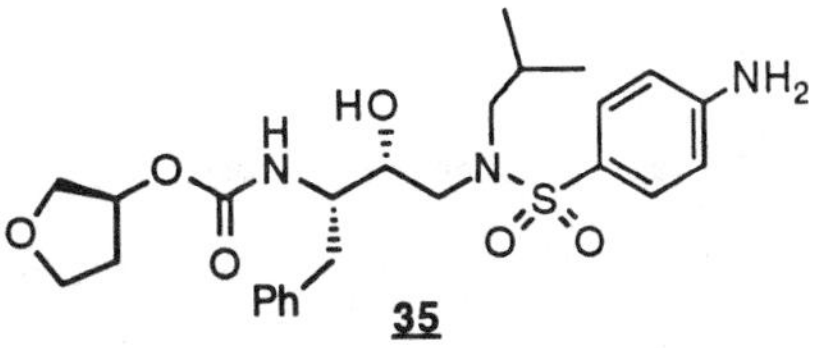

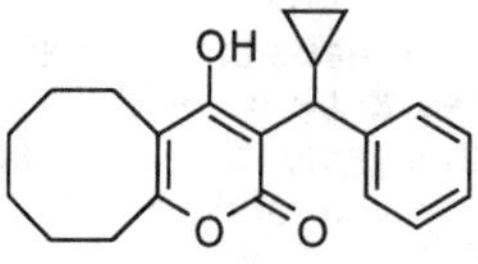

36

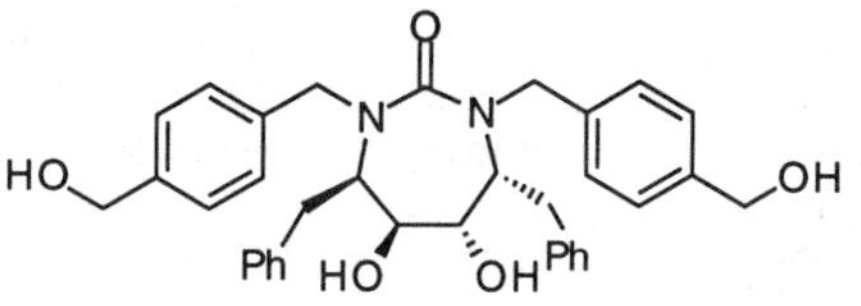

37

<u>Vascular Receptor Occupancy and Activity</u> - The interactions of drugs with receptors found on the luminal surface of vascular endothelium or on cells in the circulation are not subject to diffusional barriers. The angiotensin II (Ang II) receptor antagonist losartan (**38**) and its pharmacologically active metabolite EXP3174 (**39**) are highly bound to albumin in animals and humans, and the receptor antagonism *in vitro* can be altered (decreased) by the addition of albumin (78,79). Studies in dogs and rats have demonstrated agreement between the IC_{50} (unbound) for the blockade of exogenous Ang II vasopressor response and the IC_{50} for the blockade of Ang II binding in the absence of albumin for **38** and **39** (80,81). Other studies of the *in vitro* potency, *in vivo* receptor occupancy and *in vivo* efficacy for Ang II antagonists including **39**, DuP 532 (**40**) and L-158,809 (**41**), found a poor correlation even when HSA or RSA was present *in vitro* (82). The potency of **40** and **41** in the presence of rat plasma correlated with *in vivo* (rat) efficacy (82).

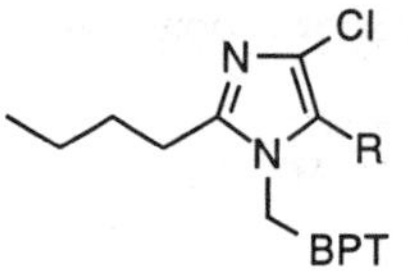

| **38** | R = CH₂OH |
| **39** | R = CO₂H |

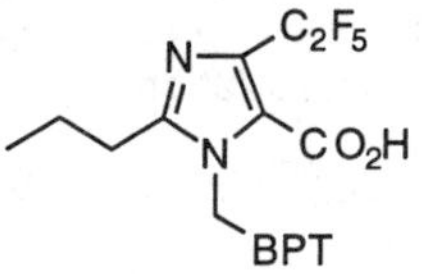

40 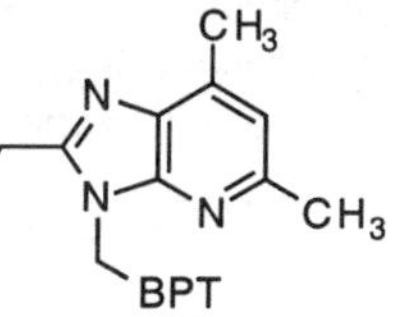   **41** 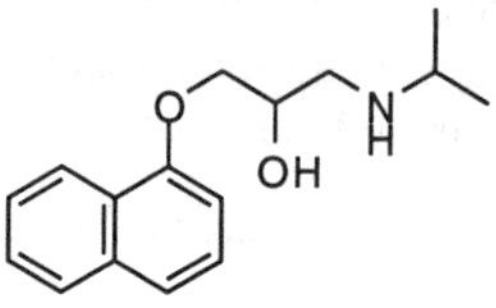   **42**

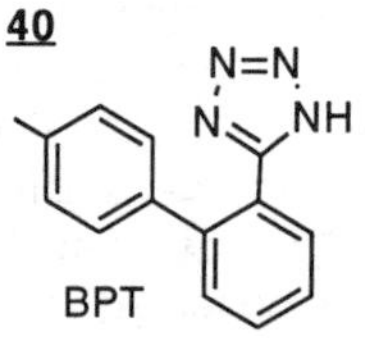

BPT

<u>CNS Penetration and Activity</u> - The blood brain barrier (BBB) effectively restricts the access of polar, hydrophilic drugs into the CNS, thus establishing a significant drug delivery impediment (83,84). The transfer of small organic molecules is believed to occur *via* passive diffusion or selective carrier-mediated uptake of the free drug which is then available for receptor occupancy and activity. Brain uptake of lipophilic drugs may in some cases exceed the apparent fu in plasma as determined by conventional techniques under equilibrium conditions. Brain extraction of ³H-propranolol (**42**) was reported to be greater than expected based on fu in the presence of varying concentrations of AAG, the major plasma protein to which propranolol is bound (85). The brain extraction for a series of benzodiazepines, drugs highly bound to albumin, also could not be described simply in terms of the fu in plasma, as determined by equilibrium dialysis (86). The mechanism postulated for these observations describes the rapid dissociation of bound drug at the capillary endothelium and the transfer of unbound drug across the membrane, reminiscent of the "nonrestrictive" binding discussed earlier. Another study of the brain transport of diazepam at steady-state using microdialysis found the uptake to be explained adequately by consideration of the fu in plasma (87). The therapeutic concentrations of neuroleptics in plasma water (unbound) and spinal fluid have been correlated with *in vitro* potencies (88).

References

1. F. Hervé, S. Urien, E. Albengres, J-C. Duché and J-P Tillement, Clin. Pharmacokinet., 26, 44 (1994).
2. M.A. Mohler, J.E. Cook and G. Baumann, in "Protein Pharmacokinetics and Metabolism," B. L. Ferraiolo, M. J. Mohler and C. A. Gloff, Eds., Plenum Press, New York, 1992, p. 35.
3. T. Peters Jr., Adv. Protein Chem., 37, 161, (1985).

4.   D.C. Carter and J.X. Ho, Adv. Protein Chem., 45, 153 (1994).
5.   X.M. He and D.C. Carter, Nature, 358, 209 (1992.)
6.   J.X. Ho, E.W. Holowachuk, E.J. Norton, P.D. Twigg and D.C. Carter, Eur. J. Biochem., 215, 205 (1993).
7.   J.R. Olivieri and A.F. Craievich, Eur. Biophys. J., 24, 77 (1995).
8.   D.C. Carter, X.-M. He, S.H. Munson, P.D. Twigg, K.M. Gernert, M.B. Broom and T.Y. Miller, Science, 244, 1195 (1989).
9.   D.C.Carter and X.M. He, Science, 249, 302, (1990).
10.   A.J. Luft and F.L. Lorscheider, Biochemistry, 22, 5978, (1983).
11.   D.C.Carter, B. Chang, J.X. Ho, K. Keeling and Z. Krishnasami, Eur. J. Biochem., 226, 1049, (1994).
12.   G. Sudlow, D.J.Birkett and D.N. Wade, Mol. Pharmacol., 12, 1052 (1976).
13.   F. Walji, A. Rosen and R.C. Hider, J. Pharm. Pharmacol., 45, 551 (1993).
14.   S. Urien, P. Nguyen, S. Berlioz, F. Brée, F. Vacherot and J.-P. Tillement, Biochem J., 302, 69 (1994).
15.   H. Aki and M. Yamamoto, J. Pharm. Sci., 83, 1712 (1994).
16.   H. Aki, M. Goto and M. Yamamoto, Thermochimica Acta, 251, 379 (1995).
17.   G. Ascoli, C. Bertucci and P. Salvadori, J. Pharm. Sci., 84, 737 (1994).
18.   S. Hashimoto, T. Yabusaki, H. Takeuchi and I. Harada, Biospectroscopy, 1, 375 (1995).
19.   F. Lapicque, N. Muller, E. Payan, N. Dubois and P. Netter, Clin. Pharmacokinet., 25, 115 (1993).
20.   M.H. Rahman, K.Yamasaki, Y.-H. Shin, C. C. Lin and M. Otagiri, Biol. Pharm. Bull., 16, 1169 (1993).
21.   T. Maruyama, C. C. Lin, K. Yamasaki, T. Miyoshi, T. Imai, M. Yamasaki and M. Otagiri, Biochem. Pharmacol., 45, 1017 (1993).
22.   T. Nomura, K. Sakamoto, T. Imai and M. Otagiri, J. Pharmacobio-Dyn., 15, 589 (1992).
23.   H. Kohita, Y. Matsushita and I. Moriguchi, Chem. Pharm. Bull., 42, 937 (1994).
24.   M.H. Rahman, T. Maruyama, T. Okada, K. Yamasaki and M. Otagiri, Biochem. Pharmacol., 46, 1721 (1993).
25.   Y. Keita, W. Wörner, G. Veile, B.G. Woodcock and U. Fuhr, Arzneim. Forsch./Drug Res., 46, 164 (1996).
26.   D.S. Hage, T.A.G. Noctor and I.W. Wainer, J. Chromatogr. A, 693, 23 (1995).
27.   S. Rahim and A.-F. Aubry, J. Pharm. Sci., 84, 949 (1995).
28.   J. Gonzàlez-Jimènez, F. Moreno and F. Garcìa Blanco, J. Pharm. Pharmacol., 47, 436 (1995).
29.   M.G. Jakoby IV, D.F. Covey and D.P. Cistola, Biochemistry, 34, 8780 (1995).
30.   R.M. Mader, G.G. Steger, B. Rizovski, R. Jakesz and H. Rainer, J. Pharm. Sci., 83, 1247 (1994).
31.   N. Takamura, M.H. Rahman, K. Yamasaki, M. Tsuruoka, M. Otagiri, Pharm. Res., 11, 1452 (1994).
32.   T. Maruyama, M.A. Furuie, S. Hibino and M. Otagiri, J. Pharm. Sci., 81, 16 (1992).
33.   R. Kaliszan, T.A.G. Noctor, I.W. Wainer, Mol. Pharm., 42, 512 (1992).
34.   F. D.-M. Péhourcq, A. Radouane, L. Labat and B. Bannwarth, Pharm. Res., 12, 1535 (1995).
35.   S. Tawara, S. Matsumoto, Y. Matsumoto, T. Kamimura and S. Goto, J. Antibiotics, 45, 1346 (1992).
36.   T. Izumi and T. Kitagawa, Chem. Pharm. Bull., 37, 742 (1989).
37.   J.M. Kremer, J. Wilting and L.H.M. Janssen, Pharmacol. Rev., 40, 1 (1988).
38.   A. Koj, in A.C. Allison, Ed., "Structure and Function of Plasma Proteins," Plenum Press, New York, 1974, p.73.
39.   K. Schmid, in P. Bauman, C.B. Eap, W.E. Muller and J.P. Tilliment, Eds., "α-Acid Glycoprotein: Genetics, Biochemistry, Physological Functions and Pharmacology," Alan R. Liss Inc., New York, 1989, p.7.
40.   R. Kaliszan, A. Nasal and M. Turowski, Biomed. Chromatogr., 9, 211 (1995).
41.   T. Muriyama, M. Otagiri and A. Takadate, Chem. Pharm. Bull., 38, 1688 (1990).
42.   M.B. Brown, J.N. Miller and N.J. Seare, J. Pharmaceut. Biomed. Analysis, 13, 1011 (1995).
43.   T. Miyoshi, R. Yamamichi, T. Maruyama and M. Otagiri, Pharm. Res., 9, 845 (1992).
44.   T. Miyoshi, K. Sukimoto and M. Otagiri, J. Pharm. Pharmacol., 44, 28 (1992).
45.   S. Urien, F. Brée, B. Testa and J.-P. Tillement, Biochem. J., 280, 277 (1991).
46.   S. Urien, F. Brée, B. Testa and J.-P. Tillement, Biochem. J., 289, 767 (1993).
47.   S. Urien, Y. Giroud, R.-S. Tsai, P.-A. Carrupt, F. Brée, B. Testa and J.-P. Tillement, Biochem. J., 306, 545 (1995).
48.   A. Nasal, A. Radwańska, K. Ośmialowski, A. Buciński, R. Kaliszan, G. E. Barker, P. Sun and R. A. Hartwick, Biomed. Chromatogr., 8, 125 (1994).
49.   R. Kaliszan, A. Nasal and M. Turowski, Biomed. Chromatogr., 9, 211 (1995).
50.   R. Kaliszan, A. Nasal and M. Turowski, J. Chromatogr. A, 722, 25 (1996).
51.   S. Urien, P. Nguyen, G. Bastian, C. Lucas and J.-P. Tillement, Investigational New Drugs, 13, 37 (1995).
52.   L.R. Pohl, H. Satoh, D.D. Christ and J.G. Kenna, Ann. Rev. Pharmacol., 28, 367 (1988).
53.   N. Dubois, F. Lapicque, M-H. Maurice, M. Pritchard, S. Fournel-Gigleux, J. Magdalou, M. Abiteboul, G. Siest and P. Netter, Drug Metab. Dispos., 21, 617 (1993).
54.   A. Ding, J.C. Ojingwa, A.F. McDonagh, A.L. Burlingame and L.Z. Benet, Proc. Natl. Acad. Sci. U.S.A., 90, 3797 (1993).

55. A. Munafo, M.L. Hyneck and L.Z. Benet, Pharmacol., 47, 309 (1993).
56. J.C. Ojingwa, H. Spahn-Langguth and L.Z. Benet, J. Pharmacokinet. Biopharmaceut., 22, 19 (1994).
57. A. Ding, P. Zia-Amirhosseini, A.F. McDonagh, A.L. Burlingame and L.Z. Benet, Drug Metab. Dispos., 23, 369 (1995).
58. J. Oravcova, B. Bohs and W. Lindner, J. Chromatogr. B., 677, 1 (1996).
59. A. Le Quellec, S. Dupin, A.E. Tufenkji, P. Genissel and G. Houin, Pharm. Res., 11, 835 (1994).
60. M. J. Alonso, A. Bruelisauer, P. Misslin and M, Lemaire, Pharm. Res., 12, 291 (1995).
61. A.S. Joshi, H.J. Pieniaszek Jr., C.Y. Quon and S.-Y.P. King, J. Pharm. Sci., 83, 1187 (1994).
62. T.A.G. Noctor, M.J. Diaz-Perez and I.W. Wainer, J. Pharm. Sci., 82, 675 (1993).
63. D.E. Epps, T.J. Raub and F.J. Kézdy, Anal. Biochem., 227, 342 (1995).
64. I. M. Klotz, Science, 217, 1247 (1982).
65. M. Gibaldi, G. Levy and P.J. McNamara, Clin. Pharmacol. Therap., 24, 1 (1978).
66. C.A. Rodriguez and D.A. Smith, Antimicrob. Agents Chemother., 35, 2395 (1991).
67. D.R. Taft and K.R. Sweeney, J. Pharmacol. Exp. Therap., 274, 752 (1995).
68. G.R. Wilkinson and D.G. Shand, Clin. Pharmacol. Therap., 18, 377 (1975).
69. B. Davies and T. Morris, Pharm. Res., 10, 1093 (1993).
70. S. Yamashita, T. Nadai, M. Sumi and Y. Suda, Int. J. Pharmaceut., 108, 241 (1994).
71. J.A. Bilello, P.A. Bilello, J.J. Kort, M.N. Dudley, J. Leonard and G.L Drusano, Antimicrob. Agents Chemotherap., 39, 2523 (1995).
72. S. Kageyama, B.D. Anderson, B.L. Hoesterey, H. Hayashi, Y. Kiso, K.P. Flora and H. Mitsuya, Antimicrob. Agents Chemother., 38, 1107 (1994).
73. M. Bryant, D. Getman, M. Smidt, J. Marr, M. Clare, R. Dillard, D. Lansky, G. DeCrescenzo, R. Heintz, K. Houseman, K. Reed, J. Stolzenbach, J. Talley, M. Vazquez and R. Mueller, Antimicrob. Agents Chemother., 39, 2229 (1995).
74. D.J. Livingston, S. Pazhanisamy, D.J.T. Porter, J.A. Partaledis, R.D. Tung and G.R. Painter, J. Infect. Dis., 172, 1238 (1995).
75. M.H. St. Clair, J. Millard, J. Rooney, M. Tisdale, N. Parry, B.M. Sadler, M.R. Blum and G. Painter, Antiviral Res., 29, 53 (1996).
76. L. Shum, C.A. Robinson and H.J. Pieniaszek Jr., Pharm. Res., 10, 366S (1993).
77. S. Erickson-Viitanen, R.M. Klabe, P.G. Cawood, P.L. O'Neal and J.L. Meek, Antimicrob. Agents Chemotherap., 38, 1628 (1994).
78. D.D. Christ, J. Clin. Pharmacol., 35, 515 (1995).
79. A.T. Chiu, D.J. Carini, J.V. Duncia, K.H. Leung, D.E. McCall, W.A. Price, P.C. Wong, R.D. Smith, R.R. Wexler and P.B.M.W.M. Timmermans, Biochem. Biophys. Res. Comm., 177, 209 (1991).
80. D.D. Christ, P.C. Wong, Y.N. Wong, S.D. Hart, C.Y. Quon and G.N. Lam, J. Pharmacol. Exp. Therap., 268, 1199 (1994).
81. P. Wong, D.D. Christ, Y.N. Wong and G.N. Lam, Pharmacol., 52, 25 (1996).
82. H.T. Beauchamp, R.S.L. Chang, P.K.S. Siegl and R.E. Gibson, J. Pharmacol. Exp. Therap., 272, 612 (1995).
83. W.M. Pardridge, Advan. Drug Deliv. Rev., 15, 1 (1995).
84. W.M. Pardridge, Drug Deliv., 1, 83 (1993).
85. W.M. Pardridge, R. Sakiyama and G. Fierer, J. Clin. Invest., 71, 900 (1983).
86. D.R. Jones, S.D. Hall, E.K. Jackson, R.A. Branch and G.R. Wilkinson, J. Pharmacol. Exp. Therap., 245, 816 (1988).
87. R.K. Dubey, C.B. McAllister, M. Inoue and G.R. Wilkinson, J. Clin. Invest., 84, 1155 (1989).
88. P. Seeman, Int. Clin. Psychopharmacol., 10 Suppl. 3, 5 (1995).

SECTION VII.   TRENDS AND PERSPECTIVES

Editor: James A. Bristol
Parke-Davis Pharmaceutical Research Division
Warner-Lambert Co., Ann Arbor, MI 48105

Chapter 34.   To Market, To Market - 1995

Xue-Min Cheng
Parke-Davis Pharmaceutical Research Division
Warner-Lambert Co., Ann Arbor, MI 48105

New chemical entities (NCEs) introduced for human therapeutic use into the world market for the first time during 1995 totaled 36 (1). This reflects a decrease from 44 in 1994 (2) and 43 in 1993 (3) but at the same level as both 1992 (4) and 1991 (5).

Japan, the perennial leader for worldwide NCE introductions, took the number one position again with 11 newly marketed drugs although its market was relatively quiet compared with past few years (2,3). The United States was the second with 8 new launches followed by the United Kingdom with 5. France and Sweden tied for the fourth place with 2 NCEs each. A similar trend was found in the category of originators for the NCEs. Japan was responsible for 10 NCEs and the United States was the originator of 8 NCEs. The United Kingdom was the third with 5. Switzerland and France followed as the fourth and the fifth places responsible for 4 and 3 new drugs, respectively. Noticeably, the European countries together accounted for 15 new launches and originated 16 of all NCEs. The European Medicine Evaluation Agency was established in 1995 and will process "European" approvals in the future through the centralized system which should have an impact on the regulatory affairs around the world.

Antineoplastic agents were the most active therapeutic category with 8 new launches. Cardiovascular drugs and antiinfectives/antivirals were other active groups with 5 NCEs each followed by 4 respiratory drugs and 3 CNS agents. Although the total number of new launches was lower in 1995 than in the previous two years, there were several innovative therapies which reflected real advances in treatment of certain serious diseases. Highlights include defeiprone, the first oral iron chelator that provides life-saving benefits to patients with thalassaemia, a disease prevalent in less developed countries; ranitidine bismuth citrate, the first therapy specifically tailored to attack *Helicobactor pylori* that appears to prevent relapse in most peptic ulcer patients. Novel NCEs with interesting mechanisms of action include saquinavir, the first HIV protease inhibitor for AIDS; tirilazad, the first peroxidation inhibitor designed to reduce tissue damage following subarachnoid hemorrhage in men; seratrodast, the first thromboxane A_2 antagonist as a therapeutic for asthma; mycophenolate mofetil, the first immunosuppressant introduced in the U.S.A. since 1983; amifostine, an effective cytoprotective for normal cells reflecting a new concept in cancer therapy; topiramate, the first of a new class of antiepileptics with a unique combined mechanism of action.

In 1995, 28 new molecular entities were approved in the United States (6,7), 6 more than in the previous year (2). Among these, 9 received "priority" reviews by the FDA resulting in an average of 17 months for approval time which was 30% less than standard reviews. Twenty three of the NCEs, compared with 14 in 1994, were assessed user fees. As usual, December was the busiest month with 9 approvals which account for 32% of the total. Antineoplastic and cardiovascular drugs dominated new approvals in 1995 with 5 each. Five of the 28 new approvals, saquinavir, lamivudine, mycophenolate mofetil, moexipril, and nalmefene reached their first worldwide market within the same year of their approval. Saquinavir, the first HIV protease inhibitor, was approved by the FDA just 97 days after NDA filing and was launched the following day.

Amifostine (Cytoprotective) (8-11)

Country of Origin: **U.S.A.**
Originator: **US Bioscience**
First Introduction: **Germany**
Introduced by: **Schering Plough**
Trade Name: **Ethyol**
CAS Registry No.:   **63717-27-1**
Molecular Weight: **232.24**
Dosage Form: **500 mg Vial**

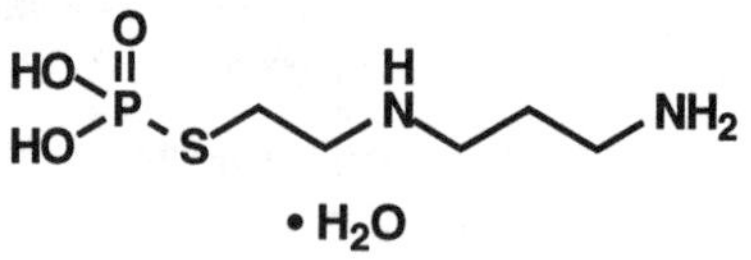

Amifostine, an organic thiophosphate, was introduced last year for the reduction of cisplatin-induced renal toxicity in patients with advanced ovarian cancer. It is also a radio-protective agent. Amifostine is a prodrug which is rapidly dephosphorylated, preferentially in non-tumor tissues to the active thiol. This agent binds to chemotherapeutic drugs and free radicals released by radiotherapy. The protective effect of amifostine was observed in a wide range of normal organs including bone marrow and gastrointestinal mucosa without interfering with the anti-tumor effect of chemo/radiotherapy, indicating an increased therapeutic index for existing cancer treatment. Amifostine is also being evaluated as a cytoprotective in other types of tumors including lung, breast, head and neck cancers. It was reported to be a potent mucolytic with potential in cystic fibrosis.

Anastrozole (Antineoplastic) (12-14)

Country of Origin: **United Kingdom**
Originator: **Zeneca**
First Introduction: **United Kingdom**
Introduced by: **Zeneca**
Trade Name: **Arimidex**
CAS Registry No.:   **120511-73-1**
Molecular Weight: **293.37**
Dosage Form: **1 mg Tablet**

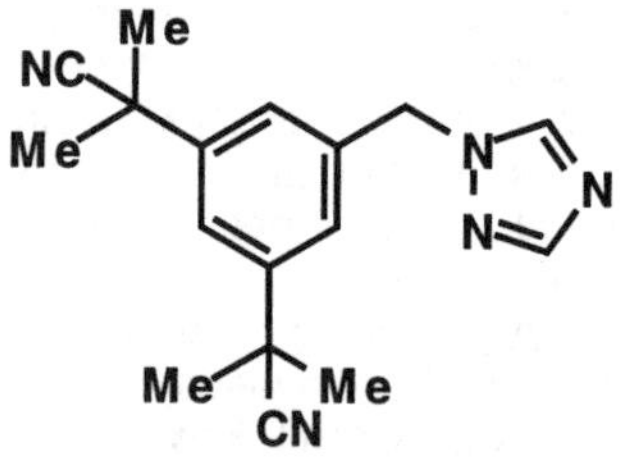

Anastrozole entered its first market in the United Kingdom for the treatment of advanced breast cancer in post-menopausal women. Anastrozole is a highly potent and selective aromatase inhibitor. It is extremely potent in lowering circulating estradiol to undetectable levels in treated patients without altering other circulating hormones. The drug is reportedly well absorbed and tolerated following oral administration.

Bicalutamide (Antineoplastic) (15-18)

Country of Origin: **United Kingdom**
Originator: **Zeneca**
First Introduction: **United Kingdom**
Introduced by: **Zeneca**
Trade Name: **Casodex**
CAS Registry No.:   **90357-06-5**
Molecular Weight: **430.37**
Dosage Form: **50 mg Tablet**

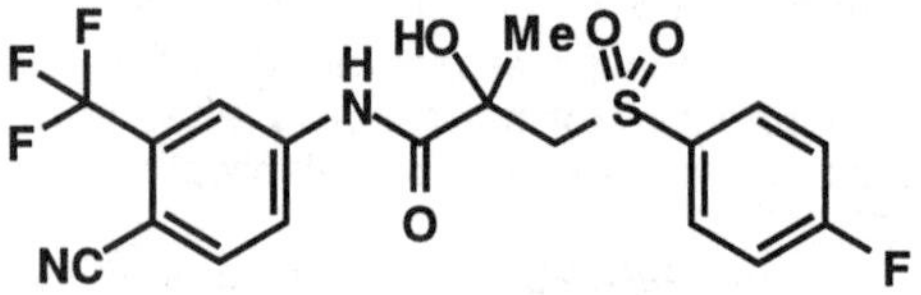

Bicalutamide was launched in the United Kingdom, its first worldwide market, for the treatment of advanced prostate cancer in combination with an LHRH analog or surgical castration. A non-steroidal, peripherally selective antiandrogen, bicalutamide inhibits the action of dihydrotestosterone and testosterone at target sites by competitive binding to the cytosolic androgen receptor. It was reportedly well tolerated with no significant cardiovascular and metabolic side effects due to the benefit of lacking any steroid activity. The efficacy of bicalutamide as a monotherapy has been demonstrated clinically. Promising response rates were also reported in treating colorectal, breast, pancreas and non-small cell lung cancers.

Carperitide (Congestive Heart Failure) (19,20)

Country of Origin:	**Japan**	Trade Name:	**Hanp**
Originator:	**Suntory**	CAS Registry No.:	**89213-87-6**
First Introduction:	**Japan**	Molecular Weight:	**3080.52**
Introduced by:	**Zeria**	Dosage Form:	**1000 μg Inj.**

Carperitide is the α-human atrial natriuretic peptide (α-hANP) produced by recombinant technology. It was introduced in Japan as a treatment for acute congestive heart failure (CHF). In dogs with CHF, carperitide significantly reduced the elevated left ventricular end-diastolic pressure and the index of myocardial oxygen consumption (systolic blood pressure x heart rate). In a clinical trial, carperitide was effective in improving hemodynamics and symptoms in 60% of patients with acute CHF. Carperitide was reported to be well tolerated with no significant adverse effects clinically.   A beneficial effect on hemodynamics has been reported in chronic heart failure patients. Carperitide is also in clinical trials for maintenance of blood pressure during surgical operations.

Cefozopran Hydrochloride (Injectable Cephalosporin) (21-25)

Country of Origin:	**Japan**
Originator:	**Takeda**
First Introduction:	**Japan**
Introduced by:	**Takeda**
Trade Name:	**Firstcin**
CAS Registry No.:	**125905-00-2**
Molecular Weight:	**551.99**
Dosage Form:	**0.5 g, 1 g Inj.**

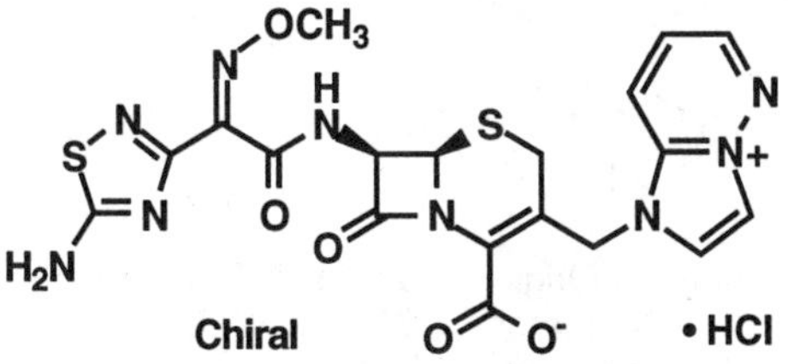

Cefozopran hydrochloride, a third-generation parenteral cephalosporin, was launched last year in Japan for the treatment of severe infections in immunocompromised patients caused by staphylococci and enterococci. Cefozopran displays a very broad antibacterial spectrum against Gram-positive and Gram-negative organisms. It is potently active against *Staphylococcus aureus, Enterococcus faecalis, Pseudomonas aeruginosa* and *Citrobacter freundii*, some of which are resistant to most other cephalosporins. The superior wide spectrum antibacterial activity is attributed to its high affinities for penicillin-binding proteins in various organisms. Cefozopran is resistant to hydrolysis by most chromosomal and plasmid-mediated β-lactamases. Cefozopran is reportedly effective in treating patients with respiratory tract, urinary tract, obstetrical, gynecological, soft tissue, and surgical infections.

Cilnidipine (Antihypertensive) (26-28)

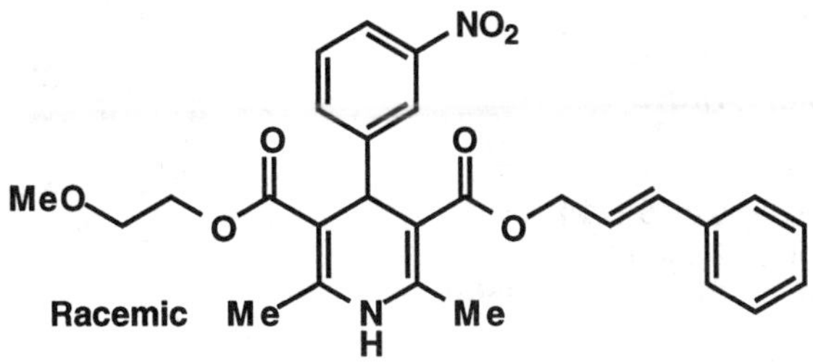

Country of Origin:	**Japan**	Trade Name:	**Cinalong;**
Originator:	**Fujirebio**		**Siscard;**
First Introduction:	**Japan**		**Atelec**
Introduced by:	**Fujirebio;**	CAS Registry No.:	**132203-70-4**
	Boehringer Ingelheim;	Molecular Weight:	**492.51**
	Roussel-Morishita	Dosage Form:	**5,10 mg Tab.**

Cilnidipine is a dihydropyridine calcium antagonist introduced last year for the treatment of essential and severe hypertension and hypertension associated with renopathy. Cilnidipine has been demonstrated to exert a potent vasodilating effect by blocking calcium influx via dihydropyridine-sensitive, voltage-dependent calcium channels. Compared to nifedipine and nicardipine, cilnidipine is superior, especially for long-term treatment, due to its characteristics of slow onset and long duration with less cardiodepressant activity. Cilnidipine is reportedly well tolerated with low toxicity. It may also be useful in ischemic heart disease.

Cisatracurium Besylate (Muscle Relaxant) (29-31)

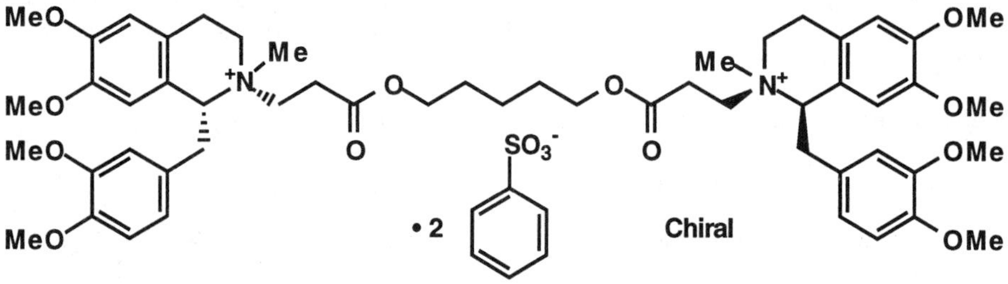

Country of Origin:	**United Kingdom**	Trade Name:	**Nimbex**
Originator:	**Glaxo Wellcome**	CAS Registry No.:	**96946-42-8**
First Introduction:	**U.S.A.**	Molecular Weight:	**1243.44**
Introduced by:	**Glaxo Wellcome**	Dosage Form:	**2 mg/ml Ampule**
			5 mg/ml Vial

Cisatracurium besylate is a new intermediate-duration non-depolarizing muscle relaxant launched last year in the U.S.A. for intubation and maintenance of muscle relaxation during surgery and intensive care. Cisatracurium besylate is the single 1*R cis*-1'*R cis* isomer of the commercial preparation of atracurium, a mixture of 10 isomers. It is 3 to 5 times more potent than the mixture with similar onset and duration of action. The single isomer was also reported to have reduced propensity to release histamine and have a stable cardiovascular profile. Similar to structurally related mivacurium, cisatracurium has distinct advantages of rapid degradation, enzymatic metabolism that is independent of liver or kidney resulting in short duration of action and fast, complete recovery.

Deferiprone (Iron Chelator) (32-35)

Country of Origin:	**India**
Originator:	**Cipla**
First Introduction:	**India**
Introduced by:	**Cipla**
Trade Name:	**Kelfer**
CAS Registry No.:	**30652-11-0**
Molecular Weight:	**139.15**
Dosage Form:	**250 mg Capsule**

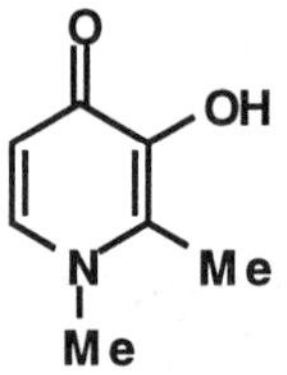

Deferiprone, the first oral iron chelator, was marketed last year in India for the management of thalassaemia. Patients with thalassaemia, a blood related genetic disorder, require life time transfusion which causes excessive deposition of iron in liver and spleen, subsequent damage to organs and eventually death unless iron is removed by a chelator. Deferiprone is a potent iron chelator that mobilizes excessive iron from iron storage proteins ferritin and hemosiderin, from iron saturated transferrin and lactoferrin, but not from hemoglobin. The deferiprone-iron complex is excreted in urine and bile. Deferiprone was reportedly well accepted by patients and no hematological toxicity was observed. Deferiprone has also been demonstrated as an effective and safe chelator in the mobilization of aluminum.

Docetaxel (Antineoplastic) (36-40)

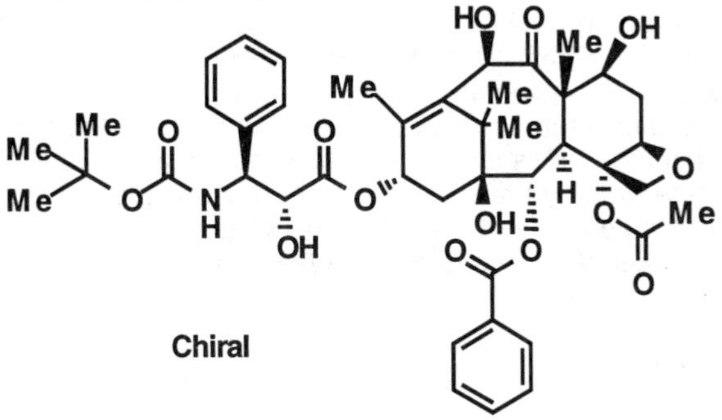

Chiral

Country of Origin:	**France**	Trade Name:	**Taxotere**
Originator:	**Rhône-Poulenc Rorer**	CAS Registry No.:	**114977-28-5**
First Introduction:	**South Africa**	Molecular Weight:	**807.86**
Introduced by:	**Rhône-Poulenc Rorer**	Dosage Form:	**20,80 mg Vial**

Docetaxel, a semi-synthetic product from the taxoid family, was launched in 1995 first in South Africa and subsequently in several other markets for the treatment of ovarian, breast and non-small cell lung cancers. Like the naturally occurring antitumor agent paclitaxel, the first marketed taxoid, docetaxel promotes both the rate and extent of tubulin assembly into stable microtubules and inhibits their depolymerization. It acts as a mitotic spindle poison and induces a mitotic block in proliferating cells. This mechanism of action for taxoids is unique from other classes of anticancer agents. Docetaxel was reported to be twice as potent as paclitaxel in several *in vitro* protocols and also exhibit higher cytotoxicity. Clinical trials are on going for docetaxel for other types of tumors including pancreatic, gastric, head and neck cancers and soft tissue sarcomas.

Dorzolamide Hydrochloride (Antiglaucoma) (41-43)

Country of Origin:	**U.S.A.**
Originator:	**Merck**
First Introduction:	**U.S.A.**
Introduced by:	**Merck**
Trade Name:	**Trusopt**
CAS Registry No.:	**130693-82-2**
Molecular Weight:	**360.89**
Dosage Form:	**5 ml Solution**

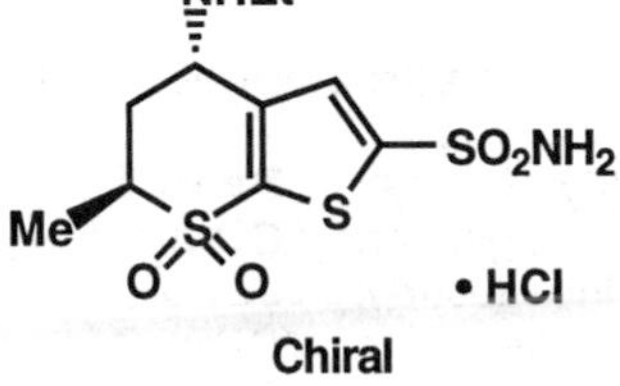

Chiral

Dorzolamide hydrochloride was introduced last year in the U.S.A. as eye drops for the treatment of open-angle glaucoma and ocular hypertension. Dorzolamide is a potent carbonic anhydrase inhibitor that, on topical administration, lowers intraocular pressure (IOP). In a clinical study in patients with bilateral primary open-angle glaucoma or ocular hypertension, dorzolamide (0.7-2%) was found to significantly lower IOP throughout the day. Dorzolamide has been shown to be effective as a single therapy or in combination with β-blockers.

Duteplase (Anticougulant) (44-46)

Country of Origin:	**Japan**	Trade Name:	**Solclot**
Originator:	**Sumitomo**	CAS Registry No.:	**120608-46-0**
First Introduction:	**Japan**	Dosage Form:	**10 MIU Inj.**
Introduced by:	**Sumitomo**		

  Duteplase, a nearly pure double-chain recombinant tissue-type plasminogen activator (rt-PA), was introduced last year in Japan for the treatment of acute myocardial infarction. Similar to other thrombolytic agents, duteplase acts by converting the proenzyme plasminogen to the active enzyme plasmin, resulting in fibrinolysis and varying degrees of depletion of circulating fibrinogen, factor V and factor VIII. In patients with acute myocardial infarction, treatment with duteplase induced patency rate of infarct-related artery at 90 minutes by up to 69%. Compared with alteplase, a single-chain rt-PA, duteplase has similar activity but a slower clearance rate and thus lower doses of duteplase have a similar therapeutic effect to alteplase.

Erdosteine (Expectorant) (47-49)

Country of Origin:	**Switzerland**
Originator:	**Refarmed**
First Introduction:	**France**
Introduced by:	**Pierre Fabre; Negma**
Trade Name:	**Edirel; Vectrine**
CAS Registry No.:	**84611-23-4**
Molecular Weight:	**249.30**
Dosage Form:	**300 mg Capsule**

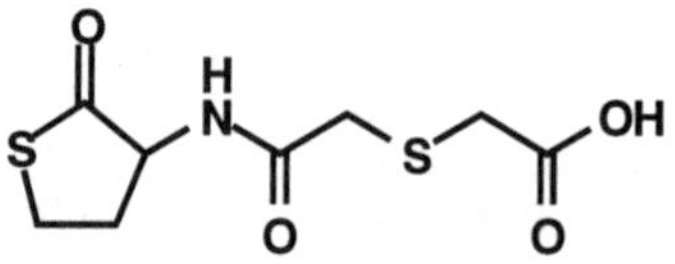

Racemic

  Erdosteine is a new expectorant introduced last year in France for the treatment of chronic bronchitis. Erdosteine possesses mucomodulator, mucolytic, mucokinetic and free radical scavenging properties. Its therapeutic efficacy has been demonstrated in patients affected by congestion and viscous mucus in higher and lower respiratory tracts as in acute and chronic bronchitis, sinusitis, otitis, tracheopharyngitis and coryza. Erdosteine acts through its active metabolites containing free sulfhydryl groups which are able to depolymerize the mucopolysaccharides of the excreatum by breaking the disulfide linkages (thereby lowing its viscosity). The prodrug approach provides unique characteristics to erdosteine by offering stable therapeutic efficacy and prevention of the gastric mucus from lythic phenomena. The agent was also reported to have local antiinflammatory and antielastase properties and enhances penetration of antibiotics into bronchial mucus.

Fadrozole Hydrochloride (Antineoplastic) (50-53)

Country of Origin:	**Switzerland**
Originator:	**Ciba-Geigy**
First Introduction:	**Japan**
Introduced by:	**Ciba-Geigy**
Trade Name:	**Afema**
CAS Registry No.:	**102676-31-3**
Molecular Weight:	**259.73**
Dosage Form:	**1 mg Tablet**

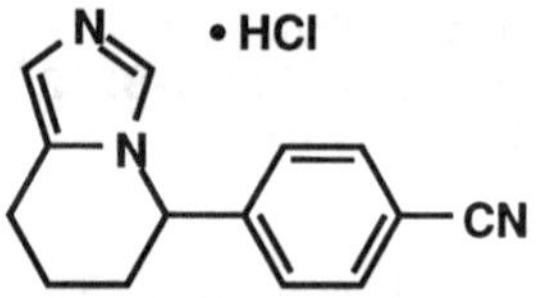

Racemic

Fadrozole hydrochloride is a non-steroidal imidazole derivative launched last year for the treatment of post-menopausal breast cancer. Fadrozole is a potent and specific aromatase inhibitor with neither androgenic nor estrogenic activities due to its structural novelty. Fadrozole exerts its action by coordinating with the iron of the porphyrin nucleus, presumably through the imidazole moiety. This strong binding competes with the binding of molecular oxygen to iron and reversibly inactivates the enzyme. An excellent clinical response rate has been reported for fadrozole with no significant side effects.

Fasudil Hydrochloride (Neuroprotective) (54-57)

Country of Origin:	**Japan**
Originator:	**Asahi Chemical**
First Introduction:	**Japan**
Introduced by:	**Asahi Chemical**
Trade Name:	**Eril**
CAS Registry No.:	**105628-07-7**
Molecular Weight:	**327.82**
Dosage Form:	**1 mg Tablet**

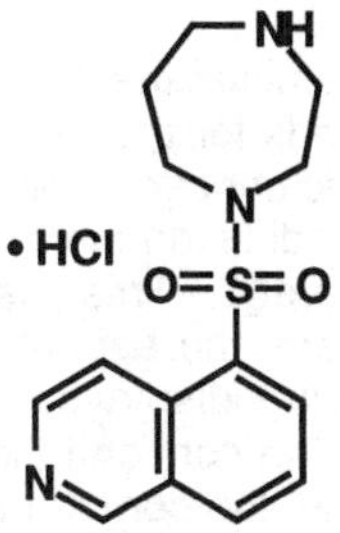

Fasudil hydrochloride, a novel calcium antagonistic vasodilator, was marketed last year for the treatment of cerebral vasospasm following subarachnoid hemorrhage (SAH). Fasudil is also a potent inhibitor of myosin light chain kinase and protein kinase C. In contrast to other calcium channel blockers, which regulate the influx of calcium ions through the cell membrane but not involved in the intracellular regulatory mechanism of the calcium, fasudil was suggested to have an intracellular mode of action in relaxing vascular smooth muscle. In patients with neurological deficits due to vasospasm, fasudil decreased the occurrence of angiographic severe and symptomatic vasospasm and cerebral infarction without decreasing systemic blood pressure. Fasudil is reportedly in clinical trials for acute ischemic stroke, sequelae of cerebral vascular diseases and angina pectoris.

Flutrimazole (Topical Antifungal) (58-60)

Country of Origin:	**Spain**
Originator:	**Uriach**
First Introduction:	**Spain**
Introduced by:	**Uriach**
Trade Name:	**Micetal**
CAS Registry No.:	**119006-77-8**
Molecular Weight:	**346.37**
Dosage Form:	**1 % Cream**

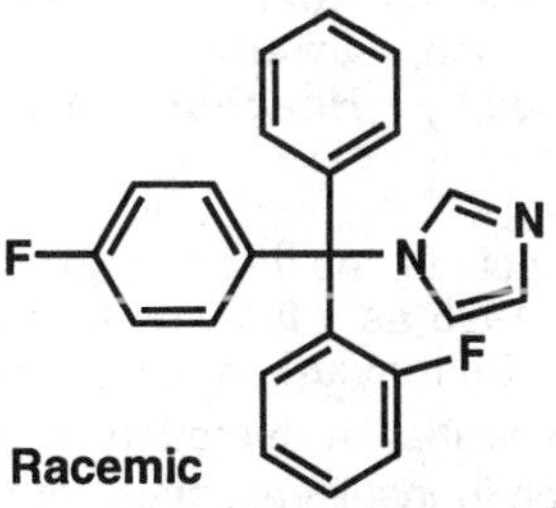

Racemic

Flutrimazole, a novel imidazole antifungal launched last year in Spain, is indicated for mycosis of the skin. Flutrimazole displays potent broad-spectrum of activity against dematophytes, filamentous fungi and yeasts, saprophytic and pathogenic to animals and humans. Similar to other imidazole and triazole antifungals, flutrimazole exerts its biological function through inhibition of fungal lanosterol 14α-demethylase. Excellent local and systemic tolerance of flutrimazole in patients has been reported.

Gemcitabine Hydrochloride (Antineoplastic) (61-63)

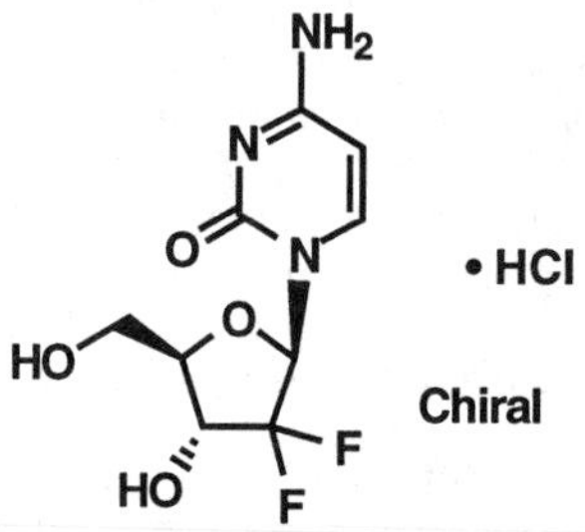

Country of Origin:	**U.S.A.**
Originator:	**Lilly**
First Introduction:	**Netherlands; Sweden**
Introduced by:	**Lilly**
Trade Name:	**Gemzar**
CAS Registry No.:	**122111-03-9**
Molecular Weight:	**299.67**
Dosage Form:	**200 mg Vial for injection**

Gemcitabine is a novel nucleoside analog that was launched in 1995 in the Netherlands for the treatment of non-small cell lung cancer (nsclc) and in Sweden for pancreatic cancer. Gemcitabine is a prodrug which is phosphorylated intracellularly by deoxycytidine kinase to its active forms, the di- and triphosphates which bind to DNA competitively. This insertion inhibits processes required for DNA synthesis and metabolism, the essential function for both cell replication and repair. Furthermore, gemcitabine displays an extraordinary array of self-potentiating mechanisms that increase the concentration and prolong the retention of its active nucleotides in tumor cells. The title compound has shown activity against a wide spectrum of human solid tumors including colon, mammary, breast, bladder cancers. Synergistic activity of gemcitabine with other anticancer agents such as cisplatin has been reported.

Glimepiride (Antidiabetic) (64,65)

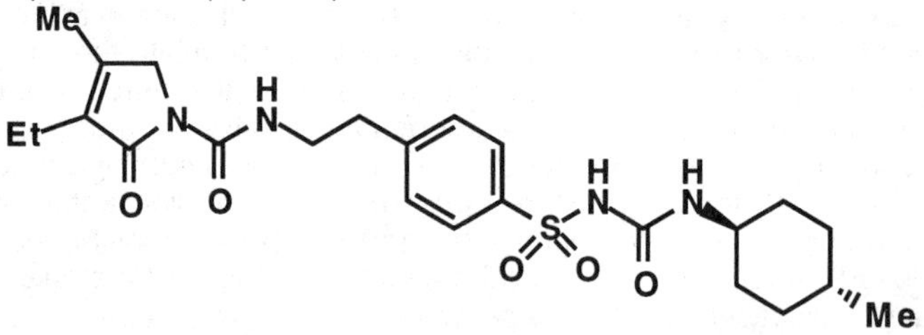

Country of Origin:	**Germany**	Trade Name:	**Amaryl**
Originator:	**Hoechst Marion Roussel**	CAS Registry No.:	**93479-97-1**
First Introduction:	**Sweden**	Molecular Weight:	**490.61**
Introduced by:	**Hoechst Marion Roussel**	Dosage Form:	**1, 2, 3, 4, 6 mg Tablet**

Glimepiride, the first of a new generation of sulfonylurea drugs, was introduced in Sweden in 1995 as a first-line therapy to lower blood glucose in patients with type II diabetes. Sulfonylureas exert their hypoglycemic function primarily by direct stimulation of insulin secretion in glucose-insensitive pancreatic β-cells and GLUT translocation in insulin-resistant fat and muscle cells. Once-daily, orally administered glimepiride in diabetes patients showed a more rapid and longer lasting glucose-lowering effect than the commonly used agent glibenclamide. Glimepiride can be used either as a monotherapy or in combination with insulin.

Itopride Hydrochloride (Gastroprokinetic) (66,67)

Country of Origin:	**Japan**
Originator:	**Hokuriku**
First Introduction:	**Japan**
Introduced by:	**Hokuriku**
Trade Name:	**Ganaton**
CAS Registry No.:	**122892-31-3**
Molecular Weight:	**394.89**
Dosage Form:	**50 mg Tablet**

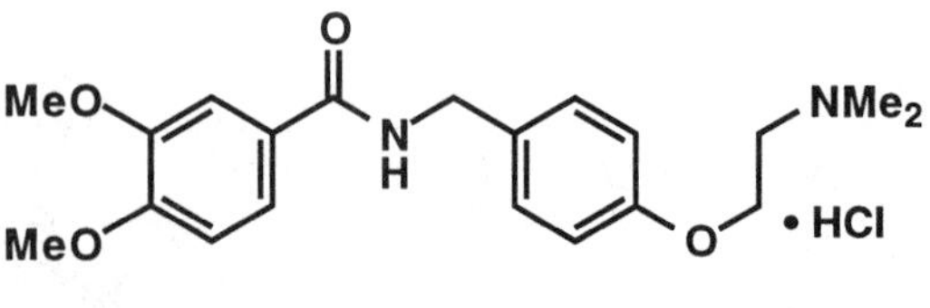

Itopride, a gastroprokinetic benzamide derivative was launched last year in Japan for the relief of gastrointestinal symptoms in patients with chronic gastritis. Itopride is a dopamine D_2-receptor antagonist that stimulates acetylcholine (Ach) release on the postganglionic cholinergic neurons to cause Ach accumulation at muscarinic receptors and, therefore, enhances Ach-induced gastric contractions. In animal models, itopride was reported to increase GI transit and gastric emptying.

Lamivudine (Antiviral) (68-71)

Country of Origin:	**Canada**
Originator:	**BioChem Pharma**
First Introduction:	**U.S.A.**
Introduced by:	**Glaxo Wellcome**
Trade Name:	**Epivir**
CAS Registry No.:	**134678-17-4**
Molecular Weight:	**229.26**
Dosage Form:	**150 mg Tablet; 10 mg/ml Sol.**

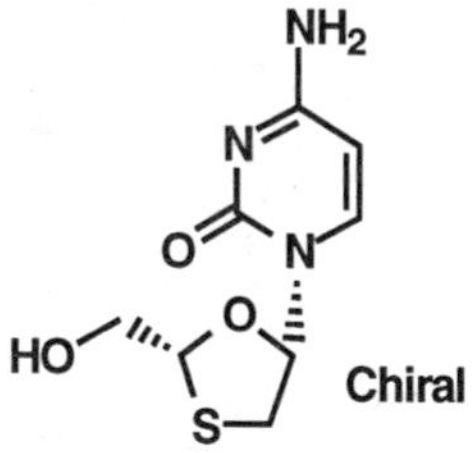

Lamivudine is a new generation orally active nucleoside analog launched last year in the U.S.A. for use in combination with zidovudine (AZT) as a first-line therapy for patients with HIV infection. Lamivudine is rapidly converted to phosphorylated metabolites in the body which act as inhibitors and chain terminators of HIV reverse transcriptase (RT), the enzyme required for the replication of the HIV genome. Lamivudine has similar inhibitory potency to RT as AZT but is 10 times less toxic and is active against AZT-resistant strains of HIV. Combination therapy of lamivudine and AZT produced a large decrease in blood-borne virus with an increase in CD4 cells, an effect that can be sustained for 2 years.   Since hepatitis B virus (HBV) also encodes a polymerase with a RT function necessary for the conversion of a RNA replicative intermediate to DNA, clinical efficacy has been reported for lamivudine in treating patients with HBV infection. It was reported that the enantiomer of lamivudine is equipotent against HIV but with considerably higher cytotoxicity.

Lanreotide Acetate (Acromegaly) (72-74)

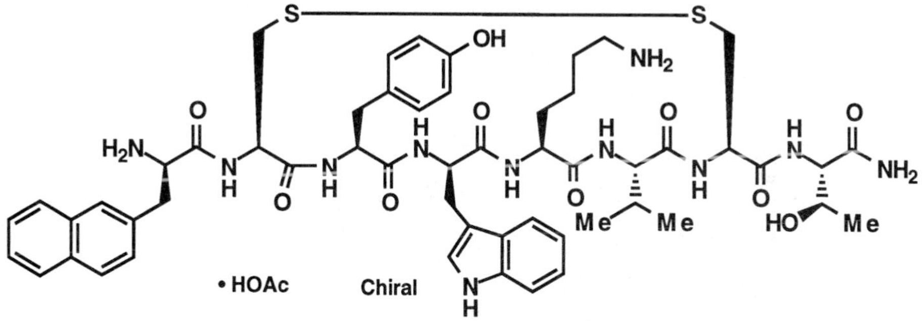

Country of Origin:	**France**	Trade Name:	**Somatuline LP**
Originator:	**Beaufour-Ipsen**	CAS Registry No.:	**127984-74-1**
First Introduction:	**France**	Molecular Weight:	**1156.37**
Introduced by:	**Beaufour-Ipsen**	Dosage Form:	**30 mg Vial + 2ml mannitol sol.**

Lanreotide acetate, an octapeptide somatostatin analog, reached its first worldwide market last year in France for acromegaly when surgery or radiotherapy have failed to restore normal growth hormone secretion. Lanreotide is a selective inhibitor of growth hormone and reduces the secretion of growth hormone, thyrotropin, motilin and pancreatic polypeptide in humans. Lanreotide has antiproliferative properties and is reportedly in clinical trials for the prevention of restenosis following coronary artery angioplasty, for diabetic retinopathy, and as a therapy for psoriasis. Its potential for neuroendocrine tumors and hormone-responsive prostate cancer has also been demonstrated.

Moexipril Hydrochloride (Antihypertensive) (75-78)

Country of Origin:	**U.S.A.**
Originator:	**Warner-Lambert**
First Introduction:	**U.S.A.**
Introduced by:	**Schwarz**
Trade Name:	**Univasc**
CAS Registry No.:	**82586-52-5**
Molecular Weight:	**535.02**
Dosage Form:	**7.5, 15 mg Tablet**

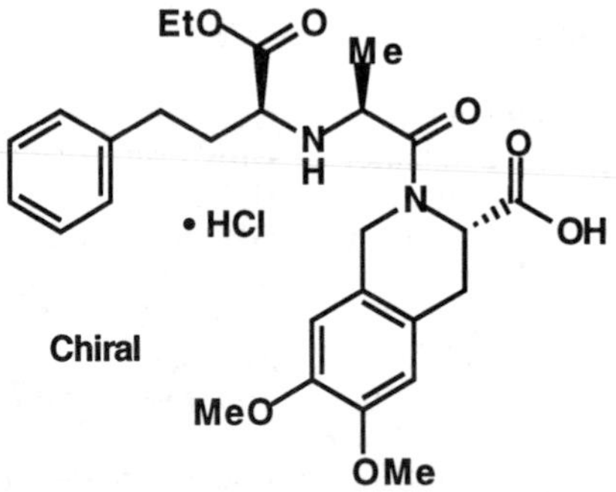

Moexipril hydrochloride is a novel ACE inhibitor first marketed in the U.S.A. for the treatment of hypertension as a monotherapy and as a second-line therapy in combination with diuretics or calcium antagonists. Like other ACE inhibitors, moexipril is a prodrug that is converted in the liver to its diacid moexiprilat which is the active agent. Moexipril displays a higher *in vitro* inhibitory potency to ACE than enalapril although its effectiveness in reduction of blood pressure in hypertensive patients is similar to that seen with enalapril. It is orally active with a rapid onset and prolonged duration of action. Excellent tolerability has been reported. Moexipril is distinguished by its lower cost than other marketed ACE inhibitors.

Mycophenolate Mofetil (Immunosuppressant) (79-82)

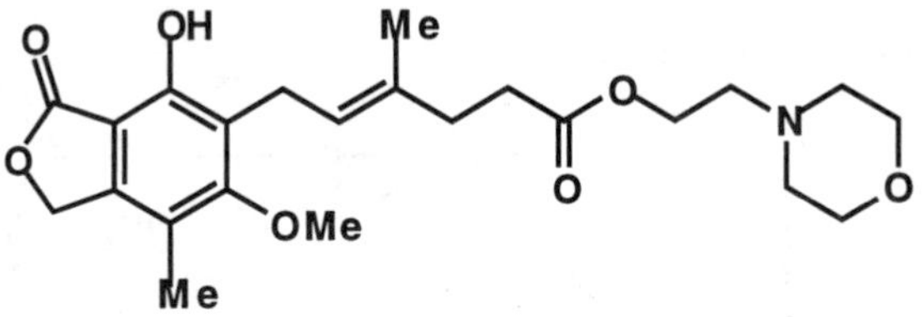

Country of Origin:	**Switzerland**	Trade Name:	**CellCept**
Originator:	**Roche**	CAS Registry No.:	**128794-94-5**
First Introduction:	**U.S.A.**	Molecular Weight:	**433.49**
Introduced by:	**Roche**	Dosage Form:	**250 mg Capsule**

Mycophenolate mofetil was launched in 1995 in the U.S.A., its first market worldwide, for the prevention of acute kidney transplant rejection in conjunction with other immunosuppressive therapy and to treat refractory acute kidney graft rejection. With improved oral absorption and bioavailability, mycophenolate mofetil is a prodrug of mycophenolic acid (MPA), a fermentation product of several *Penicillium* species. MPA is a selective, reversible, non-competitive inhibitor of inosinate dehydrogenase and guanylate synthetase. It inhibits the *de novo* pathway of purine biosynthesis. MPA was found to have more potent antiproliferative effects on T and B lymphocytes than other cell types. Compared with other immunosuppressants, mycophenolate mofetil is reportedly superior due to its unique mechanism of action and excellent safety profile for long term use. Mycophenolate mofetil is being investigated clinically in the treatment of heart and liver transplantation rejection, asthma, in preventing coronary artery restenosis, and in treating rheumatoid arthritis.

Nalmefene Hydrochloride (Dependence Treatment) (83-85)

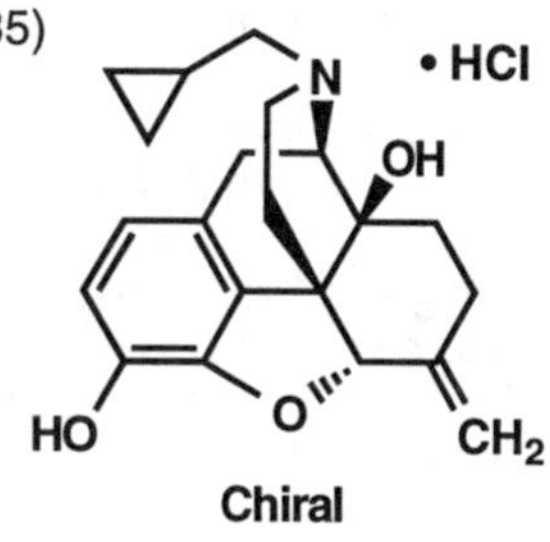

Country of Origin:	**U.S.A.**
Originator:	**IVAX**
First Introduction:	**U.S.A.**
Introduced by:	**Ohmeda**
Trade Name:	**Revex**
CAS Registry No.:	**58895-64-0**
Molecular Weight:	**375.88**
Dosage Form:	**100 µg/ml, 1 mg/ml Ampule**

• HCl

Chiral

Nalmefene hydrochloride has been introduced in the U.S.A. for opioid reversal following surgery, and in the reversal of opioid overdoses and epidurally administered narcotics. Nalmefene is an opioid antagonist that inhibits respiratory, analgesic and subjective effects of opioids. It has a higher potency, a longer duration of action, and superior bioavailability than the structurally related naltrexone. Nalmefene has a wide spectrum of biological activity and is in clinical trials for the treatment of interstitial cystitis, recalcitrant pruritus of cholestasis, stroke, rheumatoid arthritis, shock, CNS trauma, and alcoholism. Studies indicate that nalmefene does not induce morphine-like effects and has no apparent abuse potential.

Nedaplatin (Antineoplastic) (86-88)

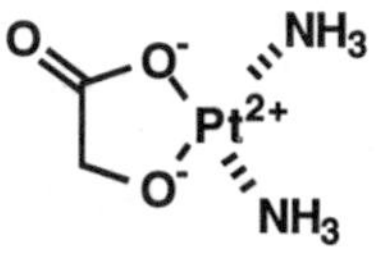

Country of Origin:	**Japan**
Originator:	**Shionogi**
First Introduction:	**Japan**
Introduced by:	**Shionogi**
Trade Name:	**Aqupla**
CAS Registry No.:	**95734-82-0**
Molecular Weight:	**303.19**
Dosage Form:	**10, 50, 100 mg Injection**

Nedaplatin, a novel second generation platinum complex, was marketed last year in Japan for the treatment of a variety of cancers including: head and neck, small-cell and non-small cell lung, oesophageal, prostatic, testicular, ovarian, cervical, bladder, and uterine cancers. Platinum anticancer agents, prototyped by cisplatin, have been reported to be hydrolyzed to the mono- or diaquated species of diamine platinum which react with nucleophilic sites on DNA to cause intrastrand and interstrand crosslinks and DNA-protein crosslinks, which result in cytotoxicity. Nedaplatin was reportedly more active than cisplatin against several solid tumors while sharing less nephro- and gastrointestinal toxicity to cisplatin *in vivo*. The minimal renal toxicity displayed by nedaplatin allows its use in patients with deteriorated renal function.

Pranlukast (Antiasthmatic) (89-91)

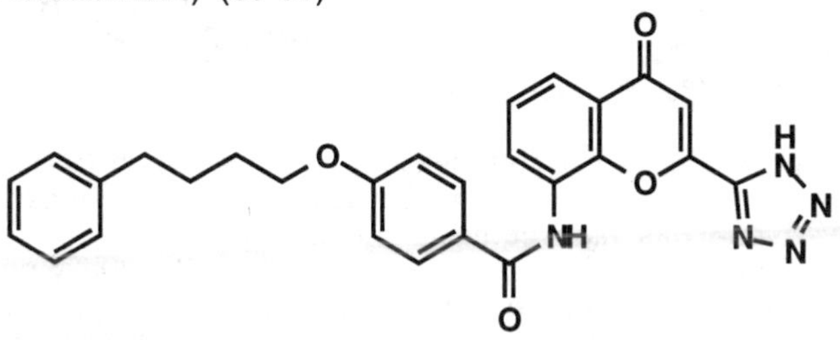

Country of Origin:	**Japan**		Trade Name:	**Onon**
Originator:	**Ono**		CAS Registry No.:	**103177-37-3**
First Introduction:	**Japan**		Molecular Weight:	**481.50**
Introduced by:	**Ono**		Dosage Form:	**112.5 mg Capsule**

Pranlukast, a novel chromone derivative, was introduced last year in Japan for the treatment of bronchial asthma and allergic diseases. Pranlukast is a highly potent, selective and competitive antagonist of peptidoleukotrienes with high affinity for the LTD_4 receptor. In patients with bronchial asthma, pranlukast was reported to induce significant improvement in both immediate and late asthmatic response induced by antigen. Pranlukast is also being evaluated clinically for the treatment of perennial allergic rhinitis, pediatric asthma, and cutaneous pruritus in dialysis patients. The therapeutic potential of pranlukast in managing irritable bowel syndrome has been suggested.

Ranitidine Bismuth Citrate (Antiulcer) (92-94)

Country of Origin:	**United Kingdom**
Originator:	**Glaxo Wellcome**
First Introduction:	**United Kingdom**
Introduced by:	**Glaxo Wellcome**
Trade Name:	**Pylorid**
CAS Registry No.:	**128345-62-0**
Molecular Weight:	**712.49**
Dosage Form:	**400 mg Tablet**

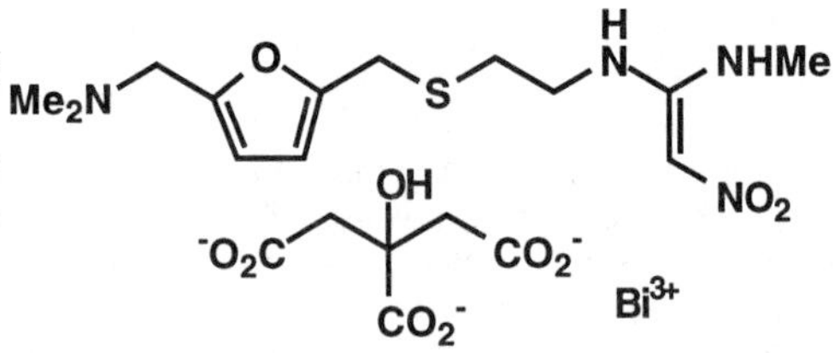

Ranitidine bismuth citrate, a novel salt formed from ranitidine and a bismuth citrate complex, is a new agent introduced last year in the United Kingdom for the treatment of duodenal and benign gastric ulcers, for eradication of *Helicobacter pylori*, and for prevention of relapse of duodenal ulcer when administered in combination with an antibiotic such as clarithromycin or amoxicillin. Ranitidine is a potent histamine H_2-receptor antagonist with ability to inhibit gastric acid secretion. Other type of agents such as tripotassium dicitrato bismuthate exert their therapeutic effects in peptic ulceration by providing a protective coating to the ulcer crater, stimulating endogenous prostaglandin production, and inhibiting pepsin activity. More significantly, they inhibit micro-organism *Helicobacter pylori*, a prevalent, specifically human pathogen associated with gastritis and gastric ulcers. The title compound possesses a novel combination of properties of the above two types of drugs and displays excellent therapeutic efficacy in clinical studies. Ranitidine bismuth citrate is being investigated for use in gastritis and non-ulcerating dyspepsia, acute healing of gastrointestinal ulcer and in prevention of the recurrence of peptic ulcer disease.

Rimexolone (Antiinflammatory) (95-97)

Country of Origin:	**Netherlands**
Originator:	**Akzo**
First Introduction:	**U.S.A.**
Introduced by:	**Alcon**
Trade Name:	**Vexol**
CAS Registry No.:	**49697-38-3**
Molecular Weight:	**370.51**
Dosage Form:	**1% Suspension**

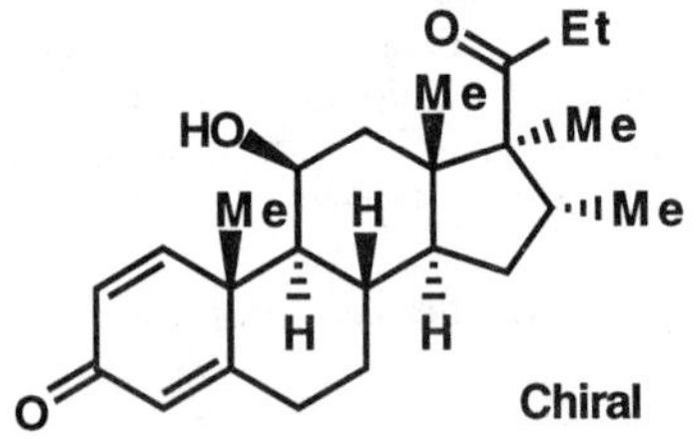

Rimexolone, an ophthalmic corticosteroid, was launched in 1995 in the U.S.A. for the treatment of postoperative inflammation following ocular surgery and anterior uveitis. Rimexolone has high corticoid receptor affinity and is a potent local antiinflammatory agent with minimal systemic effects and virtually no atrophogenic action in many animal models studied, unique among a wide range of topical steroids. Rapid onset, long duration of action plus a superior safety profile are characteristics of rimexolone. It was also approved in Europe for the treatment of rheumatoid arthritis and is currently in clinical trials for tendinitis and osteoarthritis.

Saquinavir Mesylate (Antiviral) (98-101)

Country of Origin:	**Switzerland**
Originator:	**Roche**
First Introduction:	**U.S.A.**
Introduced by:	**Roche**
Trade Name:	**Invirase**
CAS Registry No.:	**127779-20-8**
Molecular Weight:	**766.94**
Dosage Form:	**200 mg Caps.**

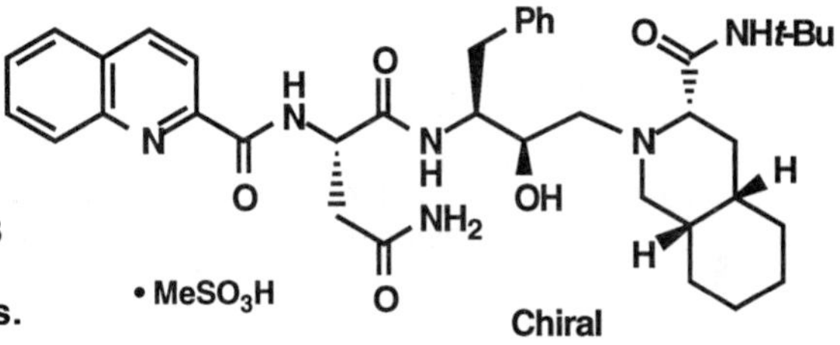

• MeSO₃H

Chiral

Saquinavir mesylate, the first HIV protease inhibitor to reach the market, was launched in the U.S.A. last year. It is indicated for use in combination with approved nucleoside analogs for the treatment of advanced HIV infection. Saquinavir, a transition state analog of Phe-Pro, is a very potent and competitive inhibitor of HIV-1 and HIV-2 proteases with high specificity. Saquinavir inhibits the last stage in the replication process of HIV and prevents virion maturation in both acute and chronically infected cells. Combination of saquinavir with the nucleoside analogs such as zidovudine (AZT) or/and zalcitabine which inhibit the enzyme reverse transcriptase and target at an earlier stage in the HIV replication process, shows a greater than additive effect in increase in CD4 cell counts and reduction in viral load, with the combination delaying the onset of resistance to either drug alone. Saquinavir is well tolerated alone and in combination with AZT.

Seratrodast (Antiasthmatic) (102-105)

Country of Origin:	**Japan**
Originator:	**Takeda**
First Introduction:	**Japan**
Introduced by:	**Takeda**
Trade Name:	**Bronica**
CAS Registry No.:	**112665-43-7**
Molecular Weight:	**354.43**
Dosage Form:	**40, 80 mg Tablet**
	100 mg/g Granule

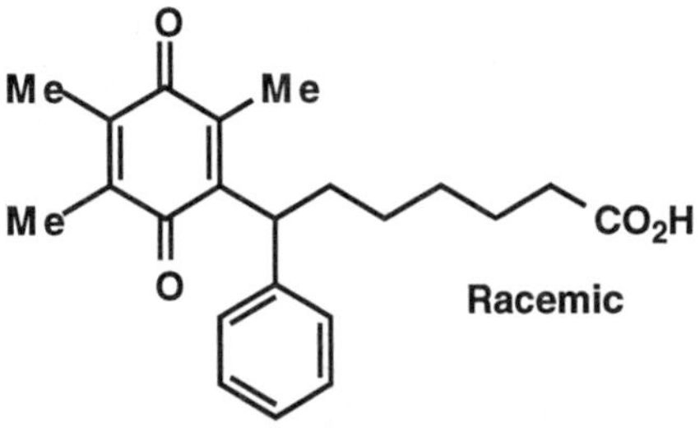

Racemic

Seratrodast, a novel benzoquinone derivative, is the first thromboxane A_2 (TxA$_2$) receptor antagonist to reach the market. It is orally active for the treatment of bronchospastic disorders such as asthma. TxA$_2$, a metabolite of the arachidonate cascade, is involved in several cardiovascular and respiratory diseases through its potent biological effects on platelet aggregation and constriction of vascular and respiratory smooth muscles. Seratrodast potently inhibits platelet aggregation and bronchoconstriction induced by a TxA$_2$ mimic and by a variety of spasmogenic prostanoids including PGF$_{2\alpha}$, PGD$_2$ and 9α,11β-PGF$_2$. Seratrodast shows excellent efficacy in asthma and has been reported to be potentially useful in hyper-responsive disorder. The R-(+)- seratrodast was reported to be the active isomer.

Spirapril Hydrochloride (Antihypertensive) (106-108)

Country of Origin:	**U.S.A.**
Originator:	**Schering-Plough**
First Introduction:	**Finland**
Introduced by:	**Sandoz**
Trade Name:	**Renpress; Renormax**
CAS Registry No.:	**94841-17-5**
Molecular Weight:	**503.06**
Dosage Form:	**3, 6, 12, 24 mg Tablet**

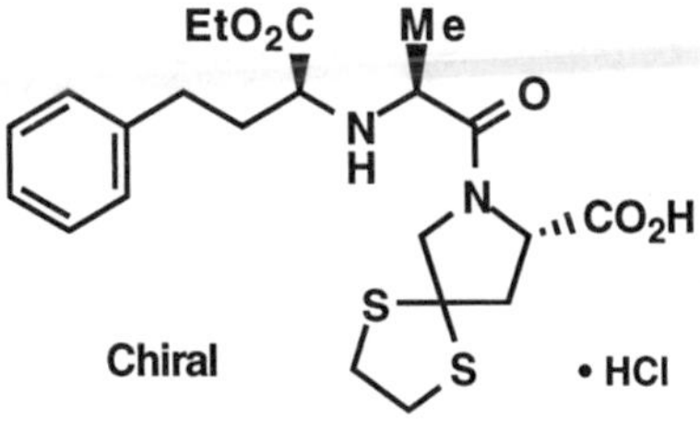

Chiral

• HCl

Spirapril hydrochloride is a new ACE inhibitor introduced last year in Finland for the treatment of hypertension. Spirapril is orally active and rapidly absorbed and converted to the active diacid spiraprilat which produces a reduction in blood pressure similar to that seen with enalapril and captopril. Spirapril has been reported to be well tolerated and exhibits an apparently lower incidence of cough compared with other ACE inhibitors. The elimination of spirapril occurs via both renal and hepatic routes so that no dosage adjustment is required for patients with renal impairment. In patients with severe congestive heart failure, spirapril has demonstrated beneficial hemodynamic effects with a duration of action >24 hours.

Suplatast Tosylate (Antiallergic) (109-111)

Country of Origin:	**Japan**
Originator:	**Taiho**
First Introduction:	**Japan**
Introduced by:	**Taiho**
Trade Name:	**IPD**
CAS Registry No.:	**94055-76-2**
Molecular Weight:	**499.62**
Dosage Form:	**50, 100 mg Capsule**

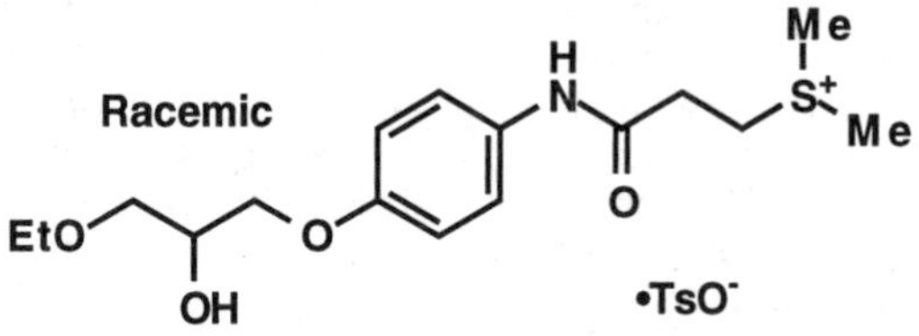

Suplatast tosylate is a unique dimethylsulfonium salt marketed last year in Japan for the treatment of bronchial asthma, atopic dermatitis and allergic rhinitis. Suplatast tosylate is a potent inhibitor of IgE synthesis without suppressing IgM and IgG. The mechanism of action for suplatast tosylate is thought to be via the inhibition of interleukin-4 and interleukin-6 production by T-cells at the gene level. In allergic patients, suplatast tosylate markedly improved clinical symptoms which correlated with a significant decrease in serum IgE antibody levels.

Tiludronate Disodium (Paget's Disease) (112-115)

Country of Origin:	**France**
Originator:	**Sanofi**
First Introduction:	**Switzerland**
Introduced by:	**Sanofi**
Trade Name:	**Skelid**
CAS Registry No.:	**149845-07-8**
Molecular Weight:	**362.56**
Dosage Form:	**240 mg Table**

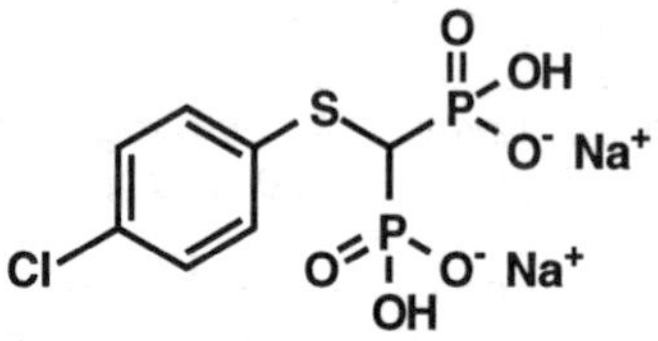

Tiludronate disodium, the first of the third generation of bisphosphonates, was introduced to the market last year in Switzerland for the treatment of Paget's disease. Bisphosphonates have high affinity for bone. When given orally, they are not metabolized but are absorbed, stored, preferentially localized to the skeleton where they inhibit bone resorption and therefore are useful in the treatment of diseases of high bone turnover such as Paget's disease and hypercalcemia of malignancy. Mechanistically, tiludronate disodium is suggested to be a specific inhibitor of functioning osteoclasts through selective incorporation into the polarized osteoclast-like multinucleated cells and direct interference with the maintenance of the cytoskeletal structure. Tiludronate disodium is reportedly in clinical trials for osteoporosis in post-menopausal women.

Tirilazad Mesylate (Subarachnoid Hemorrhage)   (116-119)

Country of Origin: **Sweden/U.S.A.**
Originator: **Pharmacia & Upjohn**
First Introduction: **Austria**
Introduced by: **Pharmacia & Upjohn**
Trade Name: **Freedox**
CAS Registry No.:   **110101-67-2**
Molecular Weight: **720.96**
Dosage Form: **100 ml Vial**

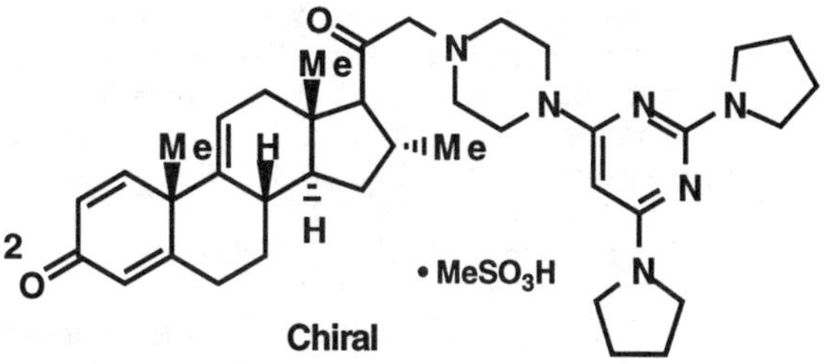

Chiral

Tirilazad mesylate, a novel 21-aminosteroid derivative, has been marketed in Austria for the intravenous treatment of subarachnoid hemorrhage (SAH) in male patients. Studies have indicated involvement of lipid peroxidation and/or lipid hydrolysis in the genesis of SAH-initiated microvascular damage and prolonged vasospasm of the large cerebral vessels. Tirilazad mesylate is a highly lipophilic compound that localizes to the cell membrane. It is a potent inhibitor of lipid peroxidation induced by oxygen free radicals, purely through a mechanism similar to vitamin E by scavenging lipid peroxyl radicals, and is a stabilizer of biological membranes. Tirilazad mesylate has been demonstrated to have a positive effect in reducing symptomatic vasospasm, cerebral infarction, and neurological dysfunction. Clinically, it is reportedly effective in reducing mortality in male SAH patients relative to the control group. The title compound is also reportedly in clinical trials for ischemic stroke, spinal cord injury, and for non-systemic ocular applications. Little or no behavioral and physiological side effects have been reported.

Topiramate (Antiepileptic)   (120-123)

Country of Origin: **U.S.A.**
Originator: **Johnson & Johnson**
First Introduction: **United Kingdom**
Introduced by: **Johnson & Johnson**
Trade Name: **Topamax**
CAS Registry No.:   **97240-79-4**
Molecular Weight: **339.36**
Dosage Form: **50, 100, 200 mg Table**

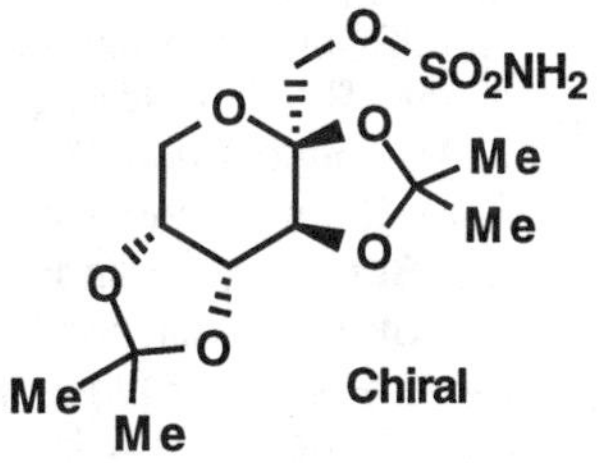

Chiral

Topiramate, a novel sulfamate-substituted D-fructose derivative, was launched last year in the United Kingdom as an adjunct therapy for use in partial seizures with or without secondary generalized seizures in adult patients inadequately controlled on conventional antiepileptics. Topiramate is structurally distinct from other available antiepileptics and functions through a unique combination of several mechanisms. It appears to act by blocking voltage-sensitive sodium channels to raise the action potential threshold and block the spread of seizure, enhancing GABA activity at postsynaptic GABA receptors and reducing glutamate activity at postsynaptic AMPA-type receptors, and is also a carbonic anhydrase inhibitor. Topiramate is orally active with rapid absorption, high bioavailability, and long duration of action. Excellent efficacy has been reported as an add-on therapy in epilepsy and it is also being evaluated as a monotherapy.

Valaciclovir Hydrochloride (Antiviral)   (124-127)

Country of Origin:	**United Kingdom**
Originator:	**Glaxo Wellcome**
First Introduction:	**United Kingdom**
Introduced by:	**Glaxo Wellcome**
Trade Name:	**Valtrex**
CAS Registry No.:	**124832-27-5**
Molecular Weight:	**360.81**
Dosage Form:	**500 mg Table**

Chiral

Valaciclovir hydrochloride, an orally active *L*-valyl ester of the potent antiviral agent aciclovir, was launched in 1995 in the United Kingdom for the treatment of herpes simplex virus (HSV) infections of the skin and mucous membranes, including initial and recurrent genital herpes. As a prodrug, valaciclovir has an improved pharmacokinetic profile to aciclovir. It is rapidly absorbed after oral administration and extensively converted to aciclovir via first-pass metabolism to achieve plasma levels of aciclovir comparable to those seen with aciclovir via i.v. route. Aciclovir is then activated selectively in virus-infected cells by viral thymidine kinase to form aciclovir triphosphate in a stepwise fashion. This active species inhibits viral DNA polymerase via irreversible binding to the active site of the enzyme. Once aciclovir is incorporated into the elongating viral DNA, it terminates replication of the viral DNA strand, an antiviral mechanism unique to aciclovir. Valaciclovir is reportedly in clinical trials for the suppression of cytomegalovirus infection and disease in renal transplant patients.

References

1. The new launches of the year is based on the combined information from the following sources:
    a. Scrip Magazine, January, 1996
    b. Pharmaprojects
    c. IMSworld Publication
    d. J.R. Prous, DN&P, $\underline{9}$, 19 (1996).
2. X.-M. Cheng, Annu. Rep. Med. Chem., $\underline{30}$, 295 (1995).
3. X.-M. Cheng, Annu. Rep. Med. Chem., $\underline{29}$, 331 (1994).
4. J.D. Strupczewski and D.B. Ellis, Annu. Rep. Med. Chem., $\underline{28}$, 325 (1993).
5. J.D. Strupczewski and D.B. Ellis, Annu. Rep. Med. Chem., $\underline{27}$, 321 (1992).
6. F-D-C Reports, 8 (January 8, 1996).
7. a). K. Rogers and W.M. Davis, Drug Topics, 84 (February 5, 1996);
    b). K. Rogers, Drug Topics, 134 (March 4, 1996).
8. J.R. Prous, ed., Drugs Future, $\underline{18}$, 18 (1993).
9. P.S. Schein, Cancer Invest., $\underline{8}$, 265 (1990).
10. M. Treskes and W.J. F. van der Vijgh, Cancer Chemother. Pharmacol., $\underline{33}$, 93 (1993).
11. R.L. Capizzi and W. Oster, Eur. J. Cancer, $\underline{31A}$ (Suppl. 1), S8 (1995).
12. P.V. Plourde, M. Dyroff, M. Dowsett, L. Demers, R. Yates, and A. Webster, J. Steroid Biochem. Mol. Biol., $\underline{53}$, 175 (1995).
13. P.V. Plourde, M. Dyroff, and M. Dukes, Breast Cancer Res. Treat., $\underline{30}$, 103 (1994).
14. P.E. Goss and K.M.E.H. Gwyn, J. Clin. Oncol., $\underline{12}$, 2460 (1994).
15. J.R. Prous, ed., Drugs Future, $\underline{20}$, 297 (1995).
16. G. Blackledge, Cancer, $\underline{72}$ (Suppl. 12), 3830 (1993).
17. C.J. Tyrrell, Prostate Suppl., $\underline{4}$, 97 (1992).
18. J. Verhelst, L. Denis, P.Van Vliet, H.Van Poppel, J. Braeckman, P.Van Cangh, J. Mattelaer, D. D'Hulster, and Ch. Mahler, Clin. Endocrinol., $\underline{41}$, 525 (1994).
19. K. Aisaka, T. Miyazaki, T. Hidaka, T. Ohno, T. Ishihara, and T. Kanai, Jpn. J.

Pharmacol., 59, 489 (1992).

20.   T. Hidaka, K. Aisaka, T. Ohno, and T. Ishihara, Folia. Pharmacol. Jpn., 101, 233 (1993).

21.   J.R. Prous, ed., Drugs Future, 19, 871 (1994).

22.   T. Iwahi, K. Okonogi, T. Yamazaki, S. Shiki, M. Kondo, A. Miyake, and A. Imada, Antimicrob. Agents Chemother., 36, 1358 (1992).

23.   Y. Iizawa, K. Okonogi, R. Hayashi, T. Iwahi, T. Yamazaki, and A. Imada, Antimicrob. Agents Chemother., 37, 100 (1993).

24.   M. Nakao, Y. Noji, T. Iwahi, and T. Yamazaki, J. Antimicrob. Chemother., 29, 509 (1992).

25.   Chemotherapy Japan, 41 (Suppl. 4), (1993), the whole issue.

26.   K. Ikeda, M. Hosino, H. Iida, and H. Ohnishi, Pharmacometrics, 44, 433 (1992).

27.   M. Hosono, H. Iida, K. Ikeda, Y. Hayashi, H. Dohmoto, Y. Hashiguchi, H. Yamamoto, N. Watanabe, and R. Yoshimoto, J. Pharmacobio. Dyn., 15, 547 (1992).

28.   T. Chibana, K. Noguchi, Y. Ojiri, and M. Sakanashi, Jpn. Heart J., 33, 239 (1992).

29.   J.M. Hunter, N. Engl. J. Med., 332, 1691 (1995).

30.   I.H. Littlejohn, K. Abhay, A. El Sayed, C.J. Broomhead, P. Duvaldestin, and P.J. Flynn, Anaesthesia, 50, 499 (1995).

31.   T.N. Calvey, Acta. Anaesthesiol. Scand., 39 (Suppl. 106), 83 (1995).

32.   J.R. Prous, ed., Drugs Future, 20, 525 (1995).

33.   G.J. Kontoghiorghes, Analyst, 120, 845 (1995).

34.   G.J. Kontoghiorghes, Ann. N. Y. Acad. Sci., 612, 339 (1990).

35.   M. Agarwal, S. Gupte, C. Viswanathan, D. Vasandani, J. Ramanathan, J. Desai, R. Puniyani, and A. Chhablani, Drugs Today, 28 (Suppl. A), 107 (1992).

36.   J.R. Prous, ed., Drugs Future, 20, 464 (1995).

37.   F. Lavelle, M.C. Bissery, C. Combeau, J.F. Riou, P. Vrignaud, and S. Andre, Semin. Oncol. 22 (2 Suppl. 4), 3 (1995).

38.   M.C. Bissery, Eur. J. Cancer, 31A (Suppl.4) S1 (1995).

39.   P.M. Ravdin and V. Valero, Semin. Oncol. 22 (2 Suppl. 4), 17 (1995).

40.   J.E. Cortes and R. Pazdur, J. Clin. Oncol., 13, 2643 (1995).

41.   J.R. Prous, ed., Drugs Future, 20, 412 (1995).

42.   E.J. Higginbotham, Chibret. Int. J. Ophthalmol., 10, 50 (1994).

43.   M.F. Sugrue, P. Mallorga, H. Schwam, J.J. Baldwin, and G.S. Ponticello, Br. J. Pharmacol. 98 (Suppl.), 820P (1989).

44.   C.B. Granger, R.M. Califf, and E.J. Topol, Drugs, 44, 293 (1992).

45.   J. Kalbfleisch, P.B. Kurnik, U. Thadani, M.A. DeWood, R. Kent, R. Magorien, A.C. Jain, L.J. Spaccavento, D.L. Morris, G. Taylor, J. Perry, M. Kutcher, H.J. Gorfinkei, and J.K. LittleJohn, Am. J. Cardiol., 71, 386 (1993).

46.   J. Kalbfleisch, U. Thadani, J.K. LittleJohn, G. Brown, R. Magorien, M. Kutcher, G. Taylor, W.T. Maddox, W.B. Campbell, J. Perry, Jr., J.F. Spann, G. Vetrovec, R. Kent, and P.W. Armstrong, Am. J. Cardiol., 69, 1120 (1992).

47.   J.R. Prous, ed., Drugs Future, 18, 859 (1993).

48.   R. Scuri, P. Giannetti, and A. Paesano, Drugs Exp. Clin. Res., 14, 693 (1988)

49.   G. Fumagalli, C. Balzarotti, P. Banfi, P. Deco, L. Ferrante, and M. Zennaro, G. Ital. Mal. Torace., 42, 299 (1988).

50.   J.R. Prous, ed., Drugs Future, 19, 874 (1994).

51.   L.J. Browne, C. Gude, H. Rodriguez, R.E. Steele, and A. Bhatnager, J. Med. Chem., 34, 725 (1991).

52.   H.V. Bossche, H. Moereels, and L.M.H. Koymans, Breast Cancer Res. Trea., 30, 43 (1994).

53.   S. Yano, M. Tanaka, and K. Nakao, Eur. J. Pharmacol. Mol. Pharmacol. Sect., 289, 217 (1995).

54.   J.R. Prous, ed., Drugs Future, 19, 1126 (1994).

55.   M. Shirotani, R. Hattori, C. Kawai, S. Sasayama, and Y. Yui, Cardiovasc. Drug Rev., 10, 333, (1992).

56.   T. Asano and H. Hidaka, Methods Find. Exp. Clin. Pharmacol., 12, 443 (1990)

57.   T. Asano, I. Ikegaki, S.I. Satoh, Y. Suzuki, M. Shibuya, M. Takayasu, and H. Hidaka, J. Pharmacol. Exp. Ther., 241, 1033 (1987).

58.   J.R. Prous, ed., Drugs Future, 18, 862 (1993).

59.   J.G. Rafanell, M.A. Dronda, M. Merlos, J. Forn, J.M. Torres, M.I. Zapatero, and N, Basi, Arzneimittelforschung, 42, 836 (1992).

60.   I. Izquierdo, M. Bayes, J. Jane, A. Alomar, and J. Forn, Arzneimittelforschung, 42, 859 (1992).

61.   J.R. Prous, ed., Drugs Future, 20, 827 (1995).

62.   W. Plunkett, P. Huang, Y.Z. Xu, V. Heinemann, R. Grunewald, and V. Gandhi, Semin. Oncol., 22 (4 Suppl. 11), 3 (1995).

63.   R.P. Abratt, W.R. Bezwoda, G. Falkson, L. Goedhals, D. Hacking, and T.A. Rugg, J. Clin. Oncol., 12, 1535 (1994).

64.   J.R. Prous, ed., Drugs Future, 19, 877 (1994).

65.   G. Mueller and S. Wied, Diabetes, 42, 1852 (1993).

66.   Y. Iwanaga, Y. Kimura, N. Miyashita, K. Morikawa, O. Nagata, Z. Itoh, and Y. Kondo, Jpn. J. Pharmacol., 66, 317 (1994).

67.   Y. Iwanaga, N. Miyashita, K. Kato, K. Morikawa, H. Kato, Y. Ito, Y. Kondo, and Z. Itoh, Eur. J. Pharmacol., 183, 2189 (1990).

68.   J.R. Prous, ed., Drugs Future, 20, 424 (1995).

69.   J.J. Eron, S.L. Benoit, J. Jemsek, R.D. MacArthur, J. Santana, J.B. Quinn, D.R. Kuritzkes, M.A. Fallon, and M. Rubin, N. Engl. J. Med., 333, 1704 (1995).

70.   E. De Clercq, Biochem. Pharmacol., 47, 155 (1994).

71.   E.H. Wiltink, Drugs Today, 31, 273 (1995).

72.   J.R. Prous, ed., Drugs Future, 19, 992 (1994).

73.   J. Marek, V. Hana, M. Krsek, V. Justova, F. Catus, and F. Thomas, Eur. J. Endocrinol., 131, 20 (1994).

74.   P. Hayry, A. Raisanen, J. Ustinov, A. Mennander, and T. Paavonen, FASEB J., 7, 1055 (1993).

75.   J.R. Prous, ed., Drugs Future, 20, 431 (1995).

76.   O. Edling, G. Bao, M. Feelisch, T. Unger, and P. Gohlke, J. Pharmacol. Exp. Ther., 275, 854 (1995).

77.   B. Persson, B.R. Widgren, A. Fox, and M. Stimpel, J. Cardiovasc. Pharmacol., 26, 73 (1995).

78.   M. Abramowicz ed., Med. Lett. Drugs Ther., 37, 75 (1995).

79.   J.R. Prous, ed., Drugs Future, 20, 356 (1995).

80.   A.C. Allison and E.M. Eugui, Agents Action, 44 (Suppl.) 165 (1993).

81.   A.C. Allison and E.M. Eugui, Immunol. Rev., 136, 5 (1993).

82.   D.O. Taylor, R.D. Ensley, S.L. Olsen, D. Dunn, and D.G. Renlund, J. Heart Lung Transplant., 13, 571 (1994).

83.   J.R. Prous, ed., Drugs Future, 20, 732 (1995).

84.   C.K. Lineberger, B. Ginsberg, R.J. Franiak, and P.S.A. Glass, Anesthesiol. Clin. North Am., 12, 65 (1994).

85.   M. De Zwaan and J.E. Mitchell, J. Clin. Pharmacol., 32, 1060 (1992).

86.   J.R. Prous, ed., Drugs Future, 18, 1085 (1993).

87.   T.C. Hamilton, P.J. O'Dwyer, and R.F. Ozols, Curr. Opin. Oncol., 5, 1010 (1993).

88.   K. Ota, T. Oguma, and K. Shimamura, Anticancer Res., 14, 1383 (1994).

89.   J.R. Prous, ed., Drugs Future, 20, 437 (1995).

90.   H. Yamamoto, M. Nagata, K. Kuramitsu, K. Tabe, H. Kiuchi, Y. Sakamoto, K. Yamamoto, and Y. Dohi, Am. J. Respir. Crit. Care Med., 150, 254 (1994).

91.   R.R. Harris, G.W. Carter, R.L. Bell, J.L. Moore, and D.W. Brooks, Int. J. Immunopharmacol., 17, 147 (1995).

92.   J.R. Prous, ed., Drugs Future, 20, 480 (1995).

93.   R. Stables, C.J. Campbell, N.M. Clayton, J.W. Clitherow, C.J. Grinham, A.A. McColm, A. McLaren, and M.A. Trevethick, Allment. Pharmacol. Ther., 7, 237 (1993).

94.   A.G. Fraser, W.M. Lam, Y.W. Luk, J. Sercombe, A.M. Sawyerr, M. Hudson, I.M. Samloff, R.E. Pounder, Gut, 34, 338 (1993).

95.   J.R. Prous, ed., Drugs Future, 20, 1075 (1995).

96.   P.K. Fox, A.J. Lewis, R.M. Rae, A.W. Sim, and G.F. Woods, Arzneim. Forsch., 30, 55 (1980).

97.   L. Joosten, M. Helsen, and W. van den Berg, Agents Action, 31, 135 (1990).

98.   J.R. Prous, ed., Drugs Future, 20, 321(1995).

99.   S. Galpin, N.A. Roberts, T. O'Connor, D.J. Jeffries, and D. Kinchington, Antiviral Chem. Chemother., 5, 43 (1994).

100.   V.A. Johnson, D.P. Merrill, T.C. Chou, and M.S. Hirsch, J. Infect. Dis., 166, 1143 (1992).

101.   J.C. Craig, I.B. Duncan, D. Hockley, C. Grief, N.A. Roberts, and J.S. Mills, Antiviral Res., 16, 295 (1991).

102.   J.R. Prous, ed., Drugs Future, 20, 847 (1995).

103.   T. Kurokawa, T. Matusmoto, Y. Ashida, R. Sasada, and S. Iwasa, Biol. Pharm. Bull., 17, 383 (1994).

104.   Y. Imura, Z. Terashita, Y. Shibouta, Y. Inada, and K. Nishikawa, Jpn. J. Pharmacol., 52, 35 (1990).

105.   Jpn. Pharmacol. Ther., 21 (Suppl. 7), 1993, the whole issue.

106.   J.R. Prous, ed., Drugs Future, 19, 887 (1994).

107.   S. Noble and E.M. Sorkin, Drugs, 49, 750 (1995).

108.   G. Reams, A. Lau, V. Knaus, and J. Bauer, J. Clin. Pharmacol., 33, 348 (1993).

109.   J.R. Prous, ed., Drugs Future, 18, 1090 (1993).

110.   S.I. Konno, M. Adachi, K. Asano, Y. Gonogami, K. Ikeda, K.I. Okamoto, and T. Takahashi, Eur. J. Pharmacol., 259, 15 (1994).

111.   Y. Yanagihara, Y. Kiniwa, K. Ikizawa, H. Yamaya, T. Shida, N. Matsuura, and A. Koda, Jpn. J. Pharmacol., 61, 23 (1991).

112.   J.R. Prous, ed., Drugs Future, 10, 482 (1985).

113.   H. Murakami, N. Takahashi, T. Sasaki, N. Udagawa, S. Tanaka, I. Nakamura, D. Zhang, A. Barbier, and T. Suda, Bone, 17, 137 (1995).

114.   R.M. Francis, Curr. Ther. Res., 56, 831, (1995).

115.   J.P. Bonjour, P. Ammann, A. Barbier, J. Caverzasio, and R. Rizzoli, Bone, 17 (Suppl.) 473S (1995).

116.   J.R. Prous, ed., Drugs Future, 20, 218 (1995).

117.   F.P. Carrea, E.J. Lesnefsky, D.G. Kaiser, and L.D. Horwitz, J. Cardiovasc. Pharmacol., 20, 230 (1992).

118.   R. McKenna, P.L. Munns, K.L. Leach, and W.R. Mathews, Methods Find. Exp. Clin. Pharmacol., 17, 279 (1995).

119.   M. Zuccarello, J.T. Marsch, G. Schmitt, J. Woodward, and D.K. Anderson, J. Neurosurg., 71, 98 (1989).

120.   J.R. Prous, ed., Drugs Future, 20, 444 (1995).

121.   R.P. Shank, J.F. Gardocki, J.L. Vaught, C.B. Davis, J.J. Schupsky, R.B. Raffa, S.J. Dodgson, S.O. Nortey, B.E. Maryanoff, Epilepsia, 35, 450 (1994).

122.   R. Fisher and D. Blum, Epilepsia, 36 (Suppl.2), S105 (1995).

123.   M.A. Rogawski and R.J. Porter, Pharmacol. Rev., 42, 223 (1990).

124.   J.R. Prous, ed., Drugs Future, 20, 744 (1995).

125.   R.J. Crooks and A. Murray, Antiviral Chem. Chemother. Suppl., 5, 31 (1994).

126.   R.J. Crooks, Antiviral Chem. Chemother. Suppl., 6, 39 (1995).

127.   A.P. Fiddian, Antiviral Chem. Chemother. Suppl., 6, 51 (1995).

Chapter 35. The Protein Structure Project, 1950-1959: First Concerted Effort of a Protein Structure Determination in the U.S.

A. Tulinsky
Department of Chemistry
Michigan State University
East Lansing, MI 48824

A scientific discipline usually develops in a small number of different laboratories, grows from there and this is more or less true of protein crystallography. In this country, however, protein crystallography started in an organized, big-bang way in 1950 with the formation of The Protein Structure Project (PSP) at the Polytechnic Institute of Brooklyn. The aim of The Project, led by David Harker, was to solve the structure of a protein molecule in ten years because no protein structure had yet been determined. The following account describes how the PSP began and proceeded through the years. It also underscores some of the monumental problems the PSP and other protein crystallographers faced and how many of them were first solved and overcome. The target protein of the PSP was ribonuclease; its structure was published by Dave Harker some seventeen years after the work started (1).

David Harker: 1906-1991

Amorphous or fiber X-ray diffraction patterns of myosin and feather keratin were originally reported by W. T. Astbury during 1931-1933. The first single crystal pattern of a soluble, globular, wet protein crystal was that of pepsin (2) recorded a year later at Cambridge University by John D. Bernal and Dorothy Crowfoot Hodgkin[1]. The work was a tactical breakthrough because it showed that such crystals are better examined in the wet state. Hodgkin went to Oxford University in 1934 where she continued her protein work on insulin and lactoglobulin (3). The next year, Max Perutz came to Bernal from Herman Mark's laboratory in Vienna, where the structure determination of gas molecules by electron diffraction began in 1929, and he started his studies on hemoglobin that were to become classic. The initial work led to a report by Bernal, Isidore Fankuchen and Perutz in 1938 of diffraction by single crystals of chymotrypsin and horse methemoglobin corroborating the importance of the use of wet crystals and offering the prophetic suggestion that a direct Fourier analysis of molecular structure may be possible once complete sets of reflections were available from different states of hydration of crystals (4). The former has now been realized in a most astonishing way although not by the application of the latter method. Thus was protein crystallography born.

The growth of single crystal protein crystallography continued in Great Britain with investigations of spherical viruses (5-7) and Bernal and Fankuchen's hallmark discovery that tobacco mosaic virus was composed of a regular substructure (8). However, further progress was soon to be adversely affected by World War II. Meanwhile in this country, pre-war successes in crystal structure studies of amino acids and peptides resulted from a program initiated by Linus Pauling at the California Institute of Technology and implemented by Robert Corey, Eddie Hughes and co-workers. The work eventually led to Pauling and Corey's monumental prediction in 1950 of the α- and γ-helical configurations that were possible for a polypeptide chain and that they were an important part of the structure of both fibrous and globular proteins (9, 10). The war years of the forties, for good reasons, were otherwise a generally subdued time for protein crystallography both here and abroad.

During those early days, David Harker was one of Pauling's students when in 1936 he developed the theory of sections and projections of Patterson diagrams (11), which are today still called Harker sections. Shortly after, he discovered with Jose Donnay a new law of crystal morphology extending Bravais' Law, to become known as the Donnay-Harker Law (12), which is said to rank with the original work of Bravais for its lasting and pervasive influence (13). Subsequently, Harker went to The General Electric (GE) Research and Development Center at Schenectady, NY. While there, he and John Kasper derived inequalities relating modified structure factors to the intensities of other X-ray reflections that became known as Harker-Kasper inequalities (14) and served for the development of modern direct methods of Hauptman and Karle relegating "small molecule" structure determination to a routine exercise.

Legend has it that in the late forties Irving Langmuir, also then at GE, asked Dave Harker what he would do if he had a million dollars. Dave said he would take a ten-year leave and solve the structure of a protein molecule. Langmuir raised the million for the ten year period. The money came from The Dean Langmuir Foundation, The Rockefeller Foundation, The Damon Runyon Memorial Fund, The New York Foundation and The American Cancer Society. Thus, in late 1950, David Harker accepted the responsibility of

---

[1] The pepsin crystals grew while a Dr. John Philpot in Svedberg's laboratory at Uppsala was on a skiing holiday, they were received by Bernal and Crowfoot four weeks before submittal of their communication, which was published about a week later on May 26, 1934.

establishing a research group that would seriously and concertedly attack the problem of finding the atomic arrangement in a protein molecule. The next thing to be done was to negotiate laboratory space and administrative assistance at a consenting institution.

The end of the war brought rapid development of punched-card techniques of computing and soon Fourier syntheses giving electron density maps and other crystallographic calculations could be done in three-dimensions with relative ease (30-40 man hours per calculation for non-protein problems). Harker recognized the importance of having suitable computational support for the protein problem. Larger computing devices, however, had not yet become commonplace in university or business settings and the Korean War tied-up many for military use. Perceiving that IBM Corp. punched-card electronic calculators would be the wave of the future, Dave arranged a favorable relationship with Thomas J. Watson whereby computations connected with the work on protein structure could be carried out on IBM computers, but by his own people, at IBM's Watson Computing Laboratory on 116th Street in New York near Columbia University. Therefore, the research group had to be located in or near New York City, which proved to be a surprising problem because most institutions in the area were unsympathetic and unenthused about providing space for establishing a full-fledged X-ray protein crystallography laboratory (15). This was not the case for the Polytechnic Institute of Brooklyn, which was already an important world center for X-ray crystallography with an outstanding contingent of crystallographers in the Department of Physics: Paul P. Ewald (Chairman), who developed the dynamical theory of diffraction as a student with Arnold Sommerfeld in Munich, Herman F. Mark (Director of Polymer Research), one of the early pioneers of X-ray crystallography and ever present for advice, Isidore (Fan) Fankuchen of early protein crystallography fame, Rudolph Brill, a research professor diffraction physicist and Ben Post, a then young, but all-around superb X-ray crystallographer. It was Fan who called the President of Brooklyn Poly and persuaded him to provide Harker space and then closed the deal with Dave to come to Poly (15). The new research group was called The Protein Structure Project and it was assigned about half of the 4th floor of 55 Johnson St. near downtown Brooklyn and the Brooklyn Bridge.

Harker gathered a small group to tackle the then known daunting aspects of protein structure determination (crystallization, intensity data collection, computation). The research space consisted of Harker's office shared with his secretary and wife Katherine, two other shared offices, a general work space, a chemical and crystallization laboratory and a X-ray diffraction room adjoining a dark room. The initial staff of postdoctorals were Beatrice S. Magdoff, to carry out crystallographic analyses, Thomas C. Furnas, Jr., an X-ray diffraction physicist to design and construct new diffraction equipment and Murray Vernon King, a chemist to grow crystals and do chemistry on proteins. Another very important, and in ways indispensible, member of the original team was William (Bridgie) Weber, instrument maker *par excellence*. Bridgie was at GE in Schenectady as a master machinist and had previously worked for Dave when Dave helped design a commercial X-ray powder diffractometer for GE X-ray Corp. in Milwaukee. Dave convinced Bridgie to come to Brooklyn because he knew that new diffraction equipment would have to be invented and built to cope with the protein data collection problem. Bridgie retired from GE, went to and remained with the PSP throughout its stay in Brooklyn sharing an apartment with Furnas until Tom was married and even moved with The Project to Buffalo in 1959. Beatrice Magdoff left about 1955 and went to the Boyce Thompson Institute for Plant Research in Yonkers to start a program on southern bean mosaic virus while Furnas and King remained until near the end of the decade when the PSP moved and continued at The Roswell Park Memorial Institute in Buffalo, NY.

The target protein of the PSP was the enzyme bovine pancreatic ribonuclease: it was available in pure form from Armour Laboratories in Chicago, it was relatively small

(MW=13.7kD) and it had been crystallized by Moses Kunitz at Rockefeller Institute (but with no diffraction studies). However, Vernon King was unable to grow ribonuclease crystals during the first year of the PSP. Late in 1951 a delegation from The Project visited Kunitz and conferred with him about his crystallization procedure. Nothing different could be found from the methods being used at the PSP so the group was understandingly discouraged. Not long after to everyone's delight, ribonuclease crystals appeared in amorphous precipitates of several of King's crystallization tubes and the PSP dodged a potentially devastating obstacle[2]. Thereafter, thirteen other different crystal forms of ribonuclease were prepared and characterized by X-ray diffraction (16, 17). To the very end, Harker suspected that during the Kunitz visit they had picked up seed crystals and carried them back to Brooklyn where they transformed incipient crystallization into a reality.

One of the overwhelming problems that faced protein structure determination was the measurement of diffraction intensities. While the number of three-dimensional reflections of small molecules were generally in the low thousands (2,500 for vitamin B_{12}), this number
was 10-20 times greater for smaller proteins. Precession photographs were the favored measurement method at the time, which gave undistorted representations of the reciprocal lattice ("diffraction space"). Protein X-ray exposures typically required 15-20 hours and since crystals suffered X-ray damage, only 1-2 exposures could be obtained from a crystal. If 30-40 layers of three-dimensional data had to be recorded, many crystals were required to complete the measurements. Add to this the time required to align each crystal, the many screen checks of crystal alignment that had to be made, the layers that had to be remeasured because crystals slipped in their wet mounts during exposure and crystals mounted in other directions to record blind regions of reciprocal space, not to mention estimating and processing all the intensities from the photographs, six months of very hard and continuous work along with some luck might possibly produce a three-dimensional data set. In practice, application of Murphy's constant π to such estimates gave more realistic times.

Harker was acutely aware of these difficulties, which were partially the reason for recruiting Tom Furnas who had just completed his Ph.D. at MIT working for Richard Bear, Martin Buerger and Bert Warren. Tom and Dave decided to measure intensities of reflections by depending solely and exclusively on calculated coordinates of reciprocal lattice points of reflections and to measure reflections only in the equatorial plane defined by the X-ray source, the crystal and the detector (18). Very importantly, all possible reflections could be measured by this procedure with only one orientation or mounting of the crystal. All that was required was a device capable of reorientating a crystal so that every diffracted beam could be made to occur in the equational plane. This instrument was designed based on a half-circle orienter mounted on a GE XRD-3 powder diffractometer with a scintillation counter detector and built by Bridgie Weber in the first few years of the PSP and the half-circle was known as an Eulerian cradle (19, 20). Tom later designed a quarter circle instrument that was marketed commercially by GE X-ray under the name

---

[2] These crystals were monoclinic, space group $P2_1$ (modification II) (16) and they were a different form from Kunitz's crystals. Discussing the crystallization procedure more carefully with Kunitz, King noted that the beef panceases of cows used for isolating ribonuclease were collected at a slaughter house in galvanized buckets. Vernon suspected that metal ions may have been leached from the pail so he tried crystallization setups in the presence of Zn^{+2} ion and various other transition metal ions like Fe^{+2}, Fe^{+3} and Ni^{+2}. It is a credit to Vernon's insight that he discovered the orthorhombic crystal form of ribonuclease grows only in the presence of Ni^{+2} ion, which are Kunitz crystals.

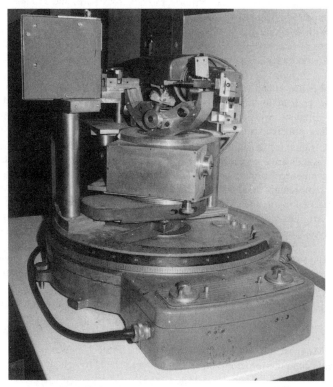

Half-circle Eulerian cradle mounted on an XRD-3

Single Crystal Orienter. He also wrote an excellent instruction manual (21), both theory and practice, to accompany the Orienter.

The Eulerian cradle was manually operated by setting three angles at pre-computed positions, measuring the intensity of a reflection and background and hard copying the results. As many as 75-100 reflections (10 second count time) could be measured in an hour, total exposure times were reduced to about 50-60 hours for about 10,000 reflections, the number of crystals to complete a data set was very small and a three-dimensional data collection could be completed in about 2 weeks. Thus, the Eulerian cradle reduced protein data collection to

manageable proportions. A second full-circle instrument using Eulerian geometry was under construction at the PSP by 1954 to incorporate complete automation through computer control by IBM punched-cards. Tom Furnas left the PSP in 1958 to go to the Picker X-ray Corp. in Cleveland where he went on to invent a whole new array of diffraction equipment, the foremost of which was the FACS I system, the first directly-coupled computer controlled single crystal X-ray diffractometer. It revolutionized intensity data collection and its offspring are still the instrument of choice thirty years later for small molecule structure determination.

Harker strengthened his group during the first five years with additional technical personnel in Dalia David, who helped with punched-card calculations on IBM plug-board machines like the 602A, 607 and 409 printer, Edith Pignataro, who was primarily concerned with measurements using the Eulerian cradle and Myra Edelman, a Ph.D.

graduate student. Illustrious visitors Vittorio Luzzati from Paris and Francis Crick from Cambridge University were also in residence during this time. Crick and Luzzati collaborated with Magdoff (1953-54) on a number of different problems ranging from shrinkage stages of ribonuclease (22) and the calculation of its three-dimensional Patterson function (23) to theoretical aspects of isomorphous replacement in protein crystals (24).

By the mid-fifties, it had become clear that study of shrinkage stages of protein crystals was not going to lead to structure. However, the new work of Max Perutz and his colleagues in England with heavy atom isomorphous replacement was showing great promise. Using p-mercuribenzoate, about 100 phase angles of centrosymmetric (h0 ℓ) reflections (± signs) of hemoglobin had been determined (25). In principle, three isomorphous crystals are a necessary and sufficient condition to determine all the phases of a non-centrosymmetrical protein crystal. The derivation and mathematics of the multiple isomorphous replacement method was first described completely by Dave Harker in a classic paper appearing as the first article, of the first issue, of *Acta Crystallographica* in 1956 (26). All the early protein structures were solved by this method (myoglobin, hemoglobin, lysozyme, ribonuclease, chymotrypsin, carboxypeptidase, cytochrome c, etc.).

I came to the PSP as a new post-doc in September 1955 from Princeton University after obtaining a degree with John G. White who was working on the "Princeton structure" of vitamin B_{12}. John's work impressed me deeply and inspired me to also go on to big structures. Little did I know what would be in store during the next 40 years. My first assignment at the PSP was to extend multiple isomorphous replacement to include anomalous scattering because Dave had recognized its potential power and many advantages. My most important contributions to The Project, however, involved writing crystallographic computer programs and computations related to structure analysis.

The plug-board IBM punched-card electronic calculators of the time (about 35 memory locations and 50 operational steps maximum) were horrendously tedious and prone to error due to the massive card handling required in even fairly simple computations. For example, when the three-dimensional Patterson map of ribonuclease was calculated it was "believed to contain only minor errors, say less than $10e^2Å^{-3}$ at any point" (23). This was not the expected error due to errors in the intensities of reflections but rather, the expected reproducibility of the calculation if it were done again. All eventually became history in 1956 with the introduction and advent of the IBM 650 computer having a 2000, ten digit word magnetic drum memory with millisecond access times. It was the first of the modern day commercial programmed computers where instructions were entered as coded numbers. Harker sent me to IBM school for the 650 at Watson Laboratory. Thereafter, programs were written for most of the computations required to solve a protein structure. Many of these were shared with Barbara Low's group (Jan Drenth, Ralph Einstein, Wolfie Traub) at Columbia University working on the structure of insulin and also using Watson Laboratory for computations.

During the final five years of the PSP (1955-59) computing underwent a tremendous transformation with the appearance of IBM 700 machines (701, 704, 709, 7090, 7094)[3]. Random access magnetic core memory was introduced along with microsecond cycle times and massive memory storage devices like magnetic tape drives with data channels and large capacity magnetic drums and disks making computational problems of protein crystallography feasible to accomplish within reasonable time frames. The crystallization, intensity data collection and computational aspects of the PSP were now in place. All that

---

[3] It was difficult keeping up with the turn-around because each new machine generally had new and quite different programmable features.

remained was to prepare several suitable heavy atom isomorphous derivatives of ribonuclease crystals to solve the protein phase angles and compute an electron density map that would reveal the structure at the atomic level.

Heavy atom isomorphous searches were a very new area of investigation and not much was known about them at the time, so success did not come easily (still somewhat true today). Heavy atom searches, in fact, rank a close second to protein crystallization as an art form. Vernon King began searching for isomorphous crystals by preparing ribonuclease derivatives attaching heavy atoms to side chains. Most of these approaches failed to produce crystals or produced crystals that were not isostructural with the native enzyme. Another approach was co-crystallizing ribonuclease with heavy atom chelate complexes or dyes[4] but with similar results. Since the expected diameter of ribonuclease was about 30Å, many of the trials probably did not succeed because this enzyme was small and the resultant changes were fairly large. The smallness would therefore also apply to the interstitial space in crystals filled with mother liquor. With time, the most expedient way of making heavy atom isomorphous derivatives proved to be by diffusing *small molecule* heavy atom containing compounds, preferably neutral or anionic, into crystals from days to weeks and examining X-ray patterns for changes in diffraction intensities to ascertain whether heavy atoms were bound in a systematic way. Many of the experiments were disappointing because: (a) the heavy atom compound did not bind, (b) intensity differences led to Patterson maps that were not interpretable because too many heavy atoms bound, (c) isomorphism was not preserved or (d) derivative formation was not reproducible. After a large number of different trials over many years, 6-7 good heavy atom isomorphous derivatives of varying degrees of phase determining quality and usefulness were eventually found that permitted the molecular structure of ribonuclease to be determined (1).

Roy Worthington joined the PSP from Adelaide, Australia about the same time as I did. Roy's thesis was on low angle diffraction of collagen for which he designed and built the first rotating anode X-ray tube. He later went on to Kings College (London) continuing low angle diffraction studies of muscle and membranes and is now at Carnegie-Mellon University. Jake Bello and Gopinath Kartha came shortly after, went to Roswell Park when The Project moved, and were the prime movers in the ultimate determination of the structure of ribonuclease (1). Jake came from Eastman Kodak Co., Rochester, and assumed the chemistry aspects of the work when Vernon King went to Childrens Hospital in Boston to work on glucagon (now at NY State Department of Health, Albany) while Gopi Kartha, a student of G.N. Ramachandran at Madras and later a postdoc at the National Research Council in Ottawa, overlapped and shared responsibilities with me until I left for a teaching and research position at Yale University in 1959. Erik von Sydow of the University of Uppsala in Sweden was also a member of the PSP about this time as a visiting scientist well-known for his crystallographic work on long chain fatty compounds.

In those early years of protein crystallography, we at the PSP (and also other protein crystallographers) worried about some seemingly insurmountable problems that many today do not know were ever a concern. Some of these included: (a) using ammonium sulfate or alcohols for crystallization, (b) allowable changes in unit cell dimensions on heavy atom derivative formation, (c) measuring mosaic spreads of X-ray reflections to choose the best unique region of reciprocal space to measure and use, (d) calculating non-centrosymmetric difference Pattersons with coefficients $(|F|_{p+h}-|F|_p)^2$ ($|F|_{p+h}=$ protein plus heavy atom, $|F|_p =$ native protein), which are not even strictly correct in the centro case, (e) referring the heavy atom positions of two or more different isomorphous derivatives to the

---

[4] Once when Vernon King was wearing a sweater with loud colors, Dave pointed out wool was a protein and dyes contained heavy atoms, which led to the use of the latter in heavy atom derivative searches at the PSP.

same crystallographic origin, (f) the absolute scattering scale of the diffraction intensities and much more. These kept us busy during the seemingly endless heavy atom isomorphous derivative searches. Ammonium sulfate and alcohols were known to be capable of causing structural changes so a protein structure of such grown crystals could be questionable. Furnas considers (c) thoroughly in his Manual (21) while Harker addressed points (b), (e) and (f) at great length in his original paper on multiple isomorphous replacement (26) without satisfactorily resolving the difficulties. The Patterson difference coefficients were eventually accepted on the faith that most of the time the heavy atom vector maps were correct. The origin problem was definitively solved by Michael Rossmann calculating the difference Pattersons between heavy atom isomorphs (27). The absolute scale, which eventually proved to be a non-issue, led Harker to commission Roy Worthington and myself to accurately solve the structure of basic beryllium acetate to be later used to experimentally place protein diffraction on an absolute scale (28-30). An unexpected by-product of the work was the very early, if not the first, observation of bonding electrons and the general acceptance of beryllium acetate crystals as an intensity standard.

The initial crystallography of ribonuclease concentrated on the orthorhombic, $P2_12_12_1$ (modification I) crystal form (16). The three centrosymmetric projections and the three Harker sections at 1/2 of this space group were distinct advantages; the crystals, however, only diffracted X-rays moderately to about 2.5-2.8Å resolution. Many potential heavy atom derivatives were measured and analyzed with high hopes over a 2-3 year period only to spawn discouragement as few appeared to be useful. When John Kendrew visited the PSP in 1958 and was shown some of the complicated difference Patterson maps, he told Kartha and me "these are the kind we discard at Cavendish". He was the person to know because their derivatives produced the structure of myoglobin (31). Most of the crystallographers of Poly's Department of Physics were also there in attendance, including Lindo Patterson from The Institute of Cancer Research in Philadelphia, examining the 6Å resolution map and model of myoglobin. All of us were somewhat surprised, if not amazed, by the unexpected irregular intestine-like folding of the myoglobin molecule and some even wondered if it were correct. Not long after, the doubts disappeared when the 2.0Å resolution map and structure of myoglobin was published (32).

Kartha and I kept trying to interpret unruly, complicated heavy atom difference Patterson maps applying many different strategies including Patterson superposition methods and minimum function principles. These led Gopi to perceive the utility of a double phased Fourier or a weighted single isomorphous replacement method to obtain structure with only one heavy atom isomorphous derivative in the non-centrocymmetrical case. The idea was to assign both the ambiguous phases (the correct phase and the incorrect one) calculated from a single isomorphous derivative to the observed amplitude and compute a double phased Fourier synthesis of the electron density (33). It was simple in its conception: the correct phases produce the structure; the incorrect ones do not systematically synthesize structure through the Fourier series and generally only contribute randomly to the background. We tested the idea on a hypothetical structure and showed that it worked satisfactorily (33). Subsequently, it was tested with a real case, the then large 50 atom (non-hydrogen) unknown antibiotic structure of isoquinocycline A that was also affected by observational errors. An unwanted complication was the 18 replaceable electrons were located at a special position introducing a spurious mirror plane of symmetry in the resulting map (34). The method worked fine regardless of the problems. Rossmann and Blow developed a similar treatment about the same time (35) (the PSP work was completed by late 1958) and even showed it could be combined with anomalous scattering data. Since then it has been applied many times in protein structure determination with B.C. Wang's solvent flattening procedure judiciously adjusting the weight of the two phase solutions.

In 1959, work at the PSP was winding down in preparation for the move to Buffalo and the Roswell Park Memorial Institute. It was easy on my part to tell Gopi as I was leaving that he should abandon the orthorhombic crystals in favor of the original monoclinic variety (modification II) (16), which diffracted X-rays to about 1.8 Å resolution. It was probably much more difficult on his part to make the decision because of all the work invested in the orthorhombic form. No matter, Gopi made the monoclinic choice so he and Jake Bello had to back-track to produce monoclinic crystals, measure a complete set of native monoclinic intensities, rewrite computer programs (because some were space group specific for speed of computation) and begin heavy atom isomorphous searches of this crystal form. The effort received a significant boost in 1963 when Smyth, Stein and Moore determined the sequence of ribonuclease and the National Science Foundation and the National Institutes of Health provided financial support. Things went fairly smoothly thereafter, the monoclinic choice was the right one, and the structure was solved at 2.0Å resolution about six years after leaving Brooklyn and rebuilding the laboratory.[5] A sad note of the time was that Dave Harker's long time dear friend Bridgie Weber (and mine also) passed away with a stroke in 1963 after they had dinner and were walking home together. Gopi Kartha stayed on at Roswell to begin structural studies of gramicidin that were cut short by his untimely death in 1984. Jake Bello also remained at Roswell and retired a few years ago. And by the early 1960s, the U.S. was well-represented in protein crystallography with many laboratories at various stages of growth and development. Among the earliest were those of Bill Lipscomb, Harvard and Lyle Jensen at the University of Washington, both also renowned for their small molecule crystallography; other labs that started were: Michael Rossmann, Purdue; Dick Dickerson and Larry Steinrauf, Illinois; Hal Wyckoff and Fred Richards, Yale; Joe Kraut, LaJolla; David Davies, NIH; Len Banazak and Scott Mathews, Washington University and myself at Michigan State.[6] Most of these crystallographers were postdocs of Kendrew and Perutz coming off the triumphant structure determinations of myoglobin and hemoglobin in Great Britain. All went on to solve protein structures and leave positive and lasting marks on the field.

It can be said safely that the tremendous progress and growth of protein crystallography worldwide over the last 30-40 years is probably stunning even to those with whom it originated. The development of ever faster computers, their time-sharing capabilities and miniaturization has proceeded at an unprecedented rate with no end in sight at corresponding lower costs so no computation is any longer "too big". However, interactive computer graphics might well be the one most crucial and important development of protein crystallography because it serves as the channel through which results of protein crystallography are passed on to the remainder of the scientific community. Early reports about myoglobin, lysozyme and hemoglobin appeared in *Scientific American* primarily because the graphical artwork was in place (36-38). Add to all this, three-dimensional intensity data collecting times reduced to hours by area detectors, high intensity mirror-focused rotating anode X-ray tubes and synchrotron X-ray sources and charged-coupled detectors on the horizon capable of making measurements an order of magnitude faster, protein crystallography is indeed wondrous and is attracting the finest intellectual talent in the world. There is reason to expect the next 40 years to be even more so. Given the opportunity, I would still not trade the experiences of my marvelous past for all the wonders that most certainly lie ahead.

---

[5] The structure of the orthorhombic form has also been recently completed in R. Parthasarathy's laboratory at Roswell.

[6] Memories are fallible. This compilation is from memory. If I overlooked anyone, I'm sorry.

David Harker became Research Professor Emeritus of the Medical Foundation of Buffalo in 1976. He was one of the most notable scientists of this century and at the Medical Foundation he continued working on his life-long passion of symmetry, studying colored space groups and infinite two-dimensional polyhedral sets. His final paper describing the symmetry of a new class of polyhedral sets appeared in the *Proceedings of the National Academy of Sciences* about a month before he passed away. Dave was a warm, friendly and unpretentious person always delighted with and concerned about others and above-all, always helpful. His joy for science and his humble demeanor made him a great teacher of many different things, not only crystallography. Herb Hauptmann put it aptly when he said "he (Dave Harker) was a tireless seeker of the truth, wherever he could find it, and in this quest he succeeded as few others have."

References

1. G. Kartha, J. Bello, D. Harker, Nature, <u>213</u>, 862 (1967).
2. J. D. Bernal and D. Crowfoot, Nature, <u>133</u>, 794 (1934).
3. D. Crowfoot, Nature, <u>135</u>, 591 (1935).
4. J. D. Bernal, I. Fankuchen, M. Perutz, Nature, <u>141</u>, 523 (1938).
5. J. D. Bernal and I. Fankuchen, Nature, <u>139</u>, 923 (1939).
6. D. Crowfoot and G. M. J. Schmidt, Nature, <u>155</u>, 504 (1945).
7. C. H. Carlisle and K. Dornberger, Acta Cryst., <u>1</u>, 194 (1948).
8. J. D. Bernal and I. Fankuchen, J. Gen. Physiol., <u>25</u>, 147 (1941).
9. L. Pauling and R. B. Corey, J. Am. Chem. Soc., <u>72</u>, 5349 (1950).
10. L. Pauling, R. B. Corey, H. R. Branson, Proc. Natl. Acad. Sci (USA), <u>37</u>, 205 (1951).
11. D. Harker, J. Chem. Phys., <u>4</u>, 381 (1936).
12. J. D. H. Donnay and D. Harker, Am. Mineral., <u>22</u>, 446 (1937).
13. C. J. Schneer in "Crystallography in North America"; D. McLachlan, Jr. and J. P. Glusker, Eds., American Crystallographic Association, Book Crafters, Chelsea, MI, 1983, p. 380.
14. D. Harker and J. S. Kasper, J. Chem. Phys., <u>15</u>, 882 (1947).
15. D. Harker in "Crystallography in North America", D. McLachlan, Jr. and J. P. Glusker, Eds., American Crystallographic Association, Book Crafters, Chelsea, MI, 1983, p. 59.
16. M. V. King, B. S. Magdoff, M. B. Adelman, D. Harker, Acta Cryst., <u>9</u>, 460 (1956).
17. M. V. King, J. Bello, E. M. Pignataro, D. Harker, Acta Cryst., <u>15</u>, 144 (1962).
18. T. C. Furnas, Jr. in "Crystallography in North America", D. McLachlan, Jr. and J. P. Glusker, Eds., American Crystallographic Association, Book Crafters, Chelsea, MI 1983, p. 220.
19. T. C. Furnas, Jr., Acta Cryst., <u>7</u>, 620 (1954).
20. T. C. Furnas, Jr. and D. Harker, Rev. Sci. Instrum., <u>26</u>, 449 (1955).
21. T. C. Furnas, Jr., Single Crystal Orienter Instruction Manual, Milwaukee: General Electric Company, 1957.
22. B. S. Magdoff and F. H. C. Crick, Acta Cryst., <u>8</u>, 461 (1955).
23. B. S. Magdoff. F. H. C. Crick, V. Luzzati, Acta Cryst., <u>9</u>, 156 (1956).
24. F. H. C. Crick and B. S. Magdoff, Acta Cryst., <u>9</u>, 901 (1956).
25. D. W. Green, V. M. Ingram, M. F. Perutz, Proc. Roy. Soc., <u>A225</u>, 287 (1954).
26. D. Harker, Acta Cryst., <u>9</u>, 1 (1956).
27. M.G. Rossmann, Acta Cryst., <u>13</u>, 221 (1960).
28. A. Tulinsky, C.R. Worthington, E. Pignataro, Acta Cryst., <u>12</u>, 623 (1959).
29. A. Tulinsky and C.R. Worthington, Acta Cryst., 12, 626 (1959).
30. A. Tulinsky, Acta Cryst., <u>12</u>, 634 (1959).
31. J.C. Kandrew, G. Bodo, H.M. Dintzis, R.G. Parrish, H.W. Wyckoff, D.C. Phillips, Nature, <u>181</u>, 662 (1958).
32. J.C. Kendrew, R.E. Dickerson, B.E. Strandberg, R.G. Hart, D.R. Davies, D.C. Phillips, V. Shore, Nature <u>185</u>, 422 (1960)..
33. G. Kartha, Acta Cryst., <u>14</u>, 680 (1961).
34. A. Tulinsky, J. Am. Chem. Soc., <u>86</u>, 5368 (1964).
35. D.M. Blow and M.G. Rossmann, Acta Cryst., <u>14</u>, 1195 (1961).
36. J.C. Kendrew, Sci. Am., <u>205</u>, 96 (1961).
37. D.C. Phillips, Sci. Am., <u>215</u>, 78 (1966).
38. M.F. Perutz, Sci., Am., <u>239</u>, 92 (1978).

Cumulative Chapter Titles Keyword Index, Vol. 1–31

GENERIC NAME	INDICATION	YEAR INTRO.	ARMC VOL., PAGE
acarbose	antidiabetic	1990	26, 297
aceclofenac	antiinflammatory	1992	28, 325
acetohydroxamic acid	hypoammonuric	1983	19, 313
acetorphan	antidiarrheal	1993	29, 332
acipimox	hypolipidemic	1985	21, 323
acitretin	antipsoriatic	1989	25, 309
acrivastine	antihistamine	1988	24, 295
actarit	antirheumatic	1994	30, 296
adamantanium bromide	antiseptic	1984	20, 315
adrafinil	psychostimulant	1986	22, 315
AF-2259	antiinflammatory	1987	23, 325
afloqualone	muscle relaxant	1983	19, 313
alacepril	antihypertensive	1988	24, 296
alclometasone dipropionate	topical antiinflammatory	1985	21, 323
alendronate sodium	osteoporosis	1993	29, 332
alfentanil HCl	analgesic	1983	19, 314
alfuzosin HCl	antihypertensive	1988	24, 296
alglucerase	enzyme	1991	27, 321
alminoprofen	analgesic	1983	19, 314
alpha-1 antitrypsin	protease inhibitor	1988	24, 297
alpidem	anxiolytic	1991	27, 322
alpiropride	antimigraine	1988	24, 296
alteplase	thrombolytic	1987	23, 326
amfenac sodium	antiinflammatory	1986	22, 315
amifostine	cytoprotective	1995	31, 338
aminoprofen	topical antiinflammatory	1990	26, 298
amisulpride	antipsychotic	1986	22, 316
amlexanox	antiasthmatic	1987	23, 327
amlodipine besylate	antihypertensive	1990	26, 298
amorolfine hydrochloride	topical antifungal	1991	27, 322
amosulalol	antihypertensive	1988	24, 297
ampiroxicam	antiinflammatory	1994	30, 296
amrinone	cardiotonic	1983	19, 314
amsacrine	antineoplastic	1987	23, 327
amtolmetin guacil	antiinflammatory	1993	29, 332
anastrozole	antineoplastic	1995	31, 338
angiotensin II	anticancer adjuvant	1994	30, 296
aniracetam	cognition enhancer	1993	29, 333
APD	calcium regulator	1987	23, 326
apraclonidine HCl	antiglaucoma	1988	24, 297
APSAC	thrombolytic	1987	23, 326
arbekacin	antibiotic	1990	26, 298
argatroban	antithromobotic	1990	26, 299
arotinolol HCl	antihypertensive	1986	22, 316
artemisinin	antimalarial	1987	23, 327
aspoxicillin	antibiotic	1987	23, 328
astemizole	antihistamine	1983	19, 314
astromycin sulfate	antibiotic	1985	21, 324
atovaquone	antiparasitic	1992	28, 326

GENERIC NAME	INDICATION	YEAR INTRO.	ARMC VOL., PAGE
auranofin	chrysotherapeutic	1983	19, 314
azelaic acid	antiacne	1989	25, 310
azelastine HCl	antihistamine	1986	22, 316
azithromycin	antibiotic	1988	24, 298
azosemide	diuretic	1986	22, 316
aztreonam	antibiotic	1984	20, 315
bambuterol	bronchodilator	1990	26, 299
barnidipine hydrochloride	antihypertensive	1992	28, 326
beclobrate	hypolipidemic	1986	22, 317
befunolol HCl	antiglaucoma	1983	19, 315
benazepril hydrochloride	antihypertensive	1990	26, 299
benexate HCl	antiulcer	1987	23, 328
benidipine hydrochloride	antihypertensive	1991	27, 322
beraprost sodium	platelet aggreg. inhibitor	1992	28, 326
betamethasone butyrate propionate	topical antiinflammatory	1994	30, 297
betaxolol HCl	antihypertensive	1983	19, 315
bevantolol HCl	antihypertensive	1987	23, 328
bicalutamide	antineoplastic	1995	31, 338
bifemelane HCl	nootropic	1987	23, 329
binfonazole	hypnotic	1983	19, 315
binifibrate	hypolipidemic	1986	22, 317
bisantrene hydrochloride	antineoplastic	1990	26, 300
bisoprolol fumarate	antihypertensive	1986	22, 317
bopindolol	antihypertensive	1985	21, 324
brodimoprin	antibiotic	1993	29, 333
brotizolam	hypnotic	1983	19, 315
brovincamine fumarate	cerebral vasodilator	1986	22, 317
bucillamine	immunomodulator	1987	23, 329
bucladesine sodium	cardiostimulant	1984	20, 316
budralazine	antihypertensive	1983	19, 315
bunazosin HCl	antihypertensive	1985	21, 324
bupropion HCl	antidepressant	1989	25, 310
buserelin acetate	hormone	1984	20, 316
buspirone HCl	anxiolytic	1985	21, 324
butenafine hydrochloride	topical antifungal	1992	28, 327
butibufen	antiinflammatory	1992	28, 327
butoconazole	topical antifungal	1986	22, 318
butoctamide	hypnotic	1984	20, 316
butyl flufenamate	topical antiinflammatory	1983	19, 316
cabergoline	antiprolactin	1993	29, 334
cadexomer iodine	wound healing agent	1983	19, 316
cadralazine	hypertensive	1988	24, 298
calcipotriol	antipsoriatic	1991	27, 323
camostat mesylate	antineoplastic	1985	21, 325
carboplatin	antibiotic	1986	22, 318
carperitide	congestive heart failure	1995	31, 339
carumonam	antibiotic	1988	24, 298
carvedilol	antihypertensive	1991	27, 323

GENERIC NAME	INDICATION	YEAR INTRO.	ARMC VOL., PAGE
clodronate disodium	calcium regulator	1986	22, 319
cloricromen	antithrombotic	1991	27, 325
clospipramine hydrochloride	neuroleptic	1991	27, 325
cyclosporine	immunosuppressant	1983	19, 317
cytarabine ocfosfate	antineoplastic	1993	29, 335
dapiprazole HCl	antiglaucoma	1987	23, 332
defeiprone	iron chelator	1995	31, 340
defibrotide	antithrombotic	1986	22, 319
deflazacort	antiinflammatory	1986	22, 319
delapril	antihypertensive	1989	25, 311
denopamine	cardiostimulant	1988	24, 300
deprodone propionate	topical antiinflammatory	1992	28, 329
desflurane	anesthetic	1992	28, 329
dexibuprofen	antiinflammatory	1994	30, 298
dexrazoxane	cardioprotective	1992	28, 330
dezocine	analgesic	1991	27, 326
diacerein	antirheumatic	1985	21, 326
didanosine	antiviral	1991	27, 326
dilevalol	antihypertensive	1989	25, 311
dirithromycin	antibiotic	1993	29, 336
disodium pamidronate	calcium regulator	1989	25, 312
divistyramine	hypocholesterolemic	1984	20, 317
docarpamine	cardiostimulant	1994	30, 298
docetaxel	antineoplastic	1995	31, 341
dopexamine	cardiostimulant	1989	25, 312
dornase alfa	cystic fibrosis	1994	30, 298
dorzolamide HCL	antiglaucoma	1995	31, 341
doxacurium chloride	muscle relaxant	1991	27, 326
doxazosin mesylate	antihypertensive	1988	24, 300
doxefazepam	hypnotic	1985	21, 326
doxifluridine	antineoplastic	1987	23, 332
doxofylline	bronchodilator	1985	21, 327
dronabinol	antinauseant	1986	22, 319
droxicam	antiinflammatory	1990	26, 302
droxidopa	antiparkinsonian	1989	25, 312
duteplase	anticougulant	1995	31, 342
ebastine	antihistamine	1990	26 302
ecabet sodium	antiulcerative	1993	29, 336
efonidipine	antihypertensive	1994	30, 299
emedastine difumarate	antiallergic/antiasthmatic	1993	29, 336
emorfazone	analgesic	1984	20, 317
enalapril maleate	antihypertensive	1984	20, 317
enalaprilat	antihypertensive	1987	23, 332
encainide HCl	antiarrhythmic	1987	23, 333
enocitabine	antineoplastic	1983	19, 318
enoxacin	antibacterial	1986	22, 320
enoxaparin	antithrombotic	1987	23, 333
enoximone	cardiostimulant	1988	24, 301
enprostil	antiulcer	1985	21, 327

GENERIC NAME	INDICATION	YEAR INTRO.	ARMC VOL., PAGE
epalrestat	antidiabetic	1992	28, 330
eperisone HCl	muscle relaxant	1983	19, 318
epidermal growth factor	wound healing agent	1987	23, 333
epinastine	antiallergic	1994	30, 299
epirubicin HCl	antineoplastic	1984	20, 318
epoprostenol sodium	platelet aggreg. inhib.	1983	19, 318
eptazocine HBr	analgesic	1987	23, 334
erdosteine	expectorant	1995	31, 342
erythromycin acistrate	antibiotic	1988	24, 301
erythropoietin	hematopoetic	1988	24, 301
esmolol HCl	antiarrhythmic	1987	23, 334
ethyl icosapentate	antithrombotic	1990	26, 303
etizolam	anxiolytic	1984	20, 318
etodolac	antiinflammatory	1985	21, 327
exifone	nootropic	1988	24, 302
factor VIII	hemostatic	1992	28, 330
fadrozole HCl	antineoplastic	1995	31, 342
famciclovir	antiviral	1994	30, 300
famotidine	antiulcer	1985	21, 327
fasudil HCl	neuroprotective	1995	31, 343
felbamate	antiepileptic	1993	29, 337
felbinac	topical antiinflammatory	1986	22, 320
felodipine	antihypertensive	1988	24, 302
fenbuprol	choleretic	1983	19, 318
fenticonazole nitrate	antifungal	1987	23, 334
filgrastim	immunostimulant	1991	27, 327
finasteride	5α-reductase inhibitor	1992	28, 331
fisalamine	intestinal antiinflammatory	1984	20, 318
fleroxacin	antibacterial	1992	28, 331
flomoxef sodium	antibiotic	1988	24, 302
flosequinan	cardiostimulant	1992	28, 331
fluconazole	antifungal	1988	24, 303
fludarabine phosphate	antineoplastic	1991	27, 327
flumazenil	benzodiazepine antag.	1987	23, 335
flunoxaprofen	antiinflammatory	1987	23, 335
fluoxetine HCl	antidepressant	1986	22, 320
flupirtine maleate	analgesic	1985	21, 328
flutamide	antineoplastic	1983	19, 318
flutazolam	anxiolytic	1984	20, 318
fluticasone propionate	antiinflammatory	1990	26, 303
flutoprazepam	anxiolytic	1986	22, 320
flutrimazole	topical antifungal	1995	31, 343
flutropium bromide	antitussive	1988	24, 303
fluvastatin	hypolipaemic	1994	30, 300
fluvoxamine maleate	antidepressant	1983	19, 319
formestane	antineoplastic	1993	29, 337
formoterol fumarate	bronchodilator	1986	22, 321
foscarnet sodium	antiviral	1989	25, 313

GENERIC NAME	INDICATION	YEAR INTRO.	ARMC VOL., PAGE
fosfosal	analgesic	1984	20, 319
fosinopril sodium	antihypertensive	1991	27, 328
fotemustine	antineoplastic	1989	25, 313
gabapentin	antiepileptic	1993	29, 338
gallium nitrate	calcium regulator	1991	27, 328
gallopamil HCl	antianginal	1983	19, 319
ganciclovir	antiviral	1988	24, 303
gemcitabine HCl	antineoplastic	1995	31, 344
gemeprost	abortifacient	1983	19, 319
gestodene	progestogen	1987	23, 335
gestrinone	antiprogestogen	1986	22, 321
glimepiride	antidiabetic	1995	31, 344
glucagon, rDNA	hypoglycemia	1993	29, 338
goserelin	hormone	1987	23, 336
granisetron hydrochloride	antiemetic	1991	27, 329
guanadrel sulfate	antihypertensive	1983	19, 319
gusperimus	immunosuppressant	1994	30, 300
halobetasol propionate	topical antiinflammatory	1991	27, 329
halofantrine	antimalarial	1988	24, 304
halometasone	topical antiinflammatory	1983	19, 320
histrelin	precocious puberty	1993	29, 338
hydrocortisone aceponate	topical antiinflammatory	1988	24, 304
hydrocortisone butyrate	topical antiinflammatory	1983	19, 320
ibopamine HCl	cardiostimulant	1984	20, 319
ibudilast	antiasthmatic	1989	25, 313
idarubicin hydrochloride	antineoplastic	1990	26, 303
idebenone	nootropic	1986	22, 321
iloprost	platelet aggreg. inhibitor	1992	28, 332
imidapril HCl	antihypertensive	1993	29, 339
imiglucerase	Gaucher's disease	1994	30, 301
imipenem/cilastatin	antibiotic	1985	21, 328
indalpine	antidepressant	1983	19, 320
indeloxazine HCl	nootropic	1988	24, 304
indobufen	antithrombotic	1984	20, 319
interferon, β-1b	multiple sclerosis	1993	29, 339
interferon, gamma	antiinflammatory	1989	25, 314
interferon, gamma-1α	antineoplastic	1992	28, 332
interferon gamma-1b	immunostimulant	1991	27, 329
interleukin-2	antineoplastic	1989	25, 314
ipriflavone	calcium regulator	1989	25, 314
irinotecan	antineoplastic	1994	30, 301
irsogladine	antiulcer	1989	25, 315
isepamicin	antibiotic	1988	24, 305
isofezolac	antiinflammatory	1984	20, 319
isoxicam	antiinflammatory	1983	19, 320
isradipine	antihypertensive	1989	25, 315
itopride HCl	gastroprokinetic	1995	31, 344
itraconazole	antifungal	1988	24, 305

GENERIC NAME	INDICATION	YEAR INTRO.	ARMC VOL., PAGE
ivermectin	antiparasitic	1987	23, 336
ketanserin	antihypertensive	1985	21, 328
ketorolac tromethamine	analgesic	1990	26, 304
lacidipine	antihypertensive	1991	27, 330
lamivudine	antiviral	1995	31, 345
lamotrigine	anticonvulsant	1990	26, 304
lanoconazole	antifungal	1994	30, 302
lanreotide acetate	acromegaly	1995	31, 345
lansoprazole	antiulcer	1992	28, 332
lenampicillin HCl	antibiotic	1987	23, 336
lentinan	immunostimulant	1986	22, 322
leuprolide acetate	hormone	1984	20, 319
levacecarnine HCl	nootropic	1986	22, 322
levobunolol HCl	antiglaucoma	1985	21, 328
levocabastine hydrochloride	antihistamine	1991	27, 330
levodropropizine	antitussive	1988	24, 305
levofloxacin	antibiotic	1993	29, 340
lidamidine HCl	antiperistaltic	1984	20, 320
limaprost	antithrombotic	1988	24, 306
lisinopril	antihypertensive	1987	23, 337
lobenzarit sodium	antiinflammatory	1986	22, 322
lodoxamide tromethamine	antiallergic ophthalmic	1992	28, 333
lomefloxacin	antibiotic	1989	25, 315
lonidamine	antineoplastic	1987	23, 337
loprazolam mesylate	hypnotic	1983	19, 321
loracarbef	antibiotic	1992	28, 333
loratadine	antihistamine	1988	24, 306
losartan	antihypertensive	1994	30, 302
lovastatin	hypocholesterolemic	1987	23, 337
loxoprofen sodium	antiinflammatory	1986	22, 322
mabuterol HCl	bronchodilator	1986	22, 323
malotilate	hepatroprotective	1985	21, 329
manidipine hydrochloride	antihypertensive	1990	26, 304
masoprocol	topical antineoplastic	1992	28, 333
medifoxamine fumarate	antidepressant	1986	22, 323
mefloquine HCl	antimalarial	1985	21, 329
meglutol	hypolipidemic	1983	19, 321
melinamide	hypocholesterolemic	1984	20, 320
mepixanox	analeptic	1984	20, 320
meptazinol HCl	analgesic	1983	19, 321
meropenem	carbapenem antibiotic	1994	30, 303
metaclazepam	anxiolytic	1987	23, 338
metapramine	antidepressant	1984	20, 320
mexazolam	anxiolytic	1984	20, 321
mifepristone	abortifacient	1988	24, 306
milrinone	cardiostimulant	1989	25, 316
miltefosine	topical antineoplastic	1993	29, 340
miokamycin	antibiotic	1985	21, 329
mirtazapine	antidepressant	1994	30, 303

GENERIC NAME	INDICATION	YEAR INTRO.	ARMC VOL., PAGE
misoprostol	antiulcer	1985	21, 329
mivacurium chloride	muscle relaxant	1992	28, 334
mitoxantrone HCl	antineoplastic	1984	20, 321
mizoribine	immunosuppressant	1984	20, 321
moclobemide	antidepressant	1990	26, 305
modafinil	idiopathic hypersomnia	1994	30, 303
moexipril HCl	antihypertensive	1995	31, 346
mofezolac	analgesic	1994	30, 304
mometasone furoate	topical antiinflammatory	1987	23, 338
moricizine hydrochloride	antiarrhythmic	1990	26, 305
moxonidine	antihypertensive	1991	27, 330
mupirocin	topical antibiotic	1985	21, 330
muromonab-CD3	immunosuppressant	1986	22, 323
muzolimine	diuretic	1983	19, 321
mycophenolate mofetil	immunosuppressant	1995	31, 346
nabumetone	antiinflammatory	1985	21, 330
nadifloxacin	topical antibiotic	1993	29, 340
nafamostat mesylate	protease inhibitor	1986	22, 323
nafarelin acetate	hormone	1990	26, 306
naftifine HCl	antifungal	1984	20, 321
nalmefene HCl	dependence treatment	1995	31, 347
naltrexone HCl	narcotic antagonist	1984	20, 322
nartograstim	leukopenia	1994	30, 304
nazasetron	antiemetic	1994	30, 305
nedaplatin	antineoplastic	1995	31, 347
nedocromil sodium	antiallergic	1986	22, 324
nefazodone	antidepressant	1994	30, 305
neltenexine	cystic fibrosis	1993	29, 341
nemonapride	neuroleptic	1991	27, 331
neticonazole HCl	topical antifungal	1993	29, 341
nicorandil	coronary vasodilator	1984	20, 322
nilutamide	antineoplastic	1987	23, 338
nilvadipine	antihypertensive	1989	25, 316
nimesulide	antiinflammatory	1985	21, 330
nimodipine	cerebral vasodilator	1985	21, 330
nipradilol	antihypertensive	1988	24, 307
nisoldipine	antihypertensive	1990	26, 306
nitrefazole	alcohol deterrent	1983	19, 322
nitrendipine	hypertensive	1985	21, 331
nizatidine	antiulcer	1987	23, 339
nizofenzone fumarate	nootropic	1988	24, 307
nomegestrol acetate	progestogen	1986	22, 324
norfloxacin	antibacterial	1983	19, 322
norgestimate	progestogen	1986	22, 324
octreotide	antisecretory	1988	24, 307
ofloxacin	antibacterial	1985	21, 331
omeprazole	antiulcer	1988	24, 308
ondansetron hydrochloride	antiemetic	1990	26, 306
ornoprostil	antiulcer	1987	23, 339

GENERIC NAME	INDICATION	YEAR INTRO.	ARMC VOL., PAGE
osalazine sodium	intestinal antinflamm.	1986	22, 324
oxaprozin	antiinflammatory	1983	19, 322
oxcarbazepine	anticonvulsant	1990	26, 307
oxiconazole nitrate	antifungal	1983	19, 322
oxiracetam	nootropic	1987	23, 339
oxitropium bromide	bronchodilator	1983	19, 323
ozagrel sodium	antithrombotic	1988	24, 308
paclitaxal	antineoplastic	1993	29, 342
parnaparin sodium	anticoagulant	1993	29, 342
panipenem/betamipron	carbapenem antibiotic	1994	30, 305
pantoprazole sodium	antiulcer	1995	30, 306
paroxetine	antidepressant	1991	27, 331
pefloxacin mesylate	antibacterial	1985	21, 331
pegademase bovine	immunostimulant	1990	26, 307
pegaspargase	antineoplastic	1994	30, 306
pemirolast potassium	antiasthmatic	1991	27, 331
pentostatin	antineoplastic	1992	28, 334
pergolide mesylate	antiparkinsonian	1988	24, 308
perindopril	antihypertensive	1988	24, 309
picotamide	antithrombotic	1987	23, 340
pidotimod	immunostimulant	1993	29, 343
piketoprofen	topical antiinflammatory	1984	20, 322
pilsicainide hydrochloride	antiarrhythmic	1991	27, 332
pimaprofen	topical antiinflammatory	1984	20, 322
pimobendan	heart failure	1994	30, 307
pinacidil	antihypertensive	1987	23, 340
pirarubicin	antineoplastic	1988	24, 309
pirmenol	antiarrhythmic	1994	30, 307
piroxicam cinnamate	antiinflammatory	1988	24, 309
plaunotol	antiulcer	1987	23, 340
polaprezinc	antiulcer	1994	30, 307
porfimer sodium	antineoplastic adjuvant	1993	29, 343
pramiracetam H_2SO_4	cognition enhancer	1993	29, 343
pranlukast	antiasthmatic	1995	31, 347
pravastatin	antilipidemic	1989	25, 316
prednicarbate	topical antiinflammatory	1986	22, 325
progabide	anticonvulsant	1985	21, 331
promegestrone	progestogen	1983	19, 323
propacetamol HCl	analgesic	1986	22, 325
propagermanium	antiviral	1994	30, 308
propentofylline propionate	cerebral vasodilator	1988	24, 310
propiverine hydrochloride	urologic	1992	28, 335
propofol	anesthetic	1986	22, 325
pumactant	lung surfactant	1994	30, 308
quazepam	hypnotic	1985	21, 332
quinagolide	hyperprolactinemia	1994	30, 309
quinapril	antihypertensive	1989	25, 317
quinfamide	amebicide	1984	20, 322

GENERIC NAME	INDICATION	YEAR INTRO.	ARMC VOL., PAGE
ramipril	antihypertensive	1989	25, 317
ranimustine	antineoplastic	1987	23, 341
ranitidine bismuth citrate	antiulcer	1995	31, 348
rebamipide	antiulcer	1990	26, 308
remoxipride hydrochloride	antipsychotic	1990	26, 308
repirinast	antiallergic	1987	23, 341
reviparin sodium	anticoagulant	1993	29, 344
rifabutin	antibacterial	1992	28, 335
rifapentine	antibacterial	1988	24, 310
rifaximin	antibiotic	1985	21, 332
rifaximin	antibiotic	1987	23, 341
rilmazafone	hypnotic	1989	25, 317
rilmenidine	antihypertensive	1988	24, 310
rimantadine HCl	antiviral	1987	23, 342
rimexolone	antiinflammatory	1995	31, 348
risperidone	neuroleptic	1993	29, 344
rocuronium bromide	neuromuscular blocker	1994	30, 309
rokitamycin	antibiotic	1986	22, 325
romurtide	immunostimulant	1991	27, 332
ronafibrate	hypolipidemic	1986	22, 326
rosaprostol	antiulcer	1985	21, 332
roxatidine acetate HCl	antiulcer	1986	22, 326
roxithromycin	antiulcer	1987	23, 342
rufloxacin hydrochloride	antibacterial	1992	28, 335
RV-11	antibiotic	1989	25, 318
salmeterol hydroxynaphthoate	bronchodilator	1990	26, 308
sapropterin hydrochloride	hyperphenylalaninemia	1992	28, 336
saquinavir mesvlate	antiviral	1995	31, 349
sargramostim	immunostimulant	1991	27, 332
sarpogrelate HCl	platelet antiaggregant	1993	29, 344
schizophyllan	immunostimulant	1985	22, 326
seratrodast	antiasthmatic	1995	31, 349
sertaconazole nitrate	topical antifungal	1992	28, 336
setastine HCl	antihistamine	1987	23, 342
setiptiline	antidepressant	1989	25, 318
setraline hydrochloride	antidepressant	1990	26, 309
sevoflurane	anesthetic	1990	26, 309
simvastatin	hypocholesterolemic	1988	24, 311
sobuzoxane	antineoplastic	1994	30, 310
sodium cellulose PO4	hypocalciuric	1983	19, 323
sofalcone	antiulcer	1984	20, 323
somatomedin-1	growth hormone insensitivity	1994	30, 310
somatotropin	growth hormone	1994	30, 310
somatropin	hormone	1987	23, 343
sorivudine	antiviral	1993	29, 345
sparfloxacin	antibiotic	1993	29, 345
spirapril HCl	antihypertensive	1995	31, 349

GENERIC NAME	INDICATION	YEAR INTRO.	ARMC VOL., PAGE
spizofurone	antiulcer	1987	23, 343
stavudine	antiviral	1994	30, 311
succimer	chelator	1991	27, 333
sufentanil	analgesic	1983	19, 323
sulbactam sodium	B-lactamase inhibitor	1986	22, 326
sulconizole nitrate	topical antifungal	1985	21, 332
sultamycillin tosylate	antibiotic	1987	23, 343
sumatriptan succinate	antimigraine	1991	27, 333
suplatast tosilate	antiallergic	1995	31, 350
suprofen	analgesic	1983	19, 324
surfactant TA	respiratory surfactant	1987	23, 344
tacalcitol	topical antipsoriatic	1993	29, 346
tacrine HCl	Alzheimer's disease	1993	29, 346
tacrolimus	immunosuppressant	1993	29, 347
tamsulosin HCl	antiprostatic hypertrophy	1993	29, 347
tazobactam sodium	β-lactamase inhibitor	1992	28, 336
tazanolast	antiallergic	1990	26, 309
teicoplanin	antibacterial	1988	24, 311
telmesteine	mucolytic	1992	28, 337
temafloxacin hydrochloride	antibacterial	1991	27, 334
temocapril	antihypertensive	1994	30, 311
temocillin disodium	antibiotic	1984	20, 323
tenoxicam	antiinflammatory	1987	23, 344
teprenone	antiulcer	1984	20, 323
terazosin HCl	antihypertensive	1984	20, 323
terbinafine hydrochloride	antifungal	1991	27, 334
terconazole	antifungal	1983	19, 324
tertatolol HCl	antihypertensive	1987	23, 344
thymopentin	immunomodulator	1985	21, 333
tiamenidine HCl	antihypertensive	1988	24, 311
tianeptine sodium	antidepressant	1983	19, 324
tibolone	anabolic	1988	24, 312
tilisolol hydrochloride	antihypertensive	1992	28, 337
tiludronate disodium	Paget's disease	1995	31, 350
timiperone	neuroleptic	1984	20, 323
tinazoline	nasal decongestant	1988	24, 312
tioconazole	antifungal	1983	19, 324
tiopronin	urolithiasis	1989	25, 318
tiquizium bromide	antispasmodic	1984	20, 324
tiracizine hydrochloride	antiarrhythmic	1990	26, 310
tirilazad mesylate	subarachnoid hemorrhage	1995	31, 351
tiropramide HCl	antispasmodic	1983	19, 324
tizanidine	muscle relaxant	1984	20, 324
toloxatone	antidepressant	1984	20, 324
tolrestat	antidiabetic	1989	25, 319
topiramate	antiepileptic	1995	31, 351
torasemide	diuretic	1993	29, 348
toremifene	antineoplastic	1989	25, 319

GENERIC NAME	INDICATION	YEAR INTRO.	ARMC VOL., PAGE
tosufloxacin tosylate	antibacterial	1990	26, 310
trandolapril	antihypertensive	1993	29, 348
tretinoin tocoferil	antiulcer	1993	29, 348
trientine HCl	chelator	1986	22, 327
trimazosin HCl	antihypertensive	1985	21, 333
trimetrexate glucuronate	*Pneumocystis carinii* pneumonia	1994	30, 312
tropisetron	antiemetic	1992	28, 337
troxipide	antiulcer	1986	22, 327
ubenimex	immunostimulant	1987	23, 345
unoprostone isopropyl ester	antiglaucoma	1994	30, 312
valaciclovir HCl	antiviral	1995	31, 352
venlafaxine	antidepressant	1994	30, 312
vesnarinone	cardiostimulant	1990	26, 310
vigabatrin	anticonvulsant	1989	25, 319
vinorelbine	antineoplastic	1989	25, 320
voglibose	antidiabetic	1994	30, 313
xamoterol fumarate	cardiotonic	1988	24, 312
zalcitabine	antiviral	1992	28, 338
zaltoprofen	antiinflammatory	1993	29, 349
zidovudine	antiviral	1987	23, 345
zinostatin stimalamer	antineoplastic	1994	30, 313
zolpidem hemitartrate	hypnotic	1988	24, 313
zonisamide	anticonvulsant	1989	25, 320
zopiclone	hypnotic	1986	22, 327
zuclopenthixol acetate	antipsychotic	1987	23, 345

CUMULATIVE NCE INTRODUCTION INDEX, 1983–1995, SORTED BY INDICATION

GENERIC NAME	INDICATION	YEAR INTRO.	ARMC VOL., PAGE
gemeprost	ABORTIFACIENT	1983	19,319
mifepristone		1988	24,306
lanreotide acetate	ACROMEGALY	1995	31,345
nitrefazole	ALCOHOL DETERRENT	1983	19,322
tacrine HCl	ALZHEIMER'S DISEASE	1993	29,346
quinfamide	AMEBICIDE	1984	20,322
tibolone	ANABOLIC	1988	24,312
mepixanox	ANALEPTIC	1984	20,320
alfentanil HCl	ANALGESIC	1983	19,314
alminoprofen		1983	19,314
dezocine		1991	27,326
emorfazone		1984	20,317
eptazocine HBr		1987	23,334
flupirtine maleate		1985	21,328
fosfosal		1984	20,319
ketorolac tromethamine		1990	26,304
meptazinol HCl		1983	19,321
mofezolac		1994	30,304
propacetamol HCl		1986	22,325
sufentanil		1983	19,323
suprofen		1983	19,324
desflurane	ANESTHETIC	1992	28,329
propofol		1986	22,325
sevoflurane		1990	26,309
azelaic acid	ANTIACNE	1989	25,310
emedastine difumarate	ANTIALLERGIC	1993	29,336
epinastine		1994	30,299
nedocromil sodium		1986	22,324
repirinast		1987	23,341
suplatast tosilate		1995	31,350
tazanolast		1990	26,309
lodoxamide tromethamine	ANTIALLERGIC OPHTHALMIC	1992	28,333
gallopamil HCl	ANTIANGINAL	1983	19,319
cibenzoline	ANTIARRHYTHMIC	1985	21,325
encainide HCl		1987	23,333
esmolol HCl		1987	23,334

GENERIC NAME	INDICATION	YEAR INTRO.	ARMC VOL. PAGE
moricizine hydrochloride		1990	26,305
pilsicainide hydrochloride		1991	27,332
pirmenol		1994	30,307
tiracizine hydrochloride		1990	26,310
amlexanox	ANTIASTHMATIC	1987	23,327
emedastine difumarate		1993	29,336
ibudilast		1989	25,313
pemirolast potassium		1991	27,331
seratrodast		1995	31,349
ciprofloxacin	ANTIBACTERIAL	1986	22,318
enoxacin		1986	22,320
fleroxacin		1992	28,331
norfloxacin		1983	19,322
ofloxacin		1985	21,331
pefloxacin mesylate		1985	21,331
pranlukast		1995	31,347
rifabutin		1992	28,335
rifapentine		1988	24,310
rufloxacin hydrochloride		1992	28,335
teicoplanin		1988	24,311
temafloxacin hydrochloride		1991	27,334
tosufloxacin tosylate		1990	26,310
arbekacin	ANTIBIOTIC	1990	26,298
aspoxicillin		1987	23,328
astromycin sulfate		1985	21,324
azithromycin		1988	24,298
aztreonam		1984	20,315
brodimoprin		1993	29,333
carboplatin		1986	22,318
carumonam		1988	24,298
cefbuperazone sodium		1985	21,325
cefdinir		1991	27,323
cefepime		1993	29,334
cefetamet pivoxil hydrochloride		1992	28,327
cefixime		1987	23,329
cefmenoxime HCl		1983	19,316
cefminox sodium		1987	23,330
cefodizime sodium		1990	26,300
cefonicid sodium		1984	20,316
ceforanide		1984	20,317
cefotetan disodium		1984	20,317
cefotiam hexetil hydrochloride		1991	27,324
cefpimizole		1987	23,330
cefpiramide sodium		1985	21,325
cefpirome sulfate		1992	28,328
cefpodoxime proxetil		1989	25,310

GENERIC NAME	INDICATION	YEAR INTRO.	ARMC VOL., PAGE
cefprozil		1992	28,328
ceftazidime		1983	19,316
cefteram pivoxil		1987	23,330
ceftibuten		1992	28,329
cefuroxime axetil		1987	23,331
cefuzonam sodium		1987	23,331
clarithromycin		1990	26,302
dirithromycin		1993	29,336
erythromycin acistrate		1988	24,301
flomoxef sodium		1988	24,302
imipenem/cilastatin		1985	21,328
isepamicin		1988	24,305
lenampicillin HCl		1987	23,336
levofloxacin		1993	29,340
lomefloxacin		1989	25,315
loracarbef		1992	28,333
miokamycin		1985	21,329
rifaximin		1985	21,332
rifaximin		1987	23,341
rokitamycin		1986	22,325
RV-11		1989	25,318
sparfloxacin		1993	29,345
sultamycillin tosylate		1987	23,343
temocillin disodium		1984	20,323
meropenem	ANTIBIOTIC, CARBAPENEM	1994	30,303
panipenem/betamipron		1994	30,305
mupirocin	ANTIBIOTIC, TOPICAL	1985	21,330
nadifloxacin		1993	29,340
angiotensin II	ANTICANCER ADJUVANT	1994	30,296
chenodiol	ANTICHOLELITHOGENIC	1983	19,317
duteplase	ANTICOAGULANT	1995	31,342
parnaparin sodium		1993	29,342
reviparin sodium		1993	29,344
lamotrigine	ANTICONVULSANT	1990	26,304
oxcarbazepine		1990	26,307
progabide		1985	21,331
vigabatrin		1989	25,319
zonisamide		1989	25,320
bupropion HCl	ANTIDEPRESSANT	1989	25,310
citalopram		1989	25,311
fluoxetine HCl		1986	22,320
fluvoxamine maleate		1983	19,319

GENERIC NAME	INDICATION	YEAR INTRO.	ARMC VOL., PAGE
indalpine		1983	19,320
medifoxamine fumarate		1986	22,323
metapramine		1984	20,320
mirtazapine		1994	30,303
moclobemide		1990	26,305
nefazodone		1994	30,305
paroxetine		1991	27,331
setiptiline		1989	25,318
sertraline hydrochloride		1990	26,309
tianeptine sodium		1983	19,324
toloxatone		1984	20,324
venlafaxine		1994	30,312
acarbose	ANTIDIABETIC	1990	26,297
epalrestat		1992	28,330
glimepiride		1995	31,344
tolrestat		1989	25,319
voglibose		1994	30,313
acetorphan	ANTIDIARRHEAL	1993	29,332
granisetron hydrochloride	ANTIEMETIC	1991	27,329
ondansetron hydrochloride		1990	26,306
nazasetron		1994	30,305
tropisetron		1992	28,337
felbamate	ANTIEPILEPTIC	1993	29,337
gabapentin		1993	29,338
topiramate		1995	31,351
centchroman	ANTIESTROGEN	1991	27,324
fenticonazole nitrate	ANTIFUNGAL	1987	23,334
fluconazole		1988	24,303
itraconazole		1988	24,305
lanoconazole		1994	30,302
naftifine HCl		1984	20,321
oxiconazole nitrate		1983	19,322
terbinafine hydrochloride		1991	27,334
terconazole		1983	19,324
tioconazole		1983	19,324
amorolfine hydrochloride	ANTIFUNGAL, TOPICAL	1991	27,322
butenafine hydrochloride		1992	28,327
butoconazole		1986	22,318
cloconazole HCl		1986	22,318
flutrimazole		1995	31,343
neticonazole HCl		1993	29,341

GENERIC NAME	INDICATION	YEAR INTRO.	ARMC VOL., PAGE
sertaconazole nitrate		1992	28,336
sulconizole nitrate		1985	21,332
apraclonidine HCl	ANTIGLAUCOMA	1988	24,297
befunolol HCl		1983	19,315
dapiprazole HCl		1987	23,332
dorzolamide HCl		1995	31,341
levobunolol HCl		1985	21,328
unoprostone isopropyl ester		1994	30,312
acrivastine	ANTIHISTAMINE	1988	24,295
astemizole		1983	19,314
azelastine HCl		1986	22,316
ebastine		1990	26,302
cetirizine HCl		1987	23,331
levocabastine hydrochloride		1991	27,330
loratadine		1988	24,306
setastine HCl		1987	23,342
alacepril	ANTIHYPERTENSIVE	1988	24,296
alfuzosin HCl		1988	24,296
amlodipine besylate		1990	26,298
amosulalol		1988	24,297
arotinolol HCl		1986	22,316
barnidipine hydrochloride		1992	28,326
benazepril hydrochloride		1990	26,299
benidipine hydrochloride		1991	27,322
betaxolol HCl		1983	19,315
bevantolol HCl		1987	23,328
bisoprolol fumarate		1986	22,317
bopindolol		1985	21,324
budralazine		1983	19,315
bunazosin HCl		1985	21,324
carvedilol		1991	27,323
celiprolol HCl		1983	19,317
cicletanine		1988	24,299
cilazapril		1990	26,301
cinildipine		1995	31,339
delapril		1989	25,311
dilevalol		1989	25,311
doxazosin mesylate		1988	24,300
efonidipine		1994	30,299
enalapril maleate		1984	20,317
enalaprilat		1987	23,332
felodipine		1988	24,302
fosinopril sodium		1991	27,328
guanadrel sulfate		1983	19,319
imidapril HCl		1993	29,339
isradipine		1989	25,315

GENERIC NAME	INDICATION	YEAR INTRO.	ARMC VOL., PAGE
ketanserin		1985	21,328
lacidipine		1991	27,330
lisinopril		1987	23,337
losartan		1994	30,302
manidipine hydrochloride		1990	26,304
moexipril HCl		1995	31,346
moxonidine		1991	27,330
nilvadipine		1989	25,316
nipradilol		1988	24,307
nisoldipine		1990	26,306
perindopril		1988	24,309
pinacidil		1987	23,340
quinapril		1989	25,317
ramipril		1989	25,317
rilmenidine		1988	24,310
spirapril HCl		1995	31,349
temocapril		1994	30,311
terazosin HCl		1984	20,323
tertatolol HCl		1987	23,344
tiamenidine HCl		1988	24,311
tilisolol hydrochloride		1992	28,337
trandolapril		1993	29,348
trimazosin HCl		1985	21,333
aceclofenac	ANTIINFLAMMATORY	1992	28,325
AF-2259		1987	23,325
amfenac sodium		1986	22,315
ampiroxicam		1994	30,296
amtolmetin guacil		1993	29,332
butibufen		1992	28,327
deflazacort		1986	22,319
dexibuprofen		1994	30,298
droxicam		1990	26,302
etodolac		1985	21,327
flunoxaprofen		1987	23,335
fluticasone propionate		1990	26,303
interferon, gamma		1989	25,314
isofezolac		1984	20,319
isoxicam		1983	19,320
lobenzarit sodium		1986	22,322
loxoprofen sodium		1986	22,322
nabumetone		1985	21,330
nimesulide		1985	21,330
oxaprozin		1983	19,322
piroxicam cinnamate		1988	24,309
rimexolone		1995	31,348
tenoxicam		1987	23,344
zaltoprofen		1993	29,349

GENERIC NAME	INDICATION	YEAR INTRO.	ARMC VOL., PAGE
fisalamine	ANTIINFLAMMATORY,	1984	20,318
osalazine sodium	INTESTINAL	1986	22,324
alclometasone dipropionate	ANTIINFLAMMATORY,	1985	21,323
aminoprofen	TOPICAL	1990	26,298
betamethasone butyrate propionate		1994	30,297
butyl flufenamate		1983	19,316
deprodone propionate		1992	28,329
felbinac		1986	22,320
halobetasol propionate		1991	27,329
halometasone		1983	19,320
hydrocortisone aceponate		1988	24,304
hydrocortisone butyrate propionate		1983	19,320
mometasone furoate		1987	23,338
piketoprofen		1984	20,322
pimaprofen		1984	20,322
prednicarbate		1986	22,325
pravastatin	ANTILIPIDEMIC	1989	25,316
artemisinin	ANTIMALARIAL	1987	23,327
halofantrine		1988	24,304
mefloquine HCl		1985	21,329
alpiropride	ANTIMIGRAINE	1988	24,296
sumatriptan succinate		1991	27,333
dronabinol	ANTINAUSEANT	1986	22,319
amsacrine	ANTINEOPLASTIC	1987	23,327
anastrozole		1995	31,338
bicalutamide		1995	31,338
bisantrene hydrochloride		1990	26,300
camostat mesylate		1985	21,325
cladribine		1993	29,335
cytarabine ocfosfate		1993	29,335
docetaxel		1995	31,341
doxifluridine		1987	23,332
enocitabine		1983	19,318
epirubicin HCl		1984	20,318
fadrozole HCl		1995	31,342
fludarabine phosphate		1991	27,327
flutamide		1983	19,318
formestane		1993	29,337
fotemustine		1989	25,313
gemcitabine HCl		1995	31,344
idarubicin hydrochloride		1990	26,303

GENERIC NAME	INDICATION	YEAR INTRO.	ARMC VOL., PAGE
interferon gamma-1α		1992	28,332
interleukin-2		1989	25,314
irinotecan		1994	30,301
lonidamine		1987	23,337
mitoxantrone HCl		1984	20,321
nedaplatin		1995	31,347
nilutamide		1987	23,338
paclitaxal		1993	29,342
pegaspargase		1994	30,306
pentostatin		1992	28,334
pirarubicin		1988	24,309
ranimustine		1987	23,341
sobuzoxane		1994	30,310
toremifene		1989	25,319
vinorelbine		1989	25,320
zinostatin stimalamer		1994	30,313
porfimer sodium	ANTINEOPLASTIC ADJUVANT	1993	29,343
masoprocol	ANTINEOPLASTIC, TOPICAL	1992	28,333
miltefosine		1993	29,340
atovaquone	ANTIPARASITIC	1992	28,326
ivermectin		1987	23,336
droxidopa	ANTIPARKINSONIAN	1989	25,312
pergolide mesylate		1988	24,308
lidamidine HCl	ANTIPERISTALTIC	1984	20,320
gestrinone	ANTIPROGESTOGEN	1986	22,321
cabergoline	ANTIPROLACTIN	1993	29,334
tamsulosin HCl	ANTIPROSTATIC HYPERTROPHY	1993	29,347
acitretin	ANTIPSORIATIC	1989	25,309
calcipotriol		1991	27,323
tacalcitol	ANTIPSORIATIC, TOPICAL	1993	29,346
amisulpride	ANTIPSYCHOTIC	1986	22,316
remoxipride hydrochloride		1990	26,308
zuclopenthixol acetate		1987	23,345
actarit	ANTIRHEUMATIC	1994	30,296
diacerein		1985	21,326

GENERIC NAME	INDICATION	YEAR INTRO.	ARMC VOL., PAGE
octreotide	ANTISECRETORY	1988	24,307
adamantanium bromide	ANTISEPTIC	1984	20,315
cimetropium bromide	ANTISPASMODIC	1985	21,326
tiquizium bromide		1984	20,324
tiropramide HCl		1983	19,324
argatroban	ANTITHROMBOTIC	1990	26,299
defibrotide		1986	22,319
cilostazol		1988	24,299
cloricromen		1991	27,325
enoxaparin		1987	23,333
ethyl icosapentate		1990	26,303
ozagrel sodium		1988	24,308
indobufen		1984	20,319
picotamide		1987	23,340
limaprost		1988	24,306
flutropium bromide	ANTITUSSIVE	1988	24,303
levodropropizine		1988	24,305
benexate HCl	ANTIULCER	1987	23,328
ecabet sodium		1993	29,336
enprostil		1985	21,327
famotidine		1985	21,327
irsogladine		1989	25,315
lansoprazole		1992	28,332
misoprostol		1985	21,329
nizatidine		1987	23,339
omeprazole		1988	24,308
ornoprostil		1987	23,339
pantoprazole sodium		1994	30,306
plaunotol		1987	23,340
polaprezinc		1994	30,307
ranitidine bismuth citrate		1995	31,348
rebamipide		1990	26,308
rosaprostol		1985	21,332
roxatidine acetate HCl		1986	22,326
roxithromycin		1987	23,342
sofalcone		1984	20,323
spizofurone		1987	23,343
teprenone		1984	20,323
tretinoin tocoferil		1993	29,348
troxipide		1986	22,327

GENERIC NAME	INDICATION	YEAR INTRO.	ARMC VOL., PAGE
didanosine	ANTIVIRAL	1991	27,326
famciclovir		1994	30,300
foscarnet sodium		1989	25,313
ganciclovir		1988	24,303
lamivudine		1995	31,345
propagermanium		1994	30,308
rimantadine HCl		1987	23,342
saquinavir mesylate		1995	31,349
sorivudine		1993	29,345
stavudine		1994	30,311
valaciclovir HCl		1995	31,352
zalcitabine		1992	28,338
zidovudine		1987	23,345
alpidem	ANXIOLYTIC	1991	27,322
buspirone HCl		1985	21,324
etizolam		1984	20,318
flutazolam		1984	20,318
flutoprazepam		1986	22,320
metaclazepam		1987	23,338
mexazolam		1984	20,321
flumazenil	BENZODIAZEPINE ANTAG.	1987	23,335
bambuterol	BRONCHODILATOR	1990	26,299
doxofylline		1985	21,327
formoterol fumarate		1986	22,321
mabuterol HCl		1986	22,323
oxitropium bromide		1983	19,323
salmeterol hydroxynaphthoate		1990	26,308
APD	CALCIUM REGULATOR	1987	23,326
clodronate disodium		1986	22,319
disodium pamidronate		1989	25,312
gallium nitrate		1991	27,328
ipriflavone		1989	25,314
dexrazoxane	CARDIOPROTECTIVE	1992	28,330
bucladesine sodium	CARDIOSTIMULANT	1984	20,316
denopamine		1988	24,300
docarpamine		1994	30,298
dopexamine		1989	25,312
enoximone		1988	24,301
flosequinan		1992	28,331
ibopamine HCl		1984	20,319
milrinone		1989	25,316
vesnarinone		1990	26,310

GENERIC NAME	INDICATION	YEAR INTRO.	ARMC VOL., PAGE
amrinone	CARDIOTONIC	1983	19,314
xamoterol fumarate		1988	24,312
cefozopran HCL	CEPHALOSPORIN, INJECTABLE	1995	31,339
cefditoren pivoxil	CEPHALOSPORIN, ORAL	1994	30,297
brovincamine fumarate	CEREBRAL VASODILATOR	1986	22,317
nimodipine		1985	21,330
propentofylline		1988	24,310
succimer	CHELATOR	1991	27,333
trientine HCl		1986	22,327
fenbuprol	CHOLERETIC	1983	19,318
auranofin	CHRYSOTHERAPEUTIC	1983	19,314
aniracetam	COGNITION ENHANCER	1993	29,333
pramiracetam H_2SO_4		1993	29,343
carperitide	CONGESTIVE HEART FAILURE	1995	31,339
nicorandil	CORONARY VASODILATOR	1984	20,322
dornase alfa	CYSTIC FIBROSIS	1994	30,298
neltenexine		1993	29,341
amifostine	CYTOPROTECTIVE	1995	31,338
nalmefene HCL	DEPENDENCE TREATMENT	1995	31,347
azosemide	DIURETIC	1986	22,316
muzolimine		1983	19,321
torasemide		1993	29,348
alglucerase	ENZYME	1991	27,321
erdosteine	EXPECTORANT	1995	31,342
cinitapride	GASTROPROKINETIC	1990	26,301
cisapride		1988	24,299
itopride HCL		1995	31,344
imiglucerase	GAUCHER'S DISEASE	1994	30,301
somatotropin	GROWTH HORMONE	1994	30,310

GENERIC NAME	INDICATION	YEAR INTRO.	ARMC VOL., PAGE
somatomedin-1	GROWTH HORMONE INSENSITIVITY	1994	30,310
pimobendan	HEART FAILURE	1994	30,307
erythropoietin	HEMATOPOETIC	1988	24,301
factor VIII	HEMOSTATIC	1992	28,330
malotilate	HEPATROPROTECTIVE	1985	21,329
buserelin acetate	HORMONE	1984	20,316
goserelin		1987	23,336
leuprolide acetate		1984	20,319
nafarelin acetate		1990	26,306
somatropin		1987	23,343
sapropterin hydrochloride	HYPERPHENYLALANINEMIA	1992	28,336
quinagolide	HYPERPROLACTINEMIA	1994	30,309
cadralazine	HYPERTENSIVE	1988	24,298
nitrendipine		1985	21,331
binfonazole	HYPNOTIC	1983	19,315
brotizolam		1983	19,315
butoctamide		1984	20,316
cinolazepam		1993	29,334
doxefazepam		1985	21,326
loprazolam mesylate		1983	19,321
quazepam		1985	21,332
rilmazafone		1989	25,317
zolpidem hemitartrate		1988	24,313
zopiclone		1986	22,327
acetohydroxamic acid	HYPOAMMONURIC	1983	19,313
sodium cellulose PO4	HYPOCALCIURIC	1983	19,323
divistyramine	HYPOCHOLESTEROLEMIC	1984	20,317
lovastatin		1987	23,337
melinamide		1984	20,320
simvastatin		1988	24,311
glucagon, rDNA	HYPOGLYCEMIA	1993	29,338
fluvastatin	HYPOLIPAEMIC	1994	30,300

GENERIC NAME	INDICATION	YEAR INTRO.	ARMC VOL., PAGE
acipimox	HYPOLIPIDEMIC	1985	21,323
beclobrate		1986	22,317
binifibrate		1986	22,317
ciprofibrate		1985	21,326
meglutol		1983	19,321
ronafibrate		1986	22,326
modafinil	IDIOPATHIC HYPERSOMNIA	1994	30,303
bucillamine	IMMUNOMODULATOR	1987	23,329
centoxin		1991	27,325
thymopentin		1985	21,333
filgrastim	IMMUNOSTIMULANT	1991	27,327
interferon gamma-1b		1991	27,329
lentinan		1986	22,322
pegademase bovine		1990	26,307
pidotimod		1993	29,343
romurtide		1991	27,332
sargramostim		1991	27,332
schizophyllan		1985	22,326
ubenimex		1987	23,345
cyclosporine	IMMUNOSUPPRESSANT	1983	19,317
gusperimus		1994	30,300
mizoribine		1984	20,321
muromonab-CD3		1986	22,323
mycophenolate mofetil		1995	31,346
tacrolimus		1993	29,347
defeiprone	IRON CHELATOR	1995	31,340
sulbactam sodium	β-LACTAMASE INHIBITOR	1986	22,326
tazobactam sodium		1992	28,336
nartograstim	LEUKOPENIA	1994	30,304
pumactant	LUNG SURFACTANT	1994	30,308
telmesteine	MUCOLYTIC	1992	28,337
cisatracurium besilate	MUSCLE RELAXANT	1995	31,340
interferon β-1B	MULTIPLE SCLEROSIS	1993	29,339
afloqualone	MUSCLE RELAXANT	1983	19,313
doxacurium chloride		1991	27,326
eperisone HCl		1983	19,318

GENERIC NAME	INDICATION	YEAR INTRO.	ARMC VOL., PAGE
mivacurium chloride		1992	28,334
tizanidine		1984	20,324
naltrexone HCl	NARCOTIC ANTAGONIST	1984	20,322
tinazoline	NASAL DECONGESTANT	1988	24,312
clospipramine hydrochloride	NEUROLEPTIC	1991	27,325
nemonapride		1991	27,331
risperidone		1993	29,344
timiperone		1984	20,323
rocuronium bromide	NEUROMUSCULAR BLOCKER	1994	30,309
fasudil HCL	NEUROPROTECTIVE	1995	31,343
bifemelane HCl	NOOTROPIC	1987	23,329
choline alfoscerate		1990	26,300
exifone		1988	24,302
idebenone		1986	22,321
indeloxazine HCl		1988	24,304
levacecarnine HCl		1986	22,322
nizofenzone fumarate		1988	24,307
oxiracetam		1987	23,339
alendronate sodium	OSTEOPOROSIS	1993	29,332
tiludronate disodium	PAGET'S DISEASE	1995	31,350
beraprost sodium	PLATELET AGGREG. INHIBITOR	1992	28,326
epoprostenol sodium		1983	19,318
iloprost		1992	28,332
sarpogrelate HCl	PLATELET ANTIAGGREGANT	1993	29,344
trimetrexate glucuronate	*PNEUMOCYSTIS CARINII* PNEUMONIA	1994	30,312
histrelin	PRECOCIOUS PUBERTY	1993	29,338
gestodene	PROGESTOGEN	1987	23,335
nomegestrol acetate		1986	22,324
norgestimate		1986	22,324
promegestrone		1983	19,323
alpha-1 antitrypsin	PROTEASE INHIBITOR	1988	24,297
nafamostat mesylate		1986	22,323

GENERIC NAME	INDICATION	YEAR INTRO.	ARMC VOL., PAGE
adrafinil	PSYCHOSTIMULANT	1986	22,315
finasteride	5α-REDUCTASE INHIBITOR	1992	28,331
surfactant TA	RESPIRATORY SURFACTANT	1987	23,344
tirilazad mesylate	SUBARACHNOID HEMORRHAGE	1995	31,351
APSAC	THROMBOLYTIC	1987	23,326
alteplase		1987	23,326
tiopronin	UROLITHIASIS	1989	25,318
propiverine hydrochloride	UROLOGIC	1992	28,335
clobenoside	VASOPROTECTIVE	1988	24,300
cadexomer iodine	WOUND HEALING AGENT	1983	19,316
epidermal growth factor		1987	23,333